Functional Plant Genomics

Functional Plant Genomics

Editors
J. F. Morot-Gaudry
P. Lea
J. F. Briat

CRC Press
Taylor & Francis Group
Boca Raton London New York

CRC Press is an imprint of the
Taylor & Francis Group, an **informa** business
A SCIENCE PUBLISHERS BOOK

CRC Press
Taylor & Francis Group
6000 Broken Sound Parkway NW, Suite 300
Boca Raton, FL 33487-2742

ISBN-13: 9781578085064 (hbk)

Visit the Taylor & Francis Web site at
http://www.taylorandfrancis.com

and the CRC Press Web site at
http://www.crcpress.com

Library of Congress Cataloging-in-Publication Data

Génomique en biologie végétale. English.
Functional plant genomics/ [edited by] J.F. Morot-Gaudry, P. Lea, J.F. Briat.
p. cm.
Includes bibliographical references and index.
ISBN-13: 978-1-57808-506-4
ISBN-10: 1-57808-506-3
1. Plant genomes. 2. Plant genetics. 3. Plant molecular genetics. I. Morot-Gaudry Jean-Francois. II. Lea, P. III. Briat, J.F. (Jean-François), 1951-IV. Title

QK981.G48513 2007
572.8'62--dc22

2006051254

Published by arrangement with INRA, Paris.

Translation of: LA GÉNOMIQUE EN BIOLOGIE VÉGÉTALE
French edition: © INRA, Paris, 2004
ISBN 2-7380-1167-5

Foreword

The publication of the sequence of the nuclear genome of the model crucifer *Arabidopsis thaliana* has been a landmark in plant sciences paving the way to innovative approaches in the new millennium. For the first time this has made possible, at least theoretically, the complete inventory of the genes operating in a plant genome, and the systematic analysis of their function. The identification of putative genes in chromosome sequences, based on annotation tools can be validated by transcription studies. This can be done on a genome scale by identifying a set of sequences specific for each of the annotated genes, and using the corresponding sequences to generate probes and build microarrays as illustrated by AFFYMETRIX and CATMA chips devised to study Arabidopsis gene transcription. However gene annotation based on prediction algorithms is not a hundred percent reliable and errors of prediction in exon-intron boundaries are frequent and complementary approaches are therefore needed to validate annotations. The production of full length cDNA sequences, and their analysis by systematic sequencing has been an efficient validation approach. Different institutions have contributed to this effort (the RIKEN in Japan, the Salk Institute in the USA and the Genoscope in France), allowing not only an improved annotation but the creation of a resource for functional studies. These have included the overexpression of a specific transcript, the intracellular localization of the corresponding protein by fusion with a reporter enzyme such a GFP, or the expression of the protein in a heterologous system to study its biochemical properties and 3D structure. Since full length cDNA inventories are always incomplete, an alternative approach has been devised by the team of J. Ecker at the Salk Institute by using an oligonucleotide-based array covering, for transcription studies, the entire genome.

Knowing the sequence of a genome paves the way to the functional analysis of the genes it contains. The inactivation of genes by random insertion of known sequences (T-DNA, transposons) in a sequenced genome, allows the identification and inventory of insertion sites. This approach has been used to perform the inventory of T-DNA insertion sites in the Arabidopsis genome, and to store the corresponding information in databases. Several programs have contributed to this effort, these have included SIGNAL, Genoplante, GABI, but also many others. Again this type of approach will not saturate the Arabidopsis genome with insertions in all annotated genes. A complementary approach is the TILLING of genes, i.e. the identification of a specific gene by screening a large number of M_2 plants. Alternatively RNAi approaches are also chemically efficient enough to allow a systematic inactivation of genes in the Arabidopsis genome as illustrated by the AGRIKOLA-EC project. These two approaches, furthermore are applicable to any plant genome, provided that a set of genes have been identified for which a functional analysis is envisaged. However, genomics can also take profit from other systematic approaches. Purifying a specific cell compartment (chloroplast, vacuole, plasma membrane) and identifying the different proteins present in the purified component by mass spectrometry analysis is facilitated by the availability of predicted protein sequences derived from the annotated genome sequences, thus allowing the massive identification of the intracellular localization of proteins. Genomic inventories are completed by the characterization of the "interactome", using the Y2H approach and more recently by TAP-TAG. Recently also regulatory motives present in the promoters of genes have been identified by comparative genomic approaches, and non-coding RNA transcripts identified systematically. Among them miRNAs are now thought to have a key regulatory function, and are the subject of intensive investigation. Genomics therefore brings together a set of molecular tools that allows the analysis of gene function, which was possible in the early days of molecular genetics, but is now renewed by the availability of additional resources. Data mining allows the development of working hypotheses for the function of unknown genes, with sometimes little work needed to validate a hypothesis. This is due to the large amount of information already existing for that specific gene scattered in various databases (mutant phenotype, pattern of gene expression, intracellular localization, similarity to other genes). Paradoxically plant genomics is expensive, but saves money. Commercial sequencing has been readily available for some time but in addition, many of the approaches which are required to study the function of a gene, can be set up to deal efficiently with large numbers. This is especially true of the inventories of T-DNA insertion sites, which can take weeks for a single line and a few hours in a systematic inventory.

Genomics of model species such as Arabidopsis has also had an important impact on the study of crop genomes. The conservation of gene function through evolution has been beautifully illustrated by the identification of the *Gai* gene involved in gibberellin signal transduction that turned out to be the functional ortholog of the green revolution gene from wheat (*Rht*). This conservation has also been established in relation to the gene order on chromosomes. The exploitation of synteny has been the rational for choosing a few species as model species, such as Arabidopsis and rice for intensive studies and their use to study other plant genomes. As for annotation, synteny predictions are not a hundred percent reliable but at least they establish a reference frame to study an unknown locus of interest in the genome of a crop, allowing the rapid construction of high density genetic maps to perform the positional cloning of genes of agronomic interest. Genomics is also central to any biodiversity study in a crop and the inventory of allelic diversity at loci contributing to agronomic traits, and ultimately to molecular breeding.

Michel Caboche

Foreword

After the Arabidopsis whole-genome sequence was achieved, many hypothesized that the conservation of gene content, function and organisation across plant species would be sufficiently tight to use knowledge on the Arabidposis genome as a generic road map of plant genome. Indeed, the Arabidopsis paradigm has proven extremely relevant as a generic model of plant gene function for many traits and pathways, but much less as a reference for gene content and order, especially for phylogenetically distant species.

Fortunately, as a counterpart to the Arabidopsis effort, numerous projects were initiated on plant species of agronomical interest such as rice, tomato and an increasing number of species first considered intractable due to their genome size and organisation such as maize, wheat, sugar cane, grapevine. This leads to gene discoveries through the sequencing of expressed sequence tags (EST), and decipher the synthenic network of genes between plant families. Nowadays, the sequencing of several whole plant genomes, the production of dense physical maps of others and draft sequencing of diverse genotypes within a species or a set of related species result in a multitude of data. Keeping in mind that "data is not knowledge" bioinformatics has emerged as a key discipline to valorise this information and phylogenetics and evolutionary approaches are becoming important currencies to understand the relationships between genes and genomes.

Just as the capacity for DNA sequencing has expanded, similar shifts have occurred in functional genomics. Reverse genetics is being used to investigate the function of candidate genes through approaches such as RNA interferences, TILLING (Targeting Induced Local Lesions In Genomes) and the use of systematically generated T-DNA populations. Analysis of natural genetic variation and the interest in distinguishing adaptative variation from genetic drift is becoming a key issue. Moreover,, technologies for high-throughput genotyping, in terms of markers and/or individuals analysed, will revolutionize our ability to fine map genes, with key impacts on genome structure understanding and mapping of loci which determine traits of agronomical interest. Its is expected that genotyping and phenotype assessment will achieve "omic" status for basic and applied plant biologists.

The rate limiting step in the genetic mapping of quantitative traits is shifting from the analysis of genotype to that of phenotype. Understanding how the environment affects the expression of quantitative traits is a key issue for both basic and applied interests. Unravelling this mystery and developing integrative models should enhance both the characterisation of yet untapped sources of genetic diversity and their use for crop improvement.

Information resulting from genomic studies is providing a wealth of new opportunities in plant biology, ranging from basic studies of evolution and molecular physiology to applied aspects of plant biology that focus on crop improvement. Many, but not all, of these topics are covered in the chapters of this book.

A Charcosset
Department of genetics of plant breeding, INRA

Preface

Life Sciences research has made a considerable amount of progress over the last fifty years due major changes in concepts and methodology. Recently, there have been numerous successes in providing the complete sequence of the genome of various eukaryotic multicellular organisms, and plant biology has participated in this enterprise. However, total genome sequencing should be seen as a start rather than an end! The openings offered by functional genomics reconciles organism biology and molecular biology, in order to define an integrative biology that should allow new insights about how a phenotype is built up from a genotype in interaction with its environment.

It was during our years spent at the head of the INRA Plant Biology Department that the idea for this book emerged. It was from this privileged position for observing the major changes occurring in Plant Biology that we decided to assemble in the same book, various topics related to genomics. These topics have had a major impact and changed the way research is carried out in Plant Biology.and beyond. Thanks to the expertise of our colleagues from INRA, CNRS, CIRAD, IRD, and the Universities, we have been able to cover a wide area of concepts and methods in genomics. These have ranged from international genome sequencing projects, to invaluable bioinformatics tools, and analytical methods of gene expression, including final metabolic products and their specific tissue and cellular compartmentalization. This new knowledge enables us to understand the integration of basic physiological functions, and developmental programs. Also, the establishment of tight relationships between genomics, and genetic and plant breeding reveals synergies for exploiting molecular markers, for analysing genetic variation, or for studying quantitative traits. The integration that is now feasible by these new approaches also effects ecophysiology, and opens up new perspectives for the use of modelling for the study of plant populations.

We are, therefore, looking at the birth of a new Plant Biology, and we hope that this book will contribute to the dissemination of these new practices.

This book is intended for advanced students in biology and agronomy, their lecturers, and for all scientists who wish to understand plant genomics rapidly, in a relatively exhaustive and synthetic form.

Jean-François Briat
Jean-François Morot-Gaudry
A. Charcosset

Contents

Plant Models in Genomics

Genomics, Genetic Variability and Plant Breeding

Genomics and Technology

Structural Genomics and *in silico* Analysis

1

CHAPTER

Physical Structure of the Plant Nuclear Genome

M. Delseny, M. Echeverria, E. Lasserre

Three general approaches are available to analyze genomes.

- *Determination of their physical organization,* which consists of describing DNA molecules in terms of physical distance, succession and distribution of various types of sequences along the chromosomes and packaging of the DNA molecule within chromatin. Physical distances are measured in length units, nanometers, or in base pair numbers.
- *Determination of their genetic organization,* which consists of elucidating how genes are distributed along the chromosome and how they segregate and/or recombine at each generation. Analysis of genetic organization results in the establishment of genetic maps on which distances are evaluated in recombination units or centiMorgans. These distances correspond to the recombinational capacity.
- *Determination of expression patterns,* which consists of analyzing transcription products (or transcriptome) and their translation products (or proteome). Most proteins are enzymes which are involved in transforming a number of organic molecule substrates. The collection of all these organic molecules is the metabolome.

This conception of genome organization is relatively recent and is the result of the development of a variety of increasingly resolutive technologies, the finest being complete sequencing of a genome.

This chapter describes the main methods which have allowed biologists to elucidate the physical organization of plant genomes. Figure 1.1 illustrates the various levels of resolution in analyzing a eukaryotic genome. The first is observation of chromosomes under the microscope, also known as cytogenetics. A second level is observation of genetic recombination in segregating populations resulting in genetic maps. Next is the

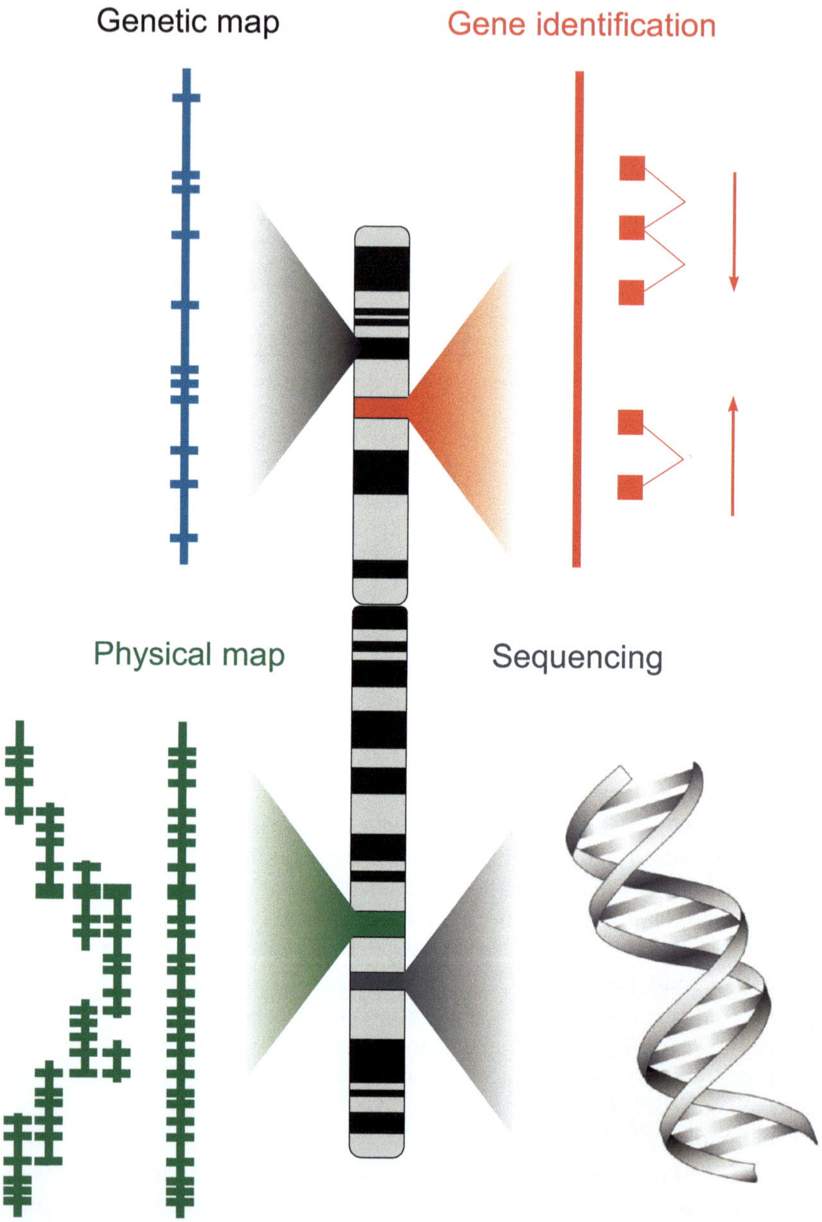

Fig. 1.1 Different levels of genome representation. In the middle, schematic representation of a chromosome after differential staining of euchromatin and heterochromatin domains. On the genetic map molecular markers are positioned. The genetic distance between markers is expressed in recombination units or centiMorgans. A physical map is represented by sets of overlapping fragments with restriction sites as landmarks. Distances are in base pairs. Each DNA fragment can be sequenced and from the sequence the next step is to identify precise gene structures with orientation and limits of their primary transcript, as well as mature transcripts.

arrangement of the DNA molecules in overlapping pieces which results in physical maps. The final step is sequencing and assembling the whole chromosome DNA molecule.

CLASSICAL CYTOGENETICS

The first level of analysis of physical structure of a genome was microscopic observation. It was carried out at stages of the cell cycle where chromosomes are individualized (the metaphase plate of mitotic nuclei) or decondensed (pachytene and diplotene stages of the first meiotic division). One could then count chromosomes, measure their sizes, observe their morphology and locate centromeric constrictions or nucleolar organizers (NOR). One can describe the various key regions of the chromosome: the telomere at each end and the centromere, whose position determines chromosome type: telocentric if the centromere is close to the telomere, acrocentric if the two arms are disymmetric and metacentric if the centromere is the middle of the chromosome. Various stains, such as Giemsa, acetic carmine or 4',6-diamidino-2-phenylindole (DAPI), often revealed regions which were more intensely colored and which corresponded to heterochromatin. It is now known that heterochromatin regions are gene poor, mostly silent and highly packed. Sometimes, Giemsa revealed a banding pattern (G-banding) which was a characteristic feature of a given chromosome and allowed its identification. These microscopic observations showed the heterogeneity of the chromosome. The whole set of individual chromosomes is the karyotype.

Chromosome number in somatic cells is always even because plant cells are diploids (or polyploids). It can be variable: five pairs in *Arabidopsis thaliana*, nine pairs in cabbage, 10 in maize, 21 in wheat and more than a hundred in sugarcane. Chromosome size can be extremely variable from one species to another and can reach several microns in species with large genomes like wheat, broad bean or lily. Chromosome size is correlated with genome size.

Differences in genome size became obvious when it was possible to measure the DNA content per nucleus, using either extraction methods and dosage or cytophotometric methods. In the latter situation, the DNA content was measured either on isolated nuclei in a flux cytometer or on microscopic slides following specific staining (DAPI, propidium iodide, ...). An internal standard, with known DNA content is used as a reference. Such experiments indicated that the content of plant nuclear DNA varies from a few hundred millions of base pairs to tens of billions (Table 1.1). Such a range of variation (over at least two orders of magnitude, was not correlated with genetic complexity (the gene number), which most likely does not vary by more than a factor of five (Flavell, 1986). The DNA content of a genome is called the "C-value" and the large discrepancy between C-value and genetic complexity is called the "C-value" paradox. In addition, the DNA content within different cells of a given plant is also variable because some cells undergo endo-replication cycles (for instance trichome cells, endosperm cells or precursors of vascular cells). However, polytene chromosomes such as those of Drosophila salivary glands are exceptional and restricted to a limited number of cells such as the suspensor cells of legume embryos.

Table 1.1 Genome sizes in Mbp

Arabidopsis thaliana	130
Oryza sativa (rice)	430
Medicago truncatula	650
Brassica. oleracea (cabbage)	700
Sorghum bicolor	800
Lycopersicum esculentum (tomato)	1,000
Brassica napus (rapeseed)	1,200
Zea mays (maize)	2,500
Hordeum vulgaris (barley)	5,200
Pisum sativum (pea)	5,000
Triticum aestivum (wheat)	16,000
Tulipa sp. (tulip)	30,000

Observation of chromosome pairing at the pachytene stage during meiosis provided information on the relationship between chromosomes. Abnormal pairing can reveal ancestral relationships or translocation events. This type of observation revealed that many plants which apparently behave as diploids are indeed polyploids, which result either from doubling of a single chromosome stock (autopolyploidy) or fusion of two distinct stocks (allopolyploidy). Classical examples are rapeseed, which is an allotetraploid resulting from fusion between a cabbage and a rape genome, and wheat which is an allohexaploid resulting from the fusion of three distinct genomes, the A, B and D genomes. These ploidy phenomena account for variations in DNA content by a factor of x 2 to x 10 but certainly not of 2 orders of magnitude. Other gross rearrangements revealed by cytogenetic methods are large deletions, inversions, duplications or translocations.

PHYSICAL METHODS TO FRACTIONATE GENOMES

CsCl gradient equilibrium ultracentrifugation was one of the first methods used to fractionate DNA. When DNA is ultracentrifuged in a CsCl solution for a long time, a concentration gradient is established along the centrifuge tube and the DNA molecule positions itself at the CsCl concentration corresponding to its buoyant density. This buoyant density is strictly correlated with the GC composition of the DNA molecule. It therefore became possible to separate DNA molecules according to their GC content. Buoyant density was usually determined after elution of the gradient by measuring the refraction index. Bacterial DNAs showed symmetrical peaks but many eukaryotic DNAs exhibited asymmetrical elution profiles. Secondary peaks or shoulders are qualified as "satellite DNA" (Fig. 1.2). Whatever the nature of the satellite DNA, such an observation indicated that DNA molecules were heterogeneously distributed with respect to average GC composition. Similar methods have used chemical drugs or dyes which specifically bind to AT-rich or GC-rich sequences to amplify the separation between the different DNA fractions (Flavell, 1986; Buchanan *et al.*, 2002).

At this stage of the analysis we have learned that the plant genome is distributed in a variable number of chromosomes, that its size can vary by at least two orders of magnitude and that the distribution of DNA molecules is heterogeneous with respect to nucleotide composition.

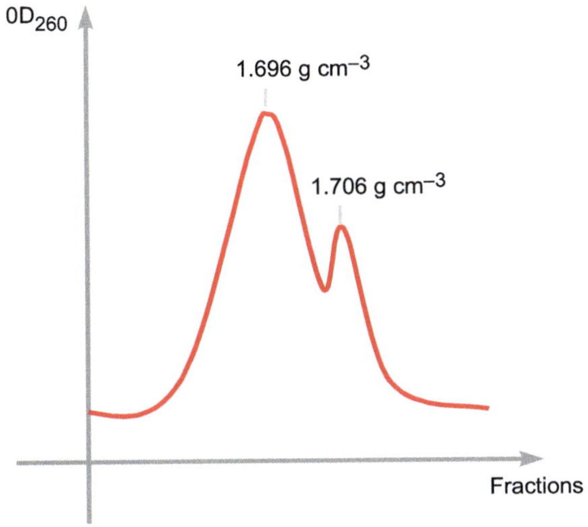

Fig. 1.2 Elution profile of melon DNA following isopycnic ultracentrifugation in a CsCl gradient. Most of the DNA is eluted at a buoyant density of 1.696 g cm^{-3} and a satellite DNA fraction at 1.706 g cm^{-3}. This fraction is heavier because it has a higher GC content.

ANALYSIS OF DNA MOLECULES USING DENATURATION RENATURATION KINETICS

DNA molecules are double-stranded. When a DNA solution in 0.18 M Na phosphate buffer is heated to 100°C, the two strands separate and the molecules are denatured. When the temperature is returned to 60°C complementary single strands pair and the DNA renatures. Alternatively, DNA molecules can be alkali denatured in the presence of NaOH and then renatured by returning to neutral pH. Denaturation is marked by an increase in optical density at 260 nm which allows one to measure the denaturation or renaturation kinetics, using a temperature controlled spectrophotometer. The temperature at which 50% of the DNA molecules in the solution are denatured is called the Tm. This value is strictly correlated with the GC content of DNA. Another way to analyze denaturation/renaturation kinetics is to physically separate single-stranded molecules from the double-stranded ones. Such fractionation can be carried out using hydroxylapatite column chromatography. At 0.18 M Na phosphate, single-stranded molecules flow through the column whereas double-stranded ones are retained. Double-stranded DNA can then be eluted at a higher Na phosphate concentration (Fig. 1.3). It is also possible to follow renaturation kinetics using S1 nuclease. This enzyme digests single-stranded molecules and mis-paired DNA in heteroduplexes. Digestion at regular intervals during renaturation kinetics allows precise measurement of the amount of correctly-paired molecules.

The first kinetic measurements of eukaryotic DNA renaturation were produced by Britten and Davidson in the early 1970s (Britten and Davidson, 1971). Their classical experiment was fairly simple: DNA was broken into small 250-300 bp fragments, using mechanical shearing by repeatedly pumping and extruding the DNA solution through a syringe needle. This solution at a concentration (Co) was distributed into identical

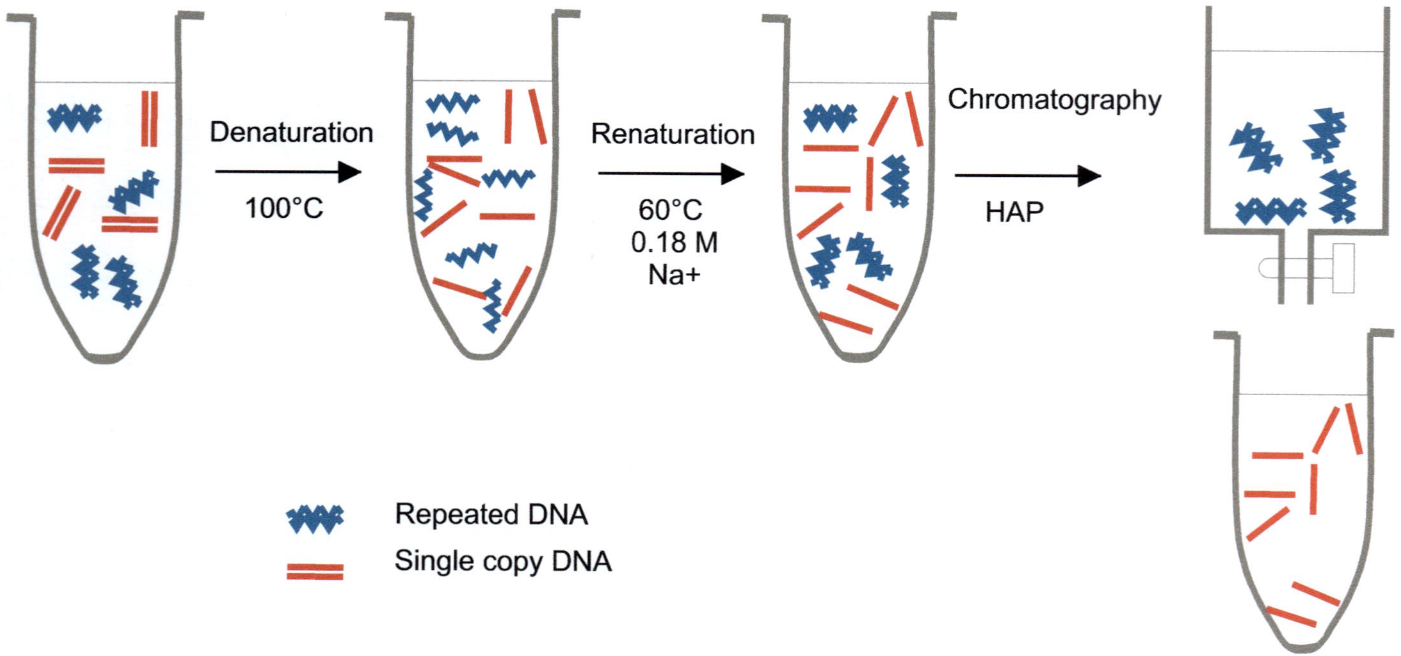

Fig. 1.3 Principle of DNA denaturation/renaturation kinetics. Sheared DNA is heterogeneous, being made of single copy sequences and repeated sequences. After heat denaturation at 100°C, it is incubated at 60°C for renaturation. After a short while, only repeated sequences are renatured. They can then be separated from non reassociated sequences by hydroxylapatite (HAP) column chromatography. Single stranded, non-reassociated unique sequences flow through the column whereas double stranded, renatured repeated sequences remain bound to the column. They can be eluted by raising the ionic strength of the elution buffer.

aliquots. They were denatured for a few minutes at 100°C and then allowed to renature for various times after cooling to 60°C. After a time t, the concentration of renatured DNA was measured, either using hydroxylapatite or S1 nuclease. The result of these measurements is represented by the Cot curve or renaturation curve in which the renatured fraction is plotted against the Co x t product. Because the renaturation time can be quite long, a logarithmic scale is used. Britten and Davidson's key observation (Britten and Davidson, 1971) was that the pattern of the renaturation curve depends on the complexity of the DNA molecules in the solution. Thus a bacteriophage DNA renatured much more rapidly than that of *E. coli*, but in both cases the curve was a simple sigmoid. However, when the same experiment was carried out with calf thymus or plant DNA, a complex curve was observed and a very long time was required to achieve almost complete renaturation. A fraction of DNA usually renatured quite rapidly, often within minutes at fairly low concentrations. This fraction corresponded to molecules with a simple nucleotide composition which were overrepresented in the solution and represented a fraction of highly repeated DNA sequences. A second fraction reassociated more slowly and corresponded to moderately-repeated sequences. More complex DNA took a longer time to reassociate, even at higher concentrations, usually within days. This fraction of DNA was interpreted as containing rare and unique sequences, so that it required a longer time to find the complementary sequence than for abundant and simple sequences. Each repeated sequence class can be characterized, for a given size, by its Cot value.

Plant DNA renaturation curves were found to be very similar to that of calf thymus DNA and indeed this pattern is characteristic of eukaryotic DNA (Fig. 1.4). DNA from different plants differs by a variable proportion of moderately and highly-repeated sequences (Cot < 1). Species like wheat contain nearly 90% of their DNA as repeated sequences (Flavell, 1986), whereas this percentage is around 50% for rice and less than 25% for *A. thaliana*. It is generally assumed that most of the protein-coding sequences are located in the "unique" sequence fraction of higher complexity (Cot > 100). It is possible to split the complex Cot curve into elementary sigmoid curves and experimentally isolate the corresponding DNA fractions. When DNA molecules are sheared into larger fragments (for instance 2,700 bp instead of 250 bp), the corresponding Cot curve is shifted to the left. This indicates that the repeated sequences are interspersed in a short range with unique sequences and that they drive the renaturation reaction to take place at a faster rate. Using this type of approach, various interspersion patterns can be distinguished amongst plant genomes. In some species, like soybean or French bean, a short interspersion pattern is observed, whereas in certain others, like rice, a long pattern is observed. The proportion of repeated sequences in plant genomes appears to be quite variable, up to 90% in the most complex genomes, such as wheat. Indeed this large range of proportion of repeated sequences accounts for the large range of genome sizes observed in plants, known as the C-value paradox

This concept of a highly heterogeneous genome with sequences of different complexity was confirmed when it became possible to isolate well-defined DNA fragments by molecular cloning in *E. coli* and analyze their distribution in a genome, using Southern blot hybridization techniques and, later on, direct sequencing of DNA molecules (Buchanan *et al.*, 2002; Lewin, 2004).

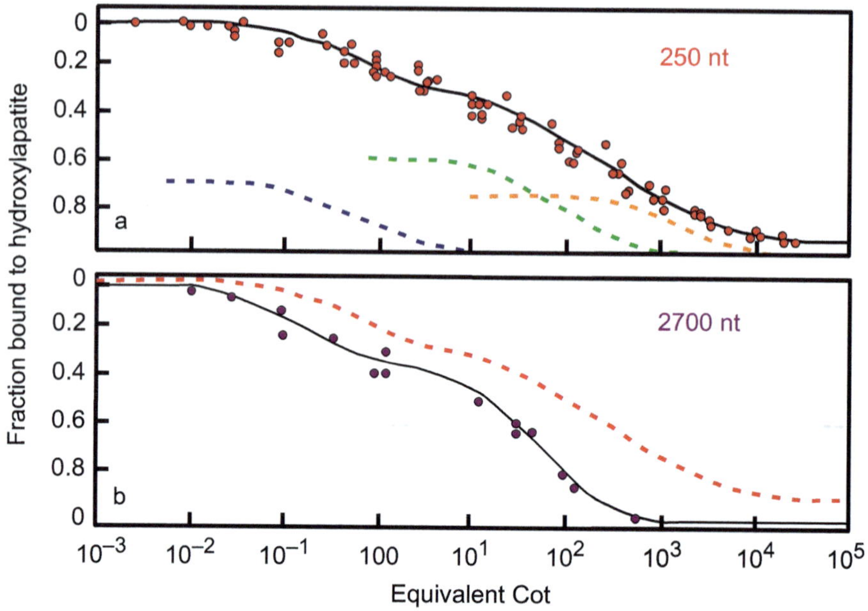

Fig. 1.4 Soybean DNA renaturation kinetics. The top panel corresponds to an experimental curve obtained with DNA sheared to 250 nucleotides (nt) . The curve can be interpreted as the result of the superimposition of three simple sigmoïd curves (dotted curves) which correspond, from left to right to renaturation of the highly repeated DNA fraction, to that of the moderately repeated sequences, and finally, to that of unique sequences. The bottom panel illustrates the effect of modifying the size of DNA fragments in the reaction mixture: the experimental points and curve correspond to DNA fragments of 2,700 nucleotides, whereas the dotted curve correspond to DNA fragments of 250 nucleotides as in the top panel.

PHYSICAL ORGANIZATION OF PLANT DNA ANALYZED BY CLONING AND HYBRIDIZATION BASED METHODS

The discovery of endonuclease restriction enzymes, by the end of the 1960s and early 1970s (Arber, 1974) offered the possibility to cut reproducibly any DNA molecule. Meanwhile, isolation of the T4 bacteriophage DNA-ligase allowed covalent ligation of restriction fragments, whether they had overhanging or blunt ends (Murray *et al.*, 1979; Sambrook *et al.*, 1989). These two discoveries made it possible to isolate a well-defined DNA fragment, to insert it at a unique site in a vector DNA molecule and to amplify it by cloning in a bacterium or phage. This major breakthrough allowed purification of any DNA fragment and its amplification *in vivo* so that large amounts of material became available. Hybridization techniques were also improved. A new technique was developed by E. Southern to transfer DNA digested by a restriction enzyme from an agarose gel in which it was fractionated according to its size, to a nitrocellulose or nylon membrane to which it is covalently bound (Southern, 1975). DNA is alkali denatured before capillary transfer, so that single-stranded molecules are fixed on the membrane in the same disposition as in the agarose gel. Such a membrane can be hybridized with a single-stranded probe corresponding to a specific DNA fragment: the probe recognizes its complementary sequence on the filter and, in appropriate saline and temperature conditions, will specifically hybridize to it. Because the probe is radioactively labeled, the DNA fragments in the digested DNA mixture which hybridize to the probe can be recognized after elimination of non-specifically hybridized probe by washing and autoradiography. This method of detecting a specific DNA fragment within a complex pattern of DNA molecules is known as Southern blotting. It has been the source of a large number of advances in unraveling the organization of all genomes.

The discovery of new restriction enzymes, along with the possibility to carry out double or triple digestions and the ability to recognize sequences according to their size and hybridization to a probe, paved the way to build restriction maps of small genomic regions with restriction sites as landmarks along the DNA molecules. The first genomes to be mapped in plants were the chloroplast genomes (Weil *et al.*, 1982) and a few DNA virus genomes (Hull, 1980) at the end of the 1970s and early 1980s. Mapping of genes became feasible after it became possible to clone cDNA molecules and identify their function at the beginning of the 1980s (Sambrook *et al.*, 1989).

Random cloning of small genomic DNA fragments from plants, and their use to probe digestion patterns of plant nuclear genomes in Southern blot experiments revealed several distinct patterns which confirmed data obtained by using denaturation/renaturation kinetics :

- The fragments revealed after hybridization were unique or in very limited numbers. To detect such fragments relatively hot probes and long exposures of the autoradiograms were necessary. Such hybridizations revealed fragments and sequences which were represented by only one or few copies in the genome. This type of hybridization was frequently obtained with cDNA probes, indicating that most genes are represented by this type of fragment.
- The number of detected fragments was still low but their detection did not require long exposures or very hot probes. This indicated that the probe recognized sequences which were relatively abundant within the genome. Because only a small number of fragments were revealed, this indicated that a limited number of restriction sites were precisely positioned with respect to the sequence. This situation was observed for example with ribosomal RNA gene fragments or with organelle DNA fragments contaminating nuclear DNA.
- There were a significant number of fragments, but they showed a regular ladder pattern, in which each band corresponded to a size fragment which was a multiple of the smallest one. This pattern was characteristic of tandemly repeated sequences from a few tens to a few hundred bp. Repeats of 175-180 bp were common in centromeric and subtelomeric regions of most plant chromosomes. The ladder pattern can be explained by the fact that, following amplification of the repeat sequence, a number of restriction sites have been lost by mutation during evolution, as shown in Fig. 1.5A. 5S RNA ribosomal genes also belong to this class of tandemly-repeated sequences.
- Sometimes one band in the ladder was more intense. This indicated reamplification of a multimer. For instance in Fig. 1.5 A the bands at 1,750 and 17,500 bp are much more intense. This demonstrates that a pentamer fragment

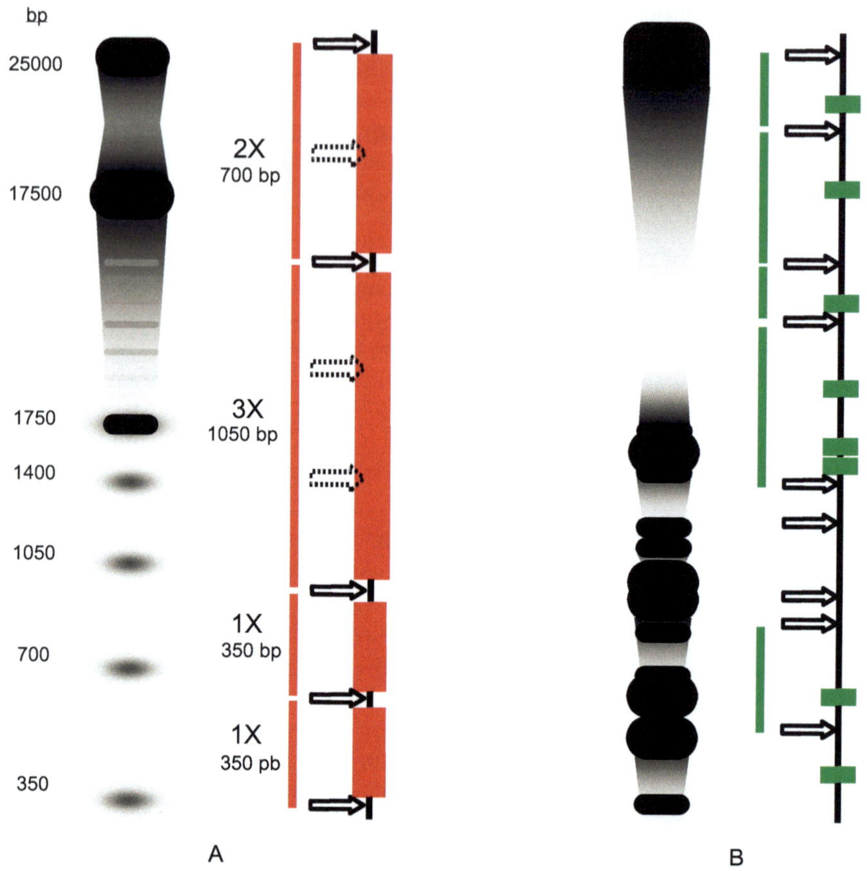

Fig. 1.5 Examples of Southern blot hybridization patterns for repeated sequences. A–Ladder pattern: the hybridizing bands correspond to multimers of 350 bp. They result from repeated units which have lost a restriction site (dotted arrows). The two bands at 1,750 bp and 17,500 bp indicate that a pentamer of the repeat unit and that another bock of 50 units have been further tandemly amplified. A non-digested fraction at 25,000 bp indicates that larger blocks exist. B–Irregular and smearing pattern: it is typical of dispersed repeated sequences. Repeated sequences are represented by small boxes. Restriction sites are randomly positioned with respect to the repeat units, so that each fragment containing a repeated sequence has a different size.

(1,750 bp) has been amplified and that a 10 x repeat of this pentamer has been further amplified (17,500 bp). In rice a 155 bp centromere-specific satellite DNA was detected on each chromosome, with a copy number varying from 60 in chromosome 8 to 1,600 in chromosome 11. Copy numbers of tandemly-repeated sequences can vary tremendously from one individual to another because of the occurrence of unequal crossing-overs. There were several levels of complexity in the tandemly-repeated sequences. When the repeated sequence was more than a few tens of bp, it was designated as satellite DNA (mostly because after characterization of satellite DNA initially isolated by CsCl gradient centrifugation, it was found that they were organized in tandemly-repeated sequences). When the sequence was about 10 bp it was designated as a minisatellite. Finally, when the complexity was only a few base pairs, the sequence was called a microsatellite or SSR (simple sequence repeat). Because they are highly variable, microsatellites and minisatellites are very useful for mapping and fingerprinting eukaryotic genomes.

- Finally, some probes revealed diffused and smearing patterns, with poorly focused bands. This indicated that these sequences are not precisely and reproducibly organized with respect to the restriction sites. Sequences corresponding to transposable elements and their derivatives usually fall in this category (Fig. 1.5B).

Analysis of larger genomic regions had to wait for more efficient analysis methods. A first improvement was the isolation of very high molecular weight DNA molecules in the range of 1-5 Mbp instead of the usual 50-100 kbp, thanks to the use of gentle extraction procedures of DNA from protoplasts or nuclei embedded in agarose beads (Creusot *et al.*, 1995). Such large DNA molecules could not be separated by classical agarose gel electrophoresis. Introduction of pulse field gel electrophoresis (Contour-clamped homogeneous electric field, CHEF) allowed one to resolve DNA molecules in the range of 100 to 2000 kbp (Schwartz and Cantor, 1984). Characterization of large molecules was also facilitated by the discovery of rare-cutting restriction enzymes with an 8 bp recognition site. Large fragments could be cloned and manipulated in new vectors such as YACs (yeast artificial chromosomes) or BACs (bacterial artificial chromosomes). With these improved methodologies a eukaryotic genome could be completely cloned and represented in a library of a few tens or hundred thousand BAC clones, whereas with previous cloning methods tens of millions of clones were required. With these new tools, it became possible to establish a physical map of a complete genome.

All these new technologies (Sambrook *et al.*, 1989), derived from cloning, allowed refinement of the physical organization of plant genomes as initially discovered using isopycnic centrifugation and denaturation/renaturation kinetics. They definitively established the occurrence of a variety of repeated sequences with heterogeneous copy numbers. When ultracentrifugation was used to fractionate DNA and when each fraction was hybridized with cDNA probes it was observed that only some of these fractions gave a positive signal. This observation suggested that genes are not homogeneously distributed in all the DNA fractions, but are concentrated in some sections, which represent the "gene space" (Barakat *et al.*, 1997). Such a situation seems particularly clear cut in large genomes such as maize where the majority of the genes occupy less than 20% of the genome. Observation of gene space is highly dependent on the size of the DNA molecules, usually in the range of 100 kb. In *A. thaliana* such a gene space could not be resolved by ultracentrifugation and was not shown by direct sequencing.

A major drawback of denaturation/renaturation kinetics, ultracentrifugation or restriction mapping is that these techniques do not provide a precise topography of the whole genome. Such an overview can be obtained by establishing long range physical maps, which are a preliminary key step before sequencing the genomes.

PHYSICAL MAPPING OF THE GENOMES

The capacity to isolate genes by molecular cloning and use them as probes in Southern blot hybridization to detect the corresponding sequence in a whole genome digested with a restriction enzyme, was a very significant breakthrough. It allowed the detection of potential molecular polymorphism between two individuals. If these two individuals could be crossed and gave rise to a progeny in which the polymorphic marker segregated, then it became possible to construct a genetic map with molecular marker landmarks. Such genetic maps are based on recombination percentages between markers (Chapter 25). These molecular markers could also be used in hybridization or PCR experiments to assign their position on a YAC or a BAC clone. PCR methods were particularly efficient when a large amount of sequence information, such as EST collections, was available. Hybridization was usually less useful because many genes are members of multigene families and cross-hybridize to some extent.

Establishing a physical map consists in assembling the YAC or BAC clones into series of overlapping clones, named "contigs". (Chapter 2 on sequencing). A first strategy to organize contigs consisted in locating as many markers as possible, usually by PCR techniques, on the various clones in the library. When a set of several markers identified different clones it could be concluded that these clones at least partially overlapped (Fig. 1.6). The limiting factor was the number of markers which could be screened in a cost-effective way. A second strategy consisted of digesting BAC clones with one or several restriction enzymes and obtaining a fingerprint of each clone. Initially the fingerprints were obtained with a single enzymes and analyzed by their agarose gels electrophoresis patterns. Now a high throughput method uses double digestions which are resolved by capillary electrophoresis on sequencing machines. Whatever the technique, when identical fragments were detected between two clones, it was concluded that the two clones partially overlapped. The likelihood of overlapping is dependent on the proportion of fragments in common. A software, FPC (Fingerprint contig), automatically builds contigs from the electrophoretic patterns (Soderlund *et al.*, 2000).

Depending on the stringency of the criteria to decide that different clones overlap (for instance 80 or 50% of the fragments in common), contig construction was more or less reliable. The two strategies of fingerprinting and PCR mapping of ESTs on BAC pools are complementary. A third useful method consisted of systematically sequencing BAC ends: with such a library of sequences one could immediately design primers and run a PCR reaction which would unambiguously validate that the corresponding BAC belonged to a well-defined contig. BAC end libraries were also extremely helpful to sequence a genomic region according to a minimum tiling path and decide which BAC should be sequenced next with the minimum overlap.

The last step in physical mapping consisted of anchoring the contigs on the genetic map. To do that, markers from the genetic map were located on the YAC or BAC clones either by hybridization or by PCR using multidimensional pools. Several important conditions were required to achieve a physical map. A deep coverage of the genome was necessary and usually 5 to 20X coverage was used, so that a region was defined by multiple overlapping clones. Such coverage could be biased by an uneven distribution of restriction sites within the genome. It proved wise to prepare several libraries with

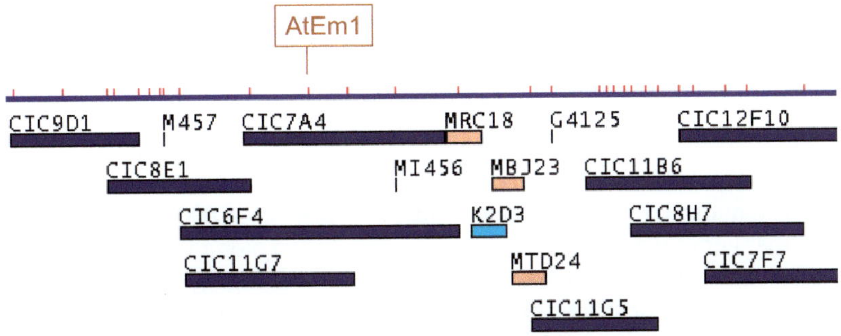

Fig. 1.6 Example of contigs around the *AtEm1* gene located on the long arm of *A. thaliana* chromosome 3. YAC clones CIC 7A4, CIC 6F4, CIC 11G7 have been identified by hybridization with *AtEm1* marker. Clone CIC 8E1 was identified with marker M457. Use of CIC 6F4 left BAC end as a probe has demonstrated that CIC 8E1 and CIC 6F4 partially overlapped. By such a walking technique it is possible to build a contig of overlapping fragments around *AtEm1*.

different enzymes. A high number of sequenced genetic markers was required on the genetic map in order to identify each contig with several markers. The result of a good physical map is the organization of a whole chromosome into just a few contigs with gaps only in heterochromatic regions such as centromeres. These regions were more difficult to organize due to the abundance of repeated sequences which brought confusion in fingerprinting and resulted in multiple very similar BAC ends. Moreover, these regions proved somewhat more difficult to clone because repeated sequences often induced instability of the BAC clones. An example of physical map is shown in Fig. 1.6.

Additional methods have been necessary to measure directly physical distance between markers. One of the most popular strategies consisted of fluorescent *in situ* hybridization (FISH) on whole chromosome mounts or on extended DNA fibers (Lysak *et al.*, 2001) . In these techniques, the probe was a BAC which has been labeled with a fluorescent dye. Several BACs were labeled with different dyes so that they could be distinguished. Using these methods it was possible to use the two last BACs in contigs flanking a gap as probes and to measure the size of the gap. The resolution was of the order of a few kb on extended DNA fibers and few tens of kb on chromosome mounts, either metaphase plates or pachytene meiotic chromosomes. Because of the abundance of repeated sequences in plant genomes, it was necessary to compete out spurious hybridization with unlabeled Cot 1 DNA.

In situ hybridization has been extensively used, initially to locate specific repeated sequences and identify each chromosome individually. Such experiments revealed the location of ribosomal RNA genes in the NOR (Nucleolar Organizer Regions) and the presence of most tandemly-repeated satellite DNA sequences within centromeric or subtelomeric regions. Finally, a variant of this method was extremely useful to detect introgression of chromosome fragments from a distinct species. Repeated DNA sequences evolve much faster than coding sequences and become specific to a given genome. When a whole genome is labeled, the sequences which hybridize faster and most efficiently are the repeated sequences. Therefore, when one hybridized chromosomes from a natural or artificial allotetraploid species with total DNA from one of its parents, usually only the chromosomes or sequences derived from this parent were detected. Alien sequences could be detected in introgression lines using the same strategy.

Detailed physical maps have been achieved for *A. thaliana* and rice in order to support sequencing projects (The Arabidopsis Genome Initiative, 2000). Such maps are presently being constructed for many species such as tomato, maize, alfalfa or poplar. The physical map of rice allowed one to anchor more than 6,000 genetic markers and it revealed that genes were clearly not evenly distributed along the chromosomes, some regions being more gene rich than others. Detailed sequencing of large maize or wheat DNA fragments demonstrated that in these species genes are usually separated from each other by large regions of junk DNA, essentially made of retroelements and their remnants. When a genome is completely sequenced, the gene number and the various types of repeated elements can be determined, at least in theory. In *A. thaliana* the gene number estimate is now close to 30,000, which are organized in roughly 12,000 families. Transposable elements represent about 14% of the genome. On average there is a gene every 4-5 kbp. This value is one gene every 7 to 10 kbp in rice, depending on the considered region.

So far, in this chapter we have described the physical structure of the DNA molecule and how to determine it. However, this is an oversimplification because within the nucleus DNA is not naked, but embedded with histones and other chromatin proteins, all of which are extensively packaged and compacted within chromatin. It has now become quite clear that coordinated genome expression, its stability and inheritance are tightly associated with chromatin structure. Thus, compactness and remodeling of chromatin are regulated during development and cell cycle.

THE DIFFERENT LEVELS OF CHROMATIN ORGANIZATION

Cytology describes two types of chromatin: heterochromatin and euchromatin. Heterochromatin was first defined by Emil Heinz in 1928. It corresponds to the chromatin fraction which is heavily stained by acetic carmine during the cell-cycle. Euchromatin, the "true" chromatin, corresponds to the fraction which is decondensed during the progression of the cell-cycle from metaphase to interphase. The majority of heterochromatin consists of highly-repeated DNA sequences, which are often highly methylated. Heterochromatin is usually associated with centromeric and pericentromeric regions as well as telomeres. In *A. thaliana*, maize and rice, tandemly-repeated sequences and transposons are concentrated within centromeres and in heterochromatic islands named "knobs" (Peacock *et al.*, 1981). These regions have a much lower gene density than euchromatin and have a reduced recombination rate. A similar situation is observed in other cereals, in which these heterochromatic islands initially described by Barbara McClintock (McClintock, 1951) in maize can represent more than 50% of the genome. They are mostly composed of retrotransposable elements. These large regions of nested retroelements are interspersed between genes and gene-rich regions.

At the molecular level, several compaction levels can be distinguished using more or less stringent methods for isolating nuclear DNA (Alberts *et al.*, 2002; Lewin, 2004). Our knowledge of this packaging organization comes mostly from mild digestion of chromatin with *Micrococcus* endonuclease, followed by sucrose gradient centrifugation fractionation, biochemical analysis and crystallography (Widom and Klug, 1985; Travers and Klug, 1987). The resulting fractions can be analyzed by extraction and characterization of proteins and DNA as well as by direct observation in the electron microscope and X-ray diffraction. Chromatin has also been reconstituted from purified DNA and histones. The first level of packaging is detected at low ionic strength and its unit is the nucleosome. A core of four distinct histones proteins, which is in fact an octamer made of two copies of each H2a, H2b, H3 and H4 histones, is wrapped with 145-146 bp of DNA. There are roughly two turns of DNA around the histone core and it is negatively twisted. Each nucleosome is separated from the next by a 10-60 bp linker which is sensitive to micrococcal nuclease digestion. When partial digests are observed in the electron microscope, they appear like a pearl necklace. The phasing of nucleosomes along the DNA molecule changes, depending on whether the chromatin region is going to be active or not and the nucleosome structure locally disappears during transcription and replication. A specific H3 variant is associated with nucleosomes of the centromeric regions (Fig. 1.7).

Adding Ca^{++} and Mg^{++} (at 1 and 2 mM respectively) induces aggregation of nucleosomes which now form a fiber similar to the 30 nm fibers observed *in vivo*. This aggregation capacity is an intrinsic property of nucleosomes but it is enhanced and stabilized by its association with another histone protein, H1 (or its variant H5) which acts as a linker between two nucleosomes.

In vivo, these H1 linkers induce nucleation centers for higher organization of chromatin and they contribute to repress the expression of large regions of chromatin.

During mitotic interphase, Ca^{++} and Mg^{++} concentrations are much higher in the nucleus than those which induce the formation of the 30 nm fiber. Accordingly, there are interactions between fibers which aggregate to produce a higher level of compaction represented by the 100 nm fibers which are visible on electron

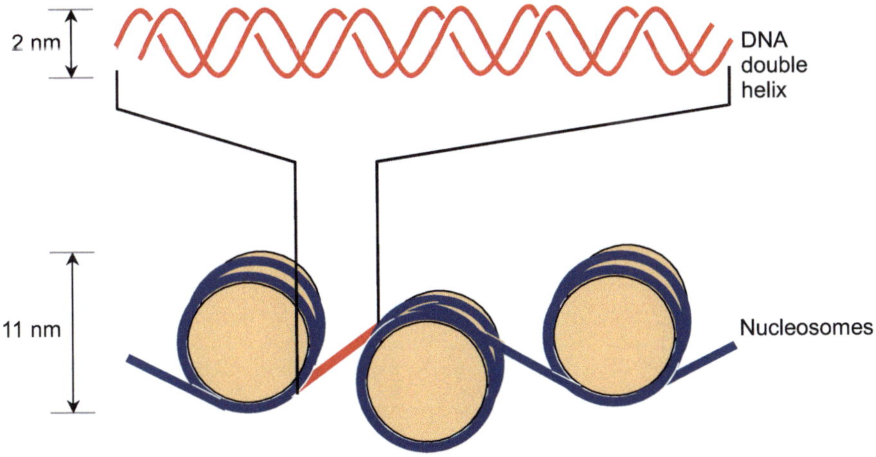

Fig. 1.7 Nucleosome organization of chromatin. The DNA double helix is wrapped (2 turns) around a core of 8 histone molecules, made of two molecules of each of histone H2A, H2B, H3 and H4. The nucleosomal pearl necklace can be supercoiled to give higher compaction levels.

micrographs. These large fibers are highly unstable and can be observed only when nuclei are isolated at specific ionic conditions.

In the nucleus, chromatin is fundamentally condensed. Transcriptionally active chromatin, which *in vivo* is the most decondensed state, is at least packaged as 30 nm fibers. However, despite more than 20 years of active research, the structure of condensed chromatin remains a subject of debate. Several models have been proposed, but none by itself can explain all the observations. In the classical model, chromatin is anchored in nuclei at periodic DNA-binding sites to a protein nuclear matrix (Franzs *et al.*, 2002). This would generate loops ranging between 3 and 45 kbp delimiting chromatin domains and functional groups of genes. These matrix attachment regions (Fig. 1.8) or MARs are defined as DNA fragments which specifically bind to the nuclear matrix *in vitro*. They are made of AT-rich sequences, showing conserved motifs and can spread over several hundred base pairs. Due to their high AT content these regions are not stable and can produce single stranded DNA molecules. In higher plants MAR have been isolated from tobacco, *A. thaliana* and several cereals. They are located in intergenic regions and are associated with gene-rich regions. MAR mapping in *A. thaliana* and maize has demonstrated that each potential loop limited by two MARs contains one or a few genes. These MARs could act as insulating elements to protect genes against the deleterious effects of condensed chromatin on their expression. Indeed this hypothesis is supported by the results of transgenic plant experiments which show that insertion of MARs in the constructs usually significantly enhance transgene expression. Although

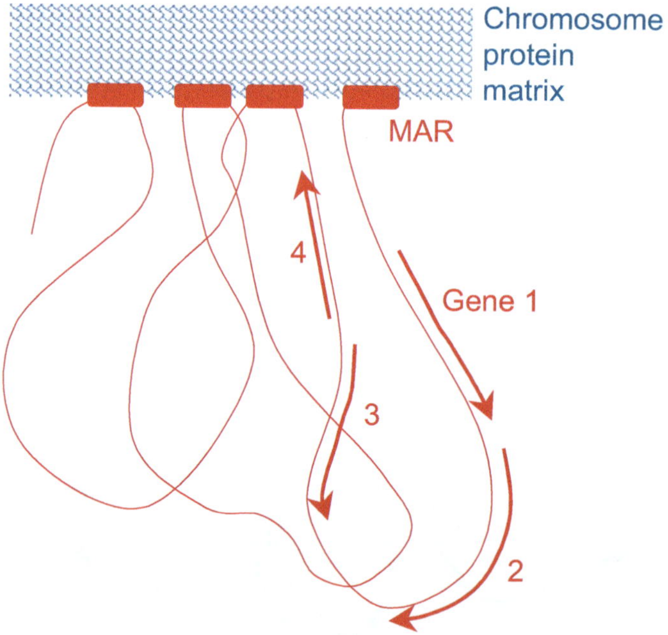

Fig. 1.8 Loop organization of decondensed chromatin. Each loop contains a few active genes and is separated (insulated) from the next loop by an anchor to the chromosome matrix throughout a MAR sequence

attractive, this model remains hypothetical and even the occurrence of a nuclear matrix is disputed, because its nature is still ill-defined. The only unambiguously-established anchoring sites for DNA are the lamins which constitute a filament network in the periphery of the nucleus with extensions inside.

HETEROGENEITY AND DYNAMICS OF CHROMATIN

Two points are particularly important. The first is the heterogeneity of chromatin. It is illustrated by the occurrence of chromatin domains with a distinct nucleoprotein composition (Alberts *et al.*, 2002; Lewin, 2004). In addition to classical nucleosomes, several other types exist with histone variants. The centromeric CENP-A protein is a variant of H3 and centromeric nucleosomes are also associated with CENP-C a kinetochore protein. Chromatin heterogeneity is increased by the fact that a large number of additional proteins are associated with it, such as histone H1 variants, HMG (High Mobility Group proteins, a group of nuclear protein with high electrophoretic mobility), transcription factors and chromatin remodeling proteins which are presents at specific developmental stages. All these types of proteins have been identified in *A. thaliana* which contains several H1 variants. Recently the LHP1 protein (like heterochromatin Protein 1), a homologue of the Drosophila Heterochromatin Protein 1 which is specific for heterochromatin, has been identified from a mutant. This mutant is altered in plant architecture and flowering time, illustrating the impact of deficiencies in heterochromatin on developmental pattern.

A second important point to keep in mind is that chromatin is a dynamic structure (Reyes *et al.*, 2002). The cytological and cytogenetic approaches which have been predominant for many years could only provide a static view of chromatin. Indeed chromatin is fluid and its structure changes throughout the cell-cycle and during the development of the plant. For instance, chromatin is condensed at interphase. This notion of chromatin fluidity can now be investigated at the molecular level, thanks to the development of microscopy and imaging techniques during the phases of the cell cycle in which chromosomes are individually visible and then decondense, associated with specific labeling of chromatin components by fluorescent antibodies. Techniques of chromatin immunoprecipitation (ChIp) allow isolation of fractions of chromatin or chromatin complexes containing a specific protein. DNA in these fractions can then be extracted, sequenced and positioned on the genome, or simply hybridized to genomic DNA chips. Modification of histone proteins has been correlated with the transcriptional activity of chromatin. Typically, in transcriptionally-active chromatin, histones are acetylated and histone H3 is methylated on the lysine residue 4 (K4). In contrast, in heterochromatin histone H3 is methylated on lysine residue 9 (K9). Antibodies against H3 mK4 or H3 mK9 have been developed and can be used to probe and isolate chromatin regions methylated on either lysine residue. Using the DNA prepared from these fractions as probes allows precise definition of the chromatin domains, at least in sequenced genomes. Mutants have been obtained in *A. thaliana* in which the chromatin domains are altered, thus giving molecular support to epigenetic phenomena. One of these mutants is the *ddm1* mutant (decrease in DNA methylation), which is altered in a chromatin remodeling protein, DDM1, involved in both DNA and histone H3 protein methylation. Recent work has demonstrated that in *ddm1* mutants, heterochromatic regions in the centromere of chromosome 4 are reactivated and transcribed.

Several additional genes have been discovered which are involved in remodeling chromatin at specific developmental stages, such as the Polycomb group of genes. These were first characterized in Drosophila and homologous genes have subsequently been independently characterized in plants because they are involved in important processes in the life of the plant. The most obvious examples are the *VRN2* gene of *A. thaliana* which controls flowering time through vernalization and the *MEDEA* or *FIS1* (*Fertilization Independent Seed*) which repress embryogenesis until fertilization (Luo *et al.*, 2000).

As a conclusion, now that the physical structure of several plant genomes is known at the sequence level, the next challenge is to understand how the interactions of specific DNA sequences with proteins in the chromatin control specific gene expression during development and adaptation to the environment. An understanding of the global changes in chromatin structure and remodeling should shed new light on the mode of control of gene expression and on epigenetic phenomena.

REFERENCES

Alberts, B., Johnson, A., Lewis, J., Raff, M., Roberts, K., Walter, P. 2002. Molecular biology of the cell. Gardland Pub. Inc. p. 1616, Textbook.

Arber, W. 1974. DNA modification and restriction. *Progr. Nucl. Acid Res. Mol. Biol.*, 14, 1-37. Review.

Barakat, A., Carels, N., Bernardi, G. 1997. The distribution of genes in the genomes of Gramineae. *Proc. Natl. Acad. Sci. USA*, 24, 6857-6861.

Britten, R.J., Davidson, E.H. 1971. Repetitive and non-repetitive DNA sequences and a speculation on the origin of evolutionary novelty. *The Quaterly Reviewof Biology*, 46, 111-138. Review.

Buchanan, B.B., Gruissem, W., Jones, R.L. 2002. Biochemistry and Molecular Biology of Plants. Wiley, J. and sons Ltd. p. 1408, Textbook.

Creusot, F., Fouilloux, E., Dron, M., Lafleuriel, J., Picard, G., Billault, A., LePaslier, D., Cohen, D., Chabouté, M.E., Durr, A. 1995. The CIC library: a large insert YAC library for genome mapping in *A. thaliana*. *Plant J.*, 85, 763-770.

Flavell, R.B. 1986. Repetitive DNA and chromosome evolution in plants. *Philos. Trans. R. Soc. London B, Biol. Sci.*, 312, 227-242.

Fransz, P., De Jong, J.H., Lysak, M., Ruffini Castiglione, M., Schubert, I. 2002. Interphase chromosomes in Arabidopsis are organized as well defined chromocenters from which euchromatine loops emanate. *Proc. Natl. Acad. Sci. USA*, 99, 14584-14589.

Hull, R. 1980. Structure of the cauliflower mosaic virus. Part 3. Restriction endonuclease mapping of 33 isolates. *Virology*, 80, 362-375.

Lewin, B. 2004. Genes VIII. Pearson Prentice Hall Pub. p. 1027, Textbook.

Luo, M., Bilodeau, P., Dennis, E.S., Peacok, W.J., Chaudhury, A. 2000. Expression and parent-of-origin effects for *FIS2*, *MEA*, and *FIE* in the endosperm and embryo of developing Arabidopsis seeds. *Proc. Natl. Acad. Sci. USA*, 97, 10637-10642.

Lysak, M.A, Fransz, P.F., Ali, H.B.M., Schubert, I. 2001. Chromosome painting in *Arabidopsis thaliana*. *Plant J.*, 28, 689-697.

McClintock, B. 1951. Chromosome organization and genic expression. *Cold Spring Harbor Symp. Quant. Biol.* 16, 13-47.

Murray, N.E., Bruce, S.A., Murray, K. 1979. Molecular cloning of the DNA-ligase from bacteriophage T4. II Amplification and preparation of gene product. *J. Mol. Biol.*, 132, 493-505.

Peacock, W.J., Dennis, E.S., Rhoades, M.M., Pryor, A.J. 1981. Highly repeated DNA sequence limited to knob heterochromatin in maize. *Proc. Natl. Acad. Sci. USA*, 78, 4490-4494.

Reyes, J.C., Hennig, L., Gruissem, W. 2002. Chromatin-remodeling and memory factors. New regulators of plant development. *Plant Physiol.*, 130, 1090-1101.

Sambrook, J., Fritsch, E.F., Maniatis, T. 1989. Molecular cloning,a laboratory manual. Cold Spring Harbor Laboratory Press. 2nd Edition.

Schwartz, D.C., Cantor, C.R. 1984. Separation of yeast chromosome sized DNAs by pulsed field gradient gel electrophoresis. *Cell*, 37, 67-75.

Soderlund, C., Humphrey, S., Dunhum, A., French, L. 2000. Contig built with fingerprints, markers and FPC V4.7. *Genome Res.*, 10, 1772-1787.

Southern, E.M. 1975. Detection of specific sequences among DNA fragments separated by gel electrophoresis. *J. Mol. Biol.*, 98, 503-517.

The Arabidopsis Genome Initiative. 2000. Analysis of the genome sequence of the flowering plant Arabidopsis thaliana. *Nature*, 408, 796-815.

Travers, A., Klug, A. 1987. Nucleoprotein complex: DNA wrapping and writhing. *Nature*, 327, 280-281.

Weil, J.H., Mumumbila, M., Kuntz, M., Keller, M., Steinmetz, A., Crouse, E.J., Burkard, G., Guillemaut, P., Selden, R., McIntosh, L., Bogorad, L., Loffelhardt, W., Mucke, H., Bonhert, H.J. 1982. Gene mapping studies and sequence determination on chloroplast transfer RNAs from various photosynthetic organisms. *Prog. Clin. Biol. Res.* 102, 321-331.

Widom, J., Klug, A. 1985. Structure of the 300 A chromatin filament: X-ray diffraction of orientated samples. *Cell*, 43, 207-213.

2

CHAPTER

The Sequencing of Plant Nuclear Genomes

N. Choisne, S. Samain, N. Demange, G. Orjeda, L. Michelet,
E. Pelletier, M. Salanoubat, J. Weissenbach, F. Quetier

THE BEGINNING

Sequencing of DNA began to be performed in laboratories in 1978. The available technology was limited to a manual approach based on the utilization of radioactive isotopes, only small regions were sequenced, about the size of a gene of several thousand bases. In 1981, the 50 kb of bacteriophage lambda were sequenced, and between 1983 and 1987, the sequences of several DNA viruses of 150 kb were obtained at Cambridge. During the same period, in 1986, Japanese scientists surprised the international community by sequencing a whole series of chloroplast genomes (about 150 kb), using only manual apparatus (Shinozaki *et al.*, 1986). To be successful, large sequencing projects (bacteria, *Saccharomyces cerevisiae*, the human genome) had to wait for the arrival of the first automatic sequencers, which could analyze 96 samples simultaneously. It was in 1995 that Fleischmann *et al.*, from TIGR (The Institute for Genomic Research, Washington, D.C.), published the sequence of the first bacterial genome, *Haemophilus influenzae* (1.8 Mb). The genome of yeast (16 Mb), sequenced by a public international consortium (Goffeau, 1997), was completed in 1996 and the first genome of a multicellular eukaryote, *Caenorhabditis elegans* was published in 1998 (The *C. elegans* Sequencing Consortium, Sanger Center, United Kingdom, and Washington University, USA).

SEQUENCING STRATEGIES FOR EUKARYOTIC NUCLEAR GENOMES

There are two types of sequencing strategies: complete or exhaustive sequencing, until now utilized by international public consortia, and more recently, partial sequencing, presently called *draft* sequences, initially performed by the Celera Genomics Company (USA), and now performed by both the public and private sectors.

Complete Sequencing

This type of sequencing aims for a very high quality, often called "Bermuda" quality, for the rules set out at the first strategic meetings for human genome sequencing which were held there from 1996 to 1998. The goal is to obtain a sequence covering the totality of the genome, if possible with no or very few gaps, and with an average error level of less than 10^{-4} (less than one wrong base in 10,000). To attain this objective, the following steps have been set up for higher eukaryotic genomes:

- construction of Bacterial Artificial Chromosome (BAC)libraries;
- establishment of a pre-sequencing physical map;
- complete sequencing of BAC selected to cover the genome entirely;
- progression along the chromosome using the presequencing map or with the help of BAC extremities;
- *ab initio* annotation.

The Yeast Artificial Chromosome (YAC) libraries with inserts of 500 kb (and up to 2 Mb later) constructed for the human genome proved to be much less useful than expected, due to production of chimeric DNA and to the instability of the cloned DNA; they have been replaced by BAC libraries, which are very stable and in which the length of the insert is more modest. The first BAC libraries were of the order of 100 to 120 kb; those currently in use are between 150 and 200 kb. Vectors of the BAC type are practically always single copy (1 molecule/bacterium), which causes problems of limiting quantity in high throughput production (culture volumes are small); this is the price to pay for stability of the insert. The presequencing mapping, which will be presented here, requires a rather high level of coverage between neighboring BACs, which necessitates the construction of BAC libraries of about 25 genome equivalents (25 x redundancies, or coverage in laboratory jargon). The libraries are then stored in 384-plaque microtiter plates.

For the complete sequencing strategy, the choice of BACs to be sequenced depends on the establishment of a presequencing physical map; this map consists of a continuum of BACs to be sequenced from one telomere to the other on each chromosome of the organism. It should be remembered that the length of the chromosomes is from about 20 to 300 Mb in plants and animals. The international public consortiums have adopted the following procedure, with a few variations. The BAC library (25 x approximately) is used to construct two supplementary resources: (i) sequences of the extremities of the BACs (BAC ends), and (ii) the Hind III restriction profile of the BACs (Fig. 2.1). These resources are essential for the establishment of the order of the BACs. For the first of these resources, each BAC DNA (purified at the "miniprep" level) is sequenced on its two extremities (single reads, often called "single pass") over a length of 500 bps to 900 bps depending on the type of automatic sequencers used (Fig. 2.1a); this leads to a database which includes all the BAC end sequences (*Bac End Sequences* or BES). The information thus generated is used later, but it should be noted that this already produces about 10% of the genome sequence in the form of very small fragments, linked two by two, dispersed throughout the genome, with no known localization at this stage. Furthermore, one can look for homologies to BAC end sequences *in silico* using a BLAST-type program (each BAC end is compared with all the others, Altschul *et*

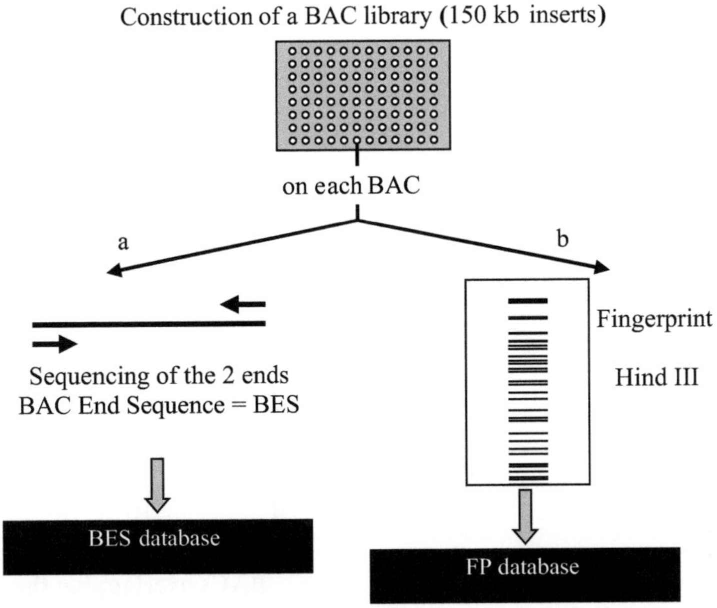

Fig. 2.1 Fig 1 : Construction of the two supplementary resources from the BAC library : a) database of the BAC end sequences ; b) database of Hind III fingerprints

al., 1990), to obtain a catalogue of repeated sequences, which is already rather complete (the selection takes the redundancy of the library into account). The second resource consists in establishing the Hind III restriction profile of each BAC (Fig. 2.1b); this restriction endonuclease produces an average of 30 bands per BAC and this operation is conducted at the rate of 40 clones per 20 cm x 20 cm electrophoresis plate, with image acquisition by CCD camera and treatment by a program to standardize the migrations. The ensemble of images, together with data on the speed of migration of the different bands for each, constitutes a database of Hind III fingerprints.

The construction of the presequencing map begins with the utilization of the FPC (Finger Print Contigs, developed at the Sanger Center, UK) program (Soderlund *et al.*, 1997). Starting from one BAC, FPC screens the database containing the fingerprints of all the BACs and selects for all BACs that show at least 50% of bands in common with the starting BAC, i.e. selects all the BACs which overlap it (Fig. 2.2). Each candidate BAC is then taken as a starting BAC and this recurrent progression, like a sliding window, establish the order of overlapping automatically, progressively constructing a coverage pathway (*"Tiling Path"*) (Fig. 2.3a and 2.3c). The deepness of the piles increases with increasing redundancy of the library. It is important to keep in mind the lack of precision of the values for distance of migration for the restriction fragments on the gel. Also the doublets (bands too close together) are poorly detected, and too small fragments (< 500 bps) can be lost. In addition, there is a not negligible probability that two non-overlapping BACs will show fortuitously, 50% of bands for which the speed of migration is not considered as significantly different (admitted margin of error). Therefore, for all these reasons, the propositions of this electronic organization of BAC

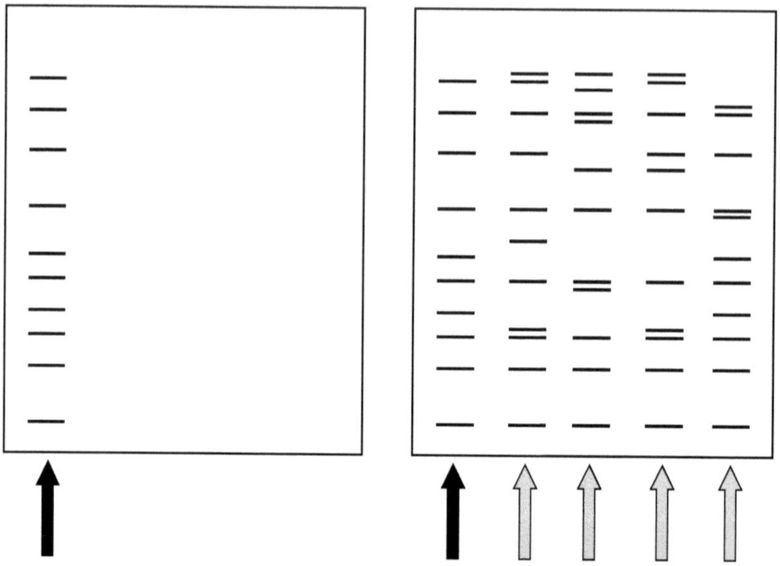

Fig. 2.2 First steps in the FPC electronic contigation of the BACs for the construction of the physical map. a) A "seed" BAC (nucleation) is selected to start the process and the Hind III restriction electrophoregram is displayed on the screen; b) the software searches in the database all the BACs whose restriction electrophoregrams show at least 50% of the bands aligned with those of the nucleation BAC. Theses BACs are therefore overlapping with the nucleation BAC. The FPC software automatically designs the overlapping scheme by using each of these candidate BACs as a new nucleation BAC; this reiterative process corresponds to a sliding window and the physical map is progressively generated (tiling path, cf Fig 2.3a).

contigs contain some regions which are not robust. Since it is not possible, for reasons of cost and time, to sequence all the BACs of the tiling path, it is necessary to determine the minimal tiling path (Fig. 2.3b); there are several strategies available for this purpose. A few groups have made a direct selection on the contig by visually choosing the BACs for their minimal overlap. However, the actual length of the overlap, which appears on the contig is not objective and remains rather blurred at this stage. It is preferable to perform a fine analysis of the Hind III restriction profiles and often to perform Southern hybridizations as controls; the STC strategy, which will be described later, is also of great value. At this stage, the fingerprints are also used to verify the integrity of the BACs chosen for sequencing.

Alternatively, when there is by chance a genetic marker of known sequence anchored on a BAC, the overlaps may be verified by PCR on neighboring (overlapping) candidate BACs in the local region. An example of a minimal tiling path 10 Mb in length is presented in Fig. 2.3d. The anchoring of markers from the genetic map on the physical map constitutes an appreciable security; this operation is generally performed by hybridization, either globally using a set of Nylon filters on which the DNA of each BAC of the library is placed (spotted), or in a targeted manner, by Southern blot or by PCR, on a region for which a tiling path, either minimal or not, has been determined.

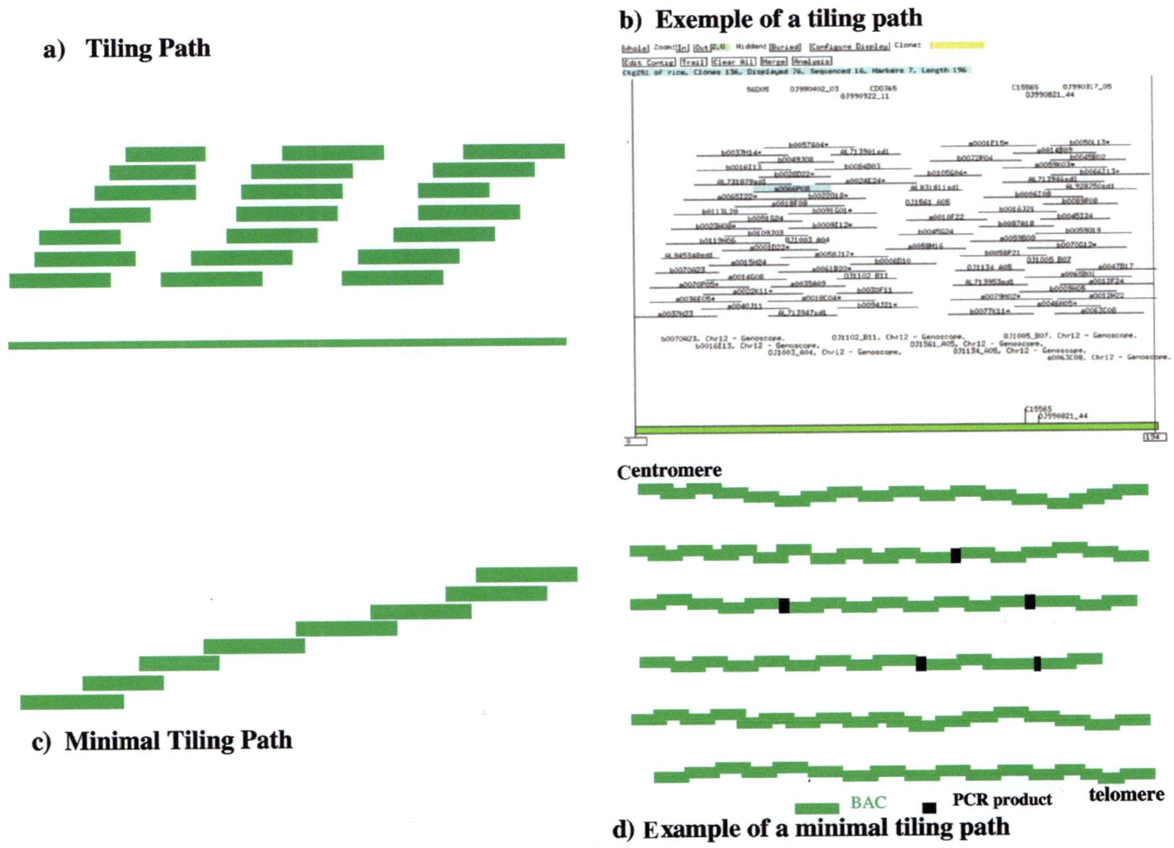

Fig. 2.3 Tiling path (TP) and minimum tiling path (MTP) : a) Basic scheme of the tiling path generated by the FPC software; b) screen display of a tiling path ; a part of the short arm of *Arabidopsis thaliana* chromosome 3; c) the minimal tiling path corresponding to the tiling path shown in a, reducing the number of BACs from 21 to 8 ; d) MTP covering completely the 10 Mb of the short arm of *A. thaliana* chromosome 3, with 102 BACs and 5 PCR products (for these later ones, the corresponding overlaping BACs were identified, but the sequencing of PCR products is cheaper and faster, while being sufficient to fill-in these short gaps).

The sequencing itself is performed BAC by BAC. The DNA of each BAC is purified, then cut mechanically, either by ultrasound (nebulization) or by shearing in a syringe (Hydroshear*) to obtain fragments of the required size (between 2 and 10 kb). These fragments are sub-cloned into a sequencing vector of the multicopy plasmid type, using lacZ selection; after electroporation into the bacteria, recombinant colonies are selected with the help of "colony-pickers" at high throughput until a coverage of 10 x is obtained. The preparation of DNA (often called matrix or template in the large sequencing centers) is done after culture in 96-well microtiter plates and purification of plasmids using a "miniprep" protocol. The two ends of the recombinant plasmids are sequenced on a length of 500 to 900 bps; the sequencing reactions use fluorescent primers ("*dye primers*") or fluorescent ddNTPs ("*dye terminators*") depending on the case; the separation is performed on automatic equipment (it is difficult to provide an exhaustive list of apparatus, and this has changed a lot during the last few years). The useful parts of the electrophoretic sequences are selected after elimination of sequences from the sequencing vector and cutoff of the sequence at the place where the reliability of the identity of the bases becomes mediocre. For this purpose, the identification of each base, or "*base calling*" is carried out by programs such as Phred, which assigns a score which becomes higher as the reliability of base calling increases; a minimum score of 20, which indicates a probability of correct identification of 99%, is required. Arranging the sequences to form a contig consists in assembling these short sequences to reconstitute the complete insert of the BAC; this operation depends on algorithms for sequence similarity on a sliding window, such as Phrap (so-called assembly programs). The presence of repeated sequences may lead to false assemblies. To remedy this inconvenience, it is possible to mask these repeated sequences by applying a counter-selection filter (a database called "*repeat masker*" which collects repeat sequences from a database of BAC end sequences, see above). The assembly of a BAC sequenced in this way produces a skeleton or "*scaffold*" which contains less than about 20 gaps on average. This BAC is then subjected to a "*finishing*" phase which consists of i) filling gaps by sequencing PCR products, or by direct sequencing of DNA from the BAC using oligonucleotides developed from the portion of the sequence which is already available, or by using specific primers on sub-clones and ii) polishing regions where the sequencing quality is inferior to the average admissible error (by additional reads by PCR). The verification of the BAC assembly is performed by comparing the theoretical digestion profile from the finished sequence, with the experimental profile with up to six different endonucleases. The up-to-date strategy is now to raise the shotgun sequencing of a BAC up to 12 x (the sum of all sequences is 12 fold the length of the BAC). Using such a high shotgun sequencing coverage for the initial assembly results in there being only two or three gaps; the finishing step, which is more expensive than shotgun, is thus dramatically reduced. The finishing of BACs encompassing repeats (microsatellites, transposable elements) is rather tedious and most large scale sequencing centers have choosen to use a transposon tagging protocol.

Sequencing progresses BAC by BAC, following the minimal tiling path previously established. In reality, many sequencing projects begin without waiting for the establishment of the minimal tiling path; these use a so-called STC (*Sequence Tagged Connectors,* also referred to as "sequence and go") strategy (Venter *et al.*, 1996); when a BAC has been completely sequenced, STC identifies the candidate BACs that overlap the former one. To do so, the end sequence of this sequenced BAC is compared to all the

BAC ends from the BAC end sequence database; this selects the BACs for which one BAC end sequence is anchored in the end of the completely sequenced BAC. It is necessary to ascertain that the query sequence, namely the end of the completely sequenced BAC, does not carry a repeated sequence that will lead to false overlaps; this check is carried out by a comparison of the query sequence against the repeat database (see above). Using the library with 25 x redundancy, there are usually several BACs which are candidates for extension. The BAC with the smallest overlap with the starting BAC, and in which the insert is longest (in the Hind III fingerprint database) is selected (Fig. 2.4). The same operation is performed on the other extremity, which permits a bi-directional progressing of sequencing from the starting BAC.

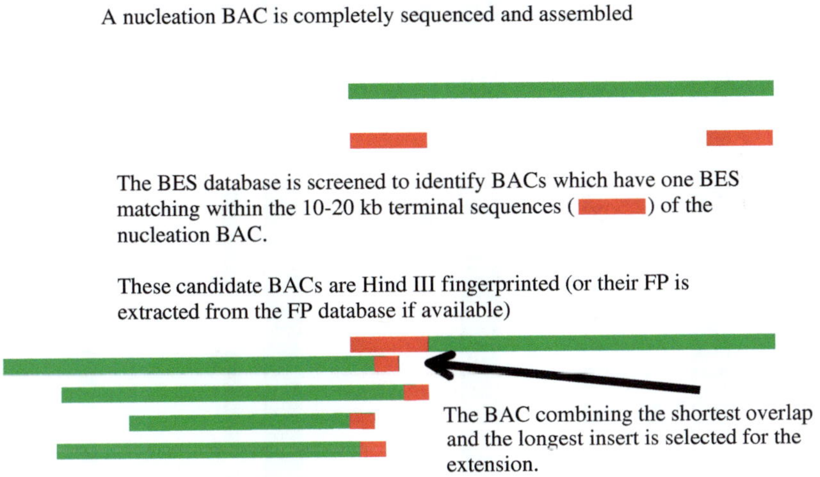

Fig. 2.4 The « sequence tag connectors » strategy (also called « sequence and go). The progression starts with "nucleation" BACs in a bi-directional process (for simplification, only the left extension is shown).

The STC strategy always begins with the development of a set of nucleation BACs (or *"seed BACs"*); these BACs should be evenly spaced along the chromosome to be sequenced, as equidistant as possible (Fig. 2.5). This is the most delicate phase, which requires a reliable anchorage of BACs on the physical map, by hybridization on a filter, by Southern blot, or, most frequently, by PCR; the use of the genetic map for anchoring of the BACs on the physical map is primordial. The number of BACs in the starting set depends on the capacity of the sequencing center. These BACs are then completely sequenced and assembled; for each of these nucleation BACs, the STC is carried out for the two ends, giving rise to a bi-directionnal progression. The process is carried out in a recurrent manner, taking then into account one end only (Fig. 2.5); there is no need (or real possibility!) to synchronize the progression for each of the nucleation points. However, if two selected nucleation BACs are too close to each other, the meeting point will occur very rapidly and two progression points will disappear (Fig. 2.5), slowing down the execution of the project. The utilization of the radiation hybrid (RH) map of the human genome has permitted the constitution of an "acceptable set of nucleation BACs in some sequencing centers: for example, the 88 Mb of human chromosome 14 sequenced by Genoscope (Heilig *et al.*, 2003) depended almost exclusively on a

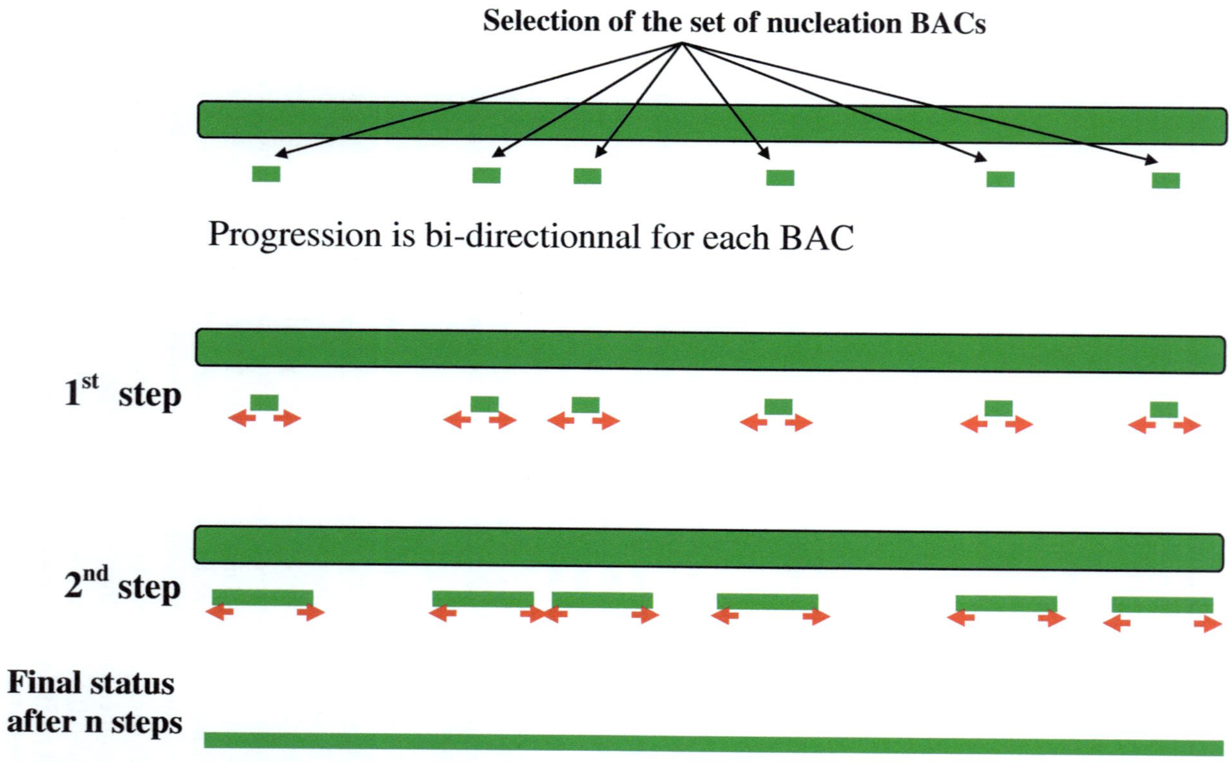

Fig. 2.5 STC progression to encompass whole chromosomes
The STC process is shown here as synchronous for the successive rounds of extension. In fact, the progression is independent and is linked to the rapidity at which each BAC is sequenced and its assembly is achieved.

"Sequence Tag Connectors" strategy starting with a set of 162 nucleation BACs. Unfortunately, this *"RH mapping"* has not yet been applied to the genomes of plants. Finally, in some regions, the local density of BACs in the tiling path is such that only one BAC is proposed, and it overlaps too much. In this case in order to avoid the complete sequencing of a BAC of 120 to 180 kb to only fill a few kb, a junctional PCR product can be produced and sequenced.

Regions which contain many repetitive rDNA units and telomeric and centromeric regions are treated separately. For the rDNA regions, the sequence of the unit has been known for a long time, and it is only necessary to ensure that no other sequences have been inserted among the concatemers. The centromeric regions are constituted of 150 to 200 bp regions, usually arranged one after the other, and when centromeric BACs are sequenced, it is usually done at low depth only, to ensure the absence of genes, without attempting to assemble completely the sequences. Telomeric sequences, on the other hand, necessitate a dedicated effort, which is difficult because of overlapping of sequences other than the short purely telomeric sequences; the sequencing progression may require using recombinant clones other than those from the BAC libraries, which are developed using special techniques, and in all cases numerous PCR products are necessary, and even this does not guarantee complete success.

When the coverage of the chromosome is considered complete, a pseudomolecule (also called *"master molecule"*) is reconstituted by placing the completely sequenced BACs end-to-end, obviously with only one of the overlapping sequences on each extremity. Furthermore, since these overlapping sequences have been sequenced twice, they are used to verify the quality of the sequencing in cases in which the BAC library was derived from a completely homozygous individual (pure line).

"Draft" Type Sequencing (WGS)

Because of the cumbersome nature of complete sequencing based on presequencing mapping, BAC-by-anchored BAC progression and a burdensome finishing phase (*cf.* above), Craig Venter and the Celera Genomics company announced in 1998 that they had developed a revolutionary method which did not require any mapping at all, or supplementary resources; genomic DNA is cloned in plasmids using the classical *shotgun* strategy, and the extremities of the inserts are sequenced at very high redundancy. The assembly of the sequence into contigs is then performed using massive computer resources. The technique itself is called WGS (*Whole Genome Shotgun*); it is rapid, less expensive and efficient according to its authors. It was tested by Celera Genomics on the genome of *Drosophila* (Adams *et al.*, 2000), then on the human genome (Venter *et al.*, 2001), and more recently by groups from the private or public sectors for rice (Goff *et al.*, 2002; Yu *et al.*, 2002) (see below), for the blow fish fugu (*Takifugu rubripes* or *Fugu rubripes)* (Aparicio *et al.*, 2002), for mouse (Mouse Genome Sequencing Consortium, 2002, and for *Ciona intestinalis* (Dehal *et al.*, 2002). All the results obtained to date have shown that this type of sequencing is very partial, because of the existence of numerous gaps: for example, initially more than one thousand gaps for the 180 Mb of *Drosophila* sequenced at 15 x, 62 000 gaps initially for the 420 Mb of rice sequenced at 6 x (reduced to 41,000 gaps at publication) and from 150,000 to 200,000 gaps for the human genome. The case of the human genome merits special attention; the assembly of the shotgun sequences without using the mapping resources, despite the claims of Celera Genomics, was not in fact completed in this way by Celera. As shown by Robert Waterson, Eric Lander and John Sulston, (Waterson *et al.*, 2002), Celera Genomics used about 2 x anchored sequences from the public databases of the Human Genome Project to assemble its 5.1 x *shotgun* sequences. No eukaryotic genome has, in fact, been completely sequenced to date using the WGS strategy. The partial sequences obtained in this way are qualified as *"drafts"*; they have the advantage of providing a general description of the genome including a first inventory of genes at reduced cost and in a short time frame, but with a quality which is much lower than by complete sequencing, and the estimates of the number of genes is only approximate. Moreover, it is extremely difficult, and in any case very onerous to later transform these drafts into a completed finished sequence, unless the BAC to anchored BAC approach is used, with a lower coverage for the sequencing of the BACs. In conclusion, the choice between complete sequencing and draft sequencing depends on the quality of information desired.

The last phase in a sequencing program is that of *in silico* annotation. This involves identifying the interesting regions along the length of the pseudo-molecules, or quite simply, the genes. Whereas the boundaries of the genes coding ribosomal RNAs are simple to define because of the conservation of consensus sequences, and that of genes coding transfer RNA are also easy to determine due to the conservation of three-dimensional structure of the encoded transcript, identification of genes encoding the small nuclear and nucleolar RNAs is more difficult to achieve. In fact, conservation is found on fragments which can adopt stem, or hairpin configurations and the detection programs still predict many false positives. However, with the genes which code for proteins, the difficulties increase dramatically. The algorithms for exon finding (Chapters 3, 5) include constraints for successions of sense codons and the GT/AG rules for the beginning and end of introns. The very existence of a whole series of programs such as Genscan, FgeneSH, GeneMark HMM, GlimmerM and Eugene to mention only those which have survived, indicates that none of them are perfect and universal; we must be careful to remember that *ab initio* annotation is predictive, and that each proposition must later be verified experimentally. Furthermore, there are two levels in the quality of the predictions: the first concerns the detection of exons, and the second groups exons together in genes (notion of gene models). Although the prediction of coding sequences is fairly satisfactory (good sensitivity and specificity), the programs are not conceived for the identification of non-coding regions such as the 5′ and 3′ UTRs and are therefore not adapted for the identification of the start and end points of the genes; this leads to predictions which fuse contiguous genes, or create artifactual breaks.

In these cases, the utilization of biological resources such as EST banks may lead to a biological validation. The construction of full-length cDNA libraries and the corresponding sequencing is actually the most useful tool for helping with the annotation, which is then no longer strictly *ab initio*. As a complement, the new Exofish method (Roest-Crollius *et al.*, 2000) for identification of exons which was developed at Genoscope, based only on the conservation of sequence similarities in exons during the course of evolution, may be applied to plant genomes (Castelli *et al.*, 2004).

THE GENOME OF *ARABIDOPSIS THALIANA*

For several years, the scientific community, following the impulsion of a nucleus composed of Chris Sommerville, David Meinke, Joe Ecker, Mike Bevan and a few others, has considered *Arabidopsis thaliana* as the model to be used for the first sequencing of a plant genome; James D. Watson was an enthusiastic supporter from the beginning (Theologis, 2001). There were numerous arguments in favour of this organism, including the smallest known genome size for a flowering plant (estimated at 130 Mb by flux cytometry), the autogamous nature of its reproduction, the rapidity of the seed-to-seed cycle (six weeks), the large number of seeds produced per plant (several thousand), the density of growth in greenhouses and constant environment chambers (the plant is small), the ability to cultivate it *in vitro* and its transformation by *Agrobacterium*. A very consensual supplementary argument was added to all these advantages: it is a plant of no direct agronomic interest, and therefore no country would derive special benefit from this choice. Pilot sequencing programs started in the United States and in Europe at the beginning of the nineties, and the project took off following the ESSA (European Scientists Sequencing *Arabidopsis*) project for the sequencing of chromosome 4 led by Mike Bevan (John Innes Center, United Kingdom), which involved a consortium of European laboratories with funding from the European Community. The National Science Foundation from USA also set up a sequencing programme in 1995. The history of the preparatory phase and the beginnings of the programme have been exhaustively described by Sakis Theologis (2001). In August 1996, the Arabidopsis Genome Initiative (AGI) was created; it included public or non-profit laboratories from the United States, Europe and Japan. The five chromosomes which constitute the *A. thaliana* genome were distributed to the various members of the consortium (Fig. 2.6). The AGI target was to obtain the complete genome in 2004; the quality should correspond to a maximum of 1 error in 10,000 bases, and the level of finishing to a single contig per chromosome arm. In the initial plans, it was agreed that neither the centromeres nor the regions with DNA repeats (localized on chromosomes 2 and 4) would be sequenced. The six sequencing groups also agreed to place their results in public databases as quickly as possible.

The resources at the beginning included two BAC libraries with inserts of 120 kb (Rod Wing, Texas A&M University, USA and Thomas Altman, IGF, Germany) for a total of 23,000 clones (18x), a physical map in YAC vectors (CIC [for CEPH/INRA/CNRS] YAC library with low chimerical clones), a genetic map including about 800 markers and an EST database which was growing constantly (90,000 entries at the end of the project). For the complementary resources described above, a collaboration between three members of the AGI (The Institute for Genomic Research [TIGR] at Rockville, Madison, USA, Genoscope at Evry in France, and the University of Pennsylvania, USA) generated 42,000 BAC end sequences (for a 91% success rate compared to the 46,000 expected), and Marco Marra *et al.* (Washington University) performed Hind III fingerprinting of more than 21,000 BACS and, using the FPC program, progressively established a tiling path. The six sequencing groups had each started to sequence several BACs to familiarize themselves with the technologies, to optimize the parameters and to evaluate the difficulties in assembly. These BACs have served as nucleation points. The methods of progression have been rather heterogeneous; for example, in the

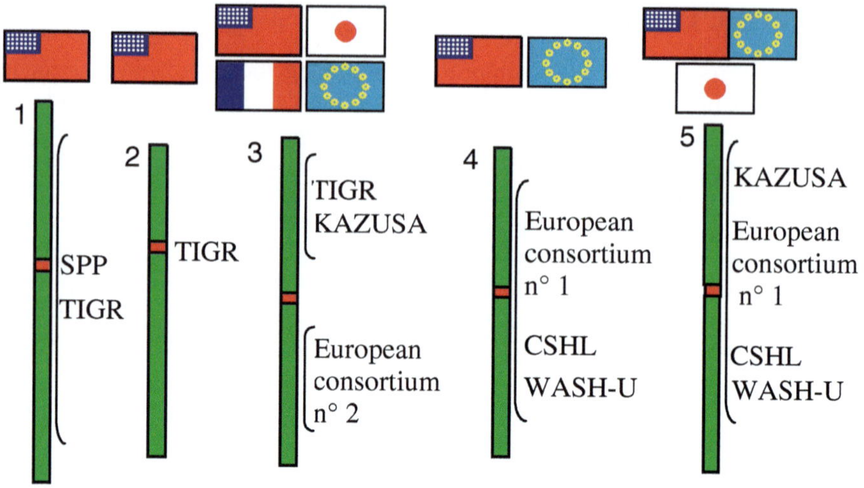

Fig. 2.6 Chromosome allocation for the *Arabidopsis thaliana* genome sequencing.
SPP Consortium
Stanford DNA Sequencing & Technology Center, Stanford University,
Arabidopsis thaliana Genome Center, University of Pennsylvannia
Plant Gene Expression Center, USDA
TIGR, The Institute for Genomic Research, Washington DC
KAZUSA DNA Research Institute, Japan
European Consortium n°1(coordinator : John Innes Center, Norwich)
European Consortium n°2 (coordinator : Genoscope, Evry)
Wash-U, Washington University, St Louis
CSHL, Cold Spring Harbor Laboratory, NY

European Consortium #2, led by Genoscope and including 14 laboratories, those with a small output (about one BAC per year) concentrated on a region with a minimum tiling path, whereas those with large output (4 BACs per year or more) advanced by the *"Sequence Tag Connector"* strategy, constructing their minimal tiling path as sequencing progressed.

Meetings of the AGI were programmed twice a year to ensure good progression of the project (Steve Rounsley then Samir Kaul, Rob Martienssen, Dick McCombie, Joe Ecker, Ron Davis and Sakis Theologis for the United States; Mike Bevan, Marcel Salanoubat and Francis Quétier for Europe; Satoshi Tabata for Japan).

During the colloquium in Melbourne in September 1999, a more rapid progression than initially anticipated made it possible to advance the target date to the end of summer 2000. Chromosomes 2 and 4 were published in December 1999 by TIGR (Lin *et al.*, 1999) and the European Consortium #1 led by Mike Bevan (Mayer *et al.*, 1999). The extra time was used to work on the centromeres: Daphne Preuss joined the AGI and coordinated the data. Several meetings were organized at the Munich Information Center for Protein Sequences (MIPS) between September and November 2000 for the analysis of results and the writing of the final article, which was published in December 2000 (Arabidopsis Genome Initiative, 2000).

The annotations of BACs were performed by each sequencing group, except for the European consortia; the European Commission entrusted this task to the MIPS (Hans

Mewes and Klaus Mayer). For a very long period, many groups annotated not only the regions which they had sequenced themselves, but also re-annotated regions which had been obtained by other groups and placed the results on their own websites; this lack of homogeneity took a long time to correct, and finally TIGR and MIPS agreed on common parameters for annotation (see, for example, the MIPS website, Fig. 2.7).

The results of the sequencing of the *A. thaliana* genome have been widely analyzed and commentated (The Arabidopsis Genome Initiative, 2000). The 115 Mb of very high quality sequence (average of less than 1 error in 20,000 bps) required the sequencing of 1569 BACs or PACs and 22 PCR products. When the 10 Mb of rDNA and centromeric regions are taken into account, the total size of the genome is 125 Mb. Important aspects of the results included the general organization of the genome; statistics on the length of the genes; exons and introns; distribution of mobile elements, etc. The most striking findings involved the number of predicted genes (a little over 25,000), which, at the moment of publication, was the largest number of genes yet described for an organism. The fact that the smallest genome in plants turned out to be highly duplicated (63% of the genome corresponded to segmental chromosome duplications) and the existence of about 50 unique genes in the centromeres, as revealed by the presence of ESTs was surprising. This project also led to a precise inventory of gene families, the identification of genes which appeared to be specific to plants and the discovery of a large range of membrane transporters. An exhaustive identification of transposable elements led to 2,000 class I and 2,000 class II elements. Furthermore, since the chloroplast (Sato *et al.*, 1999) and mitochondrial genomes (Unseld *et al.*, 1997) of *A. thaliana* had also been sequenced, the complex relationships between the three genomes of a plant could be better defined. In particular the number of nuclear encoded proteins necessary for chloroplast or mitochondrial function has been estimated at about 2,000 for each organelle and pieces of plastid and mitochondrial DNA transferred during evolution into nuclear chromosomes have been identified.

All these observations on *A. thaliana* are going to be compared with the information obtained for rice (see next section), in order to identify what is in common (such as the general machinery), and what is specific to plants in general, and also which are specific to each of these two plants.

Detailed information is displayed on http://www.arabidopsis.org

THE RICE GENOME

Since the first sequencing of a plant genome was performed on a dicotyledonous plant, it was logical to choose a monocotyledon for the next plant. Rice was an obvious choice; this cereal is in fact the most widely cultivated plant on the planet (it nourishes about half of the human population) and at the same time its genome is estimated at only 430 Mb, which makes it a "compact" model for other cereals such as maize or wheat, for which the genomes attain 3,500 and 16,000 Mb respectively. The sequencing of rice had already begun to be seriously envisaged even before the sequencing of *A. thaliana* was finished. The IRRI (International Rice Research Institute) and especially the RGP (Rice Genome Research Programme) of Japan began to develop the necessary foundations. In 1999, the RGP was joined by about 10 other countries to constitute the IRGSP (International Rice Genome Sequencing Program), a public consortium which had as its

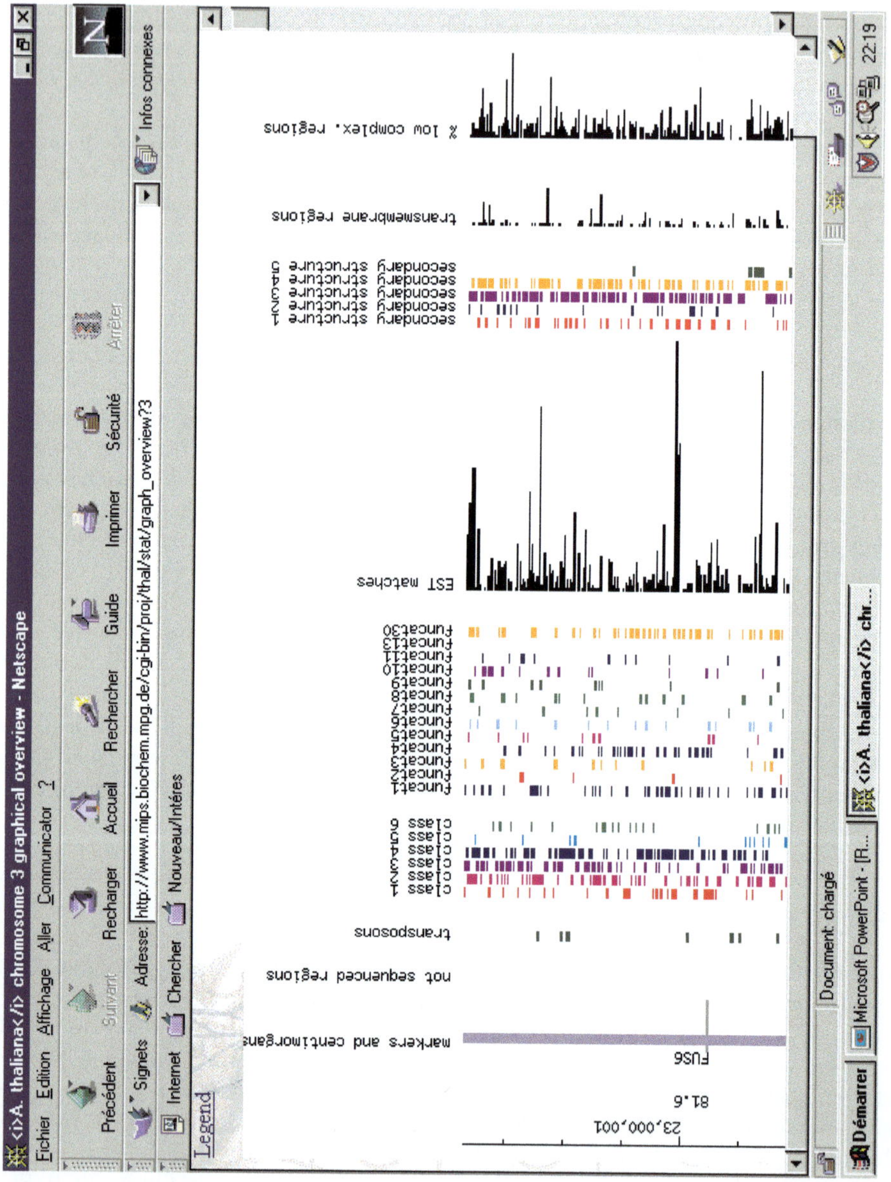

Fig. 2.7 An example in the *ab initio* annotation of *Arabidopsis thaliana* genome. Screen display (early 2000) of one end of *A. thaliana* chromosome 3 on the MIPS website (Münich, Germany).

goal the exhaustive sequencing of the Nipponbare variety of *Oryza sativa* L. *ssp japonica*, with a final quality of the "Bermuda" type, equivalent to that achieved for *A. thaliana* (a very small number of gaps, and an average of less than 1 error in 10,000 bps). The quasi-totality of the 12 chromosomes was thus distributed to the members of the IRGSP. Several modifications occurred between 2000 and 2001: the absence of financing from Canada and the European Community caused the suspension of work on chromosome 2 (Genoscope remained the only participant in Europe, it sequenced chromosome 12 with financing from the French Research Ministry) and Japan increased its participation considerably such that the ensemble of chromosomes could be covered. The final distribution of the sequencing in the member countries of IRGSP is shown in Fig. 2.8.

The resources of IRGSP were dependent on the BAC libraries constructed by Rod Wing's team (Clemson, United States, financing from the Nobel Foundation), which represented 25 genome equivalents, and the YAC/PAC libraries created by Takuji Sasaki's group (Tsukuba, Japan). The strategy was modelled on that of *A. thaliana*, with BAC end sequencing and Hind III fingerprinting of 97,000 BACs (Rod Wing, Clemson, using funds from the Novartis Foundation). The physical presequencing maps combined maps of YACs (only for some of the groups; the others considered that this approach was not sufficiently robust), and especially relied on the electronic contigs assembly of BACs by FPC 200 (Cari Soderlund, Clemson). Each IRGSP group constructed its own presequencing map to create a minimal tiling path which was more or less extensive, using approaches which varied as a function of the date of their entry into the massive sequencing phase; indeed, the fact that data from two partial sequencing efforts by the Monsanto Company and then Syngenta (Novartis) had been made available to the IRGSP increased the resources that could be used. In April 2000,

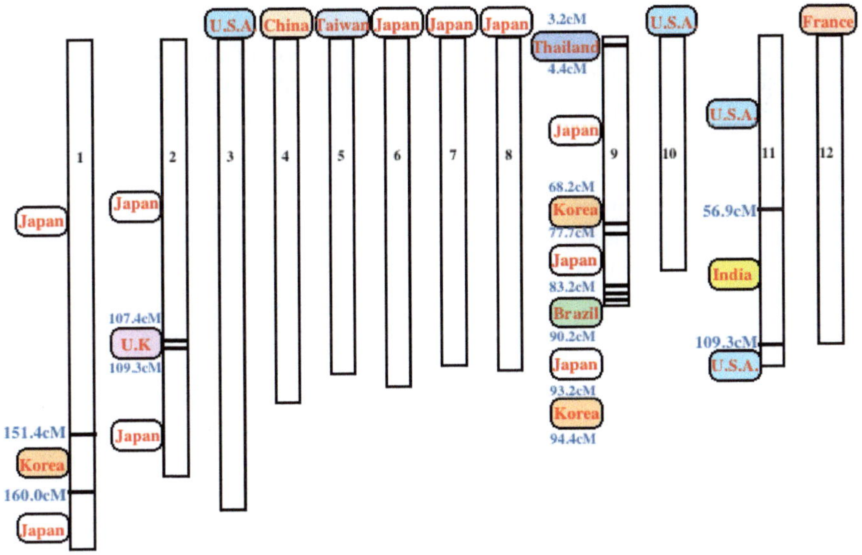

Fig. 2.8 Chromosome allocation for the IRGSP sequencing groups (November 2002)

Monsanto announced the complete sequencing, which in reality was only partial: starting with a Hind III BAC library of 76,229 clones (21 x) and 135,000 BAC end sequences, Lee Hood (Seattle) performed partial sequencing for the company at 5 x using 3434 BACs (at this stage, each BAC sequence contained an average of 30 gaps) and the data became available in 2000 and 2001 to members of IRGSP who signed agreement protocols separately. This agreement stipulated that the members of IRGSP could deposit the sequence of each Monsanto BAC in public databases as soon as complementary sequencing by an IRGSP member reached a minimum quality level (only one gap per 500 bps and known orientation of 2 contigs). IRGSP members could not divulge the Monsanto results to third parties; contrary to erroneous information published by *Nature* (Adam, 2000), Monsanto also would not have access to any IRGSP results before deposit in the public databases.

In January 2001 it was the turn of Syngenta to announce complete sequencing, which was in reality also partial, as was the preceding. This was a 6 x *shotgun* sequencing effort, which left 62,000 gaps at the time, and therefore corresponded to a *"draft"*, which was published in April 2002 (Goff *et al.*, 2002). The sequence contigs were made available to the members of IRGSP in May-June 2002, *via* an agreement protocol very similar to that concluded with Monsanto. At the same time that this draft of *Oryza sativa* L. *ssp. japonica* became available, a group from Beijing, China, published a "draft" of another subspecies of rice, *indica* (Yu *et al.*, 2002), sequenced at 4x and for which the sequences have been deposited in Genbank.

The progression of the complete sequencing directed by the IRGSP took advantage of all the available resources. For example, Genoscope in France, which was in charge of sequencing chromosome 12, used about 10 BACs for which the chromosome assignment had been verified by Michel Delseny's group (CNRS, Perpignan University, France) as initial anchoring points; these were rapidly increased by 148 BACs from Monsanto, which were anchored on chromosome 12 and elevated to "Bermuda" quality by complementary sequencing. The progression of the contigs on each side used physical mapping data (established on Clemson BACs) when they were available (there were gaps, and periodically, substantial modifications were needed). In addition complementary approaches were employed, including *"Sequence Tag Connectors"* (BAC end sequences), verification by PCR and Southern hybridization and identification of BAC borders by hybridization of probes (generated at the ends of the contigs) on filters containing DNA from the BACs. After all these possibilities were exhausted, probes generated from 6,000 ESTs anchored by Takuji Sasaki's group (Wu *et al.*, 2002) (Tsukuba, Japan) and localized outside of the contigs, permitted a few supplementary extensions. As a last step, contigs of the Syngenta sequences were used to try to fill the gaps or to reduce their size. Thus, the utilization of complementary resources generated by exterior groups was difficult and rather time-consuming.

In December 2002, IRGSP announced the achievement of a very high quality *"draft"* covering more than 90% of the genome: 3,665 BACs or PACs had been sequenced, making a total of 509.34 Mb, which was 118.2% of the estimated size of 430 Mb estimated for rice (the value of over 100% is due to overlapping sequences between adjacent BACs, which had not been removed). The sequences were either finished, or in phase 2: the orientations and order of the sequence contigs was known. It should be noted that this was not a *Whole Shotgun Sequence* type of *"draft"* because the pieces of the puzzle had already been positioned on the genetic map of rice; by using the 13,895 available markers (13,246 from the Japanese Rice Genome Program and 649 from the Cornell Rice Genome Database). 12,725 sequence markers from rice enabled *in silico* anchorage of 3,180 of the 3,665 BAC/PACs, or 87%. In fact, this type of "*draft*" is similar to that produced by the Human Genome Project consortium for the human genome in June 2000 (Fig. 2.9).

The completion of sequencing took place in early 2005. Filling the remaining gaps was helped considerably by two complementary techniques; the first one involved the hybridization of BACs on chromosomes at the beginning of the pachytene stage, a technique called BAC-FISH (Fig. 2.10), and the second one used hybridization of stretched fibers of DNA (fiber-FISH). The latter technique, which is very precise, had already been used in the United States in an expert collaboration with Jiming Jiang (University of Wisconsin, United States) to estimate the size of the small sequencing gaps. Sasaki's group carried out the sizing of the centromeres for each of the 12 chromosomes. At the end of this sequencing program, the size of the rice nuclear genome was established as 395 Mb.

Concerning the annotation, the progress achieved at the end of the *A. thaliana* program and the optimizations tested for rice indicate that the quality of the *ab initio* annotation should be robust; nevertheless, grasses have a higher percentage of GC, with intragenic heterogeneity (with a decreasing gradient from the 5′ end to the 3′ end), which requires an adaptation of the algorithms for exon identification. As previously mentioned, the Exofish approach has been adapted to the *A. thaliana*-rice pair as a complement to existing methods. Based on the lessons from the sequencing of *A.thaliana*, each major step in the annotation is performed by a single group; furthermore, the results have been collected on only two sites, that of the National Institute of Agrobiological Sciences [NIAS] at Tsukuba in Japan, and that of TIGR in the United States. An annotation jamboree, RAP1, was organized in November 2004 and led to the prediction of 37,544 protein coding genes (non-Transposable Element containing genes). The final publication appeared on August 11th 2005 (IRGSP, 2005), presenting the major traits of the second plant genome completely sequenced. A second annotation jamboree, RAP2, is scheduled for February 2006 to prepare an updated version of the gene content. The group leaded by Takuji Sasaki produced a large piece of work for the Japan emblematic plant.

Detailed information is displayed on http://rgp.dna.affrc.go.jp/IRGSP

THE MAIZE GENOME

Two unfavorable characteristics of the maize genome worked *a priori* against the sequencing of this cereal: the first was the enormous size (2,700 to 3,200 Mb depending on the strain) and the second a very high content of repeated sequences (80 to 85% of the genome), massively implicating mobile elements. Maize is certainly the largest "graveyard " of retrotransposons which constitute between 65 and 70% of its genome. On the other hand, it is undoubtedly the best known genetically of all the plants, and is extremely important economically in some countries such as the United States.

The resources available consisted of three BAC libraries, produced with three different restriction endonucleases, for a total of 400,000 clones (26 x coverage), and of

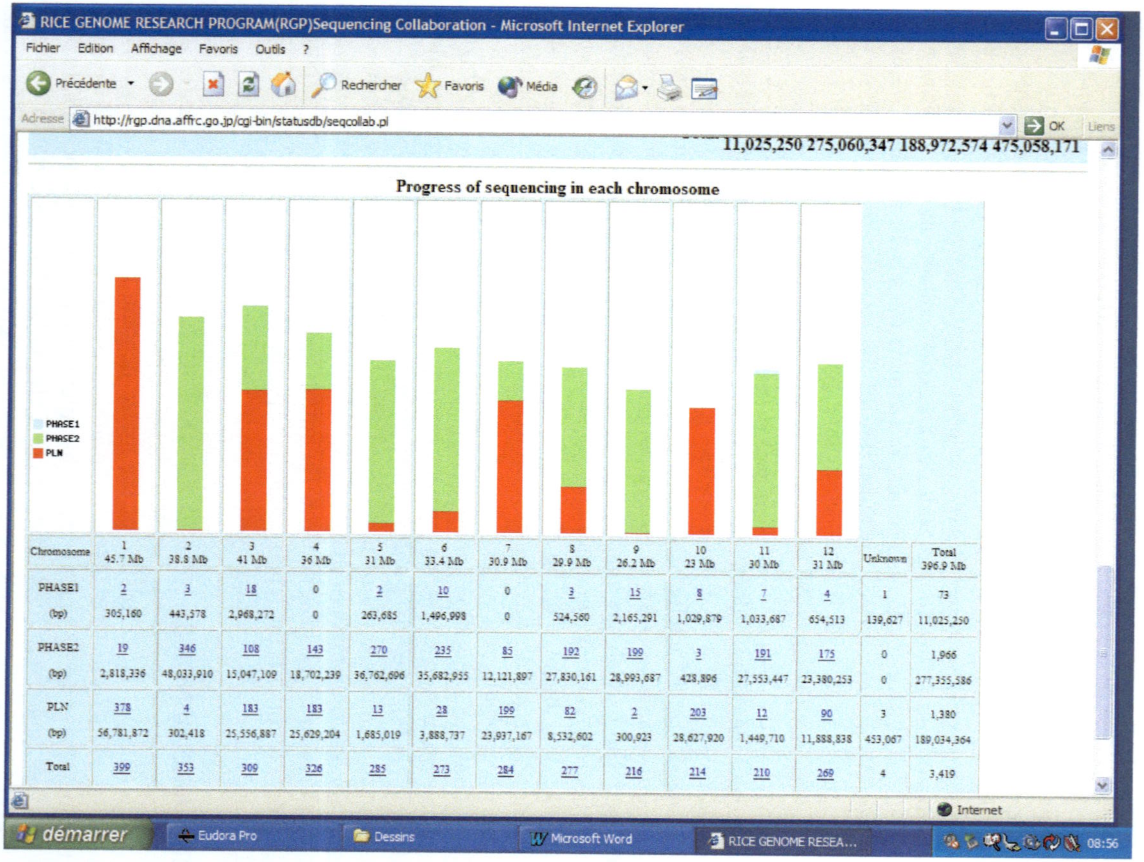

Fig. 2.9 Graphical view of the sequencing status for the Rice genome in March 2003
The sequencing status was regularly updated on the website http://rgp.dna.affrc.go.jp/cgi-bin/statusdb/irgsp-status.cgi. Phase 1 and phase 2 correspond to steps in partial sequencing, PLN is the GenBank section collecting finished sequences for plants, fungi and algae. The sequencing was completed in mid 2005.

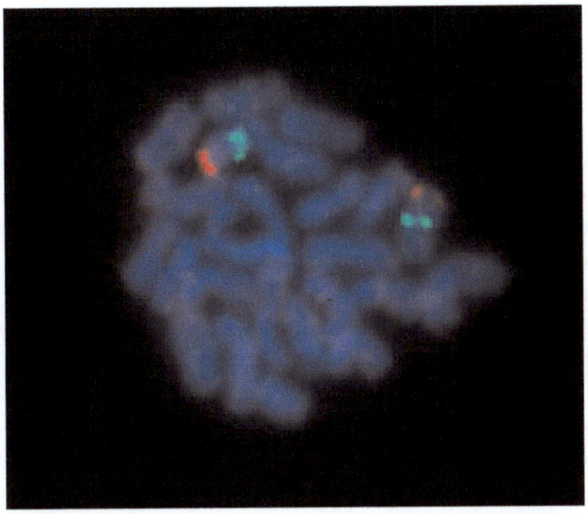

Fig. 2.10 BAC-FISH hybridizations
An example of BAC FISH hybridization on early pachytene rice chromosomes. Results confirm that the 2 candidate BACs are located on chromosome 12 and give several helpful complementary indications : which of the two arms, a rough estimate of the BAC positions, whether the two BACs are very close each other or far away, which one is the most proximal to the telomere (important information when two BACs are very close each other) (picture Angélique D'HONT, CIRAD, Montpellier, France).

course a very dense genetic map with 4,800 markers which can be anchored on a physical map; the EST libraries contained 568,000 entries.

A colloquium organized by the NSF in July 2001 in St. Louis (United States) examined possible strategies for sequencing only the gene-rich regions of the maize genome. The first possibility, proposed by Rob Martienssen's group, was based on the fact that the level of methylation is less in gene-rich regions compared with other sequences (retrotransposons, centromeric sequences, etc.). By using MCR strains of *E. coli* which do not propagate methylated DNA, genomic libraries which are very enriched for genes may be constructed; alternatively, genomic DNA digested by restriction endonucleases which are sensitive to methylation may be cloned in normal strains to obtain a similar enrichment. A second strategy is dependent on the kinetics of reassociation (Cot) and was utilized by Jeff Bennetzen's group. DNA is mechanically cut (Hydroshear®) into small fragments of 1-2 kb, which are denatured and then subjected to renaturation up to a value of Cot = 500. The single-stranded DNA is purified by chromatography on hydroxyapatite, allowed to renature to the double-stranded state (hexameric primers and DNA polymerase) and then cloned. Pilot experiments performed using these two strategies demonstrated a very substantial enrichment in genes; perfect filtering is still a long way off (one can hope to obtain about 80% of the genes).

The now-classical approach of BAC-end sequencing and electronic contig assembly by Hind III fingerprinting has been used; the former is completed and the later is about to be released (Cari Soderlund *et al.*, Arizona Genomics Computational Laboratory(AGCoL), Tucson University). The MTP will presumably contain over 19,000 BACs and the agarose fingerprints data have been implemented in complex areas by using a much more sensitive method, based on a combination of two restriction endonucleases to produce about a hundred fragments (identified by fluorescent markers and separated on capillary sequencers using a 96-format).

The sequencing program for the maize genome has been carried out exclusively by groups from the United States. Two consortia began a pilot study: one with Jo Messing (Rutgers University), Rod Wing (Arizona University), Ed Coe (University of Missouri), Mark Vaudin (Monsanto) and Steve Rounsley (Cereon); the other included Jeff Bennetzen (Purdue University), Karel Schubert and Roger Beachy (Danforth Center), Cathy Whitelaw and John Quackenbush (TIGR) and Nathan Lakey (Orion). These two pioneering programs have now been extended by a massive USA program (32 M$ from NSF, USDA and DOE) led by Rick Wilson (Washington University), which includes researchers from Cold Spring Harbor Laboratory, Iowa State University, University of Arizona, University of California, Berkeley, the Department of Energy Joint Genome Institute, University of Georgia and Stanford University. The sequencing strategy will be a hybrid between a BAC by BAC approach and a WGS: the 19,000 BACs of the MTP will sequenced at a 6 x coverage and directed reads (primer walks) will be added to improve quality and coverage of gene islands within each BAC. A complementary WGS layer will be provided by fosmid end sequences (FES, similar to BES), as well as by the methyl-filtration and high-Cot sequences obtained in the two above-mentioned pilot programmes (Rick Wilson, personal communication). The final quality will correspond to a Bermuda quality for all gene-islands and to phase II for inter gene-islands areas. According to the partial results gained so far, the maize gene content is estimated to be 50,000-60,000.

In addition to this huge programe, carried out on the popular B73 line, a second USA grant is devoted to the sequencing of one chromosome from a different cultivar.

Information: http://usinfo.state.gov/gi/Archive/2005/Nov/22-57172.html and http://mednews.wustl.edu/news/page/normal/6180.html?emailID=7210

THE GENOME OF *MEDICAGO TRUNCATULA*

At the same time, a project for the sequencing of the genome of a model legume was strongly considered. Besides the purely scientific arguments such as the importance of legumes in human and animal nutrition (they are just behind grasses) and the exhaustive study of genes implicated in symbiotic nitrogen fixation, issues relating to the ecology of the planet have been emphasized, quite logically. Indeed, the intensive use of nitrogen-containing fertilizers continues to increase and has created two serious problems. The first is linked to the mode of production of these nitrogen-containing fertilizers; the chemical process requires the burning of about two tons of fossil energy to produce one ton of fertilizer; besides the fact that this contributes to the depletion of fossil fuel reserves, it causes the release of heat and CO_2, which is a factor in the greenhouse effect. The second problem involves the intensive use of these nitrogenous fertilizers which also contributes to the pollution of ground water and has led to a critical situation in many places in different countries. Planting legumes on a larger proportion of arable land therefore appears to be an efficient alternative, which is both ecologically and politically commendable. The scientific community became rapidly focused on *Medicago truncatula*, which is a diploid ancestor of the tetraploid cultivated alfalfa. It is a model legume with a genome of moderate size, about 500 Mb spread over 8 chromosomes. In comparison, soyabean, which is the most widely cultivated legume, has a genome of more than 1,000 Mb. Moreover, a large number of legumes of agronomic interest exhibit a very high level of synteny, combined with very good sequence conservation at the protein level (up to 85% at the amino acid level). The sequencing of a model legume will thus be rapidly beneficial for many legumes of agronomic interest, from all latitudes and longitudes. The legume family is third in taxonomic importance in the plant kingdom, with 650 genera and 17,000 species; many of them store large quantities of proteins. Symbiotic fixation of atmospheric nitrogen seems quite frequent in legumes: out of 3,000 species tested, 90% are able to form root nodules.

The *M. truncatula* resources consist of three BAC libraries for a total of 23 x coverage (Doug Cook, Davis University; Jean Denarie, UMR 215 CNRS-INRA, Toulouse); a genetic map of 1,081 cM with 495 markers and 164,500 ESTs. BES of 90,000 BACs have been produced (Bruce Roe, Noble Foundation, Oklahoma; Genoscope/INRA/UNIP, France; Gilles Olroyd, Chris Lamb, John Innes Center, Norwich and Gyorgy Kiss, University of California at Davis/MBK Hungary).

Although a complete sequencing of the estimated 500-550 Mb genome of *M. truncatula* is quite feasible, based on what has been accomplished for *A. thaliana* and rice, a large portion of the scientific community favored concentrating on gene-rich regions from the beginning, taking advantage of a special characteristic of the *M. truncatula* genome: heterochromatin, which represents more than half of the 500 Mb genome, is mainly separated into large blocks localized on each side of the centromeres; the euchromatin is localized on the chromosome arms in the more telomeric regions and its size is estimated at 200 Mb by Di Aminido Phenyl lndol (DAPI) fluorescence intensity.

The sequencing programe (Nevin Young, PI) has been limited to BACs from the euchromatin at first; sequencing has been performed BAC by BAC, relying on a MTP developed by Dongjin Kim (Doug Cook's group, University of California at Davis)) or using the STC strategy, depending on the sequencing group. The hetero/euchromatic junctions have been localized by FISH hybridization of BACs on early-pachytene chromosomes (Rene Geurts, Ton Bisseling's group, University of Wageningen). In a first step, 750 BACs containing at least one EST were randomly selected and sequenced by Bruce Roe (Oklahoma University, Noble Foundation funding); a large fraction of theses BACs have been used as nucleation points by the sequencing groups. The eight chromosomes have been shared by USA and EU (Oklahoma n° 1,4,6 and 8, TIGR n°2 and 7, Sanger Centre n°3 and GENOSCOPE n°5). On August 2005, although the sequencing program seemed to be 80% completed, only 50% of the ESTs were captured in the sequenced BACs; it is likely that the euchromatin part of the genome has been initially undersized.

Based on the sequencing which has already been carried out, the density of genes in the euchromatic regions would be around one gene in every 10 kb (average value for the gene space).

Detailed information is displayed on http://www.medicago.org/genome

THE GENOME OF *LOTUS JAPONICUM*

The scientific community is also markedly interested in this other model legume which displays attractive features. The 380-400 Mb of its genome are contained in six chromosomes. The sequencing was undertaken by the group led by Satoshi Tabata at the KAZUSA DNA Research Institute near Tokyo: 11 BAC libraries have been constructed, totalling up 20.4 x and a BAC by BAC approach is followed. A genetic map containing 576 markers for 367 cM (AFLP, SSR, dCAPs, PCR products) and 98,000 ESTs are available as well as FL cDNAs. More than 30% of the genome has been sequenced and although the evolutionary distance between the two model legumes is rather large, the sequence comparison between *L. japonicum and M. truncatula* provides helpful information for the progression of the sequencing of both genomes. There is one gene every 11 kb in *L. japonicus*, a value similar to the density reported for *M. truncatula*.

Detailed information is displayed on http://www.kazusa.or.jp

THE POPLAR GENOME

The sequencing of the first tree genome was initiated in 2002 by an international collaboration, comprising DOE-JGI (Dan Rokhsar, USA) which achieved the sequencing, the Oakridge National laboratory (USA), Genome-Canada and the Genome British Columbia/Genome (Canada), the Umea Plant Science Centre (Swedaen) and the University of Ghent (Belgium). The genome size of *Populus trichocarpa*, genotype Nisqually, has been sized at 485 Mb in a set of 19 chromosomes. The sequencing strategy used WGS, only achieved at 8.5 x (7.9 x from 3 and 8 kb plasmids, completed by BES and FES). A physical map obtained by fingerprinting a BAC library followed by a FPC assembly (Marco Mara, Vancouver) has been used to position and orients the assembly contigs. Because of the WGS strategy and its moderate coverage, their present status displays numerous gaps; even if an additional WGS coverage could close some gaps and extend contigs to some extent, it is likely that non-WGS sequences will be required to bring the final quality up to a top quality.

Working with the present assembly, the different algorithms used for gene predictions gives numbers ranging from 32,000 to 50,000. Annotation is helped by 260,000 ESTs and several collections of FL-cDNAs.

Detailed information is displayed on http://www.ornl.gov/sci/ipgc/the populus genome science plan.pdf

THE TOMATO GENOME

The sequencing of the tomato genome was discussed during a workshop held in Washington in November 2003 and officially announced during the Plant and Animal Genomes- XIII meeting in San Diego in January 2005. The genome has been sized at 950 Mb and is located in 12 chromosomes. The euchromatic part of the genome has been estimated to be 220 Mb; the occurrence of very large blocks along the arms favors a sequencing strategy similar to that described for the legumes. The sequencing strategy will be a BAC by BAC progression covering the euchromatic areas. The international project includes 10 countries: USA (3 chromosomes), and China, France, India, Italy, Japan, Korea, Netherlands, Spain, and UK for one chromosome each.

Genomic resources consist of 3 BAC libraries totalling a 31 x coverage (Rod Wing, Tucson, USA and Jim Giovannoni, Cornell, USA), 400,000 BES and a BAC MTP for the 220 Mb euchromatic compartment which is under construction; the current estimation for the MTP is 2,276 BACs . BACs to be sequenced will be chosen from the MTP when locally known, or by STC otherwise. Anchoring will use the 1,825 markers of the genetic map and BAC-FISH localizations.

Annotation resources have already identified 162,000 Tomato ESTs and 77 Solanaceae cDNA libraries, of which 32 come from *Solanum lycopersicum.*

Detailed information is displayed on http://www.sgn.cornell.edu

THE GRAPEVINE GENOME

The International Grape Genomics Program has gathered together a number of countries interested in this plant. Some years ago, France and Italy began the construction of genomic resources to prepare the sequencing of this 500 Mb-long genome. Using both a French Cabernet-Sauvigon (therefore a hybrid strain) and an Italian Pinot noir, two laboratories (URGV, Michel Caboche and Anne-Françoise Adam-Blondon, Evry-France; IASMA,Ricardo Velasco, Trento-Italy) started to build both a genetical and a physical consensus map, this later one using a BAC library fingerprinted with the Snapshot technology followed by a FPC tiling. BES have been also generated.

Thanks to an INRA team (Colmar, France), a Pinot noir strain (PN 40025) brought close to homozygocity (97% from SSR assays) by retiterative back-crosses is now available for sequencing; the launching of the grapevine genome sequencing was announced in May 2005 by France and Italy. The sequencing strategy will be a WGS achieved at 12 x through a collaboration between GENOSCOPE (France) and the Universities of Padova (Giorgio Valle) and Udine (Michele Morgante) in Italy. Resources consist of a 13.6 x BAC library with BES, 3 and 5 kb plasmid libraries, and a fosmid library (GENOSCOPE); in addition to the already available Cabernet-Sauvignon / Italian Pinot noir consensus physical map. The Snapshot fingerprinting and FPC mapping of the PN 40025 strain are being carried out by Michele Morgante. The current genetic map contains some 800 SSR markers; the density will be raised to ca. 2,000 SSR markers by Anne-Françoise Adam-Blondon. Both the physical map and the SSR-based genetic map will be used to anchor the assembly contigs. As pure WGS leaves numerous gaps even at high sequence coverage, complementary sequences of fosmid inserts will be necessary to reach a high quality.

There are 140,000 grapevine ESTs in public databases and a set of various FL-cDNA libraries will be constructed for the PN 40025 strain by Italy and France; FL-cDNAs from other strains will be used as well.

THE WHEAT GENOME

Wheat is a complex plant in several respects: i) three ploidy levels occur, with AA (*Triticum monococcum*), AABB (*T. durum*) and AABBDD (*T. aestivum*) ; ii) even the lowest level of ploidy, AA, corresponds to a genome size larger than the 3 Gb displayed by the human one; iii) the repetitive sequences constitutes more than 80% of the genome. Nevertheless, unravelling the genetical information of wheat has been an long held dream, supported by a very rich genetical ancestory. ITMI (International Triticinea Mapping Initiative) gave rise to dense genetical mapping and opened the route to IGROW (International Genome Research on Wheat), which in turn is now launching the sequencing of this genome. In November 2003, an IGROW meeting hold in Washington DC decided to choose the 17 Gb hexaploid AABBDD genome of *Tricicum aestivum* var

Chinese Spring; such a goal was anticipated in the early 1980's as the biggest nightmare for future plant sequencing scientists, (Richard A. Flavell and F. Quétier, pers. comm.). Due to massive technical progress and strategic improvements, the early dreams are now coming true.

A breakthrough has been introduced by a collaboration between Jaroslav Dolezel (Inst. Exp. Bot, Oloumouc, Czech Rep.) who succeeded in the purification of the 3 B chromosome by flux cytometry and Boulos Chaloub (URGV, Evry, France) who constructed the corresponding BAC library. This pioneering work brought on the pilot project (International Wheat Genome Sequencing Consortium, May 2005) to sequence the 3B chromosome of the hexaploïd *T. aestivum* Chinese Spring: even this represents a huge project with some 900 to 1,000 Mb to decipher. A physical map is under construction through a collaboration between B. Chaloub and the team of Catherine Feuillet (INRA, Clermont-Ferrand, France) who carried out the SNapSHOT fingerprinting. The FPC physical mapping (MTP) is well advanced. The subsequent sequencing will require a coordinated international effort after some pilot local sequencing (Catherine Feuillet).

There are several genetical maps with over 1,000 markers each, 580,000 ESTs, and multiple cDNA libraries. These apparently large numbers must however be considered with the 17 Gb genome size in mind.

Information is displayed on http://www.wheatgenome.org/sitebuildercontent/sitebuilderfiles/IWGSC_White_Paper.pdf

OTHER PLANT GENOMES CURRENTLY BEING SEQUENCED

Most of these genome sequencing projects are run by the Joint Genome Institute (JGI) in the USA, either as US programs or generous Community Sequencing Programs. In 2004 and 2005, JGI started the sequencing of *Laccaria bicolor* (a mushrom, 40 Mb) and *Physcomitrella patens* (a moss, 450 Mb). This year, five more plants have been selected: whereas *Arabidopsis lyrata* (230 Mb), *Capsella rubella* (pink shepherd's-purse, 250 Mb) and *Mimulus guttatus* (monkey flower, 430 Mb) exhibit moderate genome sizes, the two other ones are large-sized genomes with *Triphysaria versicolor* (a parasitic wildflower) at 1,200 Mb and *Sorghum bicolor* (a helpful relative of maize, sugar cane and rice) at 736 Mb. All these sequencing programs follow the WGS strategy.

Information is displayed on http://www.jgi.doe.gov/sequencing/seqplans.html

SYNTENY

Never before has the word synteny been used so often or deviated more drastically from its original definition as at this moment. Today the concept refers to the conservation of the order of the genes or markers between maps or sequences of the genomes of different plants. Whereas macro-synteny concerns conservation at the level of large chromosomal blocks, micro-synteny refers to the nucleotide sequence, over shorter regions. The importance of this type of research is two-fold: first, the data is very helpful for phylogenetic studies, for which they provide a very solid base. Second, this conservation is useful for its predictive value in the assembly and annotation of sequences: if gene B is in between gene A and gene C in a certain plant, and if this ensemble belongs to a region known to be syntenic with another plant, one can expect to find gene B between genes A and C in this second plant as well. This is only a hypothesis, however, because examination at the micro-syntenic level generally reveals a non-negligible level of local rearrangements. Furthermore, differential local insertion/deletion events involving mobile elements interfere with distance relationships, and there are also local changes at the level of the genes, which may have disappeared, or become truncated or interrupted.

A particularly elegant and relatively cheap strategy for the determination of syntenic regions has been proposed by A. Fujiyama *et al.* (2002) for the human-chimpanzee pair: since the complete sequence of several human chromosomes is already available, it is possible to determine the anchorage of chimpanzee BAC-ends on the complete human sequence *in silico* (using BLAST, for instance). For example, if the ends of a chimpanzee BAC of 150 kb, for example, are found to be positioned about 150 kb apart and in the correct orientation on the human sequence, there is a certain probability that the internal sequences of the BAC will also align between human and chimpanzee (Fig. 2.11a). If a single BAC is considered, this probability is already high; when BAC libraries with significant depth are used (15-20 x is common), the ends of neighboring BACs overlap extensively and provide quasi-certainty (Fig. 2.11b). Furthermore, this coverage also makes it possible to avoid the pitfalls due to BAC ends which fall in intergenic regions or large introns, in which the sequences are very divergent compared with exons. This strategy seems promising (after the human-chimpanzee pair, it has been used with success in human-mouse pairs, especially for coding sequences); it necessitates a large informatics effort, whereas the construction of BAC libraries and the sequencing of BAC ends generally form part of the very first phases of a structural genomics program and are less cumbersome. This strategy has been successfully used by the groups of Rod Wing (AGI, USA) and Takuji Sasaki (NIAR, Japan) to construct quickly the physical map of *Oriza sativa* Kasalath by aligning the BES of the Kasalath BACs on the *Oriza sativa* Niponbare IRGSP complete sequence.

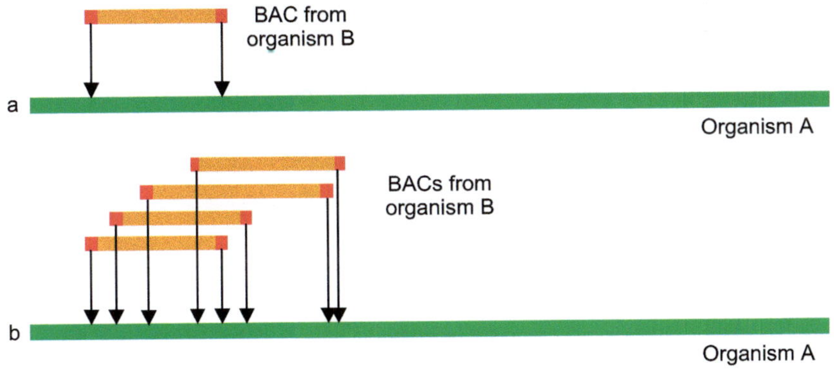

Fig. 2.11 A new strategy for the rapid construction of a physical map of a genome (when a complete sequence is available for a relative organism).
The genome of the organism A (green line) has been completely sequenced. a) the two end sequences (red blocks) of a BAC from the other organism B are positioned (BLAST alignment) on the A sequence ; b) even a medium coverage (6x) of a BAC library of the organism B allows the validation of the internal parts of this BAC from B (overlapping BACs of A).

FUTURE TECHNICAL AND BIOLOGICAL DIRECTIONS?

The sequencing of plant genomes is perhaps at a turning point. The *A. thaliana* project was accomplished by a public consortium, which produced a complete high-quality sequence; the complete sequence of rice was also undertaken by a public consortium but was aided by two drafts from private companies. James D. Watson, in his message which accompanied the December 2002 announcement of the IRGSP draft using anchored BACs before the achievement of the complete sequence, declared that: "... It is likely that your group will provide the only finished cereal genome at any time in the near future...". It is true that after the complete sequencing of a monocotyledon and a dicotyledon, the general organization of the genetic information in plants should be understood at a basic level; the cost of a complete sequence for plants in the future will have to be balanced against the portion of specific information which is the main target of the research, taking into consideration the fact that we are far from the relatively compact genome models.

It is to be anticipated that the cost of sequencing will progressively diminish, which is already happening for the conventional techniques (cf above). The new "454" technology introduces a dramatic decrease in the cost of one sequence reading; however, the reading length is limited to some 110 bp, thus limiting this cheap sequencing to re-sequencing projects, to EST and to cDNAs. Because longer readings are necessary to go through repeated sequences, eukaryotic genomes cannot benefit from such an advance. Beside, regarding the sequencing strategies confined to gene-rich regions, it is hard to imagine that this selection would have an efficiency of much greater than 80% of the genes, and the incomplete nature of the result may defeat the original purpose. The results of the several projects now in progress on maize, *Medicago truncatula*, poplar, grapevine, and wheat chromosome 3B will certainly measure this difficulty, as well as the other projects above reported.

A major breakthrough allowed by the completion of the *A. thaliana* and rice genomes is the design of whole genome tiling arrays: this new generation of arrays consist of huge sets of oligos spanning each chromosome. The most efficient pattern (Fig. 2.12) is 24 mers oligos generated in three sets which are frame-shifted by eight bases each, so that any sequence of 24 bp of the genome corresponds to one oligo (this length matches with the minimal size of a probe for hybridization). These WGA (Whole

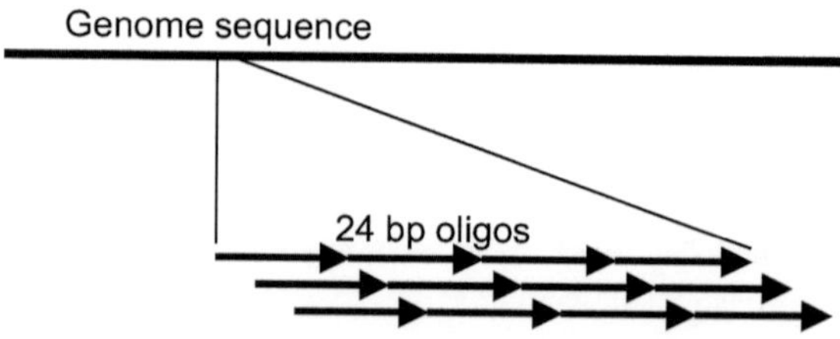

Fig. 2.12 Whole Genome Arrays (Tiling arrays) :
This newly released array technology is limited to completely sequenced genomes (or chromosomes). The spotting pattern consists of 3 sets of 24-mers oligos which scan the entire sequence, the 3 set being frameshifted for 8 bp; as a result, each genomic sequence of 24 bp is represented by one specific oligo whose length corresponds to the minimal size allowing a DNA-DNA hybridization.

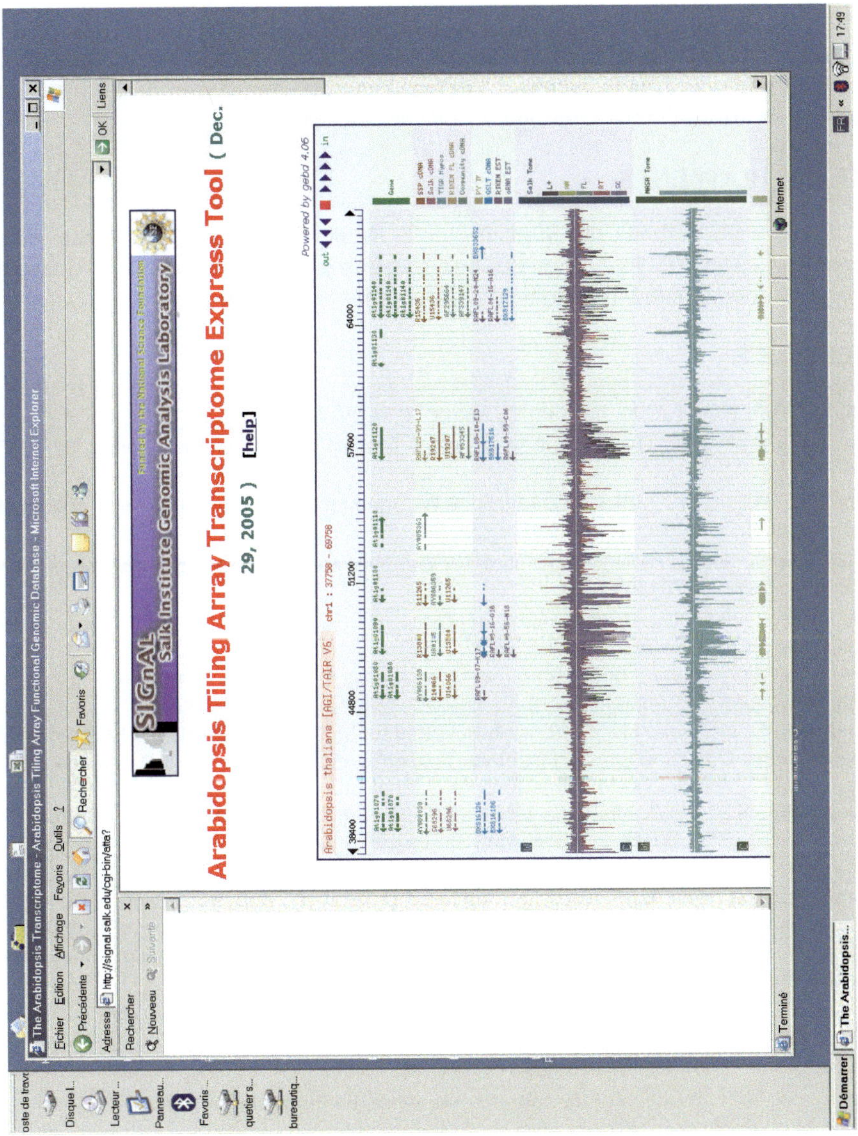
The Arabidopsis Transcriptome - Arabidopsis Tiling Array Functional Genomic Database - Microsoft Internet Explorer
http://signal.salk.edu/cgi-bin/atta?
SIGnAL
Salk Institute Genomic Analysis Laboratory
Funded by the National Science Foundation
Arabidopsis Tiling Array Transcriptome Express Tool (Dec. 29, 2005)
[help]
Arabidopsis thaliana [AGI/TAIR V5] chr1 : 37758 - 69758
38400
44800
51200
57600
64000
Internet
Terminé
Démarrer
The Arabidopsis...
17:49

Fig. 2.13 L'

Genome Arrays), better called Tiling Arrays allow the identification of all transcripts (micro-RNAs included) with an incredible efficiency and accuracy. They provide a great help in the annotation process; their only limitation is the level of completion of the collection of RNA molecules used in the hybridization (as well as their price for the time being). They are available for *A. thaliana* (http://signal.salk.edu/cgi-bin/atta?) and for several chromosomes of rice.

A general conclusion is that it is easy to speculate that the need in plant genome sequencing will continue to increase; international lobbying committees do surface regularly for various plants, hunting for financial support.

ACKNOWLEDGMENTS

The authors wish to thank the Ministère de la Recherche, l'Agence Nationale de la Recherche, Genopole® Evry (France) and the European Community for financial supports. Authors are indebted to Susan Cure for her help in translation.

REFERENCES

Adam, D., 2000. Now for the hard ones. *Nature,* 408, 792-793.

Adams, M.D. *et al.* 2000. The genome sequence of *Drosophila melanogaster. Science,* 287, 2185-2195.

Altschul, S.F., Gish, W., Miller, W., Myers, E.W., Ljpman, D.J. 1990. Basic local alignment search tool. *J. Mol. Biol,* 215, 403-410.

Aparicio, S. *et al.* 2002. Whole-genome shotgun assembly and analysis of the genome of *Fugu rubripes. Science,* 297, 1301-1310.

Dehal, P. *et al.* 2002. The draft genome of *Ciona intestinalis* : insights into chordate and vertebrate origins. *Science,* 298, 2157-2167.

Fleischmann, R.D., Adams, M.D., White, O., Clayton, R.A., Kirkness, E.F., Kerlavage, A.R., Bult, C.J., Tomb, J.K., Dougherty, B.A., Merrick, J.M., McKenney, K., Sutton, G., Fitzhugh, W., Flelds, C., Gocayne, J.D., Scott, J., Shirley, R., Liu, L., Glodek, A., Kelley, J.M., Weidman, J.K, Phillipps, C.A., Spriggs, T., Hedblom, E., Cotton, M.D., Utterback, T.R., Hanna, M.C., Nguyen, D.T., Saudek, D.M., Brandon, R.C., Fine, L.D., Fritchman, J.L., Fuhrmann, J.L., Geoghagen, N.S.M., Gnehm, C.L., McDonald, L.A., Small, K.V., Fraser, C.M., Smith, H.O., Venter, J.C. 1995. Whole-genome random sequencing and assembly of *Haemophilus influenzae* Rd. *Science,* 269, 496-512.

Fujiyama, A., Watanabe, H., Toyoda, A., Taylor, T.D., Itoh, T., Tsai, S.K., Park, H.S., Yaspo, M.L., Lehrach, H., Chen, Z., Fu, G., Saitou, N., Osoegawa, K., De Jong, P.J., Suto, Y., Hattori, M., Sakaki, Y. 2002. Construction and analysis of a human-chimpanzee comparative clone map. *Science,* 295, 131-134.

Goff, S. *et al.* 2002. A draft sequence of the Rice genome *(Oryza sativa* L. *ssp. japonica). Science,* 296, 92-100.

Goffeau, A. *et al.* 1997. The Yeast genome directory. *Nature,* 387 (suppl.), 1-105.

Heilig, R. *et al.* 2003 . Analysis of the complete sequence of human chromosome 14. *Nature,* 421, 601-607.

International Rice Genome Sequencing Project. 2005. The map-based sequence of the rice genome. *Nature,* 436, 793-800.

Lin, X. *et al.* 1999. Sequence and analysis of chromosome 2 of the plant *Arabidopsis thaliana. Nature,* 402, 761-768.

Marra, M., Kucaba, T., Sekhon, M., Hillier, L.D., Martienssen, R., Chinwalla, A., Crockett, J., Fedele, J., Grover, H., Gund, C., Mccombie, W.R., Mcdonald, K., Mcpherson, J., Mudd N., Parnell, L., Schein, J., Seim, R, Shelby, P., Waterson, R., Wilson, R. 1999. A map for sequence analysis of the *Arabidopsis thaliana* genome. *Nature Genet*, 22, 265-275.

Mayer K. *et al.* 1999. Sequence and analysis of chromosome 4 of the plant *Arabidopsis thaliana. Nature,* 402, 761-768.

Mouse Genome Sequencing Consortium. 2002. Initial sequencing and comparative genomics of the mouse genome. *Nature,* 420, 520-562.

Roest Crollius, H., Jaillon, O., Bernot, A., Dasilva, C., Bouneau, L., Fischer, C., Fizames, C., Wincker, P., Brottier, P., Quetier, F., Saurin, W., Weissenbach, J. 2000. Estimate of human gene number provided by genome-wide analysis using *Tetraodon nigroviridis* DNA sequence. *Nature Genet.,* 25, 235-238.

Sato, S., Nakamura, Y., Kaneko, T., Asamizu, E., Tabata, S. 1999. Complete structure of the chloroplast genome of *Arabidopsis thaliana. DNA Research,* 6, 283-290.

Shinozaki, K., Ohme, M., Tanaka, M., Wakasugi, T., Hayashida, N., Matsubayashi, T., Zaita, N., Chunwongse, J., Obokata, J., Yamaguchi-Shinozaki, K., Ohto, C., Torazawa, K., Meng, B. Y., Sugita, M., Deno, H., Kamogashira, T., Yamada, K., Kusuda, J., Takaiwa, F., Kato, A., Tohdoh, N., Shimada, H., Sugiura, M. 1986. The complete nucleotide sequence of the tobacco chloroplast genome: its gene organization and expression, *EMBO J.,* 5, 2043-2049.

Soderlund, C., Longden, I., Mott, R. 1997. FPC, a system for building contigs from restriction fingerprinted clones. *Comput. Appl. Biosci.,* 13, 523-535.

The Arabidopsis Genome Initiative. 2000. Analysis of the genome sequence of the flowering plant *Arabidopsis thaliana. Nature,* 408, 796-815.

The C. *elegans* Sequencing Consortium. 1998. Genome sequence of the nematode C. *elegans :* a platform for investigating biology. *Science,* 282, 2012-2018,

Theologis, A. 2001. Goodbye to «one by one» genetics. *Genome Biology,* 2(4), 1-9.

Unseld, M., Marienfeld, J., Brandt, P., Brennicke, A. 1997. The mitochondrial genome in *Arabidopsis thaliana* contains 57 Bgenes in 366,924 nucleotides. *Nature Genet.,* 15, 57-61.

Vanina Castelli, Jean-Marc Aury, Olivier Jaillon, Patrick Wincker, Christian Clepet, Manuella Menard, Corinne Cruaud, Francis Quétier, Claude Scarpelli, Vincent Schächter, Gary Temple, Michel Caboche, Jean Weissenbach and Marcel Salanoubat. 2004. Whole genome sequence comparisons and "Full-Length" cDNA sequences: a combined approach to evaluate and improve Arabidopsis genome annotation, *Genome Research,* 14, 406-413.

Venter, J.C., Smith, H.O., Hood, L. 1996. A new strategy for genome sequencing. *Nature,* 381, 364-366.

Venter J.C. *et al.* 2001.The sequence of the human genome. *Science,* 291, 1304-1351.

Waterson, R.H., Lander, E.S., Sulston, J. 2002. On the sequencing of the human genome. *Proc. Nat. Acad. Sci. USA,* 99, 3712-3716.

Wu J. *et al.* 2002. A comprehensive rice transcript map containing 6591 expressed sequences tag sites. *The Plant Cell,* 14, 525-535.

Yu J. *et al.* 2002. A draft sequence of the Rice genome *(Oryza sativa* L. *ssp. indica}. Science,* 296, 79-91.

3

CHAPTER

Introduction to Bioinformatics

H. Chiapello, F. Rodolphe

Current experiments in genomics produce a large amount of biological data. One of the primary objectives of bioinformatics is to develop methods and tools needed for the management, exploitation, and analysis of these data. But the process goes further, since such a change of dimension in biology raises new, fundamental questions on the scale of genomes and organisms. The knowledge of entire genomes is thus a veritable challenge for research in biology and bioinformatics, a challenge that continues to raise fundamental questions and generate surprises.

From a practical point of view, the organization and dissemination of these data on the scale of the international scientific community requires the use of conventional information tools (databases, Web interfaces), which are described in Chapter 4. At the same time, this massive quantity of genomic data is meaningful only within the framework of development of tools and strategies of automated (or more precisely semi-automated) annotation of genomes that are robust and reliable. In the context of bioinformatics, we have chosen to present three fundamental levels of gene annotation: gene detection (Chapter 5), which allows us to predict genes boundaries and corresponding proteins from a complete genome, annotation strategies that allow us to assign a function to the predicted proteins (Chapter 6), and finally massive sequence comparisons (Chapter 7), which are also an essential tool for annotation and functional analysis on the organism scale.

Over the course of these chapters, the reader will be provided with at least two major ideas.

First, the databases, tools, and analytical strategies are complex and varied, to the extent of the quantity and diversity of the available data. It is sometimes difficult for biologists to test and make pertinent choices among the available data (what are the appropriate sources of information?) and especially among the existing tools (what algorithm of comparison is to be used with which parameters?). Throughout these chapters, the authors have attempted to describe the existing tools as precisely and simply as possible, as well as their advantages and disadvantages.

The second major idea that emerges is that bioinformatics is faced with such problems of data interpretation that it will probably be essential to define in this context, strategies that are much stricter than those followed so far. For example, it is essential to develop procedures to obtain reliable traceability of conditions of data production, or annotation (which are presently very difficult or impossible to obtain). More striking still, the sequencing of a large quantity of complete genomes leads biologists to rethink and formalize notions as fundamental as the names of genes, or the description of the function(s) of a gene.

This chapter does not cover all the research on bioinformatics. In particular, the development of innovative statistical tools, which is another biological challenge, is not addressed: how the properties of an organism can be interpreted, analyzed, and understood through knowledge of the linking of bases of its nucleic acid sequence.

One characteristic of the approach is the need to make decisions in an uncertain environment, i.e., to resolve statistical problems. To detect genes in a genome, or homologies by sequence comparison, involves making decisions that do not have a firm basis. The decision results from the confrontation of a judiciously chosen observation with what is expected.

What is expected is described by a model. But sequence analysis, like many other domains of the application of statistics, does not have a theory that provides a mathematical representation of its object. The models are thus constructions that take into account some characteristics of the sequences analyzed that may appear naïve to the biologist.

A model is built to answer a question. It is a construction made at a moment with the knowledge at hand related to the question to be answered. Usually a model does not have a long life span, it does often not survive the question, it permitted to answer. However, one needs not be excessively pessimistic about models. In fact, experience shows that they are used to find elements that are authentic, biologically relevant discoveries. More and better information is discovered than with methods that are not based on any model. Moreover, models that are not too naïve begin to appear in the context, for example, of gene detection. For this well-defined objective, models take into account general characteristics of genomes that are linked to the function of memory of the primary structure of proteins and structural RNA. There are different textures for different types of sequences (coding, non-coding), signals (start, stop, ribosome binding site), and especially characteristics of general organization (a coding sequence begins with a start codon and ends in a stop codon, it often comprises a ribosome binding site at a certain distance upstream, a site possibly preceded by a promoter). It is true that with a more realistic model, one can hope to achieve more than with a naïve model, and some of these relatively complicated constructions, will undoubtedly benefit from a prolonged life beyond their primary objective.

REFERENCES

Baxevenis, A., Ouellette, F. 2001. *Bioinformatics: A Practical Guide to the Analysis of Genes and Proteins*. 2nd Edition. J. Wiley.

Durbin, R., Eddy, S., Krogh, A., Mitchison, G. 1999. *Biological Sequence Analysis*. Cambridge University Press.

Gusfield, D. 1997. *Algorithms in Strings, Trees and Sequences: Computer Science and Computational Biology*. Cambridge University Press.

Lesk, A.M. 2002. *Introduction to Bioinformatics*. Oxford Univ. Press.

Mount, D.W. 2001. *Bioinformatics: Sequence and Genome Analysis*. Cold Spring Harbor Laboratory Press.

Pevzner, P. 2000. *Computational Molecular Biology: An Algorithmic Approach*. MIT Press.

Setubal, J., Meidanis, J. 1997. *Introduction to Computational Molecular Biology*. PWS Pub. Co.

Young, P.G. 2002. *Exploring Genomes*. W.H. Freeman.

4

CHAPTER

Databases in Biology

H. Chiapello

Databases are today a source of information that the researcher in biology cannot ignore. This established fact is based on two major developments. The rapid progress in molecular biology and the large volume of studies generating massive quantities of data of a varied nature, which can no longer be managed manually by an individual, or a laboratory. Experiments in genomics produce potentially an enormous mass of knowledge that can be validated, only if it is organized and centralized in databases.

At the same time, in the field of informatics, two elements are critical: on the one hand, the availability of database management tools that are easy to use and inexpensive, and on the other hand the development of the Internet network, which allows dissemination of information contained in the databases across the Web, with a power and ease, that has been unequalled in history.

These two observations explain why the number of databases available on the Web has constantly increased over the past 10 years, as seen in the number of articles in the annual database issue of the *Nucleic Acids Research* journal, one reference in this field.

DIFFERENT TYPES OF DATABASES

The term *database** is easy to use but not easy to define. In the context of genomics, databases can be seen very simply as tools that allow efficient data storage, rapid data retrieval and complex data analysis. In general, it is presumed that the data volume is large and that the data have to be accessed or retrieved frequently, by several users simultaneously.

Very simply put, there are three major types of databases: flat file databases, relational databases, and object-oriented databases. This chapter will not exhaustively cover the concepts relevant to each type of database but only summarize the principal characteristics.

Flat File Databases

A flat file database is the simplest way of organizing data. Information is stored in plain text files, one database record per line or per file. Each record is divided into fields using delimiters, or fixed positions. Data retrieval may be complex in flat file databases and elaborate parsing* is often required. Access to large volumes of data can be enhanced through the creation of index files. An index in this context, is a file containing for one or several field(s), their different values and their positions in the indexed file.

This type of organization has been widely used in biology (in public *databanks*), for example by the Protein Data Bank (PDB) at the time it was created. It poses problems of efficiency of access, when the collection of data files becomes excessively large.

Relational Databases

Relational databases are presently the most common and most effective databases for managing large volumes of data. The underlying relational model represents the information in the form of a collection of relationships between entities. The data are organized in the form of tables. Each row of the table is called a tuple* and represents an occurrence of the entity described in the table. The rows are structured in fields (columns), which are the principal pieces of information and must be atomic* (non-decomposable into several sub-elements). Each table contains one or several particular field(s), defining the primary key, which identifies uniquely each line of the table. Particular tables are used to describe the nature of relationships between the base tables. To establish the scheme of the relational database thus consists in defining precisely all the base tables with their columns, then characterizing the relationships that will make it possible to form links between these tables in terms of their common attributes. The mathematical foundation of the relational database is the relational algebra, which provides all the operators that will make possible selective access to all the combinations of relations of the database. The success of relational databases relies on several factors, including the simplicity of concepts operating in the model (tables), the power of the Structured Query Language (SQL*) for access to the data, and most of all the efficiency of the associated Relational Data Base Management Systems (RDBMS).

Object-oriented Databases

In object-oriented databases, certain concepts of the object-oriented programming languages (encapsulation*, inheritance*) are applied to the field of databases. These databases contribute two interesting functionalities: the persistence* of objects and the possibility of managing complex structures of objects. Persistence is a mechanism that allows perennial storage of the state of an object and procedures that modify its state. The management of complex structures of objects particularly involves hierarchical structures, or the factorization of elements common to a set of objects. This is not easy to achieve in relational databases, in that there is no possible hierarchy between tables. In practice, object-oriented databases are rarely used, since there is presently no standard query language of object-oriented databases that is as efficient as SQL. However, several recent database management systems implement certain of these concepts to some extent.

Software and Languages Associated with Databases

Databases are used by means of specially designed software in which most of the conception functions (creation of the base and tables of the relational model, for example), query (associated SQL type of language), and administration (backups, creation of indexes, management of concurrent access) are facilitated. Only the most commonly used software is mentioned here, especially in the field of biology. The tools cited are summarized in Table 4.1.

Table 4.1 Some database and library management tools used in biology

Name	*Type*	*Site*
Tools designed for biology		
SRS	Flat Files	http://srs.ebi.ac.uk
Entrez	Flat Files	http://www.ncbi.nlm.nih.gov/Entrez
ACeDB	Object	http://www.acedb.org
EyeDB	Object	http://www.sysra.com/eyedb
General tools		
Oracle	Relational/Object	http://www.oracle.com
PostgreSQL	Relational/Object	http://www.postgresql.org
mySQL	Relational	http://www.mysql.com
Ozone	Object	http://www.ozone-db.org

In the field of flat file databases, Sequence Retrieval System (SRS) is the software most commonly used by Biologists. Initially developed by the European Molecular Biology Library (EMBL) for the management of nucleic acid banks, its functions have been extended to facilitate the integration of most biological libraries in the software. It is now marketed by the Lion Bioscience company[1] but is available free for academic research on the website of the European Bioinformatics Institute.[2]

In the field of relational database management systems, the commercial software Oracle has almost become a standard in a large number of applications. This software not only implements SQL in an extremely complete manner but also provides a large number of functions involving the administration and optimization of database performance. The latest version also provides object functionalities and a free licence for development. On the other hand, the cost of dissemination of applications under Oracle, often remains prohibitive for research laboratories.

One of the most commonly used alternatives to Oracle is the Open Source software PostgreSQL, which implements SQL in an extremely complete manner and provides a simple interface for handling a wide variety of data types. The management of transactions (sequences of operations that modify the contents of the database) and certain concepts of object programming are also implemented. Finally, the software MySQL is also an Open Source database management system, of the relational type only, which is also very easy to use and has provided full transaction support since September 2002.

[1] http://www.lionbioscience.com
[2] http://www.ebi.ac.uk

Among the object-oriented databases, the ACeDB (A *C. elegans* database) software developed in 1989 in the context of a sequencing program of the nematode *C. elegans*, is still widely used among Biologists. It provides not only a model adapted for the management of genomic data but also a user-friendly graphic interface for the management of genomic data. There are other object-oriented, database management systems in the field of Biology (see for example the EyeDB software) but they are still few in number and rarely used. The Open Source project Ozone, more general in scope, seems promising. It has been developed entirely in Java for Java developers, and it implements notably the concept of persistence (saving the states of an object).

Web Interfaces

The development of Web interfaces has been a turning point in the conception of a database in Biology. Several techniques allow the communication of Web pages with a database: the oldest is the Common Gateway Interface (CGI)* program, but more recent technologies have given rise to new concepts, for example PHP* (Programmed Hypertext Preprocessor) and XML* (extensible Markup Language).

A CGI script* is an application that is executed on the Web server and allows communication of data to an independent application (for example a database). This type of program is implemented easily by means of specially designed modules in most programming languages (for example, the CGI.pm library in Perl language).

PHP* is a language for which there is a module included in most Web servers that allows the PHP code directly included in the Web pages, to be read and interpreted. In practice, the PHP program resembles that of classical languages, such as C or Perl and makes it possible to generate dynamic Web interfaces very simply.

The eXtensible Markup Language (XML)* is a language of symbols used to describe the structure of a set of data (not the presentation of data as in HTML). An XML document is a text document describing the data in a tree form. The "grammar' of the document is described in a Document Type Definition (DTD) that is used to define a sort of dictionary of format of data. The major contribution of the XML is to make it possible to structure the "simple" text files in a standardized and uniform manner, which allows any XML interpreter to access the contents of a document structured in this manner. Several XML interpreters are already available in certain languages (Perl, Java), which greatly facilitates the dissemination and exchange of data. Moreover, working groups have tackled the definition of DTDs in the domain of biology and genomics. For example, the data of Genbank and Pubmed are today available in the XML format on the site of the National Center for Biotechnology Information (NCBI).

THE PRINCIPAL PUBLIC DATABASES IN BIOLOGY

This section provides a rapid survey of the principal libraries and databases that cannot be ignored in the framework of genomics research. We do not distinguish here between the terms *libraries* and *databases*. Most of the applications cited maintain data in two forms: simple files (or libraries) for the diffusion of data and databases for the internal management of data. The contents and addresses of all the databases cited in this section are summarized in Tables 4.2 and 4.3.

Table 4.2 Contents and addresses of the principal biological databases

Name	*Type*	*Site*
Nucleic acid sequences	Genbank	http://www.ncbi.nlm.nih.gov/Genbank
	EMBL	http://www.ebi.ac.uk/embl
	DDBJ	http://www.ddbj.nig.ac.jp
Protein sequences	PIR	http://pir.georgetown.edu
	SwissProt	http://www.expasy.ch/sprot
	TrEMBL	http://www.ebi.ac.uk/trembl
	GenPept	http://www.ncbi.nlm.nih.gov/query.fcgi?db =Protein
	Uniprot	http://www.expasy.uniprot.org
Bibliographical references	PubMed	http://www.ncbi.nlm.nih.gov/entrez/ query.fcgi?DB=pubmed
	Biosis	http://www.biosis.org
	CAB	http://www.cabi-publishing.org
	ISI Web on Knowledge	http://www.isiwebofknowledge.com
Families, motifs and protein domains	Prosite	http://www.expasy.ch/prosite
	Pfam	http://pfam.wustl.edu
	ProDom	http://protein.toulouse.inra.fr/prodom
	Prints	http://bioinf.man.ac.uk/dbbrowser/PRINTS
	InterPro	http://www.ebi.ac.uk/interpro
Protein structures	PDB	http://www.rcsb.org/pdb
Metabolic pathways and regulation pathways	Kegg	http://www.genome.ad.jp/kegg
	Metacyc	http://metacyc.org
Transcriptome	SMD	http://genome-www5.stanford.edu
	ArrayExpress	http://www.ebi.ac.uk/arrayexpress
	GEO	http://www.ncbi.nlm.nih.gov/projects/geo
	BASE	http://base.thep.lu.se
Protein interactions	DIP	http://dip.doe-mbi.ucla.edu
	BIND	http://www.bind.ca
	IntAct	http://www.ebi.ac.uk/intact
	SPiD	http://genome.jouy.inra.fr/cgi-bin/spid/ index.cgi
	STRING	http://string.embl.de

General Libraries

Libraries manage, archive, and make available to the scientific community all the nucleic acid and protein sequences published in the world and the biological annotations associated with them.

At present, three nucleic acid libraries cooperate and daily exchange their data in order to guarantee maximum standardization in the availability of sequences for the scientific community: GenBank (Benson *et al.*, 2005), the American database managed by the NCBI, a division of the National Library of Medicine; EMBL (Kanz *et al.*, 2005), the European bank maintained at EBI; and the DNA Data Bank of Japan (DDBJ) (Tateno *et al.*, 2005).

Table 4.3 Contents and addresses of principal databases devoted to an organism or group

Contents	*Source*	*Site*
Prokaryotes	GenoList	http://genolist.pasteur.fr
	EcoGene	http://bmb.med.miami.edu/EcoGene/EcoWeb
	Emglib	http://pbil.univ-lyon1.fr/emglib/emglib.html
	Micado	http://genome.jouy.inra.fr/micado
Eukaryotes	Proteome	http://www.proteome.com
	MYGD	http://mips.gsf.de/projects/fungi/yeast.html
	FlyBase	http://flybase.bio.indiana.edu
	MGD	http://www.informatics.jax.org
	Ensembl	http://www.ensembl.org
Plants	TAIR	http://arabidopsis.org
	MatDB	http://mips.gsf.de/proj/thal/db/index.html
	FLAGdb/FST	http://genoplante-info.infobiogen.fr/FLAGdb
	Gramene	http://www.gramene.org
	CropNet	http://ukcrop.net
	GrainGenes	http://www.graingenes.org

Even though each of these libraries has its own specializations, the philosophy of data structuring is identical: each "entry" corresponds to a nucleic acid sequence and is structured in four parts: (1) a heading containing the general information (identifier, accession number, definition, key word; taxonomy); (2) a part describing the bibliographic references associated with the sequence; (3) the "features/qualifiers" part, which contains all the biological annotations associated with the sequence (the associated *feature table*[3] is an annotation format common to the three libraries since 1990); and (4) the nucleic acid sequence itself.

The entries of protein libraries are structured according to the same principles and make available to researchers all the known proteins, as well as their associated biological annotations. The American library Protein Information Resource (PIR) (Wu *et al.*, 2003) distributes a set of libraries of protein sequences that attempt to disseminate all the annotated and non-redundant protein sequences (in the PIR-NREF base, very similar sequences are organized within a single entry). The European database SwissProt (Boeckmann *et al.*, 2003) is characterized by excellent quality data annotation (with manual expertise of biological annotations), at the cost of not being totaly comprehensive. Finally, the libraries TrEMBL and GenPept contain all the conceptual protein sequences obtained by automatic translation of sequences considered to be coding in EMBL and GenBank. It is important to mention that these last two libraries contain unverified sequences, in which the annotations follow from those of the nucleic acid libraries.

These generalist libraries constitute an essential source of information for the biologist. However, they must be used with care because they pose several problems. First, the quality of individual sequences varies greatly depending on the authors, and errors frequently occur in the annotation of the sequence (or sometimes even in the sequence itself). This problem becomes crucial when data from these databases are used in automated genome annotation procedures. Second, the databases are often not

[3]http://www.ebi.ac.uk/embl/Documentation/FT_definitions/feature_table.html

verified and highly redundant, although the source of this redundancy (polymorphism, duplication of genes, sequence errors) cannot be easily identified. In 2002, the Swiss-Prot + TrEMBL groups at the SIB (Swiss Institute of Bioinformatics), the EBI (European Bioinformatics Institute), the PIR (Protein Information Resource) group at Georgetown University Medical Center and National Biomedical Research Foundation joined forces to establish the UniProt resource (United Protein Databases) (Bairoch *et al.*, 2005). The UniProt database provides the scientific community with a single, centralized, authoritative resource for protein sequences and functional information.

Specialized Databases

Some essential specialized databases in the field of genomics are presented here, although this list is not exhaustive. Several of these databases are also called "derived database" according to the fact that they often integrate data derived from primary nucleotide and protein librairies.

THEMATIC BASES

In the context of a genomic project, the bibliography is naturally a major source of information. The development of the bibliographical database PubMed, which can be consulted free on the web using the Entrez software (Schuler *et al.*, 1996), has been a small revolution for biologists. However, this database, created by the National Library of Medicine, was dedicated at first to the biomedical sciences. It thus does not cover biological literature exhaustively. In particularly, the coverage of plant biology by PubMed is not satisfactory and may pose a problem. Other bibliographical bases, more complete in the plant domain, must also be used (BIOSIS, CAB, ISI Web of Knowledge).

Alongside libraries of protein sequences, libraries of motifs or protein domains and signatures are essential tools for projects of functional genomics. In this field, several complementary bases have developed in parallel: Prosite (Hulo. *et al.*, 2004), Pfam (Bateman *et al.*, 2004), ProDom (Servant *et al.*, 2002), and Prints (Attwood *et al.*, 2003). In 2000, the EBI took the initiative of combining all these databases within a single structure: InterPro (Mulder *et al.*, 2005). This resource was created to integrate the major protein signature databases. Currently, it includes PROSITE (Database of protein families and domains), Pfam (Protein families), PRINTS (Protein fingerprints), ProDom (Protein Domains), SMART (Simple Modular Architecture Research Tool), TIGRFAMs (TIGR protein FAMilies), PIRSF (PIR Super Families) and SUPERFAMILY (HMM library of proteins of known structures at the superfamily level).

In the field of protein structures, Protein Data Bank (PDB) (Westbrook *et al.*, 2002) is an exceedingly valuable source of information: it archives all the available data on protein structures.

Databases managing knowledge linked to metabolism and biological pathways have recently achieved considerable importance. In this field, the Japanese database KEGG (Kyoto Encyclopedia of Genes and Genomes) (Kanehisa *et al.*, 2004) is one of the oldest and most complete. It can be used to access in detail the metabolic pathways of a large number of model organisms in the form of clickable images. KEGG is in fact a suite of databases and associated software, integrating our current knowledge on molecular interaction networks in biological processes, the information about genes and proteins, and the information about chemical compounds and reactions.

The database MetaCyc (database of pathways) (Krieger *et al.*, 2004) was developed to centralize the metabolic data determined experimentally. It contains non-redundant, experimentally elucidated metabolic pathways from more than 300 different organisms obtained from the scientific experimental literature. This resource also provides a set of tools to edit, visualize, and analyse the metabolic networks. It also offers the possibility of interpreting genomic data in a given metabolic context.

In parallel, several projects involving the management of transcriptome data are ongoing. Among the most important is the Stanford Microarray Database (SMD) (Ball *et al.*, 2005), which manages and disseminates experimental data published on the transcriptome, as well as the corresponding filter images, free of charge. In the same spirit, the EBI has made available a database called ArrayExpress (Parkinson *et al.*, 2005), which allows transcriptome data to be submitted in a standardized format defined by the Microarray Gene Expression Database (MGED) and the performance of some analyses on these data. The NCBI also proposes a structure of storage and query of transcriptome data under the Gene Expression Omnibus (GEO) project. More recently, a free microarray database system called BASE (BioArray Software Environment) has been developed (Haas *et al.*, 2002). Designed by microarray biologists, it offers a natural workflow for microarray experiments, all the way from microarray production to the analysis of results. The system is meant to be set up as a local repository for data from a microarray laboratory, including raw images if so desired. The main part of BASE is written in PHP and works alongside a MySQL database. BASE is free software, distributed under the GNU (Gnu's Not Unix) General Public License, encouraging software distribution and people contribution. For more information, a comprehensive list of principle databases on the transcriptome is available on the SMD website.[4]

Finally, several databases were developed recently to allow management of interactions between proteins: see for example the Database of Interacting Proteins (DIP) (Salwinski *et al.*, 2004), the Biomolecular Interaction Network Database (BIND) (Alfarano *et al.*, 2005) or the IntAct database (Hermjakob *et al.*, 2004). Some of these are dedicated to a model organism, like the Subtilis Protein Interaction Database (SPiD) which focuses on the analysis of networks of protein interaction in the model bacterium *Bacillus subtilis* (Hoebeke *et al.*, 2001). Generally, their main purpose is to organize storage and display of experimental protein–protein interactions (for example from experiments of the double-hybrid type). Some also include gene co-expressions deduced from transcriptome experiments or bibliographical references. They raise interesting problems of management, visualization, and analysis of networks of complex interactions. More recent, the STRING (Search Tool for the Retrieval of Interacting Genes: proteins) database (von Mering *et al.*, 2005) provides simplified access to this information by providing a comprehensive, yet quality-controlled collection of protein–protein associations for a large number of organisms. It includes known and predicted protein-protein interactions. The interactions include direct (physical) and indirect (functional) associations; they are derived from four sources: genomic context, high-throughput experiments, conserved co-expression and previous knowledge from bibliographical resources.

[4]http://genome-www5.stanford.edu/MicroArray/SMD/databases.html

BASES DEVOTED TO AN ORGANISM OR A TAXONOMIC GROUP

In parallel with the development of general libraries, a certain number of databases dedicated to model organisms have been developed in the context of systematic sequencing programs (Table 4.3). In the prokaryotes, the database Subtilist (Moszer *et al.*, 2002) was developed to manage the sequencing data and functional annotations of the Gram-positive model bacterium *Bacillus subtilis*. Other bases, such as Colibri or EcoGene (Rudd, 2000), are dedicated to the management of the genome of the Gram-negative model bacterium *Escherichia coli*. The Pasteur Institute presently combines the databases of complete genomes of several bacteria across a common interface called Genolist (including Subtilist and Colibri). Other servers allow access to sets of complete genomes, for example Micado (Biaudet *et al.*, 1997) for microbial species or Emglib (Perriere *et al.*, 2000) for prokaryotes and yeast. The Micado base also integrates the functional analysis data of *B. subtilis* produced by a group of European laboratories. In model eukaryotic organisms, one of the references is the Yeast Proteome Database (YPD) (Hodges *et al.*, 1999), which is devoted to the functional and biochemical annotation of proteins of the model yeast *Saccharomyces cerevisiae*. The "Proteome" server of the IncyteGenomics company has subsequently integrated with the same logic all the proteins and their annotations of a large number of model organisms, including mammals, yeasts, nematodes, and parasites (Csank *et al.*, 2002). Other bases constitute important references for model eukaryotic organisms: for example the MIPS Yeast Genome Database (MYGD) for the yeast *Saccharomyces cerevisiae* (Mewes et al., 1997), FlyBase for *Drosophila melanogaster* (Drysdale *et al.*, 2005), or the Mouse Genome Database (MGD) (Eppig *et al.*, 2005) for the mouse. More recently, in the context of the human genome sequencing program, the Ensembl project (Hubbard *et al.*, 2005) was developed. Ensembl is a joint project between EMBL - European Bioinformatics Institute (EBI) and the Wellcome Trust Sanger Institute (WTSI) to develop a software system which produces and maintains automatic annotation on selected eukaryotic genomes.

SOME DATABASES SPECIFIC TO PLANTS

For plants, the primary databases are logically focused on the genomic data of the model plant *Arabidopsis thaliana*. The relational database known as The *Arabidopsis* Information Resource (TAIR) (Rhee *et al.*, 2003) centralizes most of the available information on *A. thaliana*, including systematic sequencing program data, genetic and physical maps, clones, and markers. In parallel, the MIPS *Arabidopsis thaliana* DataBase (MatDB) (Schoof *et al.*, 2002) provides a systematic analysis of all the proteins of *A. thaliana*, notably using the InterPro system. Initially designed for the management and visualization of physical maps of flanking sequences of insertion mutants of *A. thaliana*, the FLAGdb/FST database (Samson *et al.*, 2004) now provides the integration and visualization of data for high-throughput functional analysis of both *A. thaliana* and rice genomes.

Other databases centralize the knowledge accumulated in parallel on model plants and on cultivated plants. For example, the Gramene database (Ware *et al.*, 2002) is an open-source, Web-accessible data resource for comparative genome analysis of grasses and a community resource for rice. In parallel, UK CropNet has developed a set of six plant databases constructed on the same model and facilitating comparative genomics approaches. Finally, the GrainGenes database offers a compilation of phenotypic and molecular information on wheat, barley, oat, and sugarcane. It is an important tool for the management of data associated with plants of agronomic interest (e.g., sequences, cartographical data, QTL, bibliography, phenotypes).

WHAT IS THE FUTURE OF DATABASES IN BIOLOGY?

Information tools must now respond to the massive production of complex biological data, with the requirements of traceability and validation of these data. In this context, the Laboratory Information Management Systems (LIMS) are increasingly used in large genomics laboratories. These are information systems that are used to manage data at the industrial level. Simply put, LIMS can be used to manage the chain of production of genomic data from the laboratory equipment up to the interpretation of biological results. The integrated information system will thus allow coherent management and traceability of data produced in genomics experiments (e.g., sequencing, transcriptome, proteome).

Another recent field has developed in the context of major genomics projects: the concept of *data mining*. This involves the use of informatics techniques to find, interpret, and extract results from huge volumes of data, typically biological libraries and databases. The tools developed in this field are based on artificial intelligence (machine learning) and statistics (data analysis, classification). They provide tools for the biological interpretation of results but cannot in any way substitute for the experimental biologist. It is often in this context that the term *data warehouse* is used, referring to a volume of data so large that data mining tools are indispensable for the exploitation of results.

Finally, one of the requirements of databases in molecular biology is the integration of distant data in order to constantly enrich the sum of biological knowledge available on the type of data managed in a database. This advantage of integration is clearly visible when we look at the advancement of generalist database entries, which are nowadays enriched with numerous links to other bases. For example, the entries of the SwissProt database contain systematic links to a large number of distant databases (e.g., PubMed, EMBL, PDB, Interpro). These hypertext links constitute a primary level of integration of distant data. Other more complex informatics techniques allow a virtual exchange of data between applications through the Web, for example the protocols of Common Object Request Broker Architecture (CORBA)* or Simple Object Access Protocol (SOAP, recently renamed XP for XML protocol)*.[5] However, these protocols are now used only occasionally and, with the exception of the EBI, which provides a CORBA* server for nucleic acid sequences from EMBL, the biological databases are no longer considered virtual tools of data exchange. It seems in fact that in this logic of integration, other barriers exist, which are more fundamental since they involve the semantics of concepts manipulated in biology. In the context of genome annotation, the Gene Ontology project[6] has attempted to use a controlled vocabulary to describe in a

[5]http://www.corba.org/ and http://www.w3.org/TR/SOAP
[6]http://www.geneontology.org

hierarchical and precise manner, the biomolecular functions of gene products (Gene Ontology Consortium, 2001). This step is in fact an essential prerequisite to the integration of databases and the comparative analysis of functional genomic data on the scale of a set of organisms.

GLOSSARY

Atomic: an atomic string or number is a term which does not contain other terms.

CGI script: a Common Gateway Interface script is a program that processes data on a Web server. It is commonly used to process a query from the user that was entered on an HTML page (Web page).

CORBA: the Common Object Request Broker Architecture is a set of standards for building software-based interfaces that allow software modules (objects) to communicate with each other no matter where they are located (on a private network or the global Internet).

Database: a database is an organized collection of data usually created and managed by a database management system (DBMS).

Encapsulation: in object technology, the creation of self-contained modules that contain both the data and the processing.

Inheritance: in object technology, the ability of one class of objects to inherit properties from a higher class.

Parsing: the process of analysing a continuous flow of text-based input and breaks it into its constituent parts via a routine.

Persistence: in object technology, the storage of an object on a disk or other permanent storage device.

PHP: the Hypertext Preprocessor programming language is an object-oriented scripting language aimed at dynamic web content generation.

SOAP: the Simple Object Access Protocol is a standard protocol based on XML for accessing services on the Web.

SQL: the Structured Query Language is the standard language used to interrogate and process data in a relational database.

Tuple: in a relational database, a tuple is one record (one row).

XML: the eXtensible Markup Language is a language for building structured documents. XML uses a similar tag structure as HTML; however, whereas HTML defines how elements are displayed, XML defines what those elements contain.

REFERENCES

Alfarano, C., Andrade C.E., Anthony, K., Bahroos, N., Bajec, M., Bantoft, K. *et al.* 2005. The Biomolecular Interaction Network Database and related tools 2005 update. Nucleic Acids Res. 33 (Database Issue): D418-D424.

Attwood, T.K., Bradley, P., Flower, D.R., Gaulton, A., Maudling, N., Mitchell, A.L., Moulton, G., Nordle, A., Paine, K., Taylor, P., Uddin, A., Zygouri, C. 2003. PRINTS and its automatic supplement pre-PRINTS. *Nucleic Acids Res.*, 31, 400–402.

Ball, C.A. Awad, I.A., Demeter, J., Gollub, J., Hebert, J.M., Hernandez-Boussard, T., Jin H., Matese, J.C., Nitzberg, M., Wymore, F., Zachariah, Z.K., Brown, P.O., Sherlock, G. 2005. The Stanford Microarray Database accommodates additional microarray platforms and data formats. *Nucleic Acids Res.* 33(1): D580-2.

Bairoch, A., Apweiler, R., Wu, C.H., Barker, W.C., Boeckmann, B., Ferro, S., Gasteiger, E., Huang, H., Lopez, R., Magrane, M., Martin, M.J., Natale, D.A., O'Donovan, C., Redaschi, N., Yeh, L.S. 2005. The Universal Protein Resource (UniProt). *Nucleic Acids Res.* 33: D154-159.

Bateman, A., Coin, L., Durbin, R., Finn, R.D., Hollich, V., Griffiths-Jones, S., Khanna, A Marshall, M., Moxon, S., Sonnhammer, E.L., Studholme, D.J., Yeats, C., Eddy, S.R. 2004. The Pfam protein families database. *Nucleic Acids Res.*, 32, 138–141.

Benson, D.A., Karsch-Mizrachi, I., Lipman, D.J., Ostell, J., Wheeler, D.L. 2005. Genbank. *Nucleic Acids Res.* 2005 Jan 1; 33(Database issue): D34-8.

Biaudet, V., Samson, F., Bessieres, P. 1997. Micado–a network-oriented database for microbial genomes. *Comp. Appl. Biosci.*, 13(4) 431-438.

Boeckmann, B., Bairoch, A., Apweiler, R., Blatter, M.-C., Estreicher, A., Gasteiger, E., Martin, M.J., Michoud, K., O'Donovan, C., Phan, I., Pilbout, S., Schneider, M. 2003. The SWISS-PROT protein knowledgebase and its supplement TrEMBL in 2003. *Nucleic Acids Res.* 31: 365-370.

Csank, C., Costanzo, M.C., Hirschman, J., Hodges, P., Kranz, J.E., Mangan, M., O'Neill, K., Robertson, L.S., Skrzypek, M.S., Brooks, J., Garrels, J.I. 2002. Three yeast proteome databases: YPD, PombePD, and Ca1PD (MycoPathPD). *Methods Enzymol.*, 350, 347-373.

Drysdale, R.A., Crosby, M.A., The FlyBase Consortium. 2005. FlyBase: genes and gene models. *Nucleic Acids Research*, 33: D390-D395.

Eppig, J.T., Bult, C.J., Kadin, J.A., Richardson, J.E., Blake, J.A., the members of the Mouse Genome Database Group. 2005. The Mouse Genome Database (MGD): from genes to mice–a community resource for mouse biology. *Nucleic Acids Res* 2005; 33: D471-D475.

Gene Ontology Consortium. 2001. Creating the gene ontology resource: design and implementation. *Genome Res.*, 11, 1425-1433.

Hermjakob, H., Montecchi-Palazzi, L., Lewington, C., Mudali, S., Kerrien, S., Orchard, S., Vingron, M., Roechert, B., Roepstorff, P., Valencia, A. *et al.* 2004. IntAct: an open source molecular interaction database. *Nucleic Acids Res.*, 32, D452–D455.

Hoebeke, M., Chiapello, H., Noirot, P., Bessieres, P. 2001. SPiD: a subtilis protein interaction database. *Bioinformatics*, 17(12), 1209-1212.

Hodges, P.E., McKee, A.H., Davis, B.P., Payne, W.E., Garrels, J.I. 1999. The Yeast Proteome Database (YPD): a model for the organization and presentation of genome-wide functional data. *Nucleic Acids Res.* 1999 Jan 1; 27(1): 69-73.

Hubbard, T., Andrews, D., Caccamo, M., Cameron, G. *et al.* Ensembl 2005. *Nucleic Acids Res.* 2005 Jan 1; 33(Database issue): D447-53.

Hulo, N., Sigrist, C.J., Le Saux, V., Langendijk-Genevaux, P.S., Bordoli, L., Gattiker, A., De Castro, E., Bucher, P., Bairoch, A. 2004. The PROSITE database, its status in 2002. *Nucleic Acids Res.*, 30, 235–238.

Karp, P.D., Riley, M., Paley, S.M., Pellegrini-Toole, A. 2002. The MetaCyc Database. *Nucl. Acids Res.*, 3059-61.

Kanehisa, M., Goto, S., Kawashima, S., Nakaya, A. 2002. The KEGG databases at GenomeNet. *Nucl. Acids Res.*, 30, 42-46.

Kanehisa, M., Goto, S., Kawashima, S., Okuno, Y., Hattori, M. 2004. The KEGG resources for deciphering the genome. *Nucleic Acids Res.* 32, D277-D280.

Kanz, C. *et al.* 2005 The EMBL Nucleotide Sequence Database. *Nucleic Acids Res.*, 33, D29–D33.

Krieger, C.J., Zhang, P., Mueller, L.A., Wang, A., Paley, S., Arnaud, M., Pick, J., Rhee, S.Y., Karp, P.D. 2004. MetaCyc: A Multiorganism Database of Metabolic Pathways and Enzymes. *Nucleic Acids Research*, 32(1): D438-42.

Mewes, H.W., Albermann, K., Heumann, K., Liebl, S., Pfeiffer, F. 1997. MIPS: A database for protein sequences, homology data and yeast genome information. *Nucl. Acids Res.*, 25, 28-30.

Moszer, I., Jones, L.M., Moreira, S., Fabry, C., Danchin, A. 2002. SubtiList: the reference database for the Bacillus subtilis genome. *Nucl. Acids Res.*, 30, 62-65.

Mulder, N.J., Apweiler, R., Attwood, T.K., Bairoch, A., Bateman, A. *et al.* 2005. InterPro, progress and status in 2005. *Nucleic Acids Res.* 33, Database Issue: D201-5.

Parkinson, H., Sarkans, U., Shojatalab, M., Abeygunawardena, N., Contrino, S., Coulson, R., Farne, A., Garcia Lara, G., Holloway, E., Kapushesky, M., Lilja, P., Mukherjee, G., Oezcimen, A., Rayner, T., Rocca-Serra, P., Sharma, A., Sansone, S., Brazma, A. 2005. ArrayExpress–a public repository for microarray gene expression data at the EBI. *Nucl. Acids Res.* 33: D553-D555

Perriere, G., Bessieres, P., Labedan, B. 2000. EMGLib: the enhanced microbial genome library. *Nucl. Acids Res.*, 28(1), 68-71.

Rhee, S.Y., Beavis, W., Berardini, T.Z., Chen, G., Dixon, D., Doyle, A., Garcia-Hernandez, M., Huala, E., Lander, G., Montoya, M., Miller, N., Mueller, L.A., Mundodi, S., Reiser, L., Tacklind, J., Weems, D.C., Wu, Y., Xu, I., Yoo, D., Yoon, J., Zhang, P. 2003. The Arabidopsis Information Resource (TAIR): a model organism database providing a centralized, curated gateway to Arabidopsis biology, research materials and community. Nucleic Acids Research 31(1):224

Rudd, K. 2000. EcoGene: a genome sequence database for *Escherichia coli* K12. *Nucl. Acids Res.*, 28, 60-64.

Saal, L.H., Troein, C., Vallon-Christersson, J., Gruvberger, S., Borg, A., Peterson, C. 2002. BioArray Software Environment: A Platform for Comprehensive Management and Analysis of Microarray Data. Genome Biology 2002 3(8): software0003.1-0003.6

Salwinski, L., Miller, C.S., Smith, A.J., Pettit, F.K., Bowie, J.U., Eisenberg, D. 2004. The Database of Interacting Proteins: 2004 update. NAR 32 Database issue: D449-51

Samson, F., Brunaud, V., Duchene, S., De Oliveira, Y., Caboche, M., Lecharny, A., Aubourg, S. 2004. FLAGdb++: a database for the functional analysis of the *Arabidopsis* genome. *Nucleic Acids Res.* 32 Database issue: D347-50.

Schoof, H., Zaccaria, P., Gundlach, H., Lemcke, K., Rudd, S., Kolesov, G., Arnold, R., Mewes, H.W., Mayer, K.F.X. 2002. MIPS *Arabidopsis thaliana* Database (MAtDB): an integrated biological knowledge resource based on the first complete plant genome. *Nucl. Acids Res.*, 30, 91-93.

Schuler, G.D., Epstein, J.A., Ohkawa, H., Kans, J.A. 1996. Entrez: molecular biology database and retrieval system. *Methods Enzymol.*, 266, 141-162.

Servant, F., Bru, C., Carrere, S., Courcelle, E., Gouzy, J., Peyruc, D., Kahn, D. 2002. ProDom: automated clustering of homologous domains. Brief Bioinformatics, 3, 246–251.

Tateno, Y., Saitou I, N., Okubo, K., Sugawara, H., Gojobori, T. 2005. *Nucleic Acids Research.* 33 (Database issue): D25-D28

von Mering Christian, Lars J. Jensen, Berend Snel, Sean D. Hooper, Markus Krupp, Mathilde Foglierini, Nelly Jouffre, Martijn A. Huynen, Peer Bork. 2005. STRING: known and predicted protein–protein associations, integrated and transferred across organisms. *Nucleic Acids Res.* D433–D437.

Ware, D., Jaiswal, P., Ni, J., Pan, X., Chang, K., Clark, K., Teytelman, L., Schmidt, S., Zhao, W., Cartinhour, S., McCouch, S., Stein, L. 2002. Gramene: a resource for comparative grass genomics. Nucleic Acids Res. 30(1): 103-5.

Westbrook, J., Feng, Z., Jain, S., Bhat, T.N., Thanki, N., Ravichandran, V., Gilliland, G.L., Bluhm, W., Weissig, H., Greer, D.S., Bourne, P.E., Berman, H.M. 2002. The Protein Data Bank: unifying the archive. *Nucl. Acids Res.*, 30, 245-248.

Wu, C.H., Yeh, L-S.L., Huang H., Arminski, L. *et al.* 2003. The Protein Information Resource. *Nucleic Acids Research,* 31: 345-347.

5

CHAPTER

Gene Prediction

P. Nicolas, H. Chiapello

The large-scale sequencing of genomic DNA poses the problem of gene detection on a DNA sequence. An array of computer methods have been developed to address this challenge. The purpose of this chapter is to give an overview of these *in silico* methods.

Currently, the term *gene prediction* is often used in the restricted sense of prediction of sequences coding for proteins, i.e., coding sequences or CDSs. Although it is mostly used in this restricted sense in the present text, it is worth emphasizing that CDS prediction differs of gene prediction in two ways. First, protein coding genes contain upstream and downstream untranslated regions that do not belong to the CDS. Second, it is increasingly recognized that many genes do not encode protein coding sequences (Riddihough, 2005).

To predict protein coding genes is certainly one of the main reason why people sequence DNA. It is routinely the first step when interpreting genomic data and it often has deep impact on the design of functional analysis experiments such as DNA cheap experiments. To predict the genes from the DNA sequence is also a way towards identifying all the proteins, even those that are expressed only in particular conditions. The importance of these ventures makes it is useful to have an idea of the principles and limits of the different prediction methods.

Two kinds of approaches for gene prediction are usually distinguished. Extrinsic approaches rely essentially on comparisons with other related sequences. These sequences may already be in the databases, or may come from experiments designed to detect genes. Intrinsic approaches, on the other hand, are based only on the local properties of the sequence under scrutiny: nucleotide composition and sequence motifs. The prediction is thus called *ab initio* and it is the aspect of gene prediction the most thoroughly developed in this chapter.

DNA SEQUENCE MODELLING AND PREDICTION *AB INITIO*

In bacteria, a crude prediction of the CDSs can be made by searching for long open reading frames (ORFs): regions without stop codons in one of the six reading frames

(Fig. 5.1a). However, this approach does not allow the precise identification of the translation start sites, nor does it allow the detection of small genes. Moreover, because stop codons (TGA, TAA and TAG) are rich in A and T nucleotides, the poorer the genome is in A and T, the more frequent the presence of long ORFs that do not correspond to CDSs (Fig. 5.1b). In organisms where CDSs are divided up into many exons, this method is not practically applicable, since numerous exons are too short to be revealed by long ORFs (Fig. 5.1c).

More refined prediction of the CDSs are achieved by taking into account the composition of the sequence in nucleotides. Nucleotides appear with different frequencies depending on the coding or non-coding nature of the sequence due to the constraints put on the nucleic acid sequence by the protein sequence (Nicolas *et al.*, 2002). At a finer scale, properties of the genetic code makes the frequency of nucleotides also varies depending on the position in the codon. The second codon position is the most constrained by the encoded protein sequence. On the opposite, the third codon position is the least constrained and is therefore the most sensitive to the mechanisms that shape the overall G+C content of a genome. Simple linear formulas that relate the compositional correlations among codon positions have been proposed (d'Onofrio and Bernardi, 1992; Besemer and Borodovsky, 1999). Those properties where recently used to question the quality of some of the rice genome annotations (Cruveiller *et al.*, 2004).

Compositional differences between coding and non-coding regions have encouraged the emergence of prediction methods based on a probabilistic modelling of DNA sequences. Before presenting those methods, it is worth to emphasize that the probabilistic models used in this context do not attempt to reflect the way the sequence actually appears in the course of evolution. They are just means of describing what are the expected sequences and of attributing a probability to each possible sequence. This probability have to account for properties of the sequence that we want to take into account in the predictions. We start by presenting the most popular models used in gene prediction.

Modelling the DNA Sequence Composition

The simplest model of the DNA composition accounts only for mono-nucleotide frequencies. It consists in imagining that the nucleotides appear independently of one another along the sequence and with different probabilities for A, C, G and T. To account for the di-nucleotide composition, nucleotides can be modelled as appearing with probabilities that depends on the nucleotide at the preceding position. This very popular model with four different sets of probabilities for A, C, G and T is known as the Markov chain model of order 1. It generalizes easily by making the probability of appearance of a nucleotide dependent on the oligo-nucleotide of length n (n-nucleotide) that precedes it. The Markov model is then say of order n and it contains as many set of probabilities for A, C, G and T as the number of n-nucleotides (i.e. 4^n). This model accounts only the composition in $n+1$-nucleotides: the probability of a particular sequence given a Markov model of order n depends only on the number of occurrences of each $n+1$-nucleotide. In coding regions, nucleotides appear with different frequencies at three codon positions: this property can be modelled using a different Markov model for each of the three codon positions. This model known as the tri-periodic Markov

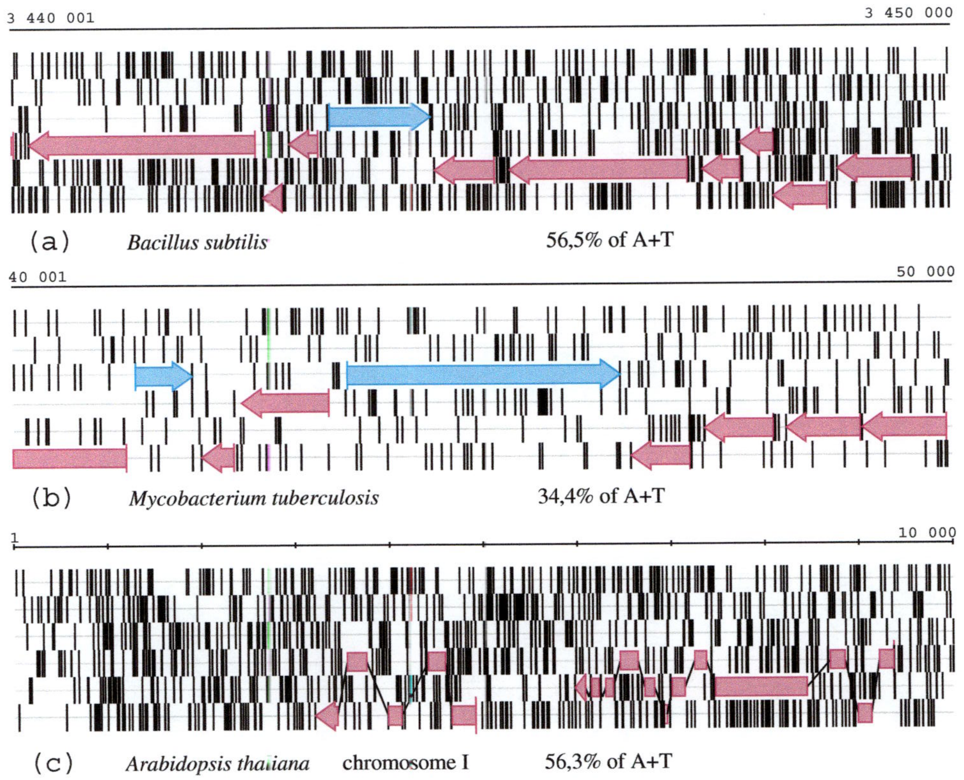

Fig. 5.1 Positions of annotated CDSs (arrows) and stop codons (black vertical bars) in the six reading frames of three sequences of 10,000 nucleotides. The coding fragments do not contain a stop in their reading frame. Sequences (a) and (b) come from bacteria that differ notably in their A+T composition. The frequency of stop codons (TAG, TGA, and TAA) is clearly higher in sequence (a), richer in A+T. Sequence (c) is a eukaryotic sequence in which the CDSs are divided into exons.

model is the basis of virtually all gene prediction softwares. It was popularized by GeneMark (Borodovsky and McIninch, 1993) which usually relies on codon position specific Markov models of order 5.

Estimating the parameters of a Markov model of order n from a set of "training" sequences is straightforward. It involves only the counting of the occurrences of each one of the four nucleotides in each possible context (after each possible n-nucleotide). As long as parameters are well estimated, biological sequences are better modelled with higher order Markov models. However, the number of parameters increases dramatically with the order of the model (3,072 parameters for a model of order 5,12,288 parameters for a model of order 6) and thus the quality of the estimates tends to drop due to the finite size of the training data. To circumvent this problem, variable order Markov models were introduced in the gene prediction software Glimmer 1.0 (Salzberg *et al.*, 1998) and later used in other programs such as EuGene'Hom (Foissac *et al.*, 2003). Variable order Markov models allow the length of the context taken into account to depend on the sequence of the context. Long context are taken into account when, first, there is enough data available for the estimation of probabilities of the next nucleotide; second, the long context can be shown to supply more information on the next nucleotide than a shorter context. A further generalization of the Markov model is found in Glimmer 2.0 (Delcher *et al.*, 1999) where context are not necessarily made of consecutive positions. These models are called *context trees*. Azad and Borodovsky (2004) compared the performances of fixed and variable order models for gene prediction but they could not show a model to systematically outperform the other. This might be due to intermingled effects of the model performance and of the estimation quality but it also probably show that the fixed order Markov model performs well and can hardly be ameliorated.

Prediction of Genes Using a Model of the Sequence Composition

There exist several approaches to predict genes using models of the DNA composition for coding and non-coding regions. The GeneMark program implements a sliding window strategy. The window slides along the sequence and at each position the program computes the probability of the sequence contained in the window under seven models: non-coding, and coding on each of the two strands in each of three reading frames (1 + 2 x 3). The introduction of a prior probability for each model makes it possible to use Bayes' formula to compute the probability of each model given the sequence. According to Bayes' formula, the probability of each model is proportional to the product of two terms: (1) the probability of appearance of the sequence under the model and (2) the prior probability for the model. The length of the window is a matter of choice: a short window gives more local information on the sequence, but information vanishes when the window becomes too small. A typical choice is a 96 nucleotides long window. Sliding the window, we obtain a probability of the locally coding nature of the sequence in each reading frame along the sequence. Results must be processed afterwards to choose the ORFs that should be predicted as CDSs and to attempt to locate start codons. A simple alternative to sliding windows is implemented by Glimmer. The program first extracts ORFs longer than a certain threshold and, second, attempts to classify them according to their coding or non-coding nature.

Sliding window and ORF classification strategies cannot be used to locate the borders between coding and non-coding regions in a simple way. From the probabilistic point of view, the natural way to tackle this problem is to model not only the composition of coding and non-coding regions but also the succession of these regions. Such a modelling can then be used to answer the question "Where are the regions of different types?", rather than "Given a region, what type is it?". These methods are discussed below.

Modelling the Sequence Heterogeneities to Predict the Borders of Genes

Hidden Markov models (HMMs) provide are a simple and flexible framework for modelling the succession of several types of regions along the DNA sequence. They model a DNA sequence as a mosaic of regions of a relatively small number of types, each type having its own properties of composition. All the composition models presented earlier can be embedded into a HMM to account for the emission of nucleotides in the different types of regions. The succession of regions along the sequence is modelled as a Markov chain. This chain is termed hidden, because it is only observed through heterogeneities of composition of the DNA sequence. Each type of region is an hidden state of the model. In bacterial gene prediction, hidden states distinguish typically the CDSs (on each strand) and the intergenic regions.

HMMs contain two layers of parameters: nucleotide emission probabilities and transition probabilities of the hidden states chain. In most models, numerous transition probabilities between hidden states are fixed as null, thus putting constraints on the path of the hidden chain along the sequence. The graph representing allowed transitions between hidden states often provides a great deal of information on the model. Figure 5.2 presents a graph of the hidden model used in GeneMark.hmm (Lukashin and Borodovsky, 1998).

The popularity of HMMs is explained by the existence of many relatively simple algorithms. In gene prediction, the most commonly used algorithm is the Viterbi algorithm. Given a model whose parameters have already been estimated, the Viterbi algorithm finds the most probable hidden path for a particular DNA sequence. This series of hidden states is thus, in a certain sense, the best prediction one can achieve with this model. The Viterbi algorithm is presently used by vast majority of the gene prediction programmes based on HMMs. After having predicted the genes with the Viterbi algorithm, Genscan (Burge and Karlin, 1997) uses another algorithm (Forward-Backward) to attribute a probability for each exon predicted.

A drawback of using HMMs that is often cited is the implicit poor modelling of the length of the coding regions. This is why numerous models used for gene prediction are in fact an extension of HMM with explicit modelling of the sojourn time in each hidden state. These models are known under various names: hidden semi-Markov model, explicit duration state HMM, generalized HMM, or simply HMM. They are used for example by GeneMark.hmm, Genscan, Genie (Kulp *et al.*, 1996), Fgenesh (Salamov and Solovyev, 2000) and GlimmerHMM (Majoros *et al.*, 2004).

Modelling of Sequence Signals and Detection of Genes with Introns

In higher eukaryotes, CDSs are often divided up on the chromosome into numerous relatively short exons separated by introns. For example, in *Arabidopsis thaliana* there

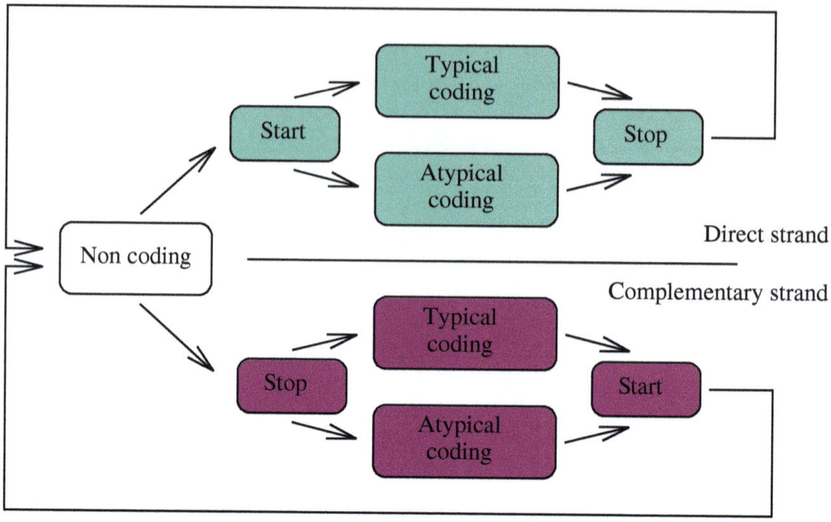

Fig. 5.2 The HMM of GeneMark.hmm (Lukashin and Borodovsky, 1998). This graph describes all the possible transitions (arrows) between hidden states (rounded rectangles). The HMM of GeneMark.hmm was ultimately enriched to account for RBS and overlapping genes.

are on average five exons per CDS of a mean length of around 250 nucleotides, the average length of introns being around 170 nucleotides (Arabidopsis Genome Initiative, 2000). The picture is even accentuated in Human where the average intron size may be around 3000 nucleotides (International Human Genome Sequencing Consortium, 2001). Introns are excised during the splicing of pre-messenger RNA, which thus gives rise to messenger RNA in which the exons are end to end in the same reading frame. The excision of introns is a complex phenomenon involving more than 50 molecules, proteins and small nuclear RNA, which constitute the *spliceosome* (Nilsen, 2003). Nearly all the introns begin with GT and finish with AG, but this information is not enough for the spliceosome to choose the splicing sites. The recognition of other sequence signals around these di-nucleotides plays an important role. The identification of splicing sites is an essential aspect of *ab initio* gene prediction in complex eukaryote genomes. It is required to locate precisely the borders of the exons that are revealed by their composition. For short exons, composition provides only weak evidence and splicing site identification can also help with their detection.

Most splicing site recognition methods are based on the evaluation of the sequences of potential sites by using a probabilistic model that describes position by position the nucleotide composition of actual splicing sites. When the appearance of nucleotides is modelled as independent between adjacent positions, the term of weighted array model (WAM) is sometimes used. An example of composition of mono-nucleotides around a splicing site is shown in Table 5.1. When not only the position in the site but also the dependency between adjacent nucleotides is taken into account, the model is sometimes called a weighted matrix model (WMM). A limitation of such models is that they do not account for dependencies between positions that are not consecutive, even so correlation between adjacent position seems predominant (O'Flanagan *et al.*, 2005).

Table 5.1 Frequencies of nucleotides position by position around the donor site of splicing in man (Burge and Karlin, 1997). These frequencies correspond to the parameters of a WAM for the recognition of the site. The frequencies of the most common nucleotides at each position are given in italics.

Nucleotide	*Position*								
	Exon			*Intron*					
	-3	*-2*	*-1*	*+1*	*+2*	*+3*	*+4*	*+5*	*+6*
A	**0.33**	**0.60**	0.08	0	0	**0.49**	**0.71**	0.06	0.15
C	**0.37**	0.13	0.04	0	0	0.03	0.07	0.05	0.19
G	0.18	0.14	**0.81**	**1**	0	**0.45**	0.12	**0.84**	0.20
T	0.12	0.13	0.07	0	**1**	0.03	0.09	0.05	**0.46**

One solution is to use a collection of models, rather than a single model. In this case, each model describes a different type of site and thereby captures some correlation between the appearances of nucleotides at non-adjacent positions within the site. An automated procedure, maximal dependence decomposition (MDD), has been proposed (Burge and Karlin, 1997) to construct such a collection of models corresponding to a splicing site. This procedure is used by Genscan and GeneSplicer (Pertea *et al.*, 2001). Other models such as Bayesian networks that account for correlations between non-adjacent positions have also been proposed (Chen *et al.*, 2005).

These models of splicing site sequences can be introduced in the form of new hidden states in a HMM (Burge and Karlin, 1997). The HMMs can then be used to make a prediction that takes into account not only differences in composition between coding and non-coding regions but also the sequence properties of splicing sites. Depending on the structure of the HMM, the predictions are of different kinds. Most often, as for example with Genscan and HMMgene (Krogh, 1997), the complete structure of the genes is predicted. However, some programs do not guarantee that the exons that compose the predicted CDS are in phase with one another; this is the case with the GeneMark.hmm version for intron genes.

Predicting without Modelling the Sequence

Broadly speaking, any prediction method requires the use of a criterion to distinguish good predictions from the set of all possible predictions. As we just saw, probabilistic modelling is a framework for constructing these criteria quite directly and has the additional advantage of providing criteria with an intuitive meaning. However, it is sometimes difficult to model all the factors that may help the prediction. Numerous methods that are not based on a complete probabilistic model of the DNA sequence have been developed for gene prediction. Because these methods bypass the complete modelling of the sequence, they are faced with the problem of the construction of a good criterion. This is often a matter of combining evidences from different measures or sensors.

Sensors for the nucleotide composition or the presence of a splicing signal often involve the probabilities of appearance of the sequence under probabilistic models. For example, the programme for predicting splicing sites GeneSplicer, combines measurements of the nucleotide composition on either side of the site (coding or non-coding texture), with a measurement of the splicing signal. In the case of GeneSplicer, it is possible to have quite a clear idea of the best way to construct the criterion to combine the different measurements. More often, the criterion is constructed by adopting a particular methodological framework that defines the type of criterion and proposes algorithms to optimize the quality of predictions on a learning set.

The following examples illustrate the diversity of methods used. To predict splicing sites, NetPlantGene (Hebsgaard *et al.*, 1996) uses an artificial neural network and makes a prediction directly from the DNA sequence. SplicePredictor (Kleffe *et al.*, 1996) combines the measurement of the splicing signal strength with two other measurements aimed at perceiving the environment of the signal: variation of the G+C and pyrimidine (C+T) composition on either side of the potential site. These measurements are combined and optimized in the framework of a logitlinear model. More recently, SpliceMachine implemented a SVM-based strategy for a related task (Degroeve, 2004). The Mzef program (Zhang, 1997) uses nine different measurements for the detection of human exons. The use of these nine measurements is optimized by quadratic discriminant analysis, while Grail (Uberbacher and Mural, 1991) uses an artificial neural network for a nearby task.

The elaboration of a criterion is not enough to allow the prediction of the structure of a gene composed of multiple exons. Among all the possible combinations of exons, therefore, it is necessary to exhibit those that best satisfy the chosen criterion. The immense number of combinations prohibits their evaluation one by one. With some conditions on the criterion chosen, dynamic programming algorithms are used to quickly explore the space of all the combinations to extract the best of them, the Viterbi algorithm for HMMs is an example of dynamic programming algorithm. Such an algorithm is used by the Glimmer version adapted for the detection of intron genes (Salzberg *et al.*, 1999) to find the gene structures that maximize the criterion corresponding to the sum of measurements of the coding texture of exons and the quality of splicing sites.

EXTRINSIC METHODS

Unlike intrinsic methods, extrinsic methods use information outside the sequence itself to predict genes. Most of the algorithms developed take into account one of three types of external information represented in Fig. 5.3: known protein sequences, messenger RNA sequences, or DNA sequences. In these three cases, there is fundamentally a problem of alignment between the DNA sequence in which the genes are sought and a sequence that resembles it more or less. The alignment is often improved by the selection of potential splicing sites.

DNA-protein Comparison

In the case of the DNA-protein similarity, the idea is to predict the CDSs that code for proteins that resemble proteins already contained in the databases. BlastX (Gish and States, 1993) and FastX (Pearson *et al.*, 1997) are used to indicate similarities by local alignment between a DNA sequence translated in six reading frames and a protein sequence library. However, to predict the structure of a CDS composed of multiple exons, the DNA sequence and the protein sequence must be aligned taking into account

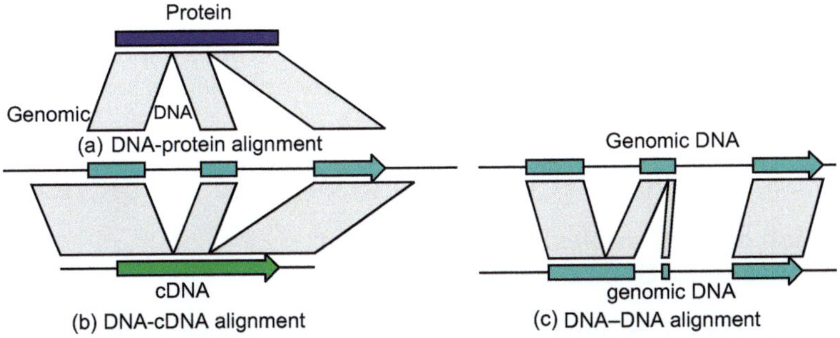

Fig. 5.3 The three major extrinsic approaches for gene prediction: (a) genomic DNA-protein alignment; (b) genomic DNA-cDNA alignment; (c) genomic DNA-genomic DNA alignment. The parallelograms represent the alignment on which the different approaches are based.

the presence of introns that constitute long fragments of the DNA sequence that do not match with the protein sequence. Procustes (Gelfand *et al.*, 1996) and PairWise (Birney *et al.*, 1996) were the first programmes developed to resolve this problem known as the spliced alignment problem.

DNA-mRNA Comparison

The sequencing of mRNA through the sequencing of cDNA (copy DNA of the mRNA) is an experimental technique to determine the sequence of genes transcribed. Since the sequencing takes place after the splicing, these sequences allow precise determination of the intron-exon structure of genes. The large-scale sequencing of mRNA produces sequences called expressed sequence tags (EST). This sequencing can be seen as an experimental technique for gene detection. However, to determine the DNA structure of genes from ESTs is not trivial, because these sequences are generally incomplete (often sequenced at the 3′ end), of poor quality (sequenced only once), and redundant (as a function of the abundance of mRNA). The problem is more difficult when the mRNA sequence has been obtained in a different organism. Algorithms of DNA-cDNA and DNA-EST alignments such as SIM4 (Florea *et al.*, 1998) and GeneSequer (Usuka *et al.*, 2000) have been developed.

DNA-DNA Comparison

There is one approach that can be called extrinsic and yet *ab initio*. It consists of predicting a pair of genes by comparing two DNA sequences. At present, the programs that predict the complete structure of genes are based essentially on the often much better conservation of coding sequences, compared to that of non-coding sequences. One example is the Rosetta program (Batzoglou *et al.*, 2000). Other criteria such as the comparison of synonymous and non-synonymous mutation rates should make it possible to distinguish, even within the conserved sequences, the coding regions from the non coding regions (Rogozin *et al.*, 1999 ; Nekrutenko *et al.*, 2002). Sophisticated HMMs were recently introduced to tackle the issue of the joint modelling of several DNA sequence in order to predict conserved gene structure. Pair-HMM are used in DOUBLESCAN and SLAM to simultaneously align two DNA sequences and perform the gene prediction (Meyer and Durbin, 2002, Alexandersson *et al.*, 2003). Phylo-HMMs were proposed to predict conserved gene structure from a set of already aligned sequences by joint modelling of the gene structure and the DNA evolution process (McAuliffe *et al.*, 2004; Siepel and Haussler, 2004).

SUCCESS AND PRESENT LIMITATIONS OF GENE PREDICTION

Illustration of the Complexity of Factors that Ensure Successful Gene Detection

It is possible to explain certain reasons why HMMs have taken time to prove their superiority in predicting bacterial genes *ab initio*. Chronologically, the EcoParse programme (Krogh *et al.*, 1994) proposed an HMM for the detection of *Escherichia coli* genes just after the first versions of the GeneMark program (Borodovsky and McIninch, 1993), which used a sliding window. However, the HMM of EcoParse lacks some of the features involved in the success of GeneMark. In particular, the GeneMark programme predicts genes on both strands in the same pass of the algorithm. On the opposite, EcoParse implements a separate prediction of the genes on each strand which sometimes results in the prediction of the fingerprint of a gene on the complementary strand as being a gene on the strand considered. GeneMark subsequently proposed to take into account the existence of a relatively large proportion of genes of atypical composition (rich in A and T), which are not predicted otherwise (Borodovsky *et al.*, 1995). GeneMark.hmm (Lukashin and Borodovsky, 1998) was the first HMM to account for these two characteristics and it proved actually better than GeneMark.

Diversity and Integration of Approaches

The approaches mentioned in this chapter tackle different aspects of the detection of genes through the prediction of coding exons, splicing sites or the complete structure of genes. For these predictions, different information is taken into account by different methods: composition, sequence similarities, experimental data. The types of algorithms and models used are also varied and do not always have a direct relationship with the quality of predictions. Comparing the performances of different approaches is a careful and difficult task. The difficulty arises partly from the diversity of information taken into account as well as the diversity of the predictions made (complete structure of CDS, exons, splicing sites). This difficulty is further increased by the diversity of species for which they have been designed. In fact, just as the quality of intrinsic methods depends clearly on their adjustment to a given species, that of extrinsic methods depends on the degree of similarity between the sequences compared. In all cases, as testified by the number of approaches developed, the difficulty of the problem means that no method so far, has proved to be completely satisfactory. Last but not least is the problem of the gold standard that could serve as a reference for the comparison of the approaches. In 2005, the EGASP competition (Guigo and Reese, 2005) addressed this issue. Gene prediction software performances were assessed on a set of regions of the human genome with very high quality gene annotations. These annotations were obtained by manual compilation multiple sources of evidence and subsequent experimental verifications as part of the ENCODE project.

Intrinsic methods *ab initio* seems to predict about 80% of coding exons correctly in *A. thaliana* (Pavy *et al.*, 1999) and at a lower rate in mammals (Guigo *et al.*, 2000; Rogic *et al.*, 2001). The figure of 80% may seem quite high but it can be deduced from it that the probability of predicting perfectly a CDS composed of 5 exons is low ($0.8^5 = 33\%$), while that of not predicting any of the exons correctly is very low: $(1 - 0.8)^5 = 0.03\%$. Thus, it must be expected that there will be a low proportion of genes that are completely ignored and a significant proportion of genes annotated incorrectly. Of course this calculation is simplistic, because the exons of a gene are rarely predicted and annotated independently of one another. Moreover, it does not take into account one of the important problems with these predictions: the difficulty of predicting the borders between genes. In the absence of an effective method to detect elements such as the promoters and signals of polyadenylation, it is in fact difficult to distinguish between intergenic regions and introns.

In real world annotation process, CDSs are usually better predicted than these small calculations would suggest. In fact, the annotation only partly relies on intrinsic *ab initio* methods, and information coming from similarities and experimental data can greatly improve these values. There are two major ways to integrate the different approaches. The first way is to use the programs separately, then carry out a post-treatment of the results. JIGSAW (Allen and Salzberg, 2005) is an example of program developed to perform this task: it uses dynamic programming algorithms to automatically combine predictions made with independent programs. The second way is to develop programs that base their predictions simultaneously on intrinsic and extrinsic criteria. For instance, Fgenesh+ (Salamov and Solovyev, 2000), GenomeScan (Yeh *et al.*, 2001) and TwinScan (Korf *et al.*, 2001) proposed to account for sequence similarity evidences by expanding the HMM framework of Genscan. Intermediate routes are imprinted by EuGene (Schiex *et al.*, 2001; Foissac *et al.*, 2003) and GAZE (Howe *et al.*, 2002).

Towards a Reliable Annotation of Genomic Sequences

Despite the use of all these sources of information to annotate CDSs, the uncertainty of annotations is often insufficiently emphasized during their publication. The annotations are then gradually improved, largely through enhanced performance of extrinsic methods resulting from the increase in the number of known sequences. The high quality and large-scale sequencing of full-length cDNA is particularly promising and several projects are now providing such experimental data for most model organisms (Carninci *et al.*, 2003, MGC Project Team, 2004; FANTOM and RIKEN, 2005). Tiling arrays are another promising tool for the experimental analysis of the transcriptional landscape of a genome. These oligonucleotide micro-arrays with probes covering the whole genomic sequence, even regions that are not known *a priori* as potentially coding, can be used to map transcription factor or DNA polymerase binding sites (Cawley *et al.*, 2004; Kim *et al.*, 2005) as well as to map RNA transcripts (Yamada *et al.*, 2003; Kampa *et al.*, 2004).

To give an example of modifications of annotations, in *A. thaliana*, out of more than 5,000 full-length cDNA sequences, 62% corresponded perfectly to the annotated CDSs, 33% to CDSs that could be corrected, and 5% to probably new genes (Haas *et al.*, 2002). In *Drosophila melanogaster*, a revision of the annotation led to the modification of 45% of

CDSs (Misra *et al.*, 2002). Besides correction of the annotation of CDSs, another advantage of cDNA sequences is that they allow annotations that cannot presently be made reliably using purely *in silico* approaches: annotation of non coding exons, location of the beginning of transcription, detection of the phenomena of alternative splicing (Modrek and Lee, 2002; Eyras *et al.*, 2004).

Because of its sensitivity, *in silico* gene prediction from a genomic sequence remains essential: it is quick and cheap as soon as the genomic sequence is available and it may be one of the only way to study poorly or rarely expressed CDSs. The performance of *in silico* approaches will probably improve with the quality and the quantity of data available for training, with the number of genome sequences available, and with better biological knowledge. Methodological advances could also be expected. Both aspects are probably essential. At present, the relative success of *ab initio* gene prediction relies largely on the tri-periodic compositional properties of the CDSs (a consequence of translation into proteins). As a consequence, the prediction of genes that do not code for proteins remains very difficult.

Additional information on gene prediction methods can be found in the reviews by Zhang (2002), Mathé *et al.* (2002) and Brent and Guigo (2004). The website maintained by Wentian Li is another valuable resource where numerous publications on the subject are listed (http://www.nslij-genetics.org/gene/).

REFERENCES

Alexandersson, M., Cawley, S., Pachter, L. 2003. SLAM: cross-species gene finding and alignment with a generalized pair hidden Markov model. Genome Res. 13, 496-502.

Allen, J.E., Salzberg, S.L. 2005. JIGSAW: integration of multiple sources of evidence for gene prediction. Bioinformatics. 21, 3596-3603.

Arabidopsis Genome Initiative. 2000. Analysis of the genome sequence of the flowering plant *Arabidopsis thaliana*. *Nature*, 408, 796-815.

Azad, R.K., Borodovsky, M. 2004. Effects of choice of DNA sequence model structure on gene identification accuracy. Bioinformatics. 20, 993-1005.

Batzoglou, S., Pachter, L., Mesirov, J.P., Berger, B., Lander, E.S. 2000. Human and mouse gene structure: comparative analysis and application to exon prediction. *Genome Res.*, 10, 950-958.

Birney, E., Thompson, J.D., Gibson, T.J. 1996. PairWise and SearchWise: finding the optimal alignment in a simultaneous comparison of a protein profile against all DNA translation frames. Nucl. Acids Res. 24, 2730-9.

Borodovsky, M., McIninch, J. 1993. GeneMark: parallel gene recognition for both DNA strands. *Computers Chem.*, 17, 123-133.

Borodovsky, M., McIninch, J., Koonin, E.V., Rudd, K.E., Medigue, C., Danchin, A. 1995. Detection of new genes in a bacterial genome using Markov models for three gene classes. *Nucl. Acids Res.*, 23, 3554-3562.

Burge, C., Karlin, S. 1997. Prediction of complete gene structures in human genomic DNA. *J. Mol. Biol.*, 268, 801-807.

Carninci, P., Waki, K., Shiraki, T., Konno, H., Shibata, K., Itoh, M., Aizawa, K., Arakawa, T., Ishii, Y., Sasaki, D. *et al.* 2003. Targeting a complex transcriptome: the construction of the mouse full-length cDNA encyclopedia. Genome Res. 13, 1273-1289.

Cawley, S., Bekiranov, S., Ng, H.H., Kapranov, P., Sekinger, E.A., Kampa, D., Piccolboni, A., Sementchenko, V., Cheng, J., Williams, A.J., Wheeler, R. Wong, B., Drenkow, J., Yamanaka,

M., Patel, S., Brubaker, S., Tammana, H., Helt, G., Struhl, K., Gingeras, T.R. 2004. Unbiased mapping of transcription factor binding sites along human chromosomes 21 and 22 points to widespread regulation of noncoding RNAs. Cell. Feb. 20; 116(4), 499-509.

Chen, T.M., Lu, C.C., Li, W.H. 2005. Prediction of splice sites with dependency graphs and their expanded bayesian networks. Bioinformatics. 21, 471-482.

Cruveiller, S., Jabbari, K., Clay, O., Bernardi, G. 2004. Incorrectly predicted genes in rice? Gene. 333, 187-188.

Degroeve, S., Saeys, Y., De Baets, B., Rouze, P., Van de Peer, Y. 2004. SpliceMachine: predicting splice sites from high-dimensional local context representations. Bioinformatics. 21, 1332-1338.

Delcher, A.L., Harmon, D., Kasif, S., White, O., Salzberg, S.L. 1999. Improved microbial gene identification with GLIMMER. *Nucl. Acids Res.*, 27, 4636-4641.

D'Onofrio, G., Bernardi, G. 1992. A universal compositional correlation among codon positions. Gene. 110, 81-88.

Eyras, E., Caccamo, M., Curwen, V., Clamp, M. 2004. ESTGenes: alternative splicing from ESTs in Ensembl. Genome Res. 2004, 14, 976-87.

FANTOM Consortium, RIKEN Genome Exploration Research Group, Genome Science Group (Genome Network Project Core Group). 2005. The transcriptional landscape of the mammalian genome. Science. 309, 1559-1563.

Florea, L., Hartzell, G., Zhang, Z., Rubin, G.G., Miller, W. 1998. A computer program for aligning a cDNA sequence with a genomic DNA sequence. *Genome Res.*, 8, 967-974.

Foissac, S., Bardou, P., Moisan, A., Cros, M.J., Schiex, T. 2003. EUGENE'HOM: A generic similarity-based gene finder using multiple homologous sequences. Nucl. Acids Res. 2003. 31, 3742-3745.

Gelfand, M.S., Mironov, A.A., Pevzner, P.A. 1996. Gene recognition via spliced sequence alignment. *Proc. Natl. Acad. Sci.* USA, 93, 9061-9066.

Gish, W., States, D.J. 1993. Identification of protein coding regions by database similarity search. *Nat. Genet.*, 3, 266-272.

Guigo, R., Agarwal, P., Abril, J.F., Burset, M., Fickett, J.W. 2000. An assessment of gene prediction accuracy in large DNA sequences. *Genome Res.*, 10, 1631-1642.

Guigo, R., Reese, M.G. 2005. EGASP: collaboration through competition to find human genes. Nat. Methods. 2, 575-577.

Haas, B.J., Volfovsky, N., Town, C.D., Troukhan, M., Alexandrov, N., Feldmann, K.A., Flavell, R.B., White, O., Salzberg, S.L. 2002. Full-length messenger RNA sequences greatly improve genome annotation. *Genome Biol.*, 3.

Hebsgaard, S.M., Korning, P.G., Tolstrup, N., Engelbrecht, J., Rouze, P., Brunak, S. 1996. Splice site prediction in *Arabidopsis thaliana* pre-mRNA by combining local and global sequence information. *Nucl. Acids Res.*, 24, 3439-3452.

Howe, K.L., Chothia, T., Durbin, R. 2002. GAZE: a generic framework for the integration of gene-prediction data by dynamic programming. Genome Res. 12, 1418-1427.

International Human Genome Sequencing Consortium. 2001. Initial sequencing and analysis of the human genome. Nature. 409, 860-921.

Kampa, D., Cheng, J., Kapranov, P., Yamanaka, M., Brubaker, S., Cawley, S., Drenkow, J., Piccolboni, A., Bekiranov, S., Helt, G., Tammana, H., Gingeras, T.R. 2004. Novel RNAs identified from an in-depth analysis of the transcriptome of human chromosomes 21 and 22. Genome Res. 14, 331-342.

Kim, T.H., Barrera, L.O., Qu, C., Van Calcar, S., Trinklein, N.D., Cooper, S.J., Luna, R.M., Glass, C.K., Rosenfeld, M.G., Myers, R.M., Ren, B. 2005. Direct isolation and identification of promoters in the human genomes. Genome Res. Jun; 15(6): 830-839.

Kleffe, J., Hermann, K., Wahrson, W., Witting, B., Brendel, V. 1996. Logitlinear models for the prediction of splice sites in plant pre-mRNA sequences. *Nucl. Acids Res.*, 24, 4709-4718.

Korf, I., Flicek, P., Duan, D., Brent, M.R. 2001. Integrating genomic homology into gene structure prediction. *Bioinformatics*, 17, S140-148.

Krogh, A. 1997. Two methods for improving performance of an HMM and their application for gene finding. *Proc. Intl. Conf. Intell. Syst. Mol. Biol.*, 5, 179-186.

Krogh, A., Mian, I.S., Haussler, D. 1994. A hidden Markov model that finds genes in *Escherichia coli* DNA. *Nucl. Acids Res.*, 22, 4768-4778.

Kulp, D., Haussler, D., Reese, M.G., Eeckman, F.H. 1996. A generalized hidden Markov model for the recognition of human genes in DNA. *Proc. Intl. Conf. Intell. Syst. Mol. Biol.*, 4, 134-142.

Likashin, A.V., Borodovsky, M. 1998. GeneMark.hmm: new solutions for gene finding. *Nucl. Acids Res.*, 26, 1107-1115.

Majoros, W.H., Pertea, M., Salzberg, S.L. 2004. TigrScan and GlimmerHMM: two open source ab initio eukaryotic gene-finders. Bioinformatics. 20, 2878-9.

Mathe, C., Sagot, M.F., Schiex, T., Rouze, P. 2002. Current methods of gene prediction, their strengths and weaknesses. *Nucl. Acids Res.*, 30, 4103-4117.

McAuliffe, J.D., Pachter, L., Jordan, M.I. 2004. Multiple-sequence functional annotation and the generalized hidden Markov phylogeny. Bioinformatics. 20, 1850-60.

Meyer, I.M., Durbin, R. 2002. Comparative ab initio prediction of gene structures using pair HMMs. Bioinformatics. 18, 1309-18.

MGC Project Team. 2004. The status, quality, and expansion of the NIH full-length cDNA project: the Mammalian Gene Collection (MGC). Genome Res. 14. 2121-2127.

Misra, S., Crosby, M.A., Mungall, C.J., Matthews, B.B., Campbell, K.S., Hradecky, P., Huang, Y., Kaminker, J.S., Millburn, G.H., Prochnik, S.E., *et al.* 2002. Annotation of the *Drosophila melanogaster* euchromatic genome: a systematic review. *Genome Biol.*, 3.

Modrek, B., Lee, C. 2002. A genomic view of alternative splicing. *Nature Genet.*, 30, 13-19.

Nekrutenko, A., Makova, K.D., Li, W.H. 2002. The Ka/Ks ratio test for assessing the protein-coding potential of genomic regions: an empirical and simulation study. *Genome Res.*, 12, 198-202.

Nicolas, P., Bize, L., Muri, F., Hoebeke, M., Rodolphe, F., Ehrlich, S.D., Prum, B., Bessières P. 2002. Mining *Bacillus subtilis* chromosome heterogeneities using hidden Markov models. *Nucl. Acids Res.* 30, 1418-1426.

Nilsen, T.W. 2003. The spliceosome: the most complex macromolecular machine in the cell? Bioessays, 25, 1147-1149

O'Flanagan, R.A., Paillard, G., Lavery, R., Sengupta, A.M. 2005. Non-additivity in protein-DNA binding. Bioinformatics. 21, 2254-2263.

Pavy, N., Rombauts, S., Dehais, P., Mathe, C., Ramana, D.V.V., Leroy, P., Rouze, P. 1999. Evaluation of gene prediction software using a genomic data set: application to *Arabidopsis thaliana* sequences. *Bioinformatics*, 15, 887-899.

Pearson, W.R., Wood, T., Zhang, Z., Miller, W. 1997. Comparison of DNA sequences with protein sequences. *Genomics*, 15, 24-36.

Pertea, M., Lin, X., Salzberg, S.L. 2001. GeneSplicer: a new computational method for site prediction. *Nucl. Acids Res.*, 29, 1185-1190.

Riddihough, G. 2005. In the forests of RNA dark matter-Science, 309, 1507.

Rogic, S., Mackworth, A.K., Ouellette, F.B.F. 2001. Evaluation of gene-finding programs on mammalian sequences. *Genome Res.*, 11, 817-832.

Rogozin, I.B., D'Angelo, D., Milanesi, L. 1999. Protein-coding regions prediction combining similarity searches and conservative evolutionary properties of protein-coding sequences. *Gene*, 226, 129-137.

Salamov, A.A., Solovyev, V.V. 2000. Ab initio gene finding in Drosophila genomic DNA. Genome Res. 10, 516-22.

Salzberg, S.L., Delcher, A.L., Kasif, S., White, O. 1998. Microbial gene identification using interpolated Markov models. *Nucl. Acids Res.*, 26, 544-548.

Salzberg, S.L., Pertea, M., Delcher, A.L., Gardner, M.J., Tettelin, H. 1999. Interpolated Markov models for eukaryotic gene finding. *Genomics*, 59, 24-31.

Schiex, T., Moisan, A., Rouze, P. 2001. EuGene: An Eukaryotic gene finder that combines several sources of evidence. In: *Computational Biology*, First Conference on Biology, Informatics and Mathematics, Jobim 2000, Montpellier, France, May 3-5, Selected papers. *Lect. Notes Comput. Sci.* 2066, 111-125.

Siepel, A., Haussler, D. 2004. Combining phylogenetic and hidden Markov models in biosequence analysis. J. Comput. Biol. 11, 413-428.

Uberbacher, E.C., Mural, R.J. 1991. Locating protein-coding regions in human DNA sequences by a multiple sensor-neural network approach. *Proc. Natl. Acad. Sci.* USA, 88, 11261-1265.

Usuka, J., Zhu, W., Brendel, V. 2000. Optimal spliced alignment of homologous cDNA to a genomic DNA template. *Bioinformatics*, 16, 203-211.

Yamada, K., Lim, J., Dale, J.M., Chen, H., Shinn, P., Palm, C.J., Southwick, A.M., Wu, H.C., Kim, C., Nguyen, M., *et al.* 2003. Empirical analysis of transcriptional activity in the Arabidopsis genome. Science. 302, 842-846.

Yeh, R.F., Lim, L.P., Burge, C.B. 2001. Computational inference of homologous gene structures in the human genome. *Genome Res.*, 803-816.

Zhang, M.Q. 1997. Identification of protein coding regions in the human genome by quadratic discriminant analysis. *Proc. Natl. Acad. Sci.* USA, 94, 565-568.

Zhang, M.Q. 2002. Computational prediction of eukaryotic protein-coding genes. *Nature*, 3, 698-709.

6

CHAPTER

Detecting Protein Function from Genome Sequences

J.F. Gibrat, A. Marin

In July 2005, there were 266 complete genomes sequenced (211 bacteria, 21 archaea, and 34 eukarya), and 1,224 ongoing sequencing projects (687 bacteria, 41 archaea, and 496 eukaria)[1]. The ultimate aim of these sequencing projects is not limited to establishing a catalogue of the genes of an organism but extended to exploring, from the knowledge of these genes, all the intermediate organizational levels, biochemical and physiological, up to the phenotype. Other techniques of integrative biology, such as transcriptome and proteome techniques, make it possible to obtain vital additional information on the overall biological properties of these organisms.

However, a primary, fundamental step of these projects is to identify the genes on the chromosomes and assign to them a function. Experimentally, it is an arduous task, even with respect to a bacterium which has quite a small number of genes. Biologists can, however, get considerable help in this regard from bioinformatics techniques. These techniques are combined under the generic term "*in silico* functional analysis". In this chapter, we will describe these techniques, as well as their strengths and weaknesses.

Functional annotation of proteins encoded by their newly sequenced genomes must meet two objectives that are rather contradictory: on the one hand, as much information as possible must be extracted from the data collected, and on the other hand it is imperative to avoid annotation errors. In fact, the genomes annotated are used for the annotation of other genomes, so annotation errors can have disastrous snowballing effects. The chapter ends with a brief description of the most frequent causes of errors and a proposed strategy to prevent them as far as possible.

There are three major classes of methods used to obtain information on protein function:

- methods using information intrinsic to the sequence;
- homology search methods;
- methods based on the context of genes.

[1]http://wit.integratedgenomics.com/GOLD

We will describe these different methods in greater detail in the following sections. They are not all equally significant: some, such as homology search techniques, provide precise information and thus occupy a central place, while others, such as certain techniques using information intrinsic to the sequence, result in only general information (for example, that the protein is located in a membrane or even that it has a peptide signal indicating that it is a secreted protein). Nevertheless, the general philosophy of *in silico* functional analysis is that no information must be neglected, no matter how insignificant it seems at first. It is necessary to "figure out" the function from a series of analyses as complete as possible and to then verify the consistency of all the information collected. Some inconsistencies in the information may indicate a problem in assigning a function to a particular protein. This primary work, of supervision, verification, and integration of data, cannot in any case be left to automated methods and constitutes the principal role of the human annotator.

PRECISION OF THE CONCEPT OF PROTEIN "FUNCTION"

In the preceding paragraphs we speak of assigning a function to a protein. This expression has the advantage of simplicity but is, in fact, highly imprecise. As often in biology, reality cannot be reduced easily into simplified categories proposed by scientists.

There are two points to consider when speaking of the "function" of a protein: the first is the modular nature of numerous proteins and the second is the exact definition of the term "function".

Modular Aspect of Proteins

Many proteins, particularly in multicellular eukaryotes, are composed of functional domains (or modules). In the rest of this chapter "module" is reserved for the functional aspect and "domain" for the structural aspect. Even though there is often a one-to-one relationship between a functional module and a structural domain, there are cases of functional modules being composed of several structural domains.

A structural domain is a part of the polypeptide chain that has a unique, compact, globular conformation, organized around a hydrophobic core. If this domain is excised from the initial protein, a new protein is obtained that must be stable.

A functional domain, as its name indicates, fulfils a particular function. For example, cytochrome P450s are heme enzymes that hydroxylate various organic compounds. The biochemical reaction requires two electrons that are, in many organisms, provided by a reductase that is docked with the cytochrome. In *Bacillus megaterium*, these two proteins are fused into a single polypeptide chain that has an enzymatic module and an integrated reducer module (Sevrioukova *et al.*, 1999). This fusion of two closely interacting proteins into a single protein is not rare. It is also the basis of the bioinformatics method allowing indication of protein-protein interactions (see: methods based on the context of genes).

Much more complex modular organizations are common, as can be seen from the site ProDom[2] (D. Kahn, INRA Toulouse). Ignoring this organization in modules can, in

[2]http://prodes.toulouse.inra.fr/prodom/doc/prodom.html

the best of cases, result in a loss of information and, at worst, be the cause of one of the most frequent and most serious errors in genome annotation (see: conventional sources of errors during annotation).

Therefore, in many cases, we cannot speak of the function of a protein but only of some functions of a protein. In fact, in annotating a genome, the appropriate level that must be considered is often the module rather than the protein.

Precise Definition of the Term "Function"

The concept of function is relatively imprecise and unsuitable for describing the complex role of a protein in a cell (Bork *et al.*, 1998). The best way to understand this function is to use a hierarchical description:

- A protein has a **molecular function**. It is, for example, an enzyme that catalyses a particular biochemical reaction, a transporter of a given chemical compound, a structural protein, a protein with a particular binding site, or a protein allowing the translation of a signal.
- A protein also has a **cellular function**. The enzyme in question is involved in a particular metabolic pathway, for example a cytochrome P450 can intervene in a process of cell detoxification or even in the steroid synthesis pathway, the structure protein is involved in the constitution of the cytoskeleton, the protein plays a role in a signaling cascade.
- Finally, a protein also has a **phenotypic function**. It plays a role in behaviour, morphology, physiology, etc. For example, serine proteases, in man, are involved in digestion, immune response regulation, and onset of spermatogenesis.

In fact, the function becomes increasingly dependent on the context as one goes up the biological hierarchy, i.e., as one goes from the genome to the whole organism.

The different methods that will be described below provide a variety of information on these three functional levels. In what follows, for the sake of simplicity we continue to speak of assigning a function to a protein, but the reader is advised to keep in mind the limitations of the term just mentioned.

METHODS USING INFORMATION INTRINSIC TO THE SEQUENCE

Methods using information intrinsic to the sequence are based on the analysis of sequences, without referring to other sequences stored in the databases. They provide relatively general information on protein function.

An important secondary effect of such methods is to isolate the regions of the sequence that have a non-standard composition of amino acids, e.g. "coiled-coils" regions. In fact, these regions, which have a biased composition of amino acids, are the cause of errors when sequence comparison methods are used. It is thus important to use methods based on intrinsic information first of all as filters. This will allow the division of the sequences into zones having a biased composition of amino acids and zones having the expected composition for a globular protein. Only the latter are subjected to homology search methods.

Detection of Transmembrane Segments

Transmembrane segments that cross the double lipidic membrane are mostly made up of hydrophobic residues. Even though there are transmembrane segments that are able to adopt a secondary structure in β sheets, for example in porin (Schirmer *et al.*, 1995), most of the transmembrane segments are made up of α helices. Around 25-30 residues are required for the polypeptide chain to cross the membrane in the form of an α helix. Transmembrane segment detection methods are thus based on the search for segments that have an appropriate size and present a marked hydrophobic character. Figure 6.1 shows a diagram of the variation of hydrophobicity along the sequence of rhodopsin. The seven transmembrane segments of the molecule, which appear highly hydrophobic, can easily be recognized. There are various ways to accomplish this segmentation, using hidden Markov models (Krogh *et al.*, 2001), neural networks (Rost *et al.*, 1996), or statistical techniques (Hirokawa *et al.*, 1998).

The principal information obtained with this type of method involves the location of the protein within the cell. The membrane proteins constitute a very important class of molecules and this type of information is thus crucial. As will be seen in the rest of the chapter, it is possible to use, in addition, homology search techniques to attempt to determine precisely the molecular function of the protein studied.

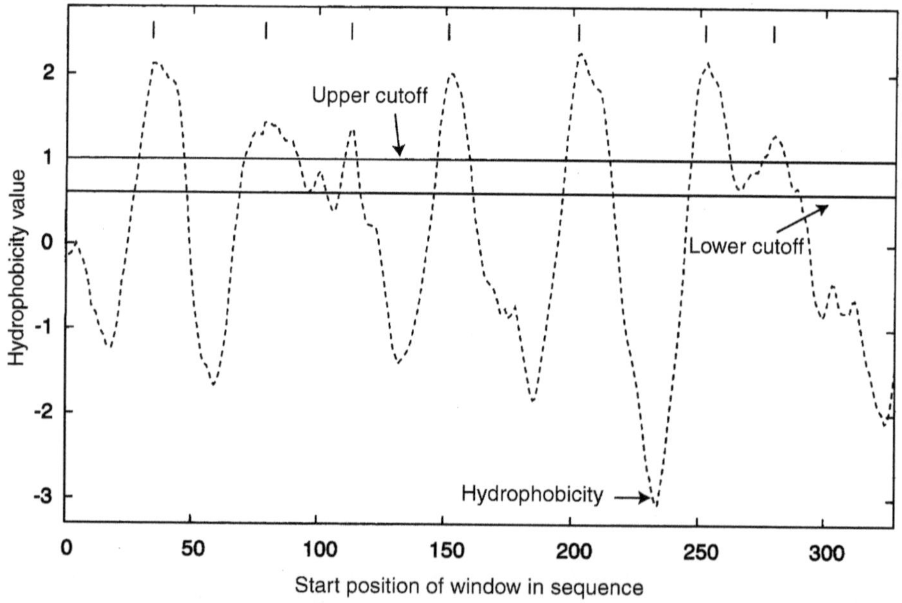

Fig. 6.1 Graph of mean hydrophobicity in a window along the rhodopsin sequence (from the Toppred program, von Heijne, 1992). The seven transmembrane segments are clearly detected with a hydrophobicity value greater than one (the center of the segment is indicated by a vertical line in the upper part of the figure).

Detection of Zones of Low Complexity

In certain segments of protein sequences there is an unusual composition of amino acids. These segments have an over-representation of a small number of amino acids (for example, zones rich in glycine or glycine and glutamic acid), or even show a more or less regular repetition of a particular peptide. Little is known about the properties of these zones, but generally it is acknowledged that they do not have a conventional three-dimensional (3D) globular structure.

Figure 6.2 shows the result of the analysis of the primary sequence of a glutenin (an important wheat protein in the process of bread making) carried out using the SEG program (Wootton and Federhen, 1996). This molecule is made up of two globular domains flanking a domain that has a more or less regular repetition of proline, glutamine, and glycine. The central domain has been studied by physical methods and it is shaped like a rod that seems to be made up of a more or less loose β spiral formed of β turns (Shewry *et al.*, 1992). It is likely that this is the part that gives the molecule the visco-elastic properties that are important in the manufacture of bread.

These zones thus supply practically no information on the function of proteins. On the other hand, they must be masked before using homology search methods, because their "abnormal" amino acid composition disturbs the statistics associated with these techniques and often results in false inference of homology. In other words, a glycine-rich region, for example, will "recruit" all the proteins having glycine-rich regions, even if those proteins have no particular relationship to one another.

Detection of "Coiled Coils"

Coiled coils zones (Lupas *et al.*, 1991) are formed of a bundle of two or three α helices in parallel that interact in an approximate angle of 20°, thus creating a coiled coil (Fig. 6.3). This coiled coils result from a periodicity of seven positions along the sequence. If the seven positions of the heptapeptide are designated *a* to *g*, then positions *a* and *d* are occupied by hydrophobic residues that constitute the interface between the helices and positions *b*, *c*, *e*, *f*, and *g* are occupied by hydrophilic residues in contact with the solvent.

Coiled coil zones are present in a certain number of proteins, for example, as leucine zippers of transcription regulation proteins, or the turning point of myosine. These zones are detected on the basis of statistical techniques that take into account the probability of observing a particular amino acid at each of seven characteristic positions (Lupas, 1996). The presence of coiled coils zones in the sequence can be used to direct the search for function towards certain types of proteins, such as regulator proteins or structural proteins. These regions are also often implicated in processes of oligomerization, this is true for leucine zippers of regulatory proteins, and it is possible that coiled coils zones could play a role in the polymerization of flagelines.

Determination of Cellular Sorting Signals

The routing of proteins towards organelles that are their final destination is a fundamental aspect of cell life. Often, this cellular sorting depends on signals present in the primary structure. For example, proteins are directed towards secretion pathways,

Low complexity fragments		'Regular' fragments
	1-120	AEGEASRQLQCERELQESSLEACRQVVDQQ LAGRLPWSTGLQMRCCQQLRDVSAKCRSVA VSQVARQYEQTVVPPKGGSFYPGETTPLQQ LQQGIFWGTSSQTVQGYYpGVTSpRQGSYY
pgqaspqqpgqgqqpgkwqepgqgqq	121-146	
	147-153	WYYPTSL
qqpgqgqqigkgqqg	154-168	
	169-174	YYPTSL
qqpgqgqqgyyp	175-186	
	187-189	TSL
qhtgqrqqpvqgqqpeqgqqpgqwqqg	190-216	
	217-222	YYPTSP
qqlgqgqqprqwqqsgqgqqg	223-243	
	244-263	HYPTSLQQPGQGQQGHYLAS
qqqpgqgqqg	264-273	
	274-278	HYPAS
qqqpgqgqqg	279-288	
	289-293	HYpAS
qqqpgqgqqg	294-303	
	304-305	HY
pasqqepgqgqqgqipasqqqpgqgqqg	306-333	
	334-339	HYPASL
qqpgqgqqg	340-348	
	349-353	HYPTS
lqqlgqgqqtgqpgqkqqpgqgqqtgqgqq peqeqqpgqgqqg	354-396	
	397-402	YYPTSL
qqpgqgqqqgqgqqg	403-417	
	418-423	YYpTSL
qqpgqgqqg	424-432	
	433-438	HYPASL
qqpgqgqpgqrqqpgqgqhpeqgkqpgqgq qgyyptspqqpgqgqqlgqgqqgyyptspq qpgqgqqpgqgqqg	439-512	
	513-517	HCPTS
pqqsgqaqqpgqgqqigqvqqpgqgqqg	518-545	
	546-551	YYPTSV
qqpgqgqqsgqgqqsgqghqpgqgqqsgqe qqg	552-584	
	585-628	YDSPYHVSAEQQAASPMVAKAQQPATQLPT VCRMEGGDALSASQ

Fig. 6.2 Division of the glutenin sequence into segments having different compositions. The column on the left lists zones of low complexity and the column on the right lists zones having the composition expected for a globular protein. In the centre is the position of these segments in the sequence (from the SEG program, Wootton and Federhen, 1996). A primary globular domain covering the first 120 residues is clearly seen, followed by a long domain having a low complexity (its composition is dominated by three amino acids: proline, glutamine, and glycine) that is known experimentally not to be globular, and finally a small globular domain in the C-terminal part.

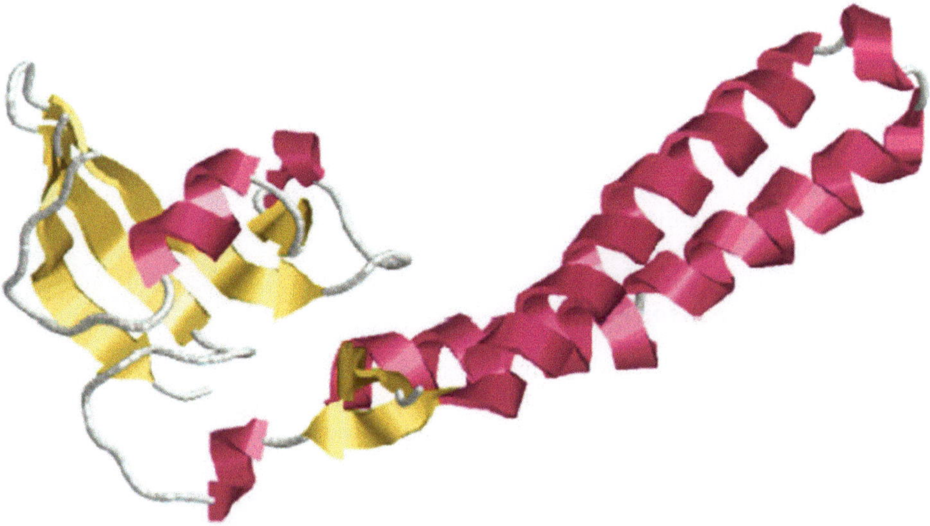

Fig. 6.3 Structure of the GREA protein (restriction factor of the transcript of *E. coli*, code PDB 1GRJ, Stebbins *et al.*, 1995) represented in a "cartoon" form: the β sheets are in light gray and the α helices in dark gray. The first 75 residues of the protein, which appear on the right side of the figure, are in the form of two α helices forming a typical coiled coil.

mitochondria or chloroplasts, or even towards a membrane anchoring by a peptide located at the N-terminal part of the sequence (Emanuelsson *et al.,* 2000). Other techniques based on the overall composition of amino acids can be used to predict the location of proteins in various organelles: e.g., the nucleus, the lysosome, the endoplasmic reticulum (Emanuelsson and von Heijne, 2001). This last type of method has an advantage, since it is based on the complete sequence and is thus not too heavily dependent on the N-terminal part of the protein. The N-terminal sequence is sometimes poorly determined by automated gene prediction techniques. Furthermore, this information can be corroborated by homology search techniques, for example by indicating a domain that is characteristic of a nuclear location.

HOMOLOGY SEARCH METHODS

Homology search methods allow the researcher to obtain the most detailed information on protein function. They thus play a central role in *in silico* functional analysis. To adopt a simplistic view, these methods consist of comparing unknown proteins with proteins stored in databases, so as to discover similar proteins in the hope that the latter possess annotations which can be transferred to the former.

Properties of Homologous Proteins

More formally, homology search techniques are based on the concept of homologous proteins. Two proteins are homologous if they evolved from a common ancestor (Fitch, 2000).

There are several types of homologous proteins, illustrated in Fig. 6.4, the two principal categories being orthologous proteins (which result from a process of speciation) and paralogous proteins (which result from a process of duplication). It is the iso-orthologues, i.e., proteins having the same function in different organisms, that are most relevant to genome annotation.

For our present purposes, homologous proteins have two interesting properties: their 3D structures are very similar and their functions are often similar. The following sections describe the different means by which a relationship of homology between two proteins can be inferred.

Sequence Comparison Methods

The most natural method, and also the oldest, of indicating a relationship of homology between two proteins, is to compare their sequences. If they are thought to have derived from a common ancestor, it can be supposed that there must remain some traces of that relationship in the sequences (it will be seen that this is not always true, if the last common ancestor is very distant). If the two sequences are sufficiently similar,

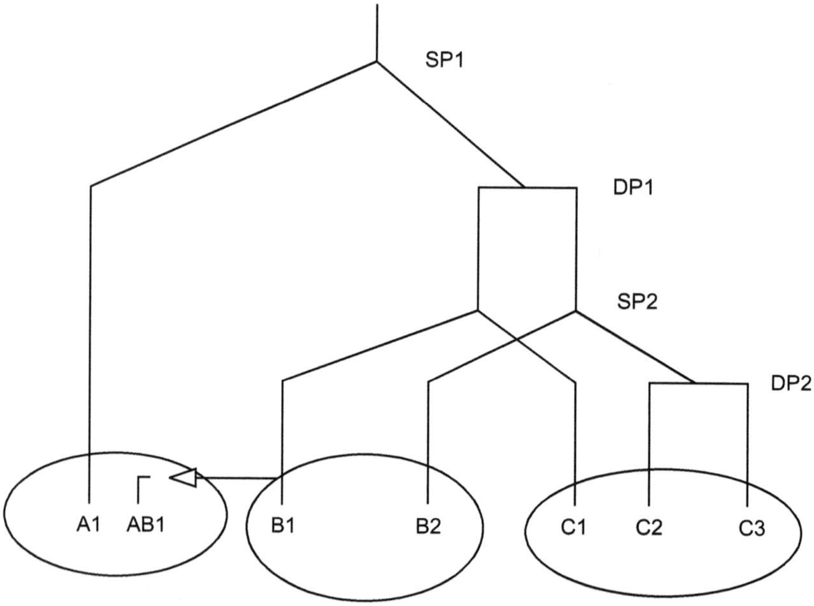

Fig. 6.4 Diagram of the notion of homology. SP represents a speciation event and DP a duplication event. In the lower part of the figure are three organisms A, B, and C, and some of their genes. Two genes are orthologous if their common ancestor is located at an SP junction. Two genes are paralogous if their ancestor is located at a DP junction. For example, C2 is paralogous with C3. C2 and C3 are orthologous with B2 but paralogous with B1. A1 is orthologous with B1, B2, C1, C2, C3. The arrow represents a horizontal transfer of the B1 gene towards organism A (where this gene is called AB1). AB1 is said to be xenologous with the six other genes represented. Given an increasing number of complete genomes of prokaryotes, the horizontal transfers of genes in these organisms can be hypothesized to be more frequent than earlier thought.

it can be inferred that they are homologous, the question is then: what is meant by "sufficiently" similar?

BLAST (Altschul *et al.*, 1990) and FASTA (Pearson, 2000) are sequence comparison methods based on this principle. The similarity between the sequences is estimated by aligning the sequences as closely as possible, each alignment being characterized by a score. BLAST is the first method to have responded rigorously to the question posed above, by allowing a statistical calculation to be made of the significance of the score. In fact, BLAST can be used to calculate the probability of obtaining by chance an alignment having a score equal to, or higher than a particular score (Karlin and Altschul, 1990). This score is expressed in the algorithm in the form of an expectation value (E-value) that also depends on the size of the sequence library used during the search. If this E-value is low, for example less than 10^{-3}, there is little chance for the similarity observed to be due to chance and thus the proteins studied are likely to be homologous.

This statistical aspect, which provides a rigorous criterion that makes it possible to state the homology of two proteins, has largely contributed to the success of BLAST. The original article has been one of the most often cited in molecular biology, during the past 15 years. The highly effective implementation of the algorithm and its widespread dissemination (through websites as well as programs locally installed in laboratories), has also helped to make it a preferred tool of molecular biologists.

In spite of these advantages, sequence comparison methods have a fundamental problem that has been mentioned briefly above: they depend critically on the conservation of sequences in homologous proteins. Brenner *et al.* (1998) have shown, for proteins that were known *a priori* to be homologous, that BLAST can reveal only around 15% of pairs of homologous proteins (with an error rate of 1%) if the percentage of residues conserved is less than 35%. In particular, when this percentage of conserved residues falls below 20%, techniques such as BLAST can no longer be used to detect relationships of homology whilst still maintaining an acceptable error rate.

What can be done to solve this problem and attempt to detect "distant" homologues?

Profile Detection Methods

A preliminary approach is to use, instead of a single sequence, a multiple alignment of similar sequences (also called profiles). In fact, there is always more information in a multiple alignment than in an isolated sequence. A multiple alignment can be used to estimate the variability of each of the positions along the sequence of a protein. This type of method will not treat an amino acid strictly conserved in several sequences (which probably has a functional or structural role), the same way as it treats the same type of amino acid at a position in the sequence where it is observed that very different amino acids could exist. Two types of techniques make use of these multiple alignments.

The first is an extension of BLAST named PSI-BLAST (Altschul *et al.*, 1997), which constructs the multiple alignment iteratively during the search. The first iteration of this method corresponds to a conventional BLAST. After this iteration, sequences having an E-value below a certain threshold are combined and used to create a profile or position specific scores matrix (PSSM). This matrix is then used to effect a new iteration that allows the recruitment of new sequences, more distant from the initial sequence. The process continues until it converges, i.e., until new sequences can no longer be recruited in this manner.

The second type of technique uses multiple alignments of families of proteins that as in the case of Pfam (Bateman *et al.*, 1999) or Smart (Schultz *et al.*, 2000) have been carefully verified and annotated manually by experts. The database Smart, for example, was created at first to describe all the "genetically mobile" domains, i.e., ubiquitous domains that are found in many different proteins. These domains generally have no enzymatic function and contribute information on, for example, the cellular role of proteins, the partners to which they are linked, the protein location in the organelles. A statistical model of these families is built with a conventional bioinformatics technique that uses hidden Markov models (HMMs). Subsequently, it is verified whether the sequence of the target protein is compatible with the statistical model of one of the different families.

These two rather similar techniques are based on two symmetrical approaches. In the first case, proteins from databases are compared with a profile created from the unknown sequence. In the second case, this unknown sequence is compared with profiles of families of proteins from databases. An additional advantage of these techniques is that they eventually allow segmentation of the unknown protein sequence into domains.

It has been demonstrated that this type of technique using multiple alignments is two to three times as sensitive, for the same rate of error, as techniques such as BLAST that use only a single sequence (Park *et al.*, 1998).

Methods of Motif Detection

Another approach to detect distant homologues, is to search for motifs (a motif is similar to a regular expression in computer science (Fig. 6.5). Motifs correspond usually to a functional signature, or even to residues necessary to maintain the correct geometry of the active site of a protein. As these residues are crucial for the function of the protein they are thoroughly conserved. Some programs, such as Scanregexp and Pfscan, developed at the Swiss Institute of Bioinformatics, can be used to search systematically for motifs characteristic of particular proteins in the Prosite library of motifs (Hofmann *et al.*, 1999).

The principal limitation with respect to the detection of motifs, is that the residues that constitute them must be more or less contiguous along the sequence. If they are far apart and separated by loops that could be of variable length, it becomes very difficult

```
C-[DE]-[LIVM](2)-[NQ]-[GTA]-D-G-[SG]-X(2)-[TA]-A
```

Fig. 6.5 Motif of ribonuclease H (code PS01277 in Prosite). This motif contains 13 contiguous positions. The amino acids are indicated by their classic single-letter code. The symbol [] signifies that any one of the residues listed may be present. The symbol (*n*) indicates that what immediately precedes is repeated *n* times, and X represents any amino acid. For example, the motif begins with a cysteine (C) followed by an acidic residue (either an aspartate (D) or a glutamate (E)), then 2 hydrophobic residues chosen among a leucine (L), an isoleucine (I), a valine (V), or a methionine (M), and so on.

to find the corresponding motif. Quite obviously, often the active site of a protein is made up of distant residues in a sequence that have been brought together by the folding of the protein.

Methods of Fold Recognition

We have seen earlier that one of the important properties of homologous proteins is that they maintain very similar 3D structures. Certain homologous proteins have highly divergent sequences, with sometimes less than 15% of conserved residues. These distant homologues cannot be detected by sequence comparison methods. There is simply not enough information in the sequence (except in the case of a functional signature, see above) for it to be a good means of indicating a relationship of homology. The 3D structure must therefore be examined.

Methods that use all the 3D structures known to indicate relationships of homology are called "threading methods".

The use of this type of technique is justified by the fact that the number of 3D structures possible is very limited. It is now estimated that this number is no more than a few thousand (Wolf *et al.*, 2000), and thus much less than the number of known protein sequences (presently 2,500,000 in non-redundant databases, but this number increases exponentially with the sequencing of an increasingly larger number of genomes). There are two reasons for this, the first is that many proteins derive from common ancestors and thus have conserved the same 3D structure, and the second is that there are a certain number of physico-chemical constraints that reduce the number of possible folds.

Threading methods are based on the alignment of a sequence on a 3D structure. The adequacy between the sequence and the structure is measured with an empirical score function. Figure 6.6 presents a simple description of this type of technique.

The method that we have developed, FROST, is twice as sensitive as PSI-BLAST with the same rate of error (Marin *et al.*, 2002a, 2002b).

This type of method has two disadvantages. One, the corresponding fold may not exist in the databases of 3D structures. Two, this technique does not directly indicate a relationship of homology between two proteins but only shows that a sequence is compatible with a 3D structure. There are proteins that have the same 3D structure but completely different functions (and for which most experts do not agree in their homologous relationship). It is thus more difficult to deduce the function with these methods than with sequence comparison techniques, where the resemblance between sequences, although it may be less significant, is unequivocally a trace of homology. With this type of technique, it is absolutely vital, if the function is to be transferred, to ensure that the residues of the active site are well conserved. It must be noted that this is an elementary precaution that must be taken even with sequence comparison methods.

METHODS BASED ON THE CONTEXT OF GENES

With the availability of an increasingly large number of complete genomes, it has recently appeared that information can be obtained on the function of genes by

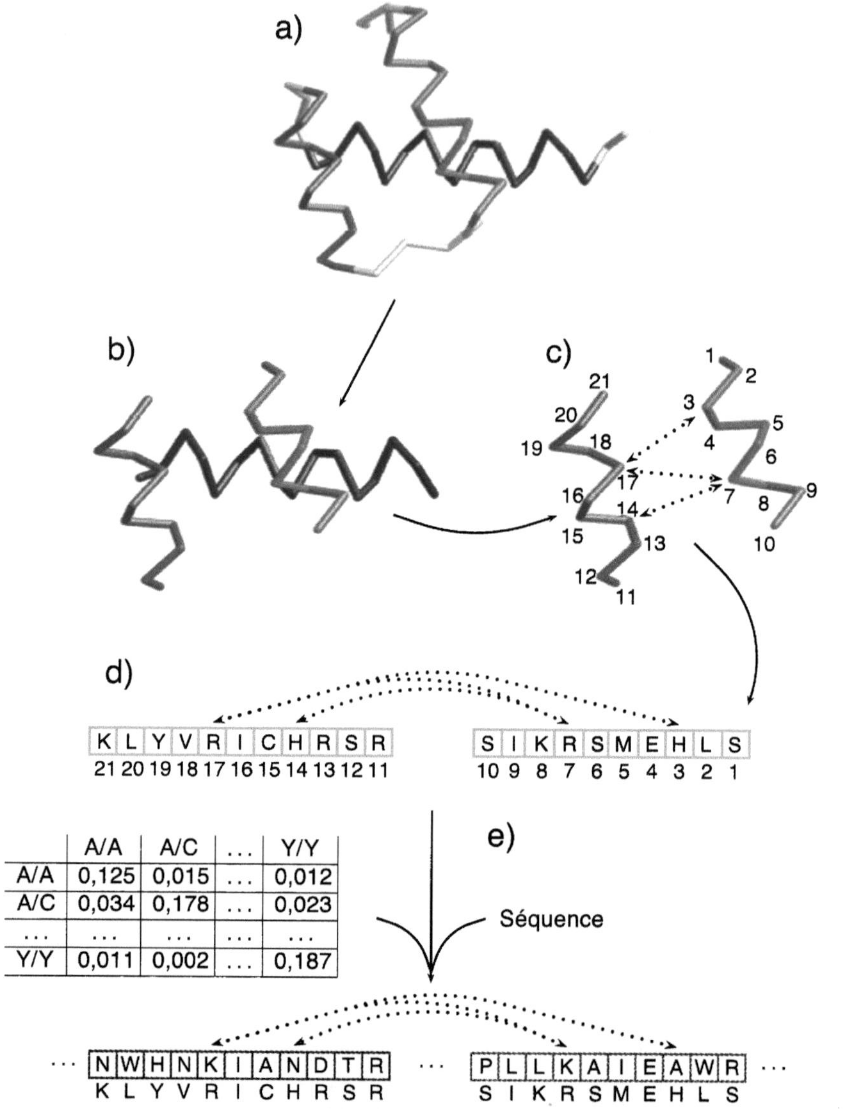

Fig. 6.6 Diagram of processes for fold recognition. (a) 3D structure composed of three α helices. (b) Only the structurally conserved elements (in the family of similar 3D structure) are taken into account (here the three α helices). That defines what is called the core of the protein. (c) The structure is coded in the form of contacts between residues (here only three contacts between two helices are represented by dotted arrows). (d) "Flat" representation of these contacts. The sequence corresponds to the actual sequence of the 3D structure studied. (e) A new sequence is aligned on the core. In the alignment represented here, for the pair in contact (3-17), the pair of original residues H-R is replaced by the pair A-K. An empirical substitution matrix provides the score when an H-R pair is replaced with an A-K pair. The total alignment score is the sum of such scores on all the pairs in contact in the 3D structure. All the possible alignments are generated and the one with the best score is retained.

studying their co-location in different genomes (Aravind, 2000). This type of approach is often called *comparative genome analysis*.

Unlike homology search techniques that often provide information on the molecular function of proteins, this type of technique generally provides information on the interactions between proteins and thus their cellular role.

The contextual information exists in various forms, more or less "diffuse", in the genomes. In the following paragraphs, the different forms are described in decreasing order of physical proximity. It should be noted that the number of genomes to be compared is as high as these constraints of proximity are low.

Gene Fusion

The process of gene fusion has been evident for some time, following the comparison of certain metabolic pathways in prokaryotes and eukaryotes. It is often observed that a metabolic pathway made up of independent enzymes in the prokaryotes is catalysed by a multi-enzymatic system in the eukaryotes (Davidson and Peterson, 1997).

The fusion of functional modules provides, *a priori*, quite precise information on the interaction between these two modules (Marcotte *et al.*, 1999). Two proteins that exist in the independent form in a genome and that are fused in another genome have a high chance, as indicated above, of being in close interaction, whether they form part of a single oligomeric complex or catalyse adjacent reactions along a particular metabolic pathway. This is the case in the example given earlier of cytochrome P450 and the reductase of *Bacillus megaterium*.

There are, however, elementary precautions to be taken before interpreting this kind of result. In fact, the fusions of functional modules can be classified in several categories that do not all give the same type of information on the functional role of the modules implicated.

The first category is that of old fusions, i.e., those that occurred in an ancestor common to most of the present lines of organisms. This is for example the case of the α subunit of bacterial polymerase III, which is composed of a phosphoesterase domain, a DNA polymerase region, and two modules linking the DNA. These are not very interesting in the prediction of interactions between modules because most of the time they are meaningful only in the particular context in which they were observed (Aravind, 2000).

The second category of fusions concerns the "genetically" mobile domains that are found in different proteins, for example the DEATH domain, which is present in proteins implicated in apoptosis, or even domains in β propeller, which allow the establishment of protein-protein interactions. Here also, this type of fusion supplies no particular information on protein-protein interactions but may provide information on location or the cellular pathways involved.

The third category is that of fusions similar to those described in the case of cytochrome P450, which provide some information on interaction between proteins. Before interpreting the results of these studies, it must be ensured that the category is relevant to extracting the pertinent information. This type of study does not require a large number of genomes to be valid; it is enough to observe a process of fusion of modules in a single genome, to be able to pronounce on the interaction of those modules. Such an analysis can prove highly useful in conjunction with double hybrid data.

Proximity of Genes

It has often been observed that functionally linked genes are co-regulated and have a tendency to be close together in the genome. This is particularly true for prokaryotes with the existence of operons.

It is known that the position of a gene in the genome may provide information about its function (Overbeek *et al.*, 1999). The presence of a non-characterized gene in a particular environment (such as an operon) in which the function of other genes is known, makes it possible to obtain information on the role of the unknown gene. This approach can be used, to determine precisely the information obtained through homology search. For example, homology search techniques may indicate that a protein is homologous with various ABC transporters, the most closely similar of which apparently is a glucose transporter. If the gene in question is in an environment in which the nearby proteins are identified as intervening in the metabolism of xylose, it may be thought that the most likely function for this unknown protein is an ABC transporter of xylose.

In practice, this approach comes up against the difficulty that the operons must be predicted precisely, as well as the large variability of operons in the bacteria. Moreover, the approach is limited to prokaryotes, since eukaryotes do not have this type of organization of genes. It is possible, however, to use a measure of more local proximity that will involve the conservation of nearby gene pairs in different genomes being compared.

This last approach is also not free of difficulties. When the genomes are too close, chromosomal rearrangements have not had time to occur, and there are few selective constraints associated with the conservation of genes. If on the other hand the genomes are too far apart, it may be difficult to indicate a relationship of orthology between the elements of the pair. As far as possible, it is necessary to avoid detecting a region in which the conservation of genes is due to recent horizontal transfer. Finally, this approach requires a larger number of complete genomes than in the case of gene fusion in order to improve the selectivity of predictions.

Co-occurrence of Genes

Co-occurrence of genes is the most diffuse contextual information. It is expressed in the form of phylogenetic profiles (Pellegrini *et al.*, 1999). A phylogenetic profile, for a given protein, is a binary vector that indicates the presence (1) or the absence (0) of iso-orthologues (proteins having the same function in different genomes), in a particular set of genomes. The idea that sustains this approach is that genes that are implicated in a particular cellular process, common to genomes of this set, share an identical phylogenetic profile.

The principal difficulty with this approach lies in detecting the iso-orthologues. Moreover, it requires a sufficient number of genomes for the resulting analysis to have a statistical meaning. An unknown protein that shares a phylogenetic profile with proteins that are known to be implicated in a particular cellular pathway, has a strong chance of playing a role in that pathway. It is, of course, advisable to test results of this type of analysis with available experimental data on the transcriptome or the proteome.

CONVENTIONAL SOURCES OF ERRORS DURING ANNOTATION

Errors in the Databases

Databases inevitably contain errors, either because of an approximate use of bioinformatics methods of sequence analysis (see below), or because of uncertain extrapolations proposed by experimenters.

Some biologists, for example, often suppose that a gene codes for an enzyme when its product complements a known mutation, even if this effect may be the consequence of other biological mechanisms. For example, the gene of *P. aeruginosa* named *thrB* complements the mutation *thrB*- in *E. coli* and it is thus supposed that it codes for a protein similar to the product of *thrB* in *E. coli*: a homoserine kinase (Clepet *et al.*, 1992). An analysis of the sequence of the corresponding protein shows no similarity with the same enzyme in related organisms or any ATP binding motif. Besides, its inactivation does not confer an auxotrophy for threonine. It is thus possible that the function assigned in this case is erroneous (Galperin and Koonin, 1998).

The main thing to remember with respect to problems of erroneous annotations in databases, is that it is absolutely necessary to obtain the largest quantity of and most diverse information on the function of a protein. The role of the annotator is then to ensure that such information is consistent and, if it is not, to attempt to resolve the inconsistencies using other types of data: e.g., bibliography, complementary experimental data, knowledge of the organism studied. If the annotator cannot resolve the contradictions it is also crucial that this be indicated clearly in the annotation of the new genome, in order to avoid perpetuating errors.

"Biased" Composition of Amino Acids

As indicated previously, regions with a biased composition of amino acids disturb the statistics of homology search methods such as BLAST or FASTA. The statistical properties of scores have been calculated for a "standard" composition of amino acids in globular proteins. A different composition invalidates the statistics. Typically, proteins rich in one type of amino acid will all seem very similar, even if they actually have no relation to each other.

It is thus very important to indicate these zones and thus to divide the protein sequence into zones of low complexity and "standard" zones. Homology search methods should be used only on the latter. Some alignment programs like BLAST do provide a filter for that type of regions (-F option), but users must pay attention that this option is often not used by default, either on the NCBI web site, or in the binary version of the program.

Modular Organization of Proteins

As we have seen earlier, it is important during the annotation to take into account the modular organization of proteins. Figure 6.7 shows a typical case of two proteins

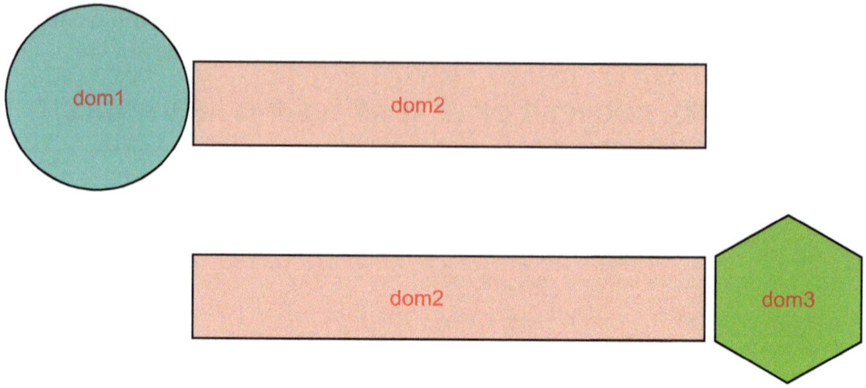

Fig. 6.7 Two proteins having a modular organization sharing a common module (the *dom2* domain). The analysis of such proteins, if it does not take into account this modular organization, is the source of errors in annotation.

sharing a common domain–here the *mod2* module–for example a regulator domain, in which the two other domains, for example enzymatic domains, are different and thus have different functions. If the first protein has been annotated with the function corresponding to the enzymatic domain *dom1*, to transfer this annotation to the second protein will result in an error that is more damaging to the extent that, it could theoretically propagate itself explosively. It is clear that the erroneous annotation could be transferred to any protein that shares the *dom3* domain with the second protein and so on, until a stage of complete chaos. This type of error is unfortunately not rare in databases!

Interpretation of the List of Homologues

Homology search methods often generate a consequent list of homologous proteins. It is necessary to remember that to annotate a genome we first look for a protein that is iso-orthologous, i.e., a protein having the same function in another organism. The list will contain, undoubtedly, a large number of paralogues or simple orthologues (Fig. 6.4). It is common in this case, but inadvisable, to attribute the function of the first protein in the list, that has the smallest E-value. This is the source of inaccuracies for at least two reasons:

(1) The protein having the best E-value may very well belong to a related genome that has just been sequenced and for which the annotation is still very rudimentary. It would be safer to run through the list to find a protein for which the annotation is better known, if such a protein exists.

(2) The E-value does not directly measure the distance between two homologous proteins in terms of evolution. The E-value measures the probability that the score observed will be due to chance. It depends significantly on the length of the two proteins compared. If we compare two small proteins having 90% of residues conserved and two others that are much larger having 30% of residues conserved, we may very well get similar E-values, even if we can assume that the

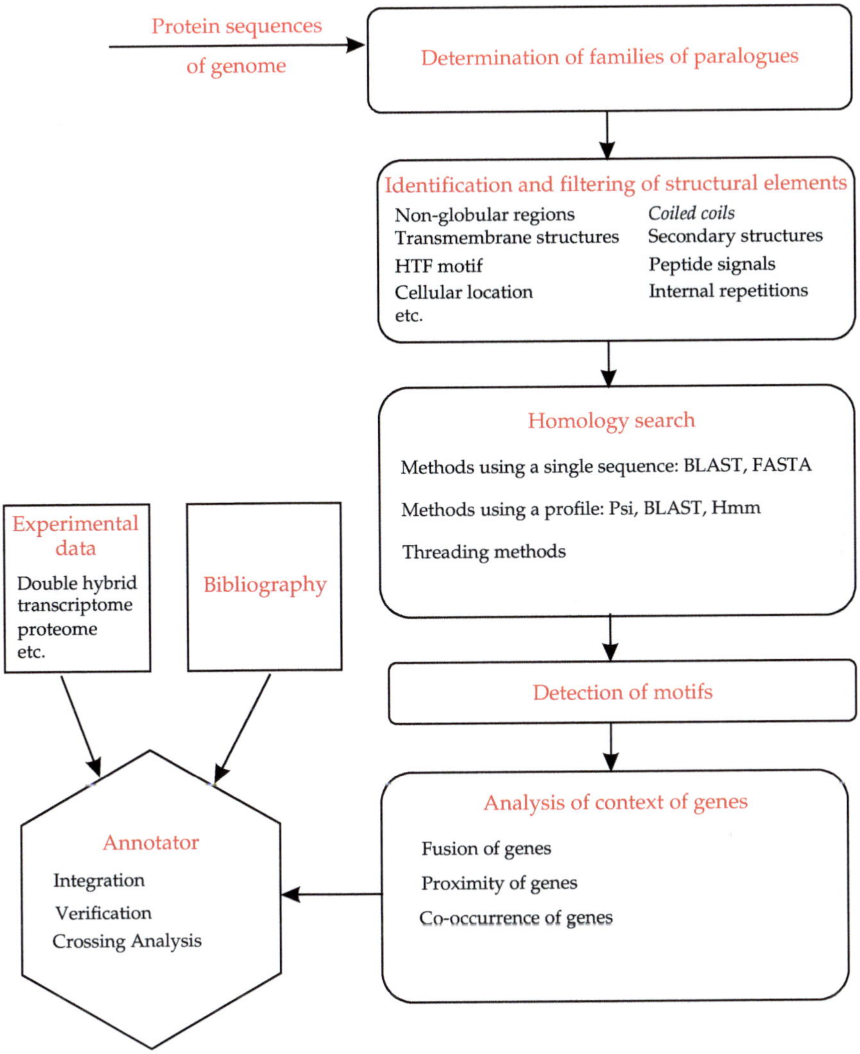

Fig. 6.8 Diagram of strategy adopted to annotate a genome

two proteins having 90% of residues conserved are closer than the two proteins having 30% of identical residues.

Often the general function of a protein can easily be detected, but its specificity (i.e., the different substrates on which it acts) is more difficult to indicate. For example, there are numerous transport proteins (such as permeases) or families of enzymes having different specificities, for which the paralogues and iso-orthologues have nearly the same level of similarity. It is thus difficult to propose a precise function without the help of other data, as we have seen elsewhere.

It is important, therefore, to analyse the results obtained by homology search methods and to cross-check them with other data.

Proximity within an Operon

As indicated in the section *Proximity of genes*, operons are not always well conserved in bacteria. A prediction of the function that is based solely on this type of data, without critical analysis, may prove false. For example, the protein RFAK-ECOLI was predicted as coding for a lipolysaccharide 1,2-N-acetylglucosamine transferase for the simple reason that it is found in the operon *raf* of *E. coli*. It was subsequently demonstrated to be, in fact, a D-glycero-D-manno-heptosyl transferase (Gibson *et al.*, 1997).

Non-orthologous Displacement of Genes

In different organisms, the same function may be accomplished by proteins that are not homologous. In general, the function is then served by a protein that has evolved from another protein by changing its specificity. For example, the gluconate kinase of *B. subtilis* (GNTK-BACSU) has no relationship with the gluconate kinase of *E. coli* (GNTK-ECOLI) but on the other hand is clearly a paralogue of the glycerol kinase of *B. subtilis* (GLPK-BACSU) (Galperin and Koonin, 1998). If this relationship had not been known through experimental data, it is likely that GNTK-BACSU would have been annotated as a glycerol kinase.

There are few solutions to remedy this type of problem, except to verify that the residues of the active site correspond closely to those of a glycerol kinase, before annotating it as such.

Knowledge of the Organism Studied

Even though it seems obvious, it is still advisable to remember that annotation must be done by, or at least supervised by, a biologist with a thorough knowledge of the organism studied. This should prevent trivial errors, such as the proposal of a function linked to a particular antibiotic for a protein of a bacterium, when experts had known very well that that antibiotic was not produced by the bacterium in question.

CONCLUSION: STRATEGY FOR ANNOTATION OF A GENOME

By way of a conclusion to this chapter on *in silico* functional analysis, we will consider a strategy based on the elements described in the preceding sections. Figure 6.8 shows a simplified diagram of the strategy.

The basic idea is to divide the protein sequences into functional modules by detecting and isolating, at the same time, the zones of the protein that may pose problems for homology search techniques (e.g., regions of low complexity, transmembrane regions).

Then, homology search methods are used on the zones corresponding to the globular proteins. These methods are applied gradually, starting with sequence comparison methods such as BLAST or FASTA. The modules for which these methods do not provide homologous protein allowing transfer of the annotation, are then subjected to techniques based on multiple alignments, such as PSI-BLAST or techniques using hidden Markov models. Then, for the most difficult cases, threading methods may be used. It must be noted that in all cases methods are used to find motifs that signify a particular function.

Finally, techniques based on the context of genes are used, by which either the results obtained earlier are more precisely determined, or other information is obtained on the interactions between proteins. All this information is provided to the annotators so that they can cross-check and integrate these data with, eventually, experimental data, bibliographic data, and their personal expertise on the subject.

Our laboratory (Mathématique, Informatique et Génome) is presently developing a semi-automated system of annotation of genomes based on this philosophy (Bryson et al., 2006). The system is semi-automated because, although it manages the automated use of different informatics methods and facilitates the annotator's task as far as possible, it leaves the entire responsibility of the final annotation to the scientist.

REFERENCES

Altshul, S., Gish, W., Miller, W., Myers, E., Lipman, D. 1990. Basic local alignment search tool. *J. Mol. Biol.*, 215, 403-410.

Altschul, S., Madden, T., Schaffer, A., Zhang, J., Zhang, Z., Miller, W., Lipman, D. 1997. Gapped BLAST and PSI-BLAST: a new generation of protein database search programs. *Nucl. Acids Res.*, 25, 3389-3402.

Aravind, L. 2000. Guilt by association: contextual information in genome analysis. *Genome Res.*, 10, 1074-1077.

Bateman, A., Birney, E., Durbin, R., Eddy, S., Finn, R., Sonnhammer, E. 1999. Pfam 3.1: 1313 multiple alignments and profile HMMs match the majority of proteins. *Nucl. Acids Res.*, 27, 260-262.

Bork, P., Dandekar, T., Diaz, L., Eisenhaber, F., Huynen, M., Yuan, Y. 1998. Predicting function: from genes to genomes and back. *J. Mol. Biol.*, 283, 707-725.

Brenner, S., Chothia, C., Hubbard, T. 1998. Assessing sequence comparison methods with reliable structurally identified distant evolutionary relationships. *Proc. Natl. Acad. Sci.* USA, 95, 6073-6078.

Bryson, K., Loux, V., Bosssy, R., Nicolas, P., Chaillou, S., van de Guchte, M., Penaud, S., Maguin, E., Hoebeke, M., Bessieres, P. and Gibrat, J.F. 2006. AGMIAL: implementing an annotation strategy for prokaryote genomes as a distributed system. Nucleic Acids Res, 34, 3533-45.

Clepet, C., Borne, F., Krishnapillai, V., Baird, C., Patte, J., Cami, B. 1992. Isolation, organization and expression of the *Pseudomonas aeruginosa* threonine genes. *Mol. Microbiol.*, 6, 3109-3119.

Davidson, J., Peterson, M. 1997. Origin of genes encoding multi-enzymatic proteins in eukaryotes. *Trends Genet.*, 13, 281-285.

Emanuelsson, O., Nielsen, H., Brunak, S., von Heijne, G. 2000. Predicting subcellular localization of proteins based on their N-terminal amino acid sequence. *J. Mol. Biol.*, 300, 1005-1016.

Emanuelsson, O., von Heijne, G. 2001. Prediction of organellar targeting signals. *Biochim. Biophys. Acta.*, 1541, 114-119.

Fitch, W. 2000. Homology: a personal view on some of the problems. *Trends Genet.*, 16, 227-231.

Galperin, M., Koonin, E. 1998. Sources of systematic error in functional annotation of genomes: domain rearrangement, non-orthologous gene displacement and operon disruption. *In Silico Biol.*, 1, 55-67.

Gibson, B., Campagnari, A., Melaugh, W., Phillips, N., Apicella, M., Grass, S., Wang, J., Palmer, K., Munson, R. 1997. Characterization of a transposon Tn916-generated mutant of

Haemophilus ducreyi 35000 defective in lipooligosaccharide biosynthesis. *J. Bacteriol.*, 179, 5062-5071.

Hirokawa, T., Boon, C., Mitaku, S. 1998. SOSUI: classification and secondary structure prediction system for membrane proteins. *Bioinformatics*, 14, 378-379.

Hofmann, K., Bucher, P., Falquet, L., Bairoch, A. 1999. The Prosite database, its status in 1999. *Nucl. Acids Res.*, 27, 215-219.

Karlin, S., Altschul, S. 1990. Methods for assessing the statistical significance of molecular sequence features by using general scoring schemes. *Proc. Natl. Acad. Sci.* USA, 87, 2264-2268.

Krogh, A., Larsson, B., von Heijne, G., Sonnhammer, E. 2001. Predicting transmembrane protein topology with a hidden Markov model: application to complete genomes. *J. Mol. Biol.*, 305, 567-580.

Lupas, A. 1996. Prediction and analysis of coiled coil structures. *Methods Enzymol.*, 266, 513-525.

Lupas, A., Van Dyke, M., Stock, J. 1991. Predicting coiled coils from protein sequences. *Science*, 252, 1162-1164.

Marcotte, E., Pellegrini, M., Ng, H., Rice, D., Yeates, T., Eisenberg, D. 1999. Detecting protein function and protein-protein interactions from genome sequences. *Science*, 285, 751-753.

Marin, A., Pothier, J., Zimmermann, K., Gibrat, J.F. 2002a. FROST: a filter based fold recognition method. *Proteins: Struct. Funct. Genet.*, 49, 493-509.

Marin, A., Pothier, J., Zimmermann, K., Gibrat, J.F. 2002b. Protein threading statistics: An attempt to assess the significance of a fold assignment to a sequence. In: *Protein Structure Prediction: Bioinformatics Approach.* Tsigelny, I.F., ed. International University Line, La Jolla, Calif., Chapter 9, 227-262.

Overbeek, R., Fonstein, M., D'Souza, M., Pusch, G., Maltsev, N. 1999. The use of gene clusters to infer functional coupling. *Proc. Natl. Acad. Sci.* USA, 96, 2896-9201.

Park, J., Karplus, K., Barrett, C., Hughey, R., Haussler, D., Hubbard, T., Chothia, C. 1998. Sequence comparisons using multiple sequences detect three times as many remote homologues as pairwise methods. *J. Mol. Biol.*, 284, 1201-1210.

Pearson, W. 2000. Flexible sequence similarity searching with the FASTA3 program package. *Methods Mol. Biol.*, 132, 185-219.

Pellegrini, M., Marcotte, E., Thompson, M., Eisenberg, D., Yeates, T. 1999. Assigning protein functions by comparative genome analysis: protein phylogenetic profiles. *Proc. Natl. Acad. Sci. USA*, 96, 4285-4288.

Rost, B., Fariselli, P., Casdio, R. 1996. Topology prediction for helical transmembrane proteins at 86. *Protein Sci.*, 1704-1718.

Schirmer, T., Keller, T., Wang, Y., Rosenbusch, J. 1995. Structural basis for sugar translocation through maltoporin channels at 3.1 Å resolution. *Science*, 267, 512-514.

Schultz, J., Copley, R., Doerks, T., Ponting, C., Bork, P. 2000. Smart: a web-based tool for the study of genetically mobile domains. *Nucl. Acids Res.*, 28, 231-234.

Sevrioukova, I., Li, H., Zhang, H., Peterson, J., Poulos, T. 1999. Structure of a cytochrome P450-redox partner electron-transfer complex. *Proc. Natl. Acad. Sci.* USA, 96, 1863-1868.

Shewry, P., Halford, N., Tatham, A. 1992. High molecular weight subunits of wheat glutenin. *J. Cereal Sci.*, 15, 105-120.

Stebbins, C., Borukhov, S., Orlova, M., Polyakov, A., Goldfarb, A., Darst, S. 1995. Crystal structure of the GreA transcript cleavage factor from *Escherichia coli. Nature*, 373, 636-640.

von Heijne, G. 1992. Membrane protein structure prediction. Hydrophobicity analysis and the positive-inside rule. *J. Mol. Biol.*, 225, 487-494.

Wolf, Y., Grishin, N., Koonin, E. 2000. Estimating the number of protein folds and families from complete genome data. *J. Mol. Biol.*, 299, 897-905.

Wootton, J., Federhen, S. 1996. Analysis of compositionally biased regions in sequence databases. *Methods Enzymol.*, 266, 554-571.

7

CHAPTER

Large-scale Protein Sequence Comparisons

J.L. Risler, S. Pasek, P. Brézellec, A. Louis

The overwhelming and exhausting avalanche of complete genome sequences (http://www.genomesonline.org) has lead to a concomitant flood of information concerning complete proteomes. In principle, knowing the nucleotide sequence of a whole genome enables the location all of its genes and therefore the sequences of all the proteins that are encoded by these genes. Several Web sites aim at collecting, annotating, organizing and distributing complete proteome sequences. As of March 2005 for example, the European Bioinformatics Institute (EBI) proteome site (http://www.ebi.ac.uk/integr8/EBI-Integr8-HomePage.do) that is tightly linked to the Uniprot database (Bairoch *et al.*, 2005) contains more than 240 proteomes, while the Munich Information Center for Protein Sequences (MIPS) Pedant database (http://pedant.gsf.de, Riley *et al.*, 2005) harbors more than 270 proteomes.

As could be expected, the availability of complete proteomes has opened the way to a new field of research : the systematic comparison of all the protein sequences from several (or many) organisms. The motivations for such studies are manifold. The most obvious application pertains to what is called "annotation by homology": if a new protein sequence happens to be very similar to another sequence of known function in another organism, then it is highly probable that the new protein will share the same function. This will remain true if the two sequences share more than 50% amino acid identities. The above inference by homology, however, is clearly dangerous if the sequence similarity is weaker – typically less than 30% over less than 150 residues (Brenner *et al.*, 1998) : the two proteins may still belong to the same structural family but their functional specificity may be different. Think, for example, of the numerous serine proteases or of the malate dehydrogenases : some genuine malate dehydrogenases look more like lactate dehydrogenases rather than malate dehydrogenases (Madern, 2002). This observation clearly puts a limit on the method of automated annotation by homology, which is known to easily produce false annotations that will propagate in the databanks. It is generally safer and more efficient to rely on the conservation of functional domains in proteins : large-scale comparisons enable the researcher to pinpoint such conserved domains, which can be compared to any new sequence once they have been stored in specialized databanks such as Prosite (Hulo *et al.*, 2004), Blocks (Henikoff *et al.*, 2000), ProDom (Bru *et al.*, 2005), Pfam (Bateman *et al.*, 2004) or Interpro (Mulder *et al.*, 2005). Another application of massive sequence comparisons consists of grouping together (clustering) into coherent families those sequences which are similar. The families are then stored in databanks that can be interrogated according to various criteria, such as Systers (Meinel *et al.*, 2005), Protomap (Yona *et al.*, 2000), Hobacgen (Perrière *et al.*, 2000), Hovergen (Duret *et al.*, 1994) or ClusTr (Kriventseva *et al.*, 2003). Large-scale comparisons are also useful in phylogenetic studies since they allow, in principle, the clustering into coherent families of homologous sequences from different species. In this case, an additional step will consist of distinguishing orthologous from paralogous sequences (Tatusov *et al.*, 1997; O'Brien *et al.*, 2005). Obviously, if two proteins are homologous, then their genes are homologous as well. It is therefore possible, through protein sequence comparisons, to locate the physical position of homologous genes in different genomes and look for chromosomal regions of conserved gene content. These so-called syntenies may prove useful in the functional annotation of genes or in the search for QTLs and for genes involved in health disorders. Finally, we have shown that the "all by all" comparisons of protein sequences may improve the annotation of the genes, particularly among eukaryotes such as *Arabidopsis thaliana* (Louis *et al.*, 2001; Mohseni-Zadeh *et al.*, 2004a). The basic idea here is to combine the annotation by homology with the search for conserved domains. A graphical representation such as that provided by Prodom (Gouzy *et al.*, 1997) enables the investigator to immediately pinpoint those sequences in which the number and/or the lengths of the domains seem inconsistent (Fig. 7.1). This generally suggests that an exon is missing, or that an in-phase intron has been incorporated into the sequence, or that a sequencing error has occurred and produced an artificial frameshift.

PROTEINS OR NUCLEIC ACIDS?

Although the raw data have come initially from genomic nucleotide sequences, we speak only of protein sequence comparisons. This is apparently contradictory but the reason is simple: sequence comparisons are more sensitive with amino acids than with nucleotides. The first obvious observation is that the nucleotide alphabet is composed of only four letters, as compared with 20 for the amino acids. Fortuitous identities are necessarily more numerous with nucleotide sequences, hence there is a loss in selectivity. In addition, the genetic code is degenerate: let us consider the nucleotide sequence of a protein-coding gene and, for each of its codons (except those for methionine and tryptophan), let us change the third base so as to obtain a synonymous codon. This will result in two nucleotide sequences sharing approximately 70% identity but coding for proteins that are 100% identical. Finally, nucleotide sequences do not allow us to take into account the physico-chemical similarities between amino acids. With similarity matrices (or scoring matrices) such as the Pam (Dayhoff *et al.*, 1978) or Blosum (Henikoff and Henikoff, 1992) series, any alignment algorithm will easily allow the substitution of a valine by an isoleucine but will put a strong penalty on the substitution of a very small cysteine by a very large tryptophan. This difference in

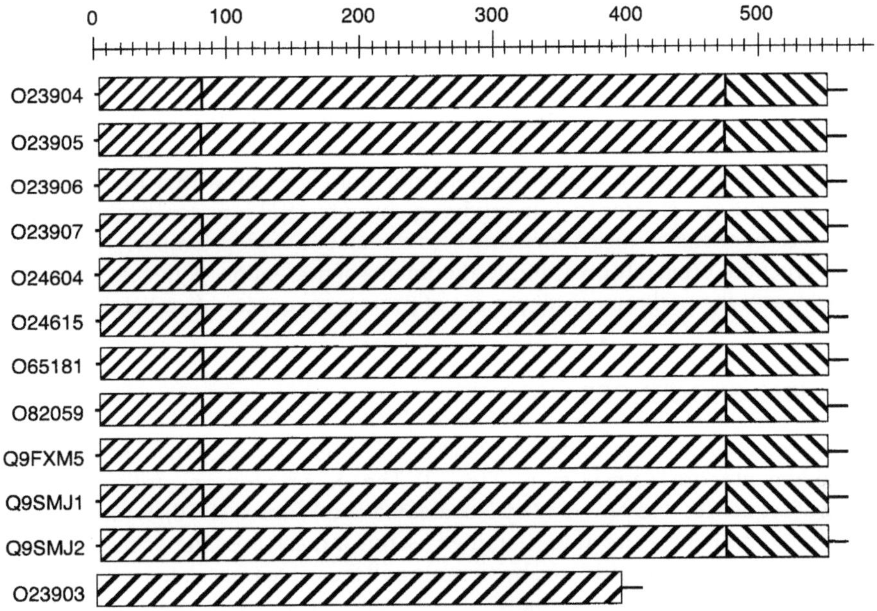


```
   2 TGTCTCAAGCAGAAGATCGATAGGATGTACAATGGAGAGAAGATTAATTG
     ||||||||||||||||||||||||||||||||||||||||||||||||||
  73 CysLeuLysGlnLysIleAspArgMetTyrAsnGlyGluLysIleAsnCy

  52 CACAGAGAATAGGTCTGTGTTGCATATAGCTCTTCGAGCTGCAAGACATA
     ||||||||||||||||||||||||||||||||||||||||||||||||||
  90 sThrGluAsnArgSerValLeuHisIleAlaLeuArgAlaAlaArgAspL

     ..................................................

1152 CTGACCATCTTATTCCTCACAAGACCTTTCCGGGCAATCGACCATCGTTG
     ||||||||||||||||||||||||||||||||||||||||||||||||||
 457 roAspHisLeuIleProHisLysThrPheProGlyAsnArgProSerLeu

1202 GAGTCTTCTATTACCTTCATTAAATGCTTACAACGTTGGACAGTTACTAG
     ||||||||||||||||||||||||||||||||||||||||||||||||||
 473 .SerLeuLeuLeuProSerLeuAsnAlaTyrAsnValGlyGlnLeuLeuA

1252 CAATATATGAGCACAGGATTGCTGTTGAAGGCTTTGTATGGGGGATCAAC
     ||||||||||||||||||||||||||||||||||||||||||||||||||
 490 laIleTyrGluHisArgIleAlaValGluGlyPheValTrpGlyIleAsn
```

Fig. 7.1 An example of a protein family, composed of glucose-6-phosphate isomerases (taken from Phytoprot). The breakdown of the sequences into domains, has been performed by the program Xdom. The last sequence is clearly different from the others: it is shorter and two domains are missing. If one aligns the nucleotide sequence of this "ugly duckling" with the closest protein sequence, it appears that the sequence of this cDNA is most probably truncated at its 5' terminus (the first ATG in the sequence is not the initiation codon). In addition, there is probably a sequencing error (G 1202 is superfluous) leading to a frameshift and a premature stop codon.

sensitivity is shown by the fact that the program Blast requires only two short similar tripeptides as anchorage points in the search for local similarity in proteins, while it requires a stretch of two identical hendecanucleotides (11 nucleotides) to align nucleotide sequences. The consequence is that the preferred method of finding homologous protein-coding genes, consists in comparing their conceptual translations into protein sequences.

Obviously the above reasoning does not hold for genes that do not code for proteins, or if one is interested in studying chromosomal intergenic regions. In such cases, a comparison of nucleotide sequences cannot be avoided. Similarly, phylogenetic reconstructions often rely on multiple alignments of nucleotide sequences, but note that in such cases the preliminary alignment of protein sequences often helps that of their genes.

ALIGNING SEQUENCES

For the purpose of "comparing" sequences, one must first align them so that, ideally, the alignment score is a maximum (Fig. 7.2). The most rigorous algorithm for aligning a pair of sequences was provided by Needleman and Wunsch (1970), which guarantees that the alignment score thus obtained is the largest possible score for the two sequences. The score is calculated by summing all the values associated with all the pairs of residues that have been brought into correspondence by the alignment (Fig. 7.2). These values are stored in a "similarity matrix" or "scoring matrix" such as Blosum62 or Pam250 (see above). If needed, insertions or deletions ("gaps") can be introduced within the sequences, in which case a "penalty" is subtracted from the alignment score. The Needleman and Wunsch algorithm is *global,* which means that it does its best to align the sequences from their beginning up to their end. The results are excellent if the two sequences are relatively close to one another, but are much less so for sequences that are divergent. In such cases, the similarities between the sequences are limited to those (short) regions that have been conserved because of their functional and/or structural importance. These fragments are generally called "domains", that can be separated along the sequences by long divergent regions containing large gaps. Here the algorithm can fail to find the correct alignment, since the penalties associated with large gaps can become too heavy. It may be more "profitable" (the score will be higher) to build a biologically false alignment containing a limited number of gaps. If we want to make massive protein sequence comparisons, it is clear that we shall have to align distantly related proteins, and that we want to spot those distant relationships. This is the reason why people do not use global alignments in massive sequence comparisons, but rather the *local* alignment algorithm provided by Smith and Waterman (1981).

The principle is very similar, but here we shall search for the most similar segments in the two sequences, even if it means that we shall align only parts of the sequences. The algorithm is no longer rigorous because there is no objective rule to define precisely the beginning and the end of most similar segments. It is this algorithm that is used in the "bestfit" program of the "Genetics Computer Group" (GCG) suite or the "Ssearch" program of the FASTA suite (see below).

Numerous studies have shown that the Smith and Waterman algorithm is effective (Brenner *et al.*, 1998). The main drawback lies in the fact that it is too demanding in

```
1 AVDHYFGRIPMASDCVWKLREMLTRERASQFKHSLMCWNAIPQS 44
  ||||::|:| |||||||             |:||||||||||||
1 AVDHFYGKIRMASDCVW.............FRHSLMCWNAIPQS 31
```

Fig. 7.2 A typical alignment by the Smith-Waterman algorithm of two short (putative) sequences. Identities are labeled with vertical bars and similarities by colons. Two amino acids i and j are considered as "similar" if the score Sij in the similarity matrix (Blosum62) is higher than an arbitrary threshold. The alignment score is computed thus: for each doublet of amino acids that have been paired (A with A, V with V, Y with F, etc.), their score is read in the similarity matrix and all the scores are summed up. If a gap is created, a penalty go (termed gap open) is subtracted. If n successive gaps are created, an extra penalty (pe) is subtracted form the score, such as pe = (n-1)ge where ge is termed gap extension and is usually about ten times smaller than go. The idea is that one must allow the creation of gaps (or insertions) but at the same time their number must be minimized, otherwise any sequence could be aligned with any totally unrelated sequence. Note that while the similarity matrices have been built upon sound theoretical backgrounds, one does not know how to model the creation of a gap (or insertion). Hence, the penalties go and ge are totally empirical.

computing time, which prevents its use in routine comparisons of one sequence against databanks containing hundreds of thousands of other sequences. It was found necessary to devise tricks (or heuristics) to speed up the process without loosing too much in selectivity and sensitivity, a goal that was reached with the programs Fasta (Lipman and Pearson, 1985; Pearson and Lipman, 1988) and Blast (Altschul *et al.*, 1990, 1997). In both cases, the principle consists of finding all the segments of similarity between two sequences. Nearby segments are joined together (gaps may be introduced here) so as to produce an alignment as long as possible. It is not within the scope of this article to give a detailed description of these two programs, the original publications are quite clear. An excellent tutorial about Fasta and protein sequence comparisons can be found at www.people.virginia.edu/~wrp/papers/mmol98f.pdf. Suffice here to say that in its most sensitive version (ktup = 1), Fasta compares favorably with Blast, but the computing time is notably longer than with Blast. As a consequence, being both rapid and sensitive, Blast is by far the most heavily used program for comparing a sequence to a databank.

THE SIMILARITY CRITERIA

When a biologist aligns two sequences, their aim is generally to try to decide whether they are biologically related or not. If the two sequences are alike, then it can be concluded that they are homologous (they share a common ancestor). But what does "are alike" mean? Is there a quantitative criterion to assess the similarity? Is there an objective means to distinguish between a "significant similarity" and a fortuitous likeness?

To any sequence alignment, there is always associated a *score* (Fig. 7.2). Whatever the alignment algorithm, the aim is always to obtain as high a score as possible. It is thus tempting to think that the higher the score, the more significant the alignment. This reasoning is correct if one deals with a homogeneous family where all the sequences have approximately the same length. In such a case, the closest sequences are those that share the highest score, which does not tell us whether the alignment score is "significant". It is clear, however, that an alignment score depends on the length of the sequences: two short sequences, even if highly similar, will necessarily produce a small score. It would be easier to obtain a higher score with two more distant sequences, provided that they are longer. Hence, in the case of massive sequence comparisons, where the lengths of the sequences are highly variable, the alignment scores cannot be used as similarity criteria. It is therefore necessary to normalize the score by dividing the alignment score by the length of the sequence. This is implemented in the Gap and Bestfit programs of the GCG suite, where the alignment score ("quality") is divided by the length of the shortest sequence, which gives the "ratio" parameter. We are now faced with another contradiction: two strictly identical sequences of length 10 and two strictly identical sequences of length 1,000 will provide the same ratio, which is not acceptable. Another idea is to use the percentage of identities between two sequences: Brenner *et al.* (1998) have shown that this criterion is catastrophic, leading to a dramatic loss in sensitivity, particularly for the shortest sequences.

The only sensible solution is to rely upon statistics, which fundamentally means comparing the real experimental score with a large number of scores obtained by aligning a large number of truly unrelated sequences. In their seminal study, Karlin and Altschul (1990) showed that the (ungapped) alignment scores between random sequences followed the "extreme values distribution", studied half a century ago by Gumbel (1958). There is no rigorous demonstration of such a behavior for gapped alignments, but simulations show that they follow the same law. Now the problem is solved. Let F(S) be the probability density of the scores (the known extreme values distribution). Then the probability P($S > S_0$) that the alignment score S between two random sequences be greater than S_0 is the integral:

$$P(S > S_0) = \int_{S_0}^{\infty} F(S)\, ds$$

Two methods are used to estimate the probability density function F(S), hence the probability P($S > S_0$):

In the case of Blast, the authors generate a great number of alignments between random sequences that possess the same composition as the databank, which allows them to estimate the parameters of the extreme values distribution. These parameters are determined once and for all and enable, for each alignment between a query sequence and a sequence belonging to the databank, the calculation of the probability to obtain by chance a score greater than the actual alignment score.

In the case of Ssearch and Fasta, the distribution of the scores is determined for each run from the actual alignment scores between the query sequence and those from the databank. The parameters of the extreme values distribution are then estimated by a least-square fit. The underlying idea is that the vast majority of the databank sequences are not related to the query sequence, hence the vast majority of the alignment scores provide a good estimate of "random" scores.

In both cases the probabilities are then transformed into "*E-values*" (expected values). Considering a sequence databank of the same size as that of the actual databank and composed of random sequences with the same composition as that of the actual databank, then E is the *number* of (random) sequences that are expected to give an alignment score with the query sequence greater than the actual alignment score. Thus, an E-value of 10^{-3} means the following : "In order to obtain an alignment between your sequence and a random sequence with a score greater than yours, then you should compare your sequence with 10^3 databanks of random sequences, each as large as your actual databank". Or, "In order to obtain an alignment between your sequence and a random sequence with a score greater than yours, then you should compare your sequence with a databank of random sequences 1,000 times larger than your actual databank". Thus, an E-value is *not* a probability. Fortunately, if E is small enough (say $E < 10^{-2}$) then $E(S_0) \approx P(S > S_0)$. Hence, if $E(S_0) = 10^{-3}$, it is correct to conclude that the probability of obtaining by chance, a score greater than S_0 is 10^{-3} (with a databank of random sequences of the same size and composition as that of the actual databank).

The previous sentence is important. The probabilities or the E-values thus calculated, depend on the size of the databank (expressed in number of amino acids or nucleotides). In other words, the same alignment between the same proteins with the same score will yield different E-values, depending on the databank. This is intuitively sensible: the probability of finding a "significant" similarity is obviously lower with a databank of 5,000 sequences than with a databank of 500,000 sequences. Thus, comparing E-values between different runs of Blast is relevant only if the databanks are of the same size. Note that this problem can be avoided through the use of the z parameter in Blast, the function of which is : "calculate the E-value as if the databank was z residues long".

The release of Blast more than 10 years ago was a major breakthrough. For the first time, the biologist was afforded a rapid, efficient and sensitive tool providing an objective estimation of the significance of alignment scores. The problem of the biologist has, however, not been fully solved. There is no definite answer to the question "which E-value threshold should I use to decide that an alignment is significant? Is it 10^{-1}, 10^{-3}, 10^{-5}?" Take for example the amino acid sequence of histidyl-tRNA synthetase from *E. coli* as a query sequence and perform a Blast search against the UNIPROT databank. Biologically relevant hits are obtained, with E-values much higher than one. For example, aspartyl-tRNA synthetase from *Geobacter metallireducens*, prolyl-tRNA synthetase from *Chlamidia pneumoniae* and threonyl-tRNA synthetase from *Clostridium thermocellum* are retrieved with E-values of 76, 99, and 99, respectively. Of course these hits are lost in an ocean of false positives but they are biologically significant since all these synthetases belong to the class II family, like the query sequence.

A totally different method (rather demanding in computing time) can be used to try to estimate the significance of an alignment score, based on a Monte-Carlo simulation (Lipman *et al.*, 1984; Comet *et al.*, 1999). It is generally used with the Smith-Waterman algorithm and proceeds in a few simple steps :

- align two sequences A and B with an alignment program, keep the score S_0
- randomly shuffle the residues in sequence B so as to generate a random sequence B_1 of the same size and composition as B. Align A with B_1, keep the score S_1.
- repeat the preceding step n times ($n = 100$ for example) to generate the random sequences $B_2 \ldots B_n$. Align A with $B_2 \ldots B_n$ and keep the scores $S_2 \ldots S_n$.
- compute the mean value Sm of $S1 \ldots Sn$ and the standard deviation σ of their distribution.
- calculate the parameter $Z = (S_0 - Sm)/\sigma$.

Z simply expresses the difference between S_0 (the actual alignment score) and Sm (the mean value of the "random" scores) in σ units (Fig. 7.3). The idea is always the same: compare the actual score with random scores. Clearly, the higher Z, the larger the difference between the actual score and the random scores (in σ units), hence the lower the probability that S_0 could be obtained by chance. Now, what is the value of Z above which a score is significant? Comet *et al.* (1999) performed a great number of protein alignments, with both real and random sequences. The two distributions of the Z values (for real and random sequences) are identical for $Z < 8$ but become divergent for $Z > 8$: the number of high Z values becomes greater for the real sequences than for the random ones. In other words, the value $Z = 8$ seems to mark the frontier between the real world and the random one, which is confirmed by common experience. The rule of thumb is the following: two sequences are probably unrelated if $Z < 6$ and probably related (homologous) if $Z > 10$. For $6 < Z < 10$ the answer is ambiguous. Another advantage of the Z values is that they are largely independent of the length of the proteins. Their main drawback is that their calculation is rather demanding in computing time.

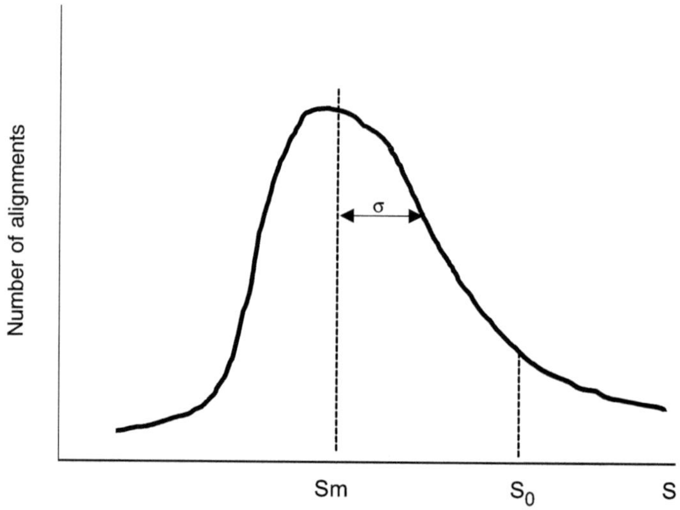

Fig. 7.3 Principle of the Z-score. Note that the distribution of the scores is not Gaussian but rather follows an "extreme values" distribution, called the Gumbel law.

"ALL AGAINST ALL" COMPARISONS: WHICH ALGORITHM TO CHOOSE?

Suppose we have a databank of n sequences, and we want to compare each sequence with all the others: this will require $n(n\text{-}1)/2$ comparisons. For "only" 50,000 sequences, this means 1.25 billion alignments. For 500,000 sequences, 125 billion! Whatever the algorithm, it is clear that massive sequence comparisons require a high computing power, far beyond the possibilities of a "standard" laboratory. Being both rapid and sensitive, Blast is largely used in such comparisons. It has enabled the construction of the following banks of protein families (clusters): COGS (Clusters of Orthologous Genes) (Tatusov *et al.*, 1997), Systers (Meinel *et al.*, 2005), Hobacgen (Perrière *et al.*, 2000), Hovergen (Duret *et al.*, 1994) and Inparanoid (O'Brien *et al.*, 2005). In the case of the Protomap databank of clusters (Yona *et al.*, 2000), the authors have even used Blast, Fasta and the Smith-Waterman algorithm. In order to achieve a greater sensitivity, the Prodom databank (Bru *et al.*, 2005) has now been built with Psi-Blast, an iterative variant of Blast (Altschul *et al.*, 1997). In spite of its long computing time, the Z-score method, as implemented in the Lassap/Biofacet package (Glémet and Codani, 1997), was the basis for constructing the databanks ClusTr (Kriventseva *et al.*, 2003) and Phytoprot (Louis *et al.*, 2001; Mohseni-Zadeh *et al.*, 2004a). The Web addresses from which these various databanks can be browsed are given in Table 7.1. We shall not give here a detailed description of all the databanks of clusters, the query interfaces and the way in which the results are displayed, as they are too diverse. The best the reader can do is to visit the web sites and make their choice according to their own preferences.

Table 7.1 Web addresses and bibliographic references of some banks of protein families that were built through massive sequence comparisons

Bank	*Web address*	*Reference*
CluSTr	http://www.ebi.ac.uk/clustr	Kriventseva *et al.*, 2003
COGs	http://www.ncbi.nlm.nih.gov/COG	Tatusov *et al.*, 1997
Hobacgen	http://pbil.univ-lyon1.fr/databases/hobacgen.html	Perrière *et al.*, 2000
Hovergen	http://pbil.univ-lyon1.fr/databases/hovergen.html	Duret *et al.*, 1994
Inparanoid	http://www.cgr.ki.se/inparanoid	Remm *et al.*, 2001
Phytoprot	http://genoplante-info.infobiogen.fr:/phytoprot	Mohseni-Zadeh *et al.*, 2004a
Prodom	http://prodes.toulouse.inra.fr/prodom/doc/prodom.html	Bru *et al.*, 2005
ProtoMap	http://protomap.cornell.edu	Yona *et al.*, 2000
Systers	http://systers.molgen.mpg.de	Meinel *et al.*, 2005

Whatever the algorithm used for building the clusters, the number of comparisons is so large, that *parallelized* versions of the programs must be used. This means that several (or many) sequence comparisons must be performed simultaneously, which is in principle not difficult since each comparison is independent of the others. Several techniques can be used for this purpose:

- There exists specialized machines, the processors of which have been explicitly "wired" for performing one particular task. This is the case for example of the "Bioccelerator" machine that was used with the Smith-Waterman algorithm for Protomap. The interest of such processors has been recognized for a long time (Guerdoux-Jamet and Lavenier, 1997), but it must be recognized that the extraordinarily fast commercial processors (Pentium and the like) are formidable competitors.
- Through the use of specialized software, it is possible to distribute the calculations to a "farm" of several local computers (PCs or workstations), each of them performing one part of the task. The master program (PVM or MPI) allocates one task (the comparison of two sequences) to each computer in the

"farm" and collects the results. To date, this principle is being extended to powerful and distant machines linked by the Internet, which is called "grid computing".

- A similar but more complicated method has been used in the frame of the Decrypthon project, which required 75,000 private PCs linked to the Internet, which were put at the disposal of the project by individual people.
- Finally, it is also possible to use multi-processor machines, where the task management is performed by the operating system itself.

BUILDING THE CLUSTERS

Once the sequence comparisons are completed, the next step consists in grouping the sequences into families (clusters), each cluster containing sequences "that are alike". We are now faced with two problems:

- Which quantitative criterion and which threshold will be used to decide that two sequences are alike?
- Practically, how shall all the related sequences be aggregated into one single cluster?

As regards the similarity criterion, we have seen that we can use either the E-value (Blast, Fasta, Ssearch) or the Z-score (Smith-Waterman). The choice of a proper threshold is less easy since we have to find a compromise between selectivity and sensitivity. With a restrictive threshold (low E-value such as 10^{-20} or high Z-score such as 40) we shall be highly selective (few false positives) but poorly sensitive (many false negatives). The inverse situation will occur with a permissive threshold. One way to answer the question consists in building several sets of clusters, each with a different threshold. All the clusters are then proposed to the user, who will make their own choice, as in the case of Protomap (E-value between 10^{-100} and 1) or ClusTr (Z-score between 10 and 200).

The building of the clusters seems to be easy, the simplest way being to use a transitive process: in one given cluster, each sequence is linked to at least *one* other sequence in the cluster, according to the principle that "the friends of my friends are my friends". Just take one sequence in the set, attach to it all its neighbors (as defined by your criterion and your threshold), then the neighbors of the neighbors, etc., until the list is exhausted. This will make one cluster. Remove these sequences from the set and start again with one other sequence to build a second cluster, and so on. It is easy, quick and efficient. Unfortunately, proteins are often made of "domains" that are not necessarily characteristic of a precise function. As a consequence and through a "chain effect", some proteins that share no similarity will be grouped together (Fig. 7.4). This phenomenon can be indeed catastrophic: After an all-by-all comparison of 23,653 sequences from the plant *A. thaliana*, we have built clusters with the transitive method. The first (largest) cluster contained 11,200 sequences, almost half of the set! This is clearly not satisfactory. This explains why a lot of efforts has been devoted to this problem, and we give thereafter some strategies that have effectively been used.

- In the case of Protomap, the first families are built with a very stringent threshold ($E = 10^{-100}$). This results in small families where the chain effect is avoided. Then

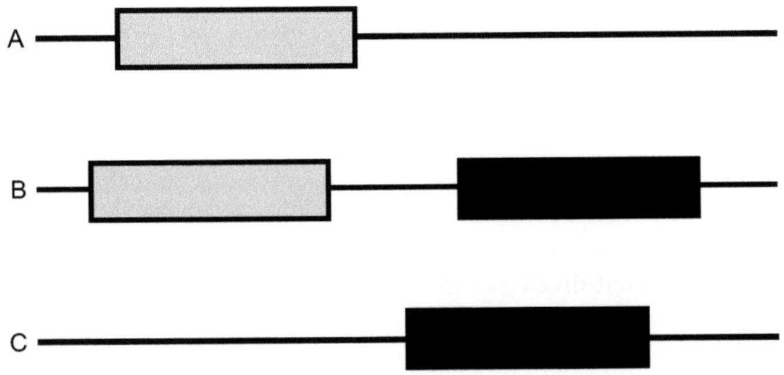

Fig. 7.4 The "chain effect" in the building of clusters by transitivity. A is similar to B through their light-gray domains and B is similar to C through their dark-gray domains, but A shares no similarity with C. They are however grouped together.

the threshold is progressively relaxed and new sequences added to the families. Some families may also be aggregated.

- The COGs are built by looking for maximal cliques of order three. In such a clique containing three sequences A, B and C from three different organisms a, b, and c, each sequence is linked to the other two sequences by a maximal link : the closest neighbor of A in organism b is sequence B and, reciprocally, the closest neighbor of B in organism a is sequence A. Each clique can thus be represented by a triangle, and two triangles will be combined if they share a common side.
- In the latest version of Phytoprot, we have searched for maximal cliques where any sequence in a clique is linked to *all* the other sequences of the clique by a Z-score greater than 14 (Mohseni-Zadeh *et al.*, 2004b). External sequences are then added to a clique if they share a link ($Z > 14$) with one sequence in the clique. As a result, the size of the largest cluster for *A. thaliana* decreased from 11,200 to 1,450.
- The way of building the clusters in the "Systers" databank is simple and ingenuous. A given sequence S having been compared to all the others, keep the set of sequences that share with S an E-value lower than 10^{-30}. Then use the last sequence of this set as a query sequence and repeat the procedure.

THE SEARCH FOR SYNTENIES

It is well known that whole chromosomal regions can be "conserved" within different species. The word "conserved" is here purposely ill-defined. "Conserved regions" generally means "local conservation of gene content and proximity across several organisms". By a misuse of language, these regions are also often called "syntenic regions". This conservation probably indicates, in many cases, a selection pressure that tends to preserve the very proximity of the genes (Overbeek *et al.*, 1999). As a consequence, the detection, across several genomes, of a local conservation of gene content and proximity, considerably helps the prediction of features of interest such as the physical interaction of proteins, or their participation in common metabolic/ regulatory networks.

Syntenic regions shared by different chromosomes mean that the chromosomal segments contain (approximately) the same *homologous* genes. Therefore, we are again faced with the same problem: to search for syntenies, one must first determine which genes (or proteins) are homologous within two (or more) species. Hence, the necessity for making massive sequence comparisons. This being done, one generally compares "manually" the list of homologous genes on the one hand, and the *ordered* list of genes on the chromosomes, on the other hand (Brachat *et al.*, 2003). Recently, we have designed an algorithm to automatically search for conserved homologous chromosomal regions within two or more chromosomes (Luc *et al.*, 2003). We have also shown that the use of the *domains* contained in the proteins, such as the Pfam domains (Bateman *et al.*, 2004), can notably improve the sensitivity of the search (Pasek *et al.*, 2005).

CONCLUSION

Since full genomic sequences are accumulating at a tremendous pace, there is no doubt as to the importance of massive sequence comparisons. The exploitation of such results depends only on the imagination of the researchers. Keep in mind, however, that if large computing resources are necessary, they are not sufficient. Different strategies of comparison and clustering are possible and, depending on preference, the resulting clusters of sequences can be markedly different, without being better or worse. It just depends on the aims of the studies. For example, clusterization by transitivity is probably the best way to find specific sequences (those that do not belong to a cluster). However, this method is certainly not the best to build biologically relevant clusters. The users must make their own enlightened choice ...

REFERENCES

Altschul, S.F., Gish, W., Miller, W., Myers, E.W., Lipman, D.J. 1990. Basic local alignment search tool. *J. Mol. Biol.* 215, 403-410.

Altschul, S.F., Madden, T.L., Schäffer, A.A., Zhang, J., Zhang, Z., Miller, W., Lipman, D. J. 1997. Gapped BLAST and PSI-BLAST : a new generation of protein database search programs. *Nucleic Acids Res.* 25, 3389-3402.

Bairoch, A., Apweiler, R., Wu, C.H., Barker, W.C, Boeckmann, B., Ferro, S., Gasteiger, E., Huang, H., Lopez, R., Magrane, M., Martin, M.J., Natale, D.A., O'Donovan, C., Redaschi, N., Yeh, L.S. 2005. The Universal Protein Resource (UniProt). *Nucleic Acids Res.* (Database Issue), 33, D154-159.

Bateman, A., Coin, L., Durbin, R., Finn, R.D., Hollich, V., Griffiths-Jones, S., Khanna, A., Marshall, M., Moxon, S., Sonnhammer, E.L., Studholme, D.J., Yeats, C., Eddy, S.R. 2004. The Pfam protein families database. *Nucleic Acids Res.* (Database issue), 32, D138-141.

Brachat, S., Dietrich, F.S., Voegeli, S., Zhang, Z., Stuart, L., Lerch, A., Gates, K., Gaffney, T., Philippsen, P. 2003. Reinvestigation of the *Saccharomyces cerevisiae* genome annotation by comparison to the genome of a related fungus: *Ashbya gossypii*. *Genome Biol.*, 4, R45.

Brenner, S.E., Chothia, C., Hubbard, T.J.P. 1998. Assessing sequence comparison methods with reliable structurally identified distant evolutionary relationships. *Proc. Natl. Acad. Sci.* USA, 95, 6073-6078.

Bru, C., Courcelle, E., Carrere, S., Beausse, Y., Dalmar, S., Kahn, D. 2005. The ProDom database of protein domain families: more emphasis on 3D. *Nucleic Acids Res.* (Database Issue), 33, D212-215.

Comet, J.P., Aude, J.C., Glemet, E., Risler, J.L., Hénaut, A., Slonimski, P.P., Codani, J.J. 1999. Significance of Z-value statistics of Smith-Waterman scores for protein alignments. *Comput. Chem.*, 23, 317-331.

Dayhoff, M.O., Schwartz, R.M., Orcutt, B.C. 1978. A model of evolutionary change in proteins. In: *Atlas of Protein sequence and Structure*, M.O. Dayhoff ed. Natl. Biomed. Res. Found., Washington, Vol. 5, 345-352.

Duret, L., Mouchiroud, D., Gouy, M. 1994. HOVERGEN: a database of homologous vertebrate genes. *Nucleic Acids Res.*, 22, 2360-2365.

Glémet, E., Codani, J.J. 1997. LASSAP, a large scale sequence comparison package. *Comput. Appl. Biosci.*, 13, 137-143.

Gouzy, J., Eugene, P., Greene, E.A., Kahn, D., Corpet, F. 1997. XDOM, a graphical tool to analyse domain arrangements in any set of protein sequences. *Comput. Appl. Biosci.*, 13, 601-608.

Guerdoux-Jamet, P., Lavenier, D. 1997. SAMBA: hazdware accelerator for biological sequence comparison. *Comput. Appl. Biosci.*, 13, 609-615.

Gumbel, E. 1958. *Statistics of extremes*. Columbia University Press, New York.

Henikoff, S., Henikoff, J.G., 1992. Amino acid substitution matrices from protein blocks. *Proc. Natl. Acad. Sci.* USA, 89, 10915-10919.

Henikoff, J.G., Greene, E.A., Pietrokovski, S., Henikoff, S. 2000. Increased coverage of protein families with the blocks database servers. *Nucleic Acids Res.*, 28, 228-230.

Hulo, N., Sigrist, C.J., Le Saux, V., Langendijk-Genevaux, P.S., Bordoli, L., Gattiker, A., De Castro, E., Bucher, P., Bairoch, A. 2004. Recent improvements to the PROSITE database. *Nucleic Acids Res.* (Database issue), 32, D134-137.

Karlin, S., Altschul, S.F. 1990. Methods for assessing the statistical significance of molecular sequence features by using general scoring schemes. *Proc. Natl. Acad. Sci.* USA, 87, 2264-2268.

Kriventseva, E.V., Servant, F., Apweiler, R. 2003. Improvements to CluSTr: the database of SWISS-PROT+TrEMBL protein clusters. *Nucleic Acids Res.*, 31, 388-389.

Lipman, D.J., Wilbur, W.J. Smith, T.F., Waterman, M.S. 1984. On the statistical significance of nucleic acid similarities. *Nucleic Acids Res.*, 12, 215-226.

Lipman, D.J., Pearson, W.R. 1985. Rapid and sentitive protein similarity. *Science*, 227, 1435-1441.

Louis, A., Ollivier, E., Aude, J.C., Risler, J.L. 2001. Massive sequence comparisons as a help in annotating genomic sequences. *Genome Res.*, 11, 1296-1303.

Luc, N., Risler, J.L., Bergeron, A., Raffinot, M. 2003. Gene teams: a new formalization of gene clusters for comparative genomics. *Comput. Biol. Chem.*, 27, 59-67.

Madern, D., 2002. Molecular evolution within the L-malate and L-lactate dehydrogenase super-family. *J. Mol. Evol.*, 54, 825-840.

Meinel, T., Krause, A., Luz, H., Vingron, M., Staub, E. 2005. The SYSTERS Protein Family Database in 2005. *Nucleic Acids Res.* (Database Issue), 33, D226-229.

Mohseni-Zadeh, S., Louis, A., Brezellec, P., Risler, J.L. 2004a. PHYTOPROT: a database of clusters of plant proteins. *Nucleic Acids Res.*, (Database issue), 32, D351-353.

Mohseni-Zadeh, S., Brézellec, P., Risler, J.L. 2004b. Cluster-C, an algorithm for the large-scale clustering of protein sequences based on the extraction of maximal cliques. *Comput. Biol. Chem.*, 28, 211-218.

Mulder, N.J., Apweiler, R., Attwood, T.K., Bairoch, A., Bateman, A., Binns, D., Bradley, P., Bork, P., Bucher, P., Cerutti, L. *et al.* 2005. InterPro, progress and status in 2005. *Nucleic Acids Res.* (Database Issue), 33, D201-205.

Needleman, S.B., Wunsch, C.D. 1970. A general method applicable to the search for similarities in the amino acid sequence of two proteins. *J. Mol. Biol.* 48, 443-453.

O'Brien, K.P., Remm, M., Sonnhammer, E.L. 2005. Inparanoid: a comprehensive database of eukaryotic orthologs. *Nucleic Acids Res.*, 33 (Database Issue), D476-480.

Overbeek, R., Fonstein, M., D'Souza, M., Pusch, G.D., Maltsev, N. 1999. The use of gene clusters to infer functional coupling. *Proc. Natl. Acad. Sci.* USA, 96, 2896-2901.

Pasek, S., Bergeron, A., Risler, J.L., Louis, A., Ollivier, E., Raffinot, M. 2005. Identification of genomic features using microsyntenies of domains : Domain teams. *Genome Res.*, 15, 867-874.

Pearson, W.R., Lipman, D.J. 1988. Improved tools for biological sequence comparison. *Proc. Natl. Acad. Sci.* USA, 85, 2444-2448.

Perrière, G., Duret, L., Gouy, M. 2000. HOBACGEN: database system for comparative genomics in bacteria. *Genome Res.*, 10, 379-385.

Riley, M.L., Schmidt, T., Wagner, C., Mewes, H.W., Frishman, D. 2005. The PEDANT genome database in 2005. *Nucleic Acids Res.* (Database Issue), 33, D308-310.

Smith, T.F., Waterman, M.S. 1981. Identification of common molecular subsequences. *J. Mol. Biol.*, 147, 195-197.

Tatusov, R.L., Koonin, E.V., Lipman, D.J. 1997. A genomic perspective on protein families. *Science*, 278, 631-637.

Yona, G., Linial, N., Linial, M. 2000. ProtoMap : automatic classification of protein sequences and hierarchy of protein families. *Nucleic Acids Res.*, 28, 49-55.

Functional Genomics: From Sequence to Function in Plants

8

CHAPTER

Gene Transfer to Plants

Y. Chupeau, M.R. Davey

INTRODUCTION

The second part of the last century witnessed the rapid development of molecular biology and molecular genetics, especially as applied to micro-organisms. Such technologies were the incentive, in the mid 1960s, to apply gene transfer procedures to plant and animal cells. Clearly, it was anticipated that DNA transfer would represent a major technology with which to elucidate the organisation and expression of genes, as well as their role in development. In fact, the idea of gene transfer technology has been one of the strongest motivations for refining tissue culture procedures to render cells *in vitro,* amenable to transformation. Gene transfer was more feasible for plant cells which were already known to express their totipotency, particularly cells from plants such as tobacco and carrot, used extensively as models in early plant transformation experiments. However, the complex wall enclosing the living cytoplasm of plant cells was seen as an obvious obstacle to DNA penetration. The idea of facilitating DNA penetration was a motivation to refine the preparation and culture of protoplasts derived from plant cells, extending the pioneering studies on such "naked" cells (Cocking, 1960). Initially, protoplasts were used to investigate the uptake of viruses and viral nucleic acids (Aoki and Takebe, 1969). However, at that time, early attemps to transfer radioactive DNA into plant cells (Ohyama *et al.*, 1972) or even entire plants (Ledoux *et al.*, 1974), were not conclusive nor accepted by the general scientific community, due to the lack of suitable and reproducible selection and characterization procedures, particularly at the molecular level.

The discovery of bacterial plasmids (Cohen *et al.*, 1970) and their major role in the natural process of gene transfer, prompted microbiologists to exploit such DNA elements to transfer into target cells, chimaeric genes constructed in the laboratory. Indeed, such advances initiated the turning point in this field of research for two main reasons. Firstly, at the technical level, marker genes, coding for resistance to various antibiotics in bacteria, were of potential value for selecting *in vitro* genetically modified (transformed) plant cells. Secondly, the discovery of the natural presence of plasmids in most bacteria, renewed interest in the suspected gene transfer to plants from pathogenic soil bacteria of the genus *Agrobacterium*. Such bacteria, discovered at the end of 19th century as the causative agent of crown gall disease (Smith and Townsend, 1907), provoke abnormal growth of plant tissues at wound sites. Agrobacteria have been studied throughout the 20th Century, since the transformed tissues of the galls that they incite are capable of growing *in vitro* without the addition of auxins and cytokinins to the culture medium. In addition, such tissues synthesize opines, unusual amino acids that are specific to crown gall tissues and that serve as a source of carbon and nitrogen for the inciting bacteria. Such characteristics led researchers to postulate possible gene transfer from *Agrobacterium* to "transformed" plant cells (Braun, 1947; Petit and Tourneur, 1972). Subsequently, this hypothesis was confirmed following the discovery of a large plasmid, the tumour-inducing (Ti) plasmid specific to *Agrobacterium tumefaciens* (Van Larebeke *et al.*, 1975), and the elucidation of its role in crown gall tumorigenesis. The sophisticated natural procedure evolved by Agrobacteria for redirecting plant metabolism to the benefit of their own nutrition and sexuality, was identified as resulting from the specific transfer into the genome of plant cells of genes located on a portion, the transfer DNA (T-DNA), of the large Ti plasmid (Zupan *et al.*, 2000; Gelvin, 2003). Later, the laboratory use of this natural gene transfer system, permitted the first experimental DNA delivery and integration into cells of tobacco, followed by the regeneration of transgenic plants (Barton *et al.*, 1983; Herrera-Estrella *et al.*, 1983).

Experimentally, the DNA transfer mechanism of *Agrobacterium* is currently the process most widely exploited for inserting genes into plants. In addition, other techniques have been established to insert DNA into plant cells and tissues. Such direct transfer procedures generally utilise *Escherichia coli* plasmids, normally used as gene cloning vectors. Importantly, the use of such vectors has extended the spectrum of plant species amenable to genetic manipulation, particularly those that were not readily amenable to transformation by *Agrobacterium*. Importantly, the techniques for gene transfer have been diversified and refined. Consequently, they represent very effective tools in many fields of genetics, enabling specific topics to be addressed, including aspects of gene regulation and the function of gene product(s). Additionally, transformation technology, associated with current progress in plant genomics, will facilitate targeted plant breeding.

Several procedures have been developed for gene transfer (Potrykus, 1990) and are summarized in Table 8.1. However, only those approaches exploited most extensively for inserting genes into plants will be discussed in this review. In short, the principle is to be able to deliver efficiently into the nucleus of recipient cells, specific fragments of DNA. The usual way to prepare large quantities of the DNA to be transferred is to have the DNA in an amplifiable bacterial genomic entity, such as a plasmid. Basic recombinant DNA technology for the construction of vectors at the molecular level may vary in detail according to specific gene transfer experiments. However, these protocols are discussed in the relevant chapters in this volume, and by Sambrook *et al.* (1989).

Table 8.1 Development of gene transfer procedures for plant cells

Process	*Reference*	*Year*
Explants + *Agrobacterium tumefaciens*	Barton *et al.*; Herrera-Estrella *et al.*,	1983
Protoplasts + Polyethylene glycol	Paszkoswski *et al.*	1984
Protoplasts + Liposomes	Caboche *et al.*,	1984
Protoplasts + Microinjection	Morikawa *et al.*,	1985
Protoplasts + Electroporation	Fromm *et al.*,	1986
Explants + Biolistics	Sandford	1988
Explants + Electroporation	d'Halluin *et al.*,	1992
Protoplasts + Sonication	Joersbo and Brunstedt	1992

DETECTION OF THE EXPRESSION OF DNA TRANSFERRED INTO PLANT CELLS

The availability of several convenient and sensitive reporter genes has facilitated the establishment of the conditions for effective gene transfer to plant cells, irrespective of the nature of the vectors and the gene transfer systems.

Transient Gene Expression Systems

The initial biochemical detection of reporter genes, such as *cat* (chloramphenicol acetyltransferase) and *npt*II (neomycin phosphotransferase), involved plant extracts and radioactivity. Such tedious and destructive procedures have been replaced by reactions involving colour changes *in situ* in living cells, such as expression of the *gus* (β-glucuronidase) gene from *E. coli*, the *luc* gene from firefly, and the *gfp* (green fluorescent protein) gene from jellyfish (Chapter 13). Indeed, the rapid exploitation of fast, non destructive and versatile detection systems, opened up a new experimental field relating to gene function and regulation through transient assays (Sheen, 2001). In addition, the range of fluorescent probes for use in molecular biology is also increasing (Shaner *et al.*, 2004). Such assays have facilitated refinement of the procedures for gene transfer, since they have permitted quantification of the number of transiently transformed cells. This figure often exceeds 50% of DNA treated cells and could be increased to 90% in those homogeneous systems that are most amenable to transformation (Sheen, 2001). Knowledge of the rates of transient expression are useful in establishing procedures for the stable transformation of plant cells, to adapt them to specific plant tissues, or to determine the strain of *Agrobacterium* most effective for gene transfer to specific cell types (Yoshioka *et al.*, 1996).

Selection Systems for Stable Gene Insertion into Nuclear Genomes

In spite of the generally high efficiency of transient gene expression in DNA treated cells, stable transformation resulting from the correct insertion of a complete sequence targeted to the DNA of plant cells remains a relatively rare event, in the order of only one transformed cell in every 10^3 or even 10^5 of cells exposed to DNA during the uptake procedure. Consequently, it is still essential to be able to favour the development and selection of rare transformed cells and their derivatives. The colour tests now used extensively in transient expression assays can also be used to recognise and, eventually, select cell/tissues from which transformed plant can be regenerated from a DNA insertion event. Several selection systems have been developed. The first ones, derived directly from studies with microorganisms, relied on bacterial genes encoding resistance to antibiotics, particularly kanamycin (*npt*II) or hygromycin (*hpt*), and herbicides such as glyphosate and phosphinothricin. These resistance genes are useful and easy to handle for experimental purposes (Potrykus, 1990). However, environmental and social reasons have resulted in strong public opposition in several countries to their use, especially in crop plants. Additionally, there are also technical difficulties owing to the necessity to use different selectable markers with different transgenes in the same recipient plant. Consequently, other means of selection have been proposed. Thus, the use of a phosphomannose isomerase gene from *E. coli,* to enable plant cells to metabolise mannose, is an excellent example of a new, positive selection approach, compared to negative selection based on antibiotics and herbicides. These positive selection systems could be of considerable benefit, especially in the transformation of hitherto recalcitrant plant species (Joersbo *et al.*, 1998).

Systems to Avoid Selectable Marker Genes in Transgenic Plants

Although screenable markers may be considered not to have potentially harmful biological activities, it is clear that, longer-term, they must be excluded from transgenic plants. One way to address this issue is to separate markers from agronomically important transgenes before transformation by having the genes on separate DNAs or plasmids. In principle, they may be integrated at different sites in the genomes of recipient cells and can be segregated by sexual crossing of transgenic plants. However, in various experimental systems, a high portion of transformed plants may carry both genes as linked copies (Stuitje *et al.*, 2003), although there are exceptions (Komari *et al.*, 1996). The use of specific nucleases from *E. coli* (*Cre*) which cleave DNA at specific sites (*lox*), was initially proposed to remove a *npt*II sequence integrated between two *lox* sites in tobacco (Dale and Ow, 1991). Several other similar systems have been developed (Puchta, 2003). Recently, a two site-specific recombination system has been proposed to achieve transgene integration and to remove sequences that are not required following DNA transfer into recipient genomes (Srivastava and Ow, 2004). However, introducing a nuclease even of bacterial origin into a plant genome, could be questionable since, for example, functional *lox* sites have been found in plastid genomes (Corneille *et al.*, 2003). Because of refinements in gene transfer techniques, it is possible, at least in some cases, to characterize transgenic plants by PCR, detecting only the transgene(s) of interest, thus avoiding the requirement for a selectable marker gene adjacent to the gene(s) of interest in the constructs used for transformation. Although a somewhat high yield of transformants is required for reproducible exploitation of such PCR-based procedures (Aziz and Machray, 2003), these approaches are workable in many cases alongside classical gene transfer techniques (de Vetten *et al.*, 2003).

Selection Systems for Stable Gene Insertion into Plastid Genomes

The plastid genome (plastome) is highly polyploid. A leaf cell of a tobacco plant contains about a hundred chloroplasts, each of which harbours about ten nucleoids, each containing several genomes. Thus, there are between a thousand and ten thousand plastomes per cell, emphasising the necessity for an efficient selection procedure. Streptomycin and lincomycin resistances have been the markers favoured for plastid transformation. However, these resistances are not cell autonomous. Consequently, the production of homoplastomic plants necessitates the growth, under selective conditions over more that 20 mitotic divisions, of cells that have received DNA. This can be achieved starting with isolated cells, and waiting for complete segregation of plastids before plant regeneration. Repeated rounds of shoot regeneration may be essential if the starting material is leaf tissue. Stable integration of the transforming DNA occurs essentially through homologous recombination, as in bacteria, with sequences in the constructs being chosen that are homologous with those in intergenic regions of the plastid genome (Maliga, 2004). Recent innovative work shows that the frequency of plastid transformation, at least in tobacco, can be increased using cell suspension cultures as the target material. Although the plastids are less numerous and much smaller than in cells of intact leaf tissues, the observed efficiency of this system could be due to faster selection of homoplastic cell lines, thus enhancing the actual rates of stable transformation (Langbecker *et al.*, 2004).

Selection Systems for Transformation of Mitochondria

Transformation of mitochondria has been reported for yeasts (Johnston *et al.*, 1988) and the single celled green alga, *Chlamydomonas* (Boyton and Gillham, 1996). In contrast, attempts to transform in higher plants have been unsuccessful. The complexity and plasticity of the genome of mitochondria are probably the reasons why selection systems will have to be more precise than those for the plastome, in order to select plants with uniformly transformed mitochondria.

THE MAIN TECHNIQUES OF GENE TRANSFER

Use of the Natural Vectors of Agrobacterium

As already indicated, *Agrobacterium tumefaciens and A. rhizogenes* have evolved a complex process to deviate the metabolism of plant cells to their advantage, by transferring a set of genes that enhance the capacity of recipient plant cells to synthesise opines. The latter are both sources of carbon and nitrogen for Agrobacteria, as well as inducers of conjugation. The various steps of T-DNA transfer into plant cells are controlled by 40 to 50 virulence genes, some of which are chromosomal, but most are on the Ti plasmid or the root inducing (Ri) plasmid in the case of *A. rhizogenes*. Gene transfer is induced naturally by phenolic compounds (*e.g.* acetosyringone) released from wounded plant cells that stimulate a two component (*virA* + *virG*) activator of the transcription of other virulence genes. Essentially, gene transfer results from the adaptation of a bacterial type IV protein secretion system (involving at least 11 proteins) rendering it effective for a nucleo-protein complex. Classically, the T-DNA is delimited by two repeated sequences of 24 bases that act as specific sites for nucleases that excise a single strand of the T-DNA. One of the nucleases remains attached to the 5' extremity of the T-DNA, to protect the latter and to target it to the nucleus of the plant cell. A protein (VirE2) is associated specifically with the single stranded T-DNA to protect it against nucleases and to maintain it in a linear form to facilitate its passage from the bacterial cell to the nucleus of the recipient plant cell (Zupan *et al.*, 2000; Gelvin, 2003).

Historically, pioneering work demonstrated that foreign DNA inserted by homologous recombination into the T-DNA of Ti or Ri plasmids was transferred and integrated into the chromosomes of plant cells, where it was expressed, provided correct regulatory sequences were associated with the chimaeric genes of interest carried on the T-DNA (Barton *et al.*, 1983; Herrera-Estrella *et al.*, 1983).

However, the large size of Ti and Ri plasmids presented difficulties with their use as DNA vectors for plant transformation. Importantly, two points were crucial for routine laboratory use of the gene transfer process from *Agrobacterium* to plant cells. First, the discovery that T-DNA transfer *per se* does not encompass genes acting in the transfer or integration processes. The only two elements required are the sites of excision of the T-DNA, specifically, the T-DNA borders each of 24 base pairs, between which it is possible to integrate foreign DNA sequences to be transferred into plant cells. Second, verification that the excision nucleases are active on those border sequences wherever the latter are located. The model proposed by Tinland (1996) still applies to T-DNA integration into plant nuclear genomes. Collectively, this information made it feasible to transfer T-DNA borders onto *E. coli* plasmids, the latter being easier to manipulate in terms of construct preparation than the much larger Ti or Ri plasmids (De Frammond *et al.*, 1983; Hoekema *et al.*, 1983), since the latter readily fragment when manipulated. These advances lead to the development and general use of binary vector systems for gene transfer to plants. In such systems, a strain of *Agrobacterium*, generally resistant to rifampicin, harbours chromosomal genes controlling bacterial attachment to plant cell walls, a resident Ti plasmid with its virulence genes but deleted of its T-DNA, and an *E. coli* derived plasmid. The latter has an origin of replication functional in both *Agrobacterium* and *E. coli* and a T-DNA from a Ti plasmid. The oncogenicity genes are removed from between the T-DNA borders and replaced with "simple" DNA, such as that from an *E. coli* plasmid. This stretch of DNA between the T-DNA borders has multiple cloning sites for transgene integration. Several binary vectors have been developed in different laboratories, targeted to specific uses with increased flexibility for gene manipulation. A useful guide to the general background of binary vector construction is provided by Hellens *et al.* (2000). It is worthwhile to mention that yeast (Bundock and Hooykaas, 1996), filamentous fungi (de Groot *et al.*, 1998) and even human cells (Kunik *et al.*, 2001), are also susceptible to *Agrobacterium*-mediated transformation.

In practice, Agrobacteria carrying binary vectors are cultured in selective media in order to maintain the binary vector within the bacterial cells. Subsequently, the bacteria are mixed with explants under axenic conditions at the start of the transformation procedure. Explants from *in vitro* propagated plants can be cocultivated directly with the bacterial inoculum, while those from glasshouse or field-grown plants must be surface sterilised by immersion in a suitable sterilant solution (*e.g.* bleach) followed by rinsing with sterile water before inoculation with Agrobacteria. Prior to transformation, explants should be evaluated for their shoot regeneration ability, with the most favoured explants being those from leaves (Horch *et al.*, 1985). Explants are incubated (co-cultured) with Agrobacteria for various periods of time depending on the strain of bacterium and the sensitivity of the explants to inoculation. This period may vary from several hours to a few days, 48 hours being the mean time reported from various laboratories (Gelvin, 2003). Subsequently, the explants are transferred to a shoot regeneration medium, the composition of which, particularly the balance of growth regulators, is adapted for the type of explant and plant species. This medium is also intended to eliminate the inciting bacteria from the plant tissues, which is achieved by the addition of antibiotics that are not toxic to plant cells. Most laboratories have adopted cefotaxime, carbenicillin and augmentin for elimination of Agrobacteria (Hellens *et al.*, 2000).

The Refinements for "Domesticated" Agrobacteria

Initially, Agrobacteria were "domesticated" without modifying their host range. Initial developments in gene transfer technology were limited to dicotyledons, as that most monocotyledons, particularly cereals, were not susceptible to transformation by Agrobacteria. This severe limitation was a strong impetus to investigate other processes of gene transfer but, in the meantime, to attempt to understand the intimacy of the mechanisms of virulence genes that are associated with the host range of Agrobacteria. The study of *Agrobacterium* carrying super-virulent Ti-plasmids and capable of transforming a wide range of plants, led to the discovery that over-expression of the *virG* gene (protein inducer for the other virulence genes), coupled with the use of acetosyringone (phenol inducer of the auto-phosphorylation of *virA*) in the coculture medium, facilitated gene transfer to hitherto recalcitrant cereal species. Cereal transformation with Agrobacteria was facilitated by incorporating onto binary vectors *virB* and *virC* genes to create "super binary" vectors (Hiei *el al.*, 1994). Plant cells already active in mitotic division, such as those in suspension culture, or cells with potential mitotic activity, for example those in young embryos cultured *in vitro* (Ishida *et al.*, 1996), provided excellent targets for transformation. A ternary system has also been described in which *Agrobacterium* carrying a disarmed Ti plasmid and a binary vector harbours a third plasmid carrying a constitutively expressed *virG* gene to facilitate transformation (Van der Fits *et al.*, 2000).

Vectors have also been developed for the introduction of large DNA fragments into plants, such gene delivery vehicles having the characteristics of both bacterial artificial chromosomes and *Agrobacterium* binary vectors (Shibata and Liu, 2000). Genomic-DNA up to 150 kb in size can be transferred, thus facilitating positional cloning, identification of dominant alleles, as well as the introduction of several genes (Cao *et al.*, 2004).

In Planta Transformation with Agrobacterium

Attempts to simplify the overall process of gene transfer through approaches that do not rely on tissue culture have been developed and continue to be refined, based upon the methodology first reported for the model plant, *Arabidopsis thaliana*. In extending the initial success obtained with *Agrobacterium* inoculation of germinating seeds (Feldmann and Marks, 1987), the procedure of "*Agrobacterium* vacuum infiltration" provided the first simple procedure of wide application. Indeed, most of the collections of T-DNA insertion mutant plants have been created by adopting the procedure of *in planta* transformation (Bechtold *et al.*, 1993). Basically, the protocol consists of the following steps, specifically:

(a) Immersing, in a suspension of Agrobacteria and under vacuum, young plants or entire inflorescences at an early flowering stage to force penetration of bacteria into the floral tissues,

(b) Growing the treated plants to maturity and harvesting the seeds,

(c) Germinating the seeds on culture medium containing an appropriate selection agent, such as an antibiotic (*e.g.* kanamycin) or herbicide.

Most transformed progeny are non-chimeric, hemizigous for the transgene(s), and somaclonal variation resulting from tissue culture is avoided. Analysis of the transformation process with a *gusA* reporter gene, revealed that the female gametophyte was the target for gene delivery, although it is still unclear whether transformation by *Agrobacterium* is facilitated by pollen tube growth, or whether Agrobacteria effectively reach the ovule unaided (Bechtold *et al.*, 2000). This method of transformation was simplified with the use of a surfactant (0.005 - 0.02% Silwet L-77) replacing vacuum infiltration, prepared in a simple medium containing sucrose (Clough and Bent, 1998). To date, the floral dip procedure has been successful with some members of the family Brassicaceae (Curtis, 2005). However, it remains to be seen the extent to which it can be applied to species from other plant families. It is interesting that, recently, specific viral vectors have been improved to exploit *Agrobacterium* transfection for efficient systemic high transient gene expression in whole plants (Marillonnet *et al.*, 2005).

Direct Gene Transfer Induced by Polyethylene Glycol

The ability of polyethylene glycol (PEG) to induce fusion of the plasma membranes of adjacent protoplasts, has been known for more than three decades (Kao and Michayluk, 1974; Wallin *et al.*, 1974). This capacity is due mainly to a strong capture of water molecules by PEG at high concentrations, thus abolishing the organization of membranes resulting from the hydrophobicity of non-polar heads of constitutive phospholipids (Lentz and Lee, 1999). However, the destabilization of plasma membranes is reversible upon dilution of the PEG, enabling viable fused cells to be obtained. Heterokaryons, the products of the fusion of protoplasts with different genomes, can be regenerated into intact somatic hybrid or cybrid plants (Davey *et al.*, 2005). PEG has been used extensively to induce uptake of DNA into plant protoplasts (Paszkoswski *et al.*, 1984). The process is still being refined and adapted because in addition to permeation, PEG has the capacity to precipitate DNA into protected complexes. Generally, protoplasts are mixed with exogenously supplied DNA, followed by treatment with concentrated (from 10 to 50%, w/v) PEG solutions, supplemented with various concentrations of divalent ions (Ca^{2+} or Mg^{2+}) at room temperature for a limited period (5 to 60 minutes). Subsequently, PEG is removed by gradual dilution and the DNA treated protoplasts cultured to tissues and regenerated plants. Various parameters need to be adapted for specific protoplast systems under investigation (Potrykus, 1990). Currently, PEG permeation is used mainly to induce DNA uptake for transient gene expression studies with isolated protoplasts as the experimental system, rather than in experiments to generate stably transformed plants (Sheen, 2001).

Direct Gene Transfer into Isolated Protoplasts by Liposomes and Electroporation

In the early 1980s, alternative ways of permeating protoplasts were proposed in order to improve the efficiency of direct transfer of DNA into recipient genomes. The first, adapted from the technology developed for cultured animal cells (Fraley *et al.*, 1981), exploited liposomes as a DNA delivery system (Caboche *et al.*, 1984). Currently, there appears to be little interest in this method, probably related to the need to liposome-encapsulate the DNA.

Electroporation was developed initially to transform maize protoplasts (Fromm *et al.*, 1986) and became one of the most popular approaches for cereal transformation involving permeation of plasma membranes by high voltage electrical discharges (Potrykus, 1990). In principle, this technique exploits temporary membrane destabilisation leading to permeation, through local perturbation of the electric potential of the plasma membrane induced by rapid electric discharge applied to the solution containing DNA and the suspended protoplasts. The permeation induced by an electric discharge is variable according the size of the target protoplasts and the characteristics of their plasma membranes. In practice, the parameters of electroporation (*e.g.* the type and intensity of the electric field, conductivity of the solution in which the protoplasts are suspended) have to be adapted for each protoplast system (Chupeau *et al.*, 1989).

Direct DNA Transfer by Biolistics

The main limitation to the use of protoplasts as target cells for transformation lies, very often, in the difficulty of regenerating plants at high efficiency from such DNA treated naked cells. The media required for plant regeneration may have to be determined empirically for each plant species under investigation. In contrast, microprojectile bombardment (Sanford, 1988) has provided the ability to deliver DNA into a range of cells, tissues and organs (*e.g.* leaf and stem explants, immature embryos, callus, embryogenic cell suspensions), especially those capable of expressing their totipotency. This bypasses any host range limitations of Agrobacteria.

The DNA to be delivered by biolistics is alcohol precipitated onto micron-sized particles of gold or tungsten (Maliga, 2004). Such metal particles are accelerated to velocities that enable them to penetrate cell walls. The rough, but effective, gunpowder charge used in the early biolistics-mediated transformation experiments to provide the energy for particle acceleration (Sanford, 1988), was replaced by a more flexible and cleaner propellant. Thus, the release of helium under pressure and its subsequent expansion, was utilised to accelerate DNA coated particles into totipotent target tissues (Johnston, 1990). Biolistic delivery has been adopted and adapted in many laboratories for both transient and stable transformation providing, a genotype-independent process. Importantly, it requires less technical experience and input of labour than the methodology associated with protoplast-based approaches (Taylor and Fauquet, 2002). However, even biolistics has been superceded, to a certain extent, by the new generation of Agrobacteria for the transformation of certain crop plants, including cereals (Afobali *et al.*, 2003). Nevertheless, biolistics remains the method favoured for delivering DNA to transform chloroplast genomes, (Maliga, 2004).

DNA INTEGRATION INTO RECIPIENT PLANT GENOMES

Irrespective of the different forms of DNA (circular, linear, double stranded or single stranded), the mechanisms by which it integrates into recipient nuclear genomes appear somewhat similar. In contrast to the situation in micro-organisms, for which homologous recombination is the rule, the most frequent events for foreign DNA integration into the nuclear genome of plant cells rely on the repair of double strand breaks, consisting of non-homologous end-joining (NHEJ), often termed illegitimate recombination. The precise mechanism by which DNA integrates into plant genomes has been difficult to determine. Initially, it was predicted that integration of a T-DNA strand (a single stranded copy of the T-DNA) occurs at a nick in the plant DNA and that both ends of the T-DNA strand anneal to the gap in the plant DNA. Subsequently, repair synthesis of the complementary strand achieves integration. The 3' end of the T strand was supposed to initiate the process, through micro-homologies of 3 or 4 bases with the plant DNA (Tinland, 1996). However, this hypothetical mechanism did not take into consideration all possible features of T-DNA transfer. More recently, it was demonstrated in tobacco with the use of transient gene expression, a very rare cutting endonuclease (I-SceI) and an elegant combination of various markers into different plasmids also harbouring I-SceI sites, that T-DNA is integrated preferentially into double strand breaks created by I-SceI in the plant genome (Tzfira *et al.*, 2003). Also, T-DNA is integrated after digestion with I-SceI, the latter being active only on double stranded DNA, indicating that the T-DNA is in a double strand form prior to integration. This experimentation confirms the rapid conversion of the single-stranded T-DNA into double strands after its transfer in the single stranded form into plant nuclei, accounting for the rapid expression of reporter genes. It is more likely that T-DNA integration proceeds as a ligation reaction between two double stranded DNAs, which generalizes the mechanism of integration to NHEJ, whatever the form of the delivered DNA, with no, or weak, inference from microhomologies. However, which component of the NHEJ complex is effective in the integration process remains to be confirmed (Friesner and Britt, 2003; Somers and Makarevitch, 2004). Most generally, transgenes tend to be inserted into gene-rich regions of plant genomic DNA. In *A. thaliana*, integration is somewhat uniform along the genome, although there is a reduction in centromeric regions (Alonso *et al.*, 2003), and an increase in promoter regions (Bako *et al.*, 2003). The general reliance of foreign DNA integration to repair double stranded breaks, so effective but so critical for the survival of organisms, has constituted the major obstacle in the characterization and use of homologous recombination in higher plants. Homologous recombination can be achieved with sophisticated selection schemes, but at low frequency (Lida and Terada, 2004). Until now, this frequency has prevented the development of a successful and universal tool for gene targeting in higher plants.

GENE TRANSFER AS A SCIENTIFIC TOOL

During the last 20 years, the techniques for gene transfer, combined with other recent developments in the genomics of higher plants, have set the stage for reconsidering general aspects of biology and, in particular, the development of organisms. The pace of gene discovery in plants has increased tremendously with the complete sequencing of the *A. thaliana* (The AGI, 2000) and rice genomes (Yuan *et al.*, 2003). Sequencing of the genomes of plants, such as tomato, will facilitate comparisons between genera. Considerable effort is still needed to characterize the role and regulation of the thousands of plant genes whose functions are only putative, or completely unknown. Following early evidence of the wide dispersion of transgene loci in the genomes of transformed plants, gene transfer has been exploited for two somewhat different goals. Because of the lack of gene targeting techniques for higher plants, mutational insertions of selectable transgenes have been used to create large collections of tagged mutants. This is the case for *A. thalina,* due to the ease of *in planta* transformation (Chapter 9), but the same principle can be extended to other schemes (Nishal *et al.*, 2005) and plants such as *Medicago* with tobacco transposons (d'Erfurth *et al.*, 2003). The function of a DNA sequence can be investigated through the analysis of its expression following insertion into a recipient genome. In this case, the insertion loci preferred are the most simple possible (one copy at one site), not creating any mutation in the recipient genome, and thus facilitating the characterization of the resulting phenotype. Basically, the transfer of a coding sequence complementing a characterised mutation provides a strong indication of the putative function of the transferred gene. Depending on the promoters (native, strong constitutive, or tissue-specific) driving transcription of the DNA sequence of interest, and of the recipient plant species, DNA transfer also provides access to the role of regulators in gene expression.

Indeed, gene transfer constitutes an essential tool in the functional analysis of individual genes. It reveals useful information in conjunction with biochemical genetics, especially in improving knowledge of the metabolic networks of plants, for example, in the case of lipid metabolism (Browse and Somerville, 1991). In addition, the biology of numerous living organisms has progressed rapidly because gene transfer may also be exploited in the functional analysis of key regulators of many developmental and physiological processes. This is especially crucial in plants, since completion of the genome sequence of *A. thaliana* revealed the presence of a large number of transcription factors (The AGI, 2000). Over-expression of a transcription factor, combined with analysis of cDNA micro-arrays, enables visualization of its effects in the expression of thousands of genes. This provides innovative experimental devices to enter in the complex metabolic network of plants (Lloyd *et al.*, 1992), or in the coordinated regulation of gene expression at key development stages, as in the onset of flowering (Levy and Dean, 2005). In *A. thalina* with its sequenced genome, such approaches have been very powerful in analysing complex adaptive capacities of plants. This is the case for freezing tolerance, at the level of transcription (Seki *et al.*, 2001), and also at the level of the metabolome (Cook *et al.*, 2004). Indeed, these approaches are beginning to unravel the complexity of the cascades of molecular networks that control plant responses to environmental stress (Vinocur and Altman, 2005) and plant-pathogen interactions (Hammond-Kosak and Parker, 2005).

Gene transfer can also improve the sophistication of various techniques of microscopy used in cell biology, for example, in following the fate of a protein, or in localising the cell type and timing of expression of a specific transcription factor (Chapter 14). In all experimental and applied uses of gene transfer, a crucial point is in mastering the expression of the transgene(s). Indeed, one possibility to reduce the variability between transgenic events is to incorporate matrix attachment regions into the constructs used for transformation, since gene activity depends on the strength with which such regions bind to the nuclear matrix (Xuc *et al.*, 2005). The silencing of genes after their transfer to plants, has opened up an entirely new vision of gene regulation, namely RNA silencing, based on sequence-specific targeting and degradation of RNA. The process was found to be active in almost all eukaryotes, probably starting as an antiviral defence and evolving in connection with cellular defence, RNA surveillance and development (Mallory and Vaucheret, 2004). Although the discovery of microRNA still raises many fundamental questions, it is also providing new tools to manipulate experimentally gene expression (Hilson *et al.*, 2004).

GENE TRANSFER AS A PLANT BREEDING TOOL

Gene transfer, combined with other techniques of genomics, has lead to entirely new approaches to plant breeding. During the last 10 years, these integrated technical developments have permitted knowledge of genome organization and functioning as well as control and adaptation of plant development and metabolism. Consequently, gene transfer has opened up entirely new experimental avenues for plant breeding that intellectually, are more exciting than the mere (and often controversial) transfer of a single bacterial gene for herbicide resistance. Nevertheless, some of the initial and simple applications of gene transfer have resulted in agronomic success, as in the case of the introduction of herbicide resistance into soybean (Carpenter, 2001) and insect resistance in other crops (Mendelsohn *et al.*, 2003). The extension to rice in China of insect resistance (Huang *et al.*, 2005) and herbicide tolerance (Cao *et al.*, 2004), should constitute further steps in the agronomic validation of these strategies.

Generally, one of the first priorities proposed for sustainable practice in crops concerns the protection of plants against pathogens to replace or to reduce the use of chemical pesticides. Improved understanding of plant-pathogen interactions (Hammond-Kosak and Parker, 2005), provides the basic knowledge essential to attempt to increase the "biological" resistance of plants. Indeed, the first exploration into this area targeted virus resistance, through the expression of viral coat protein genes in transgenic plants. Interestingly, to date, transgene technology may be the only practical option for use against Papaya Ring Spot Virus, in order to maintain the production of papaya in Hawaii (Gonsalves, 1998).

Transgenic potatoes, rendered resistant to various isolates of *Phytophthora infestans* after the transfer of a defence gene (Leucine Reach Repeat, LRR type) cloned from *Solanum bulbocastanum*, is an excellent illustration of the current possibilities of targeted breeding (Song *et al.*, 2003). These genetically engineered potatoes have been shown to be resistant to *P. infestans* for more than three years in field tests intended to access the stability of the resistance under agronomic conditions. Resistance to a broad range of pathogens (bacterial wilt, *Fusarium* wilt, grey leaf spot and bacterial spot) has been observed in transgenic tomatoes over-expressing NON EXPRESSOR of PR1 (*NPR1*, a key component of the salicylic acid defence signaling pathway), a gene from *A. thaliana* (Lin *et al.*, 2004). The NPR1 protein is a key regulator in the systemic acquired resistance response in *A. thaliana*. As the signal pathway seems to be conserved in both dicotyledons and monocotyledons (Chern *et al.*, 2001), it is easy to predict that this methodology, following validation in the field, will be extended to a range of crops.

Another priority concerns the improvement of stress tolerance in order to secure, and even to increase, crop yields. For low temperature and drought tolerance, it was logical to try to define the role of CBF/DREB (COLD BINDING FACTORS/ DEHYDRATATION RESPONSIVE ELEMENT) transcription factors (Cook *et al.*, 2004). Experiments aiming to constitutively over-express DREB1A in transgenic plants improved freezing tolerance, but also caused severe growth retardation under normal growth conditions. However, controlling the DREB1A transcription activator by a stress inducible promoter (rd29A) minimized any negative effects on plant growth whilst improving stress tolerance in transgenic tobacco (Kasuga *et al.*, 2004). Importantly, transfer of *A. thaliana* transcription factors to rice increased stress tolerance without an effect on the growth of the cereal crop (Oh *et al.*, 2005).

The utilization of transcription factors will also extend the possibilities of tailoring metabolism towards the generation of plants with improved nutritional composition for consumption by both humans and animals. Such improvements will be facilitated by increased precision of gene targeting and improved knowledge of plant metabolism and its regulation (Galili *et al.*, 2002). This is true even for the up-regulation of basic metabolism, such as nitrogen assimilation (Yanagisawa *et al.*, 2004). It is now possible to advance in terms of food safety through the precise elimination of toxic or allergenic substances. In this respect, it is worthwhile to emphasise the successful targeted suppression, by co-suppression, of the immunodominant allergen, the P34 protein, in soybean seed (Herman, 2005).

The quest to increase the nutritional quality of storage organs, seeds in the case of most crop plants, is also crucial for roots of cassava or stem tubers of potato, that contain up to 90% by dry weight of starch, but that are deficient in protein. A number of promising approaches are already being conducted with tissue-specific promoters controlling genes encoding proteins of known nutritive value, rich in essential amino acids, as in the introduction of a gene from *Amaranthus hypochondriatus* into potato tubers (Chakraborty *et al.*, 2000) and even a synthetic gene, based on zein, into cassava (Zhang *et al.*, 2003). Gene transfer is also of relevance to seed storage proteins. An example is provided by a gene construct that leads to a chimeric protein composed of phaseolin and γ-zein, named zeolin, which accumulates stably, and in large quantities in the endoplasmic reticulum of tobacco leaves. The γ-zein portion of zeolin is sufficient to induce the formation of protein bodies when fused to another protein. Such a characteristic can be used to engineer nutritionally balanced and highly stable chimeric storage protein (Maineri *et al.*, 2004). The capacity to transfer large fragments of DNA is also providing an entirely new means of improving seed storage proteins. Transferring 90 kb of a bacterial artificial chromosome clone containing the 35 kb of 10 linked genes of the α-kafirin cluster from sorghum to maize, proved that a targeted exchange of valuable multigenic traits is feasible. In addition, it revealed, surprisingly, that accumulation of kafirin in maize seeds did not repress the level of zeins, thus demonstrating that protein content may be increased in seeds (Song *et al.*, 2004). Since our understanding of the control of plant development is also progressing rapidly, it should be possible to define the architecture of crop plants (Sakamoto and Matsuoka, 2004), coupled with a fine tuning of their production capacities and ability to overcome stress. Improving storage capability has been linked to the control of cell size in seeds (Jokufu *et al.*, 2005).

CONCLUSION AND PROSPECTS

Recently, progress in plant genetics has accelerated tremendously, with gene transfer playing a crucial role in facilitating a basic understanding of this topic. At the same time, developments in "omics" technologies are, and will continue, to play a key role alongside transformation technologies. Gene transfer techniques have evolved so rapidly and in such sophisticated ways that it is now conceivable to engineer a the genome of a plant to perform specific functions without introducing foreign DNA sequences (Rommens *et al.*, 2004). Improved comprehension of the intimate process of DNA integration into plant genomes is still awaited, as this will enhance the frequency of "clean DNA transfers", and could provide insight into homologous recombination. It will also assist in refining the characteristics of *Agrobacterium* vectors (Gelvin, 2003). Recombinant Agrobacteria are by far the preferred system for delivering genes into plants, as a consequence of which they are employed for a broad range of DNA transfers, including complete viral sequences, often using *in planta* techniques. Therefore, there is an urgent requirement to develop biologically confined vectors. In addition to the usual physical containment through standard microbiological procedures and good laboratory practice, there is a need to monitor safety procedures and to upgrade these as appropriate. The use of strains of *Agrobacteria* rendered "safe" by auxotrophic mutations or targeted deletions (Collens *et al.*, 2004), should be encouraged and even rendered obligatory.

In the last decade, there has emerged a clear vision of the valuable agronomic applications of gene transfer to crops, particularly with respect to the nutritional characteristics and agronomic performance through improved tolerance to stress. Clearly, many more sustainable practices in crop production will emerge from exploitation of transgenic technologies. However, public concerns about the use of gene transfer to food crops, remains the basis of many of the current objections to the more widespread development of transgenic plants. It is of interest that gene transfer has been, and remains, a frequent event in nature for the plant genome (Huang *et al.*, 2004; Matsuo *et al.*, 2005). Such natural DNA movement can be difficult to comprehend in some cases, but an excellent example has been provided by the horizontal transfer of mitochondrial genes from diverse land plant donors to the angiosperm *Amborella* (Bergthorsson *et al.*, 2004). Undoubtedly, there is a need to improve public understanding of such naturally occurring events and to provide an unbiased opinion of the merits and limitations of the current approaches for plant genetic manipulation, since, in many ways, laboratory-based procedures are merely extensions of events that occur in nature.

REFERENCES

Afobali, A.S., Worland, B., Snape, J.W., Vain, P. 2004. A large scale study of rice plants transformed with different T-DNAs provides new insights into locus composition and T-DNA linkage configurations. *Theor. Appl. Genet.* 109, 815-826.

Alonso, J.L., Stepanova, A.N., Leisse, T.J., Kim, C.J., Chen, H., Shinn, P., Stevenson, D.K., Zimmerman, J., Barajas, P., Cheuk, R. 2003. Genome wide insertional mutagenesis of *Arabidopsis thaliana*. *Science*, 301, 653-657.

Aoki, S., Takebe, I. 1969. Infection of tobacco mesophyll protoplasts by tobacco mosaic virus ribonucleic acid. *Virology*, 39, 439-441.

Aziz, N., Machray, G.C. 2003. Efficient male germ transformation for transgenic tobacco production without selection. *Plant Mol. Biol.*, 23, 203-211.

Bako, L., Umeda, M., Tiburcio, A.F., Schell, J., Koncz, C. 2003. The VirD2 pilot protein of *Agrobacterium*-transfered DNA interacts with the TATA box-binding protein and a nuclear protein kinase in plants. *Proc. Natl. Acad. Sci.* USA, 100, 10108-10113.

Barton, K.A., Binns, A.N., Matze, A.J.M., Chilton, M.D. 1983. Regeneration of intact tobacco plants containing full lengh copies of genetically engineered T-DNA and transmission of T-DNA to R1 progeny. *Cell*, 32, 1033-1043.

Bechtold, N., Ellis, J., Pelletier, G. 1993. *In planta Agrobacterium* mediated gene transfer by infiltration of adult *Arabidopsis thaliana* plants. *CR Acad. Sci. III Vie*, 316, 1194-1199.

Bechtold, N., Jaudeau, B., Jolivet, S., Maba, B., Vezon, D., Voisin, R., Pelletier, G. 2000. The maternal chromosome set is the target of the T-DNA in the *in planta* transformation of *Arabidopsis thaliana*. *Genetics*, 155, 1875-1887.

Bergthorsson, U., Richardson, A.O., Young, G.J., Goertzen, L.R., Palmer, J.D. 2004. Massive horizontal transfer of mitochondrial genes from diverse land plant donors to the basal angiosperm *Amborella*. *Proc. Natl. Acad. Sci.* USA, 101, 17747-17752.

Boyton, J.E., Gillham, N.W. 1996. Genetics and transformation of mitochondria in the green alga *Chlamydomonas*. *Methods Enzymol.*, 264, 279-296.

Braun, A.C. 1947. Thermal studies on the factors responsible for tumor induction in crown-gall. *Am. J. Bot.*, 34, 234-240.

Browse, J., Sommerville, C. 1991. Glycerolipids synthesis, biochemistry and regulation. *Ann. Rev. Plant Mol. Biol.*, 42, 467-506.

Bundock, P., Hooykaas, P.J.J. 1996. Integration of *Agrobacterium tumefaciens* T-DNA in the *Saccharomyces cerevisiae* genome by illegitimate recombination. *Proc. Natl. Acad. Sci.* USA, 14, 3206-3214.

Caboche, M., Deshayes, A. 1984. Utilisation de liposomes pourla transformation de protoplastes de mésophylle de tabac par un plasmide de *E. coli* leur conférant la résistance à la kanamycine. *CR Acad. Sci. Paris*, 299, 663-666.

Cao, M.X., Huang, J.Q., Wei, Z.M., Yao, Q.H., Wan, C.Z., Lu, J.A. 2004. Engineering yield and herbicide resistance in rice by Agrobacterium mediated multiple gene transformation. *Crop Sci.*, 44, 2206-2213.

Carpenter, J.E. 2001. Case studies in benefits and risks of agricultural biotechnology : Round-up ready soybeans and Bt field corn. Report of the National Center for Food and Agricultural Policy, Washington DC, USA; p. 54.

Chakraborty, S., Chakraborty, N., Datta, A. 2000. Increased nutritive value of transgenic potato by expressing a non allergenic seed albumin gene from *Amaranthus hypochondriatcus*. *Proc. Natl. Acad. Sci.* USA, 97, 3724-3729.

Chern, M.S., Fitzgerald, H.A., Yadav, R.C., Canlas, P.E., Dong, X., Ronald, P.C. 2001. Evidence for a disease-resistance pathway in rice similar to the *NPR1* mediated signaling pathway in *Arabidopsis*. *Plant J.*, 27, 101-113.

Chupeau, M.C., Bellini, C., Guerche, P., Maisonneuve, B., Vastra, G., Chupeau, Y. 1989. Transgenic plants of lettuce (*Lactuca sativa*) obtained through electroporation of protoplasts. *Nature Biotechnol.*, 7, 503-507.

Clough, S.J., Bent, A.F. 1998. Floral dip: a simplified method for *Agrobacterium* mediated transformation of *Arabidopsis thaliana*. *Plant J.*, 16, 735-743.

Cocking, E.C. 1960. A method for the isolation of plant protoplasts and vacuoles. *Nature*, 187, 927-929.

Cohen, S.N., Miller, C.A. 1970. Non-chromosomal antibiotic resistance in bacteria. II Molecular nature of R-factor isolated from *Proteus mirabilis* and *Escherichia coli*. *J. Mol. Biol.*, 50, 671-687.

Collens, J.I., Lee, D.A.E., Seeman, A.M., Curtis, W.R. 2004. Development of auxotrophic *Agrobacterium tumefaciens* for gene transfer in plant tissue culture. *Biotechnol. Prog.*, 20, 890-896.

Cook, D., Fowler, S., Fiehn, O., Thomashow, F. 2004. A prominent role for the CBF cold response pathway in configuring the low temperature metabolome of *Arabidopsis*. *Proc. Natl. Acad. Sci.* USA, 101, 1543-1548.

Corneille, S., Lutz, K.A., Azhagiri, A.K., Maliga, P. 2003. Identification of a functional *Lox* site in the plastid genome. *Plant J.* 35, 753-762.

Curtis, I.S. 2005. Production of transgenic crops by the floral dip method. *Methods Mol. Biol.*, 286, 103-110.

Dale, E.C., Ow, D.W. 1991. Gene transfer with subsequent removal of the selection gene from the host genome. *Proc. Natl. Acad. Sci.* USA, 23, 10558-10562.

Davey, M.R., Anthony, P., Power, J.B., Lowe, K.C. 2005. Plant protoplasts : status and biotechnological perspectives. *Biotechnol. Adv.*, 23, 131-171.

De Frammond, A.J., Barton, K.A., Chilton M.D. 1983. Mini-Ti: a new vector strategy for plant genetic engineering. *Nature Biotechnol.*, 5, 262-269.

De Groot, M.J.A., Bundock, P., Hooykaas, P.J.J., Beijersbergen, A.G.M. 1998. *Agrobacterium tumefaciens* mediated transformation of filamentous fungi. *Nature Biotechnol.*, 16, 839-842.

De Vetten, N., Wolters, A.M., Raemakers, K., van der Meer, I., ter Stege, R., Heeres, E., Heeres, P., Visser, R. 2003. A transformation method for obtaining marker free plants of cross pollinating and vegetatitvely propagated crop. *Nature Biotechnol.*, 21, 439-442.

D'Erfurth, I., Cosson, V., Eschtruth, A., Lucas, H., Kondorosi, A., Ratet, P. 2003. Efficient transposition of the Tnt1 tobacco transposon in the model legume *Medicago truncatula*. *Plant J.*, 33, 1-12.

D'Halluin, K., Bonne, E., Bossut, M., De Beuckeller, M., Leemans, J. 1992. Transgenic maize plants by tissue electroporation. *The Plant Cell*, 4, 1495-1505.

Feldmann, K.A., Marks, M.D. 1987. *Agrobacterium* mediated transformation of germinating seeds of *Arabidopsis thaliana* : a non tissue culture approach. *Mol. Gen. Genet.*, 208, 1-9.

Fraley, R., Straubinger, R.M., Springer, E.L., Papahadjopoulos, D. 1981. Liposome mediated delivery of deoxyribonucleic acid to cells enhanced efficiency of delivery related to lipid composition and incubation conditions. *Biochemistry*, 24, 6978-6987.

Friesner, J., Britt, A.B. 2003. Ku80- and DNA ligase IV-deficient plants are sensitive to ionizing radiation and defective in T-DNA integration. *Plant J.*, 34, 427-440.

Fromm, M.E., Taylor, L.P., Walbot, V. 1986. Stable transformation of maize after gene transfer by electroporation. *Nature*, 319, 791-793.

Galili, G., Lewinsohn, E., Tadmor, Y. 2002. Genetic, molecular and genomic approaches to improve the value of plant feeds and foods. *Crit. Rev. Plant Sci.*, 21, 167-204.

Gelvin, S.B. 2003. *Agrobacterium* mediated plant transformation : the biology behind the "gene-jokeying" tool. *Microbiol. Mol. Biol. Rev.*, 67, 16-37.

Gonsalves, D. 1998. Control of papaya ringspot virus in papaya : a case study. *Annu. Rev. Phytopathol.*, 36, 415-437.

Hammond-Kosac, K.E., Parker, J. 2005. Deciphering plant-pathogen communication : fresh perspectives for molecular resistance breeding. *Curr. Opin. Biotechnol.*, 14, 177-193.

Hellens, R., Mulineaux, P., Klee, H. 2000. A guide to *Agrobacterium* binary Ti vectors. *Trends Plant Sci.*, 5, 446-451.

Herman E. 2005. Soybean allergenicity and suppression of the immunodominant allergen. *Crop Sci.*, 45, 462-467.

Hiei, Y., Ohta, S., Komari, T., Kumashiro, T. 1994. Efficient transformation of rice (*Oryza sativa*) mediated by *Agrobacterium* and sequence analysis of the boundaries of the T-DNA. *Plant J.*, 6, 271-282.

Herrera-Estrella, L., Depicker, A., Van Montagu, M., Schell, J. 1983. Expression of chimeric genes transferred into plants cells using a Ti plasmid-derived-vector. *Nature,* 303, 209-213.

Hilson, P., Allemeersch, J., Altmann, T., Aubourg, S., Avon, A., Beynon, J., Bhalerao, R.P., Bitton, F., Caboche, M., Cannoot, B., Chardakov, V., Cognet-Holliger, C., Colot, V., Crowe, M., Darimont, C., Durinck, S., Eickhoff, H., Falcon de Longevialle, A., Farmer, E. E., Grant, M., Kuiper, M.T.R., Lehrach, H., Léon, C., Leyva, A., Lundeberg, J., Lurin, C., Moreau, Y., Nietfeld, W., Paz-Ares, J., Reymond, P., Rouzé, P., Sandberg, G., Segura, M.D., Serizet, C., Tabrett, A., Taconnat, L., Thareau, V., Van Hummelen, P., Vercruysse, S., Vuylsteke, M., Weingartner, M., Weisbeek, P.J., Wirta, V., Wittink, F.R.A., Zabeau, M., Small, I. 2004. Versatile gene-specific sequence tags for *Arabidopsis* functional genomics : transcript profiling and reverse genetics applications. *Genome Research,* 14, 2176-2189.

Hoekema, A., Hirsh, P.R., Hooykaas, P.J.J., Schilperoort, R.A. 1983. A binary plant vector strategy based on separation of *vir* and T region of the *Agrobacterium tumefaciens* Ti-plasmid. *Nature,* 303, 179-180.

Horsch, R.B., Fry, J.E., Hoffman, N.L., Eicholtz, D., Rogers, S.G., Fraley, R.T. 1985. A simple and general method for transferring genes into plants. *Science,* 227, 1229-1231.

Huang, C.Y., Ayliffe, M.A., Timmis, J.N. 2004. Simple and complex nuclear loci created by newly transferred chloroplast DNA in tobacco. *Proc. Natl. Acad. Sci.* USA, 101, 9710-9715.

Huang, J., Hu, R., Rozelle, S., Pray, C. 2005. Insect-resistant rice in farmers' fields : assessing productivity and health effects in China. *Science,* 308, 688-690.

Ishida, Y., Saito, H., Ohta, S., Hiei, Y., Komari, T., Kumashiro, T. 1996. High efficiency transformation of maize (*Zea mays*) mediated by *Agrobacterium tumefaciens. Nature Biotechnol.,* 14, 745-750.

Joersbo, M., Brundstedt, J. 1992. Sonication: a new method for gene transfer to plants. *Physiol. Plant.,* 85, 230-234.

Joersbo, M., Donaldson, I., Kreiberg, J., Petresen, S.G., Brundstedt, J., Okkels, F.T. 1998. Analysis of mannose selection used for transformation of suger beet. *Mol. Breed.,* 4, 111-117.

Johnston, S.A., Anziano, P.Q., Shark, K., Sanford, J.C., Butow, R.A. 1988. Mitochondrial transformation in yeast by bombardment with microprojectiles. *Science,* 240, 1538-1541.

Johnston, S.A. 1990. Biolistic transformation: microbes to mice. *Nature,* 346, 776-777.

Jokufu, K.D., Omidyar, P.M., Gee, Z., Okamuro, J.K. 2005. Control of seed mass and yield by the floral homeotic gene *APETALA2. Proc. Natl. Acad. Sci.* USA, 102, 3117-3122.

Kao, K.N., Michayluk, M.R. 1974. A method for hight frequency intergeneric fusion of plant protoplasts. *Planta,* 115, 355-367.

Kasuga, M., Miura, S., Shinozaki, K., Yamaguchi-Shinozaki, K. 2004. A combination of the *Arabidopsis DREB1A* and *rd29A* promoter improved drought and low temperature stress tolerance in tobacco by gene transfer. *Plant Cell Physiol.,* 45, 346-350.

Komari, T., Hiei, S.Y., Murai, N., Kumashuiro, T. 1996. Vectors carrying two separate T-DNAs for co-transformation of higher plants mediated by *Agrobacterium tumefaciens* and segregation of transformants free from selection markers. *Plant J.,* 10, 165-174.

Kunik, T., Tzfira, T., Kapulnik, Y., Gafni, Y., Dingwall, C., Citovsky, V. 2001. Genetic transformation of Hela cells by *Agrobacterium. Proc. Natl. Acad. Sci.* USA, 98, 1871-1876.

Langbecker, C.L., Ye, G.N., Broyles, D.L., Duggan, L.L., Xu, C.W., Hajdukiewicz, P.T.C., Armstrong, C.L., Staub, J.M. 2004. High frequency transformation of undeveloped plastids in tobacco suspension cells. *Plant Physiol.,* 135, 39-46.

Ledoux, L., Huart, R., Jacobs, M. 1974. DNA mediated genetic correction of thiamine-less *Arabidopsis thaliana. Nature,* 249, 17-21.

Lentz, B.R., Lee, J.K. 1999. Poly(ethyleneglycol) (PEG)-mediated fusion between pure lipid bilayers : a mechanism in common with viral fusion and secretory vesicule release? *Mol. Membr. Biol.,* 16, 279-296.

Levy, Y.Y., Dean, C. 2005. The transition to flowering. *The Plant Cell,* 10, 1973-1989.

Lida, S., Terada, R. 2004. A tale of two integrations, transgene and T-DNA: gene targeting by homologous recombination in rice. *Curr. Opin. Biotechnol.,* 15, 132-138.

Lin, W.C., Lu, C.F., Wu, J.W., Cheng, M.L., Lin, Y.M., Yang, N.S., Black, L., Green, S.K., Wang, J.F., Cheng, C.P. 2004. Trangenic tomato plants expressing the *Arabidopsis NPR1* gene display enhanced resistance to a spectrum of fungal and bacterial diseases. *Transgenic Res.,* 13, 567-581.

Lloyd, A.M., Walbot, V., Davis, R.W. 1992. *Arabidopsis* and *Nicotiana* anthocyanin production activated by maize regulators R and C1. *Science,* 258, 1773-1775.

Maineri, D., Rossi, M., Archinti, M., Belluci, M., De Marchis, F., Vavassori, S., Pompa, A., Arcioni, S., Vitale, A. 2004. Zeolin. A new recombinant protein constructed using maize γ-zein and bean phaseolin. *Plant Physiol.,* 136, 3447-3456.

Maliga, P. 2004. Plastid transformation in higher plants. *Ann. Rev. Plant Biol.,* 55, 289-3313.

Mallory, A.C., Vaucheret, H. 2004. MicroRNA: something important between the genes. *Curr. Opin. Plant Biol.,* 7, 120-125.

Marrillonet, S., Thoeringer, C., Kandzia, R., Klimyuk, V., Gleba, Y. 2005. Systemic *Agrobacterium tumefaciens* mediated transfection of viral replicons for efficient transient expression in plants. *Nature Biotechnol.,* 23, 718-723.

Matsuo, M., Ito, Y., Yamauchi, R., Obokata, J. 2005. The rice nuclear genome continuously integrates, shuffles and eliminates the chloroplast genome to cause chloroplast-nuclear DNA flux. *The Plant Cell,* 17, 665-675.

Mendelsohn, M., Kough, J., Matthews, K. 2003. Are Bt crops safe? *Nature Biotechnol.,* 21, 1003-1009.

Morikawa, H., Yamada, Y. 1985. Capillary micro-injection into protoplasts and intranuclear localization of injected materials. *Plant Cell Physiol.,* 26, 229-236.

Nishal, B., Tantikanjana, T., Sundaresan, V. 2005. An inducible targeted tagging system for localized saturation mutagenesis in *Arabidopsis*. *Plant Physiol.,* 137, 3-12.

Oh, S.J., Song, S.I., Kim, Y.S., Jang, H.J., Kim, S.Y., Kim, M., Kim, Y.K., Nahm, B.H., Kim, J.K. 2005. *Arabidopsis* CFB3/DREB1A and ABF3 in transgenic rice increased tolerance to abiotic stress without stunting growth. *Plant Physiol.,* 138, 341-351.

Ohyama, K., Gamborg, O.L., Miller, R.A. 1972. Uptake of exogenous DNA by plant protoplasts. *Can. J. Bot.,* 50, 2077-2080.

Paszkowski, J., Shillito, R.D., Saul, M., Mandàk, V., Hohn, B., Potrykus, I. 1984. Direct gene transfer to plants. *EMBO J.,* 3, 2717-2722.

Petit, A., Tourneur, J. 1972. Perte de virulence associée à la perte d'une activité enzymatique chez *Agrobacterium tumefaciens. CR Acad. Sci. Paris,* 275, 137-139.

Potrykus, I. 1990. Gene transfer to plants : assessment and perspectives. *Physiol. Plant.,* 79, 125-134.

Puchta, H. 2003. Marker free transgenic plants. *Plant Cell Tissue Org. Cult.,* 74, 123-134.

Rommens, C.M., Humara, J.M., Ye, J., Yan, H., Richael, C., Zhang, L., Perry, R., Swords, K. 2004. Crop improvement through modification of the plant's own genome. *Plant Physiol.,* 135, 421-431.

Sakamoto, T., Matsuoka, M. 2004. Generating high yielding varieties by genetic manipulation of plant architecture. *Curr. Opin. Biotechnol.,* 15, 144-147.

Sambrook, J., Fritsch, E.F., Maniatis, T. 1989. Molecular cloning : A laboratory manual. Cold Spring Harbor Laboratory Press, New York, USA. Vols. 1, 2, 3.

Sanford, J.C. 1988. The biolistic process. *Trends Biotechnol.,* 6, 299-302.

Seki, M., Narusaka, M., Abe, H., Kazuga, M., Yamaguchi-Shinozaki, K., Carninci, P., Hayashizaki, Y., Shinozaki, K. 2001. Monitoring the expression pattern of 1300 *Arabidopsis* genes under drought and cold stresses by using a full-lenght cDNA microarray. *The Plant Cell*, 13, 61-72.

Shaner, N.C., Campbell, R.E., Steinbach, P.A., Giepmans, B.N.G., Palmer, A.E., Tsien, R.Y. 2004. Improved monomeric red, orange and yellow fluorescent proteins derived from *Discosoma sp.* red fluorescent protein. *Nature Biotechnol.*, 22, 1567-1572.

Sheen, J. 2001. Signal transduction in maize and *Arabidopsis* mesophyl protoplasts. *Plant Physiol.*, 127, 1466-1475.

Shibata, D., Liu, Y.C. 2000. *Agrobacterium* mediated plant transformation with large DNA fragments. *Trends Plant Sci.*, 5, 354-357.

Smith, E.F., Townsend, C.O. 1907. A plant tumor of bacterial origin. *Science*, 25, 671-673.

Somers, D.A., Makarevitch, I. 2004. Transgene integration in plants : poking or patching holes in promiscuous genomes? *Current Opin. Biotech.*, 15, 126-131.

Song, J., Bradeen, J.M., Naess, S.K., Raasch, J.A., Wielgus, S.M., Haberlach, G.T., Liu, J., Kuang, H., Austin-Phillips, S., Buell, C.R., Helgeson, J.P., Jiang, J. 2003. Gene RB cloned from *Solanum bulbocastanum* confers broad spectrum resistance to potato late blight. *Proc. Natl. Acad. Sci.* USA, 100, 9128-9133.

Song, R., Segal, G., Messing, J. 2004. Expression of the sorghum 10-members kafirin gene cluster in maize endosperm. *Nucleic Acids Res.*, 32, e189.

Srivastava, V., Ow, D.W. 2004. Marker-free site specific gene integration in plants. *Trends Biotech.*, 22, 627-629.

Stuitje, A.R., Verbree, E.C., van der Linden, K.H., Mietkiewska, E.M., Nap, J.P., Kneppers, T.J.A. 2003. Seed-expressed fluorescent proteins as versatile tools for easy (co)transformation and high throughput functional genomics in *Arabidopsis*. *Plant Biotech J.*, 1, 301-309.

Taylor, N.J., Fauquet, C.M. 2002. Microparticule bombardment as a tool in plant science and agricultural biotechnology. *DNA Cell Biol.*, 21, 936-977.

The AGI : *Arabidopsis* Genome Initiative. 2000. Analysis of the genome sequence of the flowering plant *Arabidopsis thaliana*. *Nature*, 408, 796-815.

Tinland, B. 1996. The integration of T-DNA into plant genomes. *Trends Plant Sci.*, 1, 178-184.

Tzfira, T., Frankman, L.R., Vaidya, M., Citovsky, V. 2003. Specific site integration of *Agrobacterium tumefaciens* T-DNA via double strended intermediates. *Plant Physiol.*, 133, 1011-1023.

Van der Fits, L., Deakin, E.A., Hoge, J.H., Memelink, J. 2000. The ternary transformation system : constitutuve *virG* on a compatible plasmid dramatically increases *Agrobacterium* mediated plant transformation. *Plant Mol. Biol.*, 43, 495-502.

Van Larebecke, N., Genetello, C., Schell, J., Schilperoort, R.A., Hermans, A.K., Hernalsteens, J.P., Van Montagu, M. 1975. Acquisition of tumor inducing ability of non oncogenic agrobacteria as a result of plasmid transfer. *Nature*, 255, 742-743.

Vinocur, B., Altman, A. 2005. Recent advances in engineering plant tolerance to abiotic stress : achievements and limitations. *Curr. Opin. Biotechnol.*, 16, 123-132.

Wallin, A., Glimelius, K., Eriksson, T. 1974. The induction of aggregation and fusion of *Daucus carota* protoplasts by polyethylene glycol. *Z. Pflanzenphysiol.*, 74, 64-80.

Xuc, H., Yang, Y.T., Wu Zhang, M.M., Zheng, C.C. 2005. TM2 a novel strong matrix attachment region isolated from tobacco, increases transgene expression in transgenic rice calli and plants. *Theor. Applied Genet.*, 110, 620-627.

Yanagisawa, S., Akyiama, A., Kisasa, H., Uchimiya, H., Miwa, T. 2004. Metabolic engineering with *Dof1* transcription factor in plants : improved nitrogen assimilation and growth under low-nitrogen conditions. *Proc. Natl. Acad. Sci.* USA, 101, 7833-7838.

Yoshioka, Y., Takahashi, Y., Matsuoka, K., Nakamura, K., Koizumi, J., Kojima, M., Machida, M. 1996. Transient gene expression in plant cells mediated by *Agrobacterium tumefaciens*: application for the analysis of virulence loci. *Plant Cell Physiol.*, 37, 782-789.

Yuan, Q., Ouyang, S., Liu, J., Suh, B., Cheung, F., Sultana, R., Lee, D., Quackenbush, J., Buell, C.R. 2003. The TIGR rice genome annotation resource : annotating the rice genome and creating resources for plant biologists. *Nucleic Acids Res.*, 31, 229-233.

Zhang, P., Jaynes, J.M., Potrykus, I., Gruissem, W., Puonti-Kaerlas, J. 2003. Transfer and expression of an artificial storage protein (ASP1) gene in cassava (*Manihot esculenta Crantz*). *Transgenic Res.*, 12, 243-250.

Zupan, J., Muth, T.R., Draper, O., Zambryski, P. 2000. The transfer of DNA from *Agrobacterium tumefaciens* into plants: a feast of fundamental insights. *Plant J.*, 23, 11-28.

9

CHAPTER

Mutant Libraries: A High-throughput *Arabidopsis* Reverse Genetics System

F. Granier, D. Bouchez

By the end of the year 2000, scientists had sequenced the first complete genome of a higher plant, *Arabidopsis thaliana* (Chapter 2) (Fig. 9.1). For more than 20 years, this weed, very common in temperate regions, has been the model plant for geneticists and biologists. It has many practical advantages, particularly for experiments in genetics: it is small and has a short vegetative cycle, it can be cross-fertilized after emasculation, and the plants are prolific (up to 50,000 seeds per plant). Its genome, which is particularly compact, is composed of five chromosomes, with a relatively simple organization compared to other plant genomes, especially of cultivated plants. It is made up of around 30,700 genes (version 5 annotation) coding for proteins belonging to 11,000-15,000 families. The genome is relatively small: 125 Mb. One of the most surprising pieces of information from sequencing data is the high rate of genome duplication. Less than half of the DNA of this plant is unique. Around 60% of the genome has undergone more or less ancient duplication over the evolutionary history of the species (AGI, 2000). The use of this model has contributed to great advances in plant biology through the multiplicity of tools and information it makes available to researchers. The use of mutants, particularly, has led to immense progress in our understanding of the development and physiology of plants and their interactions with pathogens. In this chapter, the principal tools of mutagenesis that are available are surveyed, especially with respect to *A. thaliana*, and current aspects of reverse genetics are discussed.

For decades, advances in conventional genetics, based on the study of variants isolated from natural populations or induced by mutagenesis, have made it possible to locate and identify genes individually, and to determine their biological function. From the isolation of mutants affected in one or another biological process, physiological and genetic studies determine the genetic architecture of these characters (number of genes, order of intervention, interactions) and often infer from them relevant physiological information (for example, the steps of a biosynthetic pathway).

Fig. 9.1 *Arabidopsis thaliana*

Until the 1990s, few tools were available to isolate genes detected by genetic approaches, especially in the absence of information about the biochemical characteristics of the process studied. The positional cloning approaches used at present are derived from numerous tools developed over the past decade, such as molecular markers (Chapter 25) or cloning of large fragments. It thus became possible to isolate genes detected by mutation solely on the basis of their genetic position, and in the absence of information on a biochemical function of the product of such a gene. However, these approaches remain delicate and particularly labour-intensive, especially because large segregating populations must be constructed in order to ensure exact genetic positioning.

This is why, since the late 1980s, the possibility of insertion mutagenesis in plants, and particularly in *A. thaliana* has offered an alternative to positional cloning approaches to isolate interesting genes. Recently, and because of the absence of techniques for effective homologous recombination in higher plants, collections of insertion mutants are also widely used in functional genomics to search for gene functions by reverse genetics.

INSERTION MUTATIONS

Insertion mutagenesis involves the insertion of an element of foreign DNA that will interrupt the gene and cause its inactivation (Fig. 9.2). This insertion of a known sequence element will "label" the mutation much more effectively than conventional mutagenesis processes (chemical mutagens, radiation), which most often induce subtle modifications in the DNA structure (e.g., base changes, microdeletions).

It thus presents a considerable advantage over conventional mutagenesis in that it induces a more drastic modification of the gene structure and "tags" genes on the molecular level, facilitating their ultimate recovery. This strategy of insertion mutagenesis was initiated in *Drosophila* with the insertion of transposable P elements and it has since been widely used in *Caenorhabditis elegans*, yeast, and plants (Azpiroz-Leehan and Feldmann, 1997; Martienssen, 1998; Krysan *et al.*, 1999).

Constitution of Collections

Insertion mutagenesis relies on the generation of collections, that is, of populations of independent plants, each of which carries one or several insertions of the mutagenic element. The objective is to generate a population that statistically contains mutants for as many genes as possible and hopefully all of them. The larger the collection, the more it tends toward "saturation", that is, a satisfying probability (more than 90% or even 95%) of finding at least one insertion in any gene. The saturation depends on the size of the gene (sequence of genomic DNA, from ATG to the stop codon), the size and organization of the genome of the plant used, and the number of insertions in the population (Bouchez and Hofte, 1998). It is important that the number of random insertions and the number of insertions per line remain low. In the case of *A. thaliana*, around 120,000 independent insertions are needed to achieve saturation close to 95% (Feldmann, 1991).

Populations of insertion mutants can be used in direct screens, to isolate genes in which inactivation gives a visible or scorable phenotype and/or a reverse screen to

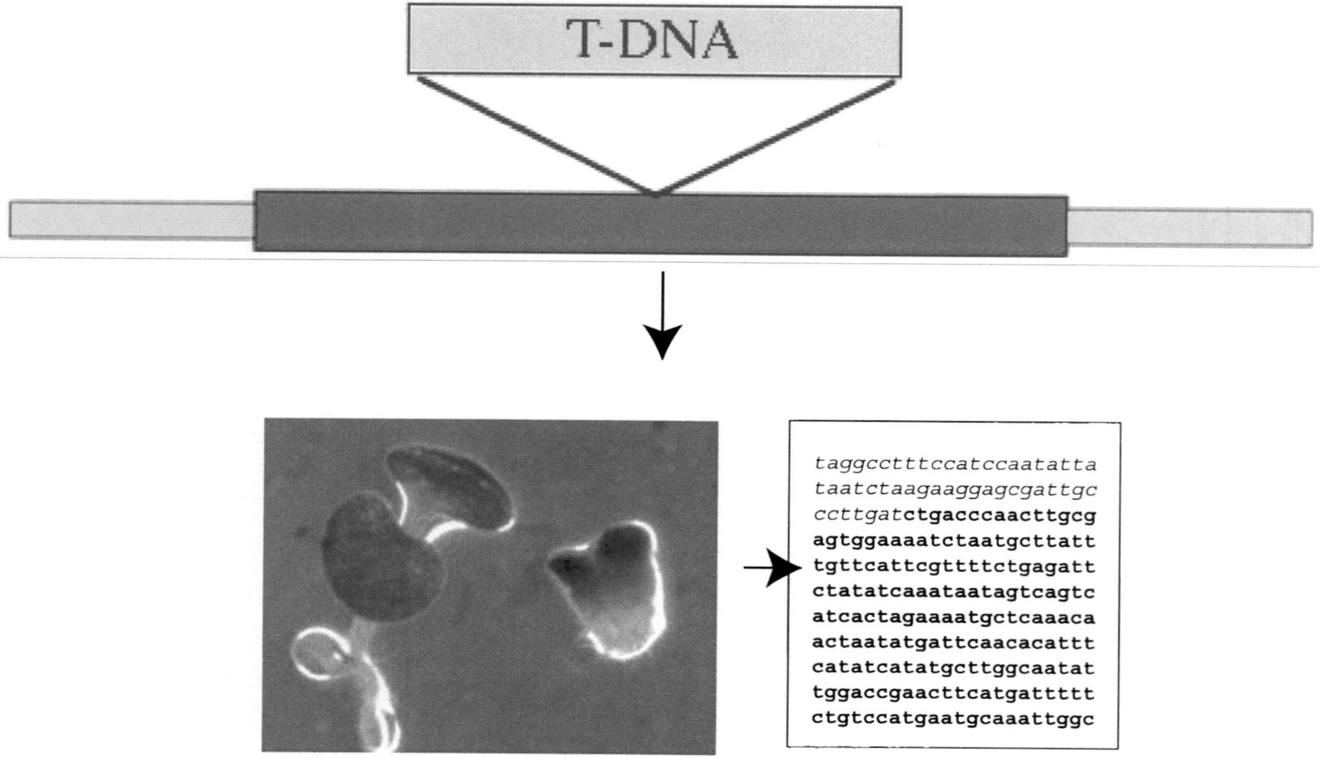

Fig. 9.2 Insertion of foreign DNA into the sequence of the *TON1* gene causes disruption of the gene and the appearance of the mutant phenotype. If the sequence of the mutagenic element (T-DNA or transposon, in italics in the sequence box) is known, the insertion site in the genome is easily located and the sort of gene affected is identified.

determine the function of a gene whose sequence is known. Large-scale reverse genetics has developed through the PCR technique, which allows detection of an insertion event in complex DNA samples using a specific primer of the target genes on the one hand, and a specific primer of the insertion element on the other (Fig. 9.3). The presence of an amplification product indicates the presence in the combination of a line that has an insertion in the gene.

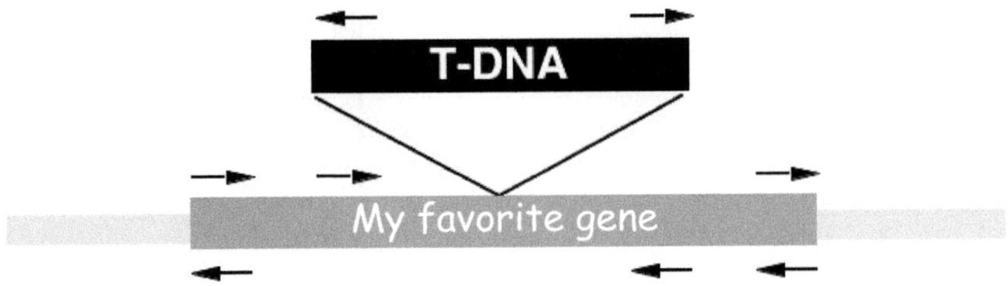

Fig. 9.3 Homologous PCR primers at the ends of an insertion element, here T-DNA, and specific PCR primers for the target gene are used to amplify the sequences at the junction of insertions. Different primer combinations are used to attempt to cover the different insertion possibilities. Only an insertion positioned in the gene studied will give a positive PCR signal.

More sophisticated methods use gene trap techniques or activation through tagging, which contributes some additional information, making it possible, for example, to obtain expression data. The expression of an integrated reporter gene will mimic that of the chromosomal gene and in certain cases make it possible to guide the search for the phenotype.

Different Types of Mutagens

Numerous collections have been generated from insertion lines in tomato, maize, rice, petunia, snapdragon, and especially *A. thaliana*. One major advance in insertion mutagenesis by T-DNA was the development of effective techniques of direct transformation that make it possible to overcome problems related to somaclonal variations not caused by a T-DNA insertion (this type of mutation commonly appears in plants grown from *in vitro* culture after regeneration).

T-DNA

T-DNA has been used as a mutagenesis tool from the late 1980s onward. In *A. thalina*, which does not have endogenous transposons with an activity allowing insertion mutagenesis, it was used in preference to transposons. The soil bacterium *Agrobacterium tumefaciens* has the capacity to transfer a plasmid fragment, T-DNA, into the nuclear genome of plant cells, where it is inserted randomly and becomes an integral part of the chromosome, transmitted to the descendants in a Mendelian manner (Fig. 9.4). This property is very useful in inactivating genes by insertion mutation particularly in *A. thaliana* , where the compactness of the genome allows such an approach.

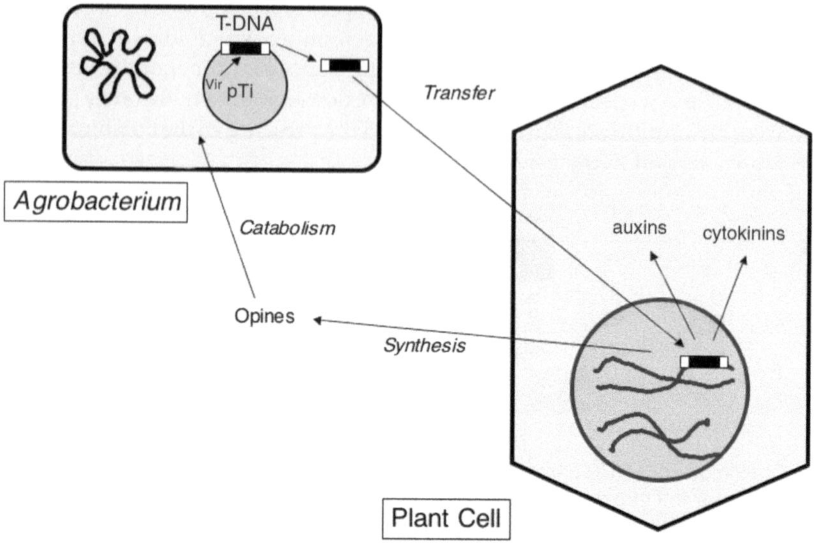

Fig. 9.4 The soil bacterium *A. tumefaciens* has the capacity to transfer a plasmidic DNA fragment, T-DNA, delimited by two repeats of 24 bp, into the DNA of a plant cell. Wild strains of *A. tumefaciens* induce diseases (crown gall or hairy root) caused by the expression of bacterial genes carried by the T-DNA into the transformed plant cell. These genes provoke the synthesis of plant hormones that induce cell proliferation (formation of galls or roots). Other genes of the T-DNA provoke the synthesis of particular compounds called "opines" that the inducing bacterial colony can use as a growth substrate. This system can be used as a biological vector for transformation of plants, after elimination of native T-DNA to introduce a synthetic genetic construction.

The only elements required for the transfer are terminal repeats of 24 bp, the flanking sequences, which delimit the region to be transferred. The increase in the efficiency of transformation methods has facilitated the development of reverse genetics.

Transformation *in planta* (Fig. 9.5), which can prevent problems linked to *in vitro* culture and regeneration of plants (somaclonal variation), was initiated by Feldman and Marks in 1987, but with a low transformation frequency. The authors transformed seeds of *A. thaliana* by soaking them in a solution of *A. tumefaciens*. Infiltration of whole plants under vacuum (Bechtold *et al.*, 1993) led to an increase in transformation efficiency (up to four to five transformations per treated plant) as well as reproducibility. The method was further simplified with the "floral dip" technique, where inflorescences were simply soaked in a solution of *A. tumefaciens* in the presence of a surfactant, Silwet L-77, and sucrose (Clough and Bent, 1998).

With these different methods, all the transformants obtained are hemizygous for T-DNA, which suggests a late transformation during floral development. The transformation target has been identified (Bechtold *et al.*, 2000; Desfeux *et al.*, 2000) as the developing ovule. Mutations that are lethal in the homozygous state are maintained in the population in the form of heterozygous plants.

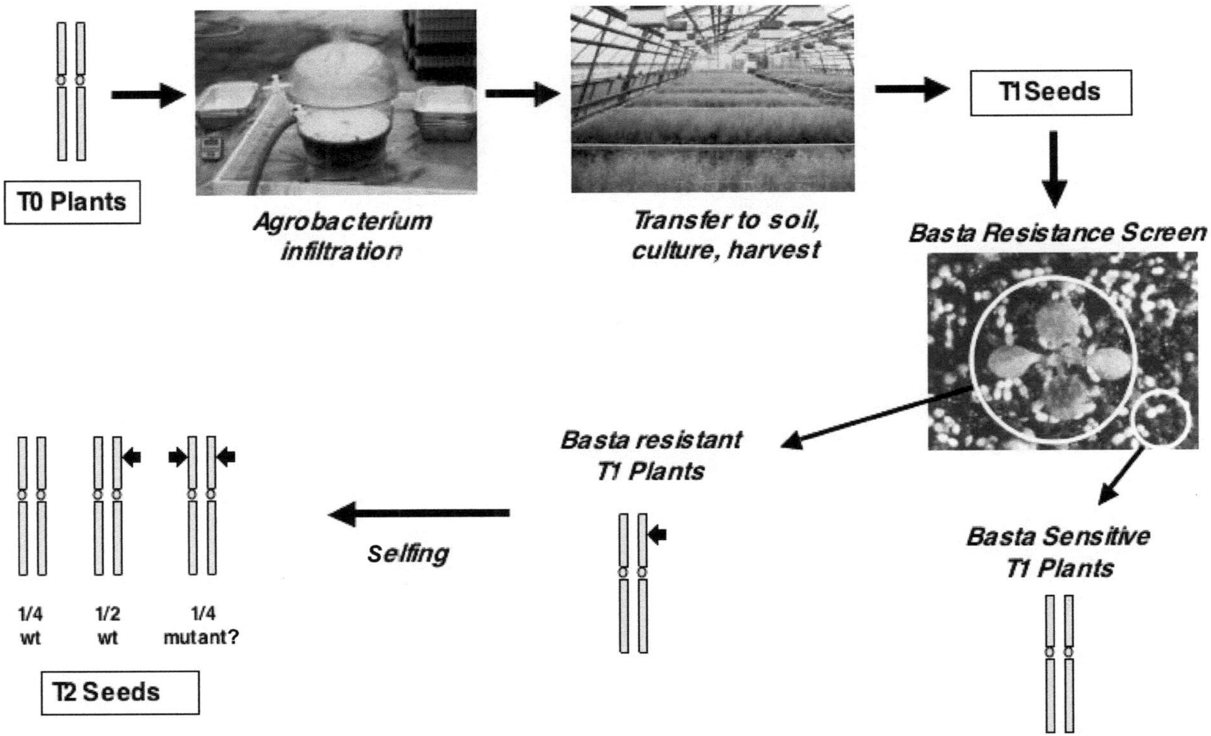

Fig. 9.5 Mutagenesis by T-DNA insertion. Wild plants T0 are infiltrated under vacuum with a colony of *A. tumefaciens* carrying a modified T-DNA containing a selection marker (resistance to an antibiotic or to a herbicide, in this case Basta). In the descendants of the infiltrated T0 plants, the transformed T1 plants are selected on Basta; the non-transformed plants are eliminated. All the resistant plants are hemizygous for the T-DNA insertion. The T1 plants are self-fertilized to yield the T2 generation, in which segregation is found: one-fourth sensitive to Basta, three-fourths resistant when there is a single insertion locus. If the insertion causes a recessive mutation that has a phenotype, this phenotype appears in one-fourth of the T2 plants, those that are homozygous for the insertion. This is called co-segregation of the phenotype and the insertion, and the mutation is thus said to be "tagged".

On an average, one or two insertion loci per primary transformant are obtained. Segregation analysis shows that 57% of the transformants contain one insertion locus, 25% contain two loci, 8% contain three loci, and 2% contain more than three loci. A molecular analysis of tagged mutants (Azpiroz-Leehan and Feldmann, 1997) showed that these insertions occur at random, are stable, and persist in the descendants, and that there are few insertion biases. There was no evidence of chromosomal preference. Insertions in the promoters or introns also revealed phenotypes.

One problem is that the T-DNA insertions are often complex, with multiple or reverse copies or copies in tandem, integration of vector sequences adjacent to the borders, and insertions of truncated or complex T-DNA that complicate the molecular analysis. The junctions between plant DNA and the left T-DNA border are prevalent over junctions with the right border (Winkler *et al.*, 1998). In the sequence, the right part of the T-DNA is frequently conserved, while the left part is much less often so and has deletions of two to several hundred nucleotides (Tinland, 1996). The insertion may induce more or less significant modifications of the genomic DNA at the insertion site: for example, deletions of tens or hundreds of nucleotides or duplications of genomic sequences flanking the T-DNA (Koncz and Redei, 1994). There may also be large chromosomal changes: large deletions, insertion of foreign genomic DNA at the insertion region or inversion or translocation of chromosomal fragments (Takano *et al.*, 1997). These chromosomal rearrangements make genetic analysis of insertions difficult (Nacry *et al.*, 1998; Laufs *et al.*, 1999). There may also be occasional mutations caused by aborted insertion events.

With the T-DNA, there is no possibility of reversion of the phenotype. It may be more difficult to reach saturation than with mobile transposons in the genome.

TRANSPOSONS

Transposons were the first agents of insertional mutation used in maize and then in petunia and the snapdragon *Antirrhinum*. In *A. thaliana*, there is no endogenous transposable element with high activity; therefore, heterologous transposable elements from maize were used, that were introduced into the genome by transformation with T-DNA. The two major transposon systems used are *Ac/Ds* (activator/dissociation) and *En/Spm* (enhancer/suppressor-mutator). Unlike T-DNA, the transposon can be excised from the interrupted gene in the presence of a transposase. The high frequency of reversion of the mutation that results makes it possible to confirm that it was induced by the transposon. The remobilization of transposons can also be used to generate some mosaics: a homozygous mutant that carries an active transposase will have somatic sectors that have lost the *Ds* transposon resulting in restored gene function. This makes it possible to determine the site of action of a gene as well as its pattern of expression. Moreover, for transposons of the *Ac/Ds* type, most of the transposition events occur at genetically linked sites. If there is a transposable element close to a target gene, it could be remobilized to be inserted nearby (Ito *et al.*, 1999). New mutant alleles can thus be isolated or a local mutation can be caused in a particular region of interest. This advantage becomes a disadvantage when the aim is to obtain a saturated collection. Some systems have thus been developed to allow negative selection of nearby transpositions, with *Ac/Ds* systems in which the two types of elements are maintained in separate lines (Parinov *et al.*, 1999) and with *En/Spm* systems that allow selection of stable insertions (Tissier *et al.*, 1999).

A great advantage of transposons is that they can activate or inactivate the transposition of defective elements, which are incapable of moving on their own, by providing a source of transposase in *trans*. This transposition activity can be introduced in a line simply by crossing, or even eliminated by genetic segregation between the defective element and the source of transposase.

There are two major types of approaches for mutagenesis with transposons:

- The use of plants with a large number of copies of very active transposons per genome. These systems with a large number of copies have the advantage of allowing coverage of the genome with small populations (mutator: 200 insertions per plant in maize). It is useful for plants with large genomes. On the other hand, thorough genetic analyses are needed to isolate the interesting insertions from others and to study the phenotypes induced.
- The use of plants with a small number of copies per genome of a given transposon. This system requires larger populations to saturate the genome but has some advantages: in particular, the phenotype of mutants can be studied much more directly, without having to carry out numerous crosses and genetic tests to purify the interesting insertion. Besides, simple copy transposons can be used to introduce reporter genes in the genome that make it possible to observe the expression of the gene. This is the most commonly used approach in *A. thaliana*.

The Ac/Ds System

The *Ac/Ds* transposons of maize have short terminal repeats that allow modifications of the element without risk of compromising the transposition function. The *Ac/Ds* transposons tend to transpose towards sites linked to the starting insertion, preferentially in a region of 500 to 1,000 kb on either side of the initial site (Bancroft and Dean, 1993). This property can be used to perform local mutagenesis to saturation, starting from the "launching pads" evenly distributed in the genome.

The *Ac* transposase gene can also be associated with a counter-selection marker such as the *iaaH* gene, which confers sensitivity to naphthalene acetamide, making it possible to select descendants without transposase activity, including stable *Ds* insertions (Parinov *et al.*, 1999).

To overcome this tendency toward local mutagenesis, which is awkward in the case of overall mutagenesis on the scale of the genome, elegant systems have been developed for the purpose of counter-selecting insertions near the starting site. For example, a small number of T-DNA "launchers" are used that contain a *Ds* insertion and a counter-selection marker (*iaaH*). The elements are mobilized by crossing with a plant containing the *Ac* transposase gene. Selection on kanamycin and naphthalene acetamide is used to obtain plants that have retained the *Ds* element but lost the donor site, and thus to select transposition events not linked to the starting site (Martienssen, 1998).

The En/Spm System

Some systems are used to saturate the genome using a small number of lines and obtain a collection of mutants from the transformation of a single plant; after self-fertilization, the primary transformant will generate a population within which copies of the starting transposon are integrated at different loci of the genome.

The *En-I* element is a particular element of *En/Spm* that maintains its mobility and remains very active when it is introduced in a heterologous host such as *A. thaliana*. This element transposes by a cut-and-paste system in which the excision of a donor site is followed by reintegration elsewhere in the genome. It is a multicopy system (collection of 8,000 lines containing 48,000 independent insertions) used to obtain saturated collections in which each gene is tagged at least once (Wisman *et al.*, 1998).

A collection of 2,592 lines of *A. thaliana* containing 20 to 25 *I* elements per line (or a total of 50,000 to 65,000 insertions) has been constituted from another system with two components based on *En-I*. This represents one insertion approximately every 2 kb thus giving a good saturation of the genome (Speulman *et al.*, 1999).

The disadvantages of lines that contain several copies of the autonomous element include events of somatic reversion that result in detection of insertions that are not transmitted. Moreover, numerous back-crosses are needed to obtain a line with a single target insertion.

A selection system for positive/negative transposition analogous to that used with the *Ac/Ds* system has been developed with *En/Spm*. The *dSpm* element carries a resistance gene to select for the integration of a T-DNA and the reinsertion of the transposon. This system can be used to rapidly obtain a large number of stable independent insertions (Tissier *et al.*, 1999). It uses a single T-DNA with positive selection for the disarmed transposon (*BAR* gene) and counter-selection for T-DNA (*SU1* gene).

GENE TRAPS AND ACTIVATION MUTAGENESIS

Some of the modifications of constructions used for insertion mutagenesis, whether they are T-DNA or transposons, may also make it possible to obtain additional information on the insertion region, especially about its transcriptional activity: this is the aim of promoter trap, enhancer trap, and other strategies. Some populations have also been created to carry out mutagenesis by activation.

GENE TRAP

A gene trap construction makes it possible to identify a gene based on its expression. These constructions contain a capture marker, often the reporter gene *GUS* or even *GFP* (*green fluorescent protein*), near the end of the insertion element. During the insertion of the element, there may be, in certain conditions of insertion and orientation, fusion of genes in which the reporter gene will express itself under the control of regulatory regions of the insertion site. Besides the indication of interesting regulatory regions, the identification of very specific expression patterns, for example in the meristem or pathogen-induced, may suggest the phenotype to be sought in the mutant line (Pereira, 2000).

This system can be used to isolate promoters or transcription activators, or to induce the expression of genes. Different types of vectors used for gene tagging have been reported in the reviews of Springer (2000) and Pereira (2000).

There are essentially three systems (Fig. 9.6), each designed to respond to regulatory elements in *cis*.

- *Promoter trap*: The reporter gene is deprived of the promoter. There is expression only if the insertion occurs in a properly oriented transcription unit and if the translation is correctly initiated at the ATG of the reporter gene. In some cases, "translational" fusions may also occur, in which there is in-frame fusion between the N-terminal end of the endogenous gene and the coding sequence of the reporter gene.
- *Enhancer trap*: The reporter gene is fused with a "minimal" promoter (containing a TATA box and a transcription initiation site) incapable of piloting the expression of the reporter gene on its own. Expression will occur only if there is activation by a nearby transcription activator. It may be difficult in this case to identify precisely the regulatory element that controls the expression of the reporter gene.
- *Gene trap*: The expression of the marker gene can be made to depend on the splicing between the endogenous gene and the reporter gene, this "cross"-splicing resulting in a translational fusion. In order to do this, a splicing acceptor site can be introduced before the reporter gene; there will be expression even if

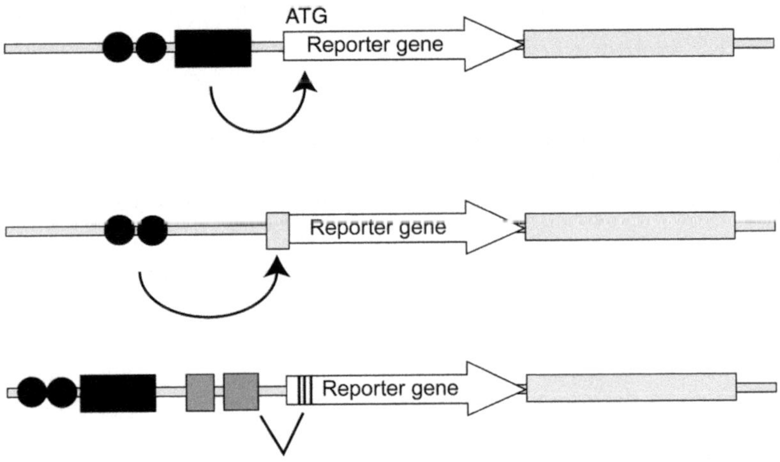

Fig. 9.6 Gene trap: After insertion in a genic region, the reporter gene carried by the insertion element (generally a color or fluorescent marker) will be expressed under the control of regulatory regions of the insertion site. The change in coloration over time and space is informative about the pattern of expression of the gene being studied.

Top to bottom: *Promoter trap*: the reporter gene is expressed if the insertion occurs upstream of an active promoter. *Enhancer trap*: The reporter gene is fused with a minimal promoter after insertion, and there is activation by a nearby transcription factor. *Gene trap*: expression depends on splicing between the endogenous gene and the reporter gene resulting from a translational fusion.

the insertion is in an intron and leads to a fusion between the exon and the reporter gene. These constructions can give information on the location of the protein as well.

"ACTIVATION TAGGING"

Insertion mutagenesis often leads to total loss of function of the genes affected. Loss-of-function, recessive mutations generally do not overcome the phenomenon of genetic redundancy (several genes controlling one function). Similarly, some mutations can lead to lethality, and strongly pleiotropic mutations may result in a complex phenotype that is difficult to approach experimentally. Gain-of-function mutations can help resolve this problem in certain cases. It is necessary in this case to stimulate the activities of genes rather than to inhibit them and disturb their regulation. For this purpose, "activation tagging" lines are used that contain an insertion element carrying a promoter or a strong transcription activator that will direct the transcription in the flanking region of the insertion, with overexpression of the plant gene located downstream of the insertion. The ectopic activation of a flanking gene in some cases allows the expression of a phenotype (Wilson *et al.*, 1996; Weigel *et al.*, 2000). These mutations are often dominant or semi-dominant and can be observed at the heterozygous stage.

T-DNA (Krysan *et al.*, 1999) or transposons (Parinov *et al.*, 1999) can be used as vectors to introduce these gene trap constructions. The large fragment of T-DNA inserted (5 to 25 kb) generally causes knock-out mutations because of the small size of introns and intergenic regions in *A. thaliana*. With a T-DNA, it may be difficult to interpret the pattern of expression of the reporter gene in the case of multiple insertions or rearrangements. T-DNA has, however, been successfully used by numerous groups to deliver gene trap constructions into plants. With transposons, two types of systems, *Ac/Ds* and *En/Spm*, are used. Systems with donor transposons allow a greater number of insertions in the adjacent regions. Two-component systems allow a reduction in the number of copies to increase the saturation of the genome in terms of insertions.

USE OF INSERTION COLLECTIONS IN REVERSE GENETICS

Reverse Genetics: Definitions and Objectives

The accumulation of sequence data changes the nature of the problem and opens up new perspectives. In order to have meaning, a gene sequence must be associated with some biological information. The present task is thus to attribute a function to these 30,700 genes, less than 10% of which have been experimentally characterized in a plant system. Putative functions can be attributed to a gene by similarity with known genes: a sequence is compared with sequence libraries in order to obtain some preliminary information, most often on its biochemical function, but this does not indicate its specific role in the organism. A putative function has been attributed to 70% of the genes in *A. thaliana*, but 30% of them still remain totally unknown either because they have no similarity with known sequences or because they are similar to unknown sequences (AGI, 2000). In any case, *in silico* analyses are not enough to define the function precisely; experimental studies are needed. Researchers will take recourse to various strategies of functional genomics, a discipline that attempts to comprehend the function of genes identified by structural genomics.

Among these strategies, the study of the spatial or temporal expression of a gene, that is, the presence of a messenger RNA and/or the protein in different cell types, during development, during the infection by a pathogen, or in different environments, may provide indirect information on gene function at the cell level or during the course of development. The subcellular location and post-translation modifications may also provide some answers. The biological characterization of mutants can give more direct information, the observation of a mutant phenotype being one of the simplest means to access the function of the mutated gene (Bouchez and Hofte, 1998). This is what is called reverse genetics, which relies on methodologies interfering with the activity of the gene.

The objective is thus to implement effective methods of targeted inactivation of genes, which makes it possible to quickly isolate mutants specifically affected in a particular gene, in the absence of other genetic modifications.

The method of choice for reverse genetics is undeniably the targeted inactivation of genes by homologous recombination that allows the precise and effective modification of genes. *In vitro* mutagenesis of the gene and reintroduction at its chromosomal location by recombination could then be carried out. In yeast, as well as in mouse, this method has been widely used with considerable success. In higher plants, despite numerous attempts, the use of gene targeting by homologous recombination remains extremely difficult, the frequencies obtained not being compatible with routine use in laboratories. Some success has been reported in *A. thaliana*, but the task remains laborious, since thousands of transgenic plants must be generated for each gene tested (Kempin *et al.*, 1997). Some recent results obtained in rice with a very strong system of selection for recombination seem more promising (Terada *et al.*, 2002).

The exception in the plant kingdom is the moss, *Physcomitrella patens*, in which homologous recombination occurs at high frequencies, close to those observed in yeast, provided a homology of a few hundreds bp exists between the transforming DNA and the genomic target. There is a correlation between the efficiency of the homologous recombination and the haploid state of the vegetative phase (Schaefer and Zryd, 1997; Schaefer, 2002).

The ectopic inactivation of genes (silencing of a gene by sense or antisense suppression) has been widely used but requires specific genetic constructions for each gene studied, requires the analysis of numerous transgenic plants for every construction, and may result in phenotypes that are difficult to analyse (e.g., partial dominance, incomplete penetration, somatic reversions). It has been suggested that the sense-antisense effects can be explained by the contamination of double-stranded RNA (Montgomery and Fire, 1998). The RNA-mediated interference technique (RNAi) is used to block the activity of a cellular gene by injection of a homologous RNA. This technique has been widely used to inhibit the function of genes in *C. elegans* (Fire *et al.*, 1998) and also has been applied successfully to other organisms, including *A. thaliana* (Chuang and Meyerowitz, 2000), in which the activities of the four genes *Agamous*, *Clavata3*, *Apetala1*, and *Perianthia* were reduced, but a large number of transgenic plants had to be constructed.

In *A. thaliana*, complete collections of RNAi lines are being constructed. For each gene, transgenic lines are constructed and made available to the public (http://www.evry.inra.fr/public/projects/agrikola/agrikola.html). On the other hand, targeted screening of chemically mutated populations is being developed. In populations obtained through conventional mutagenesis, the advantages of which have been demonstrated, research continues on point mutations specifically affecting a particular gene or even part of a gene. The undeniable advantage here is the wide variety of types of mutation potentially available in such populations (point mutations, microdeletions) leading to, for example, partial or complete loss of function or modifications of structure or regulation. This approach, called TILLING (Targeting Induced Local Lesions in Genomes) (McCallum *et al.*, 2000a, b), undoubtedly has a promising future, especially in the study of structure-function relationships for various proteins of interest. An overview of public TILLING efforts in *Arabidopsis thaliana* and others organisms is reviewed in Henikoff *et al.*, 2004. A TILLING facility for the scientific community has been developped: the Arabidopsis TILLING project (ATP) (Till *et al.*, 2003).

With respect to reverse genetics, and in order to overcome the lack of techniques based on homologous recombination, populations obtained through insertion mutagenesis have been widely used. These approaches have been mostly developed in *A. thaliana* and take on a special importance since the complete catalogue of genes of this species is available for research. First, mutated populations as complete as possible must be generated according to the methods described above, and then screening strategies have to be developed to discover the lines carrying interesting insertions in target genes from among these populations.

Screening Methods

The presence even of the insertion element within lines of the population allows for several strategies of mutant isolation. In the case of targeted approaches, intended to discover insertions in one or several genes of interest, a direct search through PCR is possible. For some years, strategies for systematic identification by sequencing of insertion sites present in the populations have tended to replace targeted approaches.

PCR SCREENING

The sensitivity and specificity of the PCR reaction make it possible to detect insertion events in the target gene from complex samples comprising several thousands of lines mixed together. Stocks of genomic DNA are first constituted from combinations of lines. These stocks constitute the primary pools. Several primary pools of DNA are then mixed together to constitute secondary pools on which the PCR experiments are carried out. The amplification is carried out by using primer pairs corresponding to the target sequence, the gene of interest, on the one hand and on the other to the insertion element. A positive signal in a reaction signifies that a line within the pool considered carries an insertion in the target gene. The primary pools constituting the secondary pool can be tested, and then the individual lines constituting the primary pool are tested until the line carrying the insertion is identified.

The strategy followed in the collection of T-DNA insertions at INRA (Versailles), for example, is described below.

— The transformation vector used is pGBK5 (Fig. 9.7), which allows tagging of genes (Bouchez *et al.*, 1993). The T-DNA region comprises the reporter gene *GUS* without a promoter joined to the right border of the T-DNA, a kanamycin resistance gene for *in vitro* selection, and a Basta resistance gene for greenhouse selection. This plasmid thus allows the search for promoters by transcriptional or translational fusion *in vivo.*

— Seed mixtures are prepared from 48 T2 or T3 families. After sterilization, the seeds are sown *in vitro* and seedlings are collected after about 15 days. The DNA from the mixtures are extracted and constitute primary pools (48 lines mixed). The DNA corresponding to eight primary pools are mixed to constitute a superpool (384 lines) and the DNA corresponding to three superpools are mixed to constitute a megapool (1,152 lines). The Versailles collection is made up of around 60,000 lines, or 90,000 insertions, which gives a saturation rate of 52% for a gene of 1 kb, 85% for a gene of 2.5 kb, and 98% for a gene of 5 kb.

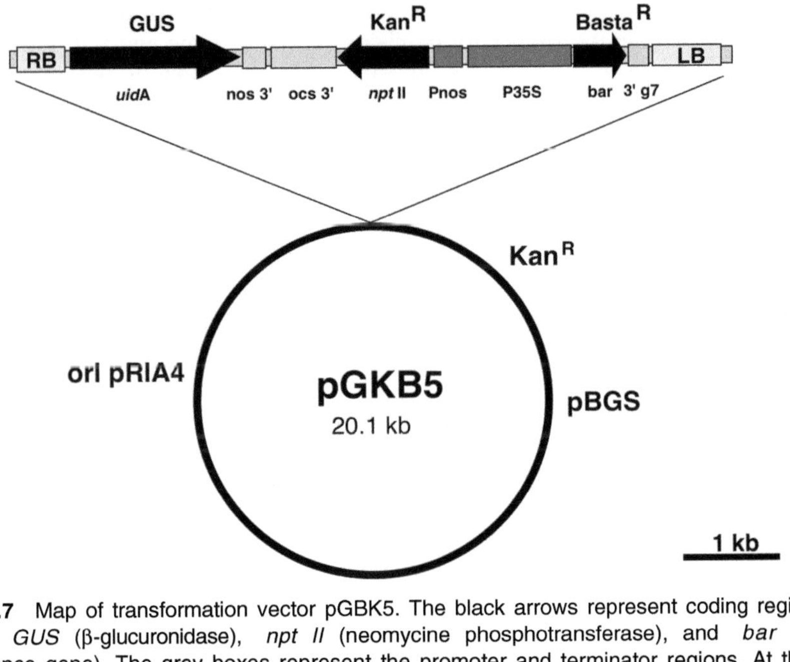

Fig. 9.7 Map of transformation vector pGBK5. The black arrows represent coding regions of genes *GUS* (β-glucuronidase), *npt II* (neomycine phosphotransferase), and *bar* (Basta resistance gene). The grey boxes represent the promoter and terminator regions. At the two ends, the LB (left border) and RB (right border) regions come from the TR-DNA of the pRiA4 plasmid of *Agrobacterium rhizogenes*. In these two regions the border sequences of 24 bp are found that serve as a T-DNA transfer signal. This binary vector is introduced in a "disarmed" colony of *Agrobacterium tumefaciens*, i.e., one in which the natural T-DNA has been deleted and which is therefore no longer pathogenic. In contrast, the bacterial functions ensuring the recognition of borders and the transfer of the T-DNA are maintained, thus allowing the transfer of synthetic T-DNA.

— PCR reactions are carried out on the available megapools and loaded on an agarose gel that is then transferred simultaneously onto two nylon membranes. One of these membranes is hybridized with a gene specific probe, and the other with a T-DNA specific probe. The potential positives give a hybridization signal with the two probes. After sequencing to confirm the presence of an insertion, new PCR reactions are carried out on the superpools and then on the 48 mixed together. The 48 lines making up the mixture are then analysed and the positive line identified by a final PCR reaction (Fig. 9.8).

This strategy is effective and quick but must begin from scratch for each gene or family of genes to be tested. This is why more systematic approaches have been developed that allow an overall economy of resources for the user laboratories.

IN SILICO SEARCH FOR INSERTIONS

In the case of *in silico* screening strategies, the insertion sites of each line of the population are characterized individually. After extraction of DNA individually from each line, an insertion site is amplified by hemi-specific PCR reaction. The band corresponding to this site is subjected to a sequence reaction and hundreds of nucleotides are obtained. The sequence is compared with the genomic sequence of *A. thaliana* to place the insertion with precision on the genome and within the genes. Several projects have created a near saturating collection of sequence-indexed knockout mutations (Parinov *et al.*, 1999; Speulman *et al.*, 1999; Tissier *et al.*, 1999, Sessions *et al.*, 2002, Alonso *et al.*, 2003, Li *et al.*, 2003, Kuromori *et al.*, 2004). As of april 2004, there are more than 330,000 flanking sequences mapped to the genome that hit around 90% of the currently known genes (Borevitz et Ecker, 2004).

After the sequences are cleaned, annotated, and validated, the databases are compiled. These sequence databases can be queried by the usual methods (Blast) or by keywords. Numerous databases are already available (see appendix to this chapter).

The chief limitation of these approaches lies in their "high-flow" character, which necessitates a compromise between efficiency and precision. Indeed, for technical reasons, the yield of such approaches is often 50-60%; about half the insertions present in the population generate no usable sequence tag. To compensate for these losses, researchers must work on populations that are large enough to obtain a sufficient number of sequenced insertions. However, on the international scale, and if we take into account all the resources of this kind, it can be estimated that almost all the mutants accessible by insertion mutagenesis are available in this type of database. Properly speaking, given that each type of mutagen has its biases and specificities, some genes cannot be mutated by insertion mutagenesis, without counting genes whose inactivation is lethal at the gametophytic level or even lethal overall and thus cannot be found in such a collection.

Characterization of Mutants

The isolation of a mutant is only the beginning of a long research task. The first step is often the genetic and molecular characterization of isolated insertion lines. At first, the number of insertions present is determined by Southern hybridization, the segregation

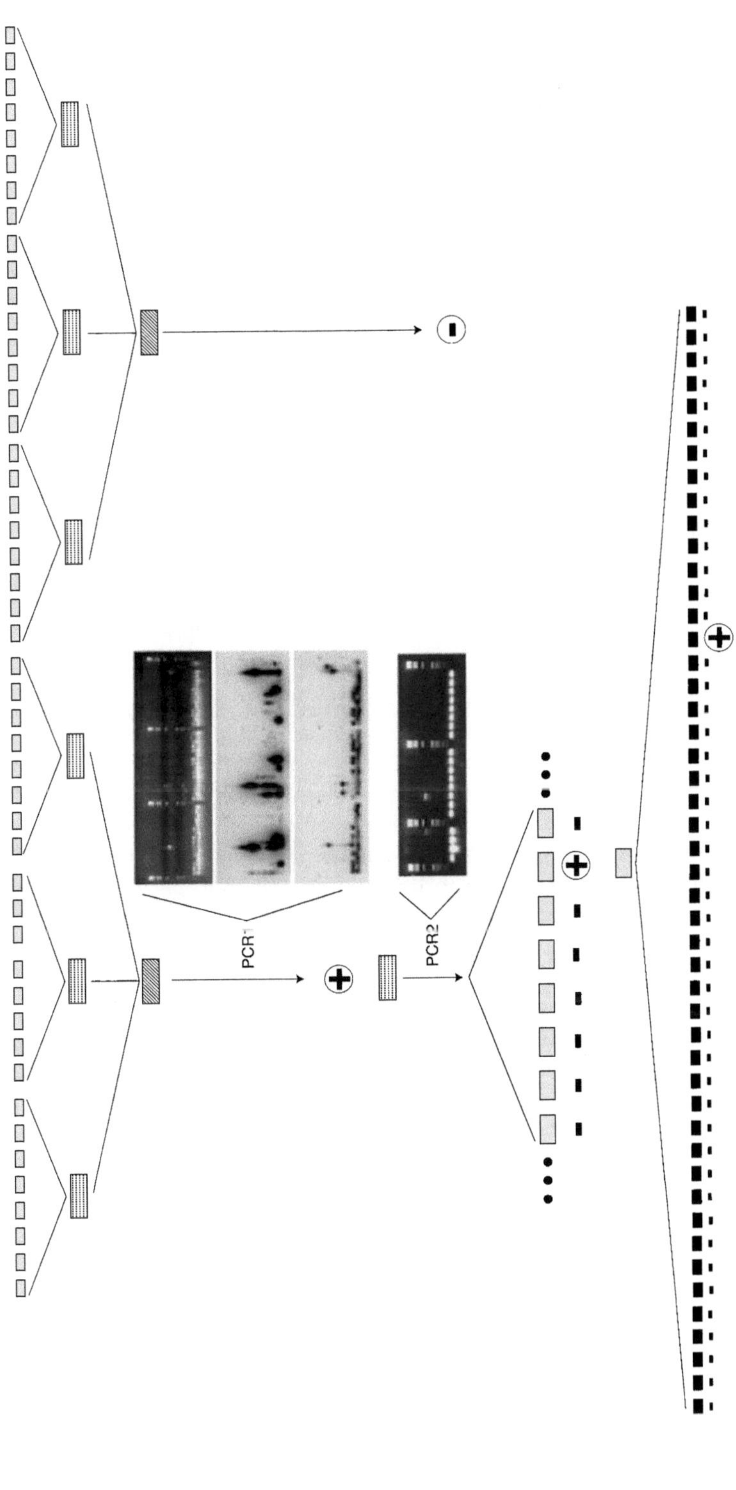

Fig. 9.8 Seeds from 48 insertion lines are combined to form a pool. Eight pools are combined into a superpool (light grey rectangles) and three superpools constitute a megapool (dark gray rectangle). A preliminary PCR is carried out on the megapools with a gene specific primer and a T-DNA specific primer. The amplification products are separated by agarose gel electrophoresis (photo at top) and then the gel is transferred simultaneously onto two nylon membranes. One membrane is hybridized with a T-DNA probe, the other with a gene specific probe (2 middle photos). The presence of a signal observed with the two probes suggests the existence of a positive. The PCRs are repeated on the superpools (PCR2, photo at bottom), then on the pools, and finally on the 48 individual plants in order to identify the plant carrying the target insertion.

of one or two selection markers must be verified, and homozygous lines must be isolated by segregation tests or PCR from a T2 or T3 family. It is always necessary to sequence the T-DNA/genomic DNA junctions in order to determine the exact structure of the mutant locus. The analysis of a mutant is greatly facilitated by prior knowledge of expression data, subcellular location, or biochemical function. In this way, a more targeted search for the phenotype can be carried out. If there is no information available, a broad search must be undertaken. The phenotype is analysed by comparing the mutant plant with the wild type under different conditions. It is useful to check the size and abundance of the messenger RNA by northern blot and the presence and/or activity of a protein using antibodies. Then, it must be demonstrated that the phenotype is closely linked to the insertion by complementing with the wild type allele or finding another mutant allele. Biochemical, physiological, and genetic studies are undertaken to link the biological function to the phenotype observed.

The Phenotype

Surprisingly, most homozygous mutant plants isolated through reverse genetic screening reveal no notable morphological modification. Only a small fraction (< 5%) show easily detected defaults in standard greenhouse or *in vitro* culture conditions.

This absence of visible effects of mutations that lead mostly to complete loss of activity of the genes affected has raised many questions and hypotheses that can be classified in two major categories, which are not mutually exclusive.

— Most mutant phenotypes are probably conditional: Plants carrying mutations that do not affect essential genes (i.e., those in which the mutation ends in lethality) can survive in favourable environmental conditions by compensating for the effect of the mutations through mechanisms of physiological adaptation, which are generally known to be highly effective in plants. Suboptimal environmental conditions, however, reveal the defect in functioning caused by the mutation, the expression of the mutant phenotype possibly being conditioned by various factors, such as the presence of pathogens, pesticides, drought, extreme temperature, salt concentrations, hormones, photoperiod. If, for example, the mutation affects a pathway of response to a particular pathogen, only the presence of that pathogen will reveal a phenotypic difference between the mutant and the wild type. On the other hand, a mutation may give rise to subtle phenotypic modifications or a loss of discrete selective value that is difficult to detect with current methods. For example, insertions in three out of eight genes of actin in *A. thaliana* (Gilliland *et al.*, 1998) led to no visible phenotype in individual plants, but studies over several generations showed that these three mutant alleles had significant deleterious effects on the selective value of *A. thaliana* plants.

— The other type of explanation relates to the high redundancy observed within the *A. thaliana* genome. The sequencing of the genome showed that 60% of the genome was duplicated, with 37% being conserved highly; 58% of the genome was distributed over 24 large duplicated fragments (100 kb or more). Only 35% of the proteins were unique in the genome and 37.5% of the proteins belonged to families with more than five members. Out of the 11,600 protein types identified,

150 are specific to plants (AGI, 2000). This significant structural redundancy, which has also been observed in other eukaryotes, may indicate functional redundancy. A loss-of-function mutation in a member of a gene family may be compensated or its effect limited by the activity of another member of the family. Different cases of functional redundancy have already been observed: three MADS-box genes (*SEPALLATA* 1/2/3) implicated in the development of different floral organs are functionally redundant, since no individual mutant has a notable phenotype, while a triple mutant, in which there is no activity of the three *SEPALLATA* genes, gives flowers in which all the organs develop into sepals (Pelaz *et al.*, 2000).

Two other MADS-box genes are implicated in fruit dehiscence in *A. thaliana*, Shatterproof1 (*SHP1*) and Shatterproof2 (*SHP2*). They control differentiation from the dehiscence zone and lignification of adjacent cells. The simple mutants present no phenotypes. The double mutant phenotype is strong, the mature fruit no longer being dehiscent. This is an example of high functional redundancy. These two genes present 87% identity at the amino acid level and have the same pattern of expression (Liljegren *et al.*, 2000).

Note that functional redundancy can also occur with two non-homologous genes (Maes *et al.*, 1999).

Gene silencing is today considered one of the major tools of functional genomics. A very large number of lines presenting insertions in known genes have been isolated and, although few insertions result in an informative phenotype (Bouche and Bouchez, 2001), reverse genetics remains an effective technique that must be included in a more global approach to determine the biological function of a gene.

APPENDIX

General *Arabidopsis* Databases

http://www.Arabidopsis.org/home.html

Available Collections

Before 1990, different collections of *A. thaliana* mutants were maintained by laboratories. They are now for the most part available from international resource centers:

ABRC: http://arabidopsis.org/abrc
NASC: http://arabidopsis.info
RIKEN BRC: http://www.brc.riken.jp/inf/en

Insertion Site Database

SAIL lines from Syngenta: http://www.arabidopsis.org/abrc/sail.jsp
SIGnAL TDNA-Express: http://signal.salk.edu/cgi-bin/tdnaexpress
FST Projects (INRA Versailles): http://flagdb-genoplante-info.infobiogen.fr/projects/fst

Biological Research Center (Hungary): http://www.szbk.u-szeged.hu/~arabidop
CSHL Arabidopsis Genetrap: http://genetrap.cshl.org/
GABI-Kat: http://www.mpiz-koeln.mpg.de/GABI-Kat/

ACKNOWLEDGMENTS

We gratefully thank Heather McKhann for her careful reading of the manuscript.

REFERENCES

Alonso, J.M., Stepanova, A.N., Leisse, T.J., Kim, C.J., Chen, H. et al. 2003. Genome wide insertional mutagenesis of *Arabidopsis thaliana*. *Science*, 301, 653-657.

The *Arabidopsis* Genome Initiative. 2000. Analysis of the genome sequence of the flowering plant *Arabidopsis thaliana*. *Nature*, 408, 796-815.

Azpiroz-Leehan, R., Feldman, K.A. 1997. T-DNA insertion mutagenesis in *Arabidopsis*: going back and forth. *Trends Genet.*, 13, 152-156.

Bancroft, I., Dean, C. 1993. Transposition pattern of the maize element Ds in *Arabidopsis thaliana*. *Genetics*, 134, 1221-1229.

Bechtold, N., Ellis, J., Pelletier, G. 1993. *In planta Agrobacterium* mediated gene transfer by infiltration of adult *Arabidopsis thaliana* plants. *C.R. Acad. Sci. Fr.*, 316, 1194-1199.

Bechtold, N., Jaudeau, B., Jolivet, S., Maba, B., Vezon, D., Voisin, R., Pelletier, G. 2000. The maternal chromosome set is the target of the T-DNA in the *in planta* transformation of *Arabidopsis thaliana*. *Genetics*, 155, 1875-1887.

Borevitz, J.O.,Ecker, J.R. 2004. Plant genomics: the third wave. *Annu. Rev. Genomics Hum. Genet.*, 5, 443-477.

Bouche, N., Bouchez, D. 2001. *Arabidopsis* gene knockout: phenotypes wanted. *Curr. Opin. Plant Biol.*, 4, 111-117.

Bouchez, D., Camilleri, C., Caboche, M. 1993. A binary vector based on Basta resistance for *in planta* transformation of *Arabidopsis thaliana*. *C.R. Acad. Sci. Fr.* 316, 1188-1193.

Bouchez, D. Hofte, H. 1998. Functional genomics in plants. *Plant Physiol.*, 118, 725-732.

Chuang, C.F., Meyerowitz, E.M. 2000. Specific and heritable genetic interference by double-stranded RNA in *Arabidopsis thaliana*. *Proc. Natl. Acad. Sci.* USA, 97, 4985-4990.

Clough, S.J., Bent, A.F. 1998. Floral dip: a simplified method for *Agrobacterium*-mediated transformation of *Arabidopsis thaliana*. *Plant J.*, 16, 735-743.

Desfeux, C., Clough, S.J., Bent, A.F. 2000. Female reproductive tissues are the primary target of *Agrobacterium*-mediated transformation by the *Arabidopsis* floral-dip method. *Plant Physiol*, 123, 895-904.

Feldmann, K.A., Marks, M.D. 1987. *Agrobacterium*-mediated transformation of germinating seeds of *Arabidopsis thaliana*. *Mol. Gen. Genet.*, 208, 1-9.

Feldmann, K.A. 1991. T-DNA insertion mutagenesis in *Arabidopsis*: mutational spectrum. *Plant J.*, 1, 71-82.

Fire, A., Xu, S., Montgomery, M.K. Kostas, S.A., Driver, S.E., Mello, C.C. 1998. Potent and specific genetic interference by double-stranded RNA in *Caenorhabditis elegans*. *Nature*, 391, 806-811.

Gilliland, L.U., McKinney, E.C., Asmussen, M.A., Meagher, R.B. 1998. Detection of deleterious genotypes in multigenerational studies. I. Disruptions in individual *Arabidopsis* actin genes. *Genetics*, 149, 717-725.

Henikoff, S., Till, B.J., Comai, J. 2004. TILLING. Traditional mutagenesis meets functional genomics. *Plant physiol.*, 135(2), 630-636.

Ito, T., Seki, M., Hayashida, N., Shibata, D., Shinozaki, K. 1999. Regional insertional mutagenesis of genes on *Arabidopsis thaliana* chromosome V using the Ac/Ds transposon in combination with a cDNA scanning method. *Plant J.*, 17, 433-444.

Kempin, S.A., Liljegren, S.J., Block, L.M., Rounsley, S.D., Yanovsky, M.F., Lam, E. 1997. Inactivation of the *Arabidopsis* AGL5 MADS-box gene by homologous recombination. *Nature*, 389, 802-803.

Koncz, C., Redei, G.P. 1994. Genetic studies in *Arabidopsis*: a historical view. In: *Arabidopsis.* Meyerowitz, E.M., Somerville, C.R., eds. Cold Spring Harbor Laboratory Press, Cold Spring Harbor, New York, 223-253.

Krysan, P.J., Young, J.C., Sussman, M.R. 1999. T-DNA as an insertional mutagen in *Arabidopsis. Plant Cell*, 11, 2283-2290.

Kuromori, T., Hirayama, T., Kiyosue, Y., Takabe, H., Mizukado, S. et al. 2004. A collection of 11 800 single-copy Ds transposon insertion lines in *Arabidopsis. Plant J.* 37, 897-905.

Laufs, P., Autran, D., Traas, J. 1999. A chromosomal paracentric inversion associated with T-DNA integration in *Arabidopsis. Plant J.*, 18, 131-139.

Li, Y., Rosso, M.G., Strizhov, N., Viehoever, P., Weisshaar, B. 2003. GABI-Kat SimpleSearch: a flanking sequence tag (FST) database for the identification of T-DNA insertion mutants in *Arabidopsis thaliana. Bioinformatics*, 19, 1441-1442.

Liljegren, S.J., Ditta, G.S., Eshed, Y., Savidge, B., Bowmann, J.L., Yanovsky, M.F. 2000. Shatterproof MADS-box genes control seed dispersal in *Arabidopsis. Nature*, 404, 766-770.

Maes, T., De Keukeleire, P., Gerats, T. 1999. Plant tagnology. *Trends in Plant Sci.*, 4(3), 90-96.

Martienssen, R.A. 1998. Functional genomics: probing plant gene function and expression with transposons. *Proc. Natl. Acad. Sci.* USA, 95, 2021-2026.

McCallum, C.M., Comai, L., Green, E.A., Henikoff, S. 2000a. Targeted screening for induced mutations. *Nature Biotech.*, 18, 455-457.

McCallum, C.M., Comai, L., Green, E.A., Henikoff, S. 2000b. High-throughput screening for induced point mutations. *Plant Physiol.*, 126, 480-484.

Nacry, P., Camilleri, C., Courtial, B., Caboche, M., Bouchez, D. 1998. Major chromosomal rearrangements induced by T-DNA transformation in *Arabidopsis. Genetics*, 149, 641-650.

Parinov, S., Sevugan, N., De, Y., Yang, W.C., Kumaran, M., Sundaresan, V. 1999. Analysis of flanking sequences from dissociation insertion lines: a database for reverse genetics in *Arabidopsis. Plant Cell*, 11, 2263-2270.

Parinov, S., Sundaresan, V. 2000. Functional genomics in *Arabidopsis*: large-scale insertional mutagenesis complements the genome sequencing project. *Curr. Opin. Biotechnol.*, 11, 157-161.

Pelaz, S., Ditta, G.S., Baumann, E., Wisman, E., Yanovsky, F. 2000. B and C floral organ identity functions require Sepallata MADS-box genes. *Nature*, 405, 200-203.

Pereira, A. 2000. A transgenic perspective on plant functional genomics. *Transgenic Res.*, 9, 245-260.

Samson, F., Brunaud, V., Balzergue, S., Dubreucq, B., Lepiniec, L. et al. 2002. FLAGdb/FST: a database of mapped flanking insertion sites (FSTs) of *Arabidopsis thaliana* T-DNA transformants. *Nucl. Acids Res.*, 30, 94-97.

Schaefer, D.G., Zyrd, J.P. 1997. Efficient gene targeting in the moss *Physcomitrella patens. Plant J.*, 11, 1195-206.

Schaefer, D.G. 2002. A new moss genetics: targeted mutagenesis in *Physcomitrella patens. Annu. Rev. Plant Biol.*, 53, 477-501.

Sessions, A., Burke, E., Presting, G., Aux, G., McElver, J. *et al.* 2002. A high-throughput *Arabidopsis* reverse genetics system. *Plant Cell*, 14, 2985-2994.

Speulman, E., Metz, P.L.J., van Arkel, G., Te Lintel Hekkert, B., Stiekma, W.J., Pereira, A. 1999. A two-component enhancer-inhibitor transposon mutagenesis system for functional analysis of the *Arabidopsis* genome. *Plant Cell*, 11, 1853-1866.

Springer, P.S. 2000. Gene traps. Tools for plant development and genomics. *Plant Cell*, 12, 1007-1020.

Takano, M., Egawa, H., Ikeda, J.E., Wakasa, K. 1997. The structures of integration sites in transgenic rice. *Plant J.*, 11, 353-361.

Terada, R., Urawa, H., Inagaki, Y., Tsugane, K., Iida, S. 2002. Efficient gene targeting by homologous recombination in rice. *Nature Biotech.*, 20, 1030-1034.

Till, B.J., Reynolds, S.H., Greene, E.A., Codomo, C.A., Enns, L.C., Johnson, J.E., Burtner, C., Odden, A.R., Young, K., Taylor, N.E., Henikoff, J.G., Comai, L., Henikoff, S. 2003. Large-scale discovery of induced point mutations with high-throughput TILLING. *Genome Res* 13: 524–530.

Tinland, B. 1996. The integration of T-DNA into plant genomes. *Trends PlantSci.*, 1(6), 178-184.

Tissier, A.F., Marillonnet, S., Klimyuk, V., Patel, K., Torres, M.A. *et al.* 1999. Multiple independent defective suppressor-mutator transposon insertions in *Arabidopsis*: a tool for functional genomics. *Plant Cell*, 11, 1841-1852.

Weigel, D., Ahn, J.H., Blazquez, M.A., Borevitz, J.O., Christensen, S.K., Fankhauser, C. *et al.* 2000. Activation tagging in *Arabidopsis. Plant Physiol.*, 122, 1003-1013.

Wilson, K., Long, D., Swinburne, J., Coupland, G. 1996. A dissociation insertion causes a semidominant mutation that increases expression of TINY, an *Arabidopsis* gene related to *Apetala2. Plant Cell*, 8, 659-671.

Winkler, R.G., Frank, M.R., Galbraigh, D.W., Feyereusen, R., Feldmann, K.A. 1998. Systematic reverse genetics of transfer-DNA-tagged lines of *Arabidopsis*. Isolation of mutations in the cytochrome p450 gene superfamily. *Plant Physiol.*, 118, 743-750.

Wisman, E., Hartmann, U., Sagasser, M., Baumann, E., Palme, K., Hahlbrock, K., Saedler, H., Weisshaar, B. 1998. Knock-out mutants from an En-1 mutagenized *Arabidopsis thaliana* population generate phenylpropanoid biosynthesis phenotypes. *Proc. Natl. Acad. Sci.* USA, 95(21), 12432-12437.

10

CHAPTER

DNA Microarrays in Plants

P. Doumas, Y. Al-Ghazi, C. Rothan, S. Robin

INTRODUCTION

The tremendous reduction in sequencing costs that has occurred in the last decade has led to a boost in total genome annotations for a wide variety of species ranging from the smallest genomes (prokaryotes, yeast) to the larger ones such as *Drosophila, Caenorhabditis,* human and an increasing number of model plants, including *Arabidopsis thaliana* and rice (Chapters 2, 9, 17, 18). Other major agronomic species like tomato or wheat have also benefited from these advanced techniques, with the creation of almost exhaustive expressed sequence tag (EST) collections. During the same period, the creation of new molecular tools has helped to unravel the mechanisms of plant growth and development. Contrary to what was previously believed, these mechanisms appear not to be distinct linear regulatory pathways but complex networks of interdependent pathways, involving a considerable number of differentially regulated genes. However, a large proportion of the annotated genes have not yet been attributed a clear function. As an example, in *Arabidopsis thaliana,* which possesses the most advanced molecular tools available for plants (Chapters 9, 17), 40% of the 25,500 annotated genes code for proteins of unknown function (Multinational coordinated *Arabidopsis thaliana* functional genomics project, June 2002). Hence, the need to associate a biological function to an annotated gene and to discover new genes within pathways, has lead to the development of large scale transcription analysis methods. Theses methods that allow genome wide expression analysis, rely either on the sequencing of expressed tags (direct measurement), or upon hybridisation of labelled targets to arrayed probes (micro and macroarrays, indirect measurement) (Table 10.1).

Direct Measurement of Gene Expression

The analysis of the data coming from the sequencing of human ESTs has revealed that gene expression could be measured by testing the occurrence of the tags

Table 10.1 The three global gene expression analysis methods have been compared and the main differences summarized. (Sequencing = DNA sequence based methods [eg., MPSS]; Fragmentation = DNA fragment based methods [differential display]; Hybridization = DNA array based methods [eg. Macro and microarrays]).

	Sequencing	*Fragmentation*	*Hybridization*
Sensitivity	Moderate / High	High	Moderate
Specificity	High	High	Low
Expression levels	Absolute	Relative	Relative
Genome sequence data required	Yes	No / Yes	Yes
Human labour cost	High	High	Low
Overall cost	High	Low	Moderate

corresponding to given sequences (Okubo *et al.*, 1992; Adams *et al.*, 1995). For most of the agronomic plant species, recent ESTs collections (TIGR http://www.tigr.org/tdb/tgi) have been built from various numbers of cDNA libraries and provide reliable references to determine the site and conditions of expression of any given gene and also to provide probes for indirect measurement with DNA arrays. These collections are however biased, in that highly expressed genes can be redundant, whereas genes weakly expressed, tissue specific, or expressed in a narrow timing window, can be completely absent. Moreover, a significant number of probes are necessary to ensure reliability of the resulting frequency data corresponding to a given gene. Theses facts elevate considerably the human and technical costs that render the method unsuitable for routine analysis. SAGE (*Serial Analysis of Gene Expression*; Velculescu *et al.*, 1995) and MPSS (*Massively Parallel Signature Sequencing*; Brenner *et al.*, 2000) have reduced the costs by analysing much shorter tag sequences ranging from 15 to 50 nucleotides and have allowed a precise quantitation of transcript abundances. The use of such techniques is, however, limited to the totally sequenced organisms, for which a link between the tags and the genes can be established (i.e., rice and *Arabidopsis*) and for the species with high coverage EST collections. In addition, SAGE and MPSS necessitate the use of high throughput sequencers or devices which are not yet commonly available to most laboratories (e.g., for MPSS). On the other hand, PCR (*Polymerase Chain Reaction*) based methods like DDRT-PCR (*Differential Display RT-PCR*; Liang et Pardee, 1992) and cDNA-AFLP (*Amplified length polymorphism*; Bachem *et al.*, 1996) have become widely used on a number of plant species. These techniques are simple, cost effective and highly sensitive. DDRT-PCR, however, generates a significant quantity of false positives which makes it unsuitable for transcript quantification. Conversely, cDNA-AFLP and derived methods like Gene Calling (Shimkets *et al.*, 1999), genEST (Qin *et al.*, 2001) and TOGA (TOtal Gene expression Analysis; Sutcliffe *et al.*, 2000), authenticate gene identification and expression profiling by cDNA sequencing, or by correlating their electrophoretic mobility with database assisted fragment sizes predictions (if data exists). Meanwhile, stepping from the isolation of a few or a few dozen genes to the screening of thousands of transcripts like the one described for tobacco BY-2 cells (Breyne and Zabeau, 2001), is inaccessible to most of laboratories due to the required financial and technical resources.

Indirect Measurement of Gene Expression

A major proportion of high scale gene expression measurement experiments are nowadays conduced in an indirect mode, e.g. involving the recognition of the target by a labelled probe. Hence, all of the methods described hereafter, which have been used on a large number of organisms (Schena *et al.*, 1995; Lockhart *et al.*, 1996; Desprez *et al.*, 1998), are derived from the classical Northern experiment. The fundamental difference with the former methods lies in the recent advances in miniaturisation techniques, sequencing capacity, scanning resolution and sensitivity, all allowing the simultaneous analysis of several thousands of genes. Using bioinformatic tools and statistical experimental design, the resulting data are able to be compared across many experiments and even between different laboratories through the creation of expression profile databases. As an example in yeast for which outcomes are the most advanced (DeRisi *et al.*, 1997), more than a thousand experimental conditions have been compared at a genomic scale. The resultant colossal amount of data is a key to unravelling the connections between regulatory networks (Devaux *et al.*, 2001). This trend is extending to more and more species or fields, as six new expression databases have been published in 2004 (http://www.nslij-genetics.org/microarray/2004.html). Considering that numerous microarray laboratories are developing their own profile database, one can expect that the number of these websites will increase as fast as the number of species being studied by microarrays. This will therefore raise, in a short time, the question of rationalizing and connecting the databases in a reduced number of comprehensive structures.

The aim of this chapter is to describe the DNA array technique, including design and realisation, but also to discuss the different strategies in which it is used to address biological questions. A bibliography of different uses of microarrays in the plant kingdom is presented.

THE DNA ARRAY PROCESS

Principle of the Method

DNA and more generally nucleic acid can pair complementary strands via non-covalent binding in a duplex; this fundamental characteristic is used in all DNA array techniques to target the transcripts of interest. Numerous other screening experiments like Northern-blotting, Southern-blotting or dot-blotting are also founded on this process, but allow the detection of only one polynucleotide species per experiment, as the target is bound to a membrane and hybridized with one labelled probe (corresponding to the specific gene being analysed). With the need to analyse several genes, these basic techniques are rapidly overwhelmed.

In DNA arrays, the numerous individual gene probes are bound in a 2D matrix on a slide or a membrane, the number of probes is then limited only by the size of the slide and the miniaturisation of the probe dots. The transcripts contained in the sample being analysed are fluorescently or radioactively labelled and then hybridised to the probes bound onto the microarray. The amount of each probed RNA species can then be measured by scanning the hybridised array for radioactivity or fluorescence. It is now collectively accepted that a probe is the sequence bound onto the DNA array and that the target is the transcriptome or sum of transcriptomes contained in a given sample to be analysed (Fig. 10. 1).

DNA Array Creation

From whatever type of slide and mode of creation one may choose to build a DNA array, the process always consists of binding a library of synthetic DNA onto a minimum surface area in a dense and homogeneous fashion. The main difference between the various sorts of DNA arrays, lies in the density of the bound probes and the way theses probes have been synthesized. Macroarrays are low density types of arrays in which dozens to hundreds of probes are bound on each square centimetre of membrane. These were the first kinds produced by laboratories at the time as miniaturisation had not been possible and are still used, as they are easier to create without advanced robotics. Microarrays are the logical evolution of macroarrays, with thousands of probe sequences being bound per square centimetre. Both Micro- and Macroarrays contain *in-vitro* synthesized probes bound on their surface. However, a third category of DNA array is now available that necessitates a higher level of technological sophistication in both design and realisation; the DNA chip, in which short probes are synthesized *in-situ* in a very high density matrix.

Printed DNA array processing

The first step in creating Macro- and Microarrays is the selection of the probe library to be bound on the slide. A few years ago, these probes were generally PCR amplification products of individual clones from a cDNA library used as templates (e.g., an EST collection) or isolated from any kind of DNA library harbouring interesting probes. As an example, a human engineered Microarray was created with probes generated by amplification of genomic sequences (Open Reading Frames) using specific primers (Penn *et al.*, 2000). In the same way, the CATMA project (Complete *Arabidopsis* Transcriptome MicroArray; http://jic-bioinfo.bbsrc.ac.uk/CATMA) allowed the building of a genome wide library of *Arabidopsis thaliana* gene specific sequences dedicated to Microarrays. Moreover, an exhaustive cDNA library can be screened with a Microarray, leading to a significant proportion of the gene expression profiles (Seki *et al.*, 2001). However, more and more, a new and attractive method based on the direct binding of specific primers on the slide as probes is preferred. The resulting printed oligo chip (Guckenberger *et al.*, 2002), is now widely used for transcriptome analysis of an increasing number of species. The recent advances in DNA synthesizers now enables the production of long (50 to 70 bases) oligonucleotides with high reliability in sequence and in reasonable amounts. These probes can then be arrayed on the slides in a cost efficient manner compared to traditional PCR amplified sequences. Designed for specific recognition of a given gene, synthetic oligonucleotides prevent cross hybridisations between homologous genes (e.g. in a multigene family), which can have a distorting effect on the resulting expression data. As an example, 60% of *Arabidopsis thaliana* genes are believed to belong to multigene families (Richmond and Somerville, 2000). Given the equivalent sensitivity of theses oligonucleotide arrays compared to PCR amplified ones, their lower cost and higher specificity, oligonucleotide microarrays are now spreading even to species for which a minimal amount of genomic sequence

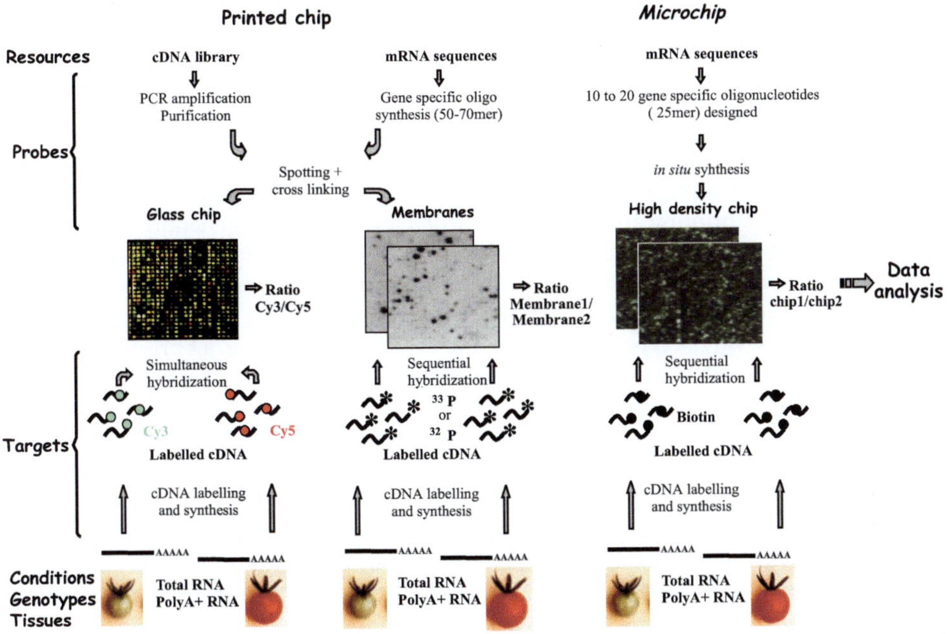

Fig. 10.1 (previous page): Scheme describing *in situ* (Affymetrix chips) and printed (micro- and macroarray) DNA array production and process.

Printed microarrays: (1) Probes are amplified from cDNA clones selected from libraries or from genomic DNA corresponding to ORFs (open reading frames) and then purified. Alternatively, 50 to 70 long oligonucleotides can be designed specifically to a given mRNA (cDNA clone or ORF) and synthesized. Probes can be then printed on the array on a slide or a nylon membrane by the use of a high density spotting robot. To prevent them from being stripped during hybridization, probes are light, heat or chemically cross linked to the support. During target processing (2), total RNA or PolyA+RNA is reverse transcribed (RT) and labeled either simultaneously or after transcription depending on the label used. Fluorochromes (Cy3 and Cy5) can be coupled to nucleotides for simultaneous labeling or bound after RT to the probe; radioactive compounds (33P or 32P) have to be incorporated during the RT only. The two fluorescent samples that are differently labeled are competitively hybridized to the slide. Each fluorescent label is excited by a specific energy laser and the fluorescence is recorded with a high resolution scanner. The relative signal intensity (ratio Cy3/Cy5) for both fluorochromes allows an estimate of the relative abundance of a given RNA species in the two samples. For radioactive labeling, targets are hybridized to different membranes or sequentially to the same membrane after stripping. The relative abundance of the transcript is estimated by comparison of the signals between the two membranes.

In situ synthesized probes (1) are 10 to 20 gene specific 25 base paired oligonucleotides synthesized directly on the chip by photolithographic methods. More than 250,000 elements can be arrayed on this kind of support (Affymetrix). Targets (2) are reverse transcribed mRNAs in which a T7 promoter is added. These cDNAs are transcribed into cRNA in the presence of biotinylated nucleotides and then fragmented before hybridization to the chips. Hybridized cRNAs are revealed upon interaction with a streptavidin/fluorophore complex. The relative abundance of the transcripts is estimated by comparison of the signals between the different hybridized chips.

data is available. As a consequence, a number of companies now advertise oligonucleotides and their associated microarrays covering the whole *Arabidopsis thaliana* expressed genome (ex: Agilent: http://www.chem.agilent.com).

The printing of the microarray probes on slides or nylon membranes can be realized by contact or via jet printing (Fig. 10. 2). When contact printed, needles carried by a printing tool, dip into probe solutions; then the DNA is dropped onto the slide surface by quick repeated contacts, printing hundreds of slides in a single session. The robotic devices for DNA printing allow the spotting of a few nanoliters, generally spreading to 60 to 250 micrometer wide spots. Spotting quality is closely tied to needle sharpness and to the precision of the robotic arm. Jet printing, on the other hand, uses much smaller sample volumes (nano- to picoliter), and was inspired directly by the inkjet printer's piezoelectric, or valvae type process. Contrary to contact printing, jet printing allows a wider range of spotted volumes and spot sizes range from 20 to 50 microns wide.

Hence, printing microarrays can be versatile and processed in any molecular biology unit that owns a spotting robot. The number of newly available spotting devices is constantly increasing as their performance is regularly improved. The new robots are faster, more precise in spotting, less probe consuming and more reliable than the earlier machines and the reduction in the overall time of spotting generally warrants their higher cost. Furthermore, the benefits of such advances extend to the hybridisation as well, needing less target in a smaller volume and hence less starting material for the experiments.

DNA probes can be spotted onto different categories of "slides", which can be glass or nylon. For glass slides, a positively charged coating (poly-L-lysine) is generally used to keep the nucleic acids bound to the surface (Protocol described in: http://nciarray.nci.nih.gov/reference/Coating_Glass.html). Although this coating is efficient in binding DNA, it can reduce the accessibility of the complements during hybridisation through steric hindrance at the slide surface. A way to avoid this is to modify the 5' end of the probes in order to create a covalent link between this position and a specific component of the coating surface. DNA probes can also be linked to the slides via molecular linkers of variable length, to facilitate the access of the targets and enhance binding capacity of the spots (Saiki *et al.*, 1989). Other types of slides and coatings (e.g. gold coating or composite materials), have been tried to create 3D structures with a higher binding capacity (polyacrylamide gels, bee nest structures, nitrocellulose and many other polymers). These alternative techniques could lead to reduced noise to signal ratios, provide higher fixation capacity, re-usability and reproducibility.

In situ Synthesis of DNA Chip Probes

DNA chips are generated via *in situ* synthesis of probes by combinatorial chemistry and photolithography based techniques (Fodor *et al.*, 1991) (Fig. 10. 2). The most accepted method is the one provided by the Affymetrix company (http://www.affymetrix.com). 20 to 25 bp long gene specific oligonucleotides are generated *in situ* on a silicon surface by combining standard DNA synthesis protocols with phosphoramidite reagents modified with photolabile 5'-protecting groups. During the process, a mercury lamp is shone through a photolithographic mask onto the chip surface, removing the photoactive group, resulting in the exposure of a 5' hydroxy group capable of reacting with another 5'-protected nucleoside. The mask therefore

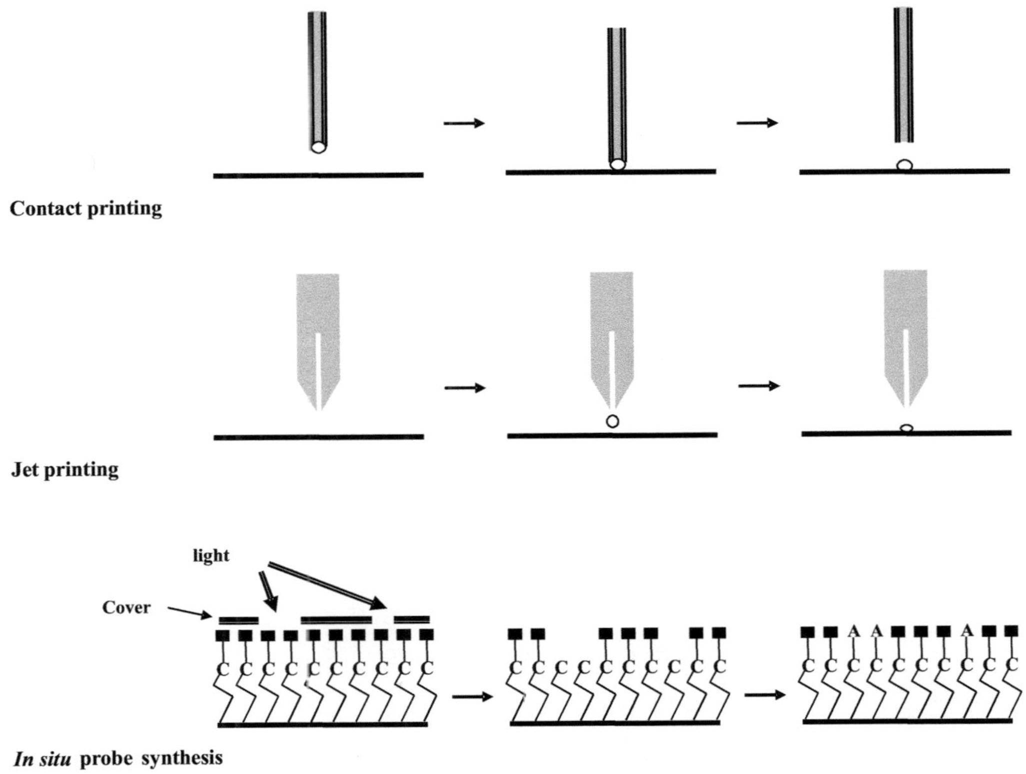

Fig. 10.2 Three different technologies to make DNA arrays. Droplets of DNA are left by a pin tool in contact printing. DNA can be also propelled like ink in a jet printer. In the in situ technique, DNA in synthesized directly on the chip with masks and photosensitive reagents.

predetermines which nucleotides are activated leading to spatially addressable synthesis (i.e., each oligo is positioned precisely at a designated spot on the chip). Successive rounds of deprotection and chemistry (using a combination of masks) allow the generation of all the desired short sequences in a combinatorial manner. This method has led to the generation of very high density chips (250,000 probes cm^{-2}). For gene expression measurements, a set of 16 oligonucleotide pairs are generally defined in highly variable regions of any given gene. Sixteen of these probes are designed as perfect matches for the gene and the 16 remaining probes are modified to produce one mismatch in the middle of the oligonucleotide sequence, in order to estimate the specificity of the hybridisation. The use of a number of probes for each gene increases the specificity and sensitivity of the data collected for each gene. Using this process, the first chips generated for *Arabidopsis thaliana* gene expression measurement contained 260,000 probes corresponding to almost 8,200 genes (Zhu *et al.*, 2001), whereas the latest Affymetrix chips allow the simultaneous analysis of 24,000 genes. Affymetrix has also created large scale chips for barley, rice, soybean, wheat, grape, tomato and yeast; and has also proposed custom made chips for any kind of species, if enough data and money is provided. Therefore, the only limit to *in situ* synthesized chip design is genome sequence data; clone library management, PCR amplifications and probe design can then be avoided. Unfortunately, the prohibitive costs of photolithographic masks to create the chips make this technique extremely expensive, inflexible and for now limited within the plant kingdom to a few well studied species.

Target Labelling and Hybridization

Before hybridization to DNA arrays or chips, one transcriptome (e.g. the totality of messenger RNAs corresponding to all the expressed genes in one particular condition) has to be labelled to allow its later detection. This labelling can be fluorescence or radioactivity mediated, depending on the nature of the chip, glass or nylon, respectively. The targets, which are cDNA counterparts of the transcriptome, are labelled during reverse transcription with modified (radioactive or fluorescent) nucleotides, but can also be processed differently depending on the experimental conditions. Hence, cDNA amplification is sometime necessary when the amount of starting material is very low; labelling is therefore performed on the resulting amplified transcriptome (Nakazono, 2003). Some have reported experimental bias induced by the differential incorporation yield of different fluorescent dyes. Cyanine 3^{tm} (Cy3) and cyanine 5^{TM} (Cy5) are large molecules that reduce reverse transcriptase efficiency of long transcripts and certain sequences. To avoid this problem, amino allyl modified dUTP nucleotides are used in reverse transcription; the resulting cDNA can then be linked by esterification to an active form of Cy3 or Cy5.

Considering radioactive labelling, ^{33}P is the most common isotope used with nylon technology. As radioactive isotopes are chemically equivalent to non radioactive ones, no bias is induced in reverse transcription efficiency. To detect the radioactive signals after hybridization, phosphor-imagers quantitatively record the amount of beta radiation produced by the labelled targets on the membrane. The disadvantage of this method remains its difficulty to be miniaturized and the tendency of many researchers to avoid the use of radioactive elements. However, the use of radiolabelling permits high sensitivity detection compared to fluorescence detection, in that less than 100,000 molecules can be detected in a few nanograms of the mRNA sample. To reach this sensitivity one would need more than one microgram of mRNA with any other technique (Bertucci *et al.*, 1999). Radioactive detection of hybridised targets is therefore an important option for gene expression analysis when sensitivity matters.

Fluorescent probes generated with Cy3 and Cy5, or sometimes fluorescein and rhodamine, are the more commonly used types of labels. Some new fluorochromes derived from rhodamine (Alexa, Molecular Bioprobes) have been created recently and seem to be spreading to a growing number of laboratories. Fluorochrome selection depends on three parameters: specificity, sensitivity and photobleaching resistance. Detection specificity is directly linked to the emission band width of the fluorochromes used and has to be sufficiently distinct. Sensitivity corresponds to the shape of the emission band, which has to be as sharp as possible. Ultimately, photobleaching is a function of fluorochrome chemical strength. All these parameters have to be taken into account during the design of an expression analysis experiment, as well as the scanning device available to detect the fluorochromes. However, the greatest advantage of fluorescence detection is its miniaturisation, that can go below 20 µm allowing significant increases in probe density on the chip.

One important feature of fluorescence detection is that it allows the simultaneous hybridization of two to several targets that have been differently labelled (Tsangaris *et al.*, 2002). Hence, comparison of two samples can be assessed in one single experiment using two non-overlapping fluorochromes. As slight differences can occur from one array to another, double labelling prevents any bias induced by slide variability. It is important to note that although slide variability should be minor, since the probes are present in saturating amounts compared to the targets, this topic is still the subject of debate. Fluorochrome diversity is increasing to the point that several condition comparisons will soon be managed with only one hybridization (Tsangaris *et al.*, 2002). Hybridizations are processed directly on the slides after target synthesis. Hybridization conditions depend on the length of the probes arrayed on the slide and has to be extensively tested before analysis. As an example, probe melting temperatures range from 42°C to 70°C depending on the nature of the buffer: the presence of formamide gives a positive effect on buffer stringency in Denhardt type buffers which are used at 42°C, whereas Sarkosyl based buffers are commonly used around 70°C. Hybridization times can also vary considerably as hybridisation follows an asymptotic curve and is therefore usually processed overnight. After signal detection, Nylon microarrays can be stripped for several hybridizations (6 to 10 times) in the case of membranes scanned while damp, or stripped after the decay of the radioactive probe in the case of membranes that are scanned dry. High resolution phosphorscreens are more efficient if the slide is in direct contact with the surface and so, the membrane has to be used dry. The analysis of a number of conditions in comparison to a control sample is then possible without slide induced bias. Although this can be time consuming since it is necessary to wait for target decay (^{33}P half life is 35 days), avoiding slide bias can be of tremendous advantage for the assessment of genes only weakly differentially expressed. Glass slides, on the other hand, are single use materials, but allow simultaneous comparison of two transcriptomes and, in the near future, maybe more.

Signal Processing and Data Extraction

Once the wet phase (e.g. slide hybridization and washing) is completed, signal detection of each of the hybridized targets can be captured. For radioactive targets, the slide signal has to be transferred for a variable time to a phosphorscreen before scanning. Radioactivity is recorded on the screen as stable high energy molecules; to be released, a light is emitted from the scanner, which frees the stored energy in a photonic process. The fluorescent signal can be recorded directly with a variety of detectors. Charge-coupled device (CCD) cameras are fast but only allow for a small area of detection (<1 square cm), whereas confocal and non-confocal laser scanners while being slower, allow several slides to be scanned at the same time. Whatever the scanner resolution, the microarray spot diameter has to be 5 to 10 times larger than the scanner resolution which can be as little as 5 μm for the most recent models.

Images recorded by the scanner correspond to target binding to the individual gene probes as well as any artefacts that can be due to mishybridization, background or dirt. To take these facts into account, image analysis software includes spot localisation, local background correction and an array of useful features to cope with image noise. Once gene signal and background values are clearly defined (this is specific to each experiment), expression data can be extracted from the image by counting the pixels within each gene and background areas and recording this in a computer readable format.

MICROARRAY EXPERIMENTAL DESIGN

Microarray Engineering

Microarray design is usually governed by the aim of the experiment and instantly raises the question of the total probe size. Considering cDNA microarrays and the high price of cDNA amplification for probe synthesis, microarray size is only limited in relation to the available budget. Although commonly the case, it is important in the first instance to understand the purpose of the experiment and the limitations imposed by using too few or the wrong types of probes. Microarray experiments are being used for two main purposes: candidate gene identification and molecular response analysis. Most of the time both goals are linked, the second being the continuation of the first. However, candidate gene searching consists of monitoring gene expression under a particular set of conditions and selecting a population of affected genes (which could be used as molecular markers of the condition, or targets for biotechnology) but without having a clear idea of their biological role. Sequence data is often poor in these experiments and the only way to access the probes is to use an anonymous cDNA array. Once a biological system is better known, however, the question of the underlying molecular events associated with the process can be asked more specifically, but this may require more accurate gene expression difference data. From here, the knowledge of gene sequences is important because it allows the design of gene specific oligonucleotide probes and permits the collection of more reliable expression data. These accurate data are the material from which statistical analyses will evaluate the potential link between conditions and genes.

The rapid developments in genome sequencing data suggests that genomic scale microarrays will spread over all the kingdoms and analysed species in the near future. Such arrays must contain a maximum of sequences with a minimum of redundancy. For model species (*Arabidopsis thaliana,* maize and rice) cloned genes and gene sequences are already available for microarray design. In other species, microarrays have to be developed from cDNA libraries. There is, however, still no fully genomic scale transcriptomic analysis yet realized in plants, although a number of projects aim to achieve it with a growing number of probes that represent 25 to 30% of the genome of *Arabidopsis thaliana.* This is about to change rapidly with the emergence of commercial printed oligoprobes (Agilent) or *in situ* synthesized chips (Affymetrix) that span the total *Arabidopsis thaliana* genome.

Enriched cDNA libraries can be used to focus the experiment on the analysis of the genes involved in the studied physiological system. As an example, Desikan*et al.* (2000) used *Differential mRNA display* to identify oxidative stress genes in a first approach and refined the results with a microarray experiment to profile expression of the previously selected genes printed to an array (Desikan *et al.*, 2001). Similarly, a subtractive cDNA library obtained from roots of tomato plants cultured in different nutrient deficiency conditions followed by nutrient resupply was used to construct a 700 probe microarray dedicated to nitrate inducible genes (Wang *et al.,* 2001). The coupling of SSH (Suppression Substraction Hybridization), which allows the creation of libraries of differentially expressed genes with DNA microarray experiments, is now widely employed as it fits in well with the discovery of new genes (Yang *et al.,* 1999). Alternatively, if expression data is already available for an array of genes of interest, one can create a microarray on the basis of previous experiments (O'Hara *et al.,* 2002; Swidzinski *et al.*, 2002), or on the basis of the genome annotation data (McGonigle *et al.*, 2000; Xu *et al.*, 2001; Chen *et al.*, 2002). On the other hand, if gene selection becomes too focussed on a particular condition, a significant component of interesting genes could be ignored leading to a misunderstanding of the phenomenon. Phosphate deprivation, for example, is responsible for an alteration of auxin inducible gene expression (Doumas pers. comm.). Without this information, a microarray dedicated to nutrient stress might have passed over auxin response genes. It is therefore important to introduce onto the microarray a number of genes related to other phenomenon, in order to assess any experimental side effects or cross regulation. It is also important to have sufficient control probes on the array, a batch of positive and negative controls should be included in order to evaluate cDNA synthesis, labelling efficiency, the hybridization process (cross hybridization effects) and to contribute to the analysis of the results. Also some experiments have used as controls, a particular mRNA isolated from a different species or a synthetic target that should not be able to cross hybridise with the microarray (e.g. human *Cot 1* or luciferase), in association with a specific probe on the array. These pairs of target/probes can also help in the assessment of the quality of the experiment and can be used in various numbers known as "spikes".

Statistical Analysis of Expression Data

Like most scientific experiments, transcriptomes requires the use of statistical tools in data analysis for two reasons:

1. The massive amount of data generated with these techniques prohibits any manual treatment.
2. Experimental variability is generally significant and has to be managed in order to exploit the data properly.

The ideal conceptual framework for statistical analysis of microarray data does not yet exist and may never appear. The reason for this is that statistics are generally framed to answer discrete questions. Hence, the more questions being asked, the more testing to be processed and experiments to be designed. The following are a few of the statistical tasks that are widely used amongst the scientific community using microarrays.

Profiling

The clustering of co-regulated genes (e.g. having similar expression profiles) under a particular set of conditions involves the analysis of huge amounts of data. Eisen *et al.* (1998) were the first to use specific mathematical tools for the analysis of gene expression data and revealed the value of such approaches (Fig. 10. 3). Among the emerging methods, hierarchical clustering algorithms were particularly praised as they allow the treatment of all the experimental data at once. These algorithms use iterative processes on the data to progressively cluster similar expression profiles, then organise similar clusters into wider groups resulting in a tree for which branch length is proportional to profile dissemblance. Hierarchical clustering algorithms are founded on three parameters: the definition of dissemblance of expression profiles, the clustering criterion that defines the clusters to be gathered at a given step that sets the resolution of the clusters and a stopping rule to avoid clustering of too unrelated elements. The choice of these parameters has a significant effect on the resulting tree and should be carefully selected during the experimental design, although most publications seem to simply ignore the existence of such effectors. One can however evaluate the reliability of a given clustered set of profiles using a randomisation technique (*bootstrap*). These clustering methods do not take data variability into account, compared to shuffle models founded on random processing of data. The former methods are anyhow much more computer-time consuming than determinist clustering methods.

Differential analysis

Monitoring gene expression profiles to understand the response of a biological system to a change in conditions, is also typical of many microarray experiments. Here, it is obvious that a number of experimental replications are necessary to record a reliable expression profile. Normalization is also of tremendous importance, as it has consequences on all the statistical analyses to be performed.

Differential analysis can be compared to population comparisons where a given gene (g) is related to a number of measurements (n) in two given conditions A and B (therefore 2n measurements). g is said to be differentially expressed in A versus B, if the difference between the two means of values in these conditions ($mA(g)$ and $mB(g)$) is significant. The key value of this analysis is therefore given by $d(g) = mA(g) - mB(g)$ and has to be significant. For more details of such analysis, the reader is referred to Dudoit *et al.* (2002).

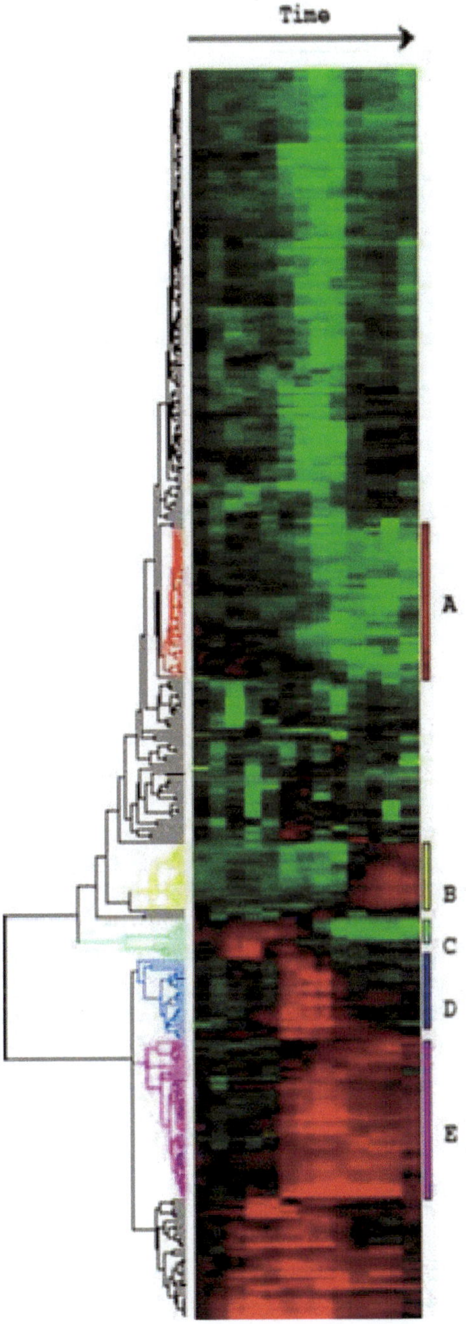

Fig. 10.3 Hierarchical clustering of time course expression profiles in yeast (Eisen *et al.*, 1998). Numeral values are coded in color (green = high expression level, red = low). Each row represents a gene and each column a condition (time). A functional interpretation of A-E groups is given in the original article.

Permutation testing: When a "sufficient" number of replicates are available, significance can be calculated without proceeding to a restrictive statistical test. The classical method consists in the calculation of all the possible values of d(*g*) over the 2n discrete values of g within the A and B conditions. The distribution of d(*g*) can therefore be tested and the gene said to be differentially expressed if the evaluated value for d(*g*) is above a chosen threshold (corresponding to the predefined risk α). This kind of test is used in SAM (Significance Analysis of Microarrays, Tusher *et al.*, 2001).

Parametric test: When the number of replicates is too low for permutation testing, the d(g) distribution can be approached using a probabilistic hypothesis. Here d(*g*) is expected to follow a normal distribution and the standard deviation $\sigma d(g)$ has to be estimated. If the number of replicates is not too low $\sigma d(g)$ can be estimated independently for each gene.

Standard deviation approximation: If the replicate number is too low, the estimation of $\sigma d(g)$ gene per gene becomes too inaccurate and a global approximation of standard deviation $\sigma d(g)$ has to be performed. The simplest way, frequently adopted, but obviously not realistic, is to assume that all genes have the same $\sigma d(g)$.

A classic approximation links $\sigma d(g)$ to the overall mean of gene intensity. Logarithmic transformation of raw data is commonly used and aims to stabilise standard deviations for highly expressed genes. Arguments on standard deviation issues in particular, and more widely on possible approximations, are presented by Rudemo *et al.* (2002). In all cases, this model is accurate if most of the genes are not differentially expressed.

Multiple comparisons: For statisticians, one other challenge for simultaneous analysis of differentially expressed genes is the detection of false positives. The threshold associated with any statistical test is the probability of identifying a gene differentially expressed when it is not, and is written as α. Hence, in a 5,000 gene set of non-differentially expressed genes, a batch of independent tests set with a threshold value of $\alpha = 5\%$, on average 250 genes would appear differentially expressed. The problem is therefore not to control the threshold for each gene, but the overall threshold of all tests. Threshold calculation is rendered extremely complex by the fact that genes are biologically linked to one another with different pathways. The SAM method (Tusher *et al.*, 2001) provides a solution founded on permutation tests were the same permutation is applied to each gene simultaneously, to avoid the loss of the existing correlation structure.

Diagnostic

Determination of differentially expressed genes is often a prerequisite to diagnostic analysis. Alizadeh *et al.* (2000) compared transcriptomes of different kinds of cancer cells harvested from a number of carcinomas. The aim was to identify genes differentially expressed between these two populations, but also to predict the type of cancer a patient may have on the basis of cellular transcriptome analysis. Although linked, these two objectives remain distinct, particularly in the selection of more discriminating groups of genes. It is clear that diagnostics have to be founded on expression levels of genes that are most strongly affected by the disease (e.g. for which the ratio $d(g)$ / $\sigma d(g)$

is the highest in the absolute). On the other hand, the subset of the most highly expressed genes may not be the best discriminating group in that most of the strongest expressed genes are correlated and therefore redundant. A second diagnostic gene should be differentially expressed but also uncorrelated with the first gene if they are to be used together diagnostically, etc. Most of the time gene selection is based on iterative algorithms that take these two facts into account at each step. Table 10.2 is a non exhaustive summary of expression data analysis softwares currently available.

Table 10.2 Software available for expression data analysis

Software	*URL address*
Acuity	http://www.axon.com/GN_Acuity.html
Bioconductor	http://www.bioconductor.org
BioMine	http://www.gnsbiotech.com/products.shtml
Cleaver	http://classify.stanford.edu
Clementine Microarray.	http://www.spssscience.com/clementine
Cluster and TreeView	http://rana.lbl.gov/EisenSoftware.htm
Cluster Identification Tool	http://www.vai.org/vari/bioinformatics.htm
Clusvafor	http://mbcr.bcm.tmc.edu/genepi
Coupled Two Way Clustering	http://ctwc.weizmann.ac.il
DNA-Chip Analyzer	http://www.dchip.org
Expression Profiler	http://ep.ebi.ac.uk/EP
GeneClust	http://odin.mdacc.tmc.edu/~kim/geneclust
GeneCluster	http://www-genome.wi.mit.edu/cancer/software
GeneLinker Gold	http://microarray.genelinker.com/products
GeneMaths	http://www.applied-maths.com/ge/ge.htm
GeneSight	http://www.biodiscovery.com/genesight.asp
Genesis	http://genome.tugraz.at/Software/GenesisCenter.html
GeneSpring	http://www.silicongenetics.com/cgi/SiG.cgi/Products
GeneTraffic	http://www.iobion.com/products/products.html
GeneViz	http://www.contentsoft.de/index.htm?geneviz.htm
J-Express Pro	http://www.molmine.com/frameset/frm_jexpress.htm
MA-ANOVA	http://www.jax.org/research/churchill/software
MatArray	http://iribhn.ulb.ac.be/microarrays/toolbox
MicroArray Explorer	http://maexplorer.sourceforge.net
MricoHelper	http://www.changbioscience.com/microhelperinfo.html
PaGE	http://www.cbil.upenn.edu/PaGE
Partisan arrayLIMS	http://www.clondiag.com/products/sw/partisan
Probality of expression	http://astor.som.jhmi.edu/poe
QuantArray	http://lifesciences.perkinelmer.com/areas/microarray
Rosetta Resolver	http://www.rosettabio.com/products/resolver
Significance Analysis of Microarrays	http://www-stat.stanford.edu/~tibs/SAM
SNOMAD	http://pevsnerlab.kennedykrieger.org/snomadinput.html
Sotarray	http://bioinfo.cnio.es/cgi-bin/tools/clustering/sotarray
Spotfire	http://www.spotfire.com
SVDMAN	http://public.lanl.gov/mewall/svdman
TIGR MultiExperiment Viewer	http://www.tigr.org/software
VERA et SAM	http://www.systemsbiology.org/VERAandSAM
XCluster	http://genome-www.stanford.edu/~sherlock/cluster.html

DNA MICROARRAY EXPERIMENTS IN PLANTS

Although DNA microarrays are a relatively recent technique (first published work by Schena *et al.*, 1995), more than a hundred publications describing a wide range of applications had emerged by 2003 (WWW Virtual Library (Botany) Plant-Arrays, http://www.univ-montp2.fr/~plant_arrays). Since then, there has been an overwhelming explosion in the use of microarrays for plant studies, e.g. more than 350 publications on this topic in 2004. The first microarrays to be carried out were on *Arabidopsis thaliana* and were used to describe different organ expression patterns (Schena *et al.*, 1995; Ruan *et al.*, 1998), lines (Kehoe *et al.*, 1999) or environmental conditions (Desprez *et al.*, 1998). Meanwhile, the DNA microarray technique was rapidly transferred to agronomic species such as strawberry (Aharoni *et al.*, 2000), wheat (Skinner *et al.*, 2005), rice (http://cdna01.dna.affrc.go.jp/RMOS/), potato (Ronning *et al.*, 2003), cotton (http://cottongenomecenter.ucdavis.edu/links.asp), maize (http://www.maizearray.org), soybean (http://psi081.ba.ars.usda.gov/SGMD/default.htm), grapevine (Water *et al.*, 2005) and flowers (http://www.floralgenome.org/fgp). The need to develop more efficient structures led to the creation of ambitious genomic programs mainly dedicated to model species like *Arabidopsis thaliana*, rice or maize (AFGC, Wisman and Ohlrogge, 2000; GARnet, Beale *et al.*, 2002; ZmdB, Gai *et al.*, 2000), that were grouped in plant transcriptome platforms (Table 10.3). The aim of these platforms was to provide methodology and expert help to research programs in DNA microarrays and to give workers the opportunity to use genome wide or dedicated microarrays for a growing number of programs that could not afford a transcriptomic experiment by themselves.

Transcriptome analyses have three main objectives:

1. Marker gene investigation, which looks for a limited number of development stage specific, or any condition specific genes, to provide tools for selection or to flag a given stage within an evolving process.
2. Characterization of a given biological stage, to describe a maximum of the expression variations occurring within this stage in order to provide an integrated view of the process.
3. Search for a function of a given gene, assuming that highly similar expression profiles are functionally linked and use the same regulation pathways. The advantage of such a scheme is to provide a way to assign a function to previously identified genes and on the other hand to allow the study of the genetic control of global regulation pathways. In most of the cases, clusters of co-regulated genes include known and unknown genes, while the conditions can be natural (development stage, day length or nutrition...) or artificial (mutations, chemicals or biotic interactions...). The resulting putative assigned function has to be confirmed with complementary experiments, such as reverse genetics.

This highly schematic view of strategies is not exhaustive, and in most cases, the use of microarrays covers more than one of the three fields described previously.

Marker Genes Investigation

A majority of the previous work using microarrays was aimed at the discovery of marker genes. Hence, arrays of constitutive or induced genes have been identified in

Table 10.3 Illustration of some academic platforms dedicated to Plant DNA microarrays

Programs	*Supports*	*Plants*	*Countries*	*URL Addresses*
Complete Arabidopsis Transcriptome MicroArray (CATMA)	PCR Amplification of genomic fragments (150 to 500 bp) 24.5 K	*Arabidopsis*	Europe	http://www.catma.org
Keck Foundation – Biotechnology Resource Laboratory – Yale University	Oligonucleotides (70 mer) 27 K	*Arabidopsis*	USA	http://keck.med.yale.edu/dnaarrays/slides.htm
Maize Oligonucleotide Array project	Oligonucleotides (70 mer) > 30 K	Maize	USA	http://www.maizearray.org/index.shtml
NSF Potato Functional Genomics	PCR Amplification of cDNA 12 K	Potato	USA	http://www.tigr.org/tdb/potato/dist2.shtml
Rice Microarray Opening Site (RMOS)	Full length cDNA 9K	Rice	Japan	http://cdna01.dna.affrc.go.jp/RMOS/index.html
NSF Rice Oligonucléotide Array Project	Oligomer (50-70 mer) 20 K	Rice	USA	http://www.ricearray.org
Cornell Center for Gene Expression (CGEP)	PCR Amplification of cDNA 13.5 K	Tomato	USA	http://bti.cornell.edu/CGEP/CGEP.html

plantlets, roots, leaves, inflorescences, flowers or siliques of *Arabidopsis thaliana* at different developmental stages (Zhu *et al.*, 2001). Conversely, organ specific genes and group of genes have also been discovered for anthers, pistil and nodules in *Lotus japonica* (Colebatch *et al.*, 2002; Endo *et al.*, 2002). Beside this descriptive way of investigation, other workers have focussed on physiological processes like embryogenesis (Lee *et al.*, 2002; O'Hara *et al.*, 2002), seed development (Girke *et al.*, 2000; Ruuska *et al.*, 2002), trichome development, programmed cell death (Swidzinski *et al.*, 2002), cell wall synthesis (Whetten *et al.*, 2001), or day length/light intensity regulated development (Harmer *et al.*, 2000; Ma *et al.*, 2001; Schaffer *et al.*, 2001). Furthermore, external constraints on plants like wounding, pathogen attack by bacteria or pests (Aharoni *et al.*, 2000; Maleck *et al.*, 2000; Reymond, 2001; Reymond *et al.*, 2000; Scheideler *et al.*, 2002; Schenk *et al.*, 2000; Stintzi *et al.*, 2001), salt stress (Bohnert *et al.*, 2001; Kawasaki *et al.*, 2001; Oztur *et al.*, 2002), water stress (Seki *et al.*, 2001; Oztur *et al.*, 2002), oxidative stress (Desikan *et al.*, 2000, 2001), iron deficit (Thimm *et al.*, 2001; Negishi *et al.*, 2002) or mercury excess (Heidenreich *et al.*, 2001), have been largely covered. Plant nutritional status modifications have also been approached in *Arabidopsis thaliana* (Wang *et al.*, 2000) and tomato (Wang *et al.*, 2001) concerning nitrate regulation pathways.

More generally, these works which benefited from the simultaneous monitoring of an important number of genes in different tissues or conditions, has led to the confirmation of previous results and the discovery of new marker genes. Furthermore, highlighting new genes or even new pathways may lead to new fields of interest and to the emergence of new working hypotheses (Gonzali *et al.*, 2005).

Characterization of a Biological State

One major purpose of DNA microarray expression profile analysis is the functional characterization of different kinds of mutant (insertion, deletion, antisense or over-expression). Hence, comparative global analysis of transgenic plant expression profiles versus wild type plants allows the clustering of the genes affected by the mutation (Osakabe *et al.*, 2002; Wu *et al.*, 2002). This analysis thus reveals regulation pathways of the mutated gene. Similarly, insertion mutant comparisons to wild type reveal a list of genes affected by the inserted DNA (Petersen *et al.*, 2000; Perez-Amador *et al.*, 2001; Brodersen *et al.*, 2002; Kim *et al.*, 2002; Tian *et al.*, 2002). DNA microarray experiments have become a complementary tool to more classical investigations (biochemistry, cell physiology or molecular physiology) for the functional characterisation of a mutant and allow the widening of the exploration towards known genes in the analysed system but also to unexpected genes. Because numerous genes can be transcriptionaly affected by different mechanisms other than the studied process or condition, the origin of the changes in expression indicated by the arrays remains uncertain. However, the strategy can be enriched by the multiplication of the samples analysed corresponding to one defined physiological process, like the light controlled development of *Arabidopsis thaliana* (Ma *et al.*, 2001).

Search for a Gene Function

Beginning with expression data of an extensive number of genes, one can use the upstream sequences of clusters of genes to look for shared cis-regulating elements. As an example, temporal analysis of *Arabidopsis thaliana* circadian rhythms has been used to isolate genes affected in a day-night shift (Harmer *et al.*, 2000). Bioinformatic analysis of these genes revealed the existence of a highly conserved regulation pattern which is necessary for the functioning of their circadian control. In another study, transcriptomic analysis of plant defence co-regulated genes in *Arabidopsis thaliana* allowed the discovery, with *in silico* analysis, of a promoter shared by all of the genes (Maleck *et al.*, 2000). One challenging goal of gene expression profile analysis is to be able to unravel regulation networks from temporal expression data (Wyrick and Young, 2002). In this, the mathematical models which have been suggested, postulate that a particular gene expression rate is simultaneously dependent on all other gene expression rates (Bucher, 1999; Genoud and Metraux, 1999; Genoud *et al.*, 2002).

Flexibility in DNA microarray construction and use, has allowed the spread of the technology to other non-model species. Hence, gene expression data through DNA microarrays have been obtained for barley (Negishi *et al.*, 2002; Oztur *et al.*, 2002), tomato (Wang *et al.*, 2001), bean (Arimura *et al.*, 2002), ectomycorhizal symbiosis *Eucalyptus globulus – Pisolithus tinctorius* (Voiblet *et al.*, 2001), strawberry (Aharoni *et al.*, 2000), *Lotus japonica* (Endo *et al.*, 2002), rice (Kawasaki *et al.*, 2001), maize (McGonigle *et al.*, 2000; Fernandes *et al.*, 2002), poplar (Hertzberg *et al.*, 2001a) and pine (Whetten *et al.*, 2001). The general strategy is to construct cDNA libraries and to recognise the amplified clones without any sequencing of the probes. Although these microarrays do not contain the entire expressed fraction of the genome (from 1,000 to 12,000 probes, Table 10.3), results are encouraging and they have been used to produce marker genes, or new working hypotheses. Finally, Girke *et al.* (2000) showed that *Arabidopsis thaliana* DNA microarrays can be used to analyse transcripts of closely related species like *Brassica,* which have strong homologies with it. As a new potential use, DNA microarrays have helped to characterize repeated sequences existing among 20 Fabacae genomic DNAs (Nouzova *et al.*, 2001).

DNA MICROARRAY LIMITS AND NEW DIRECTIONS

The transcriptome analysed by microarrays is not a real picture of what happens fully at the transcriptional level. Indeed, one of the biological processes that remains invisible to DNA microarrays is post-transcriptional regulation. Numerous response processes trigger protein interactions or enzyme catalysed post-transcriptional modifications without any transcriptional alterations. Another limit of transcriptomics concerns multigene families, for which no reliable quantitation can be done due to their highly conserved regions (Richmond and Somerville, 2000; Schenk *et al.*, 2000). One useful way to determine the difference in expression between members of a multigene family is to focus on the 5′ or 3′ untranslated regions (UTRs) of those genes where higher sequence variation occurs and design short probes within these sequences. Recording weakly expressed genes is also a requirement, if there is to be a further improvement in the sensitivity of microarray techniques. On average, DNA microarrays are able to detect transcripts present in a relative amount of 1 per 100,000 RNAs. However, data analysis methods usually remove data which has considerable variability and that is mostly below the significance threshold. The amount of RNA available for target synthesis is sometimes limiting and therefore affects microarray result quality. The yield of the experiments is improving and the required amount of RNA has been reduced from 1 to

2 µg polyA RNA to 2 µg total RNA, with the use of new technologies. However even this quantity can be hard to obtain for some tissues, specific cells, or even small organs. Gene expression in roots, for instance, is the integration of an array of expression profiles for a number of cell types (cortical, mesenchyma, xylem, phloem and pericycle). A full analysis of a single class of cell is challenging (Kehr, 2003) and the experimental design has to take this fact into account. To achieve this goal, RNA *in vitro* amplification methods have been engineered on the basis of RNA polymerase linear amplification (Byrne *et al.*, 2000), cDNA direct amplification (Endege *et al.*, 1999), or with linked adaptators (Dixon *et al.*, 1998; Hertzberg *et al.*, 2001b; Ivashuta *et al.*, 2002). The final limitation of microarrays comes from their own conception, in that they are made from cDNA libraries containing a population of the most expressed transcripts in a given situation, leading to a loss of potential data from those genes that are not represented on the array.

DNA arrays are able to measure more than expression profiles. Classically gene expression measured by microarrays is a balance of transcription activity and RNA degradation. "Run-on" experiments transiently label RNA during the targeted process, to record newly synthesized transcripts. In order to access transcription rate and RNA degradation data, run-on can be processed concurrently with microarrays. Similarly, transcription can also be monitored by purification of polysomes before RNA extraction (Diehn *et al.*, 2000).

The most stimulating perspective is the combination of transcriptomic, proteomic and metabolomic data in one situation. This new approach has already started in yeast, and has led to the production of true functional maps (Ideker *et al.*, 2001, Castrillo and Oliver, 2004). In the model plants such as *Arabidopsis*, gene expression databases obtained for a wide range of conditions are now publicly available (Genevestigator: https://www.genevestigator.ethz.ch and AthCoR: http://csbdb.mpimp-golm.mpg.de/csbdb/dbcor/ath.html#transcor). They allow the efficient *in silico* discovery of new genes or functions, e.g. the identification of whole set of genes encoding co-regulated proteins belonging to multiprotein complexes, without the need to perform time and cost-consuming transcriptome analyses. In addition, tools that display gene expression data sets onto diagrams of metabolic pathways have been created for *Arabidopsis* (MapMan at http://gabi.rzpd.de/projects/MapMan), providing new ways for studying plant systems biology.

CONCLUSION

Microarray analysis of transcriptomes can produce quantitative information for tens of thousands of genes simultaneously (Alizadeh *et al.*, 2000). Furthermore, improvements in chip miniaturization and analytical power now allow genomic scale transcript analysis, which is currently impossible for proteins or metabolites. Among the numerous perspectives opened by microarray techniques, the discovery of gene function and associated regulatory pathways appear the most promising. In most of cases, expression profiles have been shown to be closely linked to function under a particular set of conditions. Beside this, metabolic and regulatory pathway studies, will raise considerable interest and will lead to the unravelling of the global plant regulation.

Hence, like in other biological fields, DNA arrays have become one of a package of tools available to plant biologists. This technology is still in its early years, as it needs further methodological and applied improvement but is also expanding widely throughout the scientific community. The next decade will undoubtedly show how powerful microarrays can be as a tool for biologists.

REFERENCES

Adams, M.D., Kerlavage, A.R., Fleischmann, R.D., Fuldner, R.A., Bult, C.J., Lee, N.H., Kirkness, E.F., Weinstock, K.G., Gocayne, J.D., White, O., *et al.* 1995. Initial assessment of human gene diversity and expression patterns based upon 83 million nucleotides of cDNA sequence, *Nature*, 377, 3-174.

Aharoni, A., Keizer, L.C., Bouwmeester, H.J., Sun, Z., Alvarez-Huerta, M., Verhoeven, H.A., Blaas, J., van Houwelingen, A.M., De Vos, R.C., van der Voet, H. *et al.* 2000. Identification of the SAAT gene involved in strawberry flavor biogenesis by use of DNA microarrays, *Plant Cell*, 12, 647-62.

Alizadeh, A.A., Eisen, M.B., Davis, R.E., Ma, C., Lossos, I.S., Rosenwald, A., Boldrick, J.C., Sabet, H., Tran, T., Yu, X., *et al.* 2000. Distinct types of diffuse large B-cell lymphoma identified by gene expression profiling, *Nature*, 403, 503-11.

Arimura, G., Ozawa, R., Nishioka, T., Boland, W., Koch, T., Kuhnemann, F., Takabayashi, J. 2002. Herbivore-induced volatiles induce the emission of ethylene in neighboring lima bean plants, *Plant J*, 29, 87-98.

Bachem, C.W., van der Hoeven, R.S., de Bruijn, S.M., Vreugdenhil, D., Zabeau, M., Visser, R.G. 1996. Visualization of differential gene expression using a novel method of RNA fingerprinting based on AFLP: analysis of gene expression during potato tuber development, *Plant J*, 9, 745-53.

Beale, M., Dupree, P., Lilley, K., Beynon, J., Trick, M., Clarke, J., Bevan, M., Bancroft, I., Jones, J., May, S. *et al.* 2002. GARNet, the Genomic *Arabidopsis* Resource Network, *Trends Plant Sci*, 7, 145-7.

Bertucci, F., Bernard, K., Loriod, B., Chang, Y.C., Granjeaud, S., Birnbaum, D., Nguyen, C., Peck, K., Jordan, B.R. 1999. Sensitivity issues in DNA array-based expression measurements and performance of nylon microarrays for small samples, *Hum Mol Genet*, 8, 1715-22.

Bohnert, H.J., Ayoubi, P., Borchert, C., Bressan, R.A., Burnap, R.L., Cushman, J.C., Cushman, M.A., Deyholos, M., Fischer, R., Galbraith, D.W. 2001. A genomics approach towards salt stress tolerance, *Plant Physiol Bioch*, 39, 295-311.

Brenner, S., Johnson, M., Bridgham, J., Golda, G., Lloyd, D.H., Johnson, D., Luo, S., McCurdy, S., Foy, M., Ewan, M., *et al.* 2000. Gene expression analysis by massively parallel signature sequencing (MPSS) on microbead arrays, *Nat Biotechnol*, 18, 630-4.

Breyne, P., Zabeau, M. 2001. Genome-wide expression analysis of plant cell cycle modulated genes. *Curr Opin Plant Biol*, 4, 136-42.

Brodersen, P., Petersen, M., Pike, H.M., Olszak, B., Skov, S., Odum, N., Jorgensen, L.B., Brown, R.E., Mundy, J. 2002. Knockout of *Arabidopsis* accelerated-cell-death11 encoding a sphingosine transfer protein causes activation of programmed cell death and defense, *Genes Dev*, 16, 490-502.

Bucher, P. 1999. Regulatory elements and expression profiles, *Current Opin Struct Biol*, 9, 400-407.

Byrne, M.C., Whitley, M.Z., Follettie, M.T. 2000. Preparation of mRNAs for expression monitoring. In : Asubel *et al.*, eds. *Current protocols in molecular Biology*. John Wiley & Sons Inc., 22.2.1-22.2.13.

Castrillo, J.I., Oliver, S.G. 2004. Yeast as a touchstone in post-genomic research: stategies for integrative analysis in functional genomics, *J Biochem Mol Biol*, 37, 93-106.

Chen, W., Provart, N.J., Glazebrook, J., Katagiri, F., Chang, H.S., Eulgem, T., Mauch, F., Luan, S., Zou, G., Whitham, S.A., *et al.* 2002. Expression profile matrix of *Arabidopsis* transcription factor genes suggests their putative functions in response to environmental stresses, *Plant Cell,* 14, 559-74.

Colebatch, G., Kloska, S., Trevaskis, B., Freund, S., Altmann, T., Udvardi, M.K. 2002. Novel aspects of symbiotic nitrogen fixation uncovered by transcript profiling with cDNA arrays, *Mol Plant Microbe Interact,* 15, 411-20.

DeRisi, J.L., Iyer, V.R., Brown, P.O. 1997. Exploring the metabolic and genetic control of gene expression on a genomic scale, *Science,* 278, 680-6.

Desikan, R., Neill, S.J., Hancock, J.T. 2000. Hydrogen peroxide-induced gene expression in *Arabidopsis thaliana, Free Radic Biol Med,* 28, 773-8.

Desikan, R., Hancock, J.T., Neill, S.J. 2001. Regulation of the *Arabidopsis* transcriptome by oxidative stress, *Plant Physiol,* 127, 159-72.

Desprez, T., Amselem, J., Caboche, M., Hofte, H. 1998. Differential gene expression in *Arabidopsis* monitored using cDNA arrays, *Plant J,* 14, 643-52.

Devaux, F., Marc, P., Jacq, C. 2001. Transcriptomes, transcription activators and microarrays. *FEBS Letters,* 498, 140-4.

Diehn, M., Eisen, M.B., Botstein, D., Brown, P.O. 2000. Large-scale identification of secreted and membrane-associated gene products using DNA microarrays. *Nat Genet, 25,* 58-62.

Dixon, A.K., Richardson, P.J., Lee, K., Carter, N.P., Freeman, T.C. 1998. Expression profiling of single cells using 3 prime end amplification (TPEA) PCR. *Nucleic Acids Res,* 26, 4426-31.

Dudoit, S., Yang, Y., Callow, M.J., Speed, T.P. 2002. Statistical methods for identifying differentially expressed genes in replicated cDNA microarray experiments. *Statistica Sinica,* 12, 111-39.

Eisen, M.B., Spellman, P.T., Brown, P.O., Botstein, D. 1998. Cluster analysis and display of genome-wide expression patterns. *Proc Natl Acad Sci* USA, 95, 14863-8.

Endege, W.O., Steinmann, K.E., Boardman, L.A., Thibodeau, S.N., Schlegel, R. 1999. Representative cDNA libraries and their utility in gene expression profiling, *Biotechniques,* 26, 548-50.

Endo, M., Matsubara, H., Kokubun, T., Masuko, H., Takahata, Y., Tsuchiya, T., Fukuda, H., Demura, T., Watanabe, M. 2002. The advantages of cDNA microarray as an effective tool for identification of reproductive organ-specific genes in a model legume, *Lotus japonicus. FEBS Lett,* 514, 229-37.

Fernandes, J., Brendel, V., Gai, X., Lal, S., Chandler, V.L., Elumalai, R.P., Galbraith, D.W., Pierson, E.A., Walbot, V. 2002. Comparison of RNA expression profiles based on maize expressed sequence tag frequency analysis and micro-array hybridization. *Plant Physiol,* 128, 896-910.

Fodor, S.P., Read, J.L., Pirrung, M.C., Stryer, L., Lu, A.T., Solas, D. 1991. Light-directed, spatially addressable parallel chemical synthesis, *Science,* 251, 767-73.

Gai, X., Lal, S., Xing, L., Brendel, V., Walbot, V. 2000. Gene discovery using the maize genome database ZmDB, *Nucleic Acids Res,* 28, 94-6.

Genoud, T., Buchala, A.J., Chua, N.H., Metraux, J.P. 2002. Phytochrome signalling modulates the SA-perceptive pathway in *Arabidopsis, Plant J,* 31, 87-95.

Genoud, T., Metraux, J.P. 1999. Crosstalk in plant cell signaling: structure and function of the genetic network, *Trends Plant Sci,* 4, 503-507.

Girke, T., Todd, J., Ruuska, S., White, J., Benning, C., Ohlrogge, J. 2000. Microarray analysis of developing *Arabidopsis* seeds, *Plant Physiol,* 124, 1570-81.

Gonzali, S., Loreti, E., Novi, G., Poggi, A., Alpi, A., Perata, P. 2005. The use of microarrays to study the anaerobic response in *Arabidopsis, Ann Bot* 96(4): 661-668.

Guckenberger, M., Kurz, S., Aepinus, C., Theiss, S., Haller, S., Leimbach, T., Panzner, U., Weber, J., Paul, H., Unkmeir, A., *et al.* 2002. Analysis of the heat shock response of *Neisseria meningitidis* with cDNA- and oligonucleotide-based DNA microarrays, *J Bacteriol*, 184, 2546-51.

Harmer, S.L., Hogenesch, J.B., Straume, M., Chang, H.S., Han, B., Zhu, T., Wang, X., Kreps, J.A., Kay S.A. 2000. Orchestrated transcription of key pathways in *Arabidopsis* by the circadian clock, *Science*, 290, 2110-3.

Heidenreich, B., Mayer, K., Sandermann, H. Jr., Ernst, D. 2001. Mercury-induced genes in *Arabidopsis thaliana*: identification of induced genes upon long-term mercuric ion exposure, *Plant Cell Environ*, 24, 1227-34.

Hertzberg, M., Aspeborg, H., Schrader, J., Andersson, A., Erlandsson, R., Blomqvist, K., Bhalerao, R., Uhlen, M., Teeri, T.T., Lundeberg, J. *et al.* 2001a. A transcriptional roadmap to wood formation, *Proc. Natl. Acad. Sci.* USA, 98, 14732-7.

Hertzberg, M., Sievertzon, M., Aspeborg, H., Nilsson, P., Sandberg, G., Lundeberg, J. 2001b. cDNA microarray analysis of small plant tissue samples using a cDNA tag target amplification protocol, *Plant J*, 25, 585-91.

Ideker, T., Thorsson, V., Ranish, J.A., Christmas, R., Buhler, J., Eng, J.K., Bumgarner, R., Goodlett, D.R., Aebersold, R., Hood, L. 2001. Integrated genomic and proteomic analyses of a systematically perturbed metabolic network, *Science*, 292, 929-34.

Ivashuta, S., Uchiyama, K., Gau, M., Shimamoto, Y. 2002. Linear amplification coupled with controlled extension as a means of probe amplification in a cDNA array and gene expression analysis during cold acclimation in alfalfa (*Medicago sativa* L.), *J Exp Bot*, 53, 351-9.

Kawasaki, S., Borchert, C., Deyholos, M., Wang, H., Brazille, S., Kawai, K., Galbraith, D., Bohnert, H.J. 2001. Gene expression profiles during the initial phase of salt stress in rice, *Plant Cell*, 13, 889-905.

Kehoe, D.M., Villand, P., Somerville, S. 1999. DNA microarrays for studies of higher plants and other photosynthetic organisms, *Trends Plant Sci*, 4, 38-41.

Kehr, J. 2003. Single cell technology. *Curr Opin Plant Biol*, 6, 617-621.

Kim, G.T., Shoda, K., Tsuge, T., Cho, K.H., Uchimiya, H., Yokoyama, R., Nishitani, K., Tsukaya, H. 2002. The ANGUSTIFOLIA gene of *Arabidopsis*, a plant CtBP gene, regulates leaf-cell expansion, the arrangement of cortical microtubules in leaf cells and expression of a gene involved in cell-wall formation, *Embo J*, 21, 1267-79.

Lee, J.M., Williams, M.E., Tingey, S.V., Rafalski, J.A. 2002. DNA array profiling of gene expression changes during maize embryo development, *Funct Integr Genomics*, 2, 13-27.

Liang, P., Pardee, A.B. Differential display of eukaryotic messenger RNA by means of the polymerase chain reaction. *Science*, 257, 967-71.

Lockhart, D.J., Dong, H., Byrne, M.C., Follettie, M.T., Gallo, M.V., Chee, M.S., Mittmann, M., Wang, C., Kobayashi, M., Horton, H., Brown, E.L. 1996. Expression monitoring by hybridization to high-density oligonucleotide arrays, *Nat Biotechnol*, 14, 1675-80.

Ma, L., Li, J., Qu, L., Hager, J., Chen, Z., Zhao, H., Deng, X.W. 2001. Light control of Arabidopsis development entails coordinated regulation of genome expression and cellular pathways, *Plant Cell*, 13, 2589-607.

Maleck, K., Levine, A., Eulgem, T., Morgan, A., Schmid, J., Lawton, K.A., Dangl, J.L., Dietrich, R.A. 2000. The transcriptome of *Arabidopsis thaliana* during systemic acquired resistance, *Nat Genet*, 26, 403-10.

McGonigle, B., Keeler, S.J., Lau, S.M., Koeppe, M.K., O'Keefe, D.P. 2000. A genomics approach to the comprehensive analysis of the glutathione S-transferase gene family in soybean and maize, *Plant Physiol*, 124, 1105-20.

Negishi, T., Nakanishi, H., Yazaki, J., Kishimoto, N., Fujii, F., Shimbo, K., Yamamoto, K., Sakata, K., Sasaki, T., Kikuchi, S., *et al.* 2002. cDNA microarray analysis of gene expression during Fe-deficiency stress in barley suggests that polar transport of vesicles is implicated in phytosiderophore secretion in Fe-deficient barley roots, *Plant J*, 30, 83-94.

Nouzova, M., Neumann, P., Navratilova, A., Galbraith, D.W., Macas, J. 2001. Microarray-based survey of repetitive genomic sequences in *Vicia spp*, *Plant Mol Biol*, 45, 229-44.

O'Hara, P., Slabas, A.R., Fawcett, T. 2002. Fatty Acid and Lipid Biosynthetic Genes Are Expressed at Constant Molar Ratios But Different Absolute Levels during Embryogenesis, *Plant Physiol*, 129, 310-20.

Okubo, K., Hori, N., Matoba, R., Niiyama, T., Fukushima, A., Kojima, Y., Matsubara, K. 1992. Large scale cDNA sequencing for analysis of quantitative and qualitative aspects of gene expression, *Nat Genet*, 2, 173-9.

Osakabe, Y., Miyata, S., Urao, T., Seki, M., Shinozaki, K., Yamaguchi-Shinozaki, K. 2002. Overexpression of *Arabidopsis* response regulators, ARR4/ATRR1/IBC7 and ARR8/ATRR3, alters cytokinin responses differentially in the shoot and in callus formation, *Biochem Biophys Res Commun*, 293, 806-15.

Oztur, Z.N., Talame, V., Deyholos, M., Michalowski, C.B., Galbraith, D.W., Gozukirmizi, N., Tuberosa, R., Bohnert, H.J. 2002. Monitoring large-scale changes in transcript abundance in drought- and salt-stressed barley, *Plant Mol Biol*, 48, 551-73.

Penn, S.G., Rank, D.R., Hanzel, D.K., Barker, D.L. 2000. Mining the human genome using microarrays of open reading frames, *Nat Genet*, 26, 315-8.

Perez-Amador, M.A., Lidder, P., Johnson, M.A., Landgraf, J., Wisman, E., Green, P.J. 2001. New molecular phenotypes in the dst mutants of *Arabidopsis* revealed by DNA microarray analysis, *Plant Cell*, 13, 2703-17.

Petersen, M., Brodersen, P., Naested, H., Andreasson, E., Lindhart, U., Johansen, B., Nielsen, H.B., Lacy, M., Austin, M.J., Parker, J.E. *et al.* 2000. *Arabidopsis* map kinase 4 negatively regulates systemic acquired resistance, *Cell*, 103, 1111-20.

Qin, L., Prins, P., Jones, J.T., Popeijus, H., Smant, G., Bakker, J., Helder, J. 2001. GenEST, a powerful bidirectional link between cDNA sequence data and gene expression profiles generated by cDNA-AFLP. *Nucleic Acids Res*, 29, 1616-22.

Reymond, P. 2001. DNA microarrays and plant defence, *Plant Physiol Biochem*, 39, 313-21.

Reymond, P., Weber, H., Damond, M., Farmer, E.E. 2000. Differential gene expression in response to mechanical wounding and insect feeding in *Arabidopsis*, *Plant Cell*, 12, 707-20.

Richmond, T., Somerville, S. 2000. Chasing the dream: plant EST microarrays, *Curr Opin Plant Biol*, 3, 108-16.

Ronning, C.M., Stegalkina, S.S., Ascenzi, R.A., Bougri, O., Hart, A.L., Utterbach, T.R., Vanaken, S.E., Riedmuller, S.B., White, J.A., Cho, J, Pertea, G.M., Lee, Y., Karamycheva, S., Sultana, R., Tsai, J., Quackenbush, J., Griffiths, H.M., Restrepo, S., Smart, C.D., Fry, W.E., Van Der Hoeven, R., Tanksley, S., Zang, P., Jin, H., Yamamoto, M.L., Baker, B.J., Buell, C.R. 2003. Comparative analyses of potato expressed sequence tag libraries. *Plant Physiol*, 131, 419-429.

Ruan, Y., Gilmore, J., Conner, T. 1998. Towards *Arabidopsis* genome analysis: monitoring expression profiles of 1400 genes using cDNA microarrays, *Plant J*, 15, 821-33.

Rudemo, M., Lobovkina, T., Mostad, P., Scheidl, S., Nilsson, S., Lindahl, P. 2002. Variance models for microarray data. http://www.math.chalmers.se /~rudemo.

Ruuska, S.A., Girke, T., Benning, C., Ohlrogge, J.B. 2002. Contrapuntal Networks of Gene Expression during *Arabidopsis* Seed Filling, *Plant Cell*, 14, 1191-206.

Saiki, R.K., Walsh, P.S., Levenson, C.H., Erlich, H.A. 1989. Genetic analysis of amplified DNA with immobilized sequence-specific oligonucleotide probes. *Proc. Natl. Acad. Sci.* USA, 86, 6230–6234.

Schaffer, R., Landgraf, J., Accerbi, M., Simon, V.V., Larson, M., Wisman, E. 2001. Microarray Analysis of Diurnal and Circadian-Regulated Genes in *Arabidopsis, Plant Cell,* 13, 113-23.

Scheideler, M., Schlaich, N.L., Fellenberg, K., Beissbarth, T., Hauser, N.C., Vingron, M., Slusarenko, A.J., Hoheisel, J.D. 2002. Monitoring the switch from housekeeping to pathogen defense metabolism in *Arabidopsis thaliana* using cDNA arrays, *J Biol Chem,* 277, 10555-61.

Schena, M., Shalon, D., Davis, R.W., Brown, P.O. 1995. Quantitative monitoring of gene expression patterns with a complementary DNA microarray, *Science,* 270, 467-70.

Schenk, P.M., Kazan, K., Wilson, I., Anderson, J.P., Richmond, T., Somerville, S.C., Manners, J.M. 2000. Coordinated plant defense responses in *Arabidopsis* revealed by microarray analysis, *Proc. Natl. Acad. Sci.* USA, 97, 11655-60.

Shimkets, R.A., Lowe, D.G., Tai, J.T., Sehl, P., Jin, H., Yang, R., Predki, P.F., Rothberg, B.E., Murtha, M.T., Roth, M.E., Shenoy, S.G., Windemuth, A., Simpson, J.W., Simons, J.F., Daley, M.P., Gold, S.A., McKenna, M.P., Hillan, K., Went, G.T., Rothberg, J.M. 1999. Gene expression analysis by transcript profiling coupled to a gene database query. *Nat Biotechnol,* 17, 798-803.

Seki, M., Narusaka, M., Abe, H., Kasuga, M., Yamaguchi-Shinozaki, K., Carninci, P., Hayashizaki, Y., Shinozaki, K. 2001. Monitoring the Expression Pattern of 1300 *Arabidopsis* Genes under Drought and Cold Stresses by Using a Full-Length cDNA Microarray, *Plant Cell,* 13, 61-72.

Skinner, D.Z., Okubara, P.A., Baek, K.H., Call, D.R., 2005. Long oligonucleotide microarrays in wheat: evaluation of hybridization signal amplification and an oligonucleotide-design computer. *Funct Integrat. Gen.,* 5, 70-79.

Stintzi, A., Weber, H., Reymond, P., Browse, J., Farmer, E.E. 2001. Plant defense in the absence of jasmonic acid: the role of cyclopentenones, *Proc. Natl. Acad. Sci.* USA, 98, 12837-42.

Sutcliffe, J.G., Foye, P.E., Erlander, M.G., Hilbush, B.S., Bodzin, L.J., Durham, J.T., Hasel, K.W. 2000. TOGA: an automated parsing technology for analyzing expression of nearly all genes, *Proc. Natl. Acad. Sci.* USA, 97, 1976-81.

Swidzinski, J.A., Sweetlove, L.J., Leaver, C.J. 2002. A custom microarray analysis of gene expression during programmed cell death in *Arabidopsis thaliana, Plant J,* 30, 431-46.

Thimm, O., Essigmann, B., Kloska, S., Altmann, T., Buckhout, T.J. 2001. Response of *Arabidopsis* to iron deficiency stress as revealed by microarray analysis, *Plant Physiol,* 127, 1030-43.

Tian, Q., Uhlir, N.J., Reed, J.W. 2002. *Arabidopsis* SHY2/IAA3 inhibits auxin-regulated gene expression, *Plant Cell,* 14, 301-19.

Tusher, V.G., Tibshirani, R., Chu, G. 2001. Significance analysis of microarrays applied to the ionizing radiation response, *Proc. Natl. Acad. Sci.* USA, 98, 5116-21.

Tsangaris, G., Botsonis, A., Politis, I., Tzortzatou-Stathopoulou, F. 2002. Evaluation of cadmium-induced transcriptome alterations by three color cDNA labeling microarray analysis on a T-cell line, *Toxicology,* 178, 135-60.

Velculescu, V.E., Zhang, L., Vogelstein, B., Kinzler, K.W. 1995. Serial analysis of gene expression, *Science,* 270, 484-7.

Voiblet, C., Duplessis, S., Encelot, N., Martin, F. 2001. Identification of symbiosis-regulated genes in *Eucalyptus globulus-Pisolithus tinctorius* ectomycorrhiza by differential hybridization of arrayed cDNAs, *Plant J,* 25, 181-91.

Wang, R., Guegler, K., LaBrie, S.T., Crawford, N.M. 2000. Genomic analysis of a nutrient response in *Arabidopsis* reveals diverse expression patterns and novel metabolic and potential regulatory genes induced by nitrate, *Plant Cell,* 12, 1491-509.

Wang, Y.H., Garvin, D.F., Kochian, L.V. 2001. Nitrate-induced genes in tomato roots. Array analysis reveals novel genes that may play a role in nitrogen nutrition, *Plant Physiol,* 127, 345-59.

Water, D.L., Holton, T.A., Ablett, E.M., Lee, S., Henry, R.J. 2005. cDNA microarrays analysis of developing grape (Vitis vinifera cv. Shiraz) berry skin, *Funct. Intreg. Genomics,* 5, 40-58.

Whetten, R., Sun, Y.H., Zhang, Y., Sederoff, R. 2001. Functional genomics and cell wall biosynthesis in loblolly pine, *Plant Mol Biol,* 47, 275-91.

Wisman, E. and Ohlrogge, J. 2000. *Arabidopsis* microarray service facilities, *Plant Physiol,* 124, 1468-71.

Wu, K., Tian, L., Hollingworth, J., Brown, D.C., Miki, B. 2002. Functional analysis of tomato Pti4 in *Arabidopsis, Plant Physiol,* 128, 30-7.

Wyrick, J.J., Young, R.A. 2002. Deciphering gene expression regulatory networks, *Curr Opin Genet Dev,* 12, 130-6.

Xu, W., Bak, S., Decker, A., Paquette, S.M., Feyereisen, R., Galbraith, D.W. 2001 Microarray-based analysis of gene expression in very large gene families: the cytochrome P450 gene superfamily of *Arabidopsis thaliana, Gene,* 272, 61-74.

Yang, G.P., Ross, D.T., Kuang, W.W., Brown, P.O., Weigel, R.J. 1999. Combining SSH and cDNA microarrays for rapid identification of differentially expressed genes, *Nucleic Acids Res,* 27, 1517-23.

Zhu, T., Budworth, P., Han, B., Brown, D., Chang, H., Zou, G,. Wang, X. 2001. Toward elucidating the global gene expression patterns of developing *Arabidopsis*: parallel analysis of 8, 300 genes by a high-density oligonucleotide probe array. *Plant Physiol Biochem,* 39, 221-242.

11

CHAPTER

Statistical Design and the Analysis of Gene Expression Determined by Independent Component Analysis

A. Riva, A.S. Carpentier, B. Torrésani, A. Hénaut

THE DATA ANALYSIS

Expression data experiments create a large amount of quantitative information and the biologist is not necessarily used to, or trained for, such a situation. In addition, the number of methods proposed for their analysis is enormous and is increasing, still. However, all these methods can be grouped into a few, fundamental types. The first part of this chapter aims to explain their nature and the principles they are based on, whilst the second part considers the important aspects of the experimental design, based on the analytical methods discussed.

Note that we talk almost exclusively about the transcriptome. However, all that which is said can also be applied to the proteome and the metabolome, *mutatis mutandis*.

The Data Used

Throughout this chapter, we will refer to a set of experimental data, which stem from experiments on the sulphur metabolism of *Bacillus subtilis* (Sekowska et al., 2001). These data are freely available at http://195.221.65.10:1234/~carpenti

The aim of these experiments was to identify the genes differentially expressed when the bacteria are grown with methionine ("met") or methylthioribose ("mtr") as a sulphur source. The experiments followed a fully crossed factorial design with four factors (sulphur source, day of experiment, type of protocol (here: amount of RNA used) and duplicate of each spot). The data (raw levels of gene expression) were gathered in an array of 4107 rows (all *B. subtilis* genes) and 16 columns (experimental conditions). The minimum value was 213 and the maximum value 13,455, with two thirds of the data having a value below 800. Note that each factor has only two states: all factors are binary (Fig. 11.1).

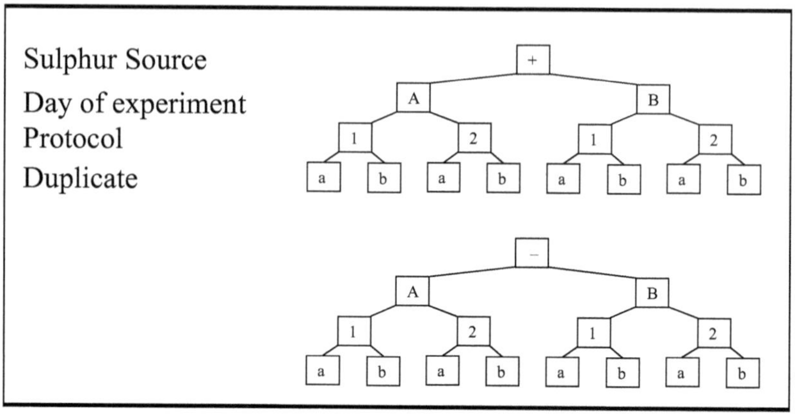

Fig. 11.1 Experimental design of the transcriptome analysis on *Bacillus subtilis* (Sekowska *et al.*, 2001).

The experimental set up follows a fully crossed factorial design. In the case of Sekowska *et al.* (2001) the quantity of RNA used for the RT-PCR differed between the two protocols. Note that changing the protocol (a different quantity of RNA or labelling with Cy3 rather than Cy5) or having duplicats for each gene on the array, are all technical factors which increase the workload without adding any biologically pertinent information. It is preferable to increase the number of states for the biological factors, in the above case an additional sulphur source or an additional experimental day. 1 NB: +/– met/metr; A/B day A/day B; 1/2 1 μg/10μg RNA, a/b spot a/spot b

The Data Table and Some Preliminary Considerations and Manipulations

The reader may be surprised to find that we do not talk about taking the ratio, a rather popular pre-processing step. We refer the reader to the excellent work of Kerr and Churchill (2001), who discuss the issue in a very clear and lucid manner.

The raw data

Surprising as it may seem, you may find yourself in the situation of not having access to the "real" or truly "raw" data. You are given a set of data, where the machine has replaced all values below a certain threshold with one or very few arbitrary values. In other words, it has subtracted the background noise for you.

This operation, a translation, does not in itself pose a problem, but will complicate things when analyzing your data. First, your data will contain a large number of very small or zero values and this is a problem, if you decide to take the log, something frequently done and described below.

Second, the data distribution will be far from being Gaussian (or just unimodal), a fundamental pre-requisite for the analysis of the microarray data. The only solution to this dilemma is to try to "restore" the Gaussian distribution by replacing the smallest values with random values (Chiappetta *et al.*, 2004).

As for the usefulness of eliminating the background noise in general, please see below.

Missing values

These pose a problem, as many data analysis methods require full sets of data. The missing values have generally two possible origins: (i) the microarray contains a defect resulting in the impossibility of taking a reading or (ii) the machine eliminates the measurement as the value is very close to the noise level (in this case it would be advisable to change the setup of the machine). The most radical solution is to eliminate the genes with missing entries, which is obviously far from ideal. A more moderate solution is to fill in the gaps with estimate values. The easiest is to use the row average; the two most common methods are:-

(a) Determining which gene the "missing gene" associates with in the other experimental conditions, i.e. determining the "neighbours" of the gene. Then, by presuming that in the missing experiment the genes are still associated, it is possible to fill in the gap with the median value (a method known under "K-nearest neighbours").

(b) Using variations around the Principle Component Analysis (PCA, described below). Examples here are the Singular Value Decomposition and Bayesian Principal Component Analysis (Oba *et al.*, 2003).

The interested reader is referred to the works of Zhou *et al.* (2003), Kim *et al.* (2004), Ouyang *et al.* (2004) for a comparison of some currently used estimation methods.

The correction of the background noise on the membranes, glass plates or silicon chips

Researchers often proceed to the correction of the background noise, before starting with the analysis of their data. A lot of effort is devoted to this, but recent studies suggest that the problem tends to be given too much importance and that the time devoted is not justified (Chen *et al.*, 2004). There is also a second aspect to be considered, namely that finding a reliable method to correct for background noise is not easy. As Lawrence *et al.* (2004) point out, the human component, difficult to quantify and correct for, plays an important role. Also, the assumption that the background level is consistent between the DNA spot and the surrounding space, frequently used for background quantification, is not correct (Konishi, 2004). Using "designated" household genes for the background determination is in itself a good idea, but finding out which the household genes are, is posing problems (Stoyanova *et al.*, 2004).

Regarding Affymetrix's GeneChips, the common practice of subtracting the mismatch (MM) probe intensities from the perfect match (PM) ones is "unjustifiable", according to Sasik *et al.* (2002), as the target sequence hybridizes not only with the PM but also with the MM probe.

A useful way to assess the utility of an anti-noise measure taken, is to check on the change of the *eigenvalue* of the first axis in a PCA (should increase) or the F value in an ANOVA (should increase), as discussed below.

N-dimensional graphs, translation and normalization

It is natural to want to plot the data in a **graph**. Each gene (*at least normally you talk about genes rather than expressed sequence tags, ESTs*) has as many coordinates as there are experimental conditions. If we have **N** experimental conditions, each gene will have N coordinates. To represent the data graphically, we will have a reference frame with N axes, each axis corresponding to one experimental condition. We will obtain one (and only one!) graph, with N dimensions, with as many points as there are genes. These points form a "cloud". However, as we are not good at coping with drawings having more than two dimensions, although three still works well on a computer screen, we are obliged to examine this cloud in little "portions" of two dimensions at a time. In this way we look at one experimental condition versus another. In other words, we choose two experimental conditions and look at the projection of the cloud on this particular plane.

Note that when you draw a graph by hand, you will automatically try to maximize the use of the paper, you look at the minimum and maximum values for both variables, and define the scale accordingly. The machine will do the same. In both cases, the data are transformed through a change of variable, 1 cm on the graph corresponds to X units of the original variable (a linear transformation).

Translation: This is an operation which in itself does not pose a problem, as one is interested in the relative position of the points to each other. The aim is to find the points that are far away relative to the main body of the cloud, which means that the reference frame used to look at the cloud does not really have much importance. However, the translation may create complications when it consists of bringing a lot of the values close to zero followed by taking the log of the data, something discussed in the next section.

Normalization: Note that drawing a graph or letting a spreadsheet (like MS-Excel) draw the graph means that one allows the data to be normalized. One has presumed that the total of the signal in each column is the same, thus total signal of column 1 = total signal of column 2. This is justified when three conditions are fulfilled: first, more than 90% of the genes do not change expression as a function of the different experimental conditions; in this case one can indeed presume that the total quantity of cDNA (and therefore of the mRNA) is the same. Second, the number of genes analyzed has to be large; this is a way to make sure that the majority of the genes does not change expression as a function of the different experimental conditions[1]. Third, the overall intensity change of up- and down-regulated genes is similar. The three conditions are fulfilled in our example, but they would not be in, say, the temporal analysis of mRNA decay. The reader is referred to the work of Stoyanova *et al.* (2004) for some interesting considerations on this subject, as well as to the work of Zhao *et al.* (2005), who propose a normalization procedure for data not fulfilling the above conditions.

Instead of just looking at the minimal and maximal values in order to best represent the graph, it is advisable to calculate the means and variance for each experimental condition. In the first case the estimates are based on two points only (minimum and maximum) per experimental condition, in the second case the estimate is made using all points. If these are numerous, the result is more stable.

[1]This maxim should be kept in mind when tempted to work with "specialized" microarrays.

Graphic Exploration

Preliminary considerations

As discussed previously, we are forced to take the columns 2 by 2, which means that we will look at PROJECTIONS OF OUR SINGLE CLOUD ON THE DIFFERENT PLANES.

What are we looking for? Presuming that the three above-mentioned conditions are fulfilled, at least 90% of the genes analyzed will not change expression under the different experimental conditions. This means that on the graph one would see them all lying on one line, if it was not for noise. The noise is responsible for making those points look more like a cigar, which is wider, the more noise there is. The remaining 10% of the genes will change expression, they have an atypical behaviour and will not lie on the line (the cigar) but will be apart. The genes that are apart from the main body of the cloud are the ones the biologist is interested in. Note that having the 90% of the points present on a line is an ideal case, the "cigar' being the reality; so one tries to find that line, which describes 90% of the genes, somehow.

How do we describe those 90% of the data? How do we determine the line? Various options are available:

(a) One can try to draw it by hand.
(b) Calculate the linear regression. This is not such a good idea as there are two lines of regression (x-axis versus y-axis and *vice versa*) and they are not identical, except when all the points lie on the same line.
(c) Use methods that are more sophisticated.

The methods all presume that the cloud follows a Gaussian distribution, or at least an unimodal and symmetrical one. They also need some preprocessing of the data, for two reasons:

(a) The fact that the data often consist of a very large amount of small values and a few, extreme points, something which affects most data analysis techniques strongly (Chiappetta *et al.*, 2004).
(b) Some effects being studied may have a multiplicative behaviour.

To solve the first of these problems, taking the log, the square (or cubic or fifth etc.) root or the hyperbolic tangent are all possible and generally accepted methods (Fig. 11.2), whilst for the second problem taking the log is preferable (Tusher *et al.*, 2001; Hoyle *et al.*, 2002; Chiappetta *et al.*, 2004; Thygesen *et al.*, 2004).

As we mentioned in the section above, the reference frame used to look at the cloud does not really matter and making a simple translation does not in itself pose a problem. One does need to be careful though: making a simple translation is indeed not a problem, nor is taking the log. However, when executing both operations, one needs to be cautious if the translation consists in bringing a lot of the values close to zero. Taking the log afterwards will create a distortion in the cloud of points, as a whole package of data will have been created with values going towards minus infinity. This means that in trying to take care of the problem of the points at the far right (few points with very large values) by taking the log, the result is worse than the starting point. Note that when executing the two operations in the inverse order (first log, then translation) the problem is not created.

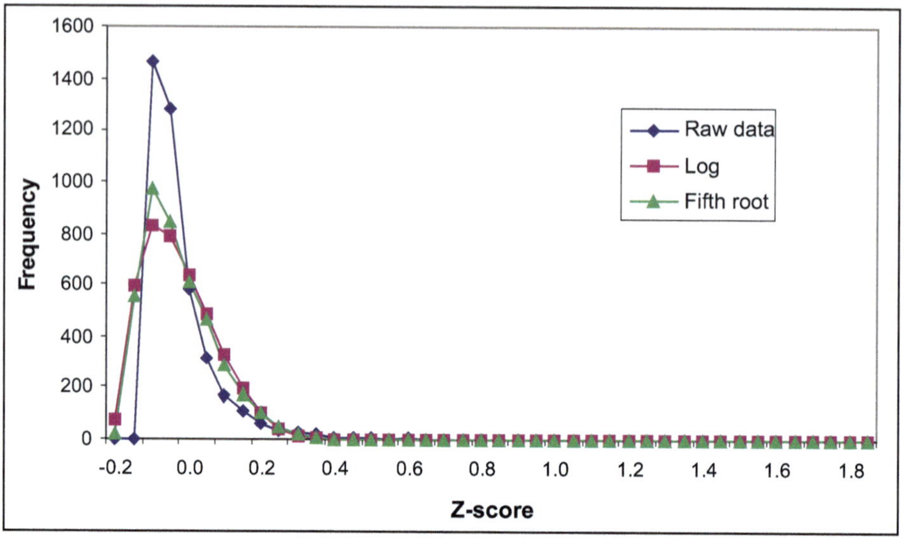

Fig. 11.2 Effect of different pre-processing methods on the data distribution.

The figure shows the effect that different pre-processing methods have on the data distribution. Shown are the distributions a) of the raw data, b) after having taken the log and c) after having taken the fifth root. As can be seen, either operation brings the distribution closer to a Gaussian one.

Some general considerations about the graphs. We come back to the graphs, which are just many projections on different planes of ONE cloud. A brief look at the general shape of each cloud projection is worthwhile. If a cloud resembles a fat cigar, a lot of genes have considerably changed expression. If, on the other hand, the cloud resembles a line, the great majority has not changed expression (Fig. 11.3 for two examples). The "cigar" may also be bent or twisted, which means that the readings were taken outside the linear range of the machine. The first and obvious recommendation is to make sure that at the moment of taking the readings, the scanning settings are correct, which they are often not (Stoyanova *et al.*, 2004). The second is to check that one is not just working at one extreme of the linear range. If this is the case, a change of concentration in the hybridization solution is a good option. If the entire linear range is taken up, two solutions can be proposed: either using two different voltage settings for the photomultiplier, or using different exposure times, when working with radioactively labelled samples. Algorithms for subsequently combining the different readings are readily available (Lyng *et al.*, 2004; Querec *et al.*, 2004). The article by Lyng *et al.* (2004) shows the relationship between the type of incorrect setting and the resulting cloud shape.

Numerous authors have proposed "remedies", if the above suggestions prove impossible to follow, but none will give back the "perfect" data, you would have had if the experiments had been executed correctly.

We can be faced with a problem, by taking the columns 2 by 2, the number or graphs increases very rapidly when increasing the number of experimental conditions. In our example we have 16 columns which means we need to look at 16 x 15/2 i.e. 120 graphs. Evaluating them all in detail, becomes a little tedious.

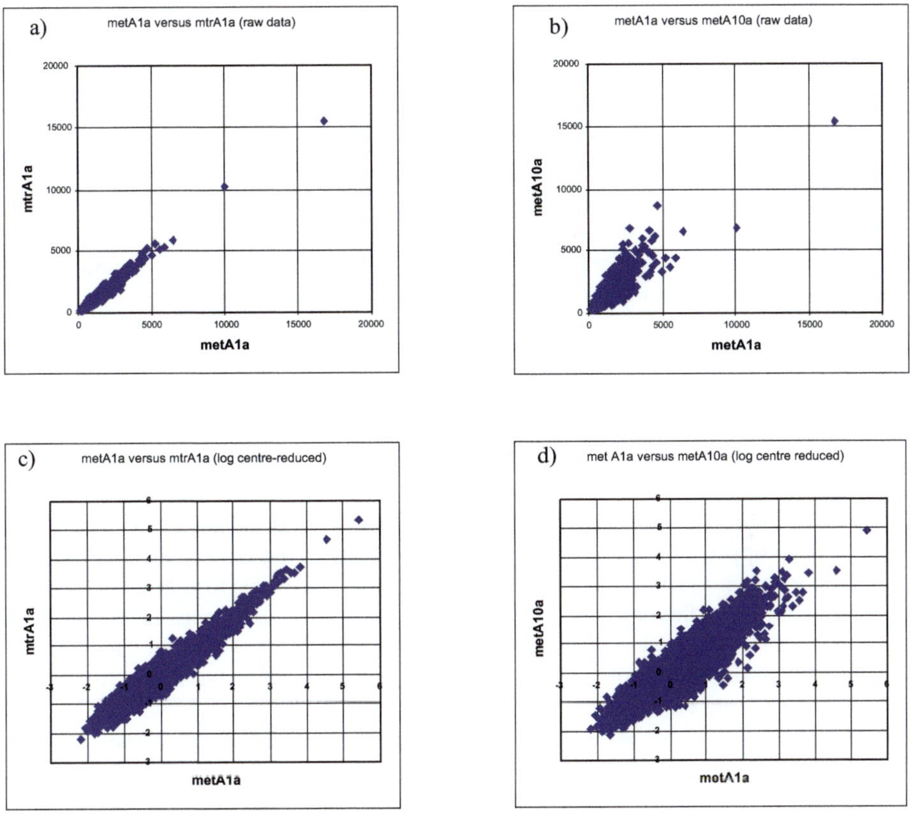

Fig. 11.3 Projections of the data on different planes.

In all four figures, each axis corresponds to an experimental condition. Figures 3a and c show metA1a versus mtrA1a, Fig. 3b and d show metA1a versus metA10a (Figure 11.1 for the nomenclature). Figures a and b show projections of the raw data, in Fig. c and d the data are log centre-reduced. Log centre-reducing the data has brought the few points which are far away from the main body in Fig. a and b closer in Fig. c and d. Note how the space is more efficiently used in Fig. c and d. The points in the two left hand pictures form a narrower "cigar", indicating that fewer genes have changed expression than on the right hand side.

Thus, we need to find ways to reduce the number of graphs we have to examine. To do this, we need to decide from which point of view we want to look at the cloud, which has to be translated into a mathematical criterion. This implies that there will be a change (rotation) of the reference frame.

By hand (with a spreadsheet)

With "by hand", we refer to the fact that the calculations are extremely simple. As the calculations have to be repeated for each gene, the number of calculations is such as to make handing the job over to a spreadsheet, a practical alternative.

The only reasonable option to reduce the 120 pictures, means concentrating on the expression changes caused by each single factor being studied, in our case four. For this we calculate the mean expression for each gene, this will be the x-value. Then, for a given factor like sulphur, we can calculate the sum of all met values and subtract from it the sum of all mtr values, which gives us the y-value. This calculation needs to be carried out for all four factors. Note that we have changed the reference frame, which may be done instinctively by hand and can be formalized via a matrix, called "mixing matrix". This allows the change from the old reference frame to the new one and is shown in Table 11.1.

We obtain four graphs, one for each factor, we then look for genes that are far away from the main body of the cloud. Figure 11.4 shows the graph obtained for the factor sulphur. Executing this operation, each experimental condition is given the same weight and the criterion chosen to look at the cloud, is "one factor per graph".

Principal component analysis (PCA)

A more sophisticated approach is PCA, first introduced in 1901 by Pearson. The reader may refer to the work by Stoyanova *et al.* (2004) for a comprehensive introduction to the subject and to Kendall *et al.* (1983) for a technical presentation.

Here, the criterion chosen to look at the cloud is to maximize the variances along the axes of the reference frame. There are numerous software packages that can do this job and which supply us with the mixing matrix, which in the case of PCA is called the *eigenvector matrix* as shown in Table 11.1. This matrix allows us to change from the old reference frame to the new one. It gives us for each of the new axes (in the table: the columns), the coefficient with which we have to multiply the value for each gene in a given experimental condition (in the table: the lines), in order to obtain the new coordinates (Table 11.1).

The *eigenvector matrix* also supplies us with other information, the variance or *eigenvalue* for each axis, expressed as a percentage. This provides an indication of the dispersion of the cloud along the axis (the bigger the value, the more the genes are dispersed along this axis). The fundamental idea is that if the dispersion is great, the image is easier to interpret than if all the points were packed together. If an experimental factor influences the expression of some genes, the factor will contribute to the dispersion of the cloud and may coincide with one of the axes determined by PCA.

The *eigenvector matrix* provides a wealth of information. Looking at our matrix, we can see that for the first axis all the sixteen coefficients have basically the same value. This means that for the first axis, all experimental conditions have the same weight, in other words, the first axis gives us the total expression of each gene, in a similar manner to a spreadsheet. This observation is generally true (Stoyanova *et al.*, 2004).

In each of the other columns (axes), the experimental conditions can be grouped together according to the sign of their coefficient (positive or negative). For some axes, this coincides with a separation of the two states of a factor. In our case, axis 2 separates the two protocols well, (1 µg RNA, all values are negative and 10 µg RNA, all values are positive). Axis 3 separates the day (A and B), axis 5 the sulphur source (met and mtr) and axis 7 the two spots (a and b). Other axes, on the other hand correspond to combinations of the experimental conditions, whose interpretation is not evident, axis 4 is one such example. Axis 4 singles out the ribosomal proteins, a biologically speaking coherent result, which is waiting for an interpretation. This is something frequently found when analyzing microarray data.

Table 11.1 The mixing matrix calculated by the spreadsheet (MS-Excel) and the *eigenvector matrix* calculated by PCA.

Experimental condition	Axis: Mean expression	Effect of protocol	Effect of day	Effect of sulphur source	Effect of duplicate
metA1a	0.250	0.250	-0.250	0.250	-0.250
metA1b	0.250	0.250	-0.250	0.250	0.250
metB1a	0.250	0.250	0.250	0.250	-0.250
metB1b	0.250	0.250	0.250	0.250	0.250
metA10a	0.250	-0.250	-0.250	0.250	-0.250
metA10b	0.250	-0.250	-0.250	0.250	0.250
metB10a	0.250	-0.250	0.250	0.250	-0.250
metB10b	0.250	-0.250	0.250	0.250	0.250
mtrA1a	0.250	0.250	-0.250	-0.250	-0.250
mtrA1b	0.250	0.250	-0.250	-0.250	0.250
mtrB1a	0.250	0.250	0.250	-0.250	-0.250
mtrB1b	0.250	0.250	0.250	-0.250	0.250
mtrA10a	0.250	-0.250	-0.250	-0.250	-0.250
mtrA10b	0.250	-0.250	-0.250	-0.250	0.250
mtrB10a	0.250	-0.250	0.250	-0.250	-0.250
mtrB10b	0.250	-0.250	0.250	-0.250	0.250

Experimental condition	Axis: 1	2	3	4	5	6	7	8	9	10	11	12	13	14	15	16
eigenvalue	94.64	2.26	1.01	0.56	0.36	0.29	0.26	0.18	0.16	0.05	0.05	0.05	0.04	0.04	0.03	0.02
metA1a	0.250	0.307	-0.164	-0.114	0.336	-0.004	-0.310	0.179	0.349	-0.146	-0.130	0.406	-0.228	0.429	-0.020	-0.017
metA1b	0.249	0.312	-0.166	-0.131	0.366	0.075	0.157	0.252	0.347	0.131	0.078	-0.429	0.262	-0.413	0.001	0.026
metB1a	0.252	0.179	0.115	0.365	0.124	0.306	-0.301	-0.280	-0.158	0.172	0.629	-0.094	-0.019	0.132	0.052	0.028
metB1b	0.251	0.181	0.107	0.369	0.153	0.425	0.211	-0.246	-0.203	-0.170	-0.592	0.072	0.003	-0.140	-0.036	-0.027
metA10a	0.249	-0.256	-0.326	0.211	0.200	-0.429	-0.183	-0.129	-0.186	0.372	-0.309	-0.254	0.193	0.264	0.082	0.028
metA10b	0.249	-0.255	-0.328	0.192	0.260	-0.303	0.315	-0.027	-0.146	-0.388	0.325	0.261	-0.198	-0.276	-0.081	-0.031
metB10a	0.250	-0.254	0.263	-0.232	0.126	0.086	-0.322	0.311	-0.322	-0.041	-0.070	-0.072	-0.146	-0.163	-0.330	0.505
metB10b	0.250	-0.253	0.259	-0.236	0.178	0.180	0.170	0.358	-0.283	0.064	0.067	0.092	0.171	0.180	0.335	-0.508
mtrA1a	0.248	0.334	-0.083	-0.365	-0.224	-0.182	-0.173	-0.256	-0.270	0.108	0.003	0.432	0.363	-0.310	0.093	0.018
mtrA1b	0.248	0.331	-0.087	-0.370	-0.189	-0.093	0.321	-0.159	-0.277	-0.089	0.029	-0.407	-0.399	0.305	-0.086	-0.028
mtrB1a	0.251	0.156	0.220	0.311	-0.360	-0.291	-0.164	0.272	0.049	-0.540	0.002	-0.211	0.315	0.102	0.061	-0.029
mtrB1b	0.252	0.149	0.214	0.322	-0.290	-0.206	0.256	0.332	0.099	0.540	-0.012	0.236	-0.301	-0.101	-0.066	0.029
mtrA10a	0.249	-0.239	-0.346	-0.031	-0.376	0.270	-0.285	0.035	0.151	-0.042	-0.089	-0.114	-0.350	-0.261	0.475	-0.045
mtrA10b	0.249	-0.240	-0.352	-0.030	-0.342	0.369	0.195	0.052	0.150	0.053	0.087	0.108	0.352	0.262	-0.471	0.045
mtrB10a	0.251	-0.226	0.335	-0.131	-0.003	-0.151	-0.196	-0.375	0.341	0.014	-0.054	-0.094	-0.118	-0.174	-0.385	-0.484
mtrB10b	0.251	-0.225	0.327	-0.140	0.041	-0.057	0.310	-0.322	0.358	-0.037	0.036	0.067	0.101	0.164	0.376	0.491

The mixing matrix at the top was calculated by the spreadsheet, the mixing matrix (or *eigenvector matrix*) at the bottom by PCA. The arrows indicate the columns which separate well the effects of the same factors. The matrices allow us to change from the old to the new reference frame: they give us for each of the new axes (the columns) the coefficient with which we have to multiply the value for each gene in a given experimental condition (the lines) in order to obtain the new coordinates. The first line in the *eigenvector matrix* contains the *eigenvalue* for each axis (in %), providing an indication of the dispersion of the cloud along that axis. Note that for the first axis all the sixteen coefficients have basically the same value; this means that for the first axis, all experimental conditions have the same weight, in other words, the first axis gives us the total expression of each gene, which is generally true (Stoyanova *et al.*, 2004). **An example** for the calculation of the new coordinates with the *eigenvector matrix*: in the original (or "old") reference frame, the gene *galK* has the coordinates (5.431; 5.432; 5.092; 5.068; 4.893; 4.744; 3.763; 3.661; 5.333; 5.265; 5.329; 5.249; 4.607; 4.444; 3.806; 3.737). To obtain coordinate of *galK* on the new axis 1, the calculations are as follows: (5.431 x 0.250) + (5.432 x 0.249) + ... + (3.737 x 0.251) = 19.0. The other coordinates are obtained accordingly.

The *eigenvector matrix* deserves a little more attention, as the values it contains can be looked at from different points of view. If we take up our example, each line represents an experimental condition and the values in the 16 columns for a given line give us the position of that particular experimental condition in the 16-dimensional space. To be precise, each value has to be multiplied with the root of the variance of that axis, in order to obtain the coordinate. This means that instead of looking at the *eigenvector matrix*, we can look at the different projections of the experimental conditions in order to figure out which axes separate well the different states of our factors. Once we have established which planes deserve to be examined in detail, we come back to the projections of the cloud on these planes and pinpoint those genes, which are far away from the main body of the cloud. Figure 11.5 shows the cloud projection on the plane formed by axis 1 versus axis 5.

Note: the normalization of the data is an integral part of PCA. To summarize, with PCA the experimental conditions are not given the same weight (contrary to a spreadsheet) and the criterion chosen to look at the cloud, is to maximize the variances along the axes.

Independent component analysis (ICA)

"ICA tries to find a linear representation of non-Gaussian data so that the components (or factors, or sources) are statistically independent, or as independent as possible" (Hyvärinen and Oja, 2000).

This search for statistical independence is generally very difficult and therefore an approximation is made: One looks for the directions that maximize the criterion of non-Gaussian distribution. As "non-Gaussian" is a "non-property", numerous possibilities exist for defining such a distribution. One criterion that seems to work quite well is to look for distributions with a positive kurtosis (distributions with "heavy tails"). ICA can be seen as a close relative of PCA. Whilst PCA looks at which directions maximize the variance, ICA approaches the question of finding genes with an "atypical behaviour" more directly, by defining "atypical" as "following a non-Gaussian distribution". The new reference frame will maximise the criterion of "non-Gaussianity". With this criterion, one can increase the weight of points that have only small deviations from the main body of the cloud and thus allows them to be detected as potentially interesting.

A latent difficulty with ICA is that there is no analytical solution (contrary to PCA), we have to look for numerical solutions. There is the danger that the algorithm finds a direction with a solution, but that this direction is not the best solution in absolute terms, thus the algorithm can get stuck with a local maximum (Chiappetta *et al.*, 2004). Launching ICA a large number of times, typically100, circumvents this problem and only those directions or solutions that have been most frequently found are kept. As with PCA we have a mixing matrix that allows us to change from the old reference frame to the new one. Again, the different experimental conditions do not have the same weight, although the weight attributed, varies from PCA. Once we have determined the axes, the procedure is the same as with PCA. Figure 11.6 shows the cloud projection on the plane that separates well the sulphur sources.

The applications of ICA in microarray analysis include the identification of groups of genes implicated in cancer, the study of the cell cycle (Liebermeister, 2002; Martoglio,

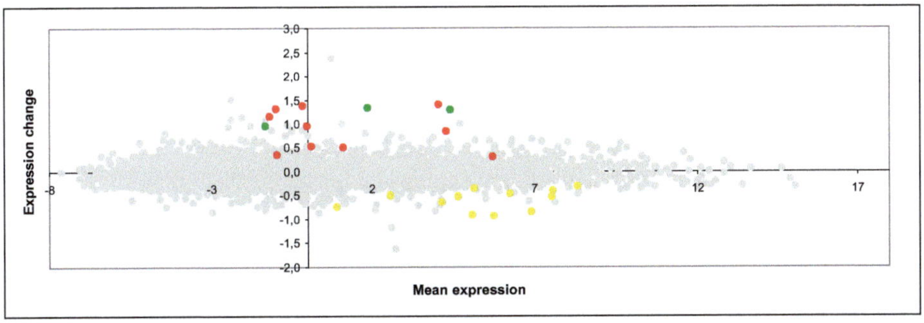

Fig. 11.4 The expression change in function of the factor sulphur as calculated by a spreadsheet (MS-Excel)

The figure shows the change of expression of the genes as a function of the sulphur source, against their mean expression. The potentially interesting genes are those away from the main body of the cloud. The highlighted genes are the ones which proved to be of particular interest for the problem investigated by Sekowska *et al.* (2001); the reader may refer to their work for a detailed discussion. Note that not all of these genes would have been detected using the spreadsheet.

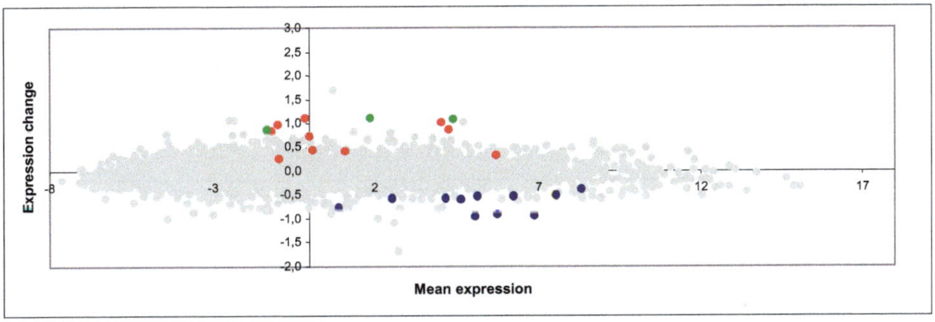

Fig. 11.5 The data cloud projected on the plane formed by axis 1 against axis 5 (PCA)

The figure shows the change of expression of the genes as a function of the sulphur source (axis 5), against their mean expression (axis 1), as calculated by PCA. The potentially interesting genes are those away from the main body of the cloud. The highlighted genes are the ones which proved to be of particular interest for the problem investigated by Sekowska *et al.* (2001); the reader may refer to their work for a detailed discussion. Note that not all of these genes would have been detected using PCA.

2002), the identification of genes that are potentially co-regulated (Chiappetta *et al.*, 2004), as well as metabolome studies (Scholz *et al.*, 2004). Carpentier et al. (2004) and Chiappetta et al. (2004) have applied both PCA and ICA to the sulphur metabolism data and have shown that the two methods perform similarly well, with ICA slightly outperforming PCA.

A brief remark

We have said that whilst a spreadsheet attributes to each experimental condition the same weight, PCA and ICA do not. Figure 11.7 shows a comparison between the three methods, from this point of view. The spreadsheet simply calculates the mean

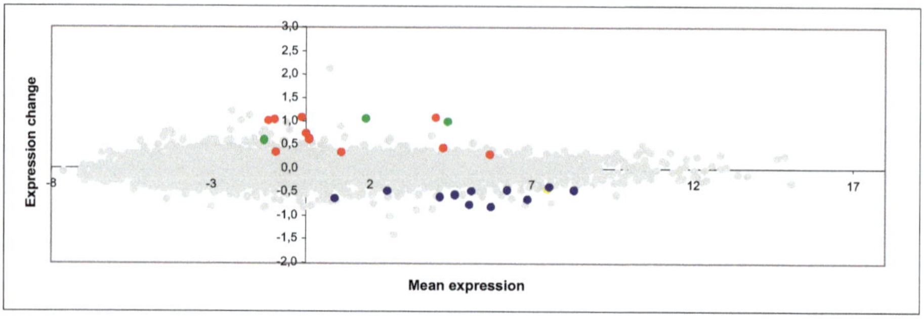

Fig. 11.6 The data cloud projected on the plane that separates well the sulphur source (ICA).

The figure shows the change of expression of the genes as a function of the sulphur source, as determined by ICA. The potentially interesting genes are those away from the main body of the cloud. The highlighted genes are the ones which proved to be of particular interest for the problem investigated by Sekowska *et al.* (2001); the reader may refer to their work for a detailed discussion. Note that not all of these genes would have been detected using ICA.

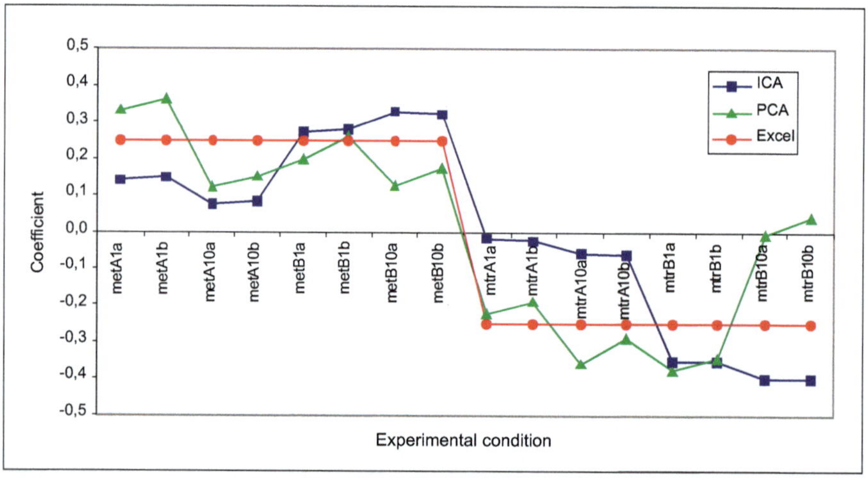

Fig. 11.7 The "weight" attributed to each experimental condition by a spreadsheet (MS-Excel), PCA and ICA.

The figure shows that a spreadsheet (MS-Excel) attributes the same "weight", or importance, to each experimental condition, whilst PCA and ICA do not.

expression. This choice is not optimal when certain experimental conditions contain more information than others. Ideally, the weight attributed should be proportional to the information contained. PCA is a good choice when the signal follows a Gaussian distribution, whilst ICA imposes itself when the distribution is non-Gaussian.

You might wonder what happens if you use, say, PCA on data that follow a non-Gaussian distribution. The answer is that you are likely to miss out on potentially interesting genes. You do not, however, risk finding "wrong" genes. Using more than one tool amounts to examining the cloud from different angles; the results obtained with the different tools are complementary.

Statistical Tests

Preliminary considerations

Our experience shows that some confusion reigns regarding the statistical tools in general and their application to microarrays in particular. Hence, this rather long introductory section.

When approaching microarray data from a statistical point of view, people seem to worry a lot about the fact that the data are "relative" and whether they should or should not take ratios.

Microarrays provide us with "relative data". The interesting information regarding a gene is "relative", as one compares the expression of a gene under condition A with that of the same gene under condition B. Microarray technology is quite recent, however, dealing with relative data is not and taking the ratio results in a reduction and a falsification of the information offered (Kerr and Churchill, 2001). It is Fisher who first tackled and solved the problem at the very beginning of the 20th century, resulting in ANOVA. For a more detailed discussion of this issue, the reader is referred to the work of Kerr and Churchill (2001). At about the same time, Gosset ("Student") came up with the *t-test*, as a solution to the problem.

Statistics help us to answer the question whether the expression differences observed are real. The answer is given indirectly, as the statistical tools give us the probability of having a false positive. A false positive is a gene whose expression difference surpasses by chance a threshold value, which has been fixed in advance. "By chance" means that if the experiment were repeated, you would not find again such a large expression change. The statistical analysis is used to evaluate the probable percentage of false positives beyond a given threshold value. 40 genes will surpass by chance the threshold value of 1% if the experiment was carried out on 4,000 genes.

The estimation of the number of false positives is only the first step. Beyond the threshold value we not only find false positives, but also genes whose expression change is "real" (we would find it again if the experiment were repeated). The key information is the proportion of false positives in the total, because it measures the risk of being on the wrong track when deciding to work on one of the genes from this group (Benjamini *et al.*, 1995). One generally chooses the threshold in order to have less than 5% of false positives in the group. Take for example an experiment carried out on 4,000 genes with 80 lying beyond the threshold of 0.1%. As there are on average 4 false positives beyond the 0.1% threshold (4000 x 0.001), the percentage of false positives is 4 / 80, or 5% of the selected genes.

The literature sometimes refers to the Bonferroni correction. This correction is not pertinent for the analysis of microarray data, as it is too restrictive.

The numerical criterion used in the statistical tests is always the ratio between the deviations observed for the factor of interest (the signal) and the deviations due to all the causes one chooses to ignore (the noise). The statistical tests differ from each other in the way they define the noise and the probability function they use to estimate the probability of false positives. In the past, the function used was the Gaussian. Nowadays one tends to employ the probability function, estimated on the data using permutations (Tusher *et al.*, 2001)

ANOVA

ANOVA is a tool that allows us to analyze simultaneously the effect of more than one factor on a variable, in our case the expression level of a gene. The method is based on the calculation of the sum of squares, degrees of freedom, mean square (short for mean square deviation from the mean) and F-statistics[2] (Zar, 1998). As we use ANOVA in a somewhat reductive manner, the reader may refer to the work of Zar (1998) for a full appreciation and pedagogic explanation of the possibilities offered.

Various quantities are used simultaneously in order to decide whether the expression of a gene varies significantly for the factor of interest.

1. V1, the variance for the total of the observations made on the gene,
2. V2, the variance for the observations made for the factor of interest,
3. V3, the variance for the observations made for those factors whose influence one wishes to subtract.

The signal is equal to V2, the noise to V1 – (V2 + V3). The possibility to calculate the term V3 is a particularity of ANOVA and it allows a finer control of the composition of the noise. In our example, V3 corresponds to the expression change caused by the day, the duplicate and the protocol used. The noise encompasses all of that which causes the difference between the actual expression level and the sum of the expression levels of the four factors.

In the case of the sulphur metabolism data, the equation used for each gene is the following:

$$Y_{ijkl} = \mu + S_i + J_j + C_k + D_l + \varepsilon_{ijkl}$$

Where

Y_{ijkl} is the total expression level measured
μ is the mean of the expression levels measured for the gene
S_i, J_j, C_k and D_l are, respectively, the effects of sulphur source i, experiment day j, protocol used k and duplicate l on the expression level
ε_{ijkl} is the residual error.

Note that the residual error ε_{ijkl} encompasses all interactions: between two factors (6), between three factors (4) and between four factors (1). The interactions are grouped together under "error" for the following reason: it is information with which we cannot work, unless we have a very precise idea of the nature of the interaction (linear, sinusoidal or other).

The *F-test* is calculated in the following manner: F = "mean square of the sulphur source"/"mean square of the residual error". We are interested in genes that posses a high *F*-value (*p-value*) for the factor sulphur source. The calculations are carried out for all genes and the results can be represented in a graphical form. The variance of the factor of interest is given on the x-axis, the *p-value* on the y-axis. The *p-value* is used to calculate how many false positives will lie below a chosen threshold value (Fig. 11.8).

Note that we are not interested, whether the expression levels of the identified genes also vary in function of the other factors. One does not preclude the other and has no impact on our analysis.

---

[2]Sometimes referred to as *F-test*

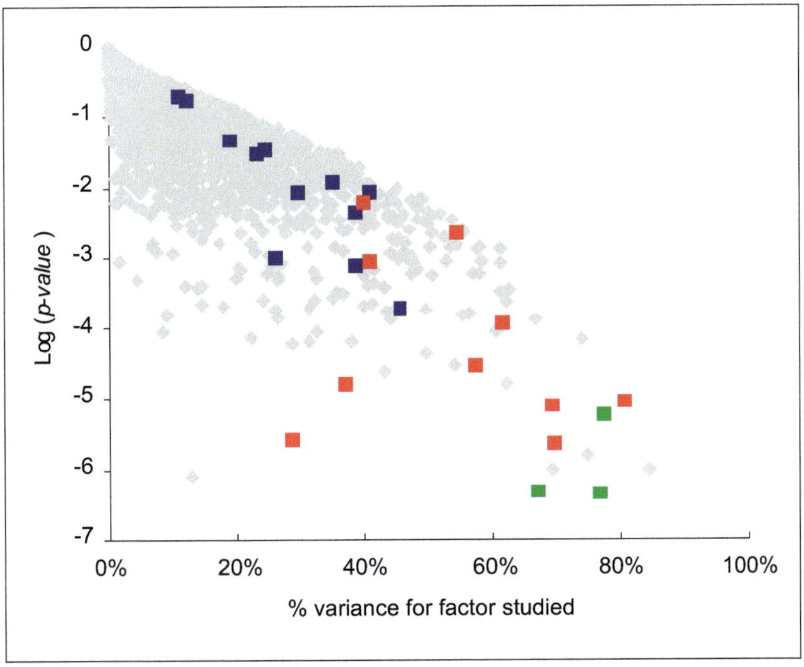

Fig. 1.8 The graphical representation of the results obtained with ANOVA for the factor sulphur.

The potentially interesting genes are those with a small *p-value* and a large variance, genes which are therefore in the bottom right part of the image. They are away from the main body of the cloud. The highlighted genes are the ones which proved to be of particular interest for the problem investigated by Sekowska *et al.* (2001); the reader may refer to their work for a detailed discussion. As can be seen, not all genes of interest would have been identified by the sole use of ANOVA.

ANOVA has some advantages when the experimental factors are not binary; in that case, it basically becomes the only tool which is easy to use.

Paired t-test

We have said that ANOVA quantifies the contribution given by each factor to the total expression of a gene, thus permitting us to isolate the contribution of our factor of interest. The paired *t-test* also allows this, but the approach is different, and we can only use it for binary factors. The paired *t-test* eliminates the influence of all the factors we are not interested in by calculating the difference between pairs of values. The members of each pair differ from each other only with respect to the factor of interest (state 1 vs. state 2), all other experimental conditions being equal.

For example, we can calculate the difference between the value obtained on met with the value obtained on mtr, both obtained on day A, with one microgram mRNA and spot a. Then we calculate the difference of met versus mtr on day B, with one microgram mRNA and spot a and so forth. This is done for each gene and we thus obtain 8 comparisons, or differences per gene. V1 is calculated on these 8 comparisons, the term V3 has disappeared.

However, as the paired *t-test* takes pairs of "similar conditions", systemic biases due to, e.g. "day" or "duplicate", are eliminated, therefore still allowing for a reasonable estimation of the error.

t-test

The *t-test* corresponds to an ANOVA with one factor and is the least favourable option. The *t-test* only considers the expression difference due to one factor, ignoring that there are pairs of measurements which have more or less in common (like the day, protocol and spot), unlike ANOVA and the paired *t-test*. Thus, we cannot separate the contribution made by our factor of interest from the contribution made by the other factors and the interaction between them , thus the expression difference due to our factor risks being drowned by the rest.

In terms of V1, V2 and V3: V1 is calculated on the total of the 16 observations made (as with ANOVA), but as the term V3 has disappeared, the noise risks being much larger.

Conclusion

The biggest difficulty is to estimate the noise with accuracy. The best solution is to repeat the experiment a large number of times. As this is not always possible, statisticians try to improve the estimation of the noise by working on groups of genes having more or less the same level of noise. A considerable amount of literature is dedicated to this effort. Numerous are the solutions proposed, no one is perfect. Generally, the grouping is done *a posteriori*, after a first estimation of the noise for all the genes separately. One speaks in this case of a Bayesian approach. The reader is referred to the work of Neuhäuser and Senske (2004) for an introduction to the subject and to the work of Kutalik *et al.* (2004) for a comparison of some of the methods proposed.

Regarding the three approaches discussed above, Table 11.2 shows the measurements obtained for *ytmJ* and the results obtained from ANOVA, the paired *t-test* and the *t-test*. It shows that although ANOVA and the paired *t-test* both identify the gene as interesting, the *t-test* results are inconclusive. The observations made on this particular example can be generalized. Table 11.3 shows a comparison of the number of genes detected by the three methods. Although ANOVA detects the highest number of genes, the paired *t-test* performs comparably well, whilst the *t-test* lags far behind.

Graphic Exploration and Statistical Tests in Comparison

We have chosen to talk about the typical representatives of the two approaches, it should be stressed that they are not the only ones proposed in the literature. The number of tools is continuously increasing and no one, definitive method has so far emerged, as is exemplified by the web-site maintained by Li, which has a steadily growing collection of articles on microarray data analysis (http://www.nslij-genetics.org/microarray). Conceptually, all these tools are based on one of the methods described above, or they fall into the category "cluster analysis", described below.

Table 11.2 Comparison between ANOVA, the paired *t-test* and the *t-test*, an example

Table 11.2a shows the measurements obtained for *ytmJ*, Tables 11.2b-d the calculations and results obtained with the *t-test*, the paired *t-test* and ANOVA, respectively. ANOVA and the paired *t-test* both identify the gene as potentially interesting, whilst the *t-test* results are inconclusive (see the relative –log (*p-value*)). DF = Degrees of Freedom; SS= Sum of Squares.

a) Measurements obtained for *ytmJ*

	met	**mtr**	**met-mtr**
A1a	1,170	1,520	-0,3494
A1b	1,176	1,580	-0,4048
B1a	0,950	1,566	-0,6158
B1b	0,891	1,541	-0,6496
A10a	1,939	2,049	-0,1096
A10b	1,565	2,048	-0,4827
B10a	0,893	1,523	-0,6296
B10b	1,007	1,485	-0,4772

b) *t-test*

Numerator	-0,465
Denominator	0,313
DF	14
t-test	**-2,971**
-log (*p-value*)	**1,99**

c) paired *t-test*

Numerator	-0,465
Denominator	0,064
DF	7
paired t-test	**-7,290**
-log (*p-value*)	**3,78**

d) ANOVA

	State 1	State 2	SS
Sulphur	9,592	13,311	0,864
Day	13,048	9,855	0,637
RNA	10,394	12,508	0,279
Spot	11,610	11,293	0,006
Residual			0,448
Total			2,235

Factor	SS	DF	Variance	F	-log (*p-value*)
Sulphur	0,864	1	0,864	**21,21**	**3,12**
Day	0,637	1	0,637	15,64	2,65
RNA	0,279	1	0,279	6,86	1,62
Spot	0,006	1	0,006	0,15	0,00
Residual			0,041		
Total			0,149		

Some methods will use the term "distance", whilst others may talk about "correlation". In mathematical terms, it boils down to the same thing, second order statistics, yielding the same type of information. As the methods all differ more or less from each other, it is normal that they do not come up with exactly the same results.

Which method is the best? Carpentier *et al.* (2004) have examined this issue and have developed a protocol that allows the comparison of the different methods, in terms of their reliability. They conclude that each of the methods analyzed gave some information not provided by the others, suggesting once more the advantage of analyzing the data with more than one statistical tool.

ANOVA, one of the methods tested, did not excel on the sulphur metabolism data. However, all factors were binary and ANOVA has the great advantage of being easily applicable in cases where the factors are non-binary. It also has another important property: ANOVA is the only method that forces the experimenter from the beginning to give the experimental set up some thought and thus to plan it carefully. It is therefore a good practice to think of an experimental set up in terms of ANOVA, even if the data are then exploited by another method (Table 11. 3).

Some Alternative Approaches

Instead of working with the raw data –our original spreadsheet–we can also decide to look at the expression similarity of the genes. For this we need to decide what kind of relationship we want to investigate. If we choose to work with second order statistics like the linear correlation or the Spearman's rank correlation, we are sure to find pairs of genes which fulfil this criterion. Genes, which are part of the same multi-enzymatic complex or which are part of the same metabolic pathway are two examples. From a biological point of view, however, these may not be the most interesting pairs to work with. It is therefore worthwhile to consider working with a more general similarity measure like the mutual information (Daub *et al.*, 2004). With this method we will not only find the genes also detected by the second order statistics but also pairs displaying a more complex behaviour, for example pairs of genes where the expression of one is only activated beyond a certain threshold expression of the other.

If you are working with N genes you end up with a table that gives you all the distances between the genes taken two by two. You can decide to represent this graphically (the same way you can express the distances between cities in a table or graphically, with a map). In other words, you can exploit your table with any of the methods described earlier. Alternatively, you can use methods, which will give a more concise representation, like the two examples described below.

Self organizing maps (SOM)

Generally speaking, only the outskirts of the cloud are visually exploitable. The internal organization is hidden by the superposition of thousands of genes on the same image. The analysis would be easier if it were possible to give a faithful representation of the density of the genes in each region of the cloud, with only k points (the magnitude of k is set by the experimenter). A rather naïve solution consists in choosing these k genes at random. This however, is unlikely to give satisfactory results. Calculating the optimal position of the k points is a difficult problem. A number of programmes exists

Table 11.3 Overall comparison between ANOVA, the paired *t-test* and the *t-test*

The table shows a comparison of the number of genes detected by the three methods. In Table 11.3a the threshold for detection was –log (*p*-value) = 3, in Table 11.3b it was equal to 4. Although ANOVA detects in both cases the highest number of genes, the paired *t-test* performs comparably well, whilst the *t-test* lags far behind.

a) Number of genes detected with a threshold of –log (*p-value*) = 3

ANOVA only	62
t-test only	5
Both	24
ANOVA only	35
Paired *t-test* only	30
Both	51
t-test only	12
Paired *t-test* only	64
Both	17
Total ANOVA	86
Total paired *t-test*	81
Total *t-test*	29

b) Number of genes detected with a threshold of –log (*p-value*) = 4

ANOVA only	16
t-test only	0
Both	9
ANOVA only	17
Paired *t-test* only	12
Both	8
t-test only	14
Paired *t-test* only	15
Both	5
Total ANOVA	25
Total paired *t-test*	20
Total *t-test*	9

proposing approximate solutions. An example is SOM (Self Organizing Maps), which chooses the *k* genes and provides a list of the genes close to the *k* genes. The interested reader may refer to the work by Kaski *et al.* (2003) for an introduction to SOM, as well as a comparison of its merits compared to some classic classification methods.

Note that all the programmes proposed necessitate the adjustment of numerous parameters for which you do not necessarily have a rational basis to make your choice. This carries the risk that you only believe those results which tell you something you already know, not the best way to discover new things!

Clustering

The principle is to group and/or to classify the genes as a function of the expression profile obtained under the various experimental conditions. The cloud is thus divided into a number of clusters, the idea being that a cluster corresponds to a functional class. Choosing a gene of unknown function, one can then look to which cluster it belongs and thus draw conclusions about its possible role. This approach poses problems from two points of views: a biological and a technical one.

From a biological point of view: we have to define what a functional class is and how many there are. These are not banal questions, as exemplified by the fact that even for well-studied organisms numerous classifications are proposed (for example SwissProt, MetaCyc, Kegg). Second, the functional classes found in the literature tend to be rather large, containing dozens or hundreds of genes, making them too large to permit their exploitation in the wet laboratory. Third, the clustering methods normally do not allow a gene to be part of more than one cluster, which goes against biological intuition and experience.

From a technical point of view: we have to choose amongst a myriad of (family of) clustering techniques. As the biological question is not clearly defined, we do not have a criterion to select the pertinent and coherent method for our needs[3]. At this point one has to make do with a data-driven attitude. This necessitates a thorough knowledge of the different families of clustering techniques, in order to make the best choice in function of the data set to be analyzed (Somorjai *et al.*, 2003), as all the clustering techniques require many prior decisions (Chiappetta *et al.*, 2004). In addition, as Somorjai *et al.* (2003) point out: "the maxim 'simpler is better' has mostly been ignored".

As clustering methods are frequently used tools (see for example the popular software proposed by Eisen *et al.*, 1998), a range of attempts have been made to circumvent the various technical problems. The reader, who would like to have a critical introduction to different families of clustering techniques, may refer to the works of Datta and Datta (2003), De Smet *et al.* (2002) and Somorjai *et al.* (2003).

Biological pertinence of these approaches

The methods described earlier allow grouping the genes into sets, which form networks. As Schäfer and Strimmer (2005) point out, these are NOT the "true model" for genetic networks. The biological interpretation of the networks obtained with these methods is complex when working with transcriptomic data. The complexity increases when working with proteomic or metabolomic data (Weckwerth, 2003).

HOW TO PLAN AN EXPERIMENT

It is quite usual to find that a rather large number of genes, typically around 10%, change expression considerably between two experimental conditions. This number is too large to be directly exploitable and it is necessary to extract a short and pertinent list of genes to work with. This task is greatly facilitated by an adequate and well thought-through experimental set up.

The following discussion does not include time series experiments, as they are rather idiosyncratic regarding the experimental set up (as well as the data analysis). The reader is referred to the work of Bar-Joseph (2004) who reviews this subject with great clarity.

The Type of Factors

An experiment is made up of three types of factors, each providing specific information.

The first factor corresponds to the phenomenon studied. The study concerns two or more states (two culture conditions, for example, or a certain number of samples taken during a time course experiment). The aim is to narrow down the target genes to a minimum, in other words to have only few genes which change expression considerably between the different experimental states. For this, the experimental states should be as close as possible, for example:

[3]An example is the definition of the distance between clusters. This is not a trivial problem. Take for example the problem of having to define the distance between two countries: do you take the two capitals? The two biggest cities? The shortest distance (0 if the countries are adjoining)?

(a) In the case of the sulphur metabolism experiments, the two sulphur sources were metabolically speaking closely related.
(b) When trying to isolate genes typical of a certain illness, one should study different subtypes, all closely related to the illness of interest

If this maxim is not observed, too many genes will change expression considerably and the identification of key target genes will become near impossible.

The second type of factor serves to verify whether the observations made hold true, if the biological parameters are changed. Do we find the same candidate genes if we work with a different ecotype, or if the experiment is carried out on a different day? Note that even repeating the experiment on a different date introduces a biological variability, as the experimental conditions will never be exactly the same (Sekowska *et al.*, 2001). This verification is extremely important as the most interesting genes are those which come up whatever the biological background. They are most likely the genes at the heart of the phenomenon studied, as their behaviour is not bound to a particular context (genetic or physiological). The reader is referred to the works of Turk *et al.* (2004) and Whitehead and Crawford (2005) who discuss this issue.

The third type of factor is a technical one. Examples are the type of protocol used, e.g. the quantity of mRNA for the RT-PCR in the case of the sulphur data, red rather than the green dye when working with fluorescence-based microarrays, or choosing to work with two spots for each gene on the array. This type of factor increases the workload without adding any biologically pertinent information. The experimental protocols have become highly reproducible and it is advisable to stick to just one protocol (with its systemic biases) and increase the number of states of the two other factors.

The Ideal Situation: A Fully Crossed Factorial Design

The best experimental setup is to follow a fully crossed experimental design (exemplified by ANOVA) as it:

(a) Allows a good exploitation of the information given.
(b) Allows a precise estimation of the error variance, see Fisher (1951) for the original discussion or Mather (1943), Zar (1998) and Kerr *et al.* (2000) for a more user-friendly approach).

Setting up a fully crossed factorial design means that each level (state) of one factor is found in combination with each level of the other factors, as shown in Fig. 11.1. Note that carrying out twice the experiment on strain 1 on day A and twice the experiment on strain 2 on day B would not be adequate, as it would be impossible to separate the effect of the day from the effect of the strain.

The Reality

A fully crossed factorial design may not be possible. The reason is often the limited availability of material. Typically, the work is carried out in a single laboratory with material originating from this source only. This material represents the learning set. To confirm the results a validation set would be needed, and to avoid finding genes specific to a particular genetic background only, one should work with different ecotypes.

This may prove to be unfeasible if not impossible. A different option is to give up on the fully crossed factorial design altogether and take a completely different approach: One can exploit all the experimental data available in the literature (freely available on the web) by pooling them together. This is not as bizarre an idea as it may seem; the aim can be to increase the number of patients (Jiang *et al.*, 2004), or to get information about the co-expression and co-regulation of genes (Lee *et al.*, 2004; Yeung *et al.*, 2004). Especially for the two latter issues, this is the only approach, as a very large amount of data is needed, which a single laboratory could not possibly come up with.

Various authors propose statistical models to help extracting the maximum information from these pooled data (Shen *et al.* [2004] and Statnikov *et al.* [2005]). Note that when working with pooled date, their analysis will have to be carried out with methods which do not need the definition of the factors *a priori*, like PCA or ICA.

The Combination of Factors

If we want to obtain useful information from our microarray experiment, we are forced to formulate precise questions. This means that we cannot combine two factors in one question, as this is equivalent to measuring the interaction between the factors, which is not separable from the error (unless we know in detail the relationship between the two factors, for example linear or sinusoidal).

This is one more reason to follow a fully crossed factorial experimental set up, as exemplified by ANOVA. It forces us to spell out in detail what we want to measure and what will be part of the error or interaction component. It has the great advantage of permitting the identification of the sources of variability and their magnitude; this allows making the improvements to the set up at the right sources, which will generally be a modification of the experimental protocol and an increase in the number of biological replications (Chen *et al.*, 2004).

The Use of Microarrays to Find Genes for an Accurate Diagnosis of a Disease

In general, the data will come from one laboratory and the observations will have been made on a limited amount of material. This material represents the learning set and the analysis of the data will always come up with *some* candidate genes. To validate the results, however, we need a validation set. It is wise to have five to six times more observations in this set than candidate genes, which may prove to be difficult, if not impossible.

To avoid finding genes that are only specific to a particular genetic background, different ecotypes should be studied. This is equivalent to increasing the states of the second type of factor. Again, this may pose a problem, from a financial as well as logistical point of view.

From a theoretical point of view, the use of microarrays for the diagnosis of a disease, poses two fundamental problems, the first one being Bellman's "curse of dimensionality" (too many features or dimensions, e.g. thousands of genes), the second one being the "curse of dataset sparsity" (too few samples) (Somorjai *et al.*, 2003), this means that we end up analyzing a space with a great number of dimensions which is nearly empty. Whatever method is applied to the analysis of the data, the result is unlikely to be statistically sound, the biological interpretation risks being inconclusive. Somorjai *et al.* (2003) discuss this problem in detail. Hwang *et al.* (2002) propose a power analysis method in order to determine the minimum sample size for the–statistically reliable–discrimination of distinct disease states.

THE ANSWER TO ALL OUR QUESTIONS?

Microarrays are sometimes seen as the miracle tool, which will give all the answers to all the questions. Paying considerable attention to the experimental set up is a necessary condition, but not a sufficient one.

The preliminary phase should already take into consideration the different analysis options available to the experimenter by consulting statisticians. There should be an exchange of ideas, not a handing over the job to a statistician, as the biological question has to stay at the front. As Vingron (2001) points out in his editorial, bioinformaticians should "go back to school and learn more statistics. Not so much with the goal of mastering all of statistics but with the goal of sufficiently educating ourselves in order to pull in statisticians."

A careful analysis of the data should follow (we suggest using more than one method, as they tend to give complementary information).

The complex nature of biological phenomena means that it is near impossible to isolate the candidate genes only through a microarray experiment (Somorjai *et al.*, 2003; Curtis and Brand, 2004; Sontag *et al.*, 2004), meaning that the list of candidate genes obtained will have to be further worked on. This may be done by further theoretical work (integration of all available biological knowledge) or additional experiments in the wet laboratory.

Sometimes the *a priori* biological knowledge about the phenomenon of interest may be very limited. In this case, it can happen that despite careful planning and execution, the genes identified as interesting are not actually the cause of the phenomenon. This was for example the case in the genome-wide analysis undertaken by Oshima *et al.* (2002). In these cases a new experimental set-up may be solution. However, the conclusion may also be that a transcriptome analysis is not the adequate tool for the study of the phenomenon (Riva *et al.*, 2004).

It is therefore important to realize that a microarray analysis will generally not be THE answer to all your questions. It is a complement to other approaches.

SOFTWARE AND DATA USED

The sulphur metabolism data from Sekowska *et al.* (2001) are freely accessible at http:// 195.221.65.10:1234/~carpenti

PCA and ANOVA were performed using GeneANOVA, freely available on request for non-commercial use. Please contact Gilles Didier at didier@iml.univ-mrs.fr

ICA was adapted to gene expression analysis by Bruno Torrésani, Marie-Christine Roubaud and Pierre Chiappetta (http://www.cmi.univ-mrs.fr/~torresan/publi.html).

ACKNOWLEDGEMENTS

The authors wish to thank J.- L. Risler for critical reading of the manuscript.

This work was supported in part by the French Ministry of the Economy, Finance and Industry (contract ASG n° 01 4 90 6093).

REFERENCES

Bar-Joseph, Z. 2004. Analyzing time series gene expression data. *Bioinformatics*, 20, 2493-2503.

Benjamini, Y., Hochberg, Y. 1995. Controlling the false discovery rate: a practical and powerful approach to multiple testing. *J. R. Stat. Soc. Ser. B* 57, 289-300.

Carpentier, A.-S., Riva, A., Tisseur, P., Didier, G., Hénaut, A. 2004. The operons, a criterion to compare the reliability of transcriptome analysis tools: ICA is more reliable than ANOVA, PLS and PCA. *Computational Biology and Chemistry*, 28, 3–10.

Chen, J., Delongchamp, R., Tsai, C.-A., Hsueh, H.-M., Sistare, F., Thompson, K.L., Desai, V.G., Fuscoe, J.C. 2004. Analysis of variance components in gene expression data. *Bioinformatics*, 20, 1436–1446.

Chiappetta, P., Roubaud, M.C., Torrésani, B. 2004. Blind Source Separation and the Analysis of Microarray Data. *J. Comput. Biology*, 11, 1090-1109.

Curtis, R.K., Brand, M.D. 2004. Analysing microarray data using modular regulation analysis. *Bioinformatics*, 20, 1272-1284.

Daub, C.O., Steure, R., Selbig, J., Kloska, S. 2004. Estimating mutual information using B-spline functions – an improved similarity measure for analysing gene expression data. *BMC Bioinformatics*, 5: 118, e-pub.

Datta, S., Datta, S. 2003. Comparisons and validation of statistical clustering techniques for microarray gene expression data. *Bioinformatics*, 19, 459–466.

De Smet, F., Mathys, J., Marchal, K., Thijs, G., De Moor, B., Moreau, Y. 2002. Adaptive quality-based clustering of gene expression profiles. *Bioinformatics*, 18, 735-746.

Eisen, M.B., Spellman, P.T., Brown, P.O., Botstein, D. 1998. Cluster analysis and display of genome-wide expression patterns. *PNAS*, 95, 14863-14868.

Fisher, R.A. 1951. The Design of Experiments, sixth ed. Oliver and Boyd, London.

Hoyle, D., Rattray, M., Jupp, R., Brass, A. 2002. Making sense of microarray data distributions. *Bioinformatics*, 18, 576–584.

Hwang, D., Schmitt, W., Stephanopoulos, G., Stephanopoulos, G. 2002. Determination of minimum sample size and discriminatory expression patterns in microarray data. *Bioinformatics*, 18, 1184-1193.

Hyvärinen, A., Oja, E. 2000. Independent component analysis: algorithms and applications. *Neural Networks*, 13, 411-430.

Jiang, H., Deng, Y., Chen, H.-S., Tao, L., Sha, Q., Chen, J., Tsai, C.-J., Zhang, S. 2004. Joint analysis of two microarray gene-expression data sets to select lung adenocarcinoma marker genes. *BMC Bioinformatics*, 5: 81, e-pub.

Kaski, S., Nikkilä, J., Oja, M., Venna, J., Törönen, P., Castrén, E. 2003. Trustworthiness and metrics in visualizing similarity of gene expression. *BMC Bioinformatics*, 4: 48, e-pub.

Kendall, M., Stuart, A., Ord, J.K. 1983. The advanced theory of statistics – Vol. 3 Design and analysis, and time-series. Charles Griffin & Co.

Kerr, K., Churchill, G. 2001. Statistical Design and the Analysis of Gene Expression Microarray Data. *Genetical Research*, 77, 123–128.

Kerr, K., Martin, M., Churchill, G. 2000. Analysis of variance for gene expression microarray data. *J. Comput. Biol.*, 7, 819–837.

Kim, K.-Y., Kim, B.-J., Yi, G.-S. 2004. Reuse of imputed data in microarray analysis increases imputation efficiency. *BMC Bioinformatics*, 5: 160, e-pub.

Konishi, T. 2004. Three-parameter lognormal distribution ubiquitously found in cDNA microarray data and its application to parametric data treatment. *BMC Bioinformatics*, 5: 5, e-pub.

Kutalik, Z., Inwald, J., Gordon, S.V., Hewinson, R.G., Bucher, P., Hinds, J., Cho, K.-H., Wokenhauer, O. 2004. Advanced significance analysis of microarray data based on weighted resampling: a comparative study and application to gene deletions in Mycobacterium bovis. *Bioinformatics*, 20, 357-363.

Lawrence, N., Milo, M., Niranjan, M., Rashbass, P., Soullier, S. 2004. Reducing the variability in cDNA microarray image processing by Bayesian inference. *Bioinformatics*, 20, 518–526.

Lee, H., Hsu, A., Sajdak, J., Qin, J., Pavlidis, P. 2004. Coexpression Analysis of Human Genes Across Many Microarray Data Sets. *Genome Research*, 14, 1085–1094.

Liebermeister, W. 2002. Linear modes of gene expression determined by independent component analysis. *Bioinformatics*, 18, 51–60.

Lyng, J., Badiee, A., Svendsrud, D., Hovig, E., Myklebost, O., Stokke, T. 2004. Profound influence of microarray scanner characteristics on gene expression ratios: analysis and procedure for correction. *BMC Genomics*, 5: 10, e-pub.

Martoglio, A.-M., Miskin, J., Smith, S., MacKay, D. 2002. A decomposition model to track gene expression signatures: preview on observer-independent classification of ovarian cancer. *Bioinformatics*, 18, 1617–1624.

Mather, K. 1943. Statistical Analysis in Biology, first ed. Methuen.

Neuhäuser, M., Senske, R. 2004. The Baumgartner-Weiß-Schindler test for the detection of differentially expressed genes in replicated microarray experiments. *Bioinformatics*, 20, 3553-3564.

Oba, S., Sato, M.-A., Takemasa, I., Monden, M., Matsubara, K.-I., Ishii, S. 2003. A Bayesian missing value estimation method for gene expression profile data. *Bioinformatics*, 19, 2088–2096.

Oshima, T., Wade, C., Kawagoe, Y., Ara, T., Maeda, M., Masuda, Y., Hiraga, S., Mori, H. 2002. Genome-wide analysis of deoxyadenosine methyltransferase-mediated control of gene expression in Escherichia coli. *Mol. Microbiol.*, 45, 673-695.

Ouyang, M., Welsh, W.J., Georgopoulos, P. 2004. Gaussian mixture clustering and imputation of microarray data. *Bioinformatics*, 20, 917-923.

Querec, T., Stoyanova, R., Ross, E., Patriotis, C. 2004. A novel approach for increasing sensitivity and correcting saturation artifacts of radioactively labeled cDNA arrays. *Bioinformatics*, 20, 1955–1961.

Riva, A., Delorme, M.-O., Chevalier, T., Guilhot, N., Henaut, C., Henaut, A. 2004. The difficult interpretation of transcriptome data: the case of the GATC regulatory network. *Computational Biology and Chemistry*, 28, 109–118.

Sasik, R., Calvo, E., Corbeil, J. 2002. Statistical analysis of high-density oligonucleotide arrays: a multiplicative noise model. *Bioinformatics*, 18, 1633–1640.

Schäfer, J., Strimmer, K. 2005. An emprirical Bayes approach to inferring large-scale gene association networks. *Bioinformatics*, 21, 754-764.

Scholz, M., Gatzek, S., Sterling, A., Fiehn, O., Selbig, J. 2004. Metabolite fingerprinting: detecting biological features by independent component analysis. *Bioinformatics*, 20, 2447-2454.

Sekowska, A., Robin, S., Daudin, J.J., Henaut, A., Danchin, A. 2001. Extracting biological information from DNA arrays: an unexpected link between arginine and methionine metabolism in Bacillus subtilis. *Genome Biology*, 2, Research 0019, e-pub.

Shen, R., Ghosh, D., Chinnaiyan, A.M. 2004. Prognostic meta-signature of breast cancer developed by two-stage mixture modeling of microarray data. *BMC Genomics*, 5: 94, e-pub.

Somorjai, R.L., Dolenko, B., Baumgartner, R. 2003. Class prediction and discovery using gene microarray and proteomics mass spectroscopy data: curses, caveats, cautions. *Bioinformatics*, 19, 1484–1491.

Sontag, E., Kiyatkin, A., Kholodenko, B. 2004. Inferring dynamic architecture of cellular networks using time series of gene expression, protein and metabolite data. *Bioinformatics*, 20, 1877–1886.

Statnikov, A., Aliferis, C.F., Tsamardinos, I., Hardin, D., Levy, S. 2005. A comprehensive evaluation of multicategory classification methods for microarray gene expression cancer diagnosis. *Bioinformatics*, 21, 631-643.

Stoyanova, R., Querec, T., Brown, T., Patriotis, C. 2004. Normalization of single-channel DNA array data by principal component analysis. *Bioinformatics*, 20, 1772–1784.

Thygesen, H., Zwinderman, A. 2004. Comparing transformation methods for DNA microarray data. *BMC Bioinformatics*, 5: 77, e-pub.

Turk., R., t'Hoen, P.A., Sterrenburg, E., de Menezes, R.X., de Meijer, E.J., Boer, J.M., van Ommen, G.-J.B., den Dunnen, J.T. 2004. Gene expression variation between mouse inbred strains. *BMC Genomics*, 5: 57, e-pub.

Tusher, V., Tibshirani, R., Chu, G. 2001. Significance analysis of microarrays applied to the ionizing radiation response. *PNAS*, 98, 5116–5121.

Vingron, M., 2001. Bioinformatics needs to adopt statistical thinking. *Bioinformatics*, 17, 389-390.

Weckwerth, W. 2003. Metabolomics in systems biology. *Annu. Rev. Plant Biol.*, 54, 669-689.

Whitehead, A., Crawford, D.L. 2005. Variation in tissue-specific gene expression among natural populations. *Genome Biology*, 6: R13, e-pub.

Yeung, K.Y., Medvedovic, M., Bumgarner, R. 2004. From co-expression to co-regulation: how many microarray experiments do we need? *Genome Biology*, 5: R48, e-pub.

Zar, J.H. 1998. Biostatistical Analysis, fourth ed. Pearson Education, Upper Saddle River, USA.

Zhao, Y., Li, M.-C., Simon, R. 2005. An adaptive method for cDNA microarray normalization. BMC *Bioinformatics*, 6: 28, e-pub.

Zhou, X., Wang, X., Dougherty, E.R. 2003. Missing-value estimation using linear and non-linear regression with Bayesian gene selection. *Bioinformatics*, 19, 2302-2307.

12

CHAPTER

Proteomics

M. Zivy, J. Joyard, M. Rossignol

The word *proteome* is derived from "protein complement expressed by a genome" (Wasinger *et al.*, 1995), and proteomics is the set of techniques used to study proteins on a large scale. The rapid development of proteomics is due together with technical advances, to the existence of increasingly rich nucleic acid sequence databases, and to the emergence of "post-genomics" questions on the function and expression of the final products of sequenced genes, the proteins.

Protein analysis can be used to tackle questions that the analysis of sequences or the transcriptome (all the mRNAs) provide at best indirect answers. These questions pertain to the amounts of proteins, their subcellular localization, their post-translation modifications, protein-protein interactions, and interactions between proteins and other molecules. An increasing number of examples show that the correlation between the quantity of protein and mRNA may be very low, which makes it risky to predict quantities of protein from quantities of mRNA (Gygi *et al.*, 1999b; Griffin *et al.*, 2002). Concerning subcellular localization and post-translation modification, only through analysis of proteins themselves can experimental data be confronted with predictions based on the recognition of motifs in the sequences. It will be demonstrated that proteomic analysis also makes it possible to gain information on the gene coding sequence.

Subcellular localization, post-translation modifications, and interactions with other molecules are information that will complement the annotation of genes and contribute to the identification of their function. This descriptive axis of the proteomics, which is developing rapidly, uses high-throughput tools for the purpose of characterizing numerous proteins, each being analysed individually. The analysis of expression, usually termed "comparative proteomics", is often more integrative. "Intra-proteome" integration is when the aim is to detect sets of co-regulated proteins or to identify proteins responding to particular treatment, and integration in the broader sense when the objective is to detect relationships between the expression of proteins and physiological or phenotypic variations. This integration is also found in transcriptome analyses but since proteins that are the results of pre- and post-translation regulation are the effectors of the cellular reactions, the proteome analysis is better adapted to the detection of the effects of the expression of genes on metabolism or phenotype.

If the genome is a static notion, the proteome can be considered to be essentially dynamic. All proteins of an organism are not expressed in all the cells or under all environmental conditions, it is not the same part of the proteome that is expressed in each sample analysed. Sometimes, several proteomes are considered for a given organism: proteome of an organ, a cell type, or an organelle. Moreover, it is generally supposed that, if the genomes contain some thousands to tens of thousands of sequences, the proteome comprises several hundreds of thousands of proteins, because of alternative splicing and post-translation modifications.

TOOLS OF PROTEOMICS

Even though there are now alternatives, proteomic analyses are conventionally carried out as follows: the proteins are extracted in denaturing conditions (quaternary, tertiary, and secondary structures eliminated) and then separated by two-dimensional (2D) gel electrophoresis (O'Farrell, 1975). The proteins are then detected on gels, and they are excised for identification by mass spectrometry (MS), or their level of expression is analysed by use of specialized software (see box).

Two-dimensional Electrophoresis of Proteins

Two-dimensional electrophoresis consists of two consecutive electrophoreses. The first separates the polypeptides as a function of their isoelectric point (pI): this is isoelectric focusing (IEF) carried out on acrylamide gels in capillary tubes or strips. Subjected to an electric field in a pH gradient, the polypeptides migrate to the position at which their charges cancel out, which is by definition the position at which the pH is equal to their pI. The pH gradient was originally created by ampholyte carrier molecules that themselves migrated in the gel to in order to establish the gradient. Immobilized pH gradients are presently used, where the molecules creating the gradient are co-polymerized within the acrylamide gel; a much better reproducibility of the position of proteins is obtained, and very narrow pH gradients are achieved (one pH unit).

After equilibration in a buffer allowing elimination of components of the gel from the first dimension, the IEF gel is placed on top of a SDS-PAGE plate gel, i.e. a polyacrylamide gel containing sodium dodecyl sulfate. The proteins, after having been separated as a function of their pI, are then separated according to an independent criterion, their molecular mass.

The proteins are then revealed by different techniques, each of which has advantages and disadvantages. The most sensitive available techniques are silver nitrate stains. However they are not very quantitative, and not favourable for MS analysis. Detection with colloidal Coomassie blue is less sensitive but is suitable for MS analysis. The SYPRO Ruby fluorescent dye is probably the best that is currently available (Patton, 2000), it is a little less sensitive than silver nitrate but much more quantitative (linear for 3 orders of magnitude) and it is compatible with MS analysis. This technique, however, requires costly reagents and equipment for the acquisition of images. Several other methods based on fluorescence have been developed recently. The DIGE (Difference Gel Electrophoresis) method allows the comparison of two samples in the same gel by using different dyes. The ProQ Diamond reagent allows the detection of phosphoproteins, which can also be revealed by

Contd.

autoradiography or by a phosphorimager, when the proteins have been labelled using radioactive isotopes.

After detection, the proteins are excised for the purpose of identification (see Appendix), or the gel is scanned and software used to detect and quantify the spots, and to match them between gels. Even though this is a matter of informatics, this part of the analysis of 2D gel electrophoresis in most cases does not have high throughput. The various parts of the treatment can be automated, but neither the problems of spot detection nor those of spot matching are trivial, and it is still necessary to inspect carefully the results of automated treatments.

The quantitative data can then be exported into statistical software that can be used to analyse the spots either one by one (comparison of averages between the different treatments, for example) or taking them all into account simultaneously (multivariable methods such as principal component analysis or clustering). Figure 12.1 is an annotated silver-stained 2D gel, representing the result of a proteomic analysis in which those proteins are highlighted for which statistical analysis indicated a significant effect of the water deficit treatment.

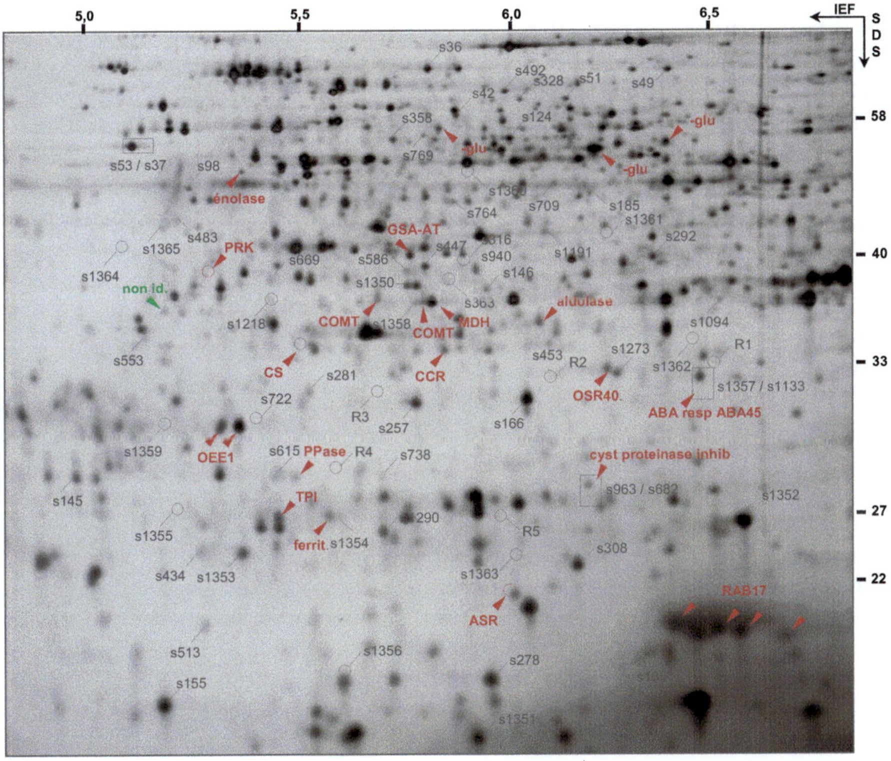

Fig. 12.1 Example of result of proteomic analysis: response to water deficit in the maize leaf base. All the proteins marked with an arrow are regulated by drought. Some are not synthesized by the genotype presented here: they are represented by circles. When allelism relationships have been identified, a rectangle surrounds the allele of the genotype presented and the position of the other allele. The pH range is indicated above, and the molecular weight is indicated at right in kDa.

Several hundreds to several thousands of polypeptides can be detected on a single 2D gel. For a long time they were difficult to identify, which led to a paradoxical situation in which the expression of thousands of proteins could be analysed as a function of various criteria, but only some were identified. During the 1990s, Edman micro-sequencing of proteins separated by 2D gel electrophoresis was developed. In order to analyse proteins at a higher rate, methods of identification by analysis of amino acid composition were then developed, but they were surpassed by MS methods (see appendix to this chapter). The principal advantages of MS methods over Edman sequencing were that they required much less protein, they could be carried out at a much higher rate, and the cost of consumables was much lower (however, the equipment costs remained very high).

The first method to appear was peptide mass fingerprinting (PMF) by matrix-assisted laser desorption/ionization – time of flight (MALDI-TOF). This is still the method with the highest yield, as the analysis itself takes only a few seconds. As for all the identifications by MS, the first step is to digest the protein contained in the 2D electrophoresis spot by an endoprotease for which the cleavage site is known (most often trypsin, which cleaves after lysine or arginine). The MALDI-TOF analysis will measure very precisely (< 0.1 dalton) the mass of peptides formed by this digestion. Since the cleavage sites are known, the digestion can be simulated by informatics, that is the masses of all the peptides produced by this digestion can be calculated for all the known sequence proteins of a given organism. These masses will depend on the length of peptides and their composition, since most amino acids have different masses. Thus, the masses predicted from sequences stored in databases can simply be compared with masses effectively measured by the MALDI-TOF equipment. The larger the number of coincidences of masses, the greater the chances that they are from the same protein. Software such as Mascot (Perkins *et al.*, 1999) also takes into account the size of databases interrogated to calculate the probability that the coincidences are due to chance.

A PMF analysis thus makes it possible to identify a protein, provided its sequence is already known and stored in a database, but there is no potential for discovering new proteins. It is therefore suitable for species for which the genome is entirely sequenced, or in which a large number of complete sequences are known. Expressed sequence tags (ESTs) can also be used (Porubleva *et al.*, 2000; Mathesius *et al.*, 2001), even if they provide information of a lower quality, e.g. short sequences that allow fewer peptides to be predicted and cannot be used to calculate the theoretical mass of the protein, or errors that cause a shift in the reading frame. For less completely sequenced species, it is possible but not recommended to compare the experimental masses of peptides with those expected from proteins sequenced in other species (Lester and Hubbard, 2002). This strategy has little chance of success, as for each peptide, if even only one amino acid is substituted, there is no longer a coincidence between the masses of the two species.

Even in an entirely sequenced species, the MALDI-TOF analyses do not always lead to identification: technical problems may swamp the signal in background noise, and the post-translation modifications, by adding different groups to the peptide, modify the mass measured and lead to loss of coincidence with the predicted mass. When the MALDI-TOF analyses have not made identification possible, or if there are good reasons to think that they will not suffice, MS/MS spectrometry is used (tandem mass spectrometry, see Addendum).

By fragmenting the peptides obtained by tryptic digestion, MS/MS provides richer information, since for each peptide analysed it produces a spectrum: one fingerprint per peptide instead of one fingerprint per protein as in MALDI-TOF. Using the same principle, the results can be compared to spectra calculated from known sequences according to rules of fragmentation using specialized software (Mascot, SEQUEST). When the spectra of several peptides are similar to those expected from a single protein, there are several independent arguments that designate this protein for identification. The protein sequences of another species can be used, since only a few peptides need to be identical for the identification to be possible. For the same reasons, ESTs can be used with greater chances of success than with MALDI-TOF.

Another possible use of MS/MS is sequencing *de novo*, that is deduction of an amino acid sequence by interpretation of a fragmentation spectrum, without the use of existing sequence databases. Although theoretically possible, sequencing *de novo* is still rarely used, but it may develop in the future with technical and informatic advances.

APPLICATIONS OF PROTEOMICS IN THE PLANT KINGDOM

Proteins as Genetic Markers

Before the techniques for protein identification appeared, 2D gel electrophoresis was already a high-throughput technique, in that it could be used to analyse several hundreds of proteins, even if for the most part their identity was unknown. One area in which the anonymity of proteins is not a hindrance is genetic mapping, where the important thing is rather the existence of a polymorphism and uniform distribution of markers on the chromosomes. It will be seen that in fact 2D electrophoresis can be used to satisfy both requirements.

Mapping of expressed genes

The isoelectric point or pI, the criterion of separation of the first dimension of 2D electrophoresis, is directly dependent on the number of amino acids carrying acid and basic groups. An amino acid substitution thus influences the position of the protein ("position shift" or PS) following IEF, if there is a gain or a loss of an acid or basic function. Around one-third of possible substitutions are of this type. An insertion/deletion polymorphism can also be detected, and some substitutions that have no effect on the pI may have an effect on the SDS migration in the second dimension. Mutations leading to the non-expression of the protein can also be detected. 2D electrophoresis thus theoretically allows detection of polymorphisms in coding sequences. It has effectively been observed in different species, and analyses of segregating populations have made it possible to map the genes coding for the proteins detected, the position shift loci (PSL). More than 170 PSL have been reported by de Vienne *et al.* (1996) in different species (maize, pine, barley, peas) and in all cases the segregations expected of a monogenic and codominant marker were observed. The only significant deviation from the proportions expected was found in a double haploid population. It was probably due to gametic selection during the *in vitro* culture process. In all cases, the PSL were uniformly dispersed along the chromosomes, intercalated with the other markers. The rate of polymorphism, which was at first believed to be less than that of the isozymes detected following 1D electrophoresis, was of the same order of magnitude, provided the 2D gels had sufficient resolution.

PSL can thus be used like other markers for mapping, however the fact that they mark only expressed genes provides additional information. In most cases, the cDNA probes used for mapping the genes expressed by restriction fragment length polymorphism (RFLP) hybridize with several other genes or pseudogenes. Several loci are thus mapped using a single probe and it is not possible to determine which ones are effectively expressed. The PSL can provide this information. For example, a cDNA of phosphoglycerate mutase (PGAM) used as an RFLP probe was able to identify four loci in maize. In parallel, the PGAM protein was identified following 2D gel electrophoresis, showing that the protein is expressed in leaves, roots, and the coleoptile. The PGAM PSL was mapped and was shown to co-localize with one of the four loci mapped by RFLP. This information indicates that at least this locus (among the four identified), corresponds to a gene expressed in the three organs, the analysis of which by RFLP, was not able to determine (de Vienne *et al.*, 1999). The genes coding for proteins can also be positioned on a map using ditelosomic lines, in which one chromosomal arm is missing. In wheat, such lines were used to locate not only the genes coding for the proteins detected, but also genes controlling their expression in leaves (Colas des Francs and Thiellement, 1985) and in the seed (Islam *et al.*, 2002).

Analysis of relationships between species and between populations

The numerous genetic markers revealed by 2D electrophoresis have also been used to establish relationships between genotypes. Burstin *et al.* (1994) studied the relationships between 21 lines of maize using different types of markers: 70 PSL, 14 isozyme loci (non-denaturing 1D electrophoresis, followed by detection of enzyme activity), and 58 RFLP markers. The polymorphism of isozymes and that of PSL were nearly identical (respectively 2.4 and 2.2 alleles on average per polymorphic locus) but lower than the rate of polymorphism revealed by RFLP (4.6 alleles on average). This higher rate of polymorphism revealed by RFLP was expected since only some of the mutations in the coding sequences influence the position of proteins following electrophoresis, and since the non-coding sequences are less conserved than the coding sequences. The distances between lines calculated as a function of these different markers were correlated among themselves, and correlated negatively to the coefficient of relationship, which shows that all these markers can be considered neutral for population genetics studies.

It is not always easy to detect PSL, in many cases the studies are based on simple presence/absence variations. Even though in some of them, variations of expression are not necessarily caused by polymorphism of the gene coding for the protein detected, this type of analysis has also proved highly effective in clarifying the phylogenetic relationships among Triticineae (Bahrman *et al.*, 1988; Thiellement *et al.*, 1989) and Brassicaceae (Marques *et al.*, 2001), as well as in analysing the relationships between different populations within a single species (Bahrman *et al.*, 1994; David *et al.*, 1997) or even between closely related varieties of durum wheat (Picard *et al.*, 1997).

Analyses of Expression

Comparative proteomics has been widely used as a means of analysing the expression of proteins as a function of different criteria, from abiotic stress, through the characterization of mutants, to differentiation and response to hormones. We will return later to the analysis of variations of expression between genotypes. In some cases, comparison of 2D gels of samples from mutant plants having undergone different treatments is sufficient to extract information of biological interest. However in most instances, the identification of proteins whose expression changes over the course of the analyses, was used to detect the metabolic pathways involved.

Characterization of mutants

Santoni *et al.* (1997) compared developmental mutants of *A. thaliana* to wild types using 2D electrophoresis. While the nine mutants studied were simple recessive point mutations, 1.8 to 7.9% of the proteins detected showed differences from the wild type. A dendrogram was constructed from distances taking into account the quantitative variations of the proteins affected. The classification thus obtained was in agreement with similarities between phenotypes of these different mutants, and with already known relationships of allelism between them. Moreover, when the developmental mutants were compared with the wild types in the presence of hormones, a mutant already known as an overproducer of auxin clustered with the wild type cultivated in the presence of an analogue of that hormone. Furthermore, predictions were made from these comparisons: a cristal (*Cri1*) mutant clustered with the wild type cultivated in the presence of cytokinin according to analysis using 2D electrophoresis, while the phenotypes was not similar. It was subsequently verified that the *Cri1* mutant overproduced cytokinin (Santoni *et al.* 1997). It should be noted that in the analysis of mutants, their classification and a prediction of the metabolism affected are based on the resemblance between the 2D electrophoresis patterns by the use of a large number of markers, and not on the identification of proteins. This type of analysis was also used by Lemarie *et al.* (1996) to show an interaction between auxin regulation and response to drought.

In other analyses of mutants, the identification of implicated proteins is the main objective. This is the case, for example, with the analysis of the pleiotropic effects of the *opaque2* (*o2*) mutation by Damerval and Le Guilloux (1998). The *o2* mutation, which renders the normally translucent maize seed opaque, also has numerous pleiotropic effects. The *O2* gene is a transcription factor expressed in the endosperm. A proteomic analysis indicated that 36 proteins were affected by the *o2* mutation in different genetic backgrounds. Among them were identified not only the proteins already known to be direct targets of *O2*, which contain *O2* recognition sites in their promoters, but also proteins implicated in the metabolism of amino acids and carbohydrates. These direct or indirect targets of the transcription factor could be responsible for the known effects of the *o2* mutation on seed carbon and nitrogen metabolism.

In the same way, proteomic analyses have been used to detect the secondary effects of mutations. For example, Herbik *et al.* (1996) showed that the increase in enzymes involved in the response to oxidative and anaerobic stress in a tomato mutant deficient in nicotianamine, were responses to abnormal Fe and Cu levels in different organs.

Generally, proteomic analyses can be considered particularly useful for analyses of secondary or pleiotropic effects of mutations on the expression of a large number of proteins.

Differentiation and interaction with the environment

In the analysis of responses to biotic and abiotic factors, as well as in the analysis of differentiation, the value of proteomic analyses is essentially to indicate the key changes in metabolism, through identification of the proteins affected.

In maritime pine subjected to constraints leading to the synthesis of compression wood, the simultaneous classification of proteomic profiles of xylem and of the growth parameters of the wood made it possible to identify the genes potentially implicated in the changes in cell wall structure associated with the formation of compression wood (Plomion *et al.*, 2000).

Hajduch *et al.* (2005) analyzed the expression profiles of soybean seed proteins at various stages after flowering. Four hundred twenty two spots were identified corresponding to 216 non-redundant proteins. They were classified into functional categories, and composite expression profiles were established, i.e. a profile of expression per category was calculated by summing the relative abundances of the proteins belonging to it. This allowed the authors to identify the principal trends in the changes in protein expression and to relate them to the metabolic changes that occured during seed filling.

Gallardo *et al.* (2001, 2002) studied the proteome of the *A. thaliana* seed at different stages of germination (strictly speaking from germination to the emergence of the radicle) under the influence of priming (pre-germination followed by dehydration, a treatment that accelerates germination). They were able to associate 84 proteins (identified by MALDI-TOF) with precise events such as imbibition, dehydration, and mobilization of reserve proteins. They then used a mutant, *ga1*, to analyse the role of gibberellin during germination. It is known that the absence of gibberellin blocks development at the stage of radicle emergence. Proteomic analysis of the mutant showed that in fact the expression of proteins was not different (with one exception) from that of the wild type during early germination, but that 11 proteins induced at the stage of emergence in the wild type were not detected in the mutants. Some of these differences were probably indirect effects of blockage of development (for example, a chloroplastic translation elongation factor), but others were very probably at least partly responsible for this blockage: two genes coding for SAM (S-adenosine methionine) synthetases figured among these 11 proteins. Genome analysis of *A. thaliana* made it possible to find sequences of the GARE type (gibberellic acid responsive element) upstream of the two genes. Given the importance of SAM as a donor of methyl groups in a very large number of reactions, it is probable that the control of the synthesis of SAM synthetases by gibberellin explains the blockage of development in the mutant.

Drought is an environmental factor that affects a very large number of plant characters and that can provoke numerous responses. Proteomic analysis is particularly suited for a study of the response to this stress. Some quantitative analyses have been carried out to study the effect of drought on needles of maritime pine (Costa *et al.*, 1998) and in the leaf elongation zone of maize (Riccardi *et al.*, 1998, 2004). Respectively 38 and 78 proteins, of which 24 and 50 were induced, were found to be regulated by drought in the two species. Different functions were identified among the proteins induced: LEA (late embryogenesis abundant) proteins, as well as proteins implicated in other functions (e.g., response to oxidative stress, chaperones, glycolysis), and even a protein similar to a transcription factor. Quantitative analysis of the response to drought in rice leaves also demonstrated that 42 proteins changed significantly, the amount of which reverted to the initial state after irrigation (Salekdeh *et al.*, 2002). Some responses were similar to those of maize and/or pine (aldolase, enzymes of response to oxidative stress), but most were specific. One of the most strongly induced proteins in roots of rice, for example, was an inactive form of a S-like RNase.

In some cases, the analysis led to the identification of genes that had not been previously identified. For example, Moons *et al.* (1997) micro-sequenced three proteins induced by abscisic acid (ABA) in rice roots. Two of them were LEAs, but the third was unknown. The authors successfully isolated the gene by screening a cDNA library with oligonucleotide probes deduced from two peptides sequenced in this protein. The gene proved to be member of a family of hydrophilic proteins that had till then been unknown. The effect of stresses can be analyzed in the proteome of subcellular compartments. Taylor *et al.* (2005) compared the quantitative responses of the mitochondrial proteome of pea to cold stress, water deficit and herbicides. Rey *et al.* (1998) found a drought-induced protein in potato chloroplasts; the three micro-sequences obtained showed no homology with any known protein. They identified the cDNA by screening a library of expression with an antiserum against the N-terminal sequence. The gene was shown to encode a new form of thioredoxin, the targets of which were identified by affinity chromatography (Rey *et al.*, 2005). Not only the induction but also the inhibition of protein accumulation can be analysed to build hypotheses related to the metabolic changes caused by variations in the environment. Thus, Vincent *et al.* (2005) predicted decreased lignification in response to drought in maize leaves, according to the observed decrease of various enzymes involved in monolignol synthesis. This hypothesis was subsequently confirmed by measurements of lignin concentration.

Over these past few years, several research teams have used the approach of symbiotic association between nitrogen-fixing bacteria and the Leguminosae. Some of their studies belong to approaches of the inventory type, indicating for example the presence of specific proteins in the peribacteroid membrane, while others were comparative studies, in which modifications of expression linked to symbiosis were detected. Proteome analysis of *Sinorhizobium meliloti*, a model bacterium of symbiosis with the Leguminosae, led to the identification of novel proteins (Guerreiro *et al.*, 1999). Panter *et al.* (2000) identified proteins in the peribacteroid membrane of the *Rhizobium* nodules formed in soybean roots. Saalbach *et al.* (2002) identified 46 proteins in the peribacteroid membrane and the space between them and the bacteroid. Natera *et al.* (2000) studied the symbiosis between *Melilotus alba* and *Sinorhyzobium meliloti*. Roots with or without nodules were compared, and bacteroids were compared to bacteria grown in culture. Variations in expression linked to symbiosis were detected in a large number of proteins, of which about 100 were identified. Bestel-Corre *et al.* (2002) studied the symbiotic interactions of *Medicago truncatula* with the fungus *Glomus mosseae* and with the bacterium *Sinorhyzobium meliloti*. Several plant proteins, which were induced specifically in one or other of the symbioses, were identified.

Genetic variations of protein expression

Just as proteomic analyses allow the study of variations of expression as a function of different physiological factors, they allow detection of variations of expression between genotypes. The causes of these variations could be a polymorphism in the gene coding for the protein (in the promoter, for example), as well as in any other gene whose expression has an influence on the quantity of the protein by modulating its synthesis (for example, a transcription factor) or its turnover (for example, a protease). The quantitative variations cannot be interpreted in the same way as the allelic variations revealed by PS: they do not correspond necessarily to a polymorphism in the gene coding for the protein in question, and also, this type of variation may not be neutral with respect to the phenotype. It was demonstrated by Burstin *et al.* (1994) that the distances between lines calculated from quantitative variations are not correlated with neutral markers or with relationship coefficients.

If these variations of expression have an effect on the phenotype, then analysis of their genetic determinism must make it possible to study the genetic determinism of phenotypic traits. This is the objective of the "candidate protein" strategy implemented for the identification of the QTL (Quantitative Trait Loci) of response to drought in maize and maritime pine (Costa and Plomion, 1999). It is in fact possible to map the QTL of proteins separated by 2D electrophoresis (Damerval *et al.*, 1994). These QTL, called PQL (protein quantity loci), no matter what their exact function, have an influence on the quantity of a protein in an organ under given environmental conditions. If the variation in the protein quantity has an influence on the variation of the character studied, then the PQL of this protein are also QTL of the character. In practice, therefore, co-localizations between the PQL of the protein and the QTL of the character of interest must be observed. To minimize the probabilities of finding chance co-localizations, only proteins for which a functional criterion has also been established, are studied. In the case of drought, this criterion is regulation by water deficit: the final candidates considered are only drought-regulated proteins for which there are co-localizations of their PQL with a QTL of response of the plant to this constraint. This strategy has been described by de Vienne *et al.* (1999).

Characterization of Proteins

Generally, inventories provide additional functional information on proteins, e.g. expression in one or other organ, tissue, or cellular compartment. In the case of subcellular compartments, the data contributed by proteomics can be tested with predictions made from nucleic acid sequences. However, proteomics can sometimes contribute information for the annotation or correction of the primary sequence itself (Mann and Pandey, 2001), even if the full amino acid sequence is not known (see appendix to this chapter). The peptide sequences themselves contain information on the sense and reading frame. If a peptide contains sequences of two consecutive exons, it provides information about the intron/exon limits, or alternative splicing. Other information on the primary structure can be obtained by MS. An amino acid acetylated at the N-terminus of a peptide indicates that it is located at the N-terminal end of the protein, a significant piece of information in the case of the cleavage of a signal peptide, or when there is ambiguity on the position of the start of translation. The absence of lysine or arginine at the C-terminal end of a peptide (in the case of a tryptic digestion) can also indicate that it is a C-terminal peptide.

Cellular fractions

Inventories have been developed of proteins present in various subcellular compartments, or in the membranes that surround them. Several membrane systems of the cell were the subject of these efforts: plasma membrane (Santoni *et al.*, 1998, 1999a), mitochondria, endoplasmic reticulum, and Golgi apparatus (Prime *et al.*, 2000). The solubilization of hydrophobic proteins in conditions compatible with two-dimensional electrophoresis is difficult, and several detergents have been developed to improve it (Santoni *et al.*, 1999b; Luche *et al.*, 2003). However, complementary techniques of separation are required to access extensively intrinsic membrane proteins (cf. below). Most subcellular studies involve mitochondria and chloroplasts. Analysis of the mitochondrial proteome by the classical method of 2D gel electrophoresis (Kruft *et al.*, 2001; Millar *et al.*, 2001; Bardel *et al.*, 2002), has resulted in a large inventory of proteins (more than 450), a large proportion of which were only predicted by systematic genome annotation until then.

Numerous proteomic studies have contributed information on the proteome of different compartments of the chloroplast, in which it is estimated that 2,500 to 3,000 different proteins are found, many of which are hydrophobic. Around 18% of the 25,498 proteins predicted in the *Arabidopsis thaliana* genome contain at least two potential transmembrane helixes, and 5 to 10% represent transport proteins (AGI, 2000). Seigneurin-Berny *et al.* (1999) and Ferro *et al.* (2000) exploited their hydrophobicity to extract proteins from the chloroplast envelope catalysing transport activities. The proteins, soluble in chloroform/methanol mixtures were analysed by electrophoresis in the presence of SDS. Micro-sequences were then obtained for numerous proteins by tandem mass spectrometry (MS/MS), and the presence of more than 40 helix-containing proteins was detected, of which about 20 had never been identified. From these studies, Ferro *et al.* (2002) defined properties common to various transport systems identified by the proteomic approach of the chloroplast envelope (hydrophobicity, charge, transit sequence). They then used these criteria to screen all the proteins predicted on the basis of the complete genome analysis of *A. thaliana,* as being located in the chloroplast envelope. Nearly 130 proteins effectively met these criteria, and 50% of those corresponded to proteins implicated in demonstrated or potential transport activities, but for the most part had not yet been identified as proteins of the chloroplast envelope.

A systematic analysis of the thylakoid proteins was undertaken by the van Wijk team (Peltier *et al.*, 2000). Protein-rich fractions of the internal space of thylakoids, the lumen, and fractions rich in peripheral proteins were analysed by 2D gel electrophoresis, and then the proteins were identified by Edman sequencing and MS. The authors thus identified 61 proteins, of which a dozen were implicated in electron transfer or carbon metabolism. Functional domains could be identified for about 30 other proteins, but no functional element could be detected for the remaining 10 proteins. The authors verified whether the nucleic acid sequences corresponding to the proteins identified in the lumen contained the known peptide signal sequences specific for this compartment. Around half of them effectively contained a lumen peptide signal sequence with the characteristic "twin arginine" motif. This observation was confirmed later by Schubert *et al.* (2002). These proteomic data of subcellular localization were also used by Peltier *et al.* (2000) to analyse the efficiency of software for prediction of location from transit peptides, and to supply information in turn that could help the software to refine predictions. Afterwards, Peltier *et al.* (2004) analyzed thylakoid membrane proteins, by using a method of phase partitioning followed by SDS-PAGE (or direct in-solution digestion) and MS. Two hundred and forty two proteins were recognized and, by combining their identification with those obtained from other compartments of the chloroplast or with various techniques of extraction or separation, they drew up a comprehensive list of the chloroplastic proteome. Gomez *et al.* (2002) analysed preparations of thylakoids rich in photosystem II by using high performance liquid chromatography (HPLC) and MS/MS. From 90 peptide tags, they detected about 40 proteins (of which 75% could be identified) and they were able to bring new information on their post-translation modifications by showing, for example, that the D1 protein of photosystem II undergoes not only phosphorylation but also acetylation and palmitoylation. Another study showed the utility of the proteomic approach of investigating subcellular fractions in characterizing the phenotype of mutants: by precipitating the proteins of thylakoid membranes of *Chlamydomonas reinhardtii* with chloroform/methanol, Hippler *et al.* (2001) successfully obtained a clear 2D electrophoresis separation of proteins containing up to 11 transmembrane domains, and were able to show the absence of most of the LHCI (light harvesting complex proteins of photosystem I) in a PSI-deficient mutant.

Macromolecular complexes

Proteins do not work in isolation in the cell but work through multiple interactions with other proteins, within pathways or complexes. Using the range of modern proteomic tools, this functional complexity can be tackled, the multi-protein groups can be purified, and the elements can be identified. In the case of the chloroplast ribosome, which can be purified by ultracentrifuge, the combination of different separation systems (2D gel electrophoresis and HPLC) with methods of characterization by MS or by chemical means (for N-terminal sequencing), has made it possible to show that the two subunits 30S and 50S contain 59 proteins in spinach, of which 6 are specific to plants, the others being orthologous to constitutive proteins of bacterial ribosomes (Yamaguchi and Subramanian, 2000; Yamaguchi *et al.*, 2000). A number of these proteins co-exist under different levels of post-translation maturation. When the multi-protein complex is not easily purified by a fractioning procedure, a "blue native" electrophoresis procedure can be used, which separates the complexes in non-denaturing conditions. The complex is then subjected to a second electrophoresis, in the presence of SDS, to separate the components. This approach was used to identify 10 Clp (Caseinolytic protease) subunits of a ClpP protease complex in the chloroplast (Peltier *et al.*, 2001) and by Heinemeyer *et al.* (2004), to investigate the supramolecular structure of photosystems in *A. thaliana*.

Post-translation modifications

An issue of major functional importance, and specific to proteomics, is the identification of states of post-translation maturation of proteins. The use of varied biochemical methods (e.g., immunological, enzymatic) has traditionally constituted the first step in analysing maturations and allows, for example, detection of original modifications such as the addition of glycosylphosphatidylinositol groups implicated in the attachment of proteins in the cell membrane (Sherrier *et al.*, 1999), or the analysis of steps of phosphorylation in signal transduction pathways (Peck *et al.*, 2001). Here also, the use of MS has cause a change in scale and led to the notion of subproteomes associated with particular types of maturation. The phosphoproteome is of course the most studied and multiple approaches have been developed to identify phosphoproteins or phosphopeptides, to identify the phosphorylated aminoacid(s) and to quantify the percentage of phosphorylation (see Appendix). Several methods have also been developed for the analysis of other post-translational modifications, many of them being based on affinity tagging or enrichment (Jensen, 2004). The regulation of the redox status of proteins by thioredoxins also plays an important role in different biological functions, and the identification of their targets has been carried out by using different methods, from the use of mutants that form a stable complex with their targets (Verdoucq *et al.*, 1999) to the use of radioactively labelled iodoacetamide followed by 2D electrophoresis (Marchand *et al.*, 2004).

NEW METHODOLOGICAL AND TECHNOLOGICAL APPROACHES

Conventional proteomic approaches (analysis of total cell extracts by 2D gel electrophoresis) have biological and technological limitations: very broadly, the conventional 2D electrophoresis allows high resolution only in the pI range of four to eight and in the molecular weight range of 10 to 150 kDa. Particular conditions are sometimes used to better separate some proteins, generally at the expense of other proteins. "Contigs" of 2D electrophoresis have sometimes been considered (series of gels of narrow and contiguous pI ranges, Wasinger *et al.*, 1997), but they involve a multiplication of 2D gel electrophoresis that is poorly suited to analyses of expression. A second constraint is the difficulty of separating the hydrophobic proteins, even if better methods have been developed to improve the resolution for some of them. Finally, only the most abundant proteins are detected by 2D gel electrophoresis in the complex mixture of total proteins. One answer is to "decomplex" the mixture by fractionating the sample, as seen earlier, but it is difficult to imagine studies of expression on the multiple subcellular fractions.

Under these conditions, and given the difficulty of automating 2D gel electrophoresis, alternative methods have been proposed. Gygi *et al.* (1999a) developed the ICAT (isotope coded affinity tag) method. ICAT is a molecule made up of three parts: a cleavable biotin group, an alkyl linker and a thiol-specific reactive group that allows fixation on protein cysteine molecules. It is manufactured in two forms differing by the replacement (heavy form) or non-replacement (light form) of nine ^{12}C carbon atoms of the linker by nine ^{13}C atoms. To compare two protein samples, one is tagged by the light form and the other by the heavy form. The samples are then mixed and digested by trypsin, and cysteine-containing tagged peptides are selected by avidin-biotin affinity chromatography. After cleavage of the biotin group, the peptides are analyzed by LC-MS. The peptide pairs separated by nine units of mass are located and quantified and then identified by MS/MS. This method is effectively quantitative because the peptides are quantified in the same spectrum and the two forms of ICAT do not modify the ionization of peptides. By selecting only the peptides containing a cysteine, a rare amino acid, this method simplifies the mixture of peptides to be analysed, so it can be analysed by LC-MS. However, this is also an obstacle to protein identification. Therefore, the identification must be certain from one or two peptides, which makes the use of this method inadvisable for organisms in which few complete sequences are known. Moreover, for the same reason, this method cannot be used to analyse post-translational modifications. Finally, the quantification is limited to the comparison of one sample with another. It has, however, attracted a great deal of interest and led to the development of alternative methods.

Methods of quantification based on stable isotope labelling of proteins have been developed. In the SILAC method (Stable Isotope Labelling by Amino acids in Cell culture), cells are cultivated in media in which one amino acid is labelled (or not) with stable isotopes: deuterium-labelled Leucine (Ong *et al.*, 2002) or ^{13}C-labelled Arginine (Gruhler *et al.*, 2005). In the iTRAQ™ method (Ross *et al.*, 2004), peptides from different samples are derivatized with different reagents in which stable isotopes are arranged in such a way that all reagents have the same mass, but different MS/MS spectra. Samples are then mixed and the relative abundance of proteins originating from the different samples is quantified by measuring the specific MS/MS peaks of the different reagents. Gerber *et al.* (2003) developed the AQUA method (Absolute QUAntification), where a peptide of the protein of interest is synthesized with stable isotopes and can be used as an internal standard. This method is not adapted to the quantification of large numbers of different proteins, since a peptide per analyzed protein must be synthesized, but it is well adapted to the quantification of few proteins in a large number of samples.

Important developments have occurred in the field of HPLC, particularly with the use of 2D-LC, that is, the in-line use of two HPLC with different separation criteria to better separate peptides before subjecting them to Electro-Spray-Ionization (ESI)-MS/MS. Wolters *et al.* (2001) developed the multidimensional protein identification technology (MudPIT) in which, after enzymatic digestion (without prior electrophoretic separation of proteins), the peptides are separated on a column comprising two compartments corresponding to an ion exchange column and a reverse phase column. Close to 1,500 proteins were thus identified in *Saccharomyces cerevisiae* (Washburn *et al.*, 2001) and nearly 2,400 in rice, by analysing proteins extracted from leaves, roots and seeds (Koller *et al.*, 2002). In-line 2D-LC-MS/MS was one of the two methods used by Carter *et al.* (2004) to identify proteins of purified vacuoles of *A. thaliana*. In the other method, total proteins or membrane proteins were separated by SDS-PAGE, then the gels were systematically cut into 2.5 mm-width gel slices, which were submitted to trypsin digestion, followed by LC-MS/MS. In total 402 proteins were identified.

Analysis of protein-protein interactions is also a rapidly developing field. Diverse methods have been used to isolate and purify protein complexes. They range from the "blue native" technique (Brookes *et al.*, 2002; Nijtmans *et al.*, 2002) to techniques based on constructions in which the protein for which the partners must be discovered is fused with one or more peptides allowing its purification by affinity chromatography, with its partners (Ho *et al.*, 2002; Tandem Affinity Purification, TAP tag: Rigaut *et al.*, 1999; Gavin *et al.*, 2002). Shevchenko *et al.* (2002) compared the results obtained from this type of analysis with those obtained using double hybrid screening. The two methods gave quite different results, with only 14% of recovery.

Chips with proteins of different types have also been developed and are already used to detect protein-protein interactions or interactions between proteins and other molecules (Zhu *et al.*, 2001, Schweitzer and Kingsmore, 2002; Bertone and Snyder, 2005). It must be noted that it is incorrect to present microarrays as an alternative to 2D electrophoresis, since they allow analysis of interactions of proteins with other molecules, which is not possible with 2D electrophoresis, and on the other hand do not allow analysis of variations of expression of a large number of proteins, which is possible with 2D electrophoresis. No systematic approach attempting to describe an "interactome" has yet been published on plants, by isolation of protein complexes or by use of chips.

Finally, MS is still an evolving field, and every year new equipment is designed that combines the high-flow capacities of one technology with the precision of a second or the sensitivity of a third.

CONCLUSION

Considerable technological advances have led to the rapid development of proteomics in the past few years. While it was only recently that most proteins detected on a 2D gel electrophoresis were unknown, and only proteins responding to one or another factor were identified, we are now close to the reverse situation, where most detected proteins are identified, or identifiable, and many analyses are directed at finding proteins that specifically do not appear in the 2D gel electrophoresis. These advances have broadened the field of proteomics, which has become a tool of characterization of proteins by supplying information on their subcellular location, their post-translation modifications, and their possible associations with other proteins within complexes. The objectives of analyses of expression are generally quite different, since they often aim to characterize not the proteins themselves but rather a given event (differentiation, phenotypic variation, interactions with the environment, metabolic modification) by means of the proteins involved in it. In this perspective, the proteins are used as detectors of the biological phenomena involved rather than as objects of analysis in themselves. Advances in MS, as well as in the various fields of 2D electrophoresis, which still remain the method of choice for quantifying numerous proteins in a large number of samples, have also transformed this aspect of proteomics. Soon it will be possible to interpret the results of an experiment not only as a function of proteins that are regulated during the phenomenon studied, but also as a function of those that do not respond. Massive identifications have also made it possible to compare expressions of the proteome and transcriptome, and to link them with metabolism.

APPENDIX: MASS SPECTROMETRY FOR PROTEIN IDENTIFICATION

The first step in protein identification by mass spectrometry (MS) is proteolysis—generally digestion by an endoprotease, most often trypsin. The digestion of a protein produces a mixture of peptides, which is the subject of the MS analysis.

In MALDI-TOF MS (matrix-assisted laser desorption-ionization–time of flight), the combination of peptides resulting from the digestion is mixed with a product (the matrix, which favours the ionization of peptides) and is deposited on a plate. Several hundreds of samples can be deposited on the same plate. Under the joint effect of a laser and an electric field, the peptides contained in the mixture will ionize (take on a positive charge: z = 1) and be precipitated at the opening of an empty tube, in which there is no electric field. All the peptides undergo the same force, conferred by their charge in the electric field, but this force does not yield the same speed since they have different masses. A detector is placed at the end of the tube to measure the time taken by each peptide to cross the tube (Fig. 12.2). It is in fact the m/z rate that is measured (mass over electric charge); it is equal to m in these conditions since z = 1. Thus, one spectrum is obtained for each protein, which should provide the measurement of the mass of all the peptides in the initial mixture. For technical reasons, only part of the peptides pass all the steps of the process in a sufficient quantity to produce a signal greater than the background noise.

In ESI-MS/MS (electro-spray ionization MS/MS), the process of ionization of peptides is different. The peptides in a liquid medium are introduced in a very fine needle, at the end of which a spray is formed in an electric field. The peptides are ionized over the course of this nebulization. The number of charges may vary, generally from 1 to 4 positive charges. The electric field drives the peptides into the apparatus, in which they will again be fragmented (Fig. 12.3).

There are different types of ESI-MS/MS apparatus. Their common characteristic is that they can be used to select just one out of the mixture of peptides penetrating the apparatus, measure its mass (or rather its m/z, first MS), subject it to fragmentation, then measure the m/z of these fragments (second MS).

In most proteomic analyses, ESI-MS/MS is coupled with a nanoHPLC (high performance liquid chromatography). The mixture of peptides resulting from the tryptic digestion of a protein is injected on a reverse phase capillary micro-column in which the eluant is perfused at a very low flow rate (100-200 nl/mn). The peptides are thus separated and concentrated before penetrating the spectrometer. This combination of techniques has numerous advantages: sensitivity, elimination of salts, quality of ionization, possibility of analysing a larger number of peptides, and also automation. The spectrometer then carries out the loop analyses (around one MS/MS analysis every 2 seconds with an ion trap, for example). Since the duration of the HPLC is around 1 hour, hundreds of MS/MS spectra can be produced during an analysis. Bioinformatics tools are essential in treating these large amounts of data. The first operations consist of testing the spectra that carry pertinent information (e.g., through elimination of redundancies, elimination of contaminants such as keratin, eliminations on the basis of signal quality). Subsequent steps consist of deciphering the spectra, that is, translating the m/z data into masses, since the ions may have multiple charges, and comparing them with spectra theoretically obtained from sequences stored in databases. These spectra are calculated as a function of fragmentation rules, which in this case include the following, to put it very simply: the peptide bonds are broken, and since peptides are mostly broken only once, each break produces a fragment containing the N-terminal part of the peptide (fragment b) and another containing the C-terminal part (fragment y).

Even though a very large number of spectra are generated, when a protein is identified the "coverage" (the proportion of the protein sequence represented by recognized peptides) is generally low. This is due partly to the large number of steps that a peptide undergoes before being detected in the spectrometry: the peptides can be

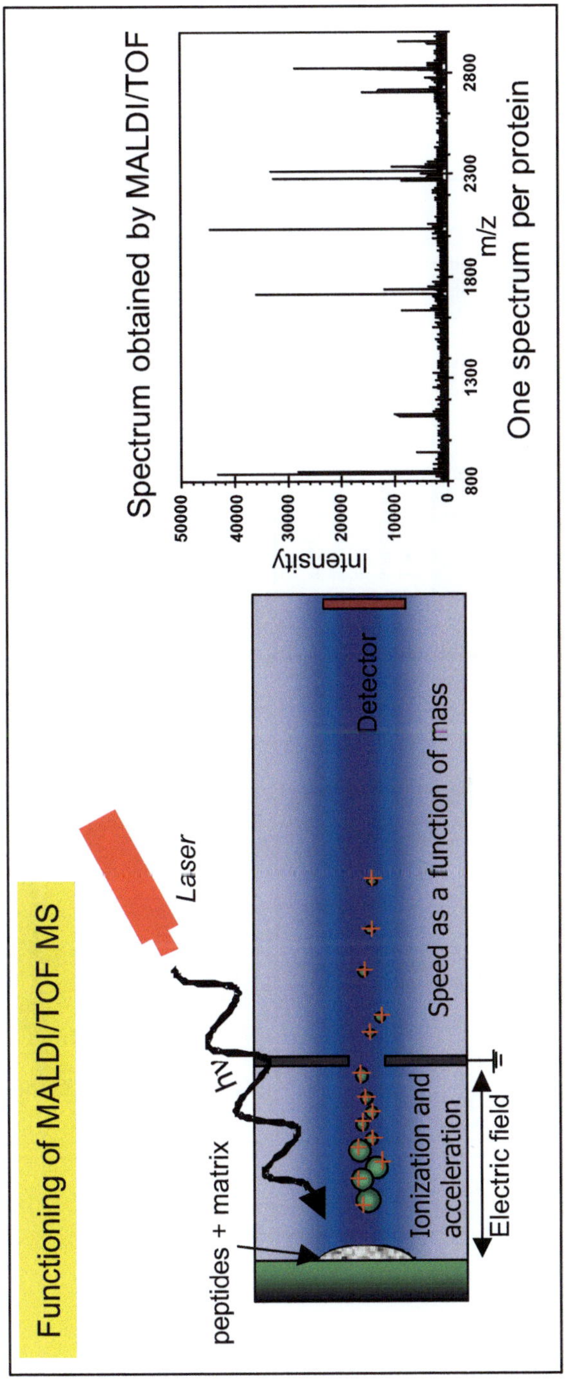

Fig. 12.2 Functioning of MALDI-TOF MS.

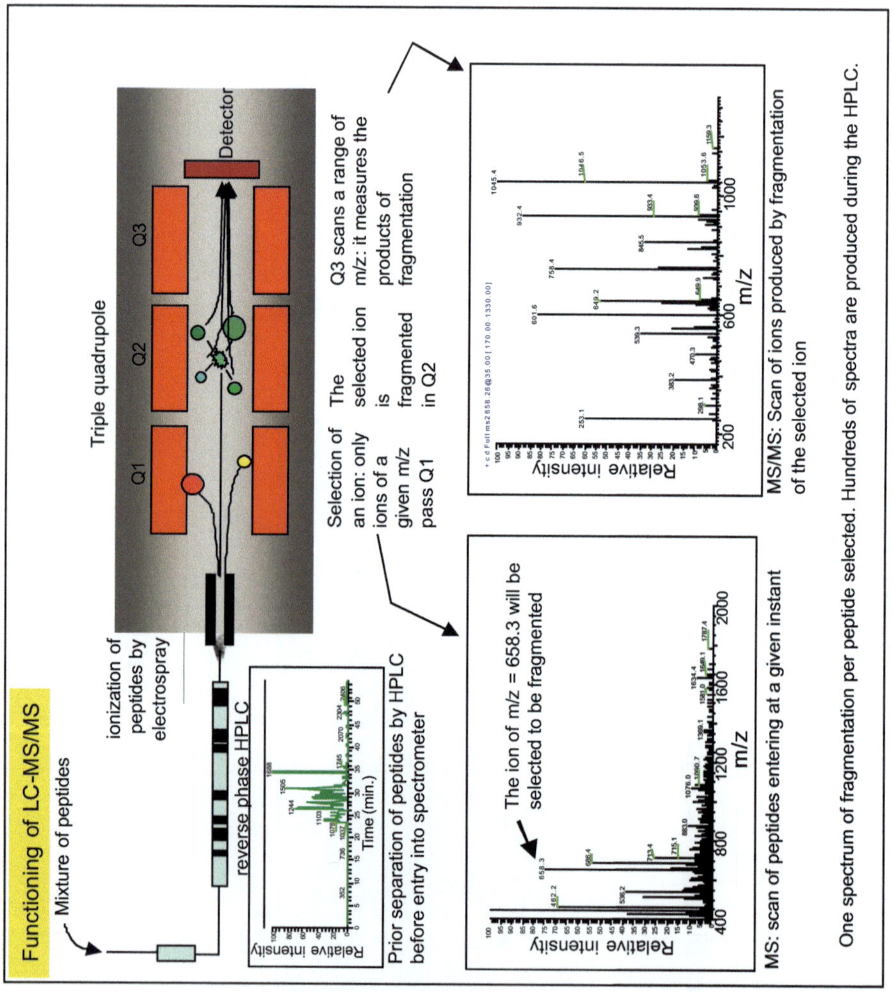

Fig. 12.3 Functioning of an LC-MS/MS combination

poorly ionized, too large, too small, too hydrophilic, and so on. But the capacity of software to analyse the spectra is also a limitation: some peptides present spectra that are too complex, and some cannot be recognized because of post-translation modifications.

In principle, a sequence can also be directly predicted from a spectrum, without taking recourse to databases, since the difference of mass between two fragments of a series b or y issued from the break at two consecutive peptide bonds is equal to the mass of the amino acid between these two bonds: this is *de novo* sequencing. Up until now this method has not been used automatically, because the complexity of spectra leads to multiple possibilities of sequences, between which it is difficult to splice. Some methods are nevertheless being developed, involving software as well as technology, to obtain spectra that are easier to interpret (better precision of spectrometers allowing certain ambiguities to be removed, chemical methods during or after the enzymatic digestion that allow b ions to be distinguished from y ions, MS^3 analyses in the case of ion traps), and it is probable that *de novo* sequencing will become increasingly common.

REFERENCES

Bahrmann, N., Zivy, M., Damerval, C., Baradat, P. 1994. Organisation of the variability of abundant proteins in seven geographical origins of maritime pine (*Pinus pinaster* Ait.). *Theor. Appl. Genet.*, 88, 407-411.

Bahrmann, N., Zivy, M., Thiellement, H. 1988. Genetic relationships in the *Sitopsis* section of *Triticum* and the origin of the B genome of the polyploidy wheats. *Heredity*, 61, 473-480.

Bardel, J., Louwagie, M., Jaquinod, M., Jourdain, A., Luche, S., Rabilloud, T., Macherel, D., Garin, J., Bourguignon, J. 2002. A survey of the plant mitochondrial proteome in relation to development. *Proteomics*, 2, 880-898.

Bertone, P., Snyder, M., 2005. Advances in functional protein microarray technology. *Febs J* 272, 5400-5411.

Bestel-Corre, G., Dumas-Gaudot, E., Poinsot, V., Dieu, M., Dierick, J.F., van Tuinen, D., Remacle, J., Gianinazzi-Pearson, V., Gianinazzi, S. 2002. Proteome analysis and identification of symbiosis-related proteins from *Medicago truncatula* Gaertn. by two-dimensional electrophoresis and mass spectrometry. *Electrophoresis*, 23, 122-137.

Brookes, P.S., Pinner, A., Ramachandran, A., Coward, L., Barnes, S., Kim, H., Darley-Usmar, V.M. 2002. High throughput two-dimensional blue-native electrophoresis: A tool for functional proteomics of mitochondria and signaling complexes. *Proteomics*, 2, 969-977.

Burstin, J., de Vienne, D., Dubreuil, P., Damerval, C. 1994. Molecular markers and protein quantities as genetic descriptors in maize. I. Genetic diversity among 21 inbred lines. *Theor. Appl. Genet.*, 89, 943-950.

Carter, C., Songqin, P., Zouhar, J., Avila, E.L., Girke, T., Raikhel, N.V. 2004. The vegetative vacuole proteome of *Arabidopsis thaliana* reveals predicted and unexpected proteins. *Plant Cell*, 16, 3285–3303.

Colas des Francs, C., Thiellement, H. 1985. Chromosomal localization of structural genes and regulators in wheat by 2D electrophoresis of ditelosomic lines. *Theor. Appl. Genet.*, 71, 31-38.

Costa, P., Bahrman, N., Frigerio, J.M., Kremer, A., Plomion, C. 1998. Water-deficit-responsive proteins in maritime pine. *Plant Mol. Biol.*, 38, 587-596.

Costa, P., Plomion, C. 1999. Genetic analysis of needle protein in Maritime pine. 2. Quantitative variation of protein accumulation. *Silvae Genet.*, 48, 146-150.

Damerval, C., Le Guilloux, M. 1998. Characterization of novel proteins affected by the *O2* mutation and expressed during maize endosperm development. *Mol. Gen. Genet.*, 257, 354-361.

Damerval, C., Maurice, A., Josse, J.M., de Vienne, D. 1994. Quantitative trait loci underlying gene product variation: a novel perspective for analyzing regulation of genome expression. *Genetics*, 137, 289-301.

David, J.L., Zivy, M., Cardin, M.L., Brabant, P. 1997. Protein evolution in dynamically managed populations of wheat: adaptive responses to macro-environmental conditions. *Theor. Appl. Genet.*, 95, 932-941.

de Vienne, D., Burstin, J., Gerber, S., Leonardi, A., Le Guilloux, M., Murigneux, A., Beckert, M., Bahrman, N., Damerval, C., Zivy, M. 1996. Two-dimensional electrophoresis of proteins as a source of monogenetic and codominant markers for population genetics and mapping the expressed genome. *Heredity*, 76, 166-177.

de Vienne, D., Leonardi, A., Damerval, C., Zivy, M. 1999. Genetics of proteome variation as a tool for QTL characterization. Application to drought stress responses in maize. *J. Exp. Bot.*, 50, 303-309.

Ferro, M., Salvi, D., Riviere-Rolland, H., Vermat, T., Seigneurin-Berny, D., Grunwald, D., Garin, J., Joyard, J., Rolland, N. 2002. Integral membrane proteins of the chloroplast envelope: identification and subcellular localization of new transporters. *Proc. Natl. Acad. Sci.* USA, 99, 11487-11492.

Ferro, M., Seigneurin-Berny, D., Rolland, N., Chapel, A., Salvi, D., Garin, J., Joyard, J. 2000. Organic solvent extraction as a versatile procedure to identify hydrophobic chloroplast membrane proteins. *Electrophoresis*, 21, 3517-3526.

Gallardo, K., Job, C., Groot, S.P., Puype, M., Demol, H., Vandekerckhove, J., Job, D. 2001. Proteomic analysis of *Arabidopsis* seed germination and priming. *Plant Physiol.*, 126, 835-848.

Gallardo, K., Job, C., Groot, S.P., Puype, M., Demol, H., Vandekerckhove, J., Job, D. 2002. Importance of methionine biosynthesis for *Arabidopsis* seed germination and seedling growth. *Physiol. Plant*, 116, 238-247.

Gavin, A.C., Bosche, M., Krause, R., Grandi, P., Marzioch, M., Bauer, A., Schultz, J., Rick, J.M., Michon, A.M., Cruciat, C.M., Remor, M., Hofert, C., Schelder, M., Brajenovic, M., Ruffner, H., Merino, A., Klein, K., Hudak, M., Dickson, D., Rudi, T., Gnau, V., Bauch, A., Bastuck, S., Huhse, B., Leutwein, C., Heurtier, M.A., Copley, R.R., Edelmann, A., Querfurth, E., Rybin, V., Drewes, G., Raida, M., Bouwmeester, T., Bork, P., Seraphin, B., Kuster, B., Neubauer, G., Superti-Furga, G. 2002. Functional organization of the yeast proteome by systematic analysis of protein complexes. *Nature*, 415, 141-147.

Gerber, S.A., Rush, J., Stemman, O., Kirschner, M.W., Gygi, S.P. 2003. Absolute quantification of proteins and phosphoproteins from cell lysates by tandem MS. *Proc. Natl. Acad. Sci.* USA 100, 6940-6945.

Gomez, S.M., Nishio, J.N., Faull, K.F., Whitelegge, J.P. 2002. The chloroplast grana proteome defined by intact mass measurements from liquid chromatography mass spectrometry. *Mol. Cell Proteomics*, 1, 46-59.

Griffin, T.J., Gygi, S.P., Ideker, T., Rist, B., Eng, J., Hood, L., Aebersold, R. 2002. Complementary profiling of gene expression at the transcriptome and proteome levels in *Saccharomyces cerevisiae*. *Mol. Cell Proteomics*, 1, 323-333.

Gruhler, A., Schulze, W.X., Matthiesen, R., Mann, M., Jensen, O.N. 2005. Stable isotope labelling of Arabidopsis thaliana cells and quantitative proteomics by mass spectrometry. *Mol. Cell Proteomics* 4, 1697-1709.

Guerreiro, N., Djordjevic, M.A., Rolfe, B.G. 1999. Proteome analysis of the model microsymbiont *Sinorhizobium meliloti*: isolation and characterization of novel proteins. *Electrophoresis*, 20, 818-825.

Gygi, S.P., Rist, B., Gerber, S.A., Turecek, F., Gelb, M.H., Aebersold, R. 1999a. Quantitative analysis of complex protein mixtures using isotope-coded affinity tags. *Nat. Biotechnol.*, 17, 994-999.

Gygi, S.P., Rochon, Y., Franza, B.R., Aebersold, R. 1999b. Correlation between protein and mRNA abundance in yeast. *Mol. Cell Biol.*, 19, 1720-1730.

Hajduch, M., Ganapathy, A., Stein, J.W., Thelen, J.J. 2005. A systematic proteomic study of seed filling in soybean. Establishment of high-resolution two-dimensional reference maps, expression profiles, and an interactive proteome database. *Plant Physiol.* 137, 1397-1419.

Heinemeyer, J., Eubel, H., Wehmhoner, D., Jansch, L., Braun, H.P. 2004. Proteomic approach to characterize the supramolecular organization of photosystems in higher plants. *Phytochemistry,* 65, 1683-1692.

Herbik, A., Giritch, A., Horstmann, C., Becker, R., Balzer, H.J., Baumlein, H., Stephan, U.W. 1996. Iron and copper nutrition-dependent changes in protein expression in a tomato wild type and the nicotianamine-free mutant chloronerva. *Plant Physiol.*, 111, 533-540.

Hippler, M., Klein, J., Fink, A., Allinger, T., Hoerth, P. 2001. Towards functional proteomics of membrane protein complexes: analysis of thylakoid membranes from *Chlamydomonas reinhardtii. Plant J.*, 28, 595-606.

Ho, Y., Gruhler, A., Heilbut, A., Bader, G.D., Moore, L., Adams, S.L., Millar, A., Taylor, P., Bennett, K., Boutilier, K., Yang, L., Wolting, C., Donaldson, I., Schandorff, S., Shewnarane, J., Vo, M., Taggart, J., Goudreault, M., Muskat, B., Alfarano, C., Dewar, D., Lin, Z., Michalickova, K., Willems, A.R., Sassi, H., Nielsen, P.A., Rasmussen, K.J., Andersen, J.R., Johansen, L.E., Hansen, L.H., Jespersen, H., Podtelejnikov, A., Nielsen, E., Crawford, J., Poulsen, V., Sorensen, B.D., Matthiesen, J., Hendrickson, R.C., Gleeson, F., Pawson, T., Moran, M.F., Durocher, D., Mann, M., Hogue, C.W., Figeys, D., Tyers, M. 2002. Systematic identification of protein complexes in *Saccharomyces cerevisiae* by mass spectrometry. *Nature,* 415, 180-183.

Islam, N., Woo, S.H., Tsujimoto, H., Kawasaki, H., Hirano, H. 2002. Proteome approaches to characterize seed storage proteins related to ditelocentric chromosomes in common wheat (*Triticum aestivum* L.). *Proteomics*, 2, 1146-1155.

Jensen, O.N. 2004. Modification-specific proteomics: characterization of post-translational modifications by mass spectrometry. *Curr. Opin. Chem. Biol.* 8, 33-41.

Koller, A., Washburn, M.P., Lange, B.M., Andon, N.L., Deciu, C., Haynes, P.A. Hays, L., Schieltz, D., Ulaszek, R., Wei, J., Wolters, D., Yates, J.R. 2002. Proteomic survey of metabolic pathways in rice. *Proc. Natl. Acad. Sci.* USA, 99, 11969-11974.

Kruft, V., Eubel, H., Jansch, L., Werhahn, W., Braun, H.P. 2001. Proteomic approach to identify novel mitochondrial proteins in *Arabidopsis. Plant Physiol.*, 127, 1694-1710.

Lester, P.J., Hubbard, S.J. 2002. Comparative bioinformatics analysis of complete proteomes and protein parameters or cross-species identification in proteomics. *Proteomics*, 2, 1392-1405.

Lemarie, J., Damerval, C., Marcotte, L., Combes, V., Vartanian, N. 1996. Two-dimensional protein patterns of *Arabidopsis* wild-type and auxin insensitive mutants, axr1, axr2, reveal interactions between drought and hormonal responses. *Plant Cell Physiol.*, 37, 966-975.

Luche, S., Santoni, V., Rabilloud, T. 2003. Evaluation of nonionic and zwitterionic detergents as membrane protein solubilizers in two-dimensional electrophoresis. *Proteomics* 3, 249-253.

Mann, M., Pandey, A. 2001. Use of mass spectrometry-derived data to annotate nucleotide and protein sequence databases. *Trends Biochem. Sci.*, 26, 54-61.

Marchand, C., Le Marechal, P., Meyer, Y., Miginiac-Maslow, M., Issakidis-Bourguet, E., Decottignies, P. 2004. New targets of Arabidopsis thioredoxins revealed by proteomic analysis. *Proteomics* 4, 2696-2706.

Marques, K., Sarazin, B., Chane-Favre, L., Zivy, M., Thiellement, H. 2001. Comparative proteomics to establish genetic relationships in the *Brassicaceae* family. *Proteomics*, 1, 1457-1462.

Mathesius, U., Keijzers, G., Natera, S.H., Weinman, J.J., Djordjevic, M.A., Rolfe, B.G. 2001. Establishment of a root proteome reference map for the model legume *Medicago truncatula* using the expressed sequence tag database for peptide mass fingerprinting. *Proteomics*, 1, 1424-1440.

Millar, A.H., Sweetlove, L.J., Giege, P., Leaver, C.J. 2001. Analysis of the *Arabidopsis* mitochondrial proteome. *Plant Physiol.*, 127, 1711-1727.

Moons, A., Gielen, J., Vandekerckhove, J., Van der Straeten, D., Gheysen, G., Van Montagu, M. 1997. An abscisic-acid- and salt-stress-responsive rice cDNA from a novel plant gene family. *Planta*, 202, 443-454.

Natera, S.H., Guerreiro, N., Djordjevic, M.A. 2000. Proteome analysis of differentially displayed proteins as a tool for the investigation of symbiosis. *Mol. Plant Microbe Interact.*, 13, 995-1009.

Nijtmans, L.G., Henderson, N.S., Holt, I.J. 2002. Blue Native electrophoresis to study mitochondrial and other protein complexes. *Methods*, 26, 327-334.

Oda, Y., Huang, K., Cross, F.R.., Cowburn, D., Chait, B.T. 1999. Accurate quantitation of protein expression and site-specific phosphorylation. *Proc. Natl. Acad. Sci.* USA, 96, 6591-6596.

O'Farrell, P.H. 1975. High resolution two-dimensional electrophoresis of proteins. *J. Biol. Chem.*, 250, 4007-4021.

Ong, S.E., Blagoev, B., Kratchmarova, I., Kristensen, D.B., Steen, H., Pandey, A., Mann, M. 2002. Stable isotope labeling by amino acids in cell culture, SILAC, as a simple and accurate approach to expression proteomics. *Mol. Cell Proteomics*, 1, 376-386.

Panter, S., Thomson, R., Bruxelles, G.D., Laver, D., Trevaskis, B., Udvardi, M. 2000. Identification with proteomics of novel proteins associated with the peribacteroid membrane of soybean root nodules. *Mol. Plant Microbe Interact.*, 13, 325-333.

Patton, W.F. 2000. A thousand points of light: the application of fluorescence detection technologies to two-dimensional gel electrophoresis and proteomics. *Electrophoresis*, 21, 1123-1144.

Peck, S.C., Nuhse, T.S., Hess, D., Iglesias, A., Meins, F., Boller, T. 2001. Directed proteomics identifies a plant-specific protein rapidly phosphorylated in response to bacterial and fungal elicitors. *Plant Cell*, 13, 1467-1475.

Peltier, J.B., Friso, G., Kalume, D.E., Roepstorff, P., Nilsson, F., Adamska, I., van Wijk, K.J. 2000. Proteomics of the chloroplast: systematic identification and targeting analysis of luminal and peripheral thylakoid proteins. *Plant Cell*, 12, 319-341.

Peltier, J.B., Ytterberg, J., Liberles, D.A., Roepstorff, P., van Wijk, K.J. 2001. Identification of a 350-kDa ClpP protease complex with 10 different Clp isoforms in chloroplasts of *Arabidopsis thaliana. J. Biol. Chem.*, 276, 16318-16327.

Peltier, J.B., Ytterberg, A.J., Sun, Q., van Wijk, K.J. 2004. New functions of the thylakoid membrane proteome of Arabidopsis thaliana revealed by a simple, fast, and versatile fractionation strategy. *J. Biol. Chem.* 279, 49367-49383.

Perkins, D.N., Pappin, D.J., Creasy, D.M., Cottrell, J.S. 1999. Probability-based protein identification by searching sequence databases using mass spectrometry data. *Electrophoresis*, 20, 3551-3567.

Picard, P., Bourgoin-Greneche, M., Zivy, M. 1997. Potential of two-dimensional electrophoresis in routine identification of closely related durum wheat lines. *Electrophoresis*, 18, 174-181.

Plomion, C., Pionneau, C., Brach, J., Costa, P., Bailleres, H. 2000. Compression wood-responsive proteins in developing xylem of maritime pine (*Pinus pinaster* Ait.). *Plant Physiol.*, 123, 959-969.

Porubleva, L., Vander Velden, K., Kothari, S., Oliver, D.J., Chitnis, P.R. 2001. The proteome of maize leaves: use of gene sequences and expressed sequence tag data for identification of proteins with peptide mass fingerprints. *Electrophoresis*, 22, 1724-1738.

Prime, T.A., Sherrier, D.J., Mahon, P., Packman, L.C., Dupree, P. 2000. A proteomic analysis of organelles from *Arabidopsis thaliana. Electrophoresis*, 21, 3488-3499.

Rey, P., Pruvot, G., Becuwe, N., Eymery, F., Rumeau, D., Peltier, G. 1998. A novel thioredoxin-like protein located in the chloroplast is induced by water deficit in *Solanum tuberosum* L. plants. *Plant J.*, 13, 97-107.

Rey, P., Cuine, S., Eymery, F., Garin, J., Court, M., Jacquot, J.P., Rouhier, N., Broin, M. 2005. Analysis of the proteins targeted by CDSP32, a plastidic thioredoxin participating in oxidative stress responses. *Plant J.* 41: 31-42.

Riccardi, F., Gazeau, P., de Vienne, D., Zivy, M. 1998. Protein changes in response to progressive water deficit in maize: quantitative variation and polypeptide identification. *Plant Physiol.*, 117, 1253-1263.

Riccardi, F., Gazeau, P., Jacquemot, M.P., Vincent, D., Zivy, M. 2004. Deciphering genetic variations of proteome responses to water deficit in maize leaves. *Plant Physiol. Biochem.*, 42, 1003-1011.

Rigaut, G., Shevchenko, A., Rutz, B., Wilm, M., Mann, M., Seraphin, B. 1999. A generic protein purification method for protein complex characterization and proteome exploration. *Nat. Biotechnol.*, 17, 1030-1032.

Ross, P.L., Huang, Y.N., Marchese, J.N., Williamson, B., Parker, K., Hattan, S., Khainovski, N., Pillai, S., Dey, S., Daniels, S., Purkayastha, S., Juhasz, P., Martin, S., Bartlet-Jones, M., He, F., Jacobson, A., Pappin, D.J. 2004. Multiplexed protein quantitation in Saccharomyces cerevisiae using amine-reactive isobaric tagging reagents. *Mol Cell Proteomics* 3, 1154-1169.

Saalbach, G., Erik, P., Wienkoop, S. 2002. Characterisation by proteomics of peribacteroid space and peribacteroid membrane preparations from pea (*Pisum sativum*) symbiosomes. *Proteosomes*, 2, 325-337.

Salekdeh, G.H., Siopongco, J., Wade, L.J., Ghareyazie, B., Bennett, J. 2002. Proteomic analysis of rice leaves during drought stress and recovery. *Proteomics*, 2, 1131-1145.

Santoni, V., Delarue, M., Caboche, M., Bellini, C. 1997. A comparison of two-dimensional electrophoresis data with phenotypical traits in *Arabidopsis* leads to the identification of a mutant (cri1) that accumulates cytokinins. *Planta*, 202, 62-69.

Santoni, V., Rouquie, D., Doumas, P., Mansion, M., Boutry, M., Degand, H., Dupree, P., Packman, L., Sherrier, J., Prime, T., Bauw, G., Posada, E., Rouze, P., Dehais, P., Sahnoun, I., Barlier, I., Rossignol, M. 1998. Use of a proteome strategy for tagging proteins present at the plasma membrane. *Plant J.*, 16, 633-641.

Santoni, V., Doumas, P., Rouquie, D., Mansion, M., Rabilloud, T., Rossignol, M. 1999a. Large scale characterization of plant plasma membrane proteins. *Biochimie.*, 81, 655-661.

Santoni, V., Rabilloud, T., Doumas, P., Rouquie, D., Mansion, M., Kieffer, S., Garin, J., Rossignol, M. 1999b. Towards the recovery of hydrophobic proteins on two-dimensional electrophoresis gels. *Electrophoresis*, 20, 705-711.

Schubert, M., Petersson, U.A., Haas, B.J., Funk, C., Schroder, W.P., Kieselbach, T. 2002. Proteome map of the chloroplast lumen of *Arabidopsis thaliana. J. Biol. Chem.*, 277, 8354-8365.

Schweitzer, B., Kingsmore, S.F. 2002. Measuringproteins on microarrays. *Curr. Opin. Biotechnol.*, 13, 14-19.

Seigneurin-Berny, D., Rolland, N., Garin, J., Joyard, J. 1999. Technical advance: Differential extraction of hydrophobic proteins from chloroplast envelope membranes: a subcellular-specific proteomic approach to identify rare intrinsic membrane proteins. *Plant J.*,19, 217-228.

Sherrier, D.J., Prime, T.A., Dupree, P. 1999. Glycosylphosphatidylinositol-anchored cell-surface proteins from *Arabidopsis. Electrophoresis*, 20, 2027-2035.

Shevchenko, A., Schaft, D., Roguev, A., Pijnappel, W.W., Stewart, A.F. 2002. Deciphering protein complexes and protein interaction networks by tandem affinity purification and mass spectrometry: analytical perspective. *Mol. Cell Proteomics*, 1, 204-212.

Taylor, N.L., Heazlewood, J.L., Day, D.A., Millar, A.H. 2005. Differential impact of environmental stresses on the pea mitochondrial proteome. *Mol Cell Proteomics*, 4, 1122-1133.

Thiellement, H., Seguin, M., Bahrman, N., Zivy, M. 1989. Homeology and phylogeny of the A, S, and D genomes of the *Triticinae. J. Mol. Evol.*, 29, 89-94.

Verdoucq, L., Vignols, F., Jacquot, J.P., Chartier, Y., Meyer, Y. 1999. In vivo characterization of a thioredoxin h target protein defines a new peroxiredoxin family. *J. Biol. Chem.* 274, 19714-19722.

Vincent, D., Lapierre, C., Pollet, B., Cornic, G., Negroni, L., Zivy, M. 2005. Water deficits affect caffeate O-methyltransferase, lignification, and related enzymes in maize leaves. A proteomic investigation. *Plant Physiol.*, 137: 949-960.

Wasinger, V.C., Bjellqvist, B., Humphery-Smith, I. 1997. Proteomic "contigs" of *Ochrobactrum anthropi*, application of extensive pH gradients. *Electrophoresis*, 18, 1373-1383.

Wasinger, V.C., Cordwell, S.J., Cerpa-Polja, A., Yan, J.X., Gooley, A.A., Wilkins, M.R., Duncan, M.W., Harris, R., Williams, K.L., Humphery-Smith, I. 1995. Progress with gene-product mapping of the Mollicutes: *Mycoplasma genitalium. Electrophoresis*, 16, 1090-1094.

Wolters, D., Washburn, M.P., Yates, J.R. 2001. An automated multidimensional protein identification technology for shotgun proteomics. *Anal. Chem.*, 73, 5683-5690.

Yamaguchi, K., Subramanian, A.R. 2000. The plastid ribosomal proteins. Identification of all the proteins in the 50 S subunit of an organelle ribosome (chloroplast). *J. Biol. Chem.*, 275, 28466-28482.

Yamaguchi, K., von Knoblauch, K., Subramanian, A.R. 2000. The plastid ribosomal proteins. Identification of all the proteins in the 30 S subunit of an organelle ribosome (chloroplast). *J. Biol. Chem.*, 275, 28455-28465.

Zhu, H., Bilgin, M., Bangham, R., Hall, D., Casamayor, A., Bertone, P., Lan, N., Jansen, R., Bidlingmaier, S., Houfek, T., Mitchell, T., Miller, P., Dean, R.A., Gerstein, M., Snyder, M. 2001. Global analysis of protein activities using proteome chips. *Science*, 293, 2101-2105.

ADDENDUM TO CHAPTER 12

12 BIS PHOSPHOPROTEOME ANALYSIS

M. Rossignol

Among the *ca* 350 types of protein modification described to date, protein phosphorylation constitutes a major post-translational modification. Actually, this reversible modification is known to be able to determine several key properties of proteins such as activity, interaction with other proteins or subcellular localization. In *A. thaliana*, the regulation of protein phosphorylation status requires more than 5% of the genes, of which *ca* 4.5% encode protein kinases whilst 0.5% encode protein phosphatases. At the protein structure level, two main features have to be taken into account. On one hand, only certain amino acids are observed to become phosphorylated and with different frequencies, pSer being much more frequent than pTyr. On another hand, proteins possess often more than one phosphorylation site that can be active in a sub-stoichiometric way. As a consequence of this diversity, the protein phosphorylation status is usually complex and its analysis requires the capacity to detect and quantify low-abundance phosphorylation sites. In recent years, several technological advances have been made and the main routes presently available to characterize phosphoproteomes are described below together with recent issues in plant phosphoproteomics. For more details, the reader is referred to various general reviews (Mc Lachlin and Chait, 2001; Mann *et al.*, 2002; Kalume *et al.*, 2003; Chalmers *et al.*, 2004; Peters *et al.*, 2004,; Zeller and König, 2004; Raggiaschi *et al.*, 2005; Reinders and Sickmann, 2005) and to reviews and protocols (Aro *et al.*, 2004; Laugesen *et al.*, 2004) in the plant area.

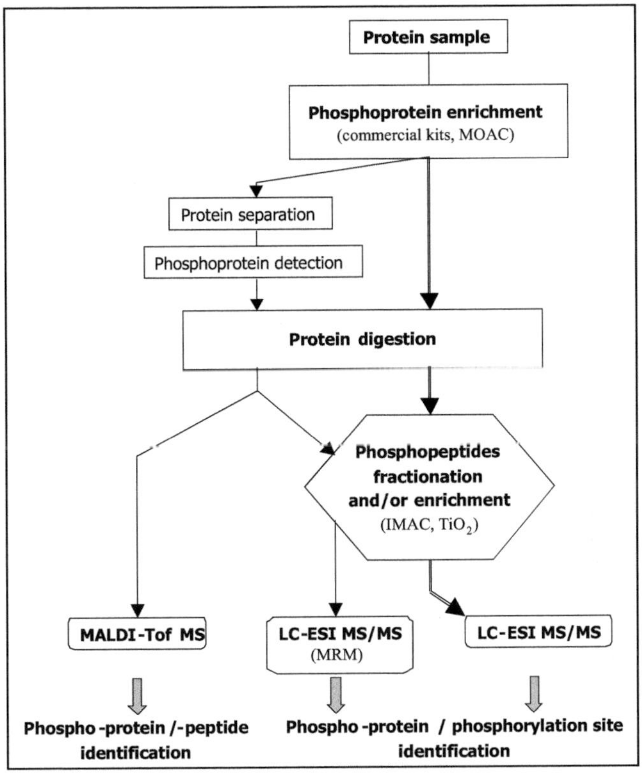

Fig. 12.4 Main routes for the characterization of protein phosphorylation

As in general proteomics, two main approaches can be used to characterize phosphoproteomes (Figure 12.4). In the first option, proteins are resolved prior to analysis. Alternatively, the analysis can be performed directly from the digest of a complex set of proteins. In both cases, current strategies include two main steps: the selection of phosphopeptides followed by their characterization by MS. In addition, prior enrichment of phosphoproteins often constitutes an additional useful step.

***Phosphoproteins enrichment*:** beside immunoprecipitation using anti-phosphoamino acid antibodies, recently available procedures involve either immobilized metal affinity chromatography (IMAC; BD™ kit), a chromatographic procedure that has not been clearly described (Qiagen kit), or metal oxide affinity chromatography (MOAC) (Wolschin *et al.*, 2005b). For proteins resolved by gel electrophoresis, novel tools for the detection of phosphoproteins were also introduced that eliminate the need for matching twin gels, such as when using the classical detection method after ^{32}P labelling, or anti-phosphoamino acid antibodies to detect phosphorylated proteins by western blotting. In this way, proteins can be successively probed on the same gel first for phosphoproteins with the specific fluorophore Pro-Q Diamond™ and then with any other MS-compatible dye for total proteins.

***Phosphopeptides selection*:** IMAC constitutes the most used tool to purify phosphopeptides and several chromatographic media are commercially available. The procedure can be easily scaled down for microcolumns. In addition IMAC can be coupled to preceding chromatography on strong anion or cation exchangers for both improved removal of non-phosphorylated peptides and phosphopeptide fractionation according to their phosphorylation level (Nühse *et al.*, 2003). Very recently, another procedure using chromatography on titanium dioxide was introduced as showing superior efficiency in selecting phosphopeptides (Larsen *et al.*, 2005).

***MS analysis of phosphopeptides*:** whereas phosphopeptides can be detected by MALDI-ToF MS (and identified provided that the protein itself could be identified), the precise location of phosphorylation sites requires MS/MS. In both types of configuration, the ability of pSer and pThr to loose phosphoric acid (- 98 Da) and of pTyr to be dephosphorylated (- 80 Da) offers further tools to identify phosphopeptides either from PSD spectra (MALDI-ToF MS) or neutral loss scan (ESI MS/MS). In addition, in the negative ion detection mode, the precursor diagnostic ion PO_3^- at *m/z* 79 can be selectively used to identify phosphopeptides in a mixture. Such approaches are convenient to investigate complex samples of (phospho)peptides. When the phosphorylation status of a known protein is to be characterized in detail, the multiple reaction monitoring strategy (MRM, on triple quadropole mass spectrometers) was recently shown to have superior sensitivity due to simultaneous selection of the parent ion and of the diagnostic ion (Unwin *et al.*, 2005).

In the plant area, the interest in a detailed characterization of the protein phosphorylation status is growing, as recently summarized (Laugesen *et al.*, 2004). Here, the focus is given specifically to very recent efforts illustrating both the classical use of current phosphoproteomics procedures and innovative approaches. Western blot analysis on 2D gels (eventually confirmed by protein dephosphorylation) still constitutes, in combination with MS identification, a simple way to decipher changes in protein phosphorylation (Yang *et al.*, 2003; Gonzalez-Camacho and Medina, 2004; Chivasa *et al.*, 2005), although providing no information about the location of phosphorylation sites. For generating such information, LC-MS/MS offers the most usual way that can be informative even in large scale approaches aimed at cataloguing proteomes, without an initial enrichment in phosphoproteins (Casati *et al.*, 2005). The recent MOAC method to select phosphoproteins was initially used to probe *A. thaliana* leaf protein samples and shown to constitute a powerful alternative procedure to the use of commercial kits (Wolschin and Weckwerth, 2005; Wolschin *et al.*, 2005b). However, even when starting from an enriched sample, the neutral loss of phosphoric acid during fragmentation may often lead to ambiguous database identification from MS/MS data. To circumvent this problem, an additional fragmentation (MS^3) of the remaining precursor was successfully introduced to identify the phosphorylation sites of *A.thaliana* sucrose-phosphate synthase (Wolschin *et al.*, 2005a). On another hand, more dedicated MRM procedures were also proven to allow for differential analysis of multiple phosphorylation sites in trehalose-6-phosphate synthase isoforms (Glinski and Weckwerth, 2005). Beside the powerful combination of refined phosphopeptide selection by IMAC and classical LC-MS/MS (Nühse *et al.*, 2003), these recent examples are likely to signal the introduction of a new generation of more sophisticated MS strategies having the capacity to decipher the phosphorylation status of proteins of interest. A last feature of recent efforts in phosphoproteomics and plant phosphoproteomics concerns the quantification of phosphorylation levels. Beside commercially available reagents to label peptides, two main routes are presently emerging. On one hand, proteins can be differentially labelled *in vivo* using stable isotopes, thus allowing direct quantification from MS data obtained with a mixture of proteins separately purified from two samples of interest. This procedure was recently shown to allow the characterization of phosphorylation-dependent protein interactions in *A.thaliana* cell suspensions (Gruhler *et al.*, 2005). Alternatively, the possibility to derive semi-quantitative data directly from MS/MS spectra of unlabelled samples was also recently explored (Glinski and Weckwerth, 2005; Wolschin *et al.*, 2005).

REFERENCES

Aro, E.M., Rokka, A., Vener, A.V. 2004. Determination of phosphoproteins in higher plant thylakoids. *Methods in Molecular Biology*, 274, 271-286.

Casati, P., Zhang, X., Burlingame, A.L., Walbot, V. 2005. Analysis of leaf proteome after UV-B irradiation in maize lines differing in sensitivity, *Mol. Cell. Proteomics*, 4, 1673-1685.

Chalmers, M.L., Kolch, W., Emmett, M.R., Marshall, A.G., Mishchak, H. 2004. Identification and analysis of phosphopeptides, *J. Chromatogr.*, B 803, 111-120.

Chivasa, S., Simon, W.J., Yu, X.L., Yalpani, N., Slabas, A.R. 2005. Pathogen elicitor-induced changes in the maize extracellular proteome, *Proteomics*, 5, DOI: 10.1002.pmic.200500047.

Glinski, M., Weckwerth, W. 2005. Differential multisite phosphorylation of the trehalose-6-phosphate synthase gene family in *Arabidopsis thaliana* – a mass spectrometry-based process for multiparallel peptide library phosphorylation analysis, *Mol. Cell. Proteomics*, 4, 1614-1625.

Gonzalez-Camacho, F., Medina, F.J. 2004. Identification of specific plant nucleolar phosphoproteins in a functional proteomics analysis, *Proteomics*, 4, 407-417.

Gruhler, A., Schulze, W.X., Matthiesen, R., Mann, M., Jensen, O.N. 2005. Stable isotope labelling of *Arabidopsis thaliana* cells and quantitative proteomics by mass spectrometry, *Mol. Cell. Proteomics*, 4, 1697-1709.

Kalume, D.E., Molina, H., Pandey, A. 2003. Tackling the phosphoproteome: tools and strategies, *Cur. Opin. Chem. Biol.*, 7, 64-69.

Larsen, M.R., Thingholm, T.E., Jensen, O.N., Roepstorff, P., Jorgensen, T.J.D. 2005. Highly selective enrichment of phosphorylated peptides from peptide mixtures using titanium dioxide microcolumns, *Mol. Cell. Proteomics*, 4 873-886.

Laugesen, S., Bergoin, A., Rossignol, M. 2004. Deciphering the plant phosphoproteome: tools and strategies for a challenging task, *Plant Physiol. Biochem.*, 42, 929-936.

Mann, M., Ong, S.E., Grønborg, M., Steen, H., Jensen, O.N., Pandey, A. 2002. Analysis of protein phosphorylation using mass spectrometry: deciphering the phosphoproteome, *Trends Biotechnol.*, 6, 261-268.

McLachlin, D.T., Chait, B.T. 2001. Analysis of phosphorylated proteins and peptides by mass spectrometry, *Cur. Opin. Chem. Biol.*, 5, 591-602.

Nühse, T.S., Stensballe, A., Jensen, O.N., Peck, S.C. 2003. Large-scale analysis of *in vivo* phosphorylated membrane proteins by immobilized metal ion affinity chromatography and mass spectrometry, *Mol. Cell. Proteomics*, 2 1234-1243.

Nühse, T.S., Stensballe, A., Jensen, O.N., Peck, S.C. 2004. Phosphoproteomics of the *Arabidopsis* plasma membrane and a new phosphorylation site database, *Plant Cell*, 16, 2394-2405.

Peters, E.C., Brock, A., Ficarro, S.B. 2004. Exploring the phosphoproteome with mass spectrometry, *Mini Rev. Med. Chem.*, 4, 313-324.

Raggiaschi, R., Gotta, S., Terstappen, G.C. 2005. Phosphoproteome analysis, *Biosci. Rep.*, 25, 33-44.

Reinders, J., Sickmann, A. 2005. State-of-the-art in phosphoproteomics, *Proteomics*, 5, 4052-4061.

Unwin, R.D., Griffiths, J.R., Leverentz, M.K., Grallert, A., Hagan, I.M., Whetton, A.D. 2005. Multiple reaction monitoring to identify sites of protein phosphorylation with high sensitivity, *Mol. Cell. Proteomics*, 4 1134-1144.

Wolschin, F., Weckwerth, W. 2005. Combining metal oxide affinity chromatography and selective mass spectrometry for robust identification of in vivo phosphorylation sites, *Plant Methods*, 1, DOI: 10.1186/1746-4811-1-9.

Wolschin, F., Lehman, U. Glinski, M., Weckwerth, W. 2005a. An integrated strategy for identification and relative quantification of site-specific protein phosphorylation using liquid chromatography coupled to MS^2/M^3, *Rapid Commun. Mass Spectrom.*, 19, 3626-3632.

Wolschin, F., Wienkoop, S., Weckwerth, W. 2005b. Enrichment of phosphorylated proteins and peptides from complex mixtures using metal oxide/hydroxide affinity chromatography (MOAC), *Proteomics*, 5, 4389-4397.

Yang, E.J., Oh, Y.A., Lee, E.S., Park, A.R., Cho, S.K., Yoo, Y.J., Park, O.K. 2003. Oxygen-evolving enhancer protein 2 is phosphorylated by glycin-rich protein 3/wall-associated kinase 1 in Arabidopsis, *Biochem. Biophys. Res. Commun.*, 305, 862-868.

Zeller, M., König, S. 2004. The impact of chromatography and mass spectrometry on the analysis of protein phosphorylation sites, *Anal. Bioanal. Chem.*, 378, 898-909.

13

CHAPTER

Tracking Gene Expression in Plant Cells: Microscopy and Associated Bio-imaging Techniques

S. Brown, S. Bolte, B. Satiat-Jeunemaître

INTRODUCTION

Light microscopy is booming as a tool for functional cell biology and genomics, due to several factors.

(a) Optical innovations have brought improved resolution notably when faced with difficult diffuse material.

(b) New light sources manipulated with opto-electronic or opto-acoustic devices facilitate optimal management of diverse configurations, offering versatility and flexibility plus the possibility to modify or to manipulate the sample with light.

(c) Probes have become more stable and bright, more specific, molecular and functional.

(d) Digital imaging is conducive to image restoration and analysis which at long last truly integrates knowledge of the image formation process.

In this chapter we shall deal with the technical aspect of cellular bio-imaging, including fluorochromes, but leaving the important theme of genetically-coded probes for the complementary Chapter 14 (Bolte *et al.*, 2007). To appreciate many of the innovations raised in this chapter, we must stress that "The image of a point taken with a microscope is not a point" (Ronot and Usson, 2001), and hence the image of an object may be very different from its true shape and size. The 3D energy distribution in the image of an infinitely small brilliant point is in reality an ovoid volume encased in a succession of dark and light ovoids produced by diffraction interference. The final image of a point has an hourglass form and the function describing this diffraction pattern is called Point-Spread Function (PSF). The PSF is an important element for imaging software; its corollary is that image information and statistics are different in the XY plane (less dispersed) than along the Z-axis (more dispersed). The sample properties, mounting medium, coverslip, immersion medium and wavelength will all affect the PSF which is in fact the complex result (the "convolution") of multiple imbricated interacting optical elements. Our chapter should encourage researchers to understand and "to feel" the overall chain of elements and processes that come into play during the light microscopy of cells.

MICROSCOPY

Wide Field Microscopy

A modern infinity-corrected optical microscope consists of an objective (bearing the ∞ symbol) and a tube lens to form the image, with an ocular to gain magnification. A condenser focuses the illuminating light onto the specimen for transmitted light microscopy.

Bright field. Most people are familiar with bright field transmission microscopy, the most basic microscopy technique and the place to start before progressing to more sophisticated imaging techniques like confocal microscopy. With an understanding of the optical components, their function and modulation, the choices of modes and of objectives, one can optimise and adapt the instrument to the specimen, to find the best balance between resolution, contrast, focus depth, speed, tissue integrity or viability. Lacey (1999) and Shaw (2001) are excellent introductory texts with basic setup protocols. The first crucial step in bright field microscopy is the adjustment of the optical components of the microscope (Köhler alignment), which the microscopist should check at each session (refer to the handbook provided with the microscope).

Dark field. Dark field microscopy uses a hollow cone of oblique (razing) light so that no signal enters the objective directly. Only light which is scattered by highly reflective or diffusive objects such as silver particles can be seen. Yasuda *et al.* (2001) used this procedure to observe the single molecular rotation of ATPase.

Phase contrast. The poor inherent contrast of certain biological structures, notably membranes or chromosomes, may be overcome by exploiting the property that refractive index gradients will shift the phase of light. These shifts are recombined with positive or negative interference amplified so that our modest human eyes can perceive them as intensity differences. Phase contrast simplifies locating cells, yeast buds or isolated chromosomes (in cytogenetics) with a large depth of field, useful when microinjecting or manipulating. The appropriate objectives carry a phase ring and the condenser a phase plate.

Differential interference contrast (DIC). In an analogous manner, Differential Interference Contrast (DIC) is based on the phenomenon that any light crossing a refractive index gradient at an angle will undergo a shift of electromagnetic polarization. The method was invented by the Nomarski brothers and is applicable with most high quality objectives. It uses two polarizers, one before the condenser and the other after the objective, and Nicol prisms to set up and analyse orthogonal regimes of incident light. It reveals edges in biological structures such as cell and organellar boundaries, the nucleus, nucleolus, chromosomes, and microorganisms. Koehler alignment is absolutely necessary. DIC can offer remarkable resolution and contrast, and the depth of field can simply be adjusted from ~2 µm (e.g. one nucleolus, with condenser diaphragm fully open) to about 8 µm (e.g. a whole nucleus, with condenser diaphragm fairly closed). DIC is commonly used for live cell studies and is a routine complement to any epifluorescence microscopy, including confocal microscopy, where cell form or condition needs to be checked. A related technique is polarised light microscopy, to detect regular or crystalline structures.

Epifluorescence. The exciting development of fluorescent molecular probes has facilitated multicolour and hence multiparametric exploration of the cell. Much as flow cytometry is the gold-standard for quantitative fluorometry, epifluorescence microscopy has become a basic tool in many laboratories. The term epi- ("above") comes from the principle that the objective contributes not only to forming the image, but also acts as the condenser providing incident light.

The optical layout of an epifluorescence microscope (Fig. 13.1A) begins with a light source: powerful high-pressure mercury or xenon lamps, or novel and convenient metal halide short-arc lamps. The filter block firstly ensures selection of the appropriate spectral window for excitation and then this light is reflected onto the sample by its dichroic mirror. The longer wavelength fluorescent emission then passes back through this dichroic mirror and the third component of the block, the emission filter, cleans up this signal. The detection of emitted fluorescence may be recorded by eye, film or digital camera.

A problem of wide field fluorescence microscopy is the blur of images acquired. This blur is caused by the capture of light originating from objects out the focal plane (Fig. 13.2): for instance a tiny fluorescent point may be a nanometric light source but its light will spread over many microns. This out of focus light might be re-assigned by mathematical methods such as the important algorithms of deconvolution (Figs. 13.2A, E), or discriminated against using specialised optical systems (Figs. 13.2A, B, C, D).

Confocal Microscopy

Confocal microscopy, using a sophisticated epifluorescence microscope, resolves two major problems: it reduces the depth of field (providing a thinner focal slice) and eliminates light from out-of-focus objects (reducing background) (Fig. 13.1B) (Pawley, 1995). Conceived in 1957 but popularised in the late 1980s due to the availability of components, confocal microscopy partially overcomes the constraint of 3D blur by optical sectioning. It thus substantially improves Z-resolution (down the observation axis) and modestly improves that on the XY plane (Fig. 13.2A). This microscopy is practised on both fixed and living material, and typically facilitates multicolour imaging of membranes, organelles, cells or tissue in complex diffusive structures such as epidermis and meristems. For instance, one can image right through an *A. thaliana* root (~130 µm) while, for reference, vesicles are similar in size to its resolution limit.

The confocal scheme blocks out from the detector all light coming from outside the focal plane. This is due to coordination ("confocality") of three points: the excitation source, the objective's focal point and a pinhole focused on the detector path. The diameter of this detection pinhole is matched to the limit of diffraction of light coming

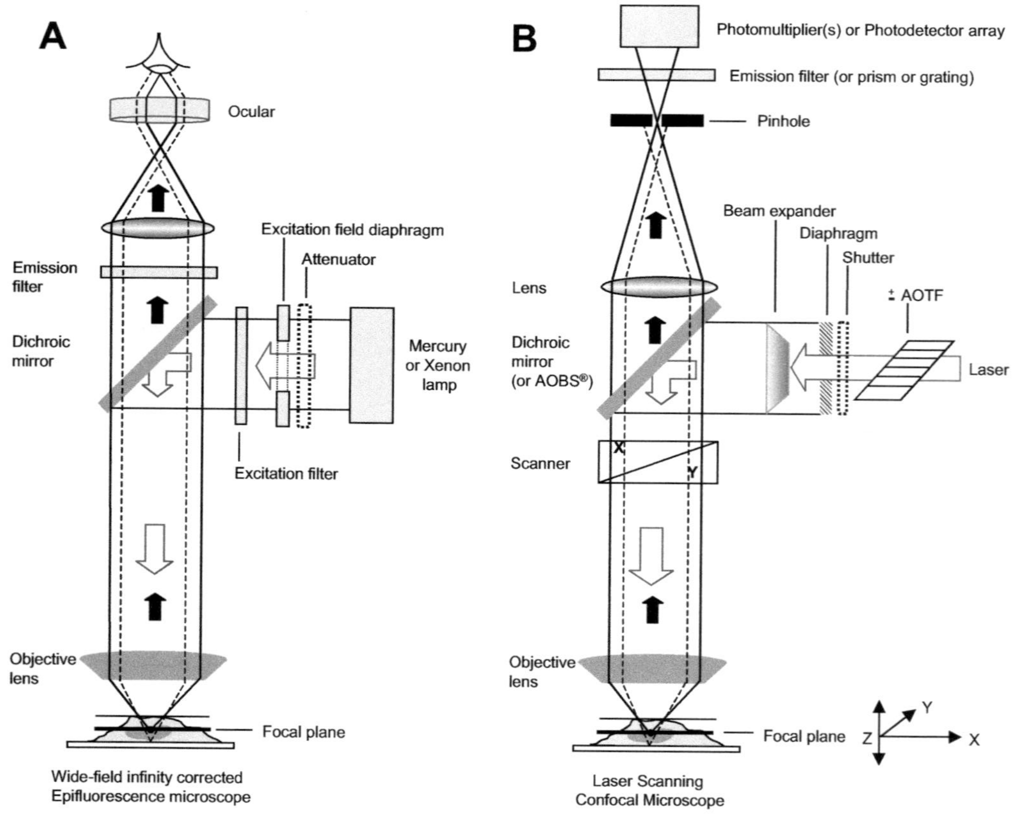

Fig. 13.1 Comparison of the optical paths of a wide-field microscope and a laser-scanning confocal microscope. In these examples, the microscopes are of infinity-corrected design, having a long sector of parallel light rather than the classical convergent light path of fixed tube-length microscopes: this parallel sector is advantageous for insertion of new optical elements without introducing aberrations. The essential principle of a confocal microscope is a pinhole ensuring confocality (coordination) of three points: a small light source (typically a laser with a diaphragm), the focal point of the objective (acting as a condenser of incident light and captor of emitted light), and the pinhole itself. Normally, the adjustable pinhole is set to accept the first Airy disc of the diffraction pattern (Fig. 13.2A), which is dependent upon wavelength and numerical aperture. The light detector (usually a photomultiplier, sometimes a camera, or array) is exposed to a signal coming from the focal plane (—) while light from higher or lower planes is mostly blocked (- - -). The result is deblurring, and improved resolution, notably along the axis of observation, the Z-axis. Galvanometric mirror(s) ensure scanning along XY-axes, and subsequent displacement of the stage or objective allows construction of the XYZ image stack. Certain temporal relations should be kept in mind: absorption is in ~fs, fluorescence ~ns, the time for an opto-electronic device (AOTF) to change laser lines or modify dichroic (AOBS®) properties is ~μs, a pixel is also ~μs and an image plane will be ~1s.

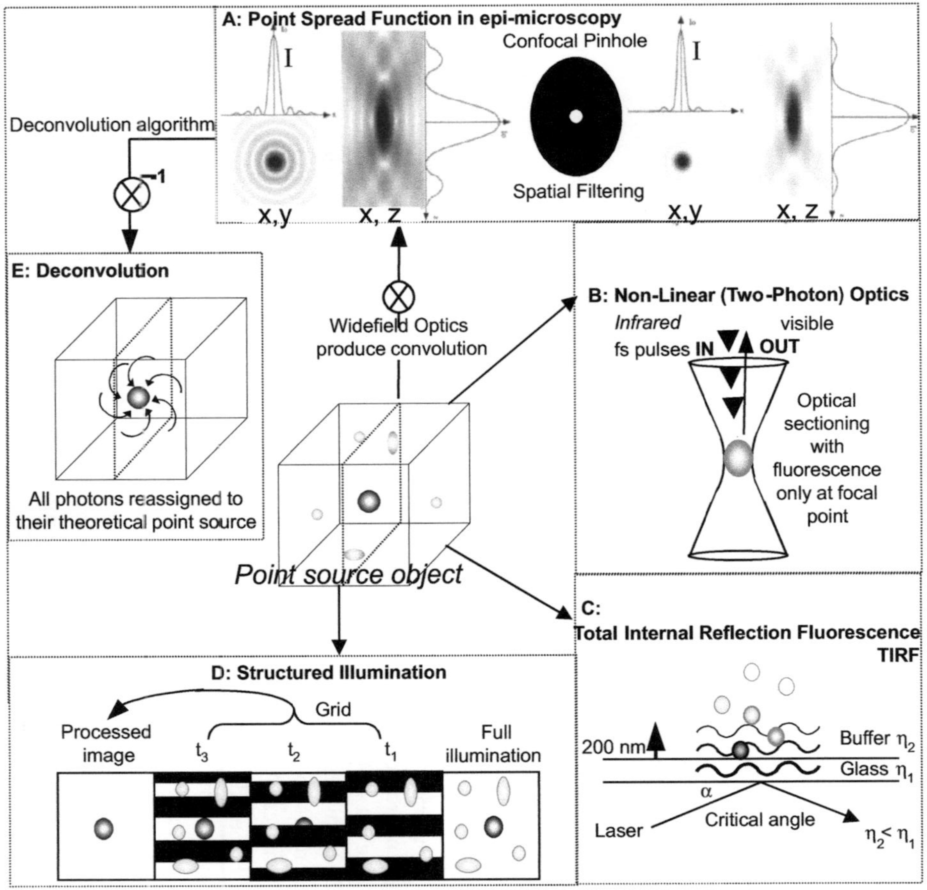

Fig. 13.2 (A, left) "The image of a point is not a point" (Ronot and Usson, 2001). Epifluorescence and epireflectance microscopy are hampered by the convolution produced in optical systems, distributing the 3D-energy as described by the Point Spread Function (PSF). (E) Mathematical deconvolution performs the inverse operation on 3D image stacks, reassigning photons and thereby improving contrast on all planes. (A, right) Confocal microscopy relies on spatial filtering to eliminate out-of-focus sources, but in so doing wastes photons. (D) Structured Illumination (e.g. Apotome® or Optigrid®) uses sequential excitation with a focussed vibrating grid to distinguish objects truly on the focal plane. (B) Non-Linear Optics (NLO) achieve optical sectioning by producing fluorescence only at the focal point, so avoiding out-of-plane interference. (C) Total Internal Reflection Fluorescence, TIRF, relies upon an evanescent electromagnetic wave (rather than transmitted photons) to excite only those objects close to the glass surface: the bulk fluorophores are not excited. (Other processes such as diffusion and readsorption also contribute to blur.)

from the object, normally accepting the first Airy disc of the diffraction pattern in order to optimise spatial resolution *versus* signal strength (Fig. 13.2A). Other solutions can also ensure confocality, and notably the spinning Nipkow disc (e.g. CARV II® from BD Biosciences, or UtraVIEW® from PerkinElmers).

The formal equation for the resolution limit, d, in the XY plane of an optical microscope is:

$$d_{xy} = 0.61\,\lambda / NA \qquad \text{Eq. 1}$$

where λ is the wavelength of light and NA is the Numerical Aperture of the objective; namely,

$$NA = \eta \sin \alpha \qquad \text{Eq. 2}$$

where η is the refractive index of the medium used with the objective and α the half solid angle of its light acceptance. In a confocal microscope,

$$d_{xy} = 0.4\,\lambda / NA \qquad \text{Eq. 3}$$

and

$$d_z = 1.4\,\lambda / (NA)^2 \qquad \text{Eq. 4}$$

The resolution in the image plane, XY, is improved by ~30% in comparison to wide field microscopy, attaining ~200 nm with an excellent objective (300 nm for wide field). There is a spectacular improvement in signal-to-noise along the axis of focus, Z, offering ~350 nm resolution and important deblurring. From these equations, resolution improves in a linear manner with wavelength, blue being better than red, and the Numerical Aperture of the objective is key.

For epifluorescence or epireflectance, the source is typically a laser, one or several wavelengths being scanned along the axes X and Y to produce an optical section in about 1 s. The fluorescence comes back up the scanning path through the pinhole and, after separation of the different coloured emissions, the sequential readout from several detectors forms several related images, pixel by pixel, scanned point by point. A typical image will be 1024 x 1024 points. The sample is then stepped along the Z axis (or exceptionally the objective can be moved rather than the stage) for another optical section and finally to create a Z-series. This image stack can be used for 3D reconstructions or, more often, the 3D information is projected onto one plane.

The spectral characteristics of the chosen fluorochromes and the available excitation windows (or laser lines) will determine the optical filters necessary for excitation and selective detection of one or several markers (Fig. 13.3). For plant biology, spectral flexibility and multicolour acquisition are essential, because the target signal may be mixed with intrinsic autofluorescence which can itself provide useful information (Meyer *et al.*, 2003), or needs to be avoided (see example of an autofluorescence spectrum in Fig. 13.3F; Cerovic *et al.*, 1999). The systems for selecting and modulating the laser lines vary from one manufacturer to another; current machines being far more rapid and convenient than those of even five years ago. Notably, acousto-optical tuneable filters (AOTF, Fig. 13.1B) facilitate the permutation of laser lines in µs, the time of one pixel. Another component can replace dichroic mirrors so that reflecting a laser line into the path only occupies several nm of the spectrum, rather than blocking a band of 15-20 nm: the acousto-optical beam splitter (AOBS®, Leica; Fig. 13.1B) and the distinctive AcroGate® (Zeiss) achieve this. Microspectroscopy is a useful addition for

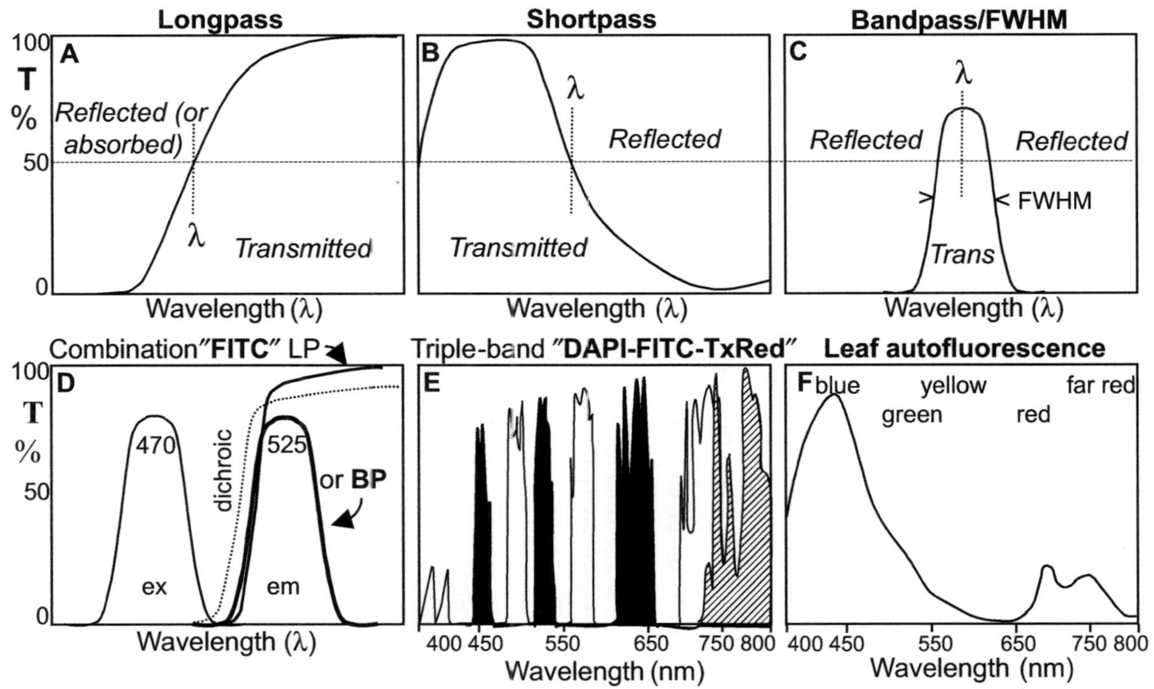

Fig. 13.3 Simplified spectra of some optical filters: (A) Longpass, (B) Shortpass, (C) Bandpass with Full Width at Half Maximum (FWHM). (D) Combination for fluorescein with either a longpass emission filter for observation (LP 515 nm) or a more specific bandpass (BP 512-542 nm). (E) A simplified representation of Omega Optical's Triple-band XF67 for simultaneous observation of DAPI, FITC and Texas Red: the excitation bands are shown as clear, the emission bands as solid (www.omegafilters.com). The supplier points out that this block is good for visual observation, but lacks protection in the infrared (hatched spectra) where cameras will still be sensitive. OD, Optical Density: data on a logarithmic scale are necessary to adequately assess the blocking power of the filter. For reference, the emission spectrum of a fresh beet leaf is shown in F (from Cerovic *et al.*, 1999).

understanding and handling autofluorescence, significant in plant material, and to resolve overlapping fluorochromes or ratiometric dyes which are popular for mapping physiological responses. The technical solution may be a prism in front of the photomultipliers or dispersion onto a multianode detector (Fig. 13.1). The aforementioned options are not luxuries, as confocal microscopy is not one great leap forward but it is panoply of useful advances. Improved spatial, temporal and spectral resolution all contribute to a better vision of the cell. There are plenty of texts on the subject that are well worth studying (Pawley 1995; Periasamy, 2001; Ronot and Usson, 2001; Benoit *et al.*, 2005).

A special application, Fluorescence Correlation Spectroscopy (FCS) uses the elementary confocal volume to measure the diffusion coefficients of molecules, complexes or particles in solution and within the cell. The entry and exit from this tiny volume is assessed in a quantum mode for one or several different markers. For instance, the entry and exit of two fluorophores will be correlated only if they are on the same mobile complex. The stoichiometry of complexes and dynamics of interaction between biomolecules can be calculated from these correlations (Guérin and Torreilles, 1998; Bachia and Schwille, 2003; Lamb *et al.*, 2005).

Non-Linear Optical Microscopy (Two-photon)

Multiphoton microscopy is based on the spatial *and* temporal focussing of light, using a special laser source, to obtain simultaneous femtosecond (fs) interaction of two or three near-infrared (IR) photons whose summed energy will be sufficient to excite a visible fluorochrome (Ronot and Usson, 2001). For example, two photons at 760 nm (or three at 1,030 nm) can excite DAPI (diamidino-phenylindole) as would a more energetic single photon at 350 nm. Femtosecond lasers excite the fluorochrome only at the focus, so there is no out-of-plane emission (Fig. 13.2B). Wide "non-descanning" detectors just behind the objective can pick up emitted photons associated with each impulsion, whether they come straight back (the "ballistic" photons) or even those which follow an irregular diffusive path. The technique is appropriate for relatively thick and diffusive material such as leaves and roots as IR penetrates much further than visible light, less bleaching is produced out-of-plane, and diffusion is less catastrophic. However, the IR can produce local heating. When using light as a tool, photo-activation, photo-switching, photo-ablation and photo-bleaching (see FRAP) effects will be confined to the focal point rather than to a long cone of incident light as with linear sources. If a confocal microscope can go to 100 μm in a certain tissue, the multiphoton mode may go twice as far. Feijo and Moreno (2004) have written an elegant review of two-photon microscopy with plant material. The feasibility of imaging in depth depends very much on the tissue. For a rose leaf with waxes, cellulose, starch, aqueous and air phases, 30 μm is already difficult (Brown, unpublished). Furthermore, delivering probes to thick material can be a limitation, so intrinsically produced probes like Green Fluorescent Protein (GFP) or autofluorescence are particularly relevant (Chapter 14). Pulsed-lasers are adapted to Fluorescence Lifetime Imaging Microscopy (FLIM). The disadvantages of a multiphoton microscope are its cost, its complexity of use, tissue destruction during adjustment, and the difficulty of multiparametric studies, since the two-photon excitation spectra tend to be flattened and non-specific.

Other Light Microscopies, Readily Accessible or Not

Microtubules are ~25 nm in diameter. Their image under light microscopy will be ~200 nm, and it is not possible to resolve two microtubules lying strictly side-by-side. Multimolecular complexes are 1-2 orders smaller than that which diffraction-limited light microscopy can resolve. However, optical nanoscopy is emerging. "Near field" techniques such as atomic force microscopy offer nanometric resolution, far below the diffraction limit of light, but are not easily applicable to living biological material. Other approaches using Spatially Modulated Illumination (SMI) allow a topological resolution of 2-20 nm (Failla *et al.*, 2002), at the scale of multimolecular complexes. This may be applied in studies of transcription machinery and macro-organisation of the nucleus, but the constraints on sample thickness preclude use on living cells. The 4π two-photon confocal microscope, with two objectives face-to-face, improves Z-resolution four-fold. By adding in the Stimulated Emission Depletion (STED) mode, resolution can be brought down to 30 nm (Egner and Hell, 2005; www.mpibpc.gwdg.de/abteilungen/200).

It is worth following such innovative *haute couture,* as the imaging of macromolecular structures in a wide field cellular environment is becoming reality. Selective Plane Illumination Microscopy (SPIM) is a high resolution optical sectioning technique for strongly diffusive live specimens such as embryos, where the sample is rotated in an aqueous chamber intercepted by a laser focussed to a knife edge (Huisken *et al.*, 2004).

Total Internal Reflection Fluorescence. Much more accessible technically (and financially) is Total Internal Reflection Fluorescence (TIRF) microscopy (Fig. 13.2C), based on the excitation of fluorochromes by the evanescent ("quickly fading") field of totally internally reflected angular light. This evanescent electromagnetic field propagates only 20-300 nm from the coverslip interface into the culture medium or cellular membrane and cortex. It thereby provides Z-super-resolution without exciting background fluorescence from fluorophores further away. As objects move down onto this surface, they become visible in high contrast. For dynamic processes such as cell adhesion and *in vitro* assembly of macromolecular structures, it is an elegant and relatively simple solution. The regulation of actin nucleation and bundling activity by the plant Formin Homology 1 domain is elegantly demonstrated *in vitro* with TIRF (Michelot *et al.*, 2005 and their Supplementary Movies).

Structured illumination. Structured illumination systems, such as Apotome® (Zeiss) and OptiGrid® (Thales and Improvision), focus the light source onto the sample plane as a grid which is then stepped laterally: that which is susceptible to the mask of the grid will "blink" as it lies on the focal plane (Fig. 13.2D). This is extracted mathematically as the essential information and displayed on a screen, suppressing the constant components of the image which are the out-of-focus signal and noise. These systems may be convenient for live studies on thick specimens and offer inexpensive excitation from 350-700 nm; but they generally demand strong labelling. They lack the complementary tools of spectroscopy and the possibility to intervene with light beams (FRAP), etc.

Examples from cell biology with various light microscopy techniques are presented in Fig. 13.4 (www.microscopyu.com and the authors' website www.ifr87.cnrs-gif.fr/

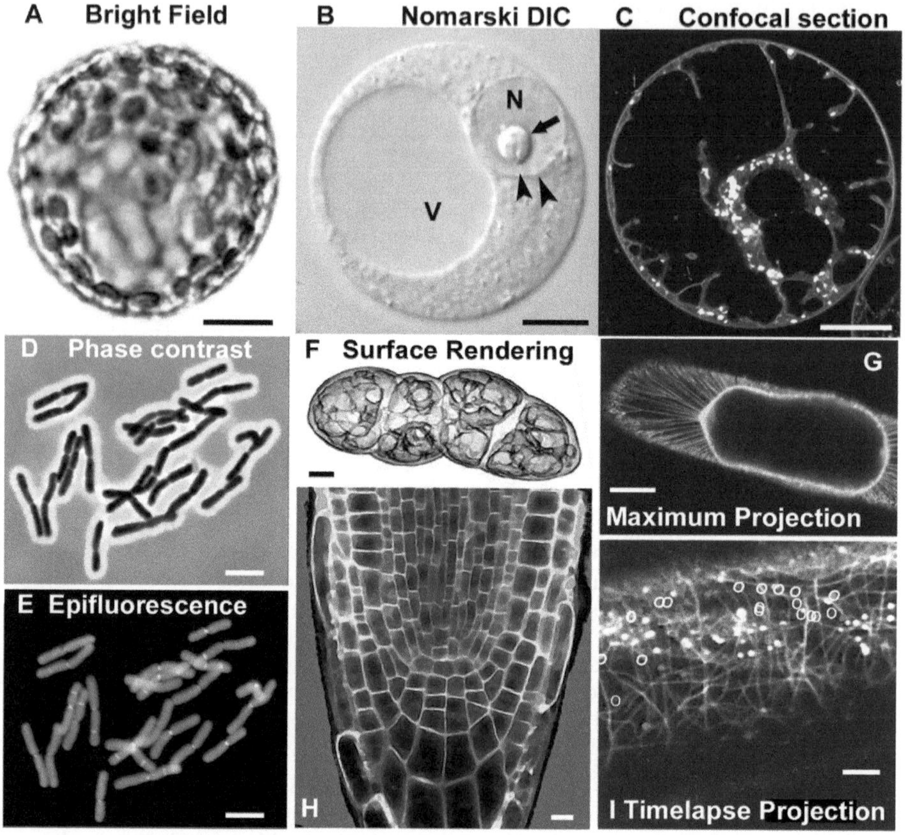

Fig. 13.4 Examples from light microscopy. (A) Bright field transmission image of a live mesophyll protoplast, with typical high field-depth. The chloroplasts dominate by their high absorptivity. (B) Differential Interference Contrast (mode Nomarski) image of a live protoplast from a tobacco BY2 culture, relying essentially upon refractive index differences. N, nucleus with its nucleolus (arrow) and nuclear membrane (arrowheads); V, vacuole. Here the condenser diaphragm is fully open, to obtain optical sectioning with a narrow depth of field. (C) A single confocal microscopy plane through the median section of such a live protoplast with dual fluorescent labelling, FM4-64 for membranes and GFP::sialyl transferase for Golgi stacks (Bolte *et al.*, 2004). (D,E) Metaphase chromosomes of *Pinus mugo* (2n = 24). Prior to laborious *in situ* hybridization or fluorochrome banding, Phase Contrast (D) has been used to locate the appropriate chromosomes on dry slides. (E) GC-banding with Chromomycin A under wide field epifluorescence microscopy. (F-I) Confocal microscopy. (F) A cluster of four live BY-2 cells with vacuolar labelling. This surface rendering is calculated from a confocal XYZ image stack. (G) A *Maximum Projection* (or *Extended Focus*) of a small confocal Z-stack through the median of a plasmolysed onion cell transiently expressing GFP, revealing the Hechtian strands and residual cytoplasmic labelling around a vacuolar zone ˙(H) An *Averaging Projection* of a Z-stack through the median of a living *A. thaliana* root labelled with FM4-64 to reveal cellular contours, at 90 μm confocal depth. (I) Tracking peroxisome movement in onion cells, superimposed upon a projection of 10 images over 10 seconds. Both the peroxisomes and microtubules were labelled with GFP fusion proteins. (D,E: Sonja Siljak-Yakovlev, CNRS, Orsay; A: Olivier Catrice, F: Vincent Fraisier, G: Marije Scholte, I: Patricia Durand and the late Hanh Trinh, ISV, Gif-sur-Yvette). Scale bars 10 μm.

pbc/contact/index.html). A confocal microscope should be equipped with a transmitted light detector to offer non-confocal bright field or DIC images, which may be overlaid upon the epifluorescence image, for it is obviously useful to record cell or tissue morphology.

Electron Microscopy

The remarkable developments concerning light microscopy should not cast doubt upon the importance of electron microscopy for subcellular exploration at 100-1,000-fold superior resolution, attaining the level of polymers and molecular complexes. "Plant Cell Biology: Structure and Function" (Gunning and Steer, 1996) is an outstanding atlas of the cell seen through both light and electron microscopes.

The transmission electron microscope images a thin specimen in a high vacuum by focussing an electron beam with electromagnetic fields. The image appears on a fluorescent screen, or is recorded with film or digital camera. Many protocols for plant material are presented in Roland and Vian (1991), Dashek (2000) and Hawes and Satiat-Jeunemaître (2001); EM immunolabelling is treated in Bolte *et al.* (Chapter 14). Preparative protocols such as ultra-rapid freeze-fixation and freeze-substitution now favour faithful preservation of cellular ultrastructure. A current development is goniometry ("angular-"), where the sample is tilted to various angles to obtain an image stack that conveys 3D information for tomo-graphy ("section-") (Segui-Simarro *et al.*, 2004).

In the scanning electron microscope, a finely focussed beam of electrons is scanned in a raster pattern (progressive lines) over the surface of the specimen. Reflected or emitted electrons form an image on a cathode ray tube, reflecting the surface topology of the specimen. Scanning electron microscopy can now handle live specimens (leaf, seed, hypocotyl, microdissected tissue, etc.) without prior metal coating, with or without a cooling stage for longer observation. This is particularly useful for rapid screening where topology is the key to understanding, such as with developmental mutants. This is generally much simpler and more instructive than 3D reconstruction from confocal microscopy.

MICROSCOPY ACCESSORIES

Objectives. Manifold objectives are available, to be considered not only for different factors of magnification, but also to provide a large light collection angle (Numerical Aperture) while ensuring planarity ("Plan") and correcting for chromatic aberrations (e.g. "Apochromatic"). The objective may be appropriate for Interference Contrast or specially fitted for phase contrast. A set of objectives will usually include some with dry lenses (convenient), oil immersion (better light collection due to the uniform refractive index between the coverslip and glass lens) and water immersion (maintaining symmetry between this water and the aqueous media of living cells). The frontal working-distance is a practical consideration with bulky plant material where some high power objectives soon hit the coverslip or sample. Water dipping lenses, used without coverslips, offer exceptional working-distances and the possibility to add biochemicals or to microinject while observing. Furthermore, a ring on the objective may control an iris, or correct for coverslip thickness, or for refractive differences in the optical path (media, glass, plastic, etc.).

Filters and Dichroic mirrors (Fig. 13.3). Optical filters use either absorbance or interference. Absorbant filters transmit certain colours and absorb the rest. Interference filters either transmit or reflect each wavelength. Dichroic filters are a particular application of interference filters, designed to transmit certain wavelengths while reflecting others, normally used at 45° to the incident beam to split it according to the wavelength. Longpass and Shortpass filters are named after their inflection point, as are dichroics; Bandpass filters are named after their maxima, with reference to their Full Width at Half Maximum (Figs. 13.3A-C). Acousto-optical devices (AOTF, AOBS) use the interference between radiowaves and crystals to achieve far superior light separation, with steeper transitions. As an example, to send a 488 nm laser beam into a microscope (Fig. 13.3B) a dichroic mirror will occupy a band of 20 nm whereas an AOBS will have a reflection band of ~4 nm, thus allowing much more of the fluorescent emission to be returned to the detectors. These devices can also be redesigned ("accorded") in several microseconds.

Fluorescein has an excitation maximum of ~489 nm and emission of ~521 nm. A generous filter combination for fluorescein in epifluorescence microscopy (Fig. 13.3D) will consist of a bandpass incident-light filter (excitor BP 450-490 nm), a dichroic splitter mirror (DS 505 nm), and a longpass emission filter for observation (LP 515 nm). To limit the observation to the green, this latter filter may be replaced with a more specific bandpass (BP 512-542 nm). The narrow bandpass filter will reject any stray reflected light or other fluorescent emissions, while a generous longpass filter would allow simultaneous observation of fluorescein and intrinsic chlorophyll fluorescence. In practice, both are needed: generous filters so that the intelligent eye can appreciate the various hues of mixed fluorescence, and highly specific filters so that captors can hopefully handle one signal at a time.

Complex multi-band filter combinations are available. The triple-pass filter in Fig. 13.3E is designed for simultaneous observation of nuclei and two immunofluorescent labels with respectively DAPI (diamidino-phenylindole) or bisbenzimide Hoechst, fluorescein and Texas Red. The XF67 block has three excitation bands, a complex dichroic mirror with three transitions, and three emission windows. Technical aids are available on the commercial sites, Omega Optical www.omegafilters.com and Chroma Optical www.chroma.com and probes.invitrogen.com (Haugland, 2005).

For versatility in fluorescence microscopy, the following filter blocks are readily justified: excitation UV (for DAPI, bisbenzimide Hoechst), violet (for plant autofluorescence, Calcofluor, Cyan Fluorescent Protein, quantum dots and other emergent probes), blue (fluorescein, GFP, etc.), green (Texas Red, Propidium Iodide, MitoTracker), red (for "Far Red" dyes like TOTO3, Cy5, DRAQ5). Whole ranges of superior fluorochromes (Alexa Fluors and Cyanines) suit these possibilities (Chapter 14; Haugland, 2005). A buyer soon invests as much in filters as in objectives.

DETECTION OF LIGHT FOR IMAGES

Of the many devices for detecting light, we shall discuss only two in detail. Cameras simultaneously capture an array of pixels and hence form an image. Photomultipliers capture a light flux as a point measurement (one pixel), and so some sequential scanning protocol is necessary to develop an image. Chapter 5 of Ronot and Usson (2001) provides excellent teaching material on light captors.

Photomultipliers

A photomultiplier tube has a photo-sensitive photocathode (0 Volts) in a vacuum behind a glass window. When incoming photons hit the photocathode, secondary electrons are emitted and attracted to the next electrode (termed a dynode) which is at a higher voltage. New secondary electrons are emitted and thus induce a cascade of perhaps 20 stages up to the anode, where the current is measured. This current is directly proportional to the incoming photons. These devices are point captors with excellent linear amplification over a wide dynamic range and with low noise. Most confocal microscopy users will be familiar with adjusting the zero voltage ("offset") and the anode voltage ("gain"), typically between 200-800 V. The spectral sensitivity of any photomultiplier should be obtained from the manufacturer. Photomultipliers are much less sensitive to red light as emitted by chlorophyll or the new far-red probes, than they are to green or blue, typically a fourfold difference. Moreover, it must be cautioned that different models may be installed in various positions on a multichannel imaging device. In most confocal microscopes, the scanner continuously sends light signals to the photomultipliers and a clock chops their output current, each interval constituting one pixel.

Cameras

In a CCD (*Charge Coupled Device*) camera, a matrix of captors convert photon energy into electrons during an integration time (exposure), the charge is then transferred to a storage area and further to a read-out register, like a "bucket brigade". The charge is finally converted to volts and digitised between Maximum and Minimum, using 8, 12 or 16 bits which corresponds to 256, 4,096 or 65,536 gray levels. The camera electronics can be adjusted so that a certain voltage is taken as black (zero offset) and another as saturated white (full scale); these hardware functions precede any subsequent contrast enhancement with software.

Plant biologists generally need color. In fluorescence imaging, colour will help distinguish the marker from background auto-fluorescence. So this is the first choice for a digital camera. Here, a single CCD chip has elements that are distinctively sensitive to Red Green Blue (RGB). If each is converted on 8 bits, the resulting image will be 24-bit and an uncompressed 1.4 M pixel image will occupy 4.2 MB. The camera electronics also need to be adjusted so that each color has its full dynamic range, black (zero) and a properly balanced slope for the three fundamental colours RGB. Associated software usually offers this function termed White Balance. One selects a truly balanced (but not saturated) white region in the microscope field and instructs the camera to readjust electronic color sensitivity to produce true white in the image.

A camera is also characterized by how much charge can accumulate on a pixel, its "full-well capacity". Many modern cameras can accumulate up to around 16,000 electrons, which only merits 12-bit Analog-to-Digital conversion. However, for convenience these files might be stored in a common 16-bit format.

One way of gaining sensitivity is "binning", where the charges on 2 x 2 capture elements (or 4 x 4) are pooled together as one bigger capture element to gain signal, speed and to have less read-out noise. Slower read-out is also cleaner, e.g. reading at 10 MHz rather than 20 MHz if time is not important. An anti-blooming function is usually available so that saturated spots do not spill over their charge into the down-stream pixels, with a smudge effect: this can be deactivated to gain sensitivity. The thermal agitation in electrical circuits randomly releases "noisy" electrons, thus affecting image quality. This can be reduced by cooling the camera with an integrated Peltier device. Even uncooled colour cameras can offer good fluorescence photomicroscopy with exposures well under a second. One word of reserve: few colour cameras achieve satisfactory colour balance and faithful reproduction of hues.

An imaging device may offer supplementary modes with more "effective pixels" than true on-chip pixels. An integrated liquid crystal optic may enable very rapid spatial resampling, micro-stepping to the east, west, north and south, to improve the average image but with a cost in overall time. Inversely, some cameras allow switching to black and white mode to gain sensitivity on the three (color) detectors, somewhat like electronic binning, while others offer this only as a way of simplifying the data and its storage. Sub-array reading (partial scan mode), where only a Region-of-Interest (ROI) is read, is a useful tool to gain reading time and to economise data-space. A "still CCD camera" is limited by slow data transfer. Modern CCD cameras can work at video rates (25 images/s) but even provisional storage of image data soon becomes a challenge.

Therefore, at long last digital cameras offer a view down a microscope almost as good as the eye, images as good as film, and far more convenient – what you see is what you get. Their detail with colours is inferior to the eye. Their sensitivity can be much better than eye and film, both for intensity and for the red to infrared spectrum. This advantage is essential when using some modern far-red markers (TOTO3, Alexa Fluor 633, Cy5, Cy7, etc.). Conversely, it also explains why some images that appear excellent to the eye can be washed out with unseen infrared in a CCD digital camera that is not fitted with an anti-IR filter!

Further Cameras

This brief commentary cannot do justice to the tremendous offer in the camera world, where more sensitive now rhymes with robustness and accessibility. Single molecule fluorescence imaging is feasible. For example, calcium imaging has a difficult temporal constraint to capture waves, spikes, sparks and flashes in both minutes and ms. Ordinary cameras may not be sufficient to image these events. A novel strategy has been applied in the quadrant-anode Time Stamped – Single Photon Counter (TS-SPC) imager (IPD 425 from Photek, UK with Institut Pasteur, Paris). The fluorochrome in a cell may be imaged firstly by transmission and epifluorescence microscopy. Then, instead of matricial imaging, single photons are detected with their X, Y, t coordinates, the time stamp being accurate to ms (Rogers *et al.*, 2005). All this is recorded in a spreadsheet, conventionally termed *Listmode* as in flow cytometry. At the end of the experiment (typically 30 minutes), the data is replayed and extracted with the appropriate resolution, using Regions-of-Interest and Times-of-Interest. A silent cell is ignored for 20 minutes, then analysed with ms resolution for a few seconds of activity. Its neighbor might have slow fluxes, analysed slowly over a longer period. Such post-analysis implemented by *Listmode* has been an essential flexibility in cytometry (Coba de la Peña and Brown, 2001), and is to be encouraged in imaging. Note that the present TS-SPC equipment has poor photon efficiency (~10%), a narrow spectral response, and poor spatial resolution.

Quite simple digital camera stations find application in growth and development studies, facilitating time-lapse recording rather than end-point measurements in processes like *in vitro* root growth, especially if they have automatic focussing devices to compensate for movement. An application developed to identify stomatal closure mutants, uses thermography to detect spatial and temporal variation of stomatal conductance over leaf surfaces (Jones, 1999). In a single image of a pot containing thousands of small *A. thaliana* plants, stomatal closure mutants may be identified thanks to their lower temperature upon incipient water stress due to the persistence of open stomata and higher evapotranspiration. This infrared thermography ($\lambda > 3$ μm) has a practical resolution of about 1°C (Merlot *et al.*, 2002). At the other end of the spectrum, Cerovic *et al.* (1999) developed whole-plant fluorescence imaging (and also biosensing, namely, without images) to assess physiological and pathological factors in the environment, using the emissions of coumarins (blue), flavins (green-yellow), chlorophylls (red) and, indirectly, the polyphenols due to their UV-screening effects (DuaLex of Force-A, Gif-sur-Yvette; www.shef.ac.uk/~ifpm/contents.html).

The CMOS (*Complementary Metal Oxide Semiconductor*) technologies have certain advantages over CCD and will increasingly appear in the laboratory (Holst, 2005; www.micron.com, www.pco.de). CMOS is cheaper, faster, has a lower energy demand, less blooming, random access to pixels, allows on-chip conversion and calculations with Very Large Scale Integration (VLSI). Essentially, the detector is just part of a semiconductor chip, costing maybe 100 €, and other functions and memories can be built into the same microcircuit to obtain an intelligent retina. Some home or cellular telephone cameras are CMOS.

Light Management

Halogen lamps generally provide the light for transmission microscopy. For fluorescence, they may be useful and economic as a soft source. However, normally high-pressure mercury and xenon arcs are used. Xenon has a more uniform spectral output and longer lifetime. Both are dangerous. Given the improved efficiency of fluorochromes, optics and captors, the current emphasis is on light quality rather than sheer energy, which had dictated the choice of mercury lamps during the last 60 years. Fortunately, alternative sources such as safer 2000 hour metal halide lamps are available. Light Emitting Diodes (LED) are also appearing on microscopes.

Light management is an essential yet neglected aspect of many microscope installations (Fig. 13.5).

- Turning on a transmission lamp wastes 4s before the color is stable for a photograph, whereas an electronic shutter allows millisecond synchronization between opening and exposure in live cell studies.
- Using fiber optics to introduce the high-power fluorescence excitation lamp will make the field more uniform, while also moving the lamp away from the work area.

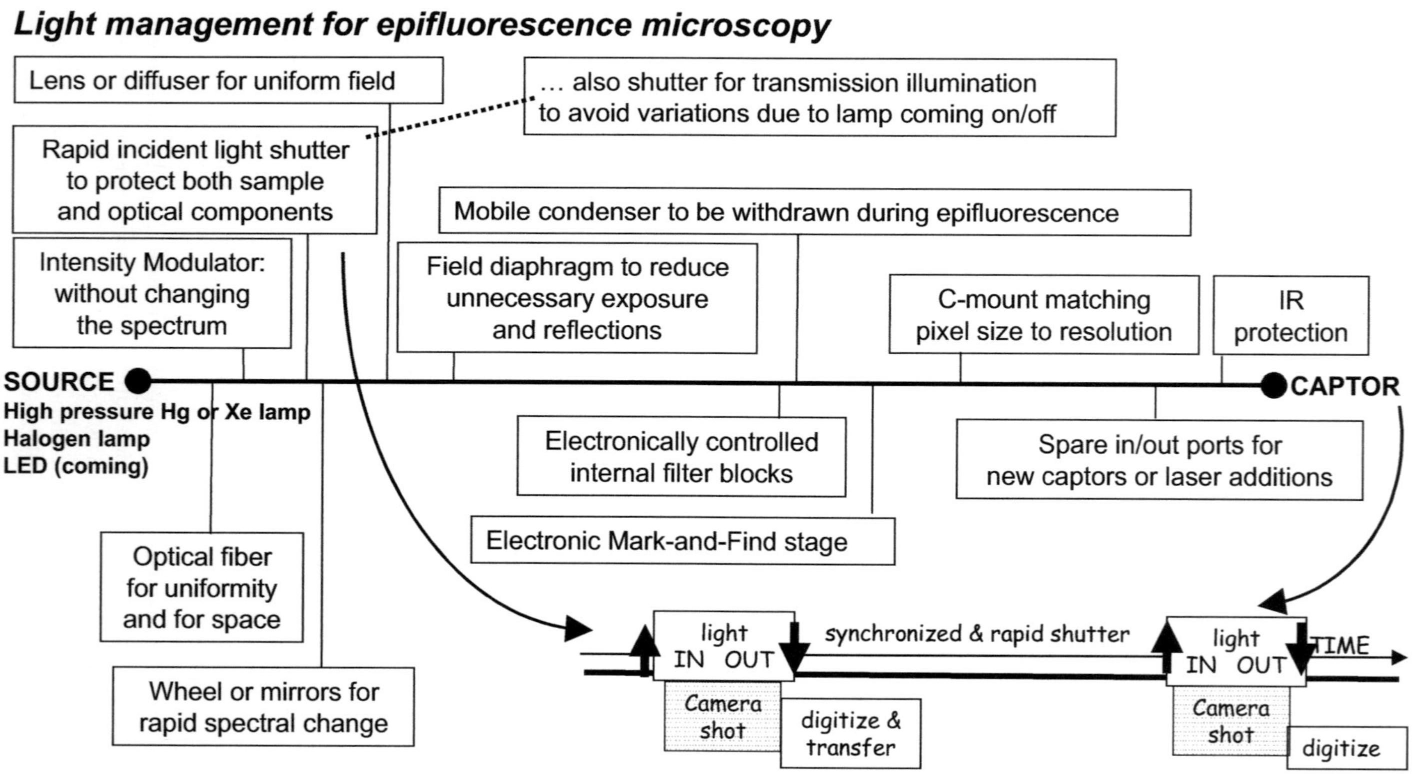

Fig. 13.5 Elements for management of light in epifluorescence microscopy. The upper part concerns essential features, the lower part optional features especially useful for videomicroscopy (including a Mark-and-Find electronic stage). The arrows refer to synchronization between camera and shutter so that if, for instance, one were to take a 50 ms image every second, the light will be blocked for the remaining 95% of the time.

- If less time is needed to capture, to digitise and transfer images, what do we do for the rest of the time in kinetic experiments? Using a fast shutter on the excitation path, the lamp may be blocked 4 ms after camera exposure while the charge transfer, digitisation and read-out are done, radically reducing sample exposure (e.g. Olympus Cell-R® module). Sharp light management is the natural complement to better cameras, even when fast imaging is not needed (Fig. 13.5).
- It is important to use field diaphragms to avoid light exposure beyond the bare essential. Otherwise, there is unnecessary peripheral cell and probe exposure, and contrast is compromised by lateral reflections.
- Furthermore, lamps and lasers must be blocked when not used, to protect filters, objectives, fibers and other optical components.

Low-light Signals

Low-light imaging requires high sensitivity cameras typically with quantum efficiencies better than 70%, a large full-well capacity at each pixel (≥18,000 electrons), low read-out noise, binning flexibility, and especially low electronic noise so that long integration times are feasible (Greer and Szalay, 2002). The latter usually means cooling the chip to 20°K below room temperature, or -20°C or even -120°C, at a price!

Bioluminescent genetic reporters such as luciferase are discussed in this volume (Bolte *et al.*, 2006). Most luminescent data are simply obtained *in vitro* with a luminometer, taking a data point every second. However, whole organisms, tissues or cell clusters can be imaged in a black box with a macrolens system (or microscope), firstly to register a bright field image and then in the dark to integrate photon emission over 1-10 minutes to obtain the luminescence images. Their spatial and temporal resolution are theoretically as good as other optical images (< 1 µm), but in practice these are limited by low light flux and hence by the long integration times (10-1,000 s). The resultant image tends to average out local dynamics during this preset integration and sometimes animals and plant tissues may move. Spikes tend to become flattened waves.

For instance, compare the temporal resolution of the following examples. Calcium fluxes in stomata have been assayed with the genetically-encoded fluorescent biosensor, Cameleon: with an experimental point every 11 seconds, one can observe rapid spikes (Allen *et al.*, 2001 for experiments lasting 3 hours). Calcium activity can also be imaged via aequorin bioluminescence in whole plants, each point necessitating ~15 minutes integration: this is sufficient to observe calcium waves of diurnal rhythms, but not spikes (Love *et al.*, 2004). Aequorin is very effective for studying calcium transients *in vitro* (Lecourieux *et al.*, 2002).

Complex plug-and-play stations are available for low-light imaging, while competition and progress in cameras are favoring price reduction. With green plant material, charge recombination can provoke fluorescence long after chloroplasts are in the dark, so background noise will be high during the first minutes of obscurity (Love *et al.*, 2004). Otherwise, the beauty of chemiluminescence is the absence of other light. The question of dynamic range is particularly important: how to handle, in parallel, both hot and cool spots and uneven background? True 12 and 16 bit cooled cameras with low chip noise are essential. However, two innovative strategies that may help solving these problems have been mentioned, the TS-SPC and the Intelligent Retina CMOS, where each captor (pixel) auto-adapts independently to local light flux.

Spectral Aspects

What is the effect of each component in terms of spectral response? With so many detectors around us – the eye, photomultipliers, diode-arrays, CCD chips, CMOS, etc. – the researcher needs to know their specifications, to take into consideration their spectral responses that can complicate signal cross-talk, auto-fluorescence, background, etc. Two photomultipliers on the same microscope may differ five-fold in their sensitivity to chlorophyll fluorescence, and both will be far more sensitive than the eye.

Fortunately, many epifluorescent microscopes offer spectral flexibility. Its simplest translation is to ensure a broad choice of excitation and emission filters, often overlooked at the time of purchase. The imaging channel may be broadened to take more signal or made narrower to be more specific. In a more complex but adaptive configuration, this can be achieved with a prism and mobile slits in front of a cascade of photomultipliers (as in the Leica confocal microscopes) or on a photomultiplier array (as in the Zeiss Meta®) or a photodiode array. On the excitation side, multiple lasers with acousto-optical tuning are the best invention since sliced-bread!

To be more precise, real-time spectroscopy can be performed on optical sections, pixel by pixel or region by region, using the devices just mentioned (Berg, 2003; Lerner and Zucker, 2004). This may serve to identify components present in a complex situation - expression of several reporter proteins (Zimmermann *et al.*, 2002), or fluorescent reporters in the presence of autofluorescence, or to characterise photoconversion of probes like Kaede (Bolte *et al.*, 2006). Indeed, because of cuticles, air-pockets and cell-wall thickening, reflectance is often a problem in confocal microscopy, so a good looking image can in fact be an unrecognised mixture of fluorescence (e.g. GFP) and reflectance from the laser. This artefact is quickly identified by spectroscopy. Moreover, spectral shifts due to the cellular environment of the probe may be usefully identified: emission of the styryl dye FM4-64 undergoes a 100 nm hypsochromic shift (= to shorter wavelengths) in protoplast membranes (Bolte *et al.*, 2004). Hutzler *et al.* (1998) provided an elegant example of spectral imaging for tissue localization of phenolic compounds in plants.

The intensity of a fluorochrome can be used as a reporter of cellular microenvironment. However, an image will also be affected by factors such as cell size, depth, rate of dye loading, etc. A key strategy to neutralise these ancillary factors in order to improve precision is ratiometric imaging, for which two images are captured under conditions where their fluorescence is inversely dependent upon the environmental factor. Spectral ratiometry is an important tool in fluorometric physiological assays of pH, calcium, etc. (Fricker *et al.*, 2001). Technically, emission spectroscopy is simpler on a microscope than excitation spectroscopy. However, one simple example is imaging of the weak acid fluorescein, excited sequentially at two wavelengths (Hoffmann and Kosegarten, 1995). The image (*img1*) obtained with deep blue excitation (~450 nm) reports fluorescein content; the image (*img2*) obtained with blue (~490 nm) reports both content and its ionization state. The pixel-by-pixel ratio image, $imgR = img2 / img1$, removes the factor "content" and reveals the pH topology, valid within a range (~pH 5.2 – 7.1) dependent upon the pKa of the probe (pKa 6.2).

POST-ACQUISITION

We have already learnt that the image of a point is not a point (Ronot and Usson, 2001) and that the 3D energy distribution in the image of an infinitely small brilliant point is characterised by the Point-Spread Function (PSF) (Fig. 13.2A). Furthermore, any optical system has its spatial resolution limit, which the user is supposed to know and to use! Indeed, the minimal justifiable sampling frequency depends upon this. The Nyquist (or Shannon) Theorem indicates that, in order to harvest information faithfully, sampling should be 0.4 times the resolution limit of the optical system used. This indicates the minimal step size when scanning, or the minimal object pixel size in cameras (Pawley, 1995; Benoit *et al.*, 2005).

Image Processing

An image acquired by an optical system, comprises the convoluted real image plus noise from the biological, optical and physical environments. The analysis, quantification and convenient presentation of scientific results based on microscopical images are crucial steps and it is therefore important to introduce some key-terms of image processing.

Image restoration uses a model of the degradative processes to mathematically reverse these and to obtain an improved image (Ronot and Usson, 2001). This may be a simple geometric correction for deformation or uneven illumination. More powerfully, deconvolution (or inverse convolution) reassigns each photon to its point source, using an experimental or theoretical Point Spread Function (Fig. 13.2E). In this way it deblurs by removing out of focal-plane light. It also improves contrast by reassigning photons to their material source. What a confocal microscope would block out or reject, deconvolution re-assigns. This can give excellent results, but image acquisition is critical, processing is laborious, and the heterogeneity of specimens often precludes application of any uniform Point Spread Function. The term deconvolution is often loosely applied to convey simple deblurring algorithms such as maximum likelihood projections that are available in standard imaging software. Real deconvolution algorithms using theoretical or measured PSFs, however, are only available in specialized imaging software packs such as Metamorph (Roper Scientific), SimplePCI (Hamamatsu) or AutoQuant Imaging (www.aqi.com). Deconvolution algorithms may be applied to an image stack (XYZ) usually obtained with a camera on a wide field microscope, but also to confocal series. For reviews on deconvolution techniques, see Wallace *et al.* (2001) and Abal *et al.* (2002).

After image restoration, image enhancement is a subjective process using numerical filters for smoothing, contrast enhancement, modifying the dynamic range, etc. in a stored data file. Image enhancement includes tools to better convey information from the image, such as modifying the relationship between data and their representation on a screen, printer or projector by modifying the Look Up Table (LUT or data Transfer Function). This includes using false colors (the eye is sensitive to many hues but to few grey levels), modifying contrast (slope) and brightness offsets, gamma (non-linearity), etc. The image displayed from these operations may or may not be stored.

These kinds of treatments are available with almost all standard imaging software delivered with your microscope or available on the www as freeware (ImageJ). Note that Photoshop which contains many of these filters is not a good example of optical software. It is very useful for image enhancement and communication, but its image statistics do not take into account any experimental limitations.

Image analysis is the process of extracting information, such as segmentation, area extraction, removal of certain structures, quantification of form, integration of intensities, granularity, etc. It includes simplification to fulfil certain aims. A useful tool is the freeware ImageJ, an image processing program based on Java and available online (rsb.info.nih.gov/ij). Numerous plugins permit the rapid analysis of images and stacks and the constant arrival of new features makes it the basis of imaging analysis software in biology laboratories.

Multidimensional Imaging

Confocal imaging and image processing allow the researcher to move to-and-fro between object and image: to extract a sharp plane from a volume, but also to constitute 2D images summarising the three dimensions of the object. In an image stack, the pixels (2D **pic**ture **el**ements) become voxels (3D, **vol**ume **el**ements). Several methods are then used to represent the 3D information. The successive images may be displayed as a gallery or over time as a sequence. Transverse sections (YZ) may be physically acquired as such, or recalculated from the stack.

A major tool is to project some or all of the 3D information onto one plane, firstly along the Z axis, but also by rotating Z to Z′. Many algorithms allow this information to be extracted under different conditions. The principle of ray casting uses virtual rays coming from far away and moving through the 3D volume to intercept the viewing screen (Ronot and Usson, 2001). The intensity displayed at the interception point is the "history" of the ray. To take one example, if 100 sections of a cell are projected onto one plane with a simple Averaging algorithm, much detail may be compromised. If small mitochondria were evident in only a few thin sections, their contrast may be destroyed by adding in the noise or information of almost 100 other images. For this case, the Maximum Projection algorithm (or Extended Focus) would retain the strong signal as the rays traverse each coordinate in the data stack, while weak values and background are abandoned (Fig. 13.4G). Another algorithm, for objects enclosed one within another like Russian dolls, uses Transparency: the mathematical rays traverse the 3D data and at each significant encounter, the intensity value of a voxel is weighted by a transparency coefficient and this is added to the previous value of the ray.

A pair of projections with slightly different positive and negative rotations (6°) can be displayed selectively for each eye to give a stereo effect. This is generally achieved with two complementary colour codes to provide a pseudo-3D perception of one channel (anaglyph). If two or more parameters have to be co-represented, other stereo techniques are available for colour such as observation of projections through cross-polarised glasses, or with intercalating videos and synchronised glasses as used in structural biology.

More delicate is the 3D use of segmentation tools to determine the surface of objects (Fig. 13.4F: subsequently enhanced by tricks of lighting and shading), or volume rendering (usually with transparency functions). Haseloff (2003) has provided an example of elegant segmentation analysis of root tissues; Grandjean *et al.* (2004) used the approach to carry out an *in vivo* analysis of meristem growth and development. Morphometry aims at counting objects, object volumes, shapes, coordinates, colocalisation and relations. (Bolte and Cordeliéres, 2007). The data will often be reduced to binary forms: black and white, in or out. On the other hand, granulometry or densitometry aims at integrating fluorescence, with statistics of density, variance, ratios, etc. Densitometry may include some binary step to extract forms as "binary masks", but these protocols then return to the intensity data in the 3D stack, with interpolation, to integrate light levels on surfaces and in volumes. Specialised software packages are noted for surface-rendering e.g. Amira (www.tgs.com), but few will handle 3D intensities for densitometry, as does Imaris (www.bitplane.com).

When processing a 3D series, the anisotropy of images should be born in mind: numerical matrices used in data filters should be different in the XY and YZ dimensions. This underlines the importance of using 3D software that has really been built around microscopy and the image formation process, and that starts with image restoration.

The scientific, technical and mathematical aspects of dynamic cell imaging are treated in Ronot and Shoëvaërt-Brossault (1999) and Periasamy (2001).

OPTIMIZATION OF FLUOROCHROMES AND PROBES

Imaging systems are increasingly sensitive, and often the real challenge is not to get signal strength but to ensure specificity by using narrow spectral windows on both excitation and emission sides. What on paper looks like mild spectral overlap can turn into violent signal cross-talk in experiments, because the intensity of labeling differs over many orders within and between cells. The answer is flexibility to maximise specificity on both the excitation and emission sides, to improve spatial discrimination, and to use spectral analysis and unmixing (= "spectral deconvolution") as a last resort. The excellent capacity of lasers for scanning systems, from the UV and violet (405 nm) through to far red (633 nm), is vitally important for specificity in plant biology.

Ruzin (1999) covers many practical aspects of plant microtechniques in general. In the context of our chapter, the choice of a fluorochrome must respond to many criteria to obtain a strong, stable, specific signal (Mason, 1999; Fricker *et al.*, 2001; Horobin and Kiernan, 2002; Haseloff 2003; Brown *et al.*, 2004; Benoit *et al.*, 2005; Haugland, 2005). In the living cell, the fluorochrome should not interfere with cellular functions. Fluorescent probes are used on living or fixed cells for the following types of information (Table 13.1)

a) To mark subcellular structures, macromolecules, complexes (DNA, cytoskeleton, etc.) or small molecules at significant concentrations. Contrast depends upon physico-chemical affinities and enhanced quantum efficiency, or spectral properties due to local environments.
b) To tag antibodies or other ligands whose affinity determines specificity and contrast (Satiat-Jeunemaître and Hawes, 2001).
c) To assess metabolic or physiological changes in living cells: probes of enzyme activity, ion fluxes, membrane permeability, etc.
d) To reveal fluxes of proteins or other compounds within and between cells *in vivo*.

Spectral properties must be coherent with the other fluorochromes present, with autofluorescence, and with the technical specifications of any instrument. One source can excite several fluorochromes with sufficiently distinct emissions: examples are blue excitation for fluorescein + propidium iodide, or GFP + chlorophyll in plastids. Alternatively, distinct spectral windows from a lamp or laser lines may be used simultaneously, or sequentially, to excite different fluorochomes: in immunolabelling, blue then green excitation for fluorescein then Texas Red, or Alexa Fluor 488 then Alexa Fluor 568. This spectral switching can be achieved in a few microseconds.

These spectral properties depend upon the environment of the fluorochrome, as in the following examples. Nile Red emits red in polar phospholipid membranes and yellow in neutral triglycerides. The emission of the carbocyanine mitochondrial dye JC1 is dependent upon membrane potential, as it is monomeric in relaxed mitochondria but, upon hyperpolaristion, the dye aggregates against the membrane (J-aggregates) and intermolecular energy transfers then shift the emission to the red. The pH will affect many fluorochromes such as Yellow Fluorescent Protein (YFP pKa 6.9) which looses quantum yield in an acidic environment and also in presence of chloride, or fluorescein (pKa 6.2) whose excitation spectrum changes with pH.

It is important to note that new fluorescent probes are now available that have unique spectral properties independent of the environment. These probes are colloidal quantum dots, composed of semiconductor nanocrystals whose size determines the spectral properties of light emitted from caged electrons. These inorganic fluorophores are incredibly bright (50-100 times fluorescein), stable, and available in sizes to offer a series of narrow emissions from the green to red, all excited conveniently in the UV to blue. Their fabrication is delicate, with layers of insulation and functionality which must be conserved in all subsequent buffers. Nevertheless, commercial supplies are available for immunolabelling (www.evidenttech.com and probes.invitrogen.com/products/qdot). Their utility is elegantly demonstrated by single neuro-receptor particle tracking on cells (Dahan *et al.*, 2003).

In summary, a fluorochrome should be used with a feeling for its physico-chemical properties: compound class, pKa, pKi, molar extinction coefficient, quantum yield, brilliance, photostability, toxicity (for cells and humans), and its environmental sensitivity. Table 13.1 presents the major mechanisms determining how and where a fluorochrome goes in a living cell.

For 3D microscopy, specimen form and structure should be conserved, even using spacers. The sample should be immobilized, with a viscous medium or surface-treated glass (poly-L-lysine, agar, Concanavalin A, etc.) or medical glue (Silicone PSA, NuSil Technology). The choice of sample mounting medium will be influenced by imaging characteristics as well as structural and optical demands. Optical coherence is a major aim (Pawley, 1995; Brown *et al.*, 2004):

- uniformity of refractive indices with mounting medium/coverslip/oil ($\eta = 1.51$), or buffer/coverslip/water ($\eta = 1.33$), or dipping lens (no coverslip);
- planarity and stability;
- transparency;
- coverslip thickness (normally 170 μm) as specified on the objective.

The fluorochrome will be protected by pH buffer, tonicity, free radical quenchers essential to reduce photodestruction, antioxidant, chelators of heavy metals, and humidifiers like glycerol, without aggravating autofluorescence. Live cell studies

Table 13.1 Examples of mechanisms determining fluorochrome distribution in cells.

Mechanism	*Properties*	*Examples*	*Some applications*
Diffusion		Highly positive Propidium iodide cannot diffuse across intact membranes, but through leaky membranes gains access to nucleic acids and fluoresces as intercalator	Discriminates dead, dying or late apoptotic cells from live cells
Covalent modification	Impermeant	Fluorescein isothiocyanate forms covalent thioureas with amines	FITC will label proteins accessible on the surface of live protoplasts
Partitioning & dissolution	Lipophilic	Nile Red partitions into all hydrophobic structures, fluorescing yellow in neutral domains and red in polar environments	Lipid bodies yellow, phospholipid membranes red
"	"	DiIC_{18}(3) crystal placed on soma will slowly dissolve and label all contiguous membrane	Axon tracer
Dissolution & anchorage	Polar lipophilic	TMA-DPH (tetra-methylammonium-DPH); FM1-43, FM4-64 (amino-styryls): all enter the outermost membrane, anchoring at the negative surface, then move with membrane mass	FM for membrane flow, endocytosis, and synaptic vesicular activity
Structural affinity	Covalent label on antibody	Antibodies labeled with fluorophores, typically 2-5 fluorophores per protein molecule	Direct or indirect immunolabeling
Structural affinity	Amphiphilic for permeation	Bisbenzimide Hoechst 33342: permeant, then high quantum yield of fluorescence in AT-rich domains	Vital labeling of DNA including AT-rich mycoplasma
"	"	nonyl-acridine orange (NAO) inserts between 4 cardiolipids specific to the mitochondrial membrane	Potential-independent mitochondrial labelling
Electric potential	Amphiphilic with delocalised charge	Small molecules with delocalised positive charge: carbocyanines DiOC_6(3), JC-1; Rhodamine123. The cation is captured inside the highly electro-negative mitochondria, according to the Nernst equation, simplified: gradient of a single-charge molecule across a charged membrane, V; then $V = -61.5 \log ([M^+]_i / [M^-]_e)$ thus -61 mV produces 10x concentration	mitochondrial labelling and assessment of electrochemical gradients

Contd.

Table 13.1 Contd.

Mechanism	*Properties*	*Examples*	*Some applications*
Proton potential	Weak base	After permeation of the neutral form, Neutral Red or Acridine Orange accumulate in an acidic compartment such as the vacuole due to protonation to the impermeant charged form, according to the Waddel & Butler equation, simplified: a permeant weak base in a two compartment system, i, inner et e, external: then $C_i / C_e = 10pH_e - pH_i$ thus Δ1 pH produces 10x concentration	Vacuole and lysosome labeling
Proton potential	Weak acid	Cytoplasmic fluorescein is a mixture of negative (anionic) and neutral fractions (pK_a 6.2). The latter are slightly permeant. Molecules reaching an alkaline compartment such as the photosynthetic chloroplast will be trapped by dissociation, then more neutral molecules enter... Δ1 pH produces 10x concentration	labeling relatively alkaline compartments
Permeation followed by hydrolysis	Permeant ester derivatives	Nonfluorescent fluorescein di-acetate (FDA) is sufficiently lipophilic to cross membranes. Unspecific cellular esterases free fluorescein, which in its anionic form accumulates behind intact membranes	Discriminating living cells according to their intact membranes. Esterification is the major strategy for loading calcium probes, etc. into cells.
Active transport	Permeant & reactive	Chloromethylfluorescein diacetate (CMFDA) penetrates, is de-esterified, the product is conjugated by reduced glutathione and transported into the vacuole	CellTracker® Green and vacuolar labeling ®Molecular Probes Invitrogen
Genetic	Tag or marker or biosensor	Green Fluorescent Protein, etc.	Chapter 14

introduce further demands such as gas and temperature regulation. Excellent commercial mounting media are available, but they need to be tested in each situation (database at www.ifr87.cnrs-gif.fr). These suffer from poor product description, including omission of their refractive index, and careless referencing in protocols – for instance, three distinctive formulations of Citifluor exist (Agar, Oxford Instruments), one (AF2) is excellent for TOPRO3 as a nuclear stain and the other incompatible due to PBS.

LIGHT AS A TOOL

Manipulating objects

There is increasing demand to manipulate what can be seen under the microscope. One beautiful example of light microscopy on plant tissue is Wildwater *et al.* (2005) where photo-ablation is used to destroy a given root cell and to follow its replacement from adjacent stem cells.

Optical tweezers use the refraction of unabsorbed light (typically Near Infra-Red, NIR) to induce an asymmetrical electromagnetic field in an object such as a molecule, vesicle, nucleolus, cell or bead whose refractive indices exceed that of their environment (Greulich, 2001). With a highly focused beam, this pulls an object to the focal point. In an inverted microscope, this levitation or trapping has an active foyer (focus) of ~1-2 μm, which is enough to hold large objects like cells. Subcellular organelles may be manipulated *in situ*, and cells may be displaced. The forces generated, typically picoNewtons, are those of an organelle moving on motors, an ATPase rotary motor or a polymerase working along a nucleic acid (Greulich, 1999). This is a tool for characterizing mechanical forces in cells.

Microinjection and micromanipulation with glass tools is a classic approach, but quite difficult with plant material (Knoblauch, 2001). Laser microdissection uses micron-size ablation due to absorption of light, typically from a UV-A laser, followed by tweezers, catapulting or pushing with light pressure. This enables recovery of the cut material on an inverted microscope (PALM Microlaser Technologies: www.palm-microlaser.com). Alternatively, on an upright microscope a robust solution is to cut and then push the fragment down with gravity into a recovery tube. Molecular biology protocols for handling minimal quantities of nucleic acids are an essential complement to laser microdissection and also to cell sorting by flow cytometry. To obtain RNA populations in the stromium cells of fixed tobacco anthers, Sanders *et al.* (2005) recovered ~500 cells in 45 cuts by laser dissection. Similar recoveries were reported from dewaxed microwave-fixed 10 μm leaf sections cut by laser (Inada and Wildermuth, 2005).

Photo-sensitive Compounds

Caged compounds are useful for functional cell studies: the principle is to load the cell with an inactive form of a modulator or probe. Typically UV or violet irradiation is then used to free the active form of the nucleotide analogue, the chelator or secondary messenger. For functional studies, a whole new range of photo-activable and photo-switchable GFP derivatives offer enhanced spatial and temporal resolution as light can be used to mark a particular pool of proteins, analogous to pulse-chase experiments

(Ando *et al.*, 2004; Arimura *et al.*, 2004; Brandizzi *et al.*, 2004; Koster *et al.*, 2005). The activation may come from a lamp or a laser. Activation is also possible with a pulsed femtosecond (fs) laser, with the advantage in this non-linear mode that the uncaging is strictly restricted to the focal point.

F-techniques

We have been heavily biased toward fluorescence, which has many useful properties in addition to intensity or quantum yield (Valeur, 2002). The electromagnetic field associated with light propagation is used in polarisation or anisotropy studies with fluorophores to appreciate viscosity (Gantet *et al.*, 1990) and molecular rotation. This is one way to discriminate free molecules from complexed forms. Another is the time constant of fluorescence, τ (tau), used in Fluorescence Lifetime Imaging Microscopy (FLIM). Yet another is the non-radiative energy transfer that may occur between two photophores in proximity, namely Förster Resonant Energy Transfer (FRET). Light can irreversibly destroy a photocentre, leaving behind a modified molecule. Fluorescence Recovery (in fact Redistribution) After Photodestruction (FRAP) and its corollary Fluorescence Loss In Photobleaching (FLIP) exploit this photodestruction to estimate diffusion or traffic coefficients in cells, artificial membranes, etc. as new fluorophores enter the darkened volume. Examples of these techniques, and many reviews, are given in the companion Chapter 14 (Bolte *et al.*, 2007).

Cytometry

Cytometry is the quantification of structure and fluorescent signals associated with individual objects. Imaging is generally precluded, although high-speed imagers (hundreds of cells per second) are now also available. For fluorometry of heterogeneous individualised cells, cytometry is unbeatable with its multiparametric quantification, with precise cross-talk correction and exceptionally high dynamic range. A cytometer can quantify fluorescence on a cell with only 50 molecules and yet another with hundreds of thousands. In cell cultures, gene expression can be quantified precisely in subpopulations and in rare cells, since millions of cells can be analysed per hour and even sorted. Unfortunately, the technical constraints are such that in plant biology we mainly use protoplasts or isolated nuclei rather than cells. We provide references concerning plant cell biology (Bergounioux *et al.*, 1992), developmental root biology (Birnbaum *et al.*, 2003), cytogenetics (Safar *et al.*, 2004), biodiversity (Doležel and Bartoš, 2004), basic methods and reviews (Coba de la Peña and Brown, 2001, Galbraith, 2003 Catrice *et al.*, 2006).

CONCLUSION AND PERSPECTIVES

Spatial, temporal, spectral resolution, and sensitivity. The number of parameters or channels necessary for imaging cells is often under-estimated, as too the needed flexibility or polyvalence and spectral coverage from violet to far red. No doubt more attention should be given to rigour in sample mounting and, in live studies, to keeping the cells in good condition which, as we have seen, includes much more careful light management.

The less expensive structured illumination modules on wide field microscopes are practical for developmental biology and screening. The non-linear optics of multiphoton microscopy have become more reliable, convenient and accessible, remain very expensive, but offer advantages in narrow fields such as deep tissue imaging or for highly diffusive tissue like leaves. But not many people could drive a Formula One car to go and fetch their daily bread!

Thanks to the physical, material and mathematical sciences, photonic devices and algorithms continue to make such significant advances that biologists would like to upgrade their tools every year to benefit from a precious gain in spatial, temporal, spectral resolution and sensitivity. Tracking few or even single molecules has become feasible in the standard laboratory, as has mapping the formation of molecular complexes. Beyond these steady improvements, there are truly revolutionary approaches underway like Stimulated Emission Depletion (STED) whose resolution is 5- to 10-fold better than diffraction-limited light microscopy, justifying the term nanoscopy.

Automation and High Throughput. High Throughput Screening of cell functions is routine in industry, based upon assays of reporter genes, signal transduction pathways, viability and apoptosis, oxidative burst, mitochondrial activity, secretion and endocytosis, etc. Automated image analysis can perform these kinetic studies on cells for the pharmaceutical industry, but plant biologists rarely have the resources or the justification for high throughput. At a simpler level, electronic stages on microscopes are the way to go in order to multiplex live-cell experiments, running through a chain of memorised positions rather than spending the day observing one or two cells, which will barely satisfy the statistician.

Molecular specificity, biosensors and Point-of-Care devices. The past two decades have seen a superb coevolution between molecular probes and biological imaging. Twelve years after the heterologous expression of GFP, it is still subject of fascinating innovations. Tools like FRAP, FRET, FLIM and Bimolecular Fluorescence Complementation (BiFC) for visualizing complexes have become routine. Molecules which report cellular or environmental changes, biosensors, are being invented by marrying the biological specificity of receptors and enzymes to optical reporters. This will be important outside the laboratory, in agriculture, industry and public health. Portable confocal microprobes and endoscopes are also available (www.optiscan.com).

Polyvalence and open doors to competence. Training for methods in cell biology will frequently include optics, image processing and tools for their communication. Much course material is available on the web (e.g. www.cyto.purdue.edu and the dictionary www.photonics.com/dictionary). Functional genetics calls upon many different imaging tools, from multicolour histology and probe design to live cell videomicroscopy and cytometry. These are expensive processes undergoing rapid change, so cooperative investment and access are more than ever justified – plus the dynamic support of competent personnel! The French network GDR2588 Microscopie Fonctionnelle du Vivant (www.gdr2588.cnrs.fr) is one such resource.

An image suggests, conveys, confirms, it shocks – and it can totally mislead. But imagine Robert Hooke in 1665, observing his slices of cork with, for the first time, the word "cell" on his lips:

"... I could exceedingly plainly perceive it to be all perforated and porous... these pores, or cells,... were indeed the first microscopical pores I ever saw, and perhaps, that were ever seen, for I had not met with any Writer or Person, that had made any mention of them before this."

ACKNOWLEDGMENTS

The Platform "Imaging and Cell Biology", part of the Gif RIO Imaging Network, has been extensively supported by the *IFR87 « La plante et son environnement »* and also by the *Conseil Général de l'Essonne* via the ASTRE program for live cell microscopy. We thank Christoph Kluge for discussions on low-light imaging, Marie-Thérèse Crosnier for technical and secretarial assistance; Olivier Catrice, Marie-Noëlle Soler and Nathalie Mansion for their support in microscopy and imaging.

REFERENCES

Abal, M., Piel, M., Bouckson-Castaing, V., Mogensen, M., Sibarita, J.B., Bornens, M. 2002. Microtubule release from the centrosome in migrating cells. *J. Cell Biology*, 159, 731-737.

Allen, G.J., Chu, S.P., Harrington, C.L., Schumacher, K., Hoffmann, T., Tang, Y.Y., Grill, E., Schroeder, J.I. 2001. A defined range of guard cell calcium oscillation parameters encodes stomatal movements. *Nature*, 411, 1053-1057.

Ando, R., Mizuno, H., Miyawaki, A. 2004. Regulated fast nucleocytoplasmic shuttling observed by reversible protein highlighting. *Science*, 306, 1370-1373.

Arimura, S., Yamamoto, J., Aida, G.P., Nakazono, M., Tsutsumi, N. 2004. Frequent fusion and fission of plant mitochondria with unequal nucleoid distribution. *Proc. Natl. Acad. Sci.* USA, 101, 7805-7808.

Bachia, K., Schwille, P. 2003. A dynamic view of cellular processes by *in vivo* fluorescence auto- and cross-correlation spectroscopy. *Methods*, 29, 74–85.

Benoit, E., Bolte, S., Brown, S., Soler, M.N. 2005. Bases de la microscopie confocale: théorie et pratique. October, Formation Permanente CNRS, Gif-sur-Yvette, France.

Berg, R.H. 2004. Evaluation of spectral imaging for plant cell analysis. *J. Microscopy*, 214, 174-181.

Bergounioux, C., Brown, S.C., Petit, P.X. 1992. Flow cytometry and plant protoplast cell biology. *Physiologia Plant*, 85, 374-386.

Birnbaum, K., Shasha, D.E., Wang, J.Y., Jung, J.W., Lambert, G.M., Galbraith, D.W., Benfey, P.N. 2003. A gene expression map of the Arabidopsis root. *Science*, 302, 1956-1960.

Bolte, S., Cordeliéres, F.P. 2007. A guided tour into sub-cellular co-localisation analysis in light microscopy. J. Microscopy, in press

Bolte, S., Boutté, Y., Kluge, C., Brown, S., Satiat-Jeunemaître, B. 2007. Tracking Gene Expression in Plant Cells: New Probes for Functional Genomics. Chapter 14 this book.

Bolte, S., Talbot, C., Boutte, Y., Catrice, O., Read, N.D., Satiat-Jeunemaître, B. 2004. FM-dyes as experimental probes for dissecting vesicle trafficking in living plant cells. *J. Microscopy*, 214, 159-173.

Brandizzi, F., Irons, S., Johansen, J., Kotzer, A., Neumann, U. 2004. GFP is the way to glow: bioimaging of the plant endomembrane system. *J. Microscopy*, 214, 138-158.

Brown, S., Poujol-Talbot, C., Kronenberger, J., Traas J., Satiat-Jeunemaître, B. 2004. L'apport de la microscopie et de l'imagerie en génomique. In: *La génomique en Biologie Végétale*. Morot-Gaudry, J.F., Briat J.F., eds., INRA Editions, Versailles, 209-236.

Catrice, O., Coba de la Peña, T., Brown, S.C. 2006. La cytométrie en flux pour la biologie végétale: contraintes, succès, espoirs. In: *La cytométrie en flux*. Ronot, X., Grunwald, D., Mayol, JF., Boutonnat, J., eds., Tec & Doc Lavoisier, Paris.

Cerovic, Z.G., Samson, G., Morales, F., Tremblay, N., Moya, I. 1999. Ultraviolet-induced fluorescence for plant monitoring: present state and prospects. *Agronomie*, 19, 543-578.

Coba de la Peña, T., Brown, S.C. 2001. Cytometry and Fluorimetry. In: *Plant Cell Biology: A Practical Approach*. 2nd. ed. Hawes, C., Satiat-Jeunemaître B., eds., IRL Press, Oxford, 85-106.

Dahan, M., Lévi, S., Luccardini, C., Rostaing, P., Riveau, B., Triller, A. 2003. Diffusion dynamics of glycine receptors revealed by single-quantum dot tracking. *Science*, 302, 442-445.

Dashek, W.V. 2000. Methods in plant electron microscopy and cytochemistry. Humana Press, Totowa, New Jersey.

Doležel, J., Bartoš, J. (2004). Plant DNA flow cytometry and estimation of nuclear genome size. *Annals Botany*, 95, 99-110.

Egner, A., Hell, S.W. 2005. Fluorescence microscopy with super-resolved optical sections. *Trends Cell Biol.*, 15, 207-215.

Failla, A.V., Spoeri, U., Albrecht, B., Kroll, A., Cremer, C. 2002. Nanosizing of fluorescent objects by spatially modulated illumination microscopy. *Appl. Opt.* , 41, 7275-7283.

Feijo, J.A., Moreno, N. 2004. Imaging plant cells by two-photon excitation. *Protoplasma*, 22, 1-32.

Fricker, M., Parsons, A., Tlaka, M., Blancaflor, E., Gilroy, S., Meyer, A., Plieth, C. 2001. Fluorescent probes for living plant cells. In: *Plant Cell Biology: A Practical Approach*. 2nd. ed. Hawes, C., Satiat-Jeunemaître B., eds., IRL Press, Oxford, 35-81.

Galbraith, D.W. 2003. Cytometry and plant sciences: a personal retrospective. *Cytometry*, 584, 37-44.

Gantet, P., Hubac, C., Brown, S.C.C 1990. Flow cytometric fluorescence anisotropy of lipophilic probes in epidermal and mesophyll protoplasts from water-stressed *Lupinus albus* L. *Plant Physiol.* 94, 729-737.

Grandjean, O., Vernoux T., Laufs, P., Belcram, K., Mizukami, Y., Traas, J. 2004. In vivo analysis of cell division, cell growth, and differentiation at the shoot apical meristem in Arabidopsis. *Plant Cell*, 16, 74-87.

Greer, L.F., Szalay, A.A. 2002. Imaging of light emission from the expression of luciferases in living cells and organisms: a review. *Luminescence*, 1, 43-74.

Greulich, K.O. 1999. Micromanipulation by light in biology and medicine: The laser microbeam and optical tweezers. Birkhaüser, Basel.

Greulich, K.O. 2001. Micromanipulation by laser microbeam and optical tweezers. In: *Plant Cell Biology: A Practical Approach*. 2nd. ed. Hawes, C., Satiat-Jeunemaître B., eds., IRL Press, Oxford, 207-233.

Guérin, M.C., Torreilles, J. 1998. Mesurer en temps réel les interactions entre ligands et biomolécules. Biofutur 177: cahier technoscope 102 (avril).

Gunning, B.E.S., Steer, M.W. 1996. Plant cell biology: Structure and Function. Jones & Bartlett Publ., Mass.

Haseloff, J. 2003. Old botanical techniques for new microscopes. *BioTechniques* 34, 1174-1182.

Haugland, R. 2005. The Handbook: A Guide to Fluorescent Probes and Labeling Technologies, 10th ed., Molecular Probes Invitrogen

Hawes, C., Satiat-Jeunemaître, B. 2001. *Plant Cell Biology*. 2nd ed. IRL Press, Oxford.

Hoffmann, B., Kosegarten, H. 1995. FITC-dextran for measuring apoplast pH and apoplastic pH gradients between various cell types in sunflower leaves. *Physiol. Plantarum*, 95, 327-335.

Holst, G. 2005. CMOS cameras rise to speed challenge. Opto Laser Europe, June, 43-45.

Horobin, R.W. Kiernan, J.A. 2002. Conn's Biological Stains (10th edition). A Handbook of dyes, stains and fluorochromes for use in biology and medecine. Garland Sciences, London.

Huisken, J., Swoger, J., Del Bene, P., Wittbrodt, J., Stelzer, E. 2004. Optical sectioning deep inside live embryos by selective plane illumination microscopy; *Science* 305, 1007-1009

Hutzler, P., Fischbach, R., Heller, W., Jungblut, T.P, Reuber, S., Schmitz, R., Veit, M., Weissenbo, G., Schnitzler, J.P. 1998. Tissue localization of phenolic compounds in plants by confocal laser scanning microscopy. *J. Exp. Bot.*, 49, 953–965.

Inada, N., Wildermuth, M.C. 2005. Novel tissue preparation method and cell-specific marker for laser microdissection of *Arabidopsis* mature leaf. *Planta*, 221, 9–16.

Jones, H.G. 1999. Use of thermography for quantitative studies of spatial and temporal variation of stomatal conductance over leaf surfaces. *Plant Cell Environ.*, 22, 1043-1055.

Knoblauch, M. 2001. Microinjection. In: *Plant Cell Biology: A Practical Approach*. 2nd. ed. Hawes, C., Satiat-Jeunemaître B., eds., IRL Press, Oxford, 207-233.

Koster, M., Frahm, T., Hauser, H. 2005. Nucleocytoplasmic shuttling revealed by FRAP and FLIP technologies. *Curr. Opin. Biotechnol.*, 16, 28-34.

Lacey, A.J. 1999. Light Microscopy in Biology: a practical approach. Oxford University Press, Oxford.

Lamb, D.C., Muller, B.K., Brauchle, C. 2005. Enhancing the sensitivity of fluorescence correlation spectroscopy by using time-correlated single photon counting. *Curr. Pharm. Biotechnol.*, 6, 405-14.

Lecourieux, D., Mazars, C., Pauly, N., Ranjeva, R., Pugin, A. 2002. Analysis and effects of cytosolic free calcium increases in response to elicitors in *Nicotiana plumbaginifolia* cells. *Plant cell*, 14, 2627-2641.

Lerner, J.M., Zucker, R.M. 2004. Calibration and validation of confocal spectral imaging systems. *Cytometry*, 62A, 8-34.

Love, J., Dodd, A.N., Webb, A.A.R. 2004. Circadian and diurnal calcium oscillations encode photoperiodic information in *Arabidopsis*. *Plant Cell*, 16, 956-966.

Mason, W.T. 1999. Fluorescent and Luminescent Probes for Biological Activity: A Practical Guide to Technology for Quantitative Real-Time Analysis. Academic Press, London.

Merlot, S., Mustilli, A., Genty, B., North, H., Lefebvre, V., Sotta, B., Vavasseur, A., Giraudat, J. 2002. Use of infrared thermal imaging to isolate *Arabidopsis* mutants defective in stomatal regulation. *Plant J.*, 30, 601-609.

Meyer, S., Cartelat, A., Moya, I., Cerovic, Z.G. 2003. UV-induced blue-green and far-red fluorescence along wheat leaves: a potential signature of leaf ageing. *J. Exp. Bot.*, 54, 757-769.

Michelot, A., Guerin, C., Huang, S., Ingouff, M., Richard, S., Rodiuc, N., Staiger, C.J., Blanchoin, L. 2005. The Formin homology 1 domain modulates the actin nucleation and bundling activity of Arabidopsis FORMIN1. *Plant Cell*, 17, 2296–2313.

Pawley, J.B. 1995. Handbook of Biological Confocal Microscopy, Plenum Press.

Periasamy, A. 2001. Methods in Cellular Imaging. Oxford University Press, Oxford.

Rogers, K., Stinnakre, J., Agulhon, C., Jublot, D., Shorte, S., Kremer, E.J., Brûlet, P. 2005. Visualization of local Ca^{2+} dynamics with genetically encoded bioluminescent reporters. *Eur. J. Neuroscience*, 21, 597-610.

Roland, J.C., Vian, B. 1991. General preparation and staining of thin sections. In: *Electron Microscopy of Plant cells*. Hall, J.L., Hawes C.H., eds., Academic Press, London, 1-66.

Ronot, X., Schoëvaërt-Brossault, D. 1999. Dynamique de la cellule vivante. Editions INSERM, Paris.

Ronot, X., Usson, Y. 2001. Imaging of nucleic acids and quantification in photonic microscopy. CRC Press, Boca Raton, USA.

Ruzin, S.E. 1999. Plant microtechnique and microscopy. Oxford University Press, Oxford.

Safar, J., Bartoš, J., Janda, J., Bellec, A., Kubalakova, M., Valarik, M., Pateyron, S., Weiserova, J., Tuskova, R., Cihalikova, J., Vrana, J., Simkova, H., Faivre-Rampant, P., Sourdille, P., Caboche, M., Bernard, M., Doležel J., Chalhoub, B. 2004. Dissecting large and complex genomes: flow sorting and BAC cloning of individual chromosomes from bread wheat. *Plant J*, 39, 960-968.

Sanders, PM., Bui, A.Q., Le, B.H., Goldberg, R.B. 2005. Differentiation and degeneration of cells that play a major role in tobacco anther dehiscence. *Sex Plant Reprod*, 17 219–241.

Satiat-Jeunemaître, B., Hawes, C. 2001. Immunocytochemistry for light microscopy. In: *Plant Cell Biology: A Practical Approach*. 2nd. ed. Hawes, C., Satiat-Jeunemaître B., eds., IRL Press, Oxford, 207-233.

Segui-Simarro, J.M., Austin, J.R., White, E.A., Staehelin. L.A. 2004. Electron tomographic analysis of somatic cell plate formation in meritematic cells of *Arabidopsis* preserved by high pressure freezing. *Plant Cell*, 16, 836-856.

Shaw, P. 2001. Introduction to optical microscopy for plant cell biology. In: *Plant Cell Biology: A Practical Approach*. 2nd. ed. Hawes, C., Satiat-Jeunemaître B., eds., IRL Press, Oxford, 1-34.

Valeur, B. 2002. Molecular Fluorescence: Principles and Applications, Editions Wiley-VCH, Weinheim, Germany.

Wallace, W., Schaefer, L.H., Swedlow, J.R. 2001. A workingperson's guide to deconvolution in light microscopy. *Biotechniques*, 3, 1076-1097.

Wildwater, M., Campilho, A., Perez-Perez, J.M., Heidstra, R., Blilou, I., Korthout, H., Chatterjee, J., Mariconti, J., Gruissem, W., Scheres B. 2005. The RETINOBLASTOMA-RELATED Gene Regulates Stem Cell Maintenance in Arabidopsis Roots. *Cell* 123, 1337–1349.

Yasuda, R., Noji, H., Yoshida, M., Kinosita, K., Itoh, H. 2001. Resolution of distinct rotational substeps by submillisecond kinetic analysis of F1-ATPase. *Nature* 410, 898-904.

Zimmermann, T., Rietdorf, J., Girod, A., Georget, V., Pepperkok, R. 2002. Spectral imaging and linear un-mixing enables improved FRET efficiency with a novel GFP2-YFP FRET pair. *FEBS Lett.*, 531, 245-249.

14

CHAPTER

Tracking Gene Expression in Plant Cells: New Probes for Functional Genomics

S. Bolte, Y. Boutté, S. Kluge, S. Brown, B. Satiat-Jeunemaître

INTRODUCTION

To understand gene function, gene products have to be characterised and their expression patterns analysed at the tissue, cell or subcellular level. In this quest, localization and quantification of RNA and proteins are essential steps. To have a full picture of the spatial and temporal pattern of gene products, strategies with reporter gene expression systems (promoter activity) and *in situ* hybridization techniques (RNA mapping) have to be integrated with immunocytochemistry and reporter protein studies (protein mapping). This field of research has largely benefited from the discovery of fluorescent proteins like the *Green Fluorescent Protein*, which is now used extensively to observe the expression of genes and the localization of proteins in living cells. These cell biology approaches have all gained from the development of bio-imaging techniques together with the development of new optical devices in microscopy and related image analysis. The biologist is now able to dissect the sequence of events occurring from gene expression to its local action in a biological model.

In this chapter, we will introduce the techniques currently used to track gene expression and proteins in cytology and histology.

Reviews on plant gene reporters, *in situ* hybridization and immunocytochemistry are already available and the basic procedures are well established (Neuhaus and Boevink, 2001; Hadlington and Denecke, 2001; Leith *et al.*, 2001; Satiat-Jeunemaitre and Hawes, 2001). The constant need for better detection leads to new developments in these fields, justifying some additional remarks on these areas. Progress in the spatial and temporal resolution to observe gene expression patterns is strongly correlated with constant development of new markers and with the recent advances in microscopy and image analysis (Brown *et al.*, 2004; Brown *et al.*, Chapter 13 this volume). Because of the increasing interest in the GFP technology and its derivatives, the major part of this manuscript will be devoted to the still rapidly evolving use of GFP methodologies in plant cells: an overview of the use of GFP for plant cell imaging will be proposed together with some guidelines to optimize the use of these fluorescent probes in plant cells.

TRACKING GENE EXPRESSION AND PRODUCTS IN CELLS AND TISSUES

To study the expression of genes *in situ*, at the cell or tissue level, two methods are generally available. The first is to couple a reporter gene construct (β-glucuronidase , luciferase or GFP) together with the regulatory sequence (promoter or enhancer region) of the gene of interest or, secondly, to detect directly the mRNA (*in situ* hybridization). Both approaches have their advantages and limits.

Reporter Genes

Once the regulatory sequence for the gene of interest has been identified, it may be coupled to a sequence coding for a reporter gene whose translated protein, if properly expressed, can be located in the tissue. Several types of reporter genes exist. The most popular are based on GFP technology, where the gene product itself is detected. Other direct fluorescent genetic tags are now available (Adams *et al.*, 2002; Hoffmann *et al.*, 2005) but their performance in plant tissues has yet to be evaluated. Alternatively to the direct visualization of gene products, the reporter gene may code for an enzyme which is able to catalyse the transition of a colourless substrate into a product which is coloured or fluorescent. These reporter enzymes can also catalyze reactions where light is produced during the process (biochemiluminescence). Although none of these enzymatic and fluorescent methods is a gold-standard for the analysis of spatial and temporal gene expression, they have become an integral part of plant molecular research and therefore the main characteristics of the most popular systems will be mentioned below.

Glucuronidase staining

The gene *gus* encodes the β-glucuronidase (GUS) enzyme from *E. coli*. It has been extensively used as a marker gene in transgenic plants because of the stability of the enzyme and the virtual absence of intrinsic GUS activity. In addition, plant metabolism remains largely unaffected by the presence of *E. coli* GUS (Jefferson *et al.*, 1987). The disadvantages of GUS-staining are the need for relatively harsh conditions for the detection procedure (incubation at 37°C for several hours, fixation in alcohol or acetone) which may disturb the tissue and change the results. The exogenous substrates of GUS will only penetrate into fixed material, therefore excluding time-lapse study of individual samples. These substrates leave a coloured precipitate in tissue or yield a soluble fluorescent product, respectively chromogenic 5-bromo-4-chloro-3-indolylβ-glucuronide, and fluorogenic 4-methylumbelliferyl-β-D-glucuronide or ELF-97 β-D-glucuronide (Zhou *et al.*, 1996; Gallagher, 1992).

The gene encoding for a β-galactosidase enzyme is also a popular reporter enzyme in mammalian cells. However, in plant cells, the high endogenous β-galactosidase activities render it difficult as extrinsic reporter. Other reporter genes such as ß-lactamase may be used, but these have not been developed with plants (Knapp *et al.*, 2003).

Bioluminescent reporter genes

Enzymatic bioluminescent reporter genes are used for analysing temporal and spatial expression in a low-light detection mode. Bioluminescence is a chemiluminescent reaction between at least two molecules. The general term for the light-emitting substrate is luciferin, while the enzyme catalysing the reaction is called luciferase. Luciferin is membrane permeate water-soluble and non-toxic and may be added regularly during prolonged experimentation. Bioluminescence measurement is done with a luminometer (typically 1 second per point) or by imaging, which is less efficient and slower. Highly sensitive imaging equipment is needed for bioluminescent assays because of the low light flux, which also implies long integration times (10-1,000 second).

Three luciferase systems have been widely adopted as reporter genes in plant cells: the coelenterate luciferase aequorin (Lin *et al.*, 2005), the Sea pansy (*Renilla reniformis*) luciferase (RUC) (Knight *et al.*, 1993; Meyerhofer *et al.*, 1993) and the firefly luciferase (LUC) (Koiwa *et al.*, 2002; Grant *et al.*, 2003). These enzyme activities are also used to monitor other subcellular events, as aequorin and RUC systems may be used as intracellular calcium sensors (Pauly *et al.*, 2000; Lecourieux *et al.*, 2002) and LUC is used to determine the presence of ATP (Schmutz and Brunhold, 1982; Phillippy, 1994; Borisjuk *et al.*, 2003; Rolletscheck *et al.*, 2004). The luciferase aequorin will also remain famous for its association with the natural GFP in the Pacific jellyfish *Aequorea victoria*, leading the way to new fields of technology.

As gene reporter, the firefly luciferase *luc* is the most sensitive and rapid for plants, allowing live macro- or micro-imaging of up- and down-regulation in the same sample (Greer and Szalay, 2002; Koiwa *et al.*, 2002; Castle, 2005). A codon-optimized form gave 40-fold higher tagging efficiency in the monocot banana (Remy *et al.*, 2005). LUC expression does not affect the metabolism of transgenic cells or organisms (Gould and Subramani, 1988; Greer *et al.*, 2002; Welsh and Kay, 2005). Since LUC is not naturally present in target cells, the assay is background-free. Hence, the *luc* reporter gene is ideal for detecting low-level gene expression.

Because LUC and RUC reporter systems emit light at quite different wavelengths (respectively 530-560 nm [yellow] and 475 nm [blue]), it is possible to use them as a dual reporter system (Horstmann *et al.*, 2004). Moreover, the luciferin emission (LUC substrate) may be strongly red-shifted at 37°C, resulting in much better light diffusion through tissue.

These bioluminescent assays are popular for measuring transcriptional activity in studies of the structural or physiological basis of regulated gene expression: gene silencing, intron splicing, *Agrobacterium* T-DNA transfer, gene regulation, and both transcriptional and post-transcriptional regulatory processes in plants. Gene-trapping experiments have used imaging of firefly LUC in *Arabidopsis* leaves (Yamamoto *et al.*, 2003). These assays have also proven useful in identifying mutations associated with novel plant phenotypes (Chinnusamy *et al.*, 2002; Grant *et al.*, 2003). Their disadvantages for imaging are compromised spatial resolution (~20 μm), the expense of highly sensitive equipment for low-light detection, and the difficulty to analyse more than two markers at a time. Moreover, the researcher has to be aware of the instability of these reactions if conditions change; oxygenation, quenchers and other factors can interfere with the lag-phase of the reaction and lead to nonlinearity, kinetic instability and non-enzymatic background emission.

Tracking mRNA

In plants the distribution of mRNA does not necessarily reflect spatial expression of genes, as mRNA can be transported from cell to cell via plasmodesmata. Tracking mRNA can be achieved on fixed cells by *in situ* hybridization, or directly on living cells.

In situ hybridization

Although delicate and labour intensive to perform with plant samples, *in situ* hybridization gives essential information on the mRNA distribution within cells and tissues at the light or electron microscopy level.

A wide variety of probes are used to reveal and image the mRNA distribution: sequences of RNA, cDNA, a PCR fragment or oligonucleotide. Visualization is direct when probes are labelled directly with radioactive nucleotides or fluorochromes. Alternatively, indirect visualisation is performed when the probes are complexed with haptens which will be subsequently detected by immunocytochemistry or affinity labelling methods. Detailed protocols are available (Jackson, 1991; Ludevid *et al.*, 1992; Kronenberger *et al.*, 1993; Feiler *et al.*, 1995; Morel and Cavalier, 1998; Leitch *et al.*, 2001). In general, the tissues are first fixed (usually by formaldehyde), then the mRNA's are made accessible to the probes, before performing the actual hybridization step. Each probe/tissue combination has to be optimised. Without presenting the details of each method, we will simply mention technical aspects important for any study where the patterns of expression of several genes are to be studied in parallel.

Probe accessibility

Most of the time, accessibility to the probe is achieved by opening the cells with tissue sectioning. Classically, three ways to prepare tissue sections are possible: from frozen material (cryosections are however difficult in highly hydrated specimens), from wax, or plastic/resin embedded specimens. By far the most common embedding medium is wax (Jackson, 1991; Kronenberger *et al.*, 1993). The occasionally poor preservation of structure in fragile tissues calls for harder embedding media like resins, which offer much superior conservation of cells (Kronenberger *et al.*, 1993), however, the preservation of mRNA is then less reliable than with paraffin. Therefore, resin may be considered useful for fragile tissues with relatively abundant mRNA.

If the specimens are small, embedding is not necessary. The intact tissue is directly incubated with the probe. The approach is appealing for its simplicity and efficient for animal tissues (Kopczynski *et al.*, 1998; Satou *et al.*, 2001). However, with plant material it is subject to major artefacts. The first problem is the highly variable accessibility of the tissue to probes; this is due to the pecto-cellulosic plant cell wall. Furthermore, certain plant tissues appear to retain probes in an unspecific manner. Early examples are the detection of mRNA in meristematic tissue and in legume nodules (Crespi *et al.*, 1994; Feiler *et al.*, 1995) (Fig. 14.1). This strategy has been extensively revisited by Friml *et al.* (2003), with novel whole mount *in situ* hybridization protocols, which are suitable for the study of mRNAs in *Arabidopsis* seedlings.

Choice of nucleic acid probe

Oligonucleotides, cDNA, PCR products and RNA probes may all be used (Leitch *et al.*, 2001). The choice to use DNA or RNA probes is empirical, based upon experience (Morel and Cavalier 1998; Leitch *et al.*, 2001). Oligonucleotide probes are not always easy to use on plant tissues. One reason is that each oligonucleotide has several labelled nucleotides, so the target mRNA is not well contrasted. This system does however work well when the messengers are abundant. *In situ* PCR appears attractive as a strategy. Here, the cDNA is directly synthesized *in situ* and then the first level product is amplified using specific oligonucleotides. An advantage of the technique is that it is no longer necessary to first obtain a cDNA clone; one simply orders the right oligonucleotides for the gene of interest. It has not been so widely adopted in plant biology, maybe because it requires detailed optimization for each probe. Nevertheless, three-colour *in situ* RT-PCR has been used in an elegant multiplex analysis enabling the simultaneous localization of several genes within a complex tissue or heterogeneous cell population, while monitoring constitutive 18S rRNA expression (Pesquet *et al.*, 2004).

In all cases, experimental controls must be carefully designed, negative to assess the specificity of the hybridization, positive to appreciate the quality of the preparation (Pesquet *et al.*, 2004). Today, enough probes work under certain standard protocols to envisage automated scale-up as in the animal domain (Kopczynski *et al.*, 1998; Satou *et al.* 2001), using bench robots (www.intavis.com). Accordingly, multiple *in situ* hybridisation probes have been applied systematically on plant tissues (Friml *et al.*, 2003).

Tracking mRNA in live plant cells

Highly sensitive methods have recently been developed for detecting RNA movement and localization in living cells. These approaches include microinjection of fluorescent RNA, *in vivo* hybridization of fluorescent oligonucleotides, and expression of fluorescent RNA Binding Particles (RBP), (Dirks *et al.*, 2001; Pederson, 2001; St Johnston, 2005). Indirect labelling using GFP/MS2 bacteriophage system to tag RNA has been used to follow the accumulation of different RNAs in plant cells, both in the nucleus and the cytoplasm. This has been combined with a DsRed tagged protein (MtRBP1) in a two colour analysis of RNA-peptide granules, to follow their complex-dependent nuclear-to-cytoplasm transfer in live cells (Cougot *et al.*, 2004; Campalans *et al.*, 2004) (Fig. 14.1B).

Tracking Protein Expression: Affinity Labelling and Immunocytochemistry

Affinity labelling is based on specific reaction sites on macromolecules detected by specific probes like lectins, enzymes and antibodies (the use of the latter probes defining the field of immunocytochemistry). These are key techniques to study protein localization with the resolution of the light or electron microscope. Since the late 1960s they have given amazing results concerning the spatio-temporal pattern of proteins

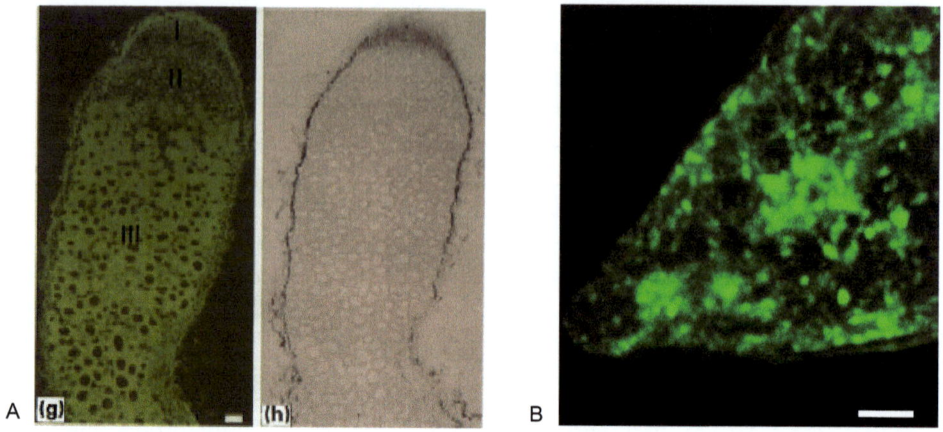

Fig. 14.1 RNA localization **A.** *In situ* hybridization with digoxigenin (DIG)-labelled probes. (g): Epifluorescence of a longitudinal section of a 20-days-old *Medicago sativa* symbiotic nodule, where characteristic zones can be seen: I: meristem; II invasion zone; III nitrogen-fixing zone enveloped by peripheral tissues. (h): Same section seen with bright field illumination indicating MsCA1 (coding for Carbonic Anhydrase) gene expression where a dark precipitate is detected with digoxigenin-labelled probes revealed by anti-DIG alkaline phosphatase antibody. For the colour reaction, nitro-blue tetrazolium salt (NTB) and 5-bromo-4-Chloro-3-indolyl-phosphate (BCTP) were used as substrate for the phosphatase (Coba de la Pena *et al.*, 1997). Bar 50 μm **B.** *In vivo* localization of enod40 RNA with GFP/MS2 bacteriophage in epidermal onion cells. Bar 20 μm (Campalans *et al.*, 2004)

within cells and tissues. However, with the "Green Revolution" of GFP (Brandizzi *et al.*, 2002a), these techniques have been slightly overshadowed, as information on dynamic events *in vivo* appears to be far more natural at first sight with GFP and closer to reality than observations on fixed cells. However, it is now obvious that GFP techniques have their constraints, GFP expression in live cells should be used in conjunction with immunocytochemistry or affinity labelling on fixed cells to limit potential artefacts (Bolte *et al.*, 2004a). Moreover, all tissues are not readily accessible to GFP techniques. Above all, the resolution of electron microscopy is often needed to properly map proteins. Affinity labelling methods remain unique tools to obtain such resolution.

Constraints of plant cell labelling

Plant cells are a challenge to cytochemists as they impose severe constraints for the success of affinity labelling experiments: to overcome cell walls or cuticles, real barriers to antibody/affinity probes penetration; to suppress the non-specific affinities of carbohydrates present in plant cells (starch, polysaccharides) for immunological probes; and, when reading the labelling at the light level, to take into account the coloration and autofluorescence of natural compounds (chlorophyll, lignin, flavins, polyphenols).

Lectins and enzymes, when linked to visible markers such as fluorochromes (lectins) or colloidal gold particles (enzymes and lectins), constitute excellent probes to detect specific cell components *in situ*, at the electron microscopy (EM) level (Roland and Vian,

1991), or light level. However, the use of antibodies as specific probes for detecting peptides or whole proteins remains the major part of the affinity labelling methods.

Immunocytochemistry techniques are based on antibody/antigen reactions within the cell. These techniques can be performed on living cells if working on plant cell surface proteins as protoplasts, but most of the time cells need to be fixed.

The fixation has to respect the antigenicity of the tissue, together with a good preservation of the structures. Chemical fixation has often been used, however cryofixation techniques (for instance by high pressure freezing) are now favoured as they may give a better protection of the antigens (Staehelin *et al.*, 1990; Hawes and Satiat-Jeunemaitre, 2001; Jolliffe *et al.*, 2004; Segui-Simarro *et al.*, 2004).

Immunolabelling is usually performed using a two step procedure (indirect immunolabelling). The antigen of interest is detected by a first specific antibody (AB1), which may be monoclonal or polyclonal (Satiat-Jeunemaitre and Hawes, 2001). This first immunoreaction is visualized by the application of a secondary antibody (anti-AB1) linked to a visual marker (enzymatic reporter, gold particles or fluorochrome) and recognizing the constant region of the first antibody. When a double labelling is needed, the immunoreactions are performed in sequential steps (AB1, Anti-AB1; AB2, Anti-AB2). Many protocols are available, for immunocytochemistry both at the light and EM levels (see protocols and references respectively in Hall and Hawes, 2000 Satiat-Jeunemaitre and Hawes, 2001). However, additional remarks regarding the two most frequently asked questions will be presented: a) What is the appropriate technique to have the best access to the proteins? b) Which secondary antibody has to be used?

Access to the antigens

To overcome the natural barriers of plant cells, antigen accessibility is handled through two main approaches: partial digestion of cell surfaces or cell opening by sectioning.

(1) *Partial digestion of cell surfaces*: it is possible to partially digest plant cell wall surfaces by an appropriate enzymatic cocktail and a modest detergent treatment. This is the basis of the "squared protoplast technique" (Lloyd *et al.*, 1979; Wick *et al.*, 1981; detailed protocols in Satiat-Jeunemaitre and Hawes, 2001; see Fig. 14.3A). The whole mount technique proposed for *Arabidopsis* roots (Friml *et al.*, 2003) is a derivative of this technique: roots are squashed lightly but without shattering, then enzymes, detergent and DMSO are still necessary for antibody penetration (Boutté *et al.*, 2006; Table 14.1).

(2) *Opening the cell by sectioning*: Sectioning renders cells accessible for immunoreactions. Usually plant material has to be embedded in a hard material like resins in order to be sectioned properly with accurate thickness. With resins like Epon or Spurr, only the surface of the subsequent section is accessible to probes as the embedding polymer cannot be easily dissolved away without a drastic treatment which may impair the antigenicity of the tissue. Acrylic resins like methacrylate or LR White resin are preferred, with the advantage of the latter one being that it is also usable for electron microscopy studies.

Polyethylene glycol embedding (Table 14.2) is also used and allows the antigenicity of the tissue to be maintained, following sectioning on a microtome at around 10 μm (Boutté *et al.*, 2006; Fig. 14.2).

Table 14.1 Protocol for immunostaining of *Arabidopsis* roots by whole mount

A. thaliana root: whole mount protocol
(1) Selection of 2-3 days old plantlets.
(2) Fixation with 3% paraformaldehyde in a modified PIPES buffer (PIPES 50 mM, EGTA 5 mM, $MgSO_4$ 5 mM, pH 6.9), 90 min.
(3) Wash 3 x 15 min with modified buffer, (possibly complemented with 0.1% Triton).
(4) Wash 3 x 15 min with distilled water (possibly complemented with 0.1% Triton).
(5) Place root tips (3 mm root apices) on poly-L-lysine coated glass slides, or on SuperFrost Plus slides (Menzel-Glaser laboratories).
(6) Gently squash the roots between slide and coverslip.
(7) Plunge the slide and coverslip into liquid nitrogen, in order to permeabilize the tissues and take out the coverslip, leaving the tissues glued to the slide.
(8) Leave the preparations dried on the slides to improve root adhesion.
(9) Rehydrate the roots by covering the roots with modified PIPES buffer, 15 min.
(10) Partial digestion of cell walls with 2% driselase in buffer, 20 min, then wash two times with the buffer.
(11) Treat the slides with a mixture of 10% DMSO and 3% NP40 (Nonidet) in modified PIPES buffer, 45 min to further increase the accessibility to the epitopes, and then washed three times in the buffer.
(12) Proceed to the immunostaining reaction.

Table 14.2 Protocol for PEG embedding prior to sectioning and immunostaining

PEG Embedding
(1) Selection of 2-3 days old Arabidopsis plantlets (or 3 mm maize root apices).
(2) Fixation with paraformaldehyde 3% in PBS (0.14 M NaCl, 2.7 mM KCl, 6.5 mM Na_2HPO_4, 1.5 mM KH_2PO_4, $2H_20$, pH 7.3), 90 min in glass vials.
(3) Wash in PBS (5 x 10 min).
(4) Dehydration at room temperature through an ethanol series, 30 min in 2 ml for each dehydrating step (30%, 50%, 70%, 90 %, 100%, 100%, 100% absolute). (For *Arabidopsis* plantlets, cut root apices at the 100% step).
(5) Maintain the last 100% ethanol step at 42°C for 10 min for the tissues to reach the embedding medium temperature.
(6) Add an equal volume of the wax solution (90 g PEG400 distearate (Sigma-Aldrich), 10g hexadecanol, mixed at 65°C but used at 42°C) to this 100% ethanol bathing medium.
(7) Incubate overnight (12 h) at 42°C in this two-phase mixture.
(8) Replace by pure wax solution (2 ml), and leave at least for 4 h at 42°C.
(9) Place the specimen in plastic embedding dishes.
(10) Leave to polymerise at room temperature (at least 1 h). Fully polymerized blocks are conserved at 4°C. Before sectioning, place them at -20°C for 10 min.
(11) Make thin sections (15 µm) with a microtome.
(12) Place the sections on glycerol-albumen (Serva) coated slides moistened with water and dry on slides overnight.
(13) Rehydrate the tissues by a decreasing series of ethanol (100%, 100%, 100%, 90%, 70%, 50%, 30% in PBS), 10 min each re-hydrating step.
(14) Wash the slides twice with PBS.
(15) Go to the immunoreaction procedure.

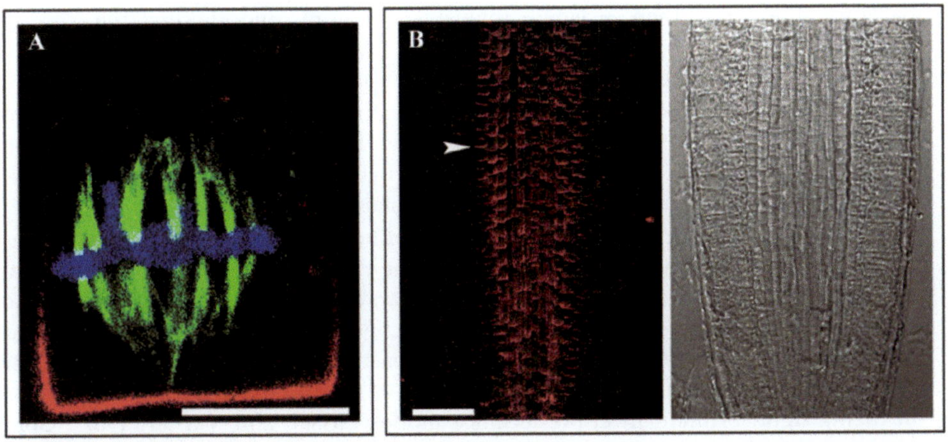

Fig. 14.2 Immunolabelling of plant cells: (A) Triple labelling in maize root cell. Immunostaining of PIN proteins i.e. plasma membrane associated proteins, facilitators of auxin transport (red channel reading Cy3 labelling), and microtubules (green channel reading the Fluorescein IsoThioCyanate (FITC) labelling). Hoechst staining of the nucleus (blue channel) (Bar: 8 µm). (B) Immunostaining of PIN proteins on thin sections of *Arabidopsis thaliana* root embedded in PEG. Left panel, PIN proteins label the basal membrane of cellular files (arrow). Right panel: differential interference contrast (Bar in B left: 40 µm). See also Boutté *et al.*, 2006.

Choice of the secondary antibody

A plethora of secondary antibodies linked to various visual markers may be found in commercial catalogues. Their choice can be critical, and the best compromise will depend on several parameters and constraints. Two aspects have to be considered: which type of Ig is needed for the immunoreaction to recognize the first antibody? Which visual marker is compatible with constraints from the tissue, techniques and complementary labels?

The secondary antibody has to recognize some conserved part of the immunoglobulins of the animal used to produce the first antibodies (often rat, and mouse for monoclonal antibodies, rabbit or chicken for polyclonal antibodies) and should be produced in an animal species (usually donkey, goat, guinea-pig) highly distinct from the first. For instance, for an antibody specific for a given protein raised in mice, one may chose a goat IgG recognizing the whole or part of the conserved chain of the mouse IgG. The question is of particular importance in the case of dual labeling of two proteins, A and B, when the two first antibodies are made in the same species, to minimize the possible cross reactions between the two sets of immunostaining. The trick in this case is to exploit the knowledge of the IgG tetrameric structure. Igs can be cleaved into fragments (F) by either papain or pepsin. Cleavage by papain releases a Fc fragment (made of the two heavy chains) and two monomeric Fab fragments, these later ones possessing the antigen binding sites. Pepsin cleaves Igs to release a dimeric fragment F(ab')2 made of the two Fab fragments and a small length of the Fc fragment. We may use an affinity purified dye-labelled Fab-fragment to reveal the first immunoreaction (Fab-fragments have a high affinity for the Fc portion of the primary antibody). After washing, this first immunolabelling event is fixed (for instance with 3%

paraformaldehyde), then washed intensively before undertaking the second immunolabelling procedure (Bolte *et al.*, 2004a). Manufacturers are now developing specific kits to minimize the risks of cross reactions in these cases.

The choice of the linked marker will depend on the microscopy mode to be used for visualisation of the immunoreaction (white light, epifluorescence, electron microscopy), and on the presence of other signals in the background (autofluorescence, chlorophyll, protein reporters, multiple immunostaining, etc.)

For *white light microscopy*, the secondary antibody can be coupled to enzymes (peroxidase, etc.): the immunolabelling will thus be visualized by treating the biological preparation with an enzyme substrate yielding an insoluble precipitate (brown), the catalysis providing massive amplification. The secondary antibody can also be coupled to gold particles (scattering rose-red light): the reaction can be enhanced with a silver reagent: black in bright field (Satiat-Jeunemaitre *et al.*, 1990), blue in polarized light, rose in razing light, brilliant in epi-reflectance mode (Cassab and Varner, 1987). Gold particles of 5-20 nm are common. One advantage is that their use is still possible in confocal microscopy in an epi-reflectance mode, despite strong tissue autofluorescence, (Satiat-Jeunemaitre and Hawes, 2001).

For *fluorescence microscopy*, the secondary antibody may be covalently coupled to a fluorochrome. The choice will depend upon the configuration of the microscope and the physico-chemical properties of the fluorochrome, notably the spectra relative to other fluorochromes present including any autofluorescence. Manufacturers are constantly improving the characteristics of the fluorescent markers linked to the antibody. For instance the family of Alexa Fluors (Molecular Probes Invitrogen) have tended to replace many classical probes including FITC. Similarly, the Cy3 dyes are brighter than Texas Red, or rhodamine. Markers fluorescing in the far red or in the blue offer important possibilities for multiple staining (three or four channels). A new type of marker has recently emerged, termed quantum dots, but as yet there have not been many applications in plant cells. These are nanocrystals of semiconductors whose dimension determines the spectral properties of light emitted from caged electrons (probes.invitrogen.com/products/qdot and www.evidenttech.com). These inorganic fluorophores are incredibly bright (50-100 times fluorescein), stable, and available in sizes to offer a series of narrow emissions from the green to red, all excited conveniently in the UV to blue. Their utility has been elegantly demonstrated by single neuro-receptor particle tracking in cells (Dahan *et al.*, 2003).

For *electron microscopy*, the immunoreaction is performed after embedding, polymerisation and sectioning. Cryo-methods for immunocytochemistry give poor tissue preservation, but permit a better protein antigenicity preservation (Vandenbosch, 1991; Hawes and Satiat-Jeunemaitre, 2001). Otherwise, LR White resins (acrylic) are preferred to Epon/Spurr (non-acrylic) resins. The immunoreaction is revealed on ultrathin sections with a secondary gold-conjugated antibody. To obtain dual labelling of several antigens, gold particles of distinctive size are used, 5 and 15 nm for instance (Hall and Hawes, 1991; Hawes and Satiat-Jeunemaitre, 2001).

Finally, one has to remember that immunocytochemistry may (and, in some cases, must) be combined with other cytochemical techniques i.e., at the light level, the use of fluorochromes or markers which will specifically stain subcellular components such as nuclei, mitochondria, cell walls, etc., (Fricker *et al.*, 2001, and Table 14.2), or, at the electron microscope level, the use of specific staining for polysaccharides for instance. Immunocytochemistry can also be combined with GFP as, fortunately, this protein remains visible after aldehyde fixation. Antibodies raised against GFP or its derivatives are commercially available, and may be used either to reinforce a low GFP signal at the light level, or to localize GFP with the resolution of electron microscopy.

Combination of Affinity Labelling and Fluorochromes as Cellular Reporters

The above examples have shown that fluorochromes are used as tags to visualize affinity labelling reactions (antibody tags). Fluorescent probes may also be used on living or fixed cells to mark subcellular structures, or small molecules at significant concentrations, or to assess metabolic or physiological changes in living cells. To chose a fluorochrome the user should consider its physico-chemical properties such as compound class, pKa, pKi, molar extinction coefficient, quantum yield, brilliance, photostability, toxicity (for cells and humans), and its environmental sensitivity. Numerous reviews and web sites on fluorochromes will help the biologist to choose the appropriate fluorochromes in response to these questions (Fricker *et al.*, 2001; Brown *et al.*, 2004).

FLUORESCENT PROTEINS AS REPORTERS FOR GENES AND PROTEINS

The study of gene expression in plant cells as in many eukaryotic organisms has highly benefited from the methodology developed around Green fluorescent protein (GFP) (Prasher *et al.*, 1992; Hanson and Koehler 2001; Brandizzi *et al.*, 2002a; Shimomura, 2005). GFP is now a well established reporter of gene expression and/or protein localisation and dynamics in living cells (Tsien, 1998). Our brief overview of GFP and its derivatives as gene or protein reporters will focus on their use for live plant cell imaging, taking into account the common pitfalls of such methodology (Zhou *et al.*, 2005). Then a guideline to help to select the most appropriate probes and the transformation methods to respond to a specific biological question, will be proposed.

An Overview on GFP Technology in Eukaryotic Cells

The gene coding for GFP has been isolated from the Pacific jellyfish *Aequorea victoria* (Prasher *et al.*, 1992). In the jellyfish, GFP fluorescence is excited by bioluminescence coming from association with aequorin. Aequorin emits blue-green light (470 nm) in the presence of oxygen and calcium (Fig. 14.3A). In the natural complex with GFP, a non-radiative transfer of energy to GFP is observed such that GFP fluoresces green (510 nm). The spectral properties of GFP permit excitation with violet to blue-green light independently of association with aequorin, or other co-factors. This makes the protein versatile for transfer into living organisms (Chalfie *et al.*, 1994; Inouye and Tsuji, 1994).

Structure

GFP is a small protein consisting of 238 amino acids (~28 kDa). The crystal structure reveals a β-can structure formed by 11 β-sheets surrounding the chromophore

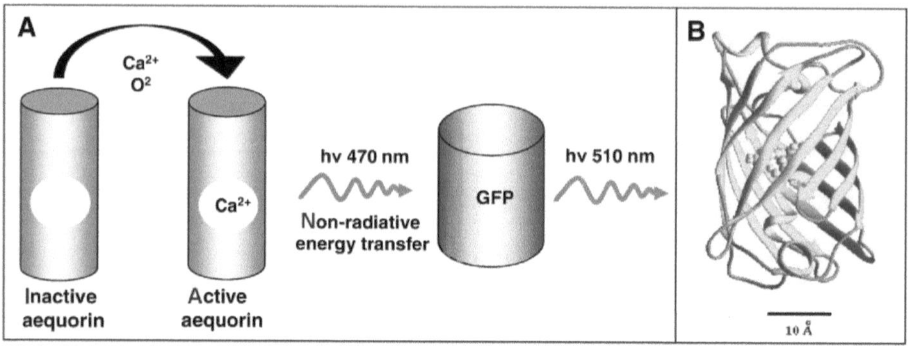

Fig. 14.3 GFP properties. (A) The bioluminescence of GFP depends on aequorin. This photoprotein, when isolated, emits blue light in a process dependent on calcium and oxygen. In complex, aequorin excites GFP by non-radiative energy transfer, which in turn emits green light. (B) Crystal structure of GFP modified from Yang *et al.*, 1996.

(Fig. 14.3B) (Ormö *et al.*, 1996; Yang *et al.*, 1996). The chromophore is formed by cyclisation and non-enzymatic oxidation of three amino acids (Ser - dehydrTyr - Gly) of the central helix, the time-limiting step of GFP protein maturation, taking 20 minutes.

Biochemical properties

GFP is a very stable protein with tendency to build dimers. In vitro, it resists high temperatures (up to 65°C), chemical treatment with guanidinium chloride (6 M), urea (8 M) and SDS (50%) and has a broad pH-tolerance (5.5 to 12). The protein is protease-resistant (trypsine, papain and thermolysine), however is usually denatured by alcohol. It has the advantage to be resistant to aldehyde fixatives, conserving its fluorescence in fixed cells.

Wild-type GFP has a principal excitation peak at 395 nm (phenol form), and a secondary peak at 475 nm (phenolate form) (Fig. 14.4). The understanding of the three-dimensional structure combined with the physico-chemical mechanisms giving rise to fluorescence has permitted the identification of particular amino acids implicated in fluorophore formation and its environment. The modification of the properties of GFP (Tsien, 1998) was performed by mutagenesis of these amino acids with the following goals: 1. Optimisation of folding and maturation properties of the protein at 37°C to permit the expression of GFP in mammalian cells (Siemering *et al.*, 1996). 2. Suppression of a cryptic intron allowing the expression of GFP in plants (Haseloff *et al.*, 1997). 3. Generation of spectral variants to obtain an excitation maximum compatible with standard laser filter sources. Numerous spectral variants of GFP such as **GFPuv** (only UV excited) (Cubitt *et al.*, 1999), **CFP** (cyan) (Heim and Tsien, 1996) and its improved variant Cerulean (Rizzo *et al.*, 2004), **YFP** (yellow) (Miyawaki *et al.*, 1997), **Venus-YFP** (yellow) (Nagai *et al.*, 2002) and **Citrine** (yellow) (Griesbeck *et al.*, 2001) have been elaborated by mutational approaches. The characteristics of these and other spectral variants are detailed in Table 14.3. A less thermo-sensitive mutation of GFP called mGFP5 (Siemering *et al.*, 1996) and a humanized version of GFP optimized for translational initiation, folding at 37°C and chromophore ionization, called EGFP (Clontech Laboratories, Palo Alto, CA), are also widely used in plants (Figs. 14.4, 14.5).

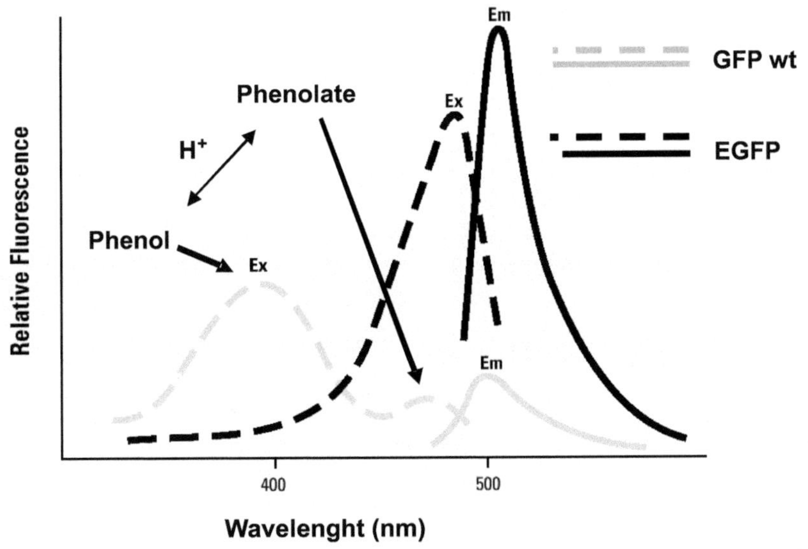

Fig. 14. 4 The dominant excitation peak of wild-type GFP is at 395 nm (phenol form) with a minor peak at 475 nm (phenolate form). In EGFP, the predominant excitation peak is shifted to 488 nm with a pKa ~4.5. This permits excitation with a standard blue argon laser.

The mGFP5 (excitation at both 395 and 475 nm) is often utilised in order to screen for GFP-positive plants with a standard UV-lamp.

Figure 14.5 presents the normalised spectra of certain Fluorescent Proteins. The expression in living cells may modify these spectra as for any fluorescent probes within cells (Bolte *et al.*, 2004b). Moreover, it is important to take into account the molar extinction values, the quantum yield and pKa (Table 14.3). For example, the pKa of the effective phenolate in EGFP and EYFP are 5.4 and 6.9: this means that the fluorescence of EYFP is more reduced in acidic compartments by protonation than that of EGFP. YFP is also very susceptible to chloride-ions. New YFP-variants such as Venus and Citrine are much less sensitive to pH and chloride; however their applicability in plant cells remains to be established.

GFP-analogues and derivatives

Besides the development of spectral variants, considerable efforts have also been made to obtain GFP-analogues from other organisms or GFP-derivatives with enhanced properties compared with the original GFP protein (Table 14.3).

GFP analogues from other organisms

PCR amplification of genes from coelenterates (Hydrozoa and Anthozoa) has given access to a variety of GFP-analogues, notably red-shifted forms (Matz *et al.*, 1999). To obtain these proteins structural information of the most striking regions of the GFP-sequence have been taken into account to design degenerated oligonucleotides. **DsRed** from the sea anemone *Discosoma striata* is one such protein (Jach *et al.*, 2001). Unfortunately, the fluorescent form of **DsRed** tends to aggregate in eukaryotic cells, forming tetramers. Moreover, the immature form fluoresces green and the emission

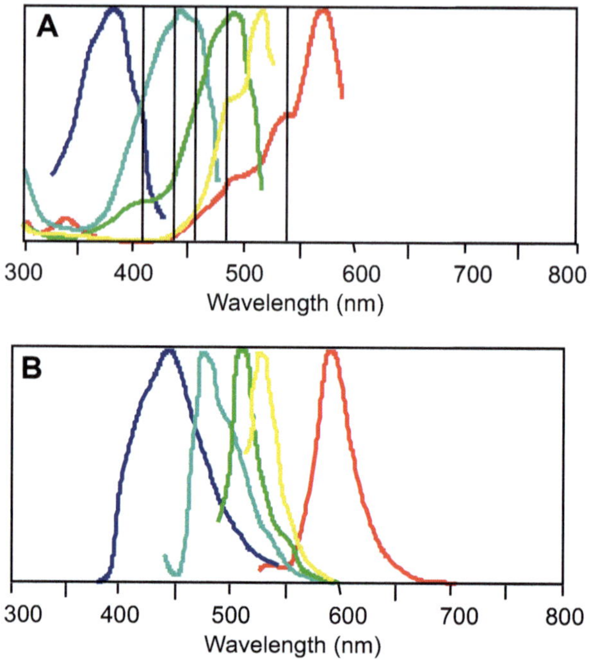

Fig. 14.5 Normalized excitation and emission spectra of EGFP and its spectral variants (modified from Clontech, Palo Alto) (From left to right: BFP, CFP, GFP, YFP, DsRed.) The positioning of laser lines often available on imaging devices is marked by vertical lines on the excitation panel (A). (B) Emission panel.

spectrum progressively shifts to higher wavelengths. This property at least made it feasible to study the time scale of gene expression by spectral shifting (Mirabella *et al.*, 2004).

A variant of DsRed, **DsRed2**, has a shorter maturation time and is more soluble. **mRFP1** is a useful monomeric DsRed-variant with a fast maturation and no green intermediary (Campbell *et al.*, 2002; Lauf *et al.*, 2002). **HcRed1** (of the reef coral *Heteractis crispa*) is a fluorescent protein with an excitation peak at 588 nm and an emission around 618 nm. To be used in green plant tissue, band-pass filters cutting the strong chlorophyll emission at 650-740 nm are needed. Another red variant with a steeper excitation spectrum is the DsRed-analogue of *Renilla reniformis* **rGFP** or **hrGFP**.

GFP variants with unique spectral properties for physiological studies

Photoactivable GFP: Photoactivation, the rapid conversion by intense irradiation of photoactivable molecules to a fluorescent state, can be used to mark and monitor selected molecules or compartments within cells. Photo-activable **GFP (PA-GFP)** is one such variant of GFP. After irradiation (< 1 s) with 413 nm light, fluorescence increases 100-fold when excited by 488 nm light and remains stable for days under aerobic conditions (Patterson and Lippincott-Schwartz, 2002). PA-GFP can be used for exploring intracellular protein dynamics by tracking photoactivated GFP-molecules which are strongly fluorescent. Photoactivable GFP has been used both as a free protein to measure protein diffusion across the nuclear envelope and as a chimera with a

Table 14.3 Properties of GFP and some of its variants or homologues

Name	*Organism*	*Origin*	*Ex (nm)*	*Em (nm)*	*pKa*	*Quantum yield* ϕ	*Extinction coefficient* ε *(*M^{-1} *x* cm^{-1}*)*	*Properties*
Derivatives of GFP								
wtGFP phenol	*Aequorea victoria*		395	508	4.5	0.79	25-30 000	
wtGFP phenolate			475	503	4.5		9.5-14 000	
EGFP		wtGFP	488	510	5.4	0.60	55 000	"Enhanced" GFP (Clontech)
mGFP5		wtGFP	395, 475	509				Elimination of a cryptic intron; excitation violet and blue
RsGFP		EGFP	505	522				Bathochromic spectra compared to GFP
PA-GFP		EGFP	400, 488	515				GFP photo-activable by violet laser
ecliptic pHluorin		EGFP	395, 475	508	7.0			optimum at pH 7.5 At pH < 6 this pHluorin is practically invisible
ratiometric pHluorin		EGFP	395 alkaline 475 acidic	508	6.6			Dynamic ~5 between pH 7.5 and pH 5.5
EBFP		EGFP	380	440		0.18	31 000	Photo-unstable
ECFP		EGFP	433	476	4.7	0.37	26 000	Reduced photo stability compared to EGFP
Cerulean		ECFP	433	475	4.7	0.62		More brilliant than ECFP, used for FRET by FLIM studies
EYFP		EGFP	515	529	6.9	0.61	80 400	-Yellow variant of GFP, -K_d for Cl^-: 110 mM
Venus-YFP		EYFP	510	530	6.0	0.57	92 000	Reinforced fluorescence at 37°C; Less sensitive to acidic pH and 100x less sensitive to Cl^- in comparison to EYFP; K_d for Cl^- >10^4 mM
Citrine		EYFP	516	529	5.7	0.76	77 000	pH and Cl^-sensitivity reduced in comparison to EYFP and Venus-YFP

Contd.

Table 14.3 Contd.

Name	*Organism*	*Origin*	*Ex (nm)*	*Em (nm)*	*pKa*	*Quantum yield* ϕ	*Extinction coefficient* ε *($M^{-1} \times cm^{-1}$)*	*Properties*
Derivatives of DsRed								
DsRed	*Discosoma striata*		558	583	4.7	0.79	57 000	
DsRed2		DsRed	561	587		0.55	57 000	Tends to oligomerize
mRFP1		DsRed	584	607	4.5	0.25	44 000	Monomeric DsRed
Fluorescent Timer		DsRed	500	580				Changes colour with time from green to red-orange
Others								
HcRed1	*Heteractis crispa*		588	618		0.04		Far red
HcRed2A		HcRed1	590	637				
MiCy	*Acropara sp.*		380, 472	495	6.6	0.9	27 200	A cyan fluorescent protein
KO	*Fungia concinna*		515, 548	561	5	0.45	110 000	orange
mKO		KO	548	561	5	0.6	51 600	Monomeric KO, orange
Kaede	*Trachyphyllia geoffroyi*		508, 572	518, 582	5.6	0.24	98 800	Photo-switchable from green to red with violet light
PS-CFP	*Aequorea coerulescens*		402 490	468 511	4.0	0.16 0.19	34 000 27 000	Photo-switchable CFP

lysosomal membrane protein to demonstrate rapid interlysosomal membrane exchange (Patterson and Lippincott-Schwartz, 2002). The validity of PA-GFP in plant cells, however, remains to be proven.

Photo-switchable XFP: A photo-switchable variant of GFP is called "Kaede" (Japanese for "maple leaf"). This tetrameric GFP-variant changes from green to red colour after photo-activation (10-60s) with a violet light or laser (Ando *et al.*, 2002) and has already been elegantly used to demonstrate mitochondrial fusions in plants (Arimura *et al.*, 2004). A new variant of photo-switchable cyan fluorescent protein is the monomeric PS-CFP. In response to intense 405 nm irradiation, PS-CFP undergoes irreversible photoconversion, with a decrease in cyan fluorescence and appearance of a 490 nm excitation peak with an emission maximum at 511 nm. After complete photoconversion of PS-CFP, a 300-fold green fluorescence increase and a fivefold cyan fluorescence decrease is observed. This adds to a 1,500-fold increase in the green-to-cyan fluorescence ratio, making PS-CFP the highest-contrast monomeric photo-switchable fluorescent protein (Chudakov *et al.*, 2004; Miyawaki, 2004).

pHluorins: pHluorins are GFP-based pH-indicators (Miesenböck *et al.*, 1998). Two different families can be distinguished: **Ecliptic pHluorin** is excited at 405 nm. Its green emission intensity is optimal at pH 7.5 and diminishes upon acidification (pKa ~6.9). **Ratiometric pHluorin** has an excitation optimum around 395 nm in an alkaline environment and an excitation optimum around 475 nm in an acidic environment. The pH is assessed by measuring the ratio of green fluorescence intensities upon excitation at 475 and 395 nm.

Application of GFP and Its Derivatives in Plant Cells

Fluorescent proteins have found numerous applications in plant biology for genomic and proteomic studies and in cell biological approaches upon living cells (Tsien, 1998; Erhardt, 2003; Brandizzi *et al.*, 2002a, 2004). Not all GFP derivatives or analogues used in mammalian and yeast cells have yet been tested in plant systems. We shall consider here the most current applications: the use of GFP either as tags for plant gene and protein location, or as indicator of distinct biological processes (protein diffusion, interactions, etc.), and propose some protocols for determining the expression of GFP systems in plant cells.

GFP as a fluorescent tag

GFP-tags have been most successfully exploited as fusion partners to study gene expression within tissue, or cellular and sub-cellular localisation of proteins. They enable the study of the fate of cells or cell lineages moving in developing organs.

A reporter of gene expression

To study gene expression, the GFP gene is cloned behind a promoter resulting in a fluorescent reporter of promoter activity. It is thus possible to study dynamics of a gene of interest at tissue or cellular level. This technique has allowed the *in vivo* analysis of cell division, growth and differentiation of the shoot apical meristem (Grandjean *et al.*, 2004), the analysis of transcription factors involved in root radial patterning (Sena *et al.*, 2004) and the sink-source transition and the vein maturation of tobacco leaves (Imlau *et al.*, 1999, Roberts *et al.*, 2001, Wright *et al.*, 2003). Cells labelled accordingly may be purified as protoplasts with the help of a cell-sorter (Birnbaum *et al.*, 2003).

Protein localization and dynamics of sub-cellular compartmentation

To visualize proteins, GFP has to be cloned in frame with the protein of interest, resulting in a chimera, which may be expressed in plant cells or tissues and even whole organisms. GFP and its spectral variants have been coupled to various peptide signals and proteins or motifs designed to mark differential locations in the plant secretory pathway. These GFP-fusions act as fluorescent reporters that allow *in vivo* the observation of the morphology and the dynamics of endomembrane compartments, membrane traffic, and protein transport pathways (Brandizzi *et al.*, 2004).

Most plant sub-cellular compartments or organelles have been labelled by a variety of GFP-fusion proteins targeting plastids, mitochondria, peroxysomes and nuclear membranes (Irons *et al.*, 2003, Kwok and Hanson, 2004; Wada and Suetsugu, 2004). The photo-switchable protein Kaede was used to follow the fusion and fission of plant mitochondria (Arimura *et al.*, 2004).

Other GFP-fusion proteins have been attributed to the secretory pathway labelling the endoplasmic reticulum, the Golgi apparatus (Boevink *et al.*, 1998), the pre-vacuolar compartment (Bolte *et al.*, 2004a), or the vacuole (Thomine *et al.*, 2003) (Fig. 14.6). The plasma membrane has been labelled by an H^+-ATPase-GFP fusion protein (Lefebvre *et al.*, 2004), a secreted form of GFP is used as a fluorescent reporter for the apoplastic space (Batoko *et al.*, 2000), and there are also reporters for the cytoskeleton (Kost *et al.*, 1998; Van Damme *et al.*, 2004) and for the nucleus (Zhou *et al.*, 2004).

GFP-fusion proteins have elucidated the function of membrane traffic regulators like monomeric Rab-GTPases (Batoko *et al.*, 2000; Sohn *et al.*, 2003; Bolte *et al.*, 2004a; Kotzer *et al.*, 2004), Sar and Arf-proteins (Brandizzi *et al.*, 2004) or SNAREs (Geelen *et al.*, 2002).

GFP technology as reporter for protein localization and dynamics has major pitfalls

The use of strong promoters has been and is still very common in order to easily visualize the GFP-fusion proteins. Over-expression of GFP-fusion proteins might, however, cause mislocation of these fusion-proteins, a shift in the dynamics and alterations of membrane structure and cell physiology. Over-expression can lead to changes in the targeting between compartments and the amount of mistargeting may also be tissue dependent (Bolte *et al.*, 2004a). Analysis of protein localization should thus be complemented by other methods: for instance, the concomitant utilisation of specific markers for subcellular compartments is recommended to identify the association of the fluorescently labelled protein within these compartments. Alternatively, after having identified the localization of an over-expressed GFP-fusion protein it is strongly recommended to check the identity and integrity of the compartment by immuno-cytochemical techniques. Immunolabelling of the endogenous protein with an antibody against the protein of interest should result in a similar pattern of distribution as for the over-expressed protein. If such an antibody is not available, it is possible to fuse a peptide epitope tag with the protein of interest. In plant cells, commonly used epitope tags include glutathione-S-transferase (GST), transcription factor c-myc, 6-histidine (6X-His) and influenza A virus haemagglutinin (HA).

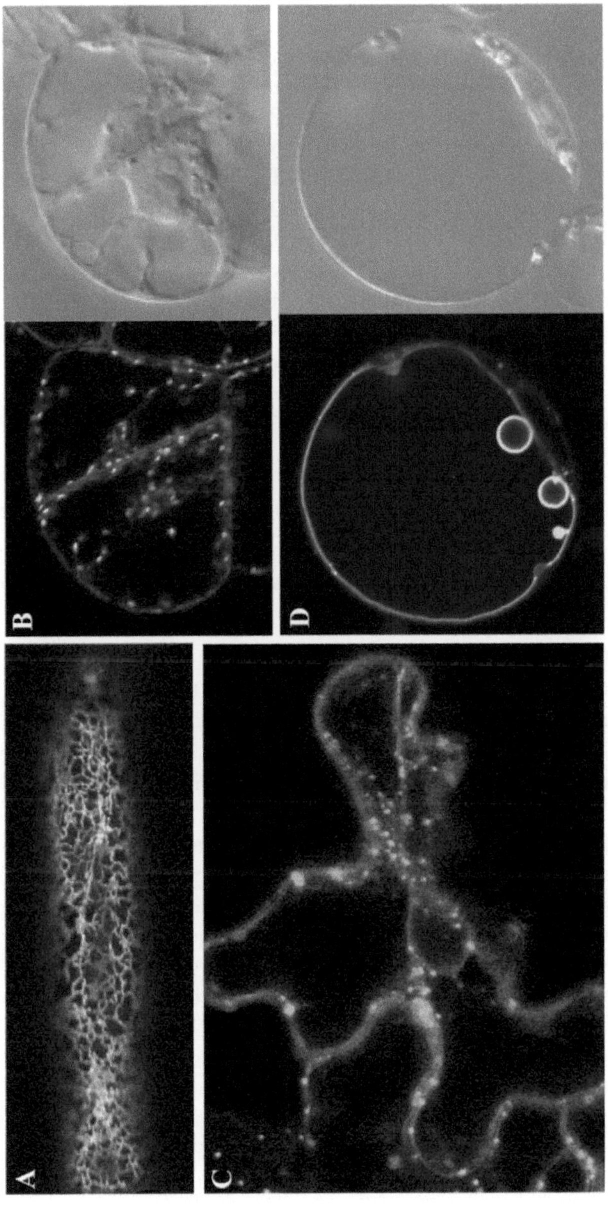

Fig. 14.6 XFP-labelling of the secretory pathway in plant cells. A, BY2-cell, ER marked with HDEL-GFP. B, left panel: BY2-cell, Golgi-stacks labelled with sialyltransferase-GFP, B, right panel: DIC. C, pre-vacuolar compartments of tobacco epidermal cells labelled with m-Rab-YFP. D, left panel: *Arabidopsis thaliana* protoplasts, vacuolar membrane marked with NRamp3-GFP, D, right panel: DIC.

GFP constructs of proteins under the control of their own promoter are also feasible if the protein is naturally expressed at a sufficient level. GFP is compatible with aldehyde fixatives like para-formaldehyde commonly used in immuno-cytochemistry (Bolte *et al.*, 2004a). It is therefore also feasible to check the localization of a GFP-labelled structure by electron microscopy (Neumann *et al.*, 2003).

It is possible that GFP-fusion proteins interact when they are co-expressed in the same cell and lead to altered morphology of membranous compartments. This has been reported for the small GTPase AtRabF2b-GFP which is localized on a sub-micron structure, the pre-vacuolar compartment. After co-expression of AtRabF2b-GFP with the vacuolar sorting receptor BP80 coupled to YFP, the pre-vacuolar compartments show tendency to aggregate (Kotzer *et al.*, 2004). It is thus important to check systematically single transformed cells before performing double transformations.

Some sub-cellular compartments may complicate the use of fluorescent reporters. Fluorescence may be quenched by highly acidic pH, chloride or metal in compartments like vacuoles and the apoplastic space (Batoko *et al.*, 2000; Tamura *et al.*, 2003). The spectra of fluorescent proteins may shift due to pH and ion content. And some compartments may exhibit a strong auto-fluorescence (chloroplasts, vacuoles, cell wall) which has to be taken into account when choosing the fluorescent reporter (Zhou *et al.*, 2005).

GFP and its spectral variants as indicators of biological processes

GFP-indicators have been successfully exploited to study distinct biological processes, such as protein diffusion and protein interactions but also to measure physiological changes in plant cells and tissues.

Protein diffusion: GFP and FRAP strategy

The advantage of GFP-tag proteins also lies in the ability to follow dynamic processes like protein movements. **FRAP** (**F**luorescence **R**ecovery **A**fter **P**hotobleaching) has allowed the study of intracellular dynamics *in vivo*, by measuring the diffusion of proteins within a sub-cellular compartment, between compartments or between cells (Zicha *et al.*, 2003, Brandizzi *et al.*, 2004; Koster *et al.*, 2005). FRAP relies on the phenomenon of photo-instability of fluorophores. A fluorescent zone is photobleached by a laser hit, the fluorescence of these target molecules being destroyed. The redistribution of fluorescence into the bleached zone and its neighbourhood can be quantified. The diffusion coefficient gives information about the nature of the transport phenomenon (e.g. lipids in artificial membranes 1 $\mu m^2/s$, lipids in biological membranes 0.01 $\mu m^2/s$, receptors 0.001 $\mu m^2/s$). A good demonstration of this technique in plant cells concerns endoplasmic reticulum-Golgi exchanges (Brandizzi *et al.*, 2002a, b). A variant of FRAP is **FLIP** (**F**luorescence **L**oss **I**n **P**hotobleaching) which also depends on the photo-instability of a fluorophore, here GFP and its derivatives. Repeated photo-destruction of a fluorescent volume leads to continued destruction of the fluorochromes. If two compartments/structures are connected, the decrease of fluorescence of the photo-bleached compartment should cause a decrease of the fluorescence in any contiguous compartment (Koster *et al.*, 2005).

Protein interactions: BiFC, FRET and BRET

Having co-localized two proteins on the same compartment it is pertinent to ask whether these proteins interact. Besides biochemical methods, it is possible to use several microscopy approaches detailed below. Light microscopy resolution is at best between 200-500 nm, whereas protein complexes will be 100-fold smaller, making it a challenge to study their nature with an optical microscope. Nevertheless, some approaches using fluorescent proteins have helped to go beyond the optical resolution limits and will be presented in the following.

Bimolecular Fluorescence Complementation (BiFC) represents an elegant exploitation of GFP to test interactions of proteins A and B (Hu *et al.*, 2002, 2003). The proteins of interest A and B are each fused to two non-fluorescent fragments of GFP (or YFP). Each fragment acts as a reporter of a distinct protein. If protein A and B interact, the two fragments of GFP (or YFP) are held in proximity to each other and may re-assemble (after several minutes), so establishing fluorescence. This strategy has been used to demonstrate interactions of transcription factors and the interaction of two pre-replication factors in plant cells (Walter *et al.*, 2004; Raynaud *et al.*, 2005).

Protein interactions may also be measured by **Förster** or **Fluorescence Resonance Energy Transfer (FRET).** FRET relies on the fact that fluorescent molecules can transfer energy by a non-radiative (electromagnetic) mechanism to an adjacent chromophore. The transfer of energy can only occur over a limited distance (< 7 nm), making FRET a powerful tool for non-invasive monitoring of protein–protein interactions. The emission spectrum of the donor fluorophore must overlap with the excitation spectrum of the acceptor for FRET to occur. Protein couples like CFP/YFP and GFP/DsRED2 (Mas *et al.*, 2000) and fluorochrome couples such as Alexa-488 and Cy3 (Achnine *et al.*, 2004) have the requisite spectral properties (Zimmermann *et al.*, 2002). FRET has enabled imaging of molecular complexes and protein interactions in plant cells (Brecht *et al.*, 2004; Kluge *et al.*, 2004; Russinova *et al.*, 2004; Bhat *et al.*, 2005), including fluorescence lifetime imaging (Russinova *et al.*, 2004).

Bioluminescence Resonant Energy Transfer (BRET) detection systems have recently been introduced in cell biology. These are based on proximity-dependent resonant energy transfer from a bioluminescent donor molecule. BRET was inspired by the natural cellular process of aequorin (Fig. 14.3) (an apo-aequorin/coelenterazine complex of the luciferase/luciferin type) which interacts rapidly with GFP in a highly specific manner to form a complex displaying very efficient energy transfer (Shimomura, 2005). A strategy to detect complex formation such as receptor cross-linking uses *Renilla* luciferase (RUC) and GFP. The luciferase and the GFP are independently attached to the proteins to be examined, either conjugated by chemical synthesis or by genetic engineering. In the latter case the gene fusion constructs are used to transfect or transform cells and to express the fusion proteins. In the presence of coelenterazine (the substrate for RUC), RUC activity is seen as a blue emission if it is not in a complex. In a complex associating the fusion proteins, part of this energy can be transferred by resonance onto GFP as energy acceptor which in turn emits green fluorescence (in absence of any external light). Excellent discrimination of the RUC and GFP signals results when Coelenterazine 400a ("DeepBlue C" BioSignal Packard) is used as the RUC substrate. The efficiency of the non-radiative energy transfer from the luciferase fusion to the GFP conjugate depends on steric effects, on the distance of the protein molecules and on their relative orientation. The method can thus be used to study protein-protein interaction as well as to screen for inhibition of interaction, e.g. in drug discovery applications or receptor-ligand studies (Boute *et al.*, 2002). Subramanian *et al.* (2004a, b) have presented protocols for BRET in plant biology with an example of mutational analysis of a protein involved in light signalling.

GFP as Biosensors

Fluorescent proteins can be engineered to obtain active physiological sensors. These sensors may then be analysed by conformation-dependent FRET analysis (Erhardt, 2003; Griesbeck, 2004). For example in the reporter Cameleon, calmodulin and the calmodulin binding peptide M13 from myosin light chain kinase have been placed in-between a cyan fluorescent protein (CFP) and a yellow fluorescent protein (YFP) (Miyawaki *et al.*, 1997). Upon calcium binding, this physiological sensor undergoes a conformational change, and enhances the energy transfer from the donor CFP to the acceptor YFP. Cameleon can thus be used to report free calcium concentrations in *Arabidopsis* tissues (Allen *et al.*, 1999, Iwano *et al.*, 2004). For non-invasive *in vivo* measurements of intra- and extracellular ion status, pHluorins can be used as pH indicators. In *Arabidopsis* they have served to measure ion dynamics in the apoplast and cytosol under various physiological conditions (Gao *et al.*, 2004) and to visualize pH gradients between different developmental compartments in intact whole-root tissues (Moseyko and Feldman, 2004).

Other examples are reporters for sugar levels in yeast cells (Fehr *et al.*, 2002). The periplasmic maltose binding protein of *Escherichia coli* has been fused to CFP and YFP. The binding of maltose to the reporter was sufficient to produce a measurable change in CFP/YFP FRET. Similar techniques were used to design a sensor for other sugars such as glucose and ribose (Fehr *et al.*, 2003, 2004), chloride sensors, sensors for kinase activity, membrane potential and cyclic AMP (Zhang *et al.*, 2002). It has yet to be established if these biosensors prove valid in plants. Other biosensors have been used to visualize the presence of galactosides on and around roots. These biosensors rely on the GFP fused to the *melA* promoter of bacteria which is induced upon exposure to galactose and galactosides (Bringhurst *et al.*, 2001).

The Right Choice of Fluorescent Protein Reporters in Plant Cells

Before using GFP, it is essential to consider the following questions:

What are the physical and optical characteristics of the specimen? Which combinations of markers (fluorescent protein and fluorochromes) are useful? Is immuno-localisation planned? What is the pH of the target compartment? What other complementary approaches like flow cytometry will be used and what are their technical constraints?

Plant cells impose some constraints

Plant cells and tissues present particular difficulties given the auto-fluorescence in blue (coumarins, phenolics), green-yellow (flavins), red (chlorophylls), not to mention the orange for algae (phycoerythrins) (Meyer *et al.*, 2003; Berg, 2004). The choice of fluorescent proteins has thus to be adapted to the tissue investigated and spectral measurements often need to be introduced in acquisition techniques. Moreover, sub-cellular compartments may have highly acidic pH which may result in a decrease, shift or even loss of fluorescence of the reporter proteins. It is possible to express several different fluorescent proteins in the same cells. Therefore, it is important to examine rigorously the spectral necessities before constructing chimeral proteins and transforming cells.

Double visualization by using double XFP probes

Figure 14.5 shows that the excitation and emission spectra of fluorescent proteins generally have large regions of overlap. Therefore, it is important to choose the right couples for double labelling experiments. They should have distinct excitation and emission spectra to avoid crosstalk. For this reason the couples CFP-YFP, GFP-DsRed and GFP-mRFP1 are mostly used. For concomitant immunolabelling studies, a large variety of fluorochromes coupled to secondary antibodies are available.

A Guideline for the Expression of Fluorescent Reporters in Plants

Fluorescent proteins may be expressed in a transient or stable manner in various plant species, such as *A. thaliana, Nicotiana tabacum, Nicotiana benthamiana, Medicago truncatula, Triticum aestivum, Hordeum vulgare, Oryza sativa, Allium cepa.* Various transformation techniques are applicable on these plants, on derived tissues, cell suspension cultures and protoplasts. The following paragraph gives an overview of some frequently used transformation techniques and plant systems and the choice of vectors.

The choice of the transformation techniques

Transient transformation techniques

The transient expression of fusion proteins can be carried out by several different approaches such as protoplast transformation by polyethyleneglycol or electroporation (Bates 1999; Datta and Datta, 1999), particle bombardment (Klein and Fitzpatrick-McElligott, 1993, Taylor and Fauquet, 2002) and infiltration of epidermal cells with *Agrobacterium tumefaciens* (Hawes *et al.*, 1999) or viruses (Boevink *et al.*, 1998). These techniques lead to a rapid temporary expression of the protein of interest and the fluorescent material can be observed 1-4 days after transformation. Transformed cells stay fluorescent for a period of 2-3 days. The advantage of transient expression techniques is that they are easy to perform and not very time consuming. Protoplast transformation and infiltration do not require sophisticated equipment and have the advantage that double expression of two different fusion proteins is simple. Particle bombardment requires specialised equipment, such as a particle gun, which is relatively expensive; it is also less efficient for double transformants. Transformation has been successful with protoplasts from *Arabidopsis* or tobacco leaves and suspension cultures (Di Sansebastiano *et al.*, 1998; Fluckiger *et al.*, 2003; Bolte *et al.*, 2004a;), leaf infiltration has given good results with tobacco and *Arabidopsis thaliana* (Batoko *et al.*, 2000; Gaire *et al.*, 1999), and particle bombardment has been performed on leaf tissue and onion epidermal tissue, BY-2 cells and pollen (Nishihara *et al.*, 1993; Thomine *et al.*, 2003).

Stable transformation techniques

The stable expression of fusion proteins has mostly been carried out with the help of *Agrobacterium tumefaciens. Agrobacterium* transformation leads to integration of the gene for the fusion protein into the genome of the host-organism and leads therefore to stable expression of the protein of interest. Stable transformation techniques are costly in time and labour. The quickest system is the infiltration of *Medicago* hairy roots, which can be observed three weeks after transformation (Rodriguez-Llorente *et al.*, 2003). Stable transformants of plants like *Arabidopsis thaliana*, *Nicotiana tabacum* or *Medicago truncatula* can be obtained in three to six months (Bent, 2000; Kamaté *et al.*, 2000). The stable transformation of whole plants depends on the regeneration time needed to obtain a F2-generation. To obtain stable transformants of BY-2 cells (Criqui *et al.*, 2000) or *Arabidopsis thaliana* suspension cells, a regeneration cycle of at least 12 weeks is recommended, transplanting the fluorescent calli on fresh culture media at least two times after transformation. The genome status of the stable line should be analysed by flow cytometry to check the ploidy-level after slow culture. BY-2 cell lines have been established which are transgenic for fluorescent markers like the Golgi resident sialyl-transferase-GFP and the ER markers ERD2-GFP (Saint-Jore *et al.*, 2002) and HDEL-GFP (Bolte *et al.*, 2004b).

Making the construct: the vector systems

The GFP-coding sequence may be coupled to several motifs and may be transferred in most biological systems. The vectors have to be chosen according to the transformation system used and there is also a choice to make between the natural gene promoters (Benkova *et al.*, 2003), strong promoters like the 35S cauliflower mosaic virus promoter (Batoko *et al.*, 2000; Bolte *et al.*, 2004a) and inducible promoters (Laporte *et al.*, 2003; Laufs *et al.*, 2004; Samalova *et al.*, 2005). The aim is normally not to alter the characteristics of the protein studied. Therefore it may be useful to couple the GFP to the N- or C-terminus (Bolte *et al.*, 2004a; Kotzer *et al.*, 2004), or even inside the protein sequence (Benkova *et al.*, 2003). Coupling to GFP may cause misfolding and thus degradation of the protein. Functional tests in knockout plants or yeast mutants may help to provide information about the influence of the fluorescent tag on the folding and function of the protein of interest. Over-expression may disturb cellular physiology and metabolism. Transient expression in protoplasts or bombarded epidermal cells can be performed with common bacterial vectors if the fusion protein is under the control of a strong promoter like the CaMV 35S-promoter. Binary vectors are also feasible in these systems, however large amounts of DNA are needed (5-15 μg per transformation), which makes some binary vectors such as pBIN19 less viable, because of their low amplification levels. All other transformation techniques require the use of binary vectors.

Perspectives

Fluorescent Proteins have proven valuable for the study of many processes in plants. They have enabled the study of tissue development and transport processes between cells (Oparka *et al.*, 1999; Gillespie *et al.*, 2002; Laufs *et al.*, 2004). Fluorescent reporters have helped to elucidate plant endomembrane traffic (Brandizzi *et al.*, 2004) and to highlight physiological processes like ion dynamics (Gao *et al.*, 2004). Furthermore these fluorescent reporters have been used to study the interactions of proteins (Raynaud *et al.*, 2005) and organelles (Arimura *et al.*, 2004).

The constant search for new variants may overcome some of the drawbacks of these fluorescent reporters such as their relatively low brightness or stability and the considerable size of the fluorescent tag, which might interfere with the function of the fusion protein. The design of new biosensors and the engineering of new variants such as fluorescent timers and photo-switchable proteins, including push-pull forms that are reversible, will further enhance the plethora of applications flowing from this extremely valuable tool to address the functional organisation of plants.

ACKNOWLEDGEMENTS

C. Kluge was recipient of a Genopole Recherche postdoctoral grant to the BioRet Company (Evry). We thank J. Daraspe (ISV, Gif sur Yvette) for providing the images of the BY2-cells expressing HDEL-GFP and ST-GFP and S. Thomine (ISV, Gif sur Yvette) for providing the Nramp3-GFP construct for *Arabidopsis thaliana* protoplast transformation. A. Campalans and F. Frugier (ISV, Gif sur Yvette) are acknowledged for producing the Figs. 14.2A, B and their comments on the *in situ* hybridation section. Thanks are due to M.T. Crosnier for her constant technical assistance. Support from the *Conseil Général de l'Essonne* via the ASTRE programme for live cell microscopy in Gif-sur-Yvette is appreciated.

REFERENCES

Achnine, L., Blancaflor, E.B., Rasmussen, S., Dixon, R.A. 2004. Colocalization of L-phenylalanine ammonia-lyase and cinnamate 4-hydroxylase for metabolic channeling in phenylpropanoid biosynthesis. *Plant Cell,* 16, 3098-3109.

Adams, S.R., Campbell, R.E., Gross, L.A., Martin, B.R., Walkup, G.K., Yao, Y., Llopis, J., Tsien, R.Y. 2002. New biarsenical ligands and tetracysteine motifs for protein labeling *in vitro* and *in vivo*: synthesis and biological applications. *J. Am. Chem. Soc.,* 124, 6063-6067.

Allen, G.J., Kwak, J.M., Chu, S.P., Llopis, J., Tsien, R.Y., Harper, J.F., Schroeder, J.I. 1999. Cameleon calcium indicator reports cytoplasmic dynamics in Arabidopsis guard cells. *Plant J.,* 19, 737-747.

Ando, R., Hama, H., Yamamoto-Hino, M., Mizuno, H., Miyawaki, A. 2002. An optical marker based on the UV-induced green-to-red photoconversion of a fluorescent protein. *Proc. Natl. Acad. Sci.* USA, 99, 12651-12656.

Arimura, S., Yamamoto, J., Aida, G.P., Nakazono, M., Tsutsumi, N. 2004. Frequent fusion and fission of plant mitochondria with unequal nucleoid distribution. *Proc. Natl. Acad. Sci.* USA, 101, 7805-7808.

Bates, G.W. 1999. Plant transformation via protoplast electroporation. *Methods Mol. Biol.,* 11, 359-366.

Batoko, H., Zheng, H.Q, Hawes, C., Moore, I. 2000. A rab1 GTPase is required for transport between the endoplasmic reticulum and Golgi apparatus and for normal Golgi movement in plants. *Plant Cell,* 12, 2201-2218.

Benkova, E., Michniewicz, M., Sauer, M., Teichmann, T., Seifertova, D., Jurgens, G., Friml, J. 2003. Local, efflux-dependent auxin gradients as a common module for plant organ formation. *Cell,* 115, 591-602.

Bent, A.F., 2000. *Arabidopsis* in Planta Transformation. Uses, mechanisms, and prospects for transformation of other species. *Plant Physiol.,* 124,1540-1547.

Berg, R.H. 2004. Evaluation of spectral imaging for plant cell analysis. *J. Microscopy,* 214, 174-181.

Bhat, R.A., Miklis, M., Schmelzer, E., Schulze-Lefert, P., Panstruga, R. 2005. Recruitment and interaction dynamics of plant penetration resistance components in a plasma membrane microdomain. *Proc. Natl. Acad. Sci.* USA, 102, 3135-3140.

Birnbaum, K., Shasha, D.E., Wang, J.Y., Jung, J.W., Lambert, G.M., Galbraith, D.W., Benfey, P.N. 2003 A gene expression map of the Arabidopsis root. *Science,* 302, 1956-1960.

Boevink, P., Oparka, K., Santa Cruz, S., Martin, B., Betteridge, A., Hawes, C. 1998. Stacks on tracks: the plant Golgi apparatus traffics on an active ER/actin network. *Plant J.*, 15, 441-447.

Bolte, S., Brown, S., Satiat-Jeunemaître, B. 2004a. The N-myristoylated Rab-GTPase m-Rabmc is involved in post-Golgi trafficking events to the lytic vacuole in plant cells. *J. Cell Sci.*, 117, 943-954.

Bolte, S., Talbot, C., Boutte, Y., Catrice, O., Read, N.D., Satiat-Jeunemaître, B. 2004b. FM-dyes as experimental probes for dissecting vesicle trafficking in living plant cells. *J. Microsc.*, 214, 159-173.

Borisjuk, L., Rolletschek, H., Walenta, S., Panitz, R., Wobus, U., Weber, H. 2003. Energy status and its control on embryogenesis of legumes: ATP distribution within Vicia faba embryos is developmentally regulated and correlated with photosynthetic capacity. *Plant J.*, 36, 318-329.

Boute, N., Jockers, R., Issad, T. 2002. The use of resonance energy transfer in high-throughput screening: BRET versus FRET. *Trends Pharmacol. Sci.*, 23, 351-354.

Boutté, Y., Crosnier, M.T., Carraro, N., Traas, J., Satiat-Jeunemaître, B. 2006. Immunocytochemistry of the plasma membrane recycling pathway and cell polarity in plants: studies on PIN proteins. *J. Cell Sci.*, 1255-1265.

Brandizzi, F., Fricker, M., Hawes, C. 2002a. A greener world: the revolution in plant bioimaging. *Nature Review Molecular Cell Biology*, 3, 520-530.

Brandizzi, F., Irons, S., Johansen, J., Kotzer, A., Neumann, U. 2004. GFP is the way to glow: bioimaging of the plant endomembrane system. *J. Microscopy*, 214, 138-158.

Brandizzi, F., Snapp, E.L., Roberts, A.G., Lippincott-Schwart, J., Hawes, C. 2002b. Membrane protein transport between the endoplasmic reticulum and the Golgi in tobacco leaves is energy dependent but cytoskeleton independent: evidence from selective photobleaching. *Plant Cell.*,14, 1293-1309.

Brecht, M., Sewald, K., Schiene, K., Keen, G., Fricke, M., Sauer, M., Niehaus, K. 2004. The use of surface plasma resonance (SPR) and fluorescence resonance energy transfer (FRET) to monitor the interaction of the plant G-proteins Ms-Rac1 and Ms-Rac4 with GTP. *J Biotechnol.*, 112, 151-164.

Bringhurst, R.M., Cardon, Z.G., Gage, D.J. 2001. Galactosides in the rhizosphere: utilization by *Sinorhizobium meliloti* and development of a biosensor. *Proc. Natl. Acad. Sci.* USA, 98, 4540-4545.

Brown, S., Bolte, S., Satiat-Jeunemaitre, B. 2007. Tracking gene expression in plant cells: Microscopy and associated Bio-imaging techniques. Chapter 13 this book.

Brown, S., Poujol-Talbot, C., Kronenberger, J., Traas J., Satiat-Jeunemaitre, B. 2004. L'apport de la Microscopie et de l'imagerie en génomique. In "La génomique en Biologie Végétale », J.F. Morot-Gaudry, J.F. Briat eds, INRA Editions, 209-236.

Campalans, A., Kondorosi, A., Crespi, M. 2004. Enod40, a Short Open Reading Frame–Containing mRNA, Induces Cytoplasmic Localization of a Nuclear RNA Binding Protein in *Medicago truncatula. Plant Cell*, 16, 1047-1059.

Campbell, R.E., Tour, O., Palmer, A.E., Steinbach, P.A., Baird, G., Zacharias, D.A, Tsien, R.Y. 2002. A monomeric red fluorescent protein. *Proc. Natl. Acad. Sci.* USA, 99, 7877-7882.

Cassab, G.I., Varner, J.E. 1987. Immunocytolocalization of extensin in developing soybean seed coats by immunogold-silver staining and by tissue printing on nitrocellulose paper. *J. Cell Biol.*, 105, 2581-2588.

Castle, J., Szekeres, M., Jenkins, G., Bishop, G.J. 2005. Unique and overlapping expression patterns of Arabidopsis CYP85 genes involved in brassinosteroid C-6 oxidation. *Plant Mol. Biol.*, 57, 129-140.

Chalfie, M., Tu, Y., Euskirchen, G., Ward, W.W., Prasher, D.C. 1994. Green fluorescent protein as a marker for gene expression. *Science*, 263, 802-805.

Chinnusamy, V., Stevenson, B., Lee, B.H., Zhu, J.K. 2002. Screening for gene regulation mutants by biolumines cence imaging. *Science STKE*, 140, PL10.

Chudakov, D.M., Verkhusha, V.V., Staroverov, D.B., Souslova, E.A, Lukyanov, S., Lukyanov, K.A. 2004. Photoswitchable cyan fluorescent protein for protein tracking. *Nat. Biotechnol.*, 22, 1435-1439.

Coba de la Peña, T., Frugier, F., McKhann, H.I., Bauer, P., Brown, S.C., Kondorosi, A., Crespi, M. 1997. A carbonic anhydrase gene is induced in the nodule promordium and its cell-specific expression is controlled by the presence of *Rhizobium* during development. *Plant J.*, 11, 407-420.

Cougot, N., Babajko, S., Séraphin, B. 2004. Cytoplasmic foci are sites of mRNA decay in human cells. *J. Cell Biology*, 165, 31-40.

Crespi, M., Jurkevitch, E., Poiret, M., d'Aubenton-Carafa, Y., Petrovics, G., Kondorosi, E., Kondorosi, A. 1994. Enod40, a gene expressed during nodule organogenesis, codes for a non-translatable RNA involved in plant growth. *EMBO Journal*, 13, 5099-5112.

Criqui, M.C., Parmentier, Y., Derevier, A., Shen, W.H., Dong, A., Genschik, P. 2000. Cell cycle-dependent proteolysis and ectopic overexpression of cyclin B1 in tobacco BY2 cells. *Plant J.*, 24, 763-773.

Cubitt, A.B., Woollenweber, L.A., Heim, R. 1999. Understanding structure-function relationships in the *Aequorea victoria* green fluorescent protein. *Methods Cell Biol.*, 58, 19-30.

Dahan, M., Lévi, S., Luccardini, C., Rostaing, P., Riveau, B., Triller, A. 2003. Diffusion dynamics of glycine receptors revealed by single-quantum dot tracking. *Science*, 302, 442-445.

Datta, K., Datta, S.K. 1999. Transformation of rice via PEG-mediated DNA uptake into protoplasts. *Methods Mol Biol.*, 111, 335-347.

Di Sansebastiano, G.P., Paris, N., Marc-Martin, S., Neuhaus, J.M. 1998. Specific accumulation of GFP in a non-acidic vacuolar compartment via a C-terminal propeptide-mediated sorting pathway. *Plant J.*, 15, 449-457.

Dirks, R.W., Molenaar, C., Tanke, H.J. 2001. Methods for visualizing RNA processing and transport pathways in living cells. *Histochem. Cell Biol.*, 115, 3–11.

Erhardt, D. 2003. GFP technology for live cell imaging, *Curr Opin Plant Biol.*, 6, 622-628.

Fehr, M., Ehrhardt, D.W., Lalonde, S., Frommer, W.B. 2003. Minimally invasive dynamic imaging of ions and metabolites in living cells. *Curr. Opin. Plant. Biol.*, 7, 345-351.

Fehr, M., Frommer, W.B., Lalonde, S. 2002. Visualization of maltose uptake in living yeast cells by fluorescent nanosensors. *Proc. Natl. Acad. Sci.* USA, 99, 9846-9851.

Fehr, M., Lalonde, S., Ehrhardt, D.W., Frommer, W.B. 2004. Live imaging of glucose homeostasis in nuclei of COS-7 cells. *J. Fluoresc.*, 14, 603-609.

Feiler, H.S., Desprez, T., Santoni, V., Kronenberger, J., Caboche, M., Traas, J. 1995. The higher plant *Arabidopsis thaliana* encodes a functional CDC48 homologue which is highly expressed in dividing and expanding cells. *EMBO Journal*, 15, 5626-5637.

Fluckiger, R., De Caroli, M., Piro, G., Dalessandro, G., Neuhaus, J.M., Di Sansebastiano, G.P. 2003. Vacuolar system distribution in Arabidopsis tissues, visualized using GFP fusion proteins. *J. Exp Bot.*, 54, 1577-1584.

Fricker, M., Parsons, A., Tlaka, M., Blancaflor, E., Gilroy, S., Meyer, A., Plieth, C. 2001. Fluorescent probes for living plant cells. In "Plant cell biology", 2nd ed., Hawes C. and Satiat-Jeunemaître, B. (eds), Oxford University Press, 35-81.

Friml, J., Benková, E., Mayer, U., Palme, K., Muster, G. 2003. Automated whole mount localisation techniques for plant seedlings. *Plant J.*, 34, 115-124.

Gallagher, S.R. (ed) 1992. Gus Protocols: Using the Gus Gene As a Reporter of Gene Expression, Academic Press;

Gao, D., Knight, M.R., Trewavas, A.J, Sattelmacher, B., Plieth, C. 2004. Self-Reporting *Arabidopsis* Expressing pH and [Ca^{2+}] Indicators Unveil Ion Dynamics in the Cytoplasm and in the Apoplast under Abiotic Stress. *Plant Physiol.*, 134, 898-908.

Geelen, D., Leyman, B., Batoko, H., Di Sansebastiano, G.P., Moore, I., Blatt, M.R. 2002. The abscisic acid-related SNARE homolog NtSyr1 contributes to secretion and growth: evidence from competition with its cytosolic domain. *Plant Cell*, 14, 387-406.

Gillespie, T., Boevink, P., Haupt, S., Roberts, A.G., Toth, R., Valentine, T., Chapman, S., Oparka, K.J. 2002. Functional analysis of a DNA-shuffled movement protein reveals that microtubules are dispensable for the cell-to-cell movement of tobacco mosaic virus. *Plant Cell*, 14, 1207-1222.

Gould, S.J., Subramani, S., 1998. Firefly luciferase as a tool in molecular and cell biology.

Grandjean, O., Vernoux T., Laufs, P., Belcram, K., Mizukami, Y., Traas, J. 2004. *In vivo* analysis of cell division, cell growth, and differentiation at the shoot apical meristem in Arabidopsis. *Plant Cell*, 16, 74-87.

Grant, J.J., Chini, A., Basu, D., Loake, G.J. 2003. Targeted activation tagging of the Arabidopsis NBS-LRR gene, ADR1, conveys resistance to virulent pathogens. *Mol. Plant Microbe Interact*, 16, 669-680.

Grant, J.J., Chini, A., Basu, D., Loake, G.J. 2003. Targeted activation tagging of the Arabidopsis NBS-LRR gene, ADR1, conveys resistance to virulent pathogens. *Mol Plant Microbe Interact.*, 16, 669-80.

Greer, L.F., Szalay, A.A. 2002. Imaging of light emission from the expression of luciferases in living cells and organisms: a review. *Luminescence*, 1, 43-74.

Griesbeck, O. 2004. Fluorescent proteins as sensors for cellular functions. *Curr. Opin. Neurobiol.*, 14, 636-641.

Griesbeck, O., Baird, G.S., Campbell, R.E., Zacharias, D.A., Tsien, R.Y. 2001. Reducing the environmental sensitivity of yellow fluorescent protein. Mechanism and applications. *J. Biol Chem.*, 276, 29188-29194.

Hadlington, J.L., Denecke, J. 2001. Transient expression, a tool to address questions in plant cell biology. In "Plant Cell Biology", Hawes, C.R. and Satiat-Jeunemaitre, B. eds, Oxford University Press, p. 107-125.

Hall, J.L., Hawes, C., 1991. Electron microscopy of plant cells. Academic Press.

Hanson, M.R., Koehler, R.H. 2001. GFP imaging: methodology and application to investigate cellular compartmentation in Plants. *J. Exp. Bot.*, 52, 529-539.

Haseloff, J., Siemering, K.R., Prasher, D.C., Hodge, S. 1997. Removal of a cryptic intron and subcellular localisation of green fluorescent proteins are required to mark transgenic Arabidopsis plants brightly. *Proc. Natl. Acad. Sci.* USA, 94, 2122-2127.

Hawes, C., Boevink, P., Moore, I. 1999. Green fluorescent protein in plants. In "Protein Localization by Fluorescence Microscopy: a Practical Approach". (Allen, V. ed) Oxford University Press, pp. 163-177.

Hawes, C., Satiat-Jeunemaître, B. 2001. Electron microscopy. In "Plant cell Biology" (Hawes CH and Satiat-Jeunemaître B eds), Oxford Univeristy Press, 235-266.

Hawes, C., Satiat-Jeunemaître, B. 2001. *Plant Cell Biology*. 2nd edition. Oxford University Press.

Heim, R., Tsien, R.Y. 1996. Engineering green fluorescent protein for improved brightness, longer wavelengths and fluorescence resonance energy transfer. *Curr. Biol.*, 6, 178-182.

Hoffmann, C., Gaiette, G., Bönemann, M., Adams, S.R., Oderdorff-Maass, S., Gehr, B., Vilardaga, J.P., Tsien, R.Y., Ellisman, M.H., Lohse, M.J. 2005. A FLASH-based approach to determine G protein-coupled receptor activation in living cells. *Nature Methods*, 2, 171-176.

Horstmann, V., Huether, C.M., Jost, W., Reski, R., Decker, E.L. 2004. Quantitative promoter analysis in Physcomitrella patens: a set of plant vectors activating gene expression within three orders of magnitude. *BMC Biotechnol.*, 4, 13.

Hu, C-D., Chinenov, Y., Kerppola, T.K. 2002. Visualization of interactions among bZIP and rel family proteins in living cells using Bimolecular Fluorescence Complementation. *Mol. Cell*, 9, 789-798.

Hu, C-D., Kerppola, T.K. 2003. Simultaneous visualization of multiple protein interactions in living cells using multicolor fluorescence complementation analysis. *Nature Biotechnology*, 21, 539-545.

Imlau, A., Truernit, E., Sauer, N. 1999. Cell-to-cell and long-distance trafficking of the green fluorescent protein in the phloem and symplastic unloading of the protein into sink tissues. *Plant Cell*, 11, 309-322.

Inouye, S., Tsuji, F.I. 1994. *Aequorea* green fluorescent protein: Expression of the gene and fluorescence characteristics of the recombinant protein. *FEBS Lett.*, 341, 277-280.

Irons, S.L., Evans, D.E., Brandizzi, F. 2003. The first 238 amino acids of the human lamin B receptor are targeted to the nuclear envelope in plants. *J. Exp. Bot.*, 54, 943-950.

Iwano, M., Shiba, H., Miwa, T., Che, F.S., Takayama, S., Nagai, T., Miyawaki, A., Isogai, A. 2004. Ca2+ dynamics in a pollen grain and papilla cell during pollination of Arabidopsis. *Plant Physiol.*, 136, 3562-3571.

Jach, G., Binot, E., Frings, S., Luxa, K., Schell, J. 2001. Use of red fluorescent protein from *Discosoma* sp. (DsRed) as a reporter for plant gene expression. *Plant J.*, 28, 483-491.

Jackson, D. 1991. *In situ* hybridization in plants. In: Molecular Plant pathology: A practical Approach". (Bowles, D. Gurr, S. McPherson, M. eds.) Oxford University Press.

Jefferson, R.A., Kavanagh, T.A., Bevan, M.W. 1987. β-glucuronidase as a sensitive and versatile gene fusion marker in higher plants. *EMBO J.*, 6, 3901-3907.

Jolliffe, N.A., Brown, J.C., Neumann, U., Vicre, M., Bachi, A., Hawes, C., Ceriotti, A., Roberts, L.M., Frigerio, L. 2004. Transport of ricin and 2S albumin precursors to the storage vacuoles of Ricinus communis endosperm involves the Golgi and VSR-like receptors. *Plant J.*, 39, 821-833.

Kamaté, K., Rodriguez-Llorente, I.D., Scholte, M., Durand, P., Ratet, P., Kondorosi, E., Kondorosi, A., Trinh, T.H. 2000. Transformation of floral organs with GFP in *Medicago truncatula*. *Plant Cell Rep*, 19, 647-653.

Klein, T.M., Fitzpatrick-McElligott, S. 1993. Particle bombardment: a universal approach for gene transfer to cells and tissues. *Curr. Opin. Biotechnol.*, 4, 583-590.

Kluge, C., Seidel, T., Bolte, S., Sharma, S.S., Hanitzsch, M., Satiat-Jeunemaître, B., Ross, J., Sauer, M., Golldack, D., Dietz, K.J. 2004. Subcellular distribution of the V-ATPase complex in plant cells, and in vivo localisation of the 100 kDa subunit VHA-a within the complex. *BMC Cell Biol.*, 5, 29.

Knapp, T., Hare, E., Feng, L., Zlokarnik, G., Negulescu, P. 2003. Detection of ß–lactamase reporter gene expression by flow cytometry. *Cytometry*, 51A, 68-78.

Knight, M.R., Read, N.D., Campbell, A.K., Trewavas, A.J. 1993. Imaging calcium dynamics in living plants using semi-synthetic recombinant aequorins. *Journal of Cell Biology*, 121, 83-90.

Koiwa, H., Barb, A.W., Xiong, L., Li, F., McCully, M.G., Lee, B.H., Sokolchik, I., Zhu, J., Gong, Z., Reddy, M., Sharkhuu, A., Manabe, Y., Yokoi, S., Zhu, J.K., Bressan, R.A., Hasegawa, P.M. 2002. C-terminal domain phosphatase-like family members (AtCPLs) differentially regulate *Arabidopsis thaliana* abiotic stress signaling, growth, and development. *Proc. Natl. Acad. Sci USA*, 99, 10893-10898.

Kopczynski, C.C., Noordermeer, J.N., Serano, T.L, Chen, W.Y., Pendleton, J.D., Lewis, S., Goodman, C.S., Rubin, G.M. 1998. A high throughput screen to identify secreted and transmembrane proteins involved in Drosophila embryogenesis. *Proc. Nat. Acad. Sci.* USA, 95, 9973-9978.

Kost, B., Spielhofer, P., Chua, N.H. 1998. A GFP-mouse talin fusion protein labels plant actin filaments in vivo and visualizes the actin cytoskeleton in growing pollen tubes. *Plant J.*, 16, 393-401.

Koster, M., Frahm, T., Hauser, H. 2005. Nucleocytoplasmic shuttling revealed by FRAP and FLIP technologies. *Curr. Opin. Biotechnol.*, 16, 28-34.

Kotzer, A.M., Brandizzi, F., Neumann, U., Paris, N., Moore, I., Hawes, C. 2004. AtRabF2b (Ara7) acts on the vacuolar trafficking pathway in tobacco leaf epidermal cells. *J. Cell Sci.*, 117, 6377-6389.

Kronenberger, J., Desprez, T., Höfte, H., Caboche, M., Traas, J. 1993. A methacrylate embedding procedure developed for immunolocalization on plant tissues is also compatible with *in situ* hybridization. *Cell Biol. Int.*, 17, 1013-1021.

Kwok, E.Y., Hanson, M.R. 2004. Plastids and stromules interact with the nucleus and cell membrane in vascular plants. *Plant Cell Rep.*, 23, 188-195.

Laporte, C., Vetter, G., Loudes, A.M., Robinson, D.G., Hillmer, S., Stussi-Garaud, C., Ritzenthaler, C. 2003. Involvement of the secretory pathway and the cytoskeleton in intracellular targeting and tubule assembly of Grapevine fanleaf virus movement protein in tobacco BY-2 cells. *Plant Cell*, 15, 2058-2075.

Lauf, U., Lopez, P., Falk, M.M. 2002. Expression of fluorescently tagged connexins: a novel approach to rescue function of oligomeric DsRed-tagged proteins. *FEBS Lett.*, 498, 11-15.

Laufs, P., Peaucelle, A., Morin, H., Traas, J. 2004. MicroRNA regulation of the CUC genes is required for boundary size control in *Arabidopsis* meristems. *Development*, 131, 4311-4322.

Lazzeri, P.A. 1995. Stable transformation of barley via Electroporation- and PEG-mediated protoplast transformation. *Methods Mol. Biol.*, 4, 95-106.

Lecourieux, D., Mazars, C., Pauly, N., Ranjeva, R., Pugin, A. 2002. Analysis and effects of cytosolic free calcium increases in response to elicitors in Nicotiana plumbaginifolia cells. *Plant cell*, 14, 2627-2641.

Lefebvre, B., Batoko, H., Duby, G., Boutry, M. 2004. Targeting of a *Nicotiana plumbaginifolia* H^+-ATPase to the plasma membrane is not by default and requires cytosolic structural determinants. *Plant Cell*, 16, 1772-1789.

Leitch, A.R., Lim, K.Y., Webb, D.R., McFadden, G.I. 2001. In situ hybridization. In "Plant Cell Biology", (Hawes, C.H. and Satiat-Jeunemaitre, B. eds) Oxford University Press, pp 267-294.

Lin, C., Yu, Y., Kadono, T., Iwata, M., Umemura, K., Furuichi, T., Kuse, M., Isobe, M., Yamamoto, Y., Matsumoto, H., Yoshizuka, K., Kawano, T. 2005. Action of aluminium, novel TPC1-type channel inhibitor, against salicylate-induced and cold-shock-induced calcium influx in tobacco BY-2 cells. *Biochem Biophys Res Commun.*, 332, 823-30.

Lloyd, C.W., Slabas, A.R., Powell, A.J., MacDonald, G., Badley, R.A. 1979. Cytoplasmic microtubules of higher plant cells visualised with anti-tubulin antibodies. *Nature*, 279, 239-241.

Ludevid, M.D., Freire, M.A., Gomez, J., Burd, C.G., Albericio, F., Giralt, E., Dreyfuss, G., Pages, M. 1992. RNA binding characteristics of a 16 KDa glycin-rich protein from maize. *Plant J.*, 2, 999-1003.

Mas, P., Devlin, P.F., Panda, S., Kay, S.A. 2000. Functional interaction of phytochrome B and cryptochrome 2. *Nature*, 408, 207-211.

Matz, M.V., Fradkov, A.F., Labas, Y.A., Savitsky, AP., Zaraisky, A.G., Markelov, M.L., Lukyanov, S.A. 1999. Fluorescent proteins from nonbioluminescent Anthozoa species. *Nature Biotech.*, 17, 969-973.

Mayerhofer, R., Langridge, W.H.R., Cormier, M.J., Szalay, A.A. 1995. Expression of recombinant Renilla luciferase in transgenic plants results in high levels of light emission. *Plant Journal*, 7, 1031-1038.

Meyer, S., Cartelat, A., Moya, I., Cerovic, Z.G. 2003. UV-induced blue-green and far-red fluorescence along wheat leaves: a potential signature of leaf ageing. *J. Exp. Bot.*, 54, 757-769.

Miesenböck, G., De Angelis, D.A., Rothman, J.E. 1998. Visualizing secretion and synaptic transmission with pH-sensitive green fluorescent proteins. *Nature*, 394, 192-195.

Mirabella, R., Franken, C., van der Krogt, G.N., Bisseling, T., Geurts, R. 2004. Use of the fluorescent timer DsRED-E5 as reporter to monitor dynamics of gene activity in plants. *Plant Physiol.*, 135, 1879-1887.

Miyawaki, A. 2004. Fluorescent proteins in a new light. *Nature Biotechnolo*, 22, 1374-1376.

Miyawaki, A., Llopis, J., Heim, R., McCaffery, J.M., Adams, J.A., Ikura, M., Tsien, R.Y. 1997. Fluorescent indicators for Ca2+ based on green fluorescent proteins and calmodulin. *Nature*, 388, 882-887.

Morel, G., Cavalier, A. 1998. Hybridation *in situ* en microscopie photonique. *Polytechnica* Paris

Moseyko, N., Feldman, L.J. 2004. Expression of pH-sensitive green fluorescent protein in Arabidopsis thaliana. *Plant Cell Environ.*, 24, 557-563.

Nagai, T., Ibata, K., Park, E.S., Kubota, M., Mikoshiba, K., Miyawaki, A. 2002. A variant of yellow fluorescent protein with fast and efficient maturation for cell-biological approaches. *Nat. Biotech.*, 20, 87-90.

Neuhaus, J.M., Boevink, P. 2001. The green fluorescent protein (GFP) as reporter in plant cell. In "Plant Cell Biology", (Hawes, C.H. and Satiat-Jeunemaitre, B. eds), Oxford University Press, pp 127-142

Neumann, U., Brandizzi, F., Hawes, C. 2003. Protein transport in plant cells: in and out of the Golgi. *Ann. Bot.*, 92, 167-180.

Nishihara, M., Ito, M., Tanaka, I., Kyo, M., Ono, K., Irifune, K., Morikawa, H. 1993. Expression of the [beta]-Glucuronidase Gene in Pollen of Lily (Lilium longiflorum), Tobacco (Nicotiana tabacum), Nicotiana rustica, and Peony (Paeonia lactiflora) by Particle Bombardment. *Plant Physiol.*, 102, 357-361.

Oparka, K.J., Roberts, A.G., Boevink, P., Santa Cruz, S., Roberts, I., Pradel, K.S., Imlau, A., Kotlizky, G., Sauer, N., Epel, B. 1999. Simple, but not branched, plasmodesmata allow the nonspecific trafficking of proteins in developing tobacco leaves. *Cell*, 97, 743-754.

Ormö, M., Cubitt, A.B., Kallio, K., Gross, L.A., Tsien, R.Y., Remington, S.J. 1996. Crystal structure of the Aequorea victoria green fluorescent protein. *Science*, 273, 1392-1395.

Patterson, G.H., Lippincott-Schwartz, J. 2002. A photoactivable GFP for selective photolabeling of proteins and cells. *Science*, 297, 1873-1877.

Pauly, N., Knight, M.R., Thuleau, P., Van der Luit, A.H., Moreau, M., Trewavas, A.J., Ranjeva, R., Mazars, C. 2000. Control of free calcium in plant cell nuclei. *Nature*, 405, 754-755.

Pederson, T. 2001. Fluorescent RNA cytochemistry: Tracking gene transcripts in living cells. *Nucleic Acids Res.*, 29, 1013–1016.

Pesquet, E., Barbier, O., Ranocha, P., Jauneau, A., Goffner, D. 2004. Multiple gene detection by *in situ* RT-PCR in isolated plant cells and tissues. *Plant J.*, 39, 947-959.

Phillippy, B.Q. 1994. Reliable luminescent detection of subpicomole amounts of ATP using a scintillation counter *Biotechniques*, 4, 596-598.

Prasher, D.C., Eckenrode, V.K., Ward, W.W., Prendergast, F.G., Cormier, M.J. 1992. Primary structure of the Aequorea victoria green-fluorescent protein, *Gene*, 111, 229-233.

Raynaud, C., Perennes, C., Reuzeau, C., Catrice, O., Brown, S., Bergounioux, C. 2005. Cell and plastid division are coordinated through the pre-replication factor AtCDT1. *Proc. Nat. Acad. Sci.* USA, 102, 8216-8221.

Remy, S., Thiry, E., Coemans, B., Windelinchx, S., Swennen, R., Sagi, L. 2005. Improved T-DNA vector for tagging plant promoters via high-throughput luciferase screening. *BioTechniques*, 38, 763-770.

Rizzo, M.A., Springer, G.H., Granada, B., Piston, D.W. 2004. An improved cyan fluorescent protein variant useful for FRET. *Nature Biotechnology*, 22, 445-449.

Roberts, I.M., Boevink, P., Roberts, A.G., Sauer, N., Reichel, C., Oparka, K.J. 2001. Dynamic changes in the frequency and architecture of plasmodesmata during the sink-source transition in tobacco leaves. *Protoplasma*, 218, 31-44.

Rodriguez-Llorente, I.D., Perez-Hormaeche, J., Dary, M., Caviedes, M.A., Kondorosi, A., Ratet, P., Palomares, A.J. 2003. Expression of MsPG3-GFP fusions in *Medicago truncatula* 'hairy roots' reveals preferential tip localization of the protein in root hairs. *Eur. J. Biochem.*, 270, 261-269.

Roland, J.C., Vian, B. 1991. General preparation and staining of thin sections. In "Electron Microscopy of Plant cells", (Hall, J.L. and Hawes, C.H. eds), Academic Press, pp 1-66.

Rolletschek, H., Weschke, W., Weber, H., Wobus, U., Borisjuk, L. 2004. Energy state and its control on seed development: starch accumulation is associated with high ATP and steep oxygen gradients within barley grains. *J. Exp. Bot.*, 55, 1351-1359.

Russinova, E., Borst, J.W., Kwaaitaal, M., Cano-Delgado, A., Yanhai, Y., Chory, J., de Vries, S. 2004. Heterodimerization and Endocytosis of Arabidopsis Brassinosteroid Receptors BRI1 and AtSERK3 (BAK1). *Plant Cell*, 16, 3216-3229.

Saint-Jore, C.M., Evins, J., Batoko, H., Brandizzi, F., Moore, I., Hawes, C. 2002. Redistribution of membrane proteins between the Golgi apparatus and endoplasmic reticulum in plants is reversible and not dependent on cytoskeletal networks. *Plant J.*, 29, 661-678.

Samalova, M., Brzobohaty, B., Moore, I. 2005. pOp6/LhGR: a stringently regulated and highly responsive dexamethasone-inducible gene expression system for tobacco. *Plant J.*, 41, 919-935.

Satiat-Jeunemaître, B. 1990. Visualization of microtubules in walled plant cells by immunogold silver-enhancement. *Cell Biol. Int. Rep.*, 14, 1109-1118.

Satiat-Jeunemaître, B., Hawes, C. 2001. Immunocytochemistry for light microscopy. In "Plant cell biology", 2nd. ed., C. Hawes and B. Satiat-Jeunemaitre (eds), Oxford University Press, 207-233

Satou, Y., Takatori, N., Yamada, L., Mochizuki, Y., Hamaguchi, M., Ishikawa, H., Chiba, S., Imai, K., Kano, S., Murakami, S.D., Nakayama, A., Nishino, A., Sasakura, Y., Satoh, G., Shimotori, T., Shin-i, T., Shoguchi, E., Suzuki, M.M., Takada, N., Utsumi, N., Yoshida, N., Saiga, H., Kohara, Y., Satoh, N. 2001. Gene expression profiles in *Ciona intestinalis* tailbud embryos. *Development*, 128, 2893-2904.

Schmutz, D., Brunold, C. 1982. Rapid and simple measurement of ATP-sulfurylase activity in crude plant extracts using an ATP meter for bioluminescence determination. *Anal Biochem.*, 121, 151-155.

Segui-Simarro, J.M., Austin, J.R., White, E.A., Staehelin. L.A. 2004. Electron tomographic analysis of somatic cell plate formation in meritematic cells of Arabidopsis preserved by high pressure freezing. *Plant Cell*, 16, 836-856.

Sena, G., Jung, J.W., Benfey, P.N. 2004. A broad competence to respond to SHORT ROOT revealed by tissue-specific ectopic expression, *Development*, 131, 2817-2826.

Shimomura, O. 2005. The discovery of aequorin and green fluorescent protein. *J. Microscop.*, 217, 3-15.

Siemering, K.R., Golbik, R., Sever, R., Haseloff, J. 1996. Mutations that suppress the thermosensitivity of green fluorescent protein. *Curr. Biol.*, 6, 1653-1663.

Sohn, E.J., Kim, E.S., Zhao, M., Kim, S.J., Kim, H., Kim, Y.W., Lee, Y.J., Hillmer, S., Sohn, U., Jiang, L., Hwang, I. 2003. Rha1, an Arabidopsis Rab5 homolog, plays a critical role in the vacuolar trafficking of soluble cargo proteins. *Plant Cell*, 15, 1057-1070.

St Johnston, D. 2005. Moving messages: the intracellular localization of mRNAs. *Nature Reviews Mol. Cell Biol.*, 6, 363-375.

Staehelin, L.A., Giddings, T.H., Kiss, J., Sack, F.D. 1990. Macromolecular differentiation of Golgi stacks in root tips of Arabidopsis and nicotiana seedlings as visualised in high pressure frozen and freeze substituted samples. *Protoplasma*, 157, 75-91.

Subramanian, C., Kim, B.H., Lyssenko, N.M., Xu, X., Johnson, C.H., von Arnim, A.G. 2004a. The Arabidopsis repressor of light signaling, COP1, is regulated by nuclear exclusions: Mutational analysis by bioluminescence resonance energy transfer. *Proc. Natl. Acad. Sci.* USA, 101, 6798-6802.

Subramanian, C., Xu, Y., Johnson, C.H,. von Arnim, A.G. 2004b. *In vivo* detection of protein-protein interaction in plant cells using BRET. *Methods Mol. Biol.*, 284, 271-286.

Tamura, K., Shimada, T., Ono, E., Tanaka, Y., Nagatani, A., Higashi, S.I., Watanabe, M., Nishimura, M., Hara-Nishimura, I. 2003. Why green fluorescent fusion proteins have not been observed in the vacuoles of higher plants. *Plant J.*, 35, 545-555.

Taylor, N.J., Fauquet, C.M. 2002. Microparticle bombardment as a tool in plant science and agricultural biotechnology: DNA. *Cell Biol.*, 21, 963-977.

Thomine, S., Lelievre, F., Debarbieux, E., Schroeder, J.I., Barbier-Brygoo, H. 2003. AtNRAMP3, a multispecific vacuolar metal transporter involved in plant responses to iron deficiency. *Plant J.*, 34, 685-695.

Tsien, R. 1998. The green fluorescent protein. *Ann. Rev. Biochem.*, 67, 509-544.

Van Damme, D., Van Poucke, K., Boutant, E., Ritzenthaler, C., Inze, D., Geelen, D. 2004. In vivo dynamics and differential microtubule-binding activities of MAP65 proteins. *Plant Physiol.*, 136, 3956-3967.

Vandenbosch, K.A. 1991. Immunogold labelling. In "Electron Microscopy of plant cells, (Hall, J.L. and Hawes C.R. eds), Academic Press, pp 181-218.

Wada, M., Suetsugu, N. 2004. Plant organelle positioning. *Curr. Opin. Plant Biol.*, 7, 626-631.

Walter, M., Chaban, C., Schütze, K., Batistic, O.l. Weckermann, K., Näke, C., Blazevic, D., Grefen, C., Schumacher, K., Oecking, C., Harter, K., Kudla, J. 2004. Visualization of protein interactions in living plant cells using bimolecular fluorescence complementation. *Plant Journal*, 40, 428-438.

Wright, K.M., Roberts, A.G., Martens, H.J., Sauer, N., Oparka, K.J. 2003. Structural and functional vein maturation in developing tobacco leaves in relation to AtSUC2 promoter activity. *Plant Physiol.*, 131, 1555-1565.

Yang, F., Moss, L.G., Phillips, J.N. Jr. 1996. The molecular structure of green fluorescent protein. *Nat Biotech*, 14, 1246-1251.

Zhang, J., Campbell, R.E., Ting, A.Y., Tsien, R.Y. 2002. Creating new fluorescent probes for cell biology. *Nature Reviews Molecular Cell Biology*, 3, 906-918.

Zhou, C., Labbe, H., Sridha,S., Wang, L., Tian, L., Latoszek-Green, M., Yang, Z., Brown, D., Miki, B., Wu, K. 2004. Expression and function of HD2-type histone deacetylases in Arabidopsis development. *Plant J.*, 38, 715-724.

Zhou, M., Upson, R.H., Diwu, Z., Haugland. R.P. 1996. A fluorogenic substrate for beta-glucuronidase: applications in fluorometric, polyacrylamide gel and histochemical assays. *J. Biochem. Biophys. Methods*, 33, 197-205.

Zhou, X., Carranco, R., Vitha, S., Hall, T.C. 2005. The dark side of green fluorescent protein. *New Phytol.*, 168, 313-321.

Zicha, D., Dobbie, I.A., Holt, M.R., Monypenny, J., Soong, D.Y.H., Gray, C., Dunn, G.A. 2003. Rapid actin transport during cell protrusion. *Science*, 300, 142-145.

Zimmermann, T., Rietdorf, J., Girod, A., Georget, V., Pepperkok, R. 2002. Spectral imaging and linear un-mixing enables improved FRET efficiency with a novel GFP2-YFP FRET pair. *FEBS Lett.*, 531, 245-249.

15

CHAPTER

Metabolomics

A. Krapp, J.F. Morot-Gaudry, S. Boutet, G. Bergot, C. Lelarge, J.L. Prioul, G. Noctor

INTRODUCTION

After DNA sequencing to establish an inventory of the genes of an organism, gene expression analysis (transcriptomics), and protein analysis (proteomics), the remaining functional genomics challenge is analysis of the metabolome. The term "metabolome" was created by analogy to transcriptome and proteome (Chapters 10 and 12). The metabolome encompasses all the metabolites that represent the catabolic and anabolic activities being performed by **proteins** at any given time (Fiehn, 2002), plus compounds that can be produced in tissues through non enzymatic reactions. Like the transcriptome and proteome, the metabolome shows specificity with respect to species, organ, developmental stage, and environmental conditions. Because metabolites are the end products of cellular functions, their levels can be viewed as the response of biological systems to environmental or genetic manipulation. Since 1998, an exponential increase of works on the metabolome has occurred and 2002 saw the cumulative number of publications in this area of research exceed 100. In April 2002, the first international congress on plant metabolomics was held in Wageningen, The Netherlands. Metabolomics denotes comprehensive, nonbiased, high-throughput analyses of the complex metabolite mixtures typical of plant extracts. The ultimate aim of these approaches is to allow a study of the entire range of chemicals of low molecular weight synthesized and metabolized by plants.

Over 50,000 biochemical compounds have already been identified in plants; (DeLuca and St Pierre, 2000). Nevertheless, 100,000 and perhaps even over 200,000 compounds are still to be discovered (Fiehn, 2002). This enormous number includes compounds belonging to a wide variety of structurally different metabolic classes which are all, in one way or another, inter-related. This multitude of chemical components, with many different carbon skeletons, functional groups and physicochemical properties, makes plants an invaluable source of food, pharmaceuticals, health-promoting compounds, flavour and fragrance compounds, biocides, and fine chemicals, but also a source of toxins and xenobiotics. Plant components have a clear determinant effect on the quality of agricultural products and a direct effect on our health and well being.

The challenges of metabolomics are, first, how to measure the maximum number of these chemicals simultaneously, and second, how to make sense of the vast datasets that are generated by such measurements (Maloney, 2004). Achieving the broadest overview of metabolic composition is very complex and entails establishing a multifaceted strategy for adequate tissue sampling, homogenisation, optimal sample extraction, metabolite separation, detection and identification, automated data gathering, handling, analysis and quantification (Hall *et al.*, 2002). All these methods must allow an unbiased analysis that ensures quantitative and qualitative measurements of the wide range of cellular metabolites, thus providing a broad view of biological status and function.

TERMINOLOGY

Metabolite analyses are used to answer a panoply of questions of varying complexity. For this reason, different analytical approaches are used and Fiehn (2002) classified them in four general approaches.

Target analysis is centred on a small number of metabolites. Very often the substrate and the product of an enzymatic reaction are analysed with the aim of characterising a given enzymatic reaction *in vivo* or *in vitro*. In this way, changes in the flux through one enzyme can be seen directly by cross-over plots. So the primary effect of a genetic modification or of a treatment can be studied, without the immediate need to analyse a larger number of metabolites. This also implies that the sample can be cleaned-up or that the compounds in question can be detected without interference from other metabolites. Target analysis is well suited for screening and it might be the only approach possible when extreme sensitivity or extraction conditions are necessary, such as for phytohormones or signalling molecules such as fructose-2,6-*bis*phosphate.

Metabolite profiling enlarges the analysis to include all metabolites of a pathway or a given chemical class of metabolites. Here too the analysis is restricted to a selected number of pre-determined metabolites, which again simplifies sample preparation and analysis. Such studies concentrate on a few, well-defined metabolites, which are often part of a well-characterised metabolic pathway. In the past, most studies on legume metabolites for example analysed a few compounds from pre-selected classes such as sugars, amino and organic acids, thiols, saponins, and phenolics, using a range of instrumentation, including HPLC, thin layer and gas chromatography (Streeter, 1987; Chen *et al.*, 2003). In plant biology, this approach has also been used for example to estimate the role of an enzymatic reaction on one branch of plant metabolism (Neuhaus and Stitt, 1990). Nevertheless, keeping in mind the highly interactive nature of metabolism, modification in one pathway can have important effects on many other pathways and indeed this is a key principle of non-specific metabolomic analysis (see below). A good example is the interaction between carbon and nitrogen metabolism, which are highly linked pathways (Scheible *et al.*, 1997a, Matt *et al.*, 2002). A further example is the effect of primary C metabolism on secondary metabolism, which can then influence plant performance by, for example, modifying resistance to pathogens

(Matt *et al.,* 2002). These considerations suggest a larger definition of metabolite profiling, focusing not only on one pathway or one class, but on key metabolites from several pathways: e.g. carbohydrates, organic acids, and amino acids in the analysis of the C/N interaction (Matt *et al.,* 2000; Dutilleul *et al.,* 2005).

Whereas metabolite profiling aims at quantifying a number of predefined targets, in contrast the goal of *metabolomic approaches* is the unbiased identification and quantification of all metabolites present in a specific biological sample from a plant grown under defined conditions. The components of the metabolome can be viewed as end products of gene expression that define the biochemical phenotype of a cell or tissue. The determination of large numbers of metabolites thus provides a broad view of the biochemical status of an organism that can be used to monitor and assess gene function (Sumner *et al.,* 2003). Differentially, expressed metabolites must be chemically identified and correlations can be sought between sets of differentially displayed metabolites. The individual as well as the correlated metabolites are used to identify metabolic pathways or networks that have been affected. These pathways are then used to determine the broader biological significance of the response or to assign gene function (Sumner *et al.,* 2003). However, any single metabolomics technique provides a less comprehensive analysis of metabolites than transcriptomics does of transcripts. In practice, a truly exhaustive analysis of the metabolome is not yet feasible. Current technologies are such that an approach can be described as 'metabolomics', if metabolite analyses are sufficiently comprehensive to cover compounds of several metabolic pathways in both primary and secondary metabolism.

Lastly, the expression *metabolic fingerprinting* is becoming more common in approaches which aim to classify samples by global and fast analysis of a large number of metabolites. Metabolic fingerprinting approaches focus on collecting and analysing data from crude metabolite mixtures to rapidly classify samples. It is a high-throughput technique which enables a snapshot of the metabolic composition at a given time, where it is not initially necessary or feasible to determine the levels of metabolites individually. Metabolic fingerprinting is the rapid classification of samples according to their original biological provenance (Fiehn, 2001). Fourier transform infrared spectroscopy (FT-IR) has been used successfully in metabolic fingerprinting technology. FT-IR is a physicochemical method that measures predominantly the vibrations of bonds within functional groups and generates a spectrum that can be regarded as a biochemical or metabolic "fingerprint" of the sample (Johnson *et al.,* 2003). NMR methods also provide metabolite fingerprints with good chemical specificity for compounds containing elements with non zero magnetic moments, such as ^{1}H, ^{13}C, ^{15}N, and ^{32}P which are commonly found in most biological metabolites. Analysis of NMR spectra of solvent extracts of plants has the potential to provide a relatively unbiased fingerprint, containing the majority of the metabolites present in the solution (Chapter 16).

ANALYTICAL TECHNIQUES

A metabolite analysis consists of several steps. Sampling has to be fast, involving efficient and rapid quenching of metabolism in order to block immediately any biochemical reaction and obtain a snapshot of the biochemical status. Metabolites then must be extracted from the matrix without modification. Next, they must be separated with a good resolution and then identified. A further step is the quantification and, last but not least, the organization of the results in a database (data processing). Automated data gathering, handling and analysis require considerable computational developments.

Sampling

When studying plant metabolism, samples must be taken in defined conditions while avoiding effects of harvesting procedures on metabolite contents. The harvest has to be rapid without changing the experimental conditions under which the plants were cultivated. For example, tissue should not be sampled by procedures which alter the temperature, gas composition or irradiance relative to growth conditions. It has been shown that metabolites central to photosynthetic and respiratory metabolism, such as ATP, 3-phosphoglycerate, and triose-phosphates, turn over extremely quickly; others undergo enormous diurnal variation, which makes it important to consider the precise point of harvesting during the light-dark cycle. The situation becomes even more complex when one considers that genetic or environmental modification can change the diurnal or nocturnal pattern of metabolites (Scheible *et al.*, 1997b). In such situations, meaningful comparison of genotypes or different treatments should be performed during a whole day/night cycle.

Rapid quenching of metabolism

This involves the arrest of all enzymatic reactions, and is very efficient in the case of tissue cultures where the liquid culture can be directly infused with extraction solvent (Gonzalez *et al.*, 1997). For thicker higher plant tissues, rapid quenching is possible by freeze clamping. Rapid freezing in liquid nitrogen is commonly used, but this technique is a lot less efficient if the sample is enclosed in a plastic tube in which freezing is not immediate, and shading can also occur if the sample is enclosed in an aluminium bag prior to freezing. Ideally, therefore the tissue should be frozen directly in liquid nitrogen under the conditions of incubation or growth. If these rapid techniques are not possible due to the experimental design, artefacts can be caused either by disturbed metabolism or by wounding or touch-induced processes. Rapid quenching can also be achieved by acid treatments (Ap Rees and Hill, 1994), but this type of treatment poses severe problems for many of the analytical methods to be used subsequently (Fiehn, 2002).

Samples should be frozen, preferably at liquid nitrogen temperatures, or lyophilized. Although such treatment may not be necessary for all metabolites, authoritative data is lacking on the stability of many metabolites in various conditions. Despite earlier studies (Galmiche, 1973), knowledge is still lacking on the stability of many metabolites under the various conditions of quenching and storage.

Extraction

Progress in bioanalytical techniques has made it possible to analyse metabolites directly in crude biological extracts. Frozen and freeze-dried sample can be extracted directly, but in the latter case, adsorption of metabolites on the walls of the equipment may cause problems. Samples should then be homogenized prior to extraction, either simply by grinding in a mortar and pestle (both chilled in liquid nitrogen) or by using a ball mill. Homogenization in the extraction solvent using Ultra-Turrax devices is also possible (Orth *et al.*, 1999). Currently, no comprehensive comparisons of extraction techniques that show high reproducibility, robustness, and recovery for all classes of components have been published. Most frequently, polar organic solvents like methanol are added directly to the ground material to extract acidic and basic ions, with an additional step using non-polar solvents to extract lipophilic components. For example, multiple components from homogenised tissues are often extracted using alcohols or water/ alcohol mixtures (Johansen *et al.*, 1996; Streeter and Strimbu, 1998). Extraction efficiency of stable structures can be enhanced by heating, pressurised liquid extraction (Benthin *et al.*, 1999), supercritical fluid extraction (Jarvis and Morgan, 1997), subcritical water extraction (Gamiz-Gracia and de Castro, 2000), pervaporation (Starmans and Nijhuis, 1996), sonication (Sargenti and Vichnewski, 2000), or microwave techniques (Namiesnik and Gorecki, 2000). Moreover, it is unclear which factors most affect robustness, which is defined as minimal analytical errors when protocols are carried out under slightly different conditions, such as subtle differences in extraction times, temperatures, solvent compositions, staff skills, tissue/solvent ratios, compounds susceptible to be oxidized at the slightest treatments, and others (Weckwerth, 2003).

Separation and Identification

As noted above, no single technique currently exists that is appropriate for measurement of all metabolites. The choice of the technique depends on the scientific question. Targeted metabolite analysis is focused on specific chemical classes of metabolites and numerous techniques have been used. True metabolomics requires more advanced techniques, and these are still in the process of development and evaluation, especially with regard to comprehensiveness, selectivity, and sensitivity. Metabolomics thus requires compromises.

Numerous techniques exist for metabolite detection. Recent developments have favoured mass spectrometry (**MS**) as a detection mode that is sensitive, relatively non-selective, and able to identify compounds with reference to databases of mass spectra obtained for authentic standards. These properties make MS the method of choice for metabolite detection in complex matrices, and this technique is particularly powerful when allied to prior chromatographic resolution of such mixtures. **Gas chromatography (GC) coupled to mass spectroscopy (GC/MS)** and more recently, **liquid chromatography (LC) coupled to mass spectroscopy (LC/MS)** are able to detect several hundred chemicals, including sugars, sugar alcohols, organic acids, amino acids, fatty acids and a wide range of diverse secondary metabolites (Weckwerth, 2003). Each technique has advantages and drawbacks.

While **GC/MS** is a low cost alternative to some metabolomic technologies and provides very fine chromatographic resolution giving high separation efficiencies of the numerous metabolites found in plant tissues, it requires that samples be both volatile and thermostable. Volatilization of intrinsically non-volatile compounds is accomplished by chemical derivatisation, but at the cost of additional time and variability. GC-MS can be performed with affordable single quadrupole mass analysers for analysis of complex mixtures. The utilization of an automated mass spectral deconvolution and identification system enhances the ability to successfully identify overlapping chromatographic peaks (Halket *et al.*, 1999). The advantage of GC-MS is that it combines high chromatographic separation power with a universal detection system to produce excellent sensitivity and selectivity. Currently, GC-TOF-MS (Gas Chromatography-Time-Of-Flight-Mass-Spectrometry) offers an attractive alternative to quadrupolar systems and provides greater m/z accuracies (Davies *et al.*, 1999). This technique increased the number of metabolites detectable by GC/MS in crude extracts from 500 to 1,000 (Fiehn, 2001; Fiehn, 2002).

The group of Fiehn in Golm, Germany, used GC/MS to compare well-characterized *Arabidopsis thaliana* mutants, two *A. thaliana* accessions and intensively studied potato transformants and demonstrated the reliability and the potential of this method (Fiehn *et al.*, 2000; Roessner *et al.*, 2001). These techniques have enabled high-throughput, nonbiased analysis of hundreds of metabolites from plants. These tools afford not only a much broader view of metabolism but also the opportunity to discover novel molecules and previously unknown aspects of metabolism (Fiehn *et al.*, 2000; Sumner *et al.*, 2003). GC-MS technologies have been used to explore plant responses to nutritional limitation, e.g., sulphur deprivation (Nikiforova *et al.*, 2004) and to take a holistic approach to defining the biochemical, molecular and physiological response of a plant to alterations and imbalances in nutrient availability.

GC/MS is known to be unsuitable for detection of certain compounds, for example, organic diphosphates or metabolites larger than tri to tetrasaccharides. In these conditions, alternative methods such as LC-MS may be appropriate. Capillary electrophoresis (CE) coupled to MS also provides a feasible method for the separation of ionic compounds and hence, represents a complementary technology to reverse phase separation focusing on lipophilic compounds (Soga *et al.*, 2002). In contrast, GC-MS and LC-MS can only poorly differentiate chemical isomers, such as those of common hexoses, since they have the same exact mass and they are difficult to resolve chromatographically (Sumner *et al.*, 2003).

LC/MS is a powerful alternative that offers high selectivity and good sensitivity (Sumner *et al.*, 2003). Unlike GC/MS a derivatisation step is not required. Fragmentation conditions are more gentle (ionization by electrospray rather than electron impact), and this allows good detection of molecular ions. Targeting can be achieved by the extraction method or the type of column/solvent used. Recently, reversed-phase LC/MS and Hydrophilic Interaction Liquid Chromatography (HILIC) coupled to MS were also used to analyse complex mixtures (Huhman and Sumner, 2002; Tolstikov and Fiehn, 2002). While MS detection is becoming increasingly popular in metabolite analyses, HPLC can also be simultaneously coupled to other detectors such as UV/PDA (Photodiode Array Detection) and 3D fluorimeters, both of which complement mass spectra by information on absorbance and fluorescence spectra. It is possible in this way to provide multiple levels of information that are useful in chemical structure elucidation (Fraser *et al.*, 2000). Such detectors are likely to be particularly useful when coupled to the new generation of HPLC systems able to function at ultra-high backpressure. These systems enable near-GC performance in liquid chromatography, and are thus better adapted than conventional HPLC for the analysis of complex mixtures.

THE CHALLENGES AND DIFFICULTIES OF METABOLOMICS

Two principal problems remain concerning exhaustive measurement of metabolites in a single extract of plant tissues. The first relates to the difficulty of finding an extraction method that is suitable for all metabolite groups. This is most evident when considering hydrophilic metabolites found in the aqueous phase of the cell and lipophilic compounds that exist in the membrane phase. Two step cold extraction (-20°C) into a chloroform/methanol/water mixture and a chloroform/methanol one followed by two-phase partitioning has been used to overcome this problem (Weckwerth *et al.*, 2004), but the key issue of 'recovery' of metabolites remains an obstacle to reliable quantification. Metabolites may be lost or modified during the extraction due to ongoing enzyme activities or chemical processes such as oxidation, and GC-MS suffers from the additional limitation that compounds to be analyzed must be volatile or made so by derivatisation. It is therefore important to establish the extent of variability of the analysis and the extraction yield of each metabolite. Extraction yield can be assessed by comparing data obtained using non-targeted techniques with established quantitative techniques that use extraction procedures and analytical methods adapted for specific metabolites or metabolite classes. **Analytical variance** is defined by Sumner *et al.* (2003) as the coefficient of variance or relative standard deviation that is directly related to the experimental approach. Biological variations typically exceed analytical variations. Roessner *et al.* (2000) reported that the biological variability exceeded the analytical variability of GC/MS by a factor of ten. Sumner *et al.* (2003) suggested that the average biological variance for *Medicago truncatula* is approximately 50%. These large biological variations represent the major limitations of the resolution of metabolomics approaches.

The second problem is that an unfractionated plant tissue extract contains compounds with concentration differences of several orders of magnitude (Sumner *et al.*, 2003). **Dynamic range** defines the concentration boundaries of an analytical determination between which the instrumental response as a function of analyte concentration is predictable and reproducible (Sumner *et al.*, 2003). The dynamic range of many techniques is known to be limited by the sample matrix or the presence of interfering and competing compounds. This is one of the most difficult issues to address in metabolomics. Although certain abundant compounds can be very useful as biomarkers in high throughput diagnosis of physiological states, their abundance can cause significant or severe chemical interferences that limit the range in which other metabolites may be successfully profiled. Interfering or competing analytes can also lower performance and/or bias MS profiling techniques. Another problem in electrospray ionisation mass spectrometry (ESI/MS) is the presence of salts (Sumner *et al.*, 2003). Both GC/MS and LC/MS are faced with the problem of dynamic range which renders metabolite quantification difficult. This problem can be partly addressed by multi-level injection of extracts, e.g., 'splitless' (injection of the total volume sampled by the GC injector) and 'split' (injection of only a fraction of the volume sampled), which enables separate analysis of minor and major metabolites, respectively (Fig. 15.1). However, this approach necessarily increases analysis time.

As noted above, lack of differentiation of isomers is a drawback of MS-based techniques. This is not necessarily problematic for the analysis of compounds for which

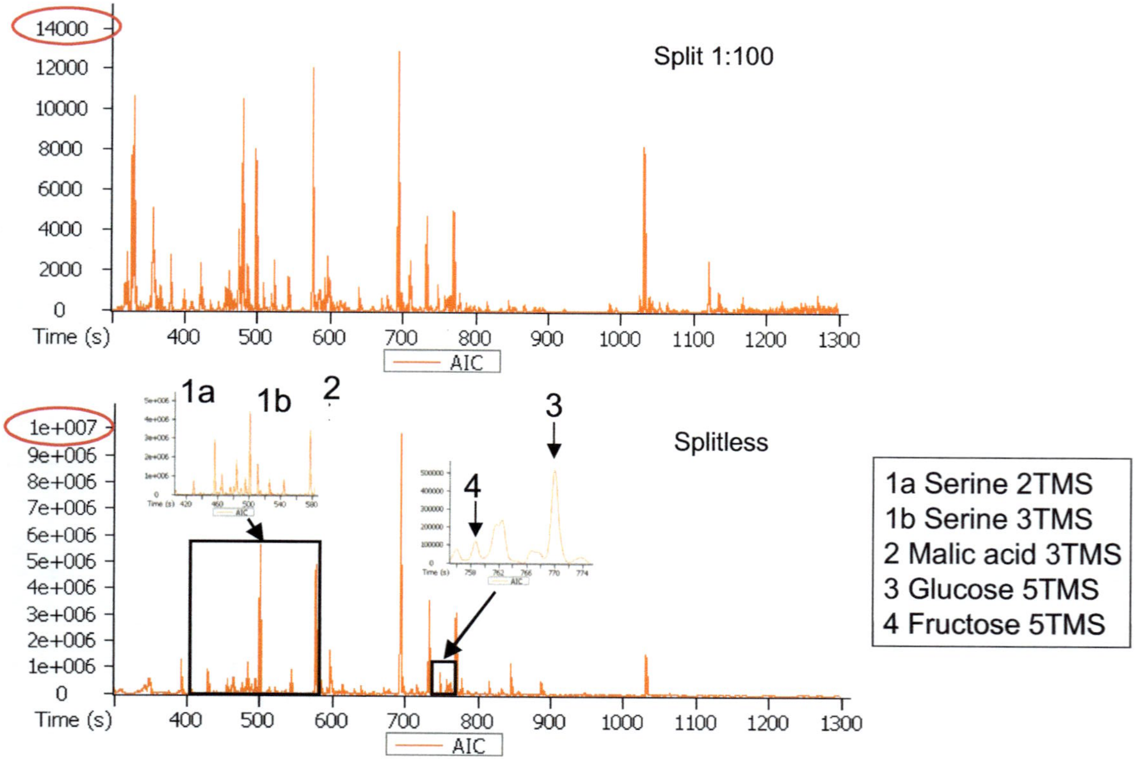

Fig. 15.1 GC-TOF-MS profile of an *Arabidopsis thaliana* leaf. Wild type (Col0) plants were grown under controlled conditions. 100 mg of leaf fresh weight was extracted using a protocol based on that of Fiehn (2002). The aqueous phase was recovered, dried, and subjected to methoxyamination (MeOX, derivatizes aldehyde and keto functions) and derivatization by trimethylsilylation (TMS, displaces labile protons). 1 µL of the derivatized mix was injected into a LECO Pegasus III GC-TOF-MS in 'splitless' mode (injection of the total volume sampled: lower panel) or 'split' 1:100 (this effectively allows a 100-fold dilution of the total volume sampled: upper panel). The y axis shows the total ion current (i.e., the signal produced by all ions detected at any given moment) in arbitrary detector units. Note the difference in y axis scaling between the two injection modes. In total 79 compounds were identified by comparing mass spectra with reference spectra available in databases and confirmed by comparison with spectra generated for authentic standards. Examples of identified compounds are shown in insets for the splitless injection. The numbered peaks correspond to 1a. serine-2-TMS. 1b. serine-3-TMS. (derivatization of serine produces both species in similar amounts). 2. malate-3-TMS. 3. glucose-1-MeOX,5-TMS. 4. fructose-1-MeOX,5-TMS. The structures and mass spectra fragmentation pattern for these derivatization products are shown in Fig. 15.2.

chromatographic retention times can be clearly defined as different using known standards that can be resolved by GC or LC. It is a major problem, however, for direct injection into the mass analyzer, for unambiguous identification of unknown compounds, or for compounds such as hexoses, for which plants may contain multiple isomers and epimers within both the D and L series. Although resolution of hexose isomers such as fructose and glucose can be achieved by GC-MS, epimers, enantiomers and diastereoisomers such as fructose and sorbose have very close retention times and identical mass spectra, making identification and quantification difficult or impossible (Fig. 15.2). In GC/MS, the use of derivatisation leads to several chemical species for many of the compounds in the chromatogram (e.g., serine, Fig. 15. 2). The presence of fragment and adduct ions also hinders analysis in MS. Our present knowledge suggests that the ultimate goal of metabolomics – to measure all metabolites present in a given, tissue or cell sample – is unlikely to be attained by any single available analytical method. These obstacles are being circumvented through the use of selective extraction and parallel analyses using a combination of technologies to obtain the most comprehensive visualisation of the metabolome (Sumner *et al.*, 2002).

STATISTICAL AND BIOINFORMATIC TOOLS

Like other "omic" approaches, metabolomics generates a large amount of data which has to be processed, analysed and stored. This can only efficiently be done through intensive data processing associated with statistics. Most of the statistical tools are the same as used in genomics (ANOVA, correlation, principal component analysis (PCA), independent component analysis (ICA), hierarchical component analysis (HCA), network...). A common prerequisite for a correct analysis is the need for careful experimental design, the best being an experimental set up which enables each level of a factor to be combined with each level of the other factors, as described in Chapter 11. However, several additional steps have to be introduced for analysing metabolomic data. A preliminary one is to be able to sort and, if possible, identify individual metabolites from the chromatographic profiles and associated mass spectra. It is only when this step is completed that multivariate analysis of the data and eventually the biological interpretation can take place.

Processing of the GC-MS Profiles and Metabolite Identification

Unfortunately, there is currently no widely used and freely available method for automated analysis of GC-MS profiles. Each laboratory has tended to construct its own program in addition to the MS deconvolution program. The general aim of automated analysis is to align profiles, then differentiate and visualize the components. This is not trivial since the number of spectra to be analysed (perhaps over one hundred) precludes a manual analysis which is very time consuming (several hours for one profile). Among the thousands of metabolites that have been described in plants, only several hundreds can be distinguished from one GS-MS run. One of the problems is that despite chromatographic separation, some metabolites co-elute prior to being subject to MS detection. A first improvement is provided by deconvolution of the elution peaks into simpler components and by examining the corresponding MS in order to find the characteristic signature of a metabolite by comparison with databases. This can be done

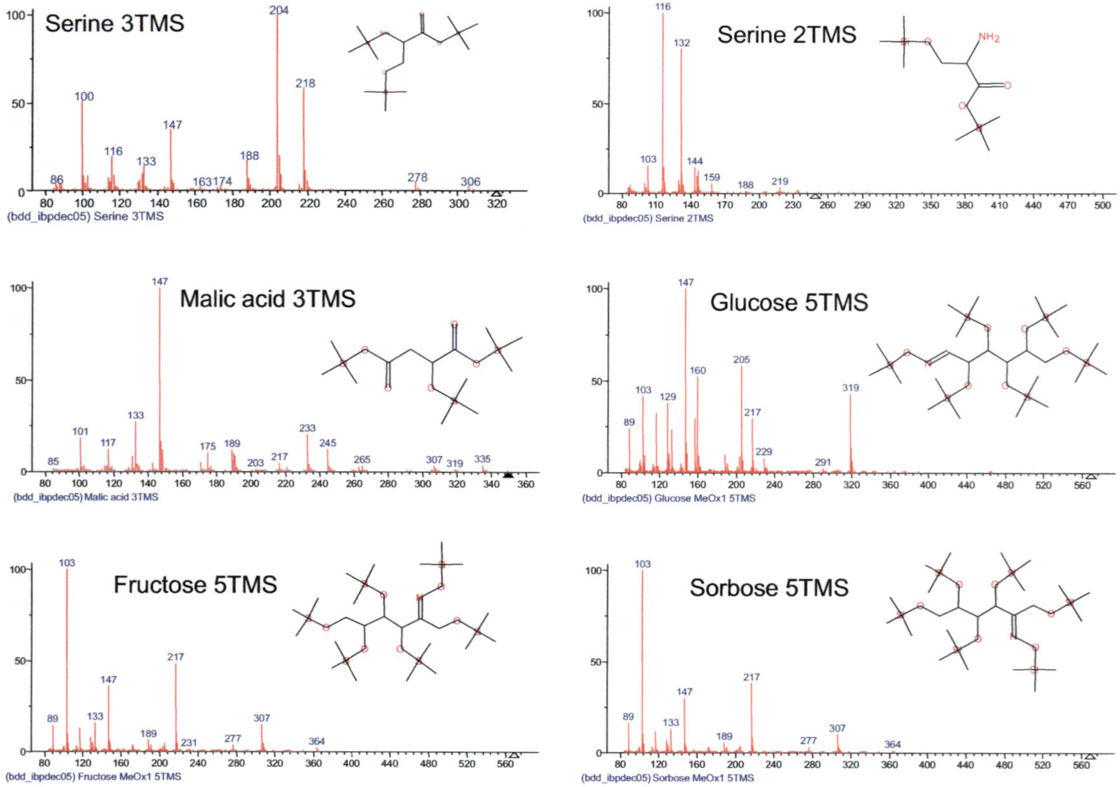

Fig. 15.2 Structures and mass spectra fragmentation patterns for authentic standards analyzed by GC-TOF-MS. Spectra are shown for all compounds indicated in Fig. 15.1. Sorbose, a diastereoisomer of fructose, is also included here. In serine-3-TMS, TMS replaces labile protons on the carboxylic acid hydroxyl group, the hydroxy-methyl hydroxyl group and the amine group; serine-2-TMS is derivatized like serine-3-TMS except that the amine group is not derivatized; malate-3-TMS is derivatized on the hydroxyl groups of the two carboxylic acid functions and the hydroxyl group of carbon 2; in glucose-1-MeOX,5-TMS the aldose oxy function is methoxyaminated while the labile protons of the five hydroxyl grouyps are all trimethylsilylated; in fructose-1-MeOX,5-TMS the derivatization is similar to glucose, i.e., TMS replaces hydroxyl protons while the keto oxy function is methoxyaminated; derivatization of sorbose is identical to fructose. Tests on authentic standards show that fructose and sorbose elute very closely. Thus, these two compounds are an example of the limitations of GC-TOF-MS: despite the analytical power of this technique, such closely related compounds as certain sugars are difficult to resolve, either chromatographically or spectrally.

on a reference spectrum, the others being treated by comparison (Taylor *et al.*, 2002). The quality of deconvolution is much improved when using GS-TOF-MS with high scanning rates since each peak is defined by a large number of data points (15-20) and even very narrow peaks can be detected. For instance, the LECO Pegasus III TOF-MS (LECO Corp, St Joseph, MI) can collect 500 mass spectra s^{-1}. The deconvolution is performed by the MS software which quantifies each peak from integration of ion current in the peak window after baseline correction, and expresses quantity in absolute terms or relative to a standard not found in plants, such as ribitol. A tentative identification is also provided by comparison to available libraries in the form of a metabolite list ranked by P-values, but the final choice must be done manually by treating spectra peak by peak following analysis of a mixture of the main metabolites of interest. Another problem, not encountered by the manufacturer, arises when comparing metabolite quantities between profiles because the software quantifies each identified metabolite from the fragment mass with the higher intensity. This may be misleading if this mass is not specific to the component and thus may be different from one profile to another, if different co-eluting components are present. Thus, in order to quantify a metabolite on the same basis between different runs, one is forced to specify manually the use of the unique mass characteristic of the compound. Since this has to be done peak by peak, it is a tedious process, but once this specific mass entered the quantification module of the software enables automatic processing of the profile.

Jonsson *et al.* (2004) proposed an automated strategy for identifying differences in large series of metabolomic samples that avoids having to deconvolute and identify all analysed samples prior to multivariate analysis of the data. The non-processed files exported from the GC-MS are baseline-corrected and the different profiles are aligned by defining time windows in which the maximum covariance between chromatograms is searched for. The components in each time window are then extracted by a multivariate curve resolution method that allows identification by comparison with MS libraries (Jonsson *et al.*, 2005). Using this technique, the time required for the processing of 70 samples was comparable to the chemical analysis time. Another method which is apparently easier to transpose was reported by Tikunov *et al.* (2005). This approach uses widely available software and classical statistical procedures, although no ready-to-use package is provided. It also removes the need for prior metabolite identification. The principle is to use several chromatograms simultaneously assuming that a given component will have the same mass fragment pattern throughout the samples. It consists of; 1) full mass spectral alignment using the MetAlign software package (sold at http://www.metalign.nl); 2) HCA on the molecular fragments for eliminating contaminants; 3) multivariate comparative analysis of the cleaned samples by HCA to control repeatability of the experiments or to allow comparison with mixture of standards and PCA for further description of the data; 4) a reconstruction of the chemical structure of metabolites from the molecular fragments in the GC-MS profile and a subsequently identification by comparison with MS libraries. The approach is based on two principles: a) the molecular fragments derived from the same metabolite will appear within a peak of a certain width at a certain retention time; b) the relative ratio between intensities of molecular fragments derived from the same metabolite is constant. Therefore, a metabolite may be defined as a group of highly correlated molecular fragments situated within a certain molecular window. Thus, clusters of molecular fragments for which HCA treatment of datasets produces very similar patterns of intensities throughout all the GC-MS analyses correspond to a unique metabolite. Starting from 198 chromatograms of tomato fruit volatiles and 20,000 molecular fragments, Tikunov *et al.* (2005) obtained 322 clusters using a threshold Pearson correlation coefficient higher than 0.8. Out of them 70 were identified from the NIST library.

Visualization Analysis and Biological Interpretation of the Data

The comparison of composition and individual quantity of metabolites in the different samples is performed using the statistical tools which are appropriate for treatment of large series of data. The aim is to find correlations between individual components and eventually to place them in metabolic networks or pathways. Within this objective, PCA reduces the dataset to orthogonal axes each of which explains a part of the total variance from a contribution of the variance of each metabolite, and is commonly used if the data are Gaussian. By this method, the concentration of a metabolite, the original variable, can be expressed as a particular linear combination of the principal components. The projection of sample value on the same axis also allows grouping of the samples by similarity to establish, for example, if mutant and wild type data are separated or if some samples are outliers or artefactual. A log transformation of data frequently helps in fulfilling the Gaussian condition. It is also advisable for each of the variables to be centered and scaled to unit variance, otherwise the result may be biased by large differences in intensity. From the graphical display of projected points on the main axes, the relationships between variables are easily represented. They may be further analysed using the correlation coefficient matrix between each metabolite. Hierarchical component analysis (HCA) is a way to group the metabolite or the sample by their similarity, using a progressive pair-wise grouping of samples by distance. The result is visualized as a dendrogram or a tree. However the problem with this method is that results are dependent on the distance metric used, such as Euclidean distance, average distance, correlation.... K-means clustering is another method of grouping data, which uses a fixed number (K) of groups (Sumner *et al.*, 2003).

In addition to these methods, which are called unsupervised since no prior information is required, supervised methods, in which information is given on data organisation, may also be useful. They are based on correlation between α priori chosen variables or factors. Partial least-squares discriminant analysis (PLS-DA) was, for example, shown by Jonsson *et al.* (2005) to discriminate the data better than PCA. A graphical display of the correlation network is obtained by using the freely available software package named Pajek (http://vlado.fmf.uni-lj.si/pub/networks/pajek). It yields a metabolite:metabolite scatter plot where each pair of correlated metabolites is joined by a line, the length of which is dependent on the strength of the correlation. From the example of the glycolytic pathway, Steuer *et al.* (2003) have shown that the emergent correlation can be interpreted as a 'fingerprint' of the underlying biochemical pathways.

One of the most promising perspectives of metabolomics is the possibility of establishing relationships between metabolites from quite different pathways. Although databases can be used to identify and visualize automatically the corresponding pathways, a significant limitation originates from the lack of appropriate databases. Sumner *et al.* (2003) have emphasized the need for databases that can store metabolomic data in a way that facilitates relevant queries and most importantly which can operate together with gene expression databases. To interpret metabolic data, it is useful to know the biochemical context, such as enzyme activities that catalyse each reaction, the proteins that show such activities, and the genes that code for them. Such data bases already exist, e.g. KEGG, EciCyc, however, Sumner et al. mentioned in 2003 that KEGG has an incomplete set of data compared with the knowledge from the literature, whereas the data set contained in EcoCyc is more complete than KEGG for *E. coli*. Although EcoCyc is being extended to other organisms including *A. thaliana* by the TAIR project, it is not clear how much information is really being recovered from the literature, i.e. experiments, as opposed to being recovered from genome annotation, i.e. BLAST. Other databases exist such as EMP, PathDB, UM-BBD and BRENDA. All these databases contain features that make them unique, but none of them alone fulfils all the requirements for a good reference for metabolomics (Mendes, 2002).

METABOLOMICS IS AN IMPORTANT TOOL FOR FUNCTIONAL GENOMICS: LINKING GENOTYPE AND PHENOTYPE

The completion of several genomes, including *A.thaliana* and rice, has focused attention on the tremendous effort still required to understand the function of expressed genes and the way in which genes and the proteins they encode interact within cells and organisms. The analysis of the genome sequence of the plant *A. thaliana* has predicted approximately 26,000 genes (AGI, July 2002). As yet the function of only a few thousand can be defined with great confidence. In addition, sequence annotation mostly denotes the biochemical activity and is often little help in elucidating the function of the protein at the whole plant level. To ensure rapid progress from genome sequence to directed experimentation and application, systematic approaches for identifying the biological functions of novel genes are needed.

Over the last few years, two main strategies have been followed in order to discover the role of a gene: **forward genetics and reverse genetics**. Forward genetics consists in searching for interesting phenotypes by exploiting all kinds of natural or induced variation of the genome. The underlying genetic modification has then to be elucidated. Reverse genetics has as its starting point a specific gene whose *in vivo* activity is deliberately modified, e.g. by antisense, mutagenesis, co-suppression, over-expression, site directed mutagenesis (SDM), RNAi approaches, etc. Transgenic or mutant plants provide an excellent means to explore changes in metabolic network connectivity through metabolomic analysis of the effect of the specific perturbation of a gene of interest.

Genetic perturbations achieved by mutations or by integration of foreign DNA sequences leading to either over-or under-expression of genes were used to modify expression products and the resultant changes have been assessed to infer gene function. Transgenic potato plants over-expressing invertase specifically in the tubers were analysed by GC/MS to establish an unbiased metabolic profiling (Roessner *et al.*, 2000). Studies of sense or antisense plants with progressively decreasing expression of a target enzyme have shown that diagnostic changes in metabolites can often be detected in lines where the alteration of enzyme activity is too small to produce any visual phenotype (Scheible *et al.*, 2000). These technologies have been developed over the past few years for the detailed profiling of various classes of metabolites specific to, or characteristic of, plants. These approaches involves the profiling of sugars, organic and amino acids (Stitt and Sonnewald, 1995; Scheible *et al.*, 1997a, Stitt, 1999; Scheible *et al.*, 2000), flavonoids and their conjugates (Lin *et al.*, 2000), carotenoids and related compounds such as tocopherols and plastoquinones (Fraser *et al.*, 2000), fatty acids (Larson and Graham, 2001), monoterpenes (Voirin and Bayet, 1996), phenylpropanoids (Muir *et al.*, 2001; Liu *et al.*, 2002), and lignin formation (Humphreys and Chapple, 2002). Application of the same profiling method to *A.thaliana* plants treated with different types of herbicides demonstrated different profiles diagnostic of the site of inhibition (Fraser *et al.*, 2000).

The use of large populations mutagenized with an insertion element (gene disruption) allows identification and study of knockout lines for virtually every *A. thaliana* gene. However, gene redundancy or silent mutations under standard growth condition (more than 80%, D. Bouchez, personal communication) prevent the functional analysis of the target gene. To overcome problems such as gene redundancy, quantitative analysis is needed rather than qualitative analysis. Moreover, many agronomic traits are under the control of multiple loci and quantitative genetic analysis is the only way to identify genes involved in these important traits. Recombinant inbred lines (RILs) are a valuable tool to detect quantitative trait loci (QTL) from allelic variations in two parental lines. The first detection of QTL for biochemical traits, such as enzyme activities and metabolite contents was carried out using maize leaves (Causse *et al.*, 1995) The use of QTL in physiological and biochemical approaches has been reviewed by Prioul *et al.* (1997). The complementary use of near isogenic inbred lines (NIL) allowed identification and cloning of a QTL for sugar content in tomato which appeared to be an invertase gene (Fridman *et al.*, 2000). The exploration of the natural variation in accessions or cultivars is another way to validate a candidate gene as a QTL, an example has been reported for hexoses and the sucrose synthase gene *Sh1* by Thevenot *et al.* (2005). Metabolite analyses enables unbiased profiling of different lines and cultivars. The use of QTL analysis in metabolic studies is more recent in *A. thaliana*, but offers an enormous potential (Loudet *et al.*, 2003).

Nevertheless, the difficulty lies in finding and studying phenotypes that can be scored easily and help to identify the role of the gene. Convenient phenotypes such as morphological alterations are not appropriate, because mutant phenotypes and natural variation are not restricted to morphological changes. Plants are also able to adjust their metabolism to perturbations resulting in unchanged growth, because the concentration of intracellular metabolites has been altered to compensate for the effect of the mutation. In some cases, therefore, mutations that are silent for fluxes (growth rate) could produce very obvious effects on metabolite concentrations. To unravel the function of all the *A. thaliana* genes, global analysis needs to be employed at all levels: genome, transcriptome, proteome, and metabolome. Compared to conventional methods of phenotype determination, which are time-consuming and not always accurate at low expression levels, metabolomic approaches allow subtle but potentially important differences to be detected (Sumner *et al.*, 2003). Thus, metabolomic analysis should be of particular value for the study of mutants with no visual phenotype.

UNDERSTANDING METABOLISM

Not least of the applications of metabolomics, will be the powerful (and sometimes unexpected) insights it will provide into the understanding of plant metabolism and its regulation. Measurement of *in vivo* metabolite levels can provide key information in several different types of approach.

Most of the recent models of plant metabolism are based around the theorem of metabolic control analysis and derive the control distribution of pathway flux from the kinetic constants of the enzymes involved. Although these kinetic models have been of some use in understanding the photosynthetic Calvin cycle, difficulties in acquiring experimental information means that they have not been used more widely and remain limited in scope. Stoichiometric analysis is an interesting alternative approach that has gained much attention recently (Schuster *et al.*, 2000). The basis of this approach is to define elementary flux modes. An elementary mode is a minimal set of enzymes that could operate at steady-state, with the enzymes weighted by the relative flux they need to carry for the mode to function. "Minimal" means that if only the enzymes belonging to this set were operating, complete inhibition of one of these would lead to cessation of any steady-state flux in the system. This allows detection of the full set of non decomposable steady-state flows that the network can support, including cyclic flow. (Schuster *et al.*, 2000) It describes all metabolic routes that are both stoichiometrically and thermodynamically feasible.

One of the first steps in elucidating important regulatory mechanisms within a network is to carry out a systematic investigation of all metabolites in a given metabolic pathway. This provides an unbiased and powerful means to identify the site(s) at which metabolic flux is regulated. This type of approach was conducted with success on potato tubers. After investigating the subcellular concentration of every metabolite in the pathway between sucrose and starch metabolism, it was concluded that ADP-glucose pyrophosphorylase is the unique site of regulation (Tiessen *et al.*, 2002). Such analyses use substrate:product ratios to provide information on changes in control strength. Thus, they provide information on important regulatory steps in metabolism and can be used to predict limiting steps for optimal production of metabolites (Noctor *et al.*, 1998). To provide a full picture of the dynamics of metabolism, it is necessary to be able to determine the metabolic fluxes. The basic approach of following the fate of the labelled atoms of a precursor was used for many years to estimate fluxes through the major pathways of primary metabolism (Morot-Gaudry and Farineau, 1978). However, the methods of determining label distribution are not used as high-throughput approaches and consequently the number of fluxes that can be measured have been severely limited. The recent emergence of more sophisticated approaches, based on a combination of steady-state stable-isotope labelling and NMR or MS-based detection systems, may revolutionise our ability to measure fluxes. This approach allows the determination of positional labelling of a range of metabolites. The position of labelling within the carbon skeleton of end products can allow retrospective evaluation of the metabolic route by which they were formed (Chapter 17). Nevertheless, the interpretation of stable-isotope labelling experiments is not a trivial task.

There is a complex interdependence between metabolite concentration and flux. While changes in flux through a pathway are often necessary for the concentration of a given metabolite to change, the reverse is also frequently true, since metabolism is finely controlled by changes in enzyme activity that occur in response to increases in substrate and by allosteric control by interacting metabolite effectors. Metabolites can potentially be used as markers for the rate of metabolic processes, although this requires that the flux:content relationship(s) be isolated and defined. By using physical measurements of gas exchange under different conditions, biochemical modelling, and assays of metabolite contents, it was shown that certain amino acid contents, or amino acid ratios show a predictable relationship with the rate of photorespiration (Novitskaya *et al.*, 2002). Surprisingly, perhaps, the amino acid that showed the best relationship with the rate of photorespiration was aspartate, which is not usually considered to be part of the glycolate pathway. This demonstrates, first, the high degree of interconnectivity of metabolic pathways and, second, the dramatic impact the rapid process of photorespiration can have on metabolism in general in C_3 leaves.

As well as providing insight into the control of given metabolic pathways, metabolite assays generate important information on the integration of different pathways. Much still remains to be discovered about how metabolic pathways interact, and metabolites are frequently measured to provide insight into the response of plants to physiological or environmental changes. This is illustrated by studies conducted on the diurnal rhythms of tobacco in response to changes in the rate of photosynthesis and photorespiration (Scheible *et al.*, 1997b). It was observed during these studies that plants use sophisticated mechanisms to maintain relatively constant levels of central metabolism. It was also shown that a remarkable consistency in the ratio between the individual minor amino acids existed, even though the overall levels varied dramatically. Coordination of minor amino acid contents was also found by analysis of wheat, barley and potato leaves (Noctor *et al.*, 2002). One interpretation of such metabolite profiles is that key enzymes involved in amino acid synthesis are under 'general control', as in yeast. Recent studies suggest that plants contain mechanisms that could act to ensure cross-pathway coordination of enzyme activities at both the level of gene expression and post-translationally by redox regulation (Halford *et al.*, 2004; Lemaire *et al.*, 2004). Indirect evidence that redox regulation occurs *in vivo* comes from data obtained on poplars transformed to accumulate glutathione in the chloroplast: these lines showed enhanced accumulation of several minor amino acids (Noctor *et al.*, 1998).

A key limitation of metabolite data is spatial resolution of compounds that are often highly mobile between different compartments. Differentiation of metabolomes of individual tissues, single cells, and subcellular organelles will allow visualisation of detailed metabolic differences. Farré and coworkers (2001) recently used a nonaqueous extraction method and GC/MS to study the compartmentalisation of glycolytic intermediates in plastids, cytosol, and vacuole of potato tubers. These preliminary results offer great hope for successful subcellular metabolomics. One drawback is the large amount of tissue needed for such an analysis. Single cell approaches are in development and will depend on more sensitive techniques for separation and identifying of metabolites.

The objective of systems biology is to simultaneously monitor all biological processes operating as an integrated system. Through the study of systems, one can visualise how individual pathways or metabolic networks are interconnected. This approach is based on solid theoretical frameworks and uses computer modelling to explain experimental observations (Chapter 18). It will be of great value for directing metabolic understanding and metabolic engineering strategies because modifications to the expression of a single gene do not always bring about the predicted or desired effects due to cross-talk between pathways (Oliver, 2002). For these models to enable predictive metabolic engineering, they must incorporate experimental information about changes in gene expression and protein abundances that underlie these fluxes (Sweetlove *et al.*, 2003). Multi-level analysis through transcriptomics, proteomics and metabolomics can potentially provide insight into whether observed regulation occurs transcriptionally, post-transcriptionally, or post-translationally. Analysis in tobacco of metabolite profiles, twinned with measurement of the expression of key enzymes and metabolic flux, showed that NADH status can affect metabolism markedly and independently of changes in the capacity of the specific enzymes involved in the pathways concerned (Dutilleul *et al.*, 2005). Finally, metabolomics is required to define the behaviour of compounds providing important information on cellular status. Such molecules include both established phytohormones and other important signal molecules (lipids, amino acids, sugars, etc.).

METABOLOMICS AS THE ONLY DEFINITIVE PROOF FOR SUBSTANTIVE BIOEQUIVALENCE?

One of the goals of agricultural research is to develop new plant varieties that have better or new qualities. Many approaches have been applied during the last decade to improve plant performance and the production of bio-manufactured goods by genetic engineering. An obvious application is the engineering of plants in order to increase flux into main storage products such as starch or oils. Recently, the engineering of the nutritional value of a given crop has been extended to the production of pharmaceuticals, nutraceuticals, food additives and food supplements in plants. However, there is public concern that genetic modification could cause the production of harmful compounds which may find their way into foodstuffs.

The gradual realization by the non-scientific community that plants contain both beneficial and harmful chemicals whose levels can be altered by genetic manipulation, has fuelled public awareness of issues relating to transgenic foodstuffs. Therefore, genetically modified food should only reach the market when substantial bioequivalence with non-engineered products has been proven. This means that the metabolic phenotypes of a number of well established cultivars have to be compared to the transgenic plants, and the transgene deemed to be 'safe', only if no larger variation occurs than that found between the non-engineered varieties. The transgene itself is chemically neutral, being composed of the same four nucleotides as any other gene. However, the transgene protein, changes in metabolites that result from its expression, or other effects might be considered undesirable, for example, if they are toxic or cause allergic reactions. In these conditions, substantial equivalence, established between transgenic and the corresponding non-transgenic plants, will be based on biochemical phenotypes that can be defined by metabolic profiling. Such an approach can only rigorously be conducted through a true global 'metabolomics' approach. Beside evaluating the composition of transgenic crops, metabolomics will also be useful in assessing risks linked to the biological impact of environmental factors on the quality of crops.

CONCLUSIONS AND PERSPECTIVES

Until recently, most metabolite measurements were targeted to specific analyses of known metabolites or metabolite classes. In contrast, a non-targeted analysis should not seek to distinguish between all metabolites in the sample. To function in an unbiased manner, the extraction, separation and the identification must be independent of the analysis. Our present understanding of plant function and regulation of metabolism, which is still rudimentary, makes it inappropriate to exclude specific metabolites from analysis on the premise that they are not important. In terms of cellular signalling and control, the most abundant metabolites may provide only limited information, and the metabolites which present the biggest variations are not necessarily the most important.

For unknown compounds, metabolite pattern matching under different conditions may be used to indicate an important role, but chemical identification is necessary for establishing physiological functions. While there is a hierarchical flow of information from DNA to RNA to protein, there is no quantitative relationship between mRNA levels and function. Furthermore, once translated a protein may or may not be enzymatically active. Ter Kuile and Westerhoff (2001) showed that regulation of glycolysis is far from being completely hierarchical, i.e. flux through the pathway did not correlate proportionally with the concentration of the corresponding biochemical enzymes for 26 out of the 29 cases studied. Due to all these factors, changes in the transcriptome or the proteome do not always correspond to alterations in biochemical phenotypes. After all, when profiling the transcriptome and the proteome, mRNA and protein are identified through sequence similarity or data base matching. Thus identification is based primarily on the quality of the match and is therefore indirect. For plant genomes, that are not well characterized, transcript and protein profiling often yield only limited information. In these conditions, profiling the metabolome may actually provide the most functional information of the 'omics' technologies since a given metabolite, once identified and precisely defined, has a unique structure. The ideal situation will be to use these techniques in parallel to integrate mRNA, protein and metabolite profiling data in order to understand the metabolic network of plants and to assign functions to all genes.

Future technical challenges relate to the chemical complexity, the biological variation and the dynamic range of metabolites in plants. It has already become clear from metabolomics studies that a key factor is a high degree of precision in experimental design and protocols. Meaningful exchange of information between laboratories will include defining and perhaps even standardizing growth and extraction procedures (Bino *et al.*, 2004). No one analytical technology will provide the single solution to complete metabolome analysis. Several complementary approaches to metabolite analysis can be favoured depending on the desired level of complexity of the analysis. Modern high-throughput assay technology will enable metabolite profiling at much higher resolution and scale than possible thus far. In some cases, where the information sought is well defined, non selective techniques may be unsuitable or even undesirable.

Such techniques may have greatest value in pinpointing changes in metabolites that would be difficult to foresee. In this way, novel effects of genotypic or environmental changes on metabolites can be identified and further investigated.

LC/MS and GC/MS are likely to be increasingly used as complementary non selective techniques. Although LC/MS is potentially more exhaustive than GC/MS, it suffers from the relatively small number of available spectra in databases. Prospects for identification of uncharacterized compounds may rely on NMR data, but could also be facilitated by high precision LC/MS analysis of molecular ions linked to predictions of formulae by computer modelling. Absorbance or fluorescence spectra provide other '3D' information that complements, or is alternative to, mass spectra in confirming compounds in complex mixtures, and use of these modes of detection will likely benefit from ongoing advances in LC technology that are greatly improving chromatographic resolution.

In conclusion, metabolite profiling is highly multidisciplinary, and strategic funding and academic networks will be essential for its further development in non industrial research. Metabolomics will not only assist in the establishment of a deeper understanding of the complex interactive nature of plant metabolic networks but will also provide unique insights into the fundamental nature of plant phenotypes in relation to development, physiology, tissue identity, resistance, biodiversity, etc. (Hall *et al.*, 2002).

REFERENCES

Ap Rees, T., Hill, S.A. 1994. Metabolite control analysis of plant metabolism. *Plant Cell Environ.*, 17, 587-599.

Benthin, B., Danz, H., Hamburger M. 1999. Pressurized liquid extraction of medicinal plants. *J. Chromatogr., A*, 837, 211-219.

Bino, R.J., Hall, R.D., Fiehn, O., Kopka, J., Saito, K., Draper, J., Nikolau, B.J., Mendes, P., Roessner-Tunali, U., Beale, M.H., Trethewey, R.N., Lange, B.M., Wurtele, E.S., Sumner, L.W. 2004. Potential of metabolomics as a functional genomics tool. *Trends Plant Sci.*, 9, 418-425.

Causse, M, Rocher, J.P., Henry, A.M., Charcosset, A., Prioul, J.L., de Vienne, D 1995. Genetic dissection of the relationship between carbon metabolism and early growth in maize, with emphasis on key-enzyme loci. *Mol. Breeding*, 1, 259-272.

Davies, S.C., Makarov, A.A., Hughes, J.D. 1999.Ultrafast gas chromatography using time-of-flight mass spectrometer. *Rapid Commun. Mass Spectra*, 13, 237-241.

Chen, F., Duran, A.L., Blount, J.W., Summer, L.W., Dixon, R.A. 2003. Profling phenolic metabolites in transgenic alfafa modified in lignin biosynthesis. *Phytochemistry*, 64, 1013-1021.

De Luca, V., St. Pierre, B. 2000. The cell and developmental biology of alkaloid biosynthesis. *Trends Plant Sci.*, 5, 168-173.

Dutilleul, C., Lelarge, C., Prioul, J.L., De Paepe, R., Foyer, C.H., Noctor, G. 2005. Mitochondria-driven changes in leaf NAD status exert a crucial influence on the control of nitrate assimilation and the integration of carbon and nitrogen metabolism. *Plant Physiol*, 139, 64-78.

Farré, E.M., Tiessen, A., Roessner, U., Geigenberger, P., Trethewey, R.N., Willmitzer, L. 2001. Analysis of compartmentation of glycolytic intermediates, nucleotides, sugars, organic acids, amino acids and sugar alcohols in potato tubers using a non-aqueous fractionation method. *Plant Physiol.*, 127, 685-700.

Fiehn, O. 2001. Combining genomics, metabolome analysis, and biochemical modelling to understand metabolic networks. *Comp. Funct. Genomics,* 2, 155-168.

Fiehn, O. 2002 Metabolomics-the link between genotypes and phenotypes. *Plant Mol. Biol.,* 48, 155-171.

Fiehn, O., Kopka, J., Dörmann, P., Altmann, T., Trethewey, R.N., Willmitzer, L. 2000. Metabolic profiling for plant functional genomics. *Nature Biotechnol.,* 18, 1157-1161.

Fraser, P.D., Pinto, M.E., Holloway, D.E., Bramley, P.M. 2000. Application of high-performance liquid chromatography with photodiode array detection to the metabolic profiling of plant isoprenoids. *Plant J.,* 24, 551–558.

Fridman, E., Pleban, T., Zamir, D. 2000. A recombination hotspot delimits a wild-species quantitative trait locus for tomato sugar content to 484 bp within an invertase gene. *Proc. Nat. Acad. Sci.* USA, 97, 4718-4723.

Galmiche, J.M. 1973. Carboxylation primaire chez les végétaux supérieurs : étude cinétique à l'aide du $^{14}CO_2$. Paris XI., pp. 177.

Gamiz-Gracia, L., de Castro, M.D.L. 2000. Continuous subcritical water extraction of medicinal plant essential oil: comparison with conventional techniques. *Talanta,* 51, 1179-1185.

Gonzalez, B., François, J., Renaud, M. 1997. A rapid and reliable method for metabolite extraction in yeast using boiling buffered ethanol. *Yeast,* 13, 1347-1356.

Halford, N.G., Hey, S., Jhurreea, D., Laurie, S., McKibbin, R.S., Zhang, Y., Paul, M.J. 2004. Highly conserved protein kinases involved in the regulation of carbon and amino acid metabolism. *J. Exp. Bot.,* 55: 35-42.

Halket, J.M., Przyborowska, A., Stein, S.E., Mallard, W.G., Down, S. and Chalmers, R.A. 1999. Deconvolution gas chromatography/mass spectrometry of urinary organic acids—potential for pattern recognition and automated identification of metabolic disorders. *Rapid Commun. Mass Spectro.,* 13, 279–284.

Hall, R., Beale, M., Fiehn, O., Hardy, N., Sumner, L., Bino, R. 2002. Plant metabolomics: the missing link in functional genomics strategies. *Plant Cell,* 14, 1437-1440.

Huhman, D.V., Sumner, L.W. 2002. Metabolic profiling of saponins in *Medicago sativa* and *Medicago truncatula* using HPLC coupled to an electrospray ion-trap mass spectometer. *Phytochemistry,* 59, 347–360.

Humphreys, J.M., Chapple, C. 2002. Rewriting the lignin roadmap. *Curr. Opin. Plant Biol.,* 5, 224–229.

Jarvis, A.P., Morgan, D. 1997. Isolation of plant products by supercritical-fluid extraction. *Phytochem. Anal.,* 8, 217-222.

Johansen, H.N., Glitso, V., Knudsen, K.E.B. 1996. Influence of extraction solvent and temperature on the quantitative determination of oligosaccharides from plant materials by high-performance liquid chromatography. *J. Agric. Food. Chem.,* 44, 1470-1474.

Johnson, H.E., Broadhurst, D., Goodacre, R., Smith, A.R. 2003. Metabolic fingerprinting of salt-stressed tomatoes. *Phytochemistry,* 62, 919-928.

Jonsson, P., Gullberg, J., Nordstrom, A., Kusano, M., Kowalczyk, M., Sjostrom, M., Moritz, T. 2004. A strategy for identifying differences in large series of metabolomic samples analyzed by GC/MS. *Anal. Chem.,* 76, 1738-1745.

Jonsson, P., Johansson, A.I., Gullberg, J., Trygg, J.A.J., Grung, B., Marklund, S., Sjostrom, M., Antti, H., Moritz, T. 2005. High-throughput data analysis for detecting and identifying differences between samples in GC/MS-based metabolomic analyses. *Anal. Chem.,* 17, 5635-5642.

Larson, T.R., Graham, I.A. 2001. A novel technique for the sensitive quantification of acyl CoA esters from plant tissues. *Plant J.,* 25, 115–125.

Lemaire, S.D., Guillon, B., Le Maréchal, P., Keryer, E., Miginiac-Maslow, M., Decottignies, P. 2004. New thioredoxin targets in the unicellular photosynthetic eukaryote *Chlamydomonas reinhardtii*. *Proc. Natl. Acad. Sci.* USA, 101, 7575-7480.

Lin, L.Z., He, X.G., Lindenmaier, M., Yang, J., Cleary, M., Qiu, S.X., Cordell, G.A. 2000. LC-ESI-MS study of the flavonoid glycoside malonates of red clover (*Trifolium pratense*). *J. Agri. Food. Chem.*, 48, 354–365.

Liu, C.J., Blount, J.W., Steele, C.L., Dixon, R.A. 2002. Bottlenecks for metabolic engineering of isoflavone glycoconjugates in *Arabidopsis*. *Proc. Natl. Acad. Sci.* USA, 99, 14578–14583.

Loudet, O., Chaillou, S., Krapp, A., Daniel-Vedele, F. 2003. Quantitative trait loci analysis of water and anion contents in interaction with nitrogen availability in *Arabidopsis thaliana*. *Genetics*, 163, 711-22.

Maloney, V. 2004. Plant metabolomics. *Biol. Teach J.*, 2, 92-99.

Matt, P., Krapp, A., Haake, V., Mock, H.P., Stitt, M. 2002. Decreased Rubisco activity leads to dramatic changes of nitrate metabolism, amino acid metabolism and the levels of phenylpropanoids and nicotine in tobacco antisense RBCS transformants. *Plant J.*, 6, 663-677.

Mendes, P. 2002. Emerging bioinformatics for the metabolome. *Briefings in Bioinformatics* , 3, 134–145.

Morot-Gaudry, J.F., Farineau, J. 1978. Etude comparée des reactions de carboxylations photosynthétiques chez un maïs normal (W64A) et chez un maïs mutant opaque 2(W64o2) : mise en évidence de déviations métaboliques chez le mutant. *Physiol.Vég.*, 16, 451-467.

Muir, S.R., Collins, G.J., Robinson, S., Hughes, S., Bovy, A., de Vos, C.H.R., van Tunen, A.J., Verhoeyen, M.E. 2001. Overexpression of petunia chalcone isomerase in tomato results in fruit containing increased levels of flavonols. *Nature Biotechnol.*, 19, 470–474.

Namiesnik, J., Gorecki, T. 2000. Sample preparation for chromatographic analysis of plant material. *JPC-J. Planar Chromatogr.-Mod. TLC*, 13, 404-413.

Neuhaus, H.E., Stitt, M. 1990. Control analysis of photosynthate partitioning: impact of reduced activity of ADP glucose pyrophosphorylase on plastid phosphoglucomutase on the fluxes to starch and sucrose in *Arabidopsis thaliana* L. Heynh,. *Planta*, 184, 445-454.

Nikiforova, V.J., Gakière, B., Kempa, S., Adamik, M., Willmitzer, L., Hesse, H., Hoefgen, R. 2004. Towards dissecting nutrient metabolism in plants: a systems biology case study on sulphur metabolism. *J. Exp. Bot.*, 55, 1861-1870.

Noctor, G., Arisi, A.C.M., Jouanin, L., Foyer, C.H. 1998. Manipulation of glutathione and amino acid biosynthesis in the chloroplast. *Plant Physiol.*, 118, 471-482.

Noctor, G., Novitskaya, L., Lea, P.J., Foyer, C.H. 2002. Co-ordination of leaf minor amino acid contents in crop species: significance and interpretation. *J. Exp. Bot.*, 53, 939-945.

Novitskaya, L., Trevanion, S., Driscoll, S.D., Foyer, C.H., Noctor, G. 2002. How does photorespiration modulate leaf amino acid contents. A dual approach through modelling and metabolite analysis. *Plant Cell Environ.*, 25, 821-836.

Oliver, S. 2002. Demand management in cells. *Nature*, 418, 33–34.

Orth, H.C.J., Rentel, C., Schmidt, P.C. 1999. Isolation purity analysis and stability of hyperforin as a standard material from *Hypericum perforatum* LJ. *Pharm. Pharmacol. (Commun.)*, 51, 193-200.

Prioul, J.L., Quarrie, S. Causse, M., Devienne, D. 1997. Dissecting complex physiological functions through the use of molecular quantitative genetics. *J. Exp. Bot.*, 48, 1151-1163.

Roessner, U., Luedemann, A., Brust, D., Fiehn, O., Linke, T., Willmitzer, L., Fernie, A.R. 2001. Metabolic profiling allows comprehensive phenotyping of genetically modified plant systems. *Plant Cell*, 13, 11-29.

Roessner, U., Wagner, C., Kopka, J., Trethewey, R.N., Willmitzer, L. 2000. Simultaneous analysis of metabolites in potato tuber by gas chromatography-mass spectrometry. *Plant J.*, 23, 131-142.

Sargenti, S.R, Vichnewski, W. 2000. Sonication and liquid chromatography as a rapid technique for extraction and fractionation of plant material. *Phytochem. Anal.*, 11, 69-73.

Scheible, W.R., Gonzalez-Fontes, A., Lauerer, M., Muller-Rober, B., Caboche, M., Stitt, M. 1997. Nitrate acts as a signal to induce organic acid metabolism and repress starch metabolism in tobacco. *Plant Cell*, 5, 783-798.

Scheible, W.R., Gonzalez-Fontes, A., Morcuende, R., Lauerer, M., Geiger, M., Glaab, J., Gojon, A., Schulze, E.D., Stitt, M., 1997. Tobacco mutants with a decreased number of functional nia genes compensate by modifying the diurnal regulation of transcription, post-translational modification and turnover of nitrate reductase. *Planta*, 203, 304-319.

Scheible, W.R., Krapp, A., Stitt, M. 2000. Reciprocal diurnal changes of phosphoenolpyruvate carboxylase expression and cytosolic pyruvate kinase, citrate synthase and NADP-isocitrate dehydrogenase expression regulate organic acid metabolism during nitrate assimilation in tobacco leaves. *Plant Cell Env.*, 23, 1155-1167.

Schuster, S., Fell, D.A., Dandekar, T. 2000. A general definition of metabolic pathways useful for systematic organization and analysis of complex metabolic networks. *Nat. Biotechnol.*, 18, 326-332.

Soga, T., Ueno, Y., Naraoka, H., Ohashi, Y., Tomita, M., Nishioka, T. 2002. Simultaneous determination of anionic intermediates for Bacillus subtilis metabolic pathways by capillary electrophoresis electrospray ionization mass spectrometry. *Anal. Chem.*, 74, 2233-2239.

Starmans, D.A.J., Nijhuis, H.H. 1996. Extraction of secondary metabolites from plant material: a review. *Trends Food Sci. Technol.*, 7, 191-97.

Steuer, R., Kurths, J., Fiehn, O., Weckwerth, M. 2003. Observing and interpreting correlation in metabolic networks. *Bioinformatics*, 8, 1019-1026.

Stitt, M. 1999. Nitrate regulation of metabolism and growth. *Cur. Opin. Plant Biol.*, 2, 178-186.

Stitt, M., Sonnewald, U. 1995. Regulation of metabolism in transgenic plants. *Annu. Rev. Plant Phys.*, 46, 341-368.

Streeter, J.G. 1987. Carbohydrates, organic acid and amino acid composition of bacteroids and cytosol from soybean nodules. *Plant Physiol.*, 85, 768-773.

Streeter, J.G., Strimbu, C.E. 1998. Simultaneous extraction and derivatization of carbohydrates from green plant tissues for analysis by gas-liquid chromatography. *Anal. Biochem.*, 259, 253-57.

Sumner, L.W., Mendes, P., Dixon, R.A. 2003. Plant metabolomics: large-scale phytochemistry in the functional genomics era. *Phytochemistry*, 62, 817-836.

Sweetlove, L.J., Last, R.L., Fernie, A.R. 2003. Predictive metabolic engineering: a goal for systems biology. *Plant Physiol.*, 132, 420-425.

Taylor, J., King, R.D., Altmann, T., Fiehn, O. 2002. Application of metabolomics to plant genotype discrimination using statistics and machine learning. *Bioinformatics*, 18, S241-248.

Ter Kuile, B.H., Westerhoff, H.V. 2001. Transcriptome meets metabolome: hierarchical and metabolic regulation of the glycolytic pathway. *FEBS Lett.*, 500, 169-71.

Thevenot, C., Simond-Cote, E., Reyss, A., Manicacci, D., Trouverie, J., Le Guilloux, M., Ginhoux, V., Sidicina, F., Prioul, J.L. 2005. QTLs for enzyme activities and soluble carbohydrates involved in starch accumulation during grain filling in maize. *J. Exp. Bot.*, 56, 945-958.

Tiessen, A., Hendriks, J.H.M., Stitt, M., Branscheid, A., Gibon, Y., Farré, E.M., Geigenberger; P. 2002. Starch Synthesis in Potato Tubers Is Regulated by Post-Translational Redox Modification of ADP-Glucose Pyrophosphorylase: a Novel Regulatory Mechanism Linking Starch Synthesis to the Sucrose Supply. *Plant Cell*, 14, 2191-2213.

Tikunov, Y., Lommen, A., de Vos, C.H., Verhoeven, H.A., Bino, R.J., Hall, R.D., Bovy, A.G. 2005. A novel approach for nontargeted data analysis for metabolomics. Large-scale profiling of tomato fruit volatiles. *Plant Physiol.*, 139, 1125-37.

Tolstikov, V.V., Fiehn, O. 2002. Analysis of highly polar compounds of plant origin: combination of hydrophilic interaction chromatography and electrospray ion trap mass spectrometry. *Anal. Biochem.*, 301 298–307.

Voirin, B., Bayet, C. 1996. Developmental changes in the monoterpene composition of *Mentha×piperita* leaves from individual peltate trichomes. *Phytochemistry*, 43, 573–580.

Weckwerth, W. 2003. Metabolomics in system biology. *Annu. Rev. Plant. Biol.*, 54, 669-689.

Weckwerth, W., Wenzel, K., Fiehn, O. 2004. Process for the integrated extraction, identification and quantification of metabolites, proteins and RNA to reveal their co-regulation in biochemical networks. *Proteomics*, 4, 78-83

16

CHAPTER

Metabolic Fingerprinting and Profiling by Proton NMR

A. Moing, C. Deborde, D. Rolin

Plants are the most prolific factories for the production of small molecules. To date, more than 100,000 metabolites have been identified in the plant kingdom and this may well be only 10% of the total (Trethewey, 2004). Metabolomics seeks to identify the whole set of metabolites in living organisms. To achieve this objective, or at least to approach it, a large range of analytical methods have been tested. Many chromatographic methods such as gas chromatography (GC) or liquid chromatography (LC) coupled to mass spectrometry (MS), have been used as the methods of choice (Chapter 15), but complementary information from other techniques, particularly NMR spectroscopy is potentially useful in the study of the metabolome (Krishnan *et al.*, 2005; Ratcliffe and Shachar-Hill, 2005; Terskikh *et al.*, 2005). Moreover, NMR is a versatile analytical tool and the possible combination of ^{1}H, ^{2}H, ^{13}C and ^{15}N and ^{31}P NMR *in vitro* and *in vivo* allows detailed analysis of plant metabolism (Bligny and Douce, 2001; Mesnard and Ratcliffe, 2005).

NMR SPECTROSCOPY TECHNIQUE

Nuclear magnetic resonance (NMR) is a spectroscopic technique that exploits the magnetic properties of the atomic nucleus (Macomber, 1998). In NMR, the sample is immersed in a strong external magnetic field and transitions between the nuclear magnetic energy levels are induced by a suitably oriented radiofrequency field. In theory, any molecule containing one atom with a non-zero nuclear spin (I) is potentially visible by NMR. Considering the isotopes with a non-zero nuclear spin such as ^{1}H, ^{13}C, ^{14}N, ^{15}N and ^{31}P, all biological molecules have at least one NMR signal. Obviously, there is a very wide variation in the sensitivity of the experiment for different nuclei, hence ^{1}H NMR remains the best choice for metabolite profiling by NMR mainly due to its natural abundance (99.8%) and its sensitivity. The NMR spectrum generally consists of a series of discrete lines (resonances), which are characterized not only by the familiar spectroscopic quantities of frequency (chemical shift), intensity and line shape, but also by relaxation times.

The great strength of NMR as an analytical tool lies in the fact that the precise frequency at which a nucleus resonates depends on the local magnetic field that the nucleus experiences. This local field depends not only on the strong external magnetic field but also on the magnetic effects associated with the electron density and with the neighbour magnetic nuclei. Thus, similar nuclei in different chemical environments are generally magnetically non-equivalent and absorb energy at different frequencies. An important practical aspect of NMR is the resolution of the corresponding resonances and their assignment to specific nuclei in the sample. It is convenient to express the energy of a transition in a field-independent form, by dividing the frequency separation (in Hz) between the resonance of interest and some standard resonance, by the operating frequency (in MHz). This relative energy scale is called the chemical shift and is expressed in parts per million (ppm).

The intensity of a resonance is related to the concentration of the nuclei in the sample, but there are some restrictions on the use of NMR as a method for quantitative analysis. NMR is an insensitive technique because of the small extinction coefficients for NMR transitions. The most favourable nucleus for quantitative analysis is ^{1}H. However, even with the best spectrometers available, the lower limit for detection is a concentration between 10 and 100 μM (Bales *et al.*, 1984; Wevers *et al.*, 1994; Krishnan *et al.*, 2005). The ^{1}H NMR detection limit or sensitivity for metabolite depends on the number of equivalent protons that contribute to the resonance and on resonance multiplicity (Wevers *et al.*, 1994). In practice, the achievable sensitivity is strongly dependent on the field strength of the magnet, and on the design of the probe head that allows the signals to be detected. Today NMR spectrometers are available with field strengths of up to 22.3 Tesla, corresponding to a ^{1}H NMR frequency of 950 MHz, but most metabolite analyses are achieved on the more commonly available instruments that operate in the range 300-600 MHz. Since increasing field strength also increases spectral resolution, reducing the number of overlapping signals in the spectrum, the spectrometers at the upper end of this frequency range are the most effective ones for metabolite fingerprinting or profiling by ^{1}H NMR. On the other hand, the increasing availability of cryogenic probe heads (Keun *et al.*, 2002; Spraul *et al.*, 2003), in which the sensitivity is increased by cooling the detection system with liquid helium, offers the opportunity of a substantial improvement in the detection of signals.

In the usual pulse or Fourier Transform NMR technique, the final spectrum is the sum of a series of spectra obtained by repeatedly exciting ("pulsing") and recording the transitions, named free induction decay (FID). This procedure increases the signal-to-noise ratio in the final spectrum and helps to overcome the problem of low sensitivity. Unfortunately, the delay between excitation pulses, or repetition delay, should be long enough to allow sufficient relaxation and accurate measurement of peak intensities. Sufficient relaxation means that the magnetization returns fully to equilibrium between pulses. This return to equilibrium is a relaxation process, and in simple cases it follows an inverse exponential time course, which is characterized by the longitudinal (spin-lattice) relaxation time T_1. Values of T_1 for nuclei in solution vary typically from 0.01 to 100 s

(Macomber, 1998). For proton nuclei in solution, T_1 varies typically from 0.01 to 20 sec, but it should be noted that these T_1 values are solvent, paramagnetic impurities (like oxygen), temperature and concentration dependent. With a 90° pulse, this repetition delay should be at least five times the longest T_1 of the sample (Pauli *et al.*, 2005). With smaller pulse angles, a more rapid repetition delay can be used. However, T_1 measurements on a relevant plant extract under the conditions used should be taken and saturation for all proton resonances in the spectrum should be measured. Saturation should not exceed 6%, to allow direct quantitative analysis of a metabolite from the spectrum (Wevers *et al.*, 1995). When saturation is avoided or controlled, the NMR signal intensity remains linear over a wide concentration range ($> 10^4$).

SAMPLE PREPARATION PREREQUISITE

The sample volume in the NMR spectrometer should be standardized. The pH of the sample in the spectrometer is critical and therefore should be closely standardized to obtain good analytical reproducibility and a correct assignment of resonances. Resonances for compounds with pKa values within the pH range of plant extracts (pH 4-8) can shift significantly. For example the chemical shift values of the methylene proton of citrate (pKa: 3.13, 4.76 and 6.40 at 25°C) and of malate (pKa: 3.46 and 5.10 at 25°C), and of the methyne proton of tartrate (meso-form pKa: 3.17 and 4.91 at 25°C) are critically dependent upon pH within this range (Bales *et al.*, 1984; Dawson *et al.*, 1986; Moore and Sillerud, 1994).

Exchange of a proton with deuterium from deuterated water as described for the OH, SH and COOH proton, but also "acidic" protons such as CH_2 from acetatoacetate, CH_3 from diacetyle may broaden the resonance of a compound, so that it becomes NMR-invisible. Resonances may also become severely broadened, and in extreme cases, NMR-invisible, if for example, a small molecule is bound to a macromolecule or forms a complex with a paramagnetic ion such as Mn^{2+}, Cu^{2+}, Fe^{3+}. TSP (3-[trimethylsilyl] [2,2,3,3-D_4] propionic acid sodium salt) is often used as an internal chemical shift reference ($\delta = 0.00$) and quantitative ^{1}H NMR standard. However, TSP is able to bind to proteins which broadens its resonance and may adversely affect quantification, and in some cases TSP may become NMR-invisible. An overview of several quantitative ^{1}H-NMR standards used for natural product analysis was published recently by Pauli *et al.* (2005). Addition of EDTA (ethylenediaminetetraacetic acid) to plant extracts is often used to make citrate and malate NMR-visible (Le Gall *et al.*, 2003). In addition, it should be kept in mind that by ^{1}H-NMR spectroscopy, metabolites with no contributing proton in their chemical structure are NMR-invisible. Moreover, correct interpretation of spectra could not be achieved without a database of spectra of relevant metabolites, acquired under the same NMR and chemical conditions.

Among the isotopes with non-zero nuclear spin, ^{1}H remains the most sensitive and this is inevitably the preferred nucleus for most metabolite fingerprinting and profiling applications of NMR. In plants, metabolite fingerprinting or profiling by NMR has been used to examine the unintended effects resulting from genetic modifications (Le Gall *et al.*, 2003; Charlton *et al.*, 2004), to identify traits in ecotypes or populations (Ward *et al.*, 2003; Moing *et al.*, 2004) or to quantify the metabolite changes due to biotic (Choi *et al.*, 2004a; André *et al.*, 2005) or abiotic stresses (Bailey *et al.*, 2003).

METABOLIC FINGERPRINTING BY PROTON NMR

High resolution ^{1}H NMR methods are generally less sensitive than MS-based methods but have been developed for rapid metabolic fingerprinting (Belton *et al.*, 1998; Noteborn *et al.*, 2000; Ward *et al.*, 2003). Proton NMR provides an overview of the most important proton-containing low-molecular mass metabolites (soluble sugars, organic and amino acids) in a tissue crude extract above a minimum threshold level, in a non-targeted way (Fan, 1996). Threshold values have been discussed by Krishnan *et al.* (2005). The suitability of ^{1}H NMR spectroscopy of tissues and body fluids for preclinical screening of xenobiotic toxicity in animals was recently reviewed (Lindon *et al.*, 2003). Generally, liquid ^{1}H NMR methods are performed on crude extracts (Charlton *et al.*, 2004), or on polar and apolar phases (Sobolev *et al.*, 2005). In such biological extracts, one-dimensional spectra contain many peaks leading to possible overlapping signals, thus making the profiles complex (Fig. 16.1).

One-dimensional ^{1}H-NMR chemical fingerprinting has been widely developed for biomedical applications (Antti *et al.*, 2002; Brindle *et al.*, 2002; Lindon *et al.*, 2001). It has also been proposed and used for food analysis and authentication (Vogels *et al.*, 1996; Belton *et al.*, 1998; Le Gall *et al.*, 2001), for the study of the biochemical mode of action of herbicides (Ott *et al.*, 2003), the effects of metal toxicity in plant cell cultures (Bailey *et al.*, 2003) and for the study of compositional changes during fruit ripening (Gil *et al.*, 2000). ^{1}H-NMR spectroscopy, either coupled or not with liquid chromatography, was used also for cultivar comparison (Sobolev *et al.*, 2003; Choi *et al.*, 2004b), for the demonstration of substantial equivalence of transgenic plants (Noteborn *et al.*, 2000) and for the detection of unintended effects following a genetic modification (Le Gall *et al.*, 2003; 2005; Choi *et al.*, 2004c; Defernez *et al.*, 2004).

Although signal intensity can be used for absolute quantification of metabolites, as it is proportional to the number of nuclei, the studies cited above only use relative quantification after elimination of the water signal using a presaturation sequence. Spectra acquisition under relative quantification conditions has the advantage of allowing high throughput analysis, since the acquisition of one spectrum may require from 2 to 5 min only.

With careful sample preparation and spectra analysis, these complex profiles can be used for metabolite fingerprinting (Defernez and Colquhoun, 2003). Moreover, these profiles can be used to monitor soluble sugars, organic and amino acids, and phenolics of a complex plant extract in a single spectrum (Gil *et al.*, 2000) if signal overlap is not significant, by comparing complex profiles with reference spectra acquired under similar conditions (Fig. 16.1). Manipulating the NMR signals to produce a two-dimensional (2D) experiment takes a longer time than a simple one-dimensional (1D) experiment, but 2D homonuclear correlation experiments are particularly useful after metabolic fingerprinting, allowing linked subsets of peaks in the conventional 1D ^{1}H NMR spectrum to be identified and assigned to particular compounds (Macomber, 1998).

High field ^{1}H-NMR spectroscopy combined with chemometric analysis offers a potentially powerful means to analyze metabolic changes in plants submitted to different stresses or between control and genetically modified plants. The combination of the non-selective yet highly specific NMR analysis with multivariate data analysis allows discrimination between differing tissue or cell populations, and elucidation of

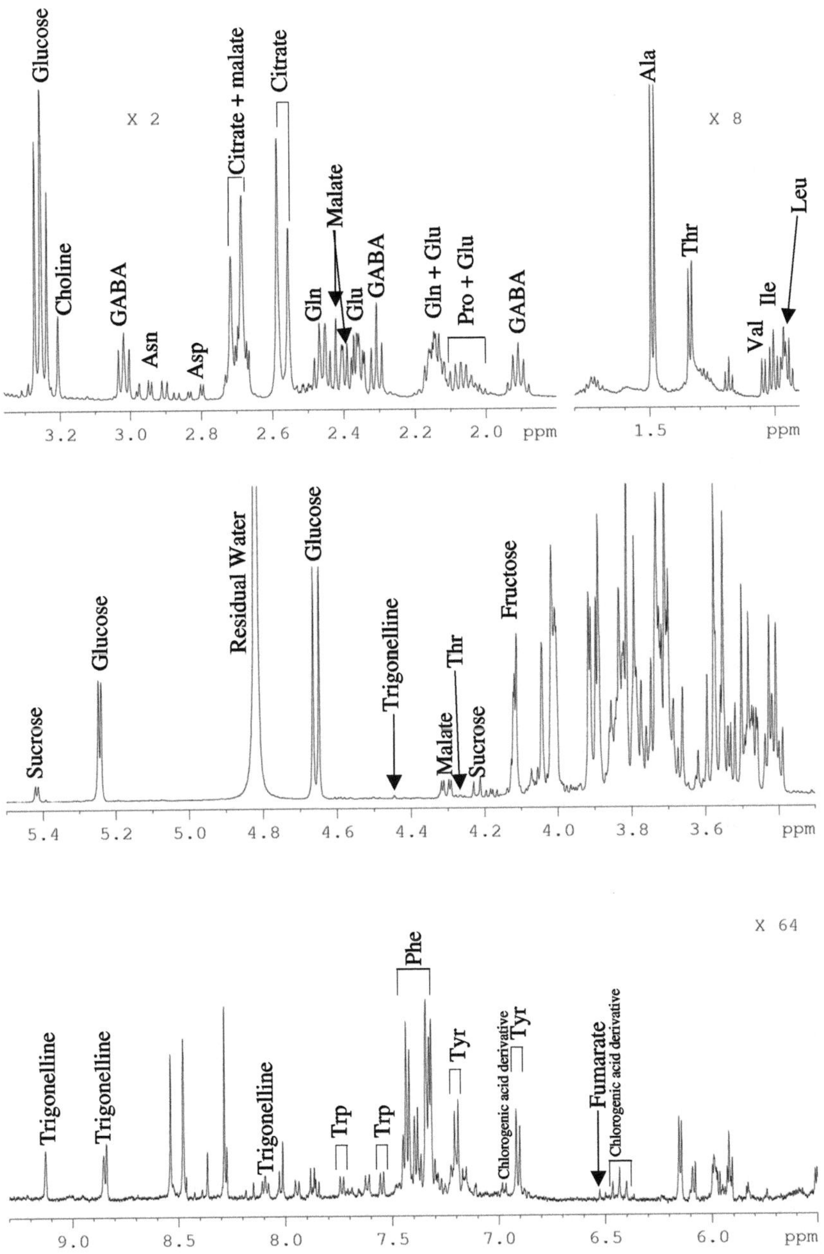

Fig. 16.1 Details of representative ^{1}H NMR spectrum of an extract of mature tomato fruit pericarp. Key: Ala, alanine; Asn, asparagine; Asp, aspartic acid; GABA, γ-aminobutyric acid; Gln, glutamine; Glu, glutamic acid; Ile, isoleucine; Leu, leucine; Phe, phenylalanine; Pro, proline; Thr, threonine; Trp, tryptophan; Tyr, tyrosine; Val, valine. (Contributed by F. Mounet, UMR Biologie du Fruit, Bordeaux).

biomarkers responsible for such differences. The comparison and classification of spectral signatures can be carried out using data reduction routines and other statistical methods (Lindon *et al.*, 2001) and adaptation and application of the clustering methods developed for transcriptome analysis (Daub *et al.*, 2003). The first step of fingerprint analyses is checking for the absence of peak shift (possibly due to small pH differences), or alignment of the regions exhibiting slight peak shifts (Vogels *et al.*, 1996). Secondly, the signal intensity is integrated over small regions of similar size (0.04 or 0.03 ppm) throughout the spectra (after elimination of the residual water zone), giving several hundred variables called spectral domains, bins or "buckets" (Lindon *et al.*, 2001). Third, principal component analysis (PCA) is performed on the "buckets" after normalization, by dividing them by the sum of the intensities (Lindon *et al.*, 2001; Defernez and Colquhoun, 2003). The examination of the loading scores (set of weights given to the original measurements) of the first principal components then indicates the spectra regions explaining most sample variability. A comparison of these regions with reference spectra indicates possible discriminatory compounds. Univariate analysis, such as analysis of variance, can be carried out after PCA, to check for significant differences between groups of samples for the relative area corresponding to these compounds in the spectra (Defernez *et al.*, 2004). Since NMR measurements are non-destructive, the extracts can be further analyzed with complementary methods dedicated to absolute quantification of the discriminatory compounds.

QUANTITATIVE METABOLIC PROFILING BY PROTON NMR

Among the different methods used for the simultaneous analysis of a large number of metabolites, GC- or LC- mass spectrometry allows the separation and identification of several hundred metabolites (Roessner *et al.*, 2001; Urbanczyk-Wochniak and Fernie, 2004; Overy *et al.*, 2005). These include minor compounds of secondary metabolism and intermediary metabolism (isoprenoids, flavonoids, phenolic compounds...), but only after mono or double-derivatization for GC-MS. In addition, absolute quantification is problematic and most data remain semi-quantitative. However, as mentioned recently by Nielsen and Oliver (2005), future metabolomic studies must focus on the accurate measurement of the concentration of identified metabolites. Although less sensitive than GC or LC-mass spectrometry, proton NMR spectroscopy is a powerful complementary technique for the identification and quantitative analysis of plant metabolites either *in vivo* or in tissue extracts (Krishnan *et al.*, 2005), which requires little or no pre-treatment of the sample. Proton NMR spectroscopy allows identification (Fan, 1996), as indicated above, and relative (Le Gall *et al.*, 2003; Le Gall *et al.*, 2004) or absolute quantification (Silvestre *et al.*, 2001; de Graaf and Behar, 2003; Moing *et al.*, 2004) of the major primary metabolites (soluble sugars, amino acids, organic acids) and other major metabolites (such as some phenolic acids and quaternary ammonium compounds). Typically, 20 to 40 metabolites have been identified in metabolite profiling of plant extracts (Krishnan *et al.*, 2005). Therefore ^{1}H-NMR profiling provides information which is intermediate between metabolic profiling (Fiehn, 2002), such as a targeted analysis usually of one family of molecules, and a limited metabolome approach.

Easy absolute quantification of metabolites from ^{1}H NMR spectra is possible following elimination of water in samples, choosing unsaturating repetition delay and using the ERETIC (Electronic Reference To access *In vivo* Concentrations) method for the quantification of metabolites (Akoka *et al.*, 1999). Recently this method was tested and used on polar extracts for offspring phenotyping for quantitative trait loci detection in strawberry, for transformant comparison in tomato and *Arabidopsis thaliana* (Moing *et al.*, 2004) and to quantify the metabolite changes due to biotic stress (André *et al.*, 2005). With such a method 20 to 30 metabolites can be quantified individually in a polar extract of 50 mg DW *A. thaliana* tissues, using a 30 min NMR acquisition time (M. Maucourt, pers. comm.). The detection limit was about 1 to 9 μg in the 5 mm NMR tube, depending on the compound. Before the development of quantification methods using an electronic reference, internal references were also used (Barantin and Akoka, 1997; Maniara *et al.*, 1998).

The number of metabolites quantified is limited by NMR sensitivity and signal overlapping. As indicated by Krishnan *et al.*, (2005), the number of metabolites quantified can be increased with higher field strength (increasing spectral resolution), and using microprobes for small quantity samples and cryogenic probe heads (increasing sensitivity). Another option is to combine proton NMR analysis with an online chromatographic separation, thus requiring extra time but eliminating the overlap problem.

In conclusion, metabolite fingerprinting and profiling using ^{1}H-NMR spectroscopy for data acquisition, combined with multivariate statistical methods for data analysis are promising methods in the emergent post-genomic technologies. One of the main advantages of ^{1}H-NMR is that structural and quantitative information can be obtained on numerous chemical species with a wide range of concentrations in a single NMR experiment, with excellent reproducibility. The speed of sample preparation, and spectra acquisition and processing under automation, makes ^{1}H-NMR suitable for applications where large numbers of samples need to be analyzed, in studies of environmental effects or functional genomics.

REFERENCES

Akoka, S., Barantin, L., Trierweiler, M. 1999. Concentration measurement by proton NMR using the ERETIC method. *Anal. Chem.*, 71, 2554-2557.

André, A., Maucourt, M., Moing, A., Rolin, D., Renaudin, J. 2005. Sugar import and phytopathogenicity of *Spiroplasma citri*: glucose and fructose play distinct roles. *Mol. Plant-Microbe Interact.*, 18, 33-42.

Antti, H., Bollard, M.E., Ebbels, T., Keun, H., Lindon, J.C., Nicholson, J.K., Holmes, E. 2002. Batch statistical processing of H-1 NMR-derived urinary spectral data. *J. Chemom.*, 16, 461-468.

Bailey, N.J.C., Oven, M., Holmes, E., Nicholson, J.K., Zenk, M.H. 2003. Metabolomic analysis of the consequences of cadmium exposure in *Silene cucubalus* cell cultures via ^{1}H-NMR spectroscopy and chemometrics. *Phytochemistry*, 62, 851-858.

Bales, J.R., Higham, D.P., Howe, I., Nicholson, J.K., Sadler, P. 1984. Use of high-resolution proton nuclear magnetic resonance spectroscopy for rapid multi-component analysis of urine. *Clin. Chem.*, 30, 426-432.

Barantin, L., Akoka, S. 1997. An overview of absolute quantification methods for *in vivo* MRS. *J. Magn. Reson. Anal.*, 3, 17-23.

Belton, P.S., Colquhoun, I.J., Kemsley, E.K., Delgadillo, I., Roma, P., Dennis, M.J., Sharman, M., Holmes, E., Nicholson, J.K., Spraul, M. 1998. Application of chemometrics to the

^{1}H-NMR spectra of apple juices: discrimination between apple varieties. *Food Chem.*, 61, 207-213.

Bligny, R., Douce, R. 2001. NMR and plant metabolism. *Curr. Opin. Plant Biol.*, 4, 191-196.

Brindle, J.T., Antti, H., Holmes, E., Tranter, G., Nicholson, J.K., Bethell, H.W.L., Clarke, S., Schofield, P.M., Mckilligin, E., Mosedale, D.E., Grainger, D.J. 2002. Rapid and noninvasive diagnosis of the presence and severity of coronary heart disease using ^{1}H-NMR-based metabonomics. *Nat. Med.*, 8, 1439-1445.

Charlton, A., Allnutt, T., Holmes, E., Chisholm, J., Bean, S., Ellis, N., Mullineaux, P., Oehlschlager, S. 2004. NMR profiling of transgenic peas. *Plant Biotechnol. J.*, 2, 27-35.

Choi, Y.H., Casas Tapias, E., Kim, H.K., Lefeber, A.W.M., Erkelens, C., Verhoeven, J.T.J., Brzin, J., Zel J., Verpoorte, R. 2004a. Metabolic discrimination of *Catharanthus roseus* leaves infected by phytoplasma using ^{1}H-NMR spectroscopy and multivariate analysis. *Plant Physiol.*, 135, 1-13.

Choi, Y.H., Kim, H.K., Hazekamp, A., Erkelens, C., Lefeber, A.W., Verpoorte, R. 2004b. Metabolomic differentiation of *Cannabis sativa* cultivars using ^{1}H-NMR and principal component analysis. *J. Nat. Prod.*, 67, 953-957.

Choi, H.K., Choi, Y.H., Verberne, M., Lefeber, A.W.M., Erkelens, C., Verpoorte, R. 2004c. Metabolic fingerprinting of wild type and transgenic tobacco plants by ^{1}H-NMR and multivariate analysis technique. *Phytochemistry*, 65, 857-864.

Daub, C.O., Kloska, S., Selbig, J. 2003. MetaGeneAlyse: analysis of integrated transcriptional and metabolite data. *Bioinformatics*, 19, 2332-2333.

Dawson, R.M.C., Elliott, D.C., Elliott, W.H., Jones, K.M. 1986. *Data for biomedical research*, Clarendon press, Oxford, UK, p. 580.

De Graaf, R.A., Behar, K.L. 2003. Quantitative ^{1}H- NMR spectroscopy of blood plasma metabolites. *Anal. Chem.*, 75, 2100-2104.

Defernez, M., Colquhoun, I.J. 2003. Factors affecting the robustness of metabolite fingerprinting using H-1 NMR spectra. *Phytochemistry*, 62, 1009-1017.

Defernez, M., Gunning, Y.M., Parr, A.J., Shepherd, L.W.T., Davies, H.V., Colquhoun, I.J. 2004. NMR and HPLC-UV profiling of potatoes with genetic modifications to metabolic pathways. *J. Agric. Food Chem.*, 52, 6075-6085.

Fan, T.W.M. 1996. Metabolite profiling by one- and two-dimensional NMR analysis of complex mixtures. *Prog. Nucl. Magn. Reson. Spectrosc.*, 28, 161-219.

Fiehn, O. 2002. Metabolomics - the link between genotypes and phenotypes. *Plant Mol. Biol.*, 48, 155-171.

Gil, A.M., Duarte, I.F., Delgadillo, I., Colquhoun, I.J., Casuscelli, F., Humpfer, E., Spraul, M. 2000. Study of the compositional changes of mango during ripening by use of nuclear magnetic resonance spectroscopy. *J. Agric. Food Chem.*, 48, 1524-1536.

Keun, H.C., Beckonert, O., Griffin, J.L., Richter, C., Moskau, D., Lindon, J.C., Nicholson, J.K. 2002. Cryogenic probe C-13 NMR spectroscopy of urine for metabonomic studies. *Anal. Chem.*, 74, 4588-4593.

Krapp, A., Morot-Gaudry, J.F., Boutet, S., Bergot, G., Lelarge, C., Prioul, J.L., Noctor, G. 2006. Metabolomics. In: Plant Genomics. Morot-Gaudry J.F., Briat J.F., eds, INRA Editions, Paris. pp 311-344.

Krishnan, P., Kruger, N.J., Ratcliffe, R.G. 2005. Metabolite fingerprinting and profiling in plants using NMR. *J. Exp. Bot.*, 56, 255-265.

Le Gall, G., Colquhoun, I.J., Davis, A.L., Collins, G.J., Verhoeyen, M.E. 2003. Metabolite profiling of tomato (*Lycopersicon esculentum*) using ^{1}H- NMR spectroscopy as a tool to detect potential unintended effects following a genetic modification. *J. Agric. Food Chem.*, 51, 2447-2456.

Le Gall, G., Colquhoun, I.J., Defernez, M. 2004. Metabolite profiling using ^{1}H-NMR spectroscopy for quality assessment of green tea, *Camellia sinensis* (L.). *J. Agric. Food Chem.*, 52, 692-700.

Le Gall, G., Metzdorff, S.B., Pedersen, J., Bennett, P.N., Colquhoun, I.J. 2005. Metabolite profiling of *Arabidopsis thaliana* (L.) plants transformed with an antisense chalcone synthase gene. *Metabolomics*, 1, 181-198.

Le Gall, G., Puaud, M., Colquhoun, I.J. 2001. Discrimination between orange juice and pulp wash by ^{1}H-nuclear magnetic resonance spectroscopy: Identification of marker compounds. *J. Agric. Food Chem.*, 49, 580-588.

Lindon, J.C., Holmes, E., Nicholson, J.K. 2001. Pattern recognition methods and applications in biomedical magnetic resonance. *Prog. Nucl. Magn. Reson. Spectrosc.*, 39, 1-40.

Lindon, J.C., Holmes, E., Nicholson, J.K. 2003. So whats the deal with metabonomics? Metabonomics measures the fingerprint of biochemical perturbations caused by disease, drugs, and toxins. *Anal. Chem.*, 75, 384A-391A.

Macomber, R.S. 1998. *A complete introduction to modern NMR spectroscopy*, Wiley, New York, 382 pp.

Maniara, G., Rajamoorthi, K., Rajan, S., Stockton, G.W. 1998. Method performance and validation for quantitative analysis by H-1 and P-31 NMR spectroscopy. Applications to analytical standards and agricultural chemicals. *Anal. Chem.*, 70, 4921-4928.

Mesnard, F., Ratcliffe, R.G. 2005. NMR analysis of plant nitrogen metabolism. *Photosyn. Res.*, 83, 163-180.

Moing, A., Maucourt, M., Renaud, C., Gaudillère, M., Brouquisse, R., Lebouteiller, B., Gousset-Dupont, A., Vidal, J., Granot, D., Denoyes-Rothan, B., Lerceteau-Kohler, E., Rolin, D. 2004. Quantitative metabolic profiling through one-dimensional ^{1}H-NMR analyses: application to plant genetics and functional genomics. *Funct. Plant Biol.*, 31, 889-902.

Moore, G.J., Sillerud, L.O. 1994. The pH dependence of chemical shift and spin-spin coupling for citrate. *J. Magn. Reson. B*, 130, 87-88.

Nielsen, J., Oliver, S. 2005. The next wave in metabolome analysis. *Trends Biotechnol.*, 23, 544-546.

Noteborn, H.P., Lommen, A., Van Der Jagt, R.C., Weseman, J.M. 2000. Chemical fingerprinting for the evaluation of unintended secondary metabolic changes in transgenic food crops. *J. Biotechnol.*, 77, 103-114.

Ott, K.H., Aranibar, N., Singh, B.J., Stockton, G.W. 2003. Metabonomics classifies pathways affected by bioactive compounds. Artificial neural network classification of NMR spectra of plant extracts. *Phytochemistry*, 62, 971-985.

Overy, S.A., Walker, H.J., Malone, S., Howard, T.P., Baxter, C.J., Sweetlove, L.J., Hill, S.A., Quick, W.P. 2005. Application of metabolite profiling to the identification of traits in a population of tomato introgression lines. *J. Exp. Bot.*, 56, 287-296.

Pauli, G.F., Jaki, B.U., Lankin, D.C. 2005. Quantitative ^{1}H-NMR: development and potential of a method for natural products analysis. *J. Nat. Prod.*, 68, 133-149.

Ratcliffe, R.G., Shachar-Hill, Y. 2005. Revealing metabolic phenotypes in plants: inputs from NMR analysis. *Biol. Rev.*, 80, 27-43.

Roessner, U., Willmitzer, L., Fernie, A.R. 2001. High-resolution metabolic phenotyping of genetically and environmentally diverse potato tuber systems. Identification of phenocopies. *Plant Physiol.*, 127, 749-764.

Silvestre, V., Goupry, S., Trierweiler, M., Robins, R., Akoka, S. 2001. Determination of substrate and product concentrations in lactic acid bacterial fermentations by proton NMR using the ERETIC method. *Anal. Chem.*, 73, 1862-1868.

Sobolev, A.P., Brosio, E., Gianferri, R., Segre, A. 2005. Metabolic profile of lettuce leaves by high-field NMR spectra. *Magn. Reson. Chem.*, 43, 625-638.

Sobolev, A.P., Segre, A., Lamanna, R. 2003. Proton high-field NMR study of tomato juice. *Magn. Reson. Chem.,* 41, 237-245.

Spraul, M., Freund, A.S., Nast, R.E., Withers, R.S., Maas, W.E., Corcoran, O. 2003. Advancing NMR sensitivity for LC-NMR-MS using a cryoflow probe: Application to the analysis of acetaminophen metabolites in urine. *Anal. Chem.,* 75, 1536-1541.

Terskikh, V.V., Feurtado, J.A., Borchardt, S., Giblin, M., Abrams, S.R., Kermode, A.R. 2005. *In vivo* ^{13}C NMR metabolite profiling: potential for understanding and assessing conifer seed quality. *J. Exp. Bot.,* 56, 2253-2265.

Trethewey, R.N. 2004. Metabolite profiling as an aid to metabolic engineering in plants. *Curr. Opin. Plant Biol.,* 7, 196-201.

Urbanczyk-Wochniak, E., Fernie, A.R. 2004. Metabolic profiling reveals altered nitrogen nutrient regimes have diverse effects on the metabolism of hydroponically-grown tomato (*Solanum lycopersicum*) plants. *J. Exp. Bot.,* 56, 309-321.

Vogels, J.T.W.E., Tas, A.C., Venekamp, J., Van Der Greef, J. 1996. Partial linear fit: a new NMR spectroscopy pre-processing tool for pattern recognition applications. *J. Chemom.,* 10, 425-438.

Ward, J.L., Harris, C., Lewis, J., Beale, M.H. 2003. Assessment of H-1 NMR spectroscopy and multivariate analysis as a technique for metabolite fingerprinting of *Arabidopsis thaliana. Phytochemistry,* 62, 949-957.

Wevers, R.A., Engelke, U., Heerschap, A. 1994. High-resolution ^{1}H-NMR spectroscopy of blood plasma for metabolic studies. *Clin. Chem.,* 40, 1245-1250.

Wevers, R.A., Engelke, U., Wendel, U., De Jong, J.G.N., Gabreels, F.J.M., Heerschap, A. 1995. Standardized method for high-resolution ^{1}H-NMR of cerebrospinal fluid. *Clin. Chem.,* 41, 744-751.

17

CHAPTER

Methods for the Qualitative and Quantitative Analysis of Metabolic Fluxes

M. Dieuaide-Noubhani, A.P. Alonso, C. Roby, P. Raymond

THE DATA ANALYSIS

The chief objectives of research in plant biology at present are to improve or diversify plant production. At the cellular level, improvement involves a change in the production or accumulation of metabolites, proteins, or RNA, whether it is a matter of desired products, regulators of their production, or agents of resistance to environmental factors. Improvements at the plant level are obtained through the choice of genotypes and the establishment of environmental conditions favourable to the expression of desired characters. In all cases, understanding the mechanisms that determine the change in production or accumulation of a given compound is based on studies of metabolism. The increased yield of a compound either results from increasing the rate of biosynthesis or decreasing the rate of degradation, or favouring storage; either the general activity is increased, or certain activities are reoriented. The unexpected effects of the modified expression of enzymes of sugar phosphate metabolism on sucrose or starch accumulation (Trethewey *et al.*, 1999), or of the tricarboxylic acid cycle on photosynthesis and, more generally, on plant development, (Carrari *et al.*, 2004; Nunes-Nesi *et al.*, 2005), show that there remains a great deal to be understood regarding the regulation of these already intensively studied metabolic pathways. Difficulties arise from the fact that metabolism and general cellular function form a closely integrated network. Consequently, the local modification of a metabolite concentration can affect the activity of one or several enzymes (Loef *et al.*, 2001), as well as the expression of a number of genes (Koch, 2004). These two mechanisms can influence the developmental programme of the cell, the organ, or the plant itself, well beyond the desired modifications.

This chapter presents the methods currently used for the measurements of fluxes in metabolic pathways and illustrates the fact that they cannot be developed independently of other methods of biochemistry or cellular biology. Their use has lead to a perception of the complexity of metabolic networks with partitioning, cycling, turnover. Integrated with metabolome, proteome, and transcriptome data, they provide a solid basis for the identification of targets for genetic or biotechnological improvements.

NET FLUX OF TERMINAL PRODUCT ACCUMULATION VERSUS UNIDIRECTIONAL FLUXES

The first step in the quantification of metabolic fluxes is to take overall measurements such as the accumulation of a final product like the increase in dry matter or any other growth indicators, or the accumulation of starch, soluble sugars, proteins etc, or the disappearance of a substrate like the consumption of oxygen or glucose in the culture medium, etc. These data (variation of quantity = ΔQ) are converted into net rates ($\Delta Q/\Delta t$), or net fluxes, that can be used to characterize the metabolism provided these rates are effectively constant in the time interval considered (Δt), that is, provided the system is in a metabolic steady state. Metabolic steady state is also characterized by a constant level of intermediary metabolites (Fig. 17.1). When the metabolic network is known, these data can be used to quantify the net fluxes through certain metabolic pathways (Marx *et al.*, 1996; Gokarn *et al.*, 2000).

It must be noted that net fluxes are the balance between synthesis and degradation: they may be very different from the unidirectional fluxes through the pathways. It is now well established that compounds such as proteins, sucrose, or starch, long considered to be terminal products of their biosynthetic pathways, are turning over actively (Dancer *et al.*, 1990; Hill and ap Rees, 1994; Dieuaide-Noubhani *et al.*, 1995). Taking net rate as the equivalent of rate of biosynthesis would lead to underestimating the activity of the biosynthetic pathway and to neglecting a potentially significant factor of net flux regulation, the degradation pathway. As an example, it was found that the accumulation of lycopene, the red pigment of mature tomatoes, is associated with a decrease in the expression of the lycopene degrading activity (Ronen *et al.*, 1999).

Finally, this overall approach considers a metabolic pathway as a homogeneous whole, not taking into account the reversible segments. For example, because of the reversibility of glycolysis between hexose-P and triose-P, the flux of hexose-P towards the triose-P is much higher than the glycolytic flux measured at the enolase step, which converts 2-phosphoglycerate to phosphoenolpyruvate. The structure of metabolic pathways in terms of fluxes can be elucidated by isotopic labelling, followed by the analysis of label redistribution by either radioactivity, nuclear magnetic resonance (NMR), or mass spectrometry (MS) methods, or through spin labelling by NMR.

ISOTOPIC LABELLING

Choice of a Tracer

Isotopic labelling is carried out by using radioactive isotopes (^{14}C, ^{35}S, ...) or stable isotopes (^{13}C, ^{15}N). Normally, the first accessible data is total labelling, as the quantity of

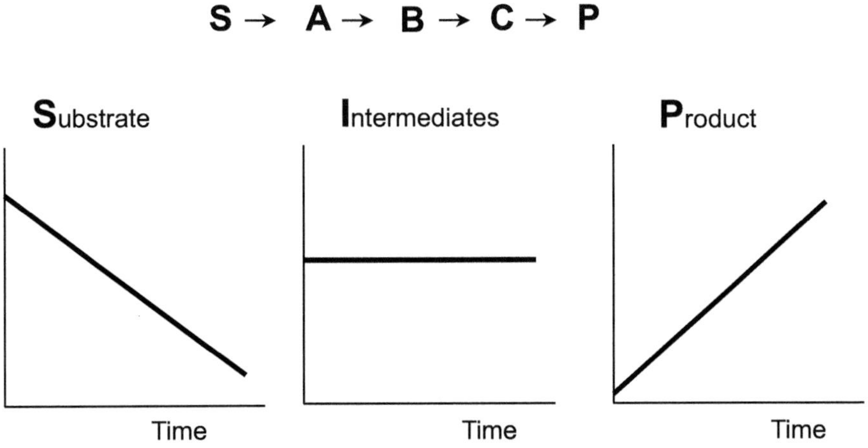

Fig. 17.1 Definition of metabolic steady state. The substrates are consumed and the products accumulate at constant rates, while the concentrations of intermediates do not change. According to Fell (1997), regulations help maintain a steady sate, while control allows changes from a steady state to another for which the fluxes and concentrations of certain intermediates, are different.

tracer incorporated, either in the whole organism, organ, etc., or in one of the compounds studied. Specific labelling can then be determined, and is expressed either as specific radioactivity in the case of radioactive markers, or as enrichments in the case of stable tracers (see box). When stable isotopes are analysed through NMR, the researcher also has access to the population of isotopomers, that tell whether labelled atoms are part of the same molecule (see below, *Inputs from isotopomer analysis*).

> The specific labelling of a pool is expressed as specific radioactivity or as enrichment. Specific radioactivity is the quantity of radioactivity per mole of compound, or per particular atom of a compound; it is expressed in Becquerel per mole, or per atom, of a compound. Enrichment is expressed, for an atom of a molecule, as the percentage of labelled atom (^{13}C, in the case of carbon) in relation to the total (labelled + unlabelled).

The choice of the tracer is in part linked to the methods of analysis available and in part to the nature of the metabolic pathway studied and to properties of the metabolic intermediates to be analysed (abundance, stability, physicochemical properties). Radioactive isotopes can be detected with a high sensitivity, compared to stable isotopes. They can thus be used in very low amounts, which offers the advantage of limiting the disturbance of the biological system. An advantage of stable isotopes is that their analysis by NMR gives direct access to the enrichment of individual atoms of the molecule. This is much more difficult to obtain with radioactive labelling, since this requires the controlled degradation of the purified molecule. The complete degradation of citrate has been described (Kent, 1972). For glucose, the radioactivity incorporated in the carbon-1 (C1) is obtained quite easily after enzymatic decarboxylation by the G6PDH-6PGDH couple: the labelling of C1 can then be easily compared to that of the rest of the molecule (Dieuaide-Noubhani *et al.*, 1995). The glucose molecule can also be degraded into its triose moieties after incubation with enzymes of glycolysis: this method was used to evaluate the randomization of label between the carbons C1 and C6 (Hatzfeld and Stitt, 1990). Similarly, the glutamate C1 can be analyzed after reaction with glutamate decarboxylase, and glutamate C5 following a delicate chemical reaction (Salon *et al.*, 1988). Access to the other carbons of glucose or glutamate is difficult.

Labelling with stable isotopes requires analysis by NMR or mass spectrometry. NMR analysis of ^{1}H and ^{13}C gives the enrichment of each of the carbon atoms of a molecule (Fig. 17.2). It may be useful to purify the compound to be analysed to avoid ambiguity that could be caused by contaminating peaks, but the difficult steps of

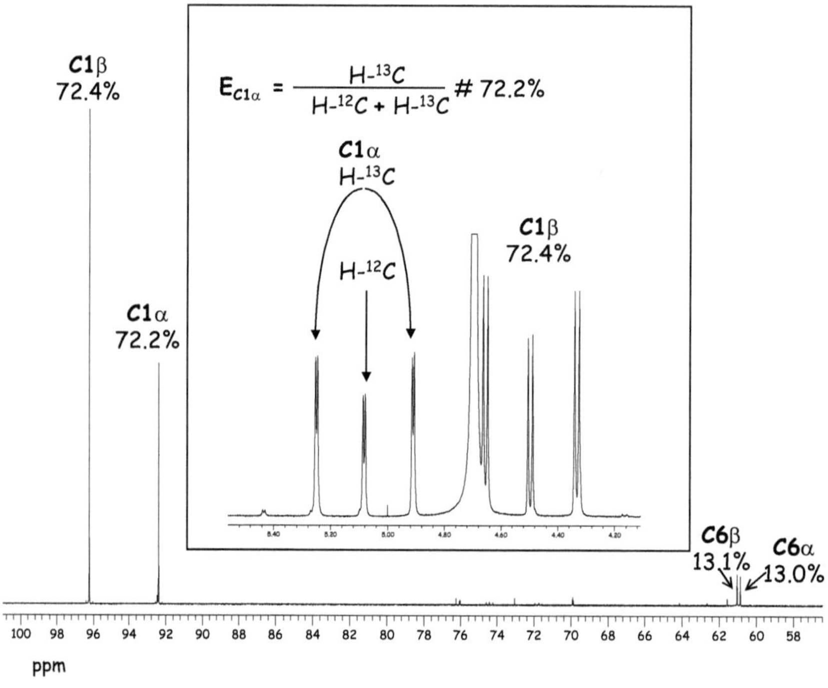

Fig. 17.2 Spectrum of glucose by ^{13}C and ^{1}H NMR. Maize root tips were incubated for 25 h with [1-^{13} C] glucose. The intracellular glucose was purified and analysed by ^{13}C NMR and ^{1}H NMR (inset). In the ^{13}C NMR spectrum, the α and β anomers of glucose appear as two clearly separated peaks for each of the carbon 1 (C1) and 6 (C6). In the ^{1}H NMR spectrum, the protons of each anomer produce a triplet of doublets, the truncated central peak being due to residual water. For each triplet, the protons linked to ^{12}C form the central doublet, while the protons linked to ^{13}C form the external doublets. The ratio of surfaces in the C1 triplet gives the absolute enrichment of C1, as indicated. In the ^{13}C spectrum, the ratio of the peak surfaces of each of the other carbons to that of C1, gives the relative enrichment of these carbons. The enrichment of carbons C2 to C5 of glucose are close to 1%, the natural level of ^{13}C in natural carbon, while that of C6 is close to 13.0%. The labelling of C6 from C1 is produced in the metabolism of hexose-phosphates; the appearance of label in free intracellular glucose is explained by the hydrolysis of oligo- and polysaccharides and of hexose-P, as discussed in the text.

selective degradation are avoided. Given its low sensitivity, NMR analysis involves the use of large quantities of biological material (about 1 g FW) and even then, only the most abundant metabolites can be studied. The appearance of cryoprobes and high field magnets has allowed improvement in the quality of spectra obtained: with proton NMR, the compound detection limit is 25 nanomoles in an analysed volume of 0.5 ml (Moing *et al.*, 2004). However, in the case of ^{13}C labelling, peak splitting due to $^{1}H/^{13}C$ coupling appears and a quantitative analysis requires as much as 0.5 µmole of compound. Mass spectrometry is an alternative: the sensitivity is high, but the quantification of metabolites and data interpretation are more difficult than with NMR (Kruger *et al.*, 2003; Roessner-Tunali *et al.*, 2004).

Labelling Kinetics

Figure 17.3 shows that, in a linear metabolic pathway, the metabolic intermediates are marked one after another. For atoms or groups of atoms derived from one another in the metabolic pathway, both total labelling per pool and specific labelling increase during the pre-stationary phase, as a function of the order of compounds in the metabolic pathway. After a sufficiently long period, the labelling of intermediates in the metabolic pathway no longer varies. This is called isotopic steady state. In quantitative terms, for groups of atoms derived directly from a precursor, the specific labelling is identical to that of the precursor and the amount of label incorporated is proportionate to the molar quantities of compounds (Fig. 17.3): these data therefore provide no information on fluxes in the metabolic pathway. However, the specific labelling is different from that of the precursor in the case of cycling or after the junction with another pathways, two situations that are illustrated below.

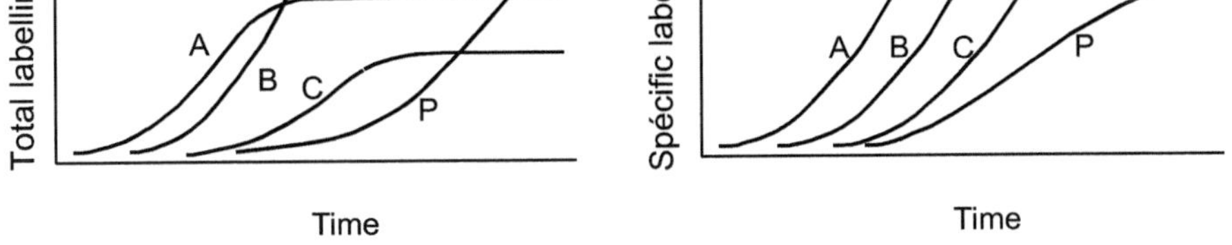

Fig. 17.3 Kinetics of labelling of a series of intermediates on a linear pathway

Isotopic pre-steady state and isotopic steady state in a linear metabolic pathway, where A, B, and C are simply derived from one another. The amount of labelled isotope in each of the intermediates increases in the course of pre-steady state. When isotopic steady state is reached, both the amount of label in each pool of intermediates and their specific labelling become constant. Enrichment is equal to that of the precursor. In the terminal product, the amount of label increases linearly whereas the enrichment continues to increase, as long as the amount of unlabelled product initially present remains significant.

The time needed to reach the isotopic steady state from labelled glucose is highly variable, from a few minutes in bacteria to several hours and sometimes several days in plants (Rontein *et al.*, 2002). This time decreases as the renewal rate, or "turnover", of compounds of the metabolic pathway increases.

Turnover of metabolic intermediates is caused by their simultaneous synthesis and degradation. Isotopic steady state is approached more quickly when the synthesis and degradation rates are high in relation to compound pool sizes. In a metabolic pathway where intermediate pool sizes are highly variable, it is the largest pools that slow down the progression of the marker in the pathway. In plants, the large pools of sugars, obviously contribute to the long labelling durations needed to reach the isotopic steady state. Vacuolar sugars in slow exchange with the cytosolic pools, as well as some cell wall polysaccharides may also intervene. A useful parameter for evaluating turnover is the turnover time, calculated for a given compound as the ratio of the quantity of compound to its unidirectional rate of synthesis: this parameter corresponds to the time required for the formation of a quantity of compound equal to that present in the tissue. It gives an idea of labelling durations compatible with each of the types of experiments described below. A formal definition of turnover with a critical discussion of the use of "turnover time" can be found in early articles, for example that of Reiner (1953).

The principles of the quantitative interpretation of labelling data are given in an article focusing on the use of mathematical models of metabolic networks (Salon and Raymond, 1988). Those methods are presented in this chapter through recent data on plant metabolism, with an emphasis on the type of information they can provide, as well as the limitations and constraints they impose.

Short-term Labelling

Short-term labelling can be used qualitatively in order to determine the sequence of intermediates of a metabolic pathway. As illustrated in Fig. 17.3, after a short time of labelling, the most labelled compound, A, is the one closest to the precursor. At the same time, the specific radioactivity of intermediates decreasing from B to C, it can be deduced that B precedes C. The name of photosynthetic types C_3 or C_4 has been attributed following short labelling experiments, according to the nature of compounds labelled most strongly after a few seconds following the introduction of radioactive CO_2. An example of this method and additional references are provided by Morot-Gaudry and Farineau (1978).

Short-term labelling experiments of around a few minutes for plant organs, can also be used to quantify metabolic fluxes. In a first step, the rate of label incorporation (dA*/dt) is measured in the compounds of interest (A). It must be noted that this rate of label incorporation is not the rate of reaction, or flux, dA/dt. It depends on both the rate of reaction and the specific labelling of the precursor. The quantification of flux (dA/dt) from radioactivity incorporation requires the specific labelling of the precursor (see box); to measure the flux to P, the precursor would be C.

> Calculation of the accumulation rate of a reaction product by short-term labelling requires knowledge of the amount of label accumulated and the specific labelling (either specific radioactivity or enrichment, E), of its precursor.
>
> $$dA/dt = dA^*/dt \times 1/E_{(S^*)}$$

Compartmentalization is a source of error in the interpretation of short-term labelling experiments. If the tracer used is distributed in different compartments, the apparent sequence may include intermediates of different pathways; in contrast, if a compound is present in several compartments, and only one is labelled, this will lead to underestimating its enrichment and thus to misplacing it in the sequence. This is the case for compounds such as sugars, organic acids, or phosphate, stored in abundance in the vacuole and for which the cytosolic or plastidial pools are more strongly labelled than the average cellular pool. The complexity of metabolic reactions with branches and cycles is another source of error in the case of labelling experiments with non-uniformly labelled precursors: the label may be incorporated at several positions in the molecules studied, which makes it risky to compare the enrichments of whole molecules. This occurs after feeding glucose labelled in the carbon 1 (C1), that leads to the formation of glutamic acid labelled in four of its five carbons (Dieuaide-Noubhani *et al.*, 1995). The interpretation of enrichments is reliable only for chosen atoms for which the metabolic origin is known, at least in the form of a hypothesis. The use of uniformly labelled precursors avoids this problem, but less information is obtained.

Finally, to accurately estimate the fluxes towards the chosen intermediate, label should not be lost, or cycled, significantly during the labelling period: this is obtained only with relatively short labelling durations which lead to linear labelling kinetics. In the case of short-term labelling, a cycle can be treated as a linear pathway. This method has been applied (Macnicol and Raymond, 1998) to the comparison of fluxes of the terminal reactions of glycolysis, those of pyruvate kinase (PK) and PEP carboxylase (PEPC), which use the same pool of PEP to form pyruvate and oxaloacetate, respectively. These two products are in rapid exchange, with alanine or aspartate and malate, respectively, and they also can enter the tricarboxylic acid cycle where they are converted to citrate and glutamate. In a first step, it was carefully verified that little radioactivity was transferred to the latter products. The PK and PEPC fluxes were then compared from the transfer of label to alanine and malate + aspartate. This comparison showed that, during maturation of the barley seed, the secretion of malic acid from the integument, which strongly acidifies the endosperm, is associated with a five-fold increase in the PEPC/PK flux ratio. Absolute values of these two fluxes were not calculated because the specific radioactivity of PEP could not be determined; in addition, due to the dual cytosolic and plastidial location of PEP, the measurement would have provided an average value not necessarily representative of the cytosolic pool.

The short-term labelling method was also used to measure the rates of sucrose synthesis and degradation in potato tubers, by labelling with radioactive glucose over a period of 4 h (Trethewey *et al.*, 1999). The sucrose synthesis rate was deduced from the rate of radioactivity incorporation into the sucrose pool and from the specific radioactivity (SRA) of glucose 6-P. The reliability of this result depends on two hypotheses: the SRA of glucose 6-P is identical to that of UDP-glucose, the actual precursor of sucrose, and the labelling time is short in relation to the turnover time of the sucrose pool. This apparently was verified in the materials used, but it may be noted that in numerous biological materials, including plants, a period of 4 h may be long enough for a significant quantity of radioactivity to be lost in the substrate cycling reactions, which could lead to underestimating the flux of synthesis.

Establishment of the Steady State

This approach is based on following the change in enrichments of one or more compounds over time, until the establishment of a steady state. The interpretation of results is based on modelling of metabolic pathways by a large number of differential equations. In fact, an equation is required to take into account the changes in the concentration of each of the intermediates involved, and in the labelling of each of its labelled atoms, in the different cell compartments. To use these equations, the quantity and specific labelling of these intermediates must be measured at different times. A major limit of this method for the description of complex networks, is that the number of unknowns often exceeds the measuring capacity. One way to simplify the model is to carry out isotopic labelling once the metabolic steady state has been reached. Despite this simplification, 200 equations were still necessary to describe the distribution of label in the tricarboxylic acid cycle (Chance *et al.*, 1983). Because of their complexity, these models are applicable to simple systems only. Addressing pool sizes and enrichments of compartmented metabolites is one of the major limitations of this approach.

Metabolic and Isotopic Steady State

At metabolic and isotopic steady state, the isotopic equations are simplified because the specific labelling and quantity of metabolic intermediates remain constant. This is the method of modelling fluxes most often used at present to study complex networks in bacteria (Marx *et al.*, 1996), animals (Katz and Grunnet, 1979; Malloy *et al.*, 1996) or plants (Salon *et al.*, 1988; Hill and ap Rees, 1994; Dieuaide-Noubhani *et al.*, 1995, 1997; Rontein *et al.*, 2002). The advantages of steady state labelling were long underestimated because earlier theoretical models, based on labelling of photosynthetic cells by $^{14}CO_2$, indicated (Morot-Gaudry *et al.*, 1993) that stationary labelling was difficult to achieve. In addition it was considered that the amount of label in a metabolite indicates nothing more than the molar quantity of metabolite, which has no relation with flux measurement. This point of view is valid for the case cited but cannot be generalized for two reasons:

(1) Isotopic steady state is relatively easy to achieve since it does not concern the entire organism but only the metabolic pathway studied. It is even useful, for the purposes of this method, that related compounds remain almost unlabelled, which may happen if they have a slower turnover. This makes it possible to measure their inflow into the pathway by comparison of enrichments before and after the junction. This situation of labelling has made it possible to evaluate the respective contribution of fatty acids and sugars in the supply of the Krebs cycle in germinating oil-rich seeds (Salon *et al.*, 1988) and in tissues subjected to sugar starvation (Dieuaide-Noubhani *et al.*, 1997), two situations in which feeding labelled glucose lead to labelling of cellular sugars whereas lipids remained unlabelled.

(2) Valuable information on intermediary metabolism reactions can be obtained by labelling to isotopic steady state provided that non-uniformly labelled molecules are used as precursors, as shown in the studies of Katz and colleagues since the 1960s (Katz and Grunnet, 1979; and examples below).

Until the 1980s, markers were essentially radioactive isotopes, but most studies now use stable isotopes, like ^{13}C, and NMR is used for the determination of enrichments (Dieuaide *et al.*, 1995; Portais *et al.*, 1996; Rontein *et al.*, 2002) as well as for the detection of isotopomers (Malloy *et al.*, 1988; de Graaf *et al.*, 2001; Lu *et al.*, 2002).

PRINCIPLES OF THE METHOD

The first step is to establish the conditions (duration, but also substrate concentration, aeration, temperature...) for labelling to metabolic and isotopic steady state. The second is to define the metabolic network from a close examination of the labelling of intermediates. The third step is to write the mathematical model that relates fluxes to enrichments or specific radioactivities, and to fit the fluxes with the measured enrichments.

The metabolic and isotopic steady state is characterized in two ways: when it is reached, the enrichments of intermediates become constant, and, as a consequence, the rate of label accumulation in the terminal products of the pathways becomes constant.

The metabolic scheme is established so as to take into account published data on metabolic pathways and the qualitative and quantitative results of labelling. The quantitative data provided by this first version of the model may eventually lead to modifications of the metabolic scheme and thus of the model itself. With respect to intermediary metabolism, the schemes produced by Dieuaide-Noubhani *et al.* (1995) and Rontein *et al.* (2002) seem to reliably represent the central metabolism of non-photosynthetic plant cells.

Two types of steady state equations allow the relationship between enrichments (which can be measured) and fluxes to be established:

The metabolic steady state equations state that for each pool, the total rate of synthesis is equal to the rate of degradation or: $\Sigma Vs = \Sigma Vd$

The isotopic steady state equation state that for each pool, the rate of label input is equal to the rate of label output, or: $\Sigma\ (Vs.E_{Pa}) = \Sigma\ (Vd.E_{Xb})$

where:

—P is a precursor of X, and atom *a* of compound P becomes atom *b* of X;

—Vs and Vd are the rates of synthesis and degradation of X;

—E_{Pa} and E_{Xb} are specific labelling (= Enrichments) of atoms *a* of P, and *b* of X.

Equations of this type must be written at each metabolic junction, for each of the atoms of the compounds represented. For example, in the case of the network presented in Fig. 15.4, the following equations are written:

Isotopic Steady State Equations

for intracellular glucose C1:

$$Ve.E_{Gel} + Vsuc.E_{suc1} = Vhk.E_{Gi1}$$

for cytosolic hexose 6-P C1:

$$Vhk.E_{Gi1} + Vhpc.E_{Hp1} + Vth.E_{T3} = (Vald + Vhcp + Vsuc).E_{Hc1}$$

for plastidial hexose 6-P C1:

$$Vhcp.\ E_{Hc1} + 2/3\ Vppp.E_{Hp1} = (Vhpc + Vppp + Va).E_{Hp1}$$

for Triose 3-P:

$$\text{Vald.}(E_{Hc1} + E_{Hc6}) + 2/3\,\text{Vppp.}E_{Hp6} = (2.\text{Vgly} + 2.\text{Vth} + 1/3\,\text{Vppp}).E_{T3}$$

where Ge = external glucose, labelled on its carbon 1 (C1) only; Gi = intracellular glucose; Hc = cytosolic hexose phosphates; Hp = plastidial hexose phosphates; T = triose-phosphates, considered to form a single pool. The number that follows indicates the position of the carbon atom in the molecule (T3 is carbon 3 of the triose phosphates and comes from carbons 1 or 6 of hexose phosphates). E_{Hc1} represents the enrichment of carbon 1 of Hc = cytosolic hexose phosphate, etc. Fluxes are designated by letter V followed by an abbreviation of the name of the reaction or set of reactions, as indicated in the legend of Fig. 17.4.

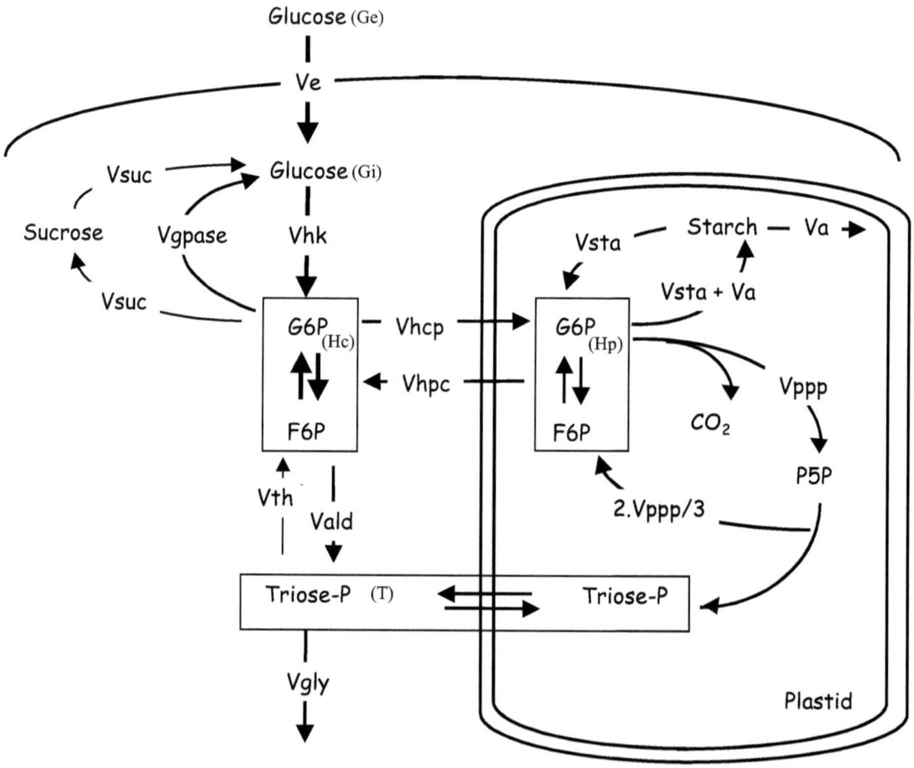

Fig. 17.4 Metabolic network, from glucose to triose-phosphates, in a plant cell supplied with glucose. The diagram illustrates the metabolic steps and their duplication (sometimes incomplete, depending on the species) in the cytosol and plastid. This diagram is the result and the basis of models of labelling at the steady state described in the text. Reaction rates are designated as follows: Ve, input of external glucose; Vhk, hexokinase; Vppp, oxidative pentose phosphate pathway (where 3 glucose 6-P are converted to 2 fructose 6-P (2/3 Vppp), 3 CO_2, and 1 triose-P); Vgly, glycolysis at the enolase (2-phosphoglycerate to phospho*enol*pyruvate conversion) step; Vsuc, sucrose synthesis by sucrose-P synthase and sucrose degradation by an invertase; Vgpase, glucose phosphatase reactions (Dieuaide-Noubhani *et al.*, 1995; Rontein *et al.*, 2002; Alonso *et al.*, 2005).

Metabolic Steady State Equations

$$\begin{aligned} Vhk &= Ve + Vsuc \\ Vhk + Vhpc + Vth &= Vald + Vhcp + Vsuc \\ Vhcp &= (Vhpc + 1/3\,Vppp + Va) \\ Vald + 1/6.Vppp &= Vgly + Vth \end{aligned}$$

These two sets of equations express the fact that the specific labelling of a molecule or carbon atom formed from two (or more) precursors depends on the relative value of the different fluxes of synthesis and of the specific labelling of the precursors. Therefore, it is not absolute rates but relative rates that are determined by the isotopic steady state method. All absolute rates in a network can be determined from relative rates if a reference flux can be determined: the rate of exogenous glucose consumption was used (Dieuaide-Noubhani *et al.*, 1995; Rontein *et al.*, 2002). Other independent measurements such as the rate of oxygen consumption or the rate CO_2 production can also be used to validate the results.

ADVANTAGES, CONSTRAINTS AND APPLICATIONS

One advantage of the metabolic and isotopic steady state approach in relation to the pre-steady state approach is that it simplifies the mathematical model and the experimentation, since the equations are linear and independent of time. Moreover, since all the molecules located between two metabolic junctions present identical labelling, it suffices to analyse the molecules situated at metabolic junctions. If a compound is too scarce for the spectra to be useful, it can be analysed through a more abundant molecule where it is faithfully represented. In plants, alanine is analysed for pyruvate, aspartate or malate are analysed for oxaloacetate, and glutamate for α-ketoglutarate. Similarly, it is possible to study the glucosyl of sucrose in place of cytosolic glucose 6-P, as sucrose glucosyl is produced from glucose 6-P only via glucose 1-P and UDP-glucose. In the case of glucose 6-P, which is present in the cytosol and the plastid, the labelling of each of these two pools was deduced from the labelling of glucosyl units of sucrose and starch respectively; the large differences in the labelling of sucrose and starch in maize root tips indicated that the two glucose 6-P pools have different fates, with the oxidative pentose phosphate pathway localized essentially in the plastids (Dieuaide-Noubhani *et al.*, 1995). This was confirmed using glucose labelled on carbon 2: the decarboxylation of [2-^{14}C]6-phosphogluconate to [1-^{14}C]ribulose-5-P results in the transfer of label from C2 towards the C1 of the fructose-6-P formed in the non-oxidative part of the pathway; the hexose-P molecules are then converted to glucosyl units of either sucrose, in the cytosol, or starch, in the plastids. In maize root tips, the C1-labelled glucosyls were found essentially in starch; model fitting was obtained with a nil flux of the pentose-phosphate pathway in the cytosol (Dieuaide-Noubhani *et al.*, 1995). In contrast, in tomato cells, no difference was observed in the labelling of sucrose and starch, which suggested that hexose-P exchange between the cytosol and the plastid was very active; alternatively, the partition between glycolysis and the pentose-P pathway may be similar in the two compartments. Using the steady state labelling approach it was possible to obtain the information needed to calculate about 20 fluxes of central metabolism, including hexose-phosphates-sugars interconversions, glycolysis, the pentose-P pathway and the tricarboxylic acid cycle (Dieuaide-Noubhani *et al.*, 1995; Rontein *et al.*, 2002).

This approach has some limitations. First, the long time required to establish isotopic steady state induces two difficulties; one is that transients, or short-term changes, like those induced by a transfer to anoxia for example, cannot be observed. Conversely, the tissues may undergo undetected modifications during incubation, so that the results of the analysis will not be representative of the initial physiological state. In the case of maize root tip labelling with [^{13}C]glucose, the time required to reach isotopic steady state was reduced by a sugar-starvation pretreatment of 4 h, which reduced the size of the intracellular pools of sugars. However this treatment itself may have led to disturbances in metabolism (Dieuaide-Noubhani *et al.*, 1995). A second limitation of the steady state labelling method is that it is not always possible to distinguish between different metabolic pathways that lead to the same label distribution. For example, when labelling with [1-^{13}C]glucose, the interconversions between hexose-phosphates and triose-phosphates lead to a transfer of label to the C-6 of hexose-phosphates. It has been observed that intracellular glucose is also labelled on C6. This indicates that glucose-P is converted to free glucose, and the steady state labelling data allow calculation of the flux. However, the data give no indication of which is, or which are, the pathway(s) involved. The hydrolysis of sucrose by an invertase activity (sucrose futile cycle) was considered to be the most likely; other well documented possibilities are the turnover of starch and cell wall polysaccharides. These three possibilities were evaluated by short-term labelling experiments (Alonso *et al.*, 2005): the flux through these three pathways was found to be much lower than that measured by steady state labelling, thus suggesting the hypothesis that a glucose-phosphatase plays this role. The nature of the pathway, or pathways involved, significantly affects the real ATP cost of this conversion, although it appears to remain high in all cases (Hill and ap Rees, 1994; Dieuaide-Noubhani *et al.*, 1995; Alonso *et al.*, 2005).

INPUTS FROM ^{13}C ISOTOPOMER ANALYSIS

The study of populations of ^{13}C isotopomers, after labelling with a ^{13}C labelled precursor also provides information on the metabolic networks (Szyperski, 1995; de Graaf, 1996). The isotopomers of a compound carry labelled carbons on different positions of the molecule. They can be distinguished by ^{13}C NMR, because the spectrum of a given ^{13}C atom is modified by the presence of another ^{13}C in its vicinity: the simple peaks are transformed into characteristic multiplets (Malloy *et al.*, 1988).

Analysis of glutamate isotopomers by [^{13}C]NMR was extensively used to study the carbon supply of the Krebs cycle after labelling with enriched substrates in animal organs (Malloy *et al.*, 1988), and in yeast (Tran Dihn *et al.*, 1998). This method has never been used on a plant, but in principle this could be done. For example, Fig. 17.5 shows how the glutamate C3 singlet changes as a function of the PEPC flux in the case of labelling with [1-13-C]glucose. A complex series of isotopomer studies was carried out by Szyperski (1995) in *E. coli*: after labelling with glucose enriched with 10% ^{13}C on carbon 6, the isotopomers of amino acids derived from the hydrolysate of a protein were analysed by 2D NMR (2D proton spectra [^{13}C,^{1}H]-COSY). Based on the known biosynthetic pathways of different amino acids, it was possible to quantify the fluxes through glycolysis, the pentose phosphate pathway and the tricarboxylic acid cycle. An

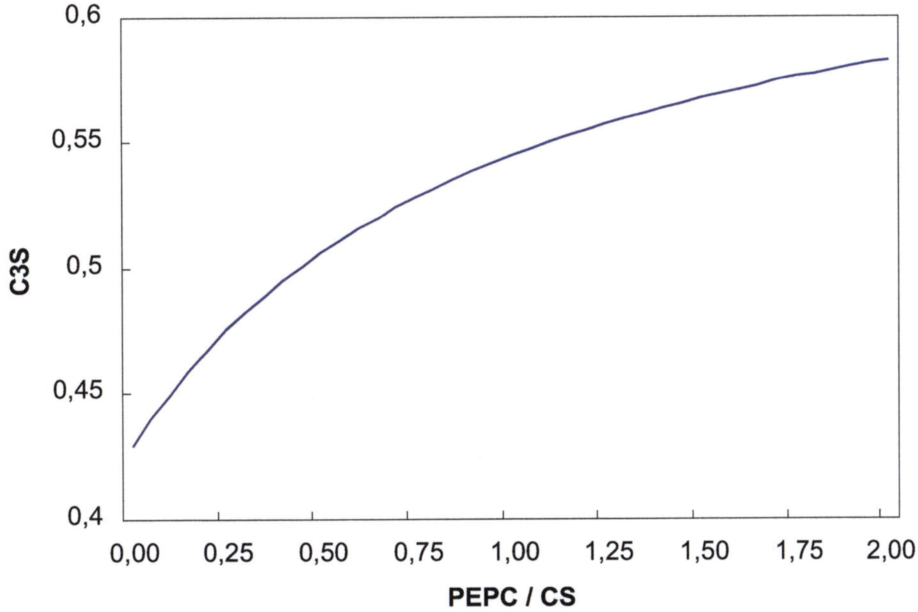

Fig. 17.5 Variation of the relative size of the singlet of glutamate carbon 3 (C3S) as a function of x= PEPC/CS, the ratio of fluxes through the reactions of phospho*enol*pyruvate carboxylase and citrate synthase, respectively. This curve was traced according to the equations defined by Malloy *et al.* (1988), with C3S = $Fc0\{(Fc0+x) / (x+1)\}$, using results obtained from maize root tips after labelling with [1-^{13}C]glucose (Dieuaide-Noubhani *et al.*, 1995). Fc*0* corresponds to the non-enriched fraction of acetyl-CoA. After incubation of maize root tips with [1-^{13}C]glucose, the C2 of acetyl-CoA is either enriched ($Fc2 = 0.34$) or not enriched ($Fc0 = 1\text{-}Fc2 = 0.66$). *Fc2* is measured as the enrichment of glutamate C4 which derives directly from acetyl-CoA C2. As the PEPC flux increases, the enrichment of glutamate C3 decreases: therefore, the relative size of the singlet, which reflects the abundance of ^{12}C, increases, as shown in the graph. The function of PEPC in plants is anaplerotic: it replenishes the carbon intermediates taken from the tricarboxylic acid cycle for various biosynthetic reactions, the most important quantitatively being the glutamate and aspartate amino acids families.

interesting property of this method, is that the study of all the amino acids provides some redundancy, because several amino acids entirely or partly derive from the same biosynthetic pathway. Recently, this method, called the retrobiosynthetic method, was applied to the pathways of isoprenoid biosynthesis (Eisenreich *et al.*, 1999) and to central carbohydrate metabolism in plants (Ettenhuber *et al.*, 2005). The very long time needed to label proteins to the steady state seems to be a major curb on the use of this approach in plants.

SPIN LABELLING

In a living cell, the distribution of metabolic fluxes is a molecular phenotype which characterizes the function of the integrated system. The *in vivo* determination of metabolic fluxes is a promising means of studying the function of subtle biochemical structures such as metabolic cycles. These coordinated enzymatic processes reflect the complex systemic organization of life. NMR techniques developed to study chemical exchange phenomena provide a unique way to approach this subject. At steady state, unidirectional enzymatic rate constants and reaction rates of a biochemical system can be measured directly *in situ* using dynamic NMR techniques (Alger and Shulman, 1984; Brindle, 1988; Hills and Belton, 1989; Rudin and Sauter, 1992). Transmembrane transport can be investigated in this way as well as chemical reactions. These NMR measurements rely on the fact that the transfer of a nuclear spin is linked to the jump of an atom between two distinct chemical sites related one to the other through the enzymatic process. These sites are part of the substrate and product molecules of an enzymatic reaction; they relate to the species located outside and inside the cell in the case of transport. At the macroscopic scale, the spin transfer leads to a magnetization transfer. These measurements are direct in that biochemical information can usually be obtained using a simplified model of the metabolic network restricted to the enzymes of interest. Here, the chief goal is to characterize a working enzyme in its natural environment. In particular, the measurement of unidirectional reaction rates makes it possible to know how far from equilibrium the enzymatic reaction is : $\Gamma/K_{eq} = v_r/v_f$ [1]. The metabolic flux through the observed enzyme is equal to the difference between the unidirectional rates. In terms of metabolic control, the number of points of measurement is very small with regards to the numerous control coefficients defined in metabolic control theories (Kacser, 1987). These NMR methods are thus restricted to non-branched elements of a metabolic network in which all the enzymes catalyse the same net fluxes.

Principles of the Method

The NMR observables (chemical shift, scalar coupling, relaxation constants) integrate the effect of atomic and molecular movements. These movements can be described precisely from relaxation studies because atomic and molecular movements generate fluctuations of local magnetic fields. These fluctuations modulate the time-dependent magnetic energy of the nuclear spins and induce a loss of coherence in the magnetic properties of the spin population which are observed. The analysis of magnetic relaxation measurements requires knowledge of the relaxation mechanisms and generally makes use of complex dynamic models (Delpuech, 1995). Two major classes of movement can be studied directly using specific techniques; these are translational diffusion and chemical exchange, the latter being defined as the periodic modification of the chemical environment of an atom. The exchange between two sites is the most commonly studied, but it is theoretically possible to study spin exchange between any number of sites (Gesmar and Led, 1986; Rudin and Sauter, 1992). The associated NMR techniques aim at measuring the spectral effects arising from the modulation of magnetic interactions by chemical exchange. Specific manipulations of the spin system under scrutiny with radiofrequency pulses, are used to achieve this goal.

In metabolic studies, the most commonly explored situation is slow exchange, where the exchange rate constant of an atom, characteristic of the biological system, is lower than the difference of its nuclear resonance frequencies intrinsic to each chemical site. In this case, the NMR spectrum of the observed spins (^{13}C, ^{31}P, ^{1}H...) displays the

[1]Γ is the mass action ratio *in vivo*, vr and vf are respectively "reverse" and "forward" reaction rates.

characteristic resonance of each site. This is, for example, the situation observed with a ^{31}P NMR spectrum of living plant cells (Baguet, 1994; Roscher, 1997). The phosphorous resonances arising from cytosolic Pi and γATP are clearly distinguished, even though the ^{31}P atoms, embedded in the phosphorylated group, pass permanently from one chemical structure to another. In the case of slow exchange, it is possible to determine the exchange rate constant of an atom between two sites if that constant is of the same order of magnitude as the magnetic relaxation rate constants of the spins involved (Rudin and Sauter, 1992). To give an order of magnitude, these kind of NMR studies are relevant when the lifetimes of the exchanging species in a living system are in the 0.1 to 1 s range. The techniques used to study slow chemical exchange combine spin labelling and longitudinal relaxation. They are based on the principle of conservation of the spin state during the exchange (Alger and Shulman, 1984). They use the fact that the magnetic effects induced by chemical exchange and relaxation can be observed if the spin system is out of magnetic thermodynamic equilibrium. Following the selective perturbation of one of the exchanging magnetizations, the magnetization components associated with the two chemical sites are measured (Fig. 17.6). Quantification is much simpler if the metabolite concentrations are constant during the NMR experiment, i.e. when the living system is in a biochemical steady state. Two major classes of experiments are distinguished, whether or not a selective saturation of the magnetizations in exchange is used as a labelling technique (Alger and Shulman, 1984). Saturation transfer, the most commonly used class of magnetization transfer experiments, is used in multiple forms. The other class of experiments is well represented by inversion transfer and two-dimensional chemical exchange spectroscopy, the latter being described in the next section.

Applications in Plant Biology

The two-steps saturation transfer technique at equilibrium was used to study ATP turnover in excised maize root tips (Roberts *et al.*, 1984) and excised pea leaves (Baguet, 1994). With this kind of experiment, one can measure directly the P/O ratio *in vivo*. Two-dimensional chemical exchange spectroscopy, detecting simultaneously all the chemical exchanges of a given isotope in a system (Jeener *et al.*, 1979), appears naturally adapted to the study of metabolic networks. This technique was optimized to study central metabolism in heterotrophic plant cells (Roscher, 1997). It was used to characterize simultaneously the main steps of sugar respiration in excised maize root tips perfused with a controlled nutrient medium (Roscher *et al.*, 1998). Apart from ATP hydrolysis and synthesis rates, reaction rates of four enzymes of sugar metabolism and glycolysis were measured. These were UDP-glucose pyrophosphorylase (UDPG-PPase), phosphoglucomutase, hexosephosphate isomerase, and enolase. The correlation peaks of the 2D chemical exchange spectrum (Fig. 17.7) arise from the magnetizations transferred during the delay characterizing the NMR experiment, called mixing time (the diagonal peaks are due to the magnetizations not transferred during this time). The intensity of a correlation peak is a function of the average value of the reaction rate catalyzing this particular exchange during the time interval required to obtain a spectrum. The net metabolic flux through an enzyme is equal to the difference between the unidirectional rates at the level of the observed enzyme. With a partial oxygen

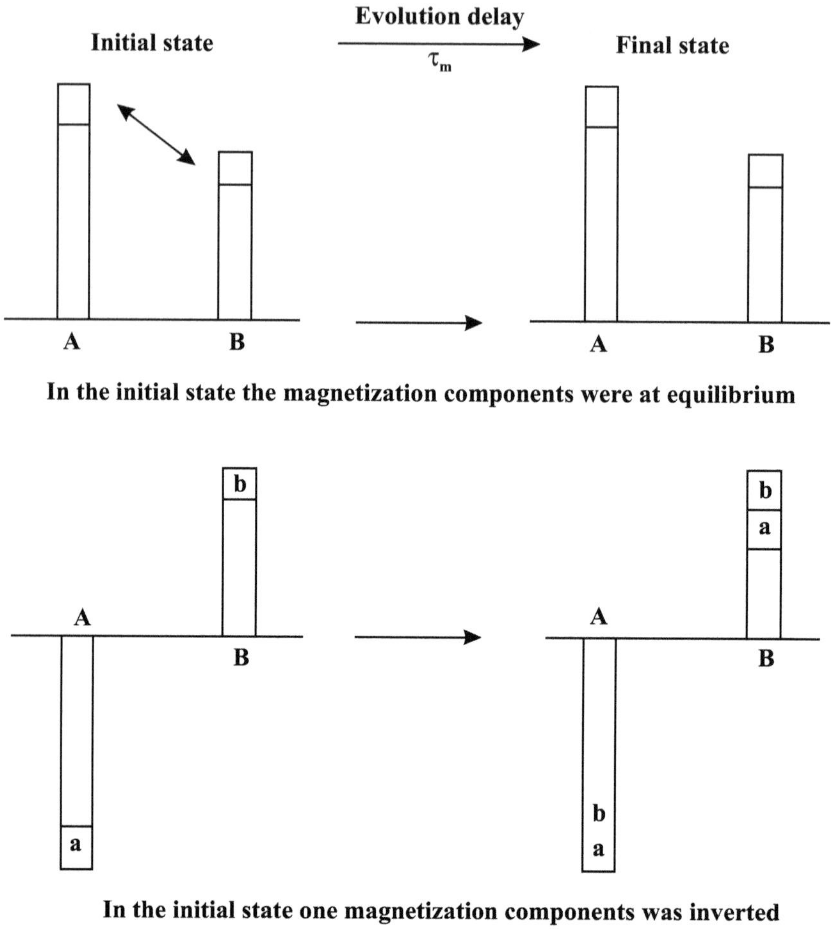

Fig. 17.6 Principle of spin labelling techniques.
A and B represent the components of the detected nuclear spin magnetization (for example ^{31}P) associated with the two sites A and B occupied by the spin (cytosolic Pi and γATP, for example). The delay τ_m separates an initial state and a final state of the spin system. The initial state is defined by a preparation of the spin system and the final state is observed with the NMR spectrum. When the two magnetization components are at thermodynamic equilibrium in the initial state, the magnetization transfer induced by the chemical exchange between the sites A and B during the delay τ_m is not observed in the final state (a standard NMR spectrum is recorded). Conversely, the magnetization transfer between A and B can be measured if one of the components is labelled (here inverted) before the delay τ_m. A increases because there is loss of *a* (negative) and gain of *b* (positive), while B diminishes because there is loss of *b* (positive) and gain of *a* (negative). The shaded areas represent the resonance intensities which would be (initial state) or are (final state) measured The longitudinal relaxation of magnetization, which tends to restore the intensities at equilibrium during τ_m, opposes this effect. In the 2D chemical exchange technique, the preparation phase aims at labelling the components A and B in many different ways by introducing a variable delay before the fixed delay τ_m. The NMR signal is recorded after τ_m during a third interval. A double Fourier transform on the two variable intervals leads to the spectrum illustrated in Fig. 17.7 (Roscher A., 1997).

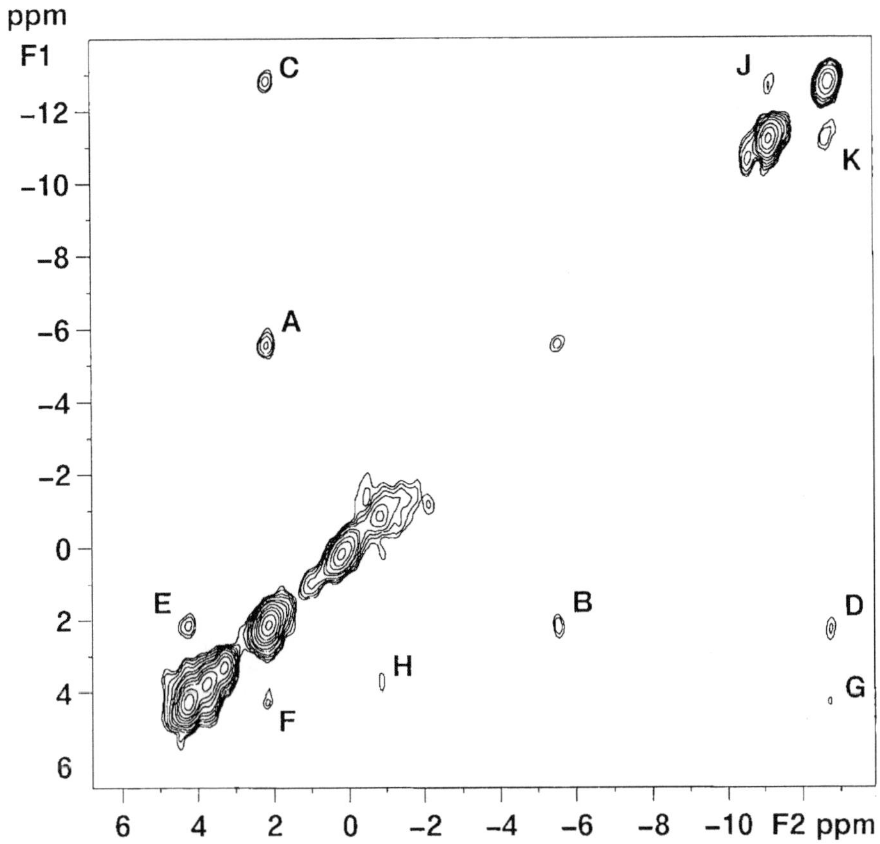

Fig. 17.7 ^{31}P 2D NMR chemical exchange spectrum of maize root tips.

Tho physiological conditions of the living tissue were continuously controlled during the NMR experiment using a perifusion device which circulates the nutrient medium through the sample. The spectrum was obtained at 161 MHz for 29 h and 16 min using a mixing time τ_m = 1 s. This spectrum is the sum of 20 elementary 2D spectra recorded for 36 min and made up of 64 pseudo-FIDs formed by adding 16 transients. The detected correlation peaks correspond to the following chemical exchange processes:

A and B: γNTP $\rightarrow$ cyt-Pi and reverse;

C and D: UDPG $\rightarrow$ G1P and reverse;

E and F: G1P $\rightarrow$ G6P and reverse;

G: G6P $\rightarrow$ G1P $\rightarrow$ UDPG;

H: 3-PGA $\rightarrow$ 2-PGA $\rightarrow$ PEP.

(J and K are attributed to dipolar coupling between the α and β ^{31}P spins of UDPG). The intensities are proportional to the unidirectionnal enzymatic reaction rates but must not be directly compared on this figure. The experimental intensities are indeed modulated by relaxation constants and the acquisition conditions of the spectrum. The absolute values of the reaction rates were obtained after due correction using the sum of all the intensities of the spectrum (including the diagonal) which is proportional to the total amount of phosphorus in the sample (Roscher *et al.,* 1998; © *J. Biol. Chem.*)

pressure of 5% in the external environment, the measurement of these fluxes at 11, 16, and 21°C lead to the following observations (Roscher *et al.*, 1998): the glycolytic flux increased as a function of the temperature, while the ATP turnover diminished because of increasing hypoxia. The UDPG-PPase and phosphoglucomutase enzymes were working close to equilibrium (Γ/K_{eq} values were respectively 2/3 and 1/2). At 21°C, the net metabolic flux through these enzymes was clearly directed towards UDPG, indicating that UDPG-PPase was a continuous source of pyrophosphate (PPi). In hypoxia, PPi is a source of energy for the proton pumps of the tonoplast and a substrate of the PPi-dependent phosphofructokinase, which is then acting as a glycolytic enzyme (Plaxton, 1996). Sucrose did not accumulate and cell wall biosynthesis could not account for the significant metabolic flux towards UDPG which was interpreted as a sucrose cycling flux. This illustrates how dynamic NMR experiments of this kind can be used to study limited but complex metabolic networks such as substrate cycles. These techniques are part of the NMR toolbox used to study the dynamic response of the intact plant cell to variations of environmental factors, at the molecular scale.

Discussion

The enzymatic properties of living systems can be studied at the molecular scale using nuclear magnetization transfer measurements: an enzyme activity can be characterized and a metabolic flux can be determined *in situ*. However, these measurements of reaction rates using spin labelling are affected by the sensitivity limits of NMR. Here, the sensitivity of the measurement is determined by the intensity of the apparent exchange rate, which is equal to the product of the rate constant by the number of atoms involved in the exchange. The second variable is severely restricted by the cellular composition itself and by the volume of the sample used for the measurements. The first variable must be of the order of magnitude of the relaxation rate constants which are barely under control in a living system. In addition, to increase the signal/noise ratio, the signals of a large number of elementary measurements (FID's) are added, which leads to a long duration of effective measurement (several hours). Therefore, the relevance of the study must be discussed in the context of the temporal hierarchy of the processes of metabolic control and regulation (Heinrich and Schuster, 1996; de la Fuente *et al.*, 2002). In theory, the metabolic fluxes are variables that characterize a given metabolic network. For a measurement of flux to be usable in terms of metabolic control, the whole biochemical system must be in a quasi-steady state during the entire duration of the measurement. Otherwise, the flux values measured represent a temporal average of the fluxes under consideration. The variations of the system parameters (quantity and activity of enzymes that define the network) over time depend upon the physiological conditions fixed for the experiment. The deviation from a reference state determines the biochemical response (even the adaptation) of the cell and it is advisable to avoid cases where the metabolic network is reorganized significantly between the beginning and end of an experiment. It is therefore necessary to measure enzyme amounts and certain enzymatic activities at the beginning and end of the experiment when the cells are placed in extreme physiological conditions (e.g., nutrient deficiency, anoxia...).

Two other facts must be addressed as potential limitations. First, *in vivo* NMR provides observable values which are averaged over a whole population of cells.

Second, the extensive compartmentation of plant cells induces an important control of metabolism. This compartmentation is not reflected directly in an NMR spectrum, except when the chemical shift is a function of the local environment (pH, ionic strength), but the relaxation constants are sensitive to the local environment of spins and that environment can be revealed by the combined use of several NMR techniques and modelling. Thus, cellular heterogeneity has been taken into account to determine metabolic fluxes in the isolated perfused heart to study the energy transfer processes mediated by creatine kinase (Joubert *et al.,* 2001). By analogy, it might be possible to address the enzymatic compartmentation of plant cells, that of glycolysis, and that involving sucrose metabolism in particular. The use of magnetization transfer in cell biology to study metabolic networks has not developed as expected since the early studies (Brindle, 1988), because NMR techniques are global and not highly sensitive. However, new NMR spectrometers and advances in modelling have considerably changed the situation and spin labelling techniques may contribute significantly to the development of enzymology *in situ*.

CONCLUSION

To best describe cell metabolism in terms of flux, it is important to combine several of the methods that we have described, such as measurement of end products, short-term or steady state labelling, and direct measurement of reaction rates. Labelling methods are based on well-established principles (Reiner, 1953; Katz and Grunnet, 1979) that are sometimes forgotten. Their results need to be interpreted in association with established knowledge of cell biology and the results of functional studies of genes encoding enzymes or transporters. During the past few years, the use NMR for the analysis of intramolecular distribution of markers both qualitatively (isotopomers) and quantitatively (enrichments) has decisively facilitated the identification of pathways and the measurement of fluxes in an increasing number of reactions in intermediary and secondary metabolism. The measurement of fluxes is a key to clarifying the relationships between modifications in the metabolome and transcriptome data that are provided by the analysis of natural or artificial mutant plants (Fernie *et al.,* 2005). Advances in the field of metabolism will then contribute to the functional analysis of genes.

REFERENCES

Alger, J.R., Shulman, R.G. 1984. NMR studies of enzymatic rates *in vitro* and *in vivo* by magnetization transfer. *Q. Rev. Biophys.,* 17, 83-124.

Alonso, A.P., Vigeolas, H., Raymond, P., Rolin, D., Dieuaide-Noubhani, M. 2005. A new substrate cycle in plants. Evidence for a high glucose-phosphate-to-glucose turnover from in vivo steady-state and pulse-labeling experiments with [C-13] glucose and [C-14] glucose. *Plant Physiol* 138, 2220-2232.

Baguet, E. 1994. Renouvellement de l'ATP dans la cellule végétale: méthodologie du transfert de saturation en RMN ^{31}P. University thesis, Grenoble I.

Brindle, K.M. 1988. NMR methods for measuring enzyme kinetics *in vivo. Prog. NMR Spectrosc.,* 20, 257-293.

Carrari, F., Nunes-Nesi, A., Gibon, Y., Lytovchenko, A., Loureiro, M.E., Fernie, A.R. 2003. Reduced expression of aconitase results in an enhanced rate of photosynthesis and marked shifts in carbon partitioning in illuminated leaves of wild species tomato. *Plant Physiol* 133, 1322-1335.

Chance, E.M., Seeholzer, S.H., Kobayashi, K., Williamson, J.R. 1983. Mathematical analysis of isotope labeling in the citric acid cycle with application to 1C NMR studies in pefused rat hearts. *J. Biol. Chem.*, 258, 13785-13794.

Dancer, J., Hatzfeld, W.D., Stitt, M. 1990. Cytosolic cycles regulate the turnover of sucrose in heterotrophic cell-suspension cultures of *Chenopodium rubrum* L. *Planta*, 182, 223-231.

de Graaf, A.A., Eggeling, L., Sahm, H. 2001. Metabolic engineering for L-lysine production by *Corynebacterium glutamicum. Adv. Biochem. Eng.-Bitechnol.*, 73, 9-29.

de la Fuente, A., Snoep, J.L., Westerhoff, H.V., Mendes, P. 2002. Metabolic control in integrated biochemical systems. *Eur. J. Biochem.*, 269, 4399.

Delpuech, J.J. 1995. Timescales in NMR: nuclear site exchange and dynamic NMR. In: *Dynamics of Solutions and Fluid Mixtures by NMR.* Delpuech, J.J. ed, Wiley, Canada, 73-172.

Dieuaide-Noubhani, M., Raffard, G., Canioni, P., Pradet, A., Raymond, P. 1995. Quantification of compartmented metabolic fluxes in maize root tips using isotope distribution from (^{13}C) or (^{14}C) labeled glucose. *J. Biol. Chem.*, 270, 13147-13159.

Dieuaide-Noubhani, M., Canioni, P., Raymond, P. 1997. Sugar-starvation-induced changes of carbon metabolism in excised maize root tips. *Plant Physiol.*, 115, 1505-1513.

Eisenreich, W., Rieder, C., Grammes, C., Hessler, G., Adam, K.P., Becker, H., Arigoni, D., Bacher, A. 1999. Biosynthesis of a Neo-epi-verrucosane diterpene in the liverwort Fossombronia alaskana – A retrobiosynthetic NMR study. *J. Biol. Chem.* 274, 36312-36320.

Ettenhuber, C., Spielbauer, G., Margl, L., Hannah, L., Gierl, A., Bacher, A., Genschel, U., Eisenreich, W. 2005. Changes in flux pattern of the central carbohydrate metabolism during kernel development in maize,. *Phytochemistry* 66, 2632-2642.

Fell, D. 1997. *Understanding the Control of Metabolism.* Portland Press Ltd., London.

Fernie, A.R., Geigenberger, P., Stitt, M. 2005. Flux an important, but neglected, component of functional genomics. *Curr. Opin. Plant. Biol.* 8, 174-182.

Gesmar, H., Led, J. J. 1986. Optimizing the multisite magnetization-transfer experiment. *J. Magn. Reson.*, 68, 95-101.

Gokarn, R.R., Evans, J.D., Walker, J.R., Martin, S.A., Eiteman, M.A., Altman, E. 2001. The physiological effects and metabolic alterations caused by the expression of *Rhizobium etli* pyruvate carboxylase in *Escherichia coli. Appl. Microbiol. Biotechnol.*, 56, 188-195.

Gokarn, R.R., Eiteman, M.A., Altman, E. 2000. Metabolic analysis of *Escherichia coli* in the presence and absence of the carboxylating enzymes phosphoenolpyruvate carboxylase and pyruvate carboxylase. *App. Environ. Microbiol.*, 66, 1844-1850.

Hatzfeld, W.D., Stitt, M. 1990. A study of the rate of recycling of trioses phosphates in heterotrophic *Chenopodium rubrum* cells, potato tubers, and maize endosperm. *Planta*, 180, 198-204.

Heinrich, R., Schuster, S. 1996. Time hierarchy in metabolism. In: *The Regulation of Cellular Systems*, Chapman & Hall, New York, 112-137.

Hill, S.A., Ap Rees, T. 1994. Fluxes of carbohydrate metabolism in ripening bananas. *Planta*, 192, 52-60.

Hills, B.P., Belton, P.S. 1989. NMR studies of membrane transport. *Annu. Rep. NMR Spectrosc.*, 21, 99-159.

Jeener, J., Meier, B.H., Bachmann, P., Ernst, R.R. 1979. Investigation of exchange processes by two-dimensional NMR spectroscopy. *J. Chem. Phys.*, 71, 4546-4553.

Joubert, F., Hoerter, J.A., Mazet, J.L. 2001. Discrimination of cardiac subcellular creatine kinase fluxes by NMR spectroscopy: a new method of analysis. *Biophys. J.*, 81, 2995-3004.

Kaczer, H., 1987. Control of metabolism. In: *The Biochemistry of Plants*, Vol 11: *Biochemistry of metabolism*. Davies, D.D. ed. Academic Press, New York, 39-67.

Katz, K., Grunnet, N. 1979. Estimation of metabolic pathways in steady state *in vitro*. Rates of tricarboxylic acid and pentose cycle. In: *Techniques in the Life Sciences*, Vol 208: *Techniques in Metabolic Research* Kornberg, H. ed. Elsevier Scientific Publishing Co., New York, 1-18.

Kent, S.S. 1972. Complete stereochemical distribution of 14C-isotope in citrate. *Anal. Biochem.*, 49, 393-406.

Koch, K. 2004. Sucrose metabolism: regulatory mechanisms and pivotal roles in sugar sensing and plant development. *Curr. Opin. Plant. Biol.* 7, 235-246.

Loef, I., Stitt, M., Geigenberger, P. 2001. Increased levels of adenine nucleotides modify the interaction between starch synthesis and respiration when adenine is supplied to discs from growing potato tubers. *Planta*, 212, 782-791.

Lu, D.H., Mulder, H., Zhao, P.Y., Burgess, S.C., Hensen, M.V., Kamzolova, S., Newgard, C.B., Sherry, A.D. 2002. C-13 NMR isotopomer analysis reveals a connection between pyruvate cycling and glucose-stimulated insulin secretion (GSIS). *Proc. Natl. Acad. Sci.* USA, 99, 2708-2713.

Macnicol, P.K., Raymond, P. 1998. Role of phosphoenolpyruvate carboxylase in malate production by the developing barley aleurone layer. *Physiol. Plant*, 103, 132-138.

Malloy, C.R., Sherry, A.D., Jeffrey, F.M.H. 1988. Evaluation of carbon flux and substrate selection through alternate pathways involving the citric acid cycle of the heart by 13C NMR spectroscopy. *J. Biol. Chem.*, 263, 6964-6971.

Marx, A., de Graaf, A.A., Wiechert, W., Egeling, L., Sahm, H. 1996. Determination of the fluxes in the central metabolism of *Corynebacterium glutamicum* by nuclear magnetic resonance spectroscopy combined with metabolite balancing. *Biotechnol. Bioeng.*, 49, 111-129.

Matern, D., Magera, M.J. 2001. Mass spectrometry methods for metabolic and health assessment. *J. Nutr.*, 131, 1615-1620.

Moing, A., Maucourt, M., Renaud, C., Gaudillere, M., Brouquisse, R., Lebouteiller, B., Gousset-Dupont, A., Vidal, J., Granot, D., Denoyes-Rothan, B., Lerceteau-Kohler, E., Rolin, D. 2004. Quantitative metabolic profiling by 1-dimensional ^{1}H-NMR analyses: application to plant genetics and functional genomics. *Functional Plant Biology* 31, 889-902

Morot-Gaudry, J.F., Farineau, J. 1978. Etude comparee de carboxylations photosynthetiques chez un mais normal (W64A) et chez un mais mutant opaque 2 (W64A o2): mise en evidence de deviations metaboliques chez le mutant. *Physiol. Veg.*, 3, 451-467.

Morot-Gaudry, J.F., Deleens, E., Prioul, J. 1993. Etude des flux de carbone et de l'azote chez le mais: tracage par les isotopes radioactifs ou les isotopes stables? In: *Utilisation des Isotopes Stables pour l'Etude du Functionnement des Plantes*. Les Colloques, no. 70, INRA, Paris.

Nunes-Nesi, A., Carrari, F., Lytovchenko, A., Smith, A.M.O., Loureiro, M.E., Ratcliffe, R.G., Sweetlove, L.J., Fernie, A.R. 2005. Enhanced photosynthetic performance and growth as a consequence of decreasing mitochondrial malate dehydrogenase activity in transgenic tomato plants. *Plant Physiol* 137, 611-622

Portais, J.C., Voisin, P., Merle, M., Canioni, P. 1996. Glucose and glutamine metabolism in C6 glioma cells studied by carbon 13 NMR. *Biochimie*, 78, 155-164.

Plaxton, W.C. 1996. The organization and regulation of plant glycolysis. *Annu. Rev. Plant Physiol. Plant Mol. Biol.*, 47, 185.

Reiner, J. 1953. The study of metabolic turnover rates by means of isotopic tracers. I. fundamental relations. *Arch. Biochem. Biophys.*, 46, 53-81.

Roberts, J.K.M., Wemmer, D., Jardetzky, O. 1984. Measurement of mitochondrial ATPase activity in maize root tips by saturation transfer 31P nuclear magnetic resonance. *Plant Physiol.*, 74, 632.

Roessner, U., Wagner, C., Kopka, J., Trethewey, R.N., Willmitzer, L. 2000. Simultaneous analysis of metabolites in potato tuber by gas chromatography-mass spectrometry. *Plant J.*, 23, 131-142.

Roessner, U., Luedemann, A., Brust, D., Fiehn, O., Linke, T., Willmitzer, L., Fernie, A.R. 2001. Metabolic profiling allows comprehensive phenotyping of genetically or environmentally modified plant systems. *Plant Cell*, 13, 11-29.

Roessner-Tunali, U., Liu, J, Leisse, A., Balbo, I., Perez-Melis, A., Willmitzer, L., Fernie, A.R. 2004. Kinetics of labelling of organic and amino acids in potato tubers by gas chromatography-mass spectrometry following incubation in 13C labelled isotopes. *Plant J* 39, 668-679.

Ronen, G., Cohen, M., Zamir, D., Hirschberg, J. 1999. Regulation of carotenoid biosynthesis during tomato fruit development: Expression of the gene for lycopene epsilon-cyclase is down-regulated during ripening and is elevated in the mutant Delta. *Plant J.* 17, 341-351

Rontein, D., Dieuaide-Noubhani, M., Dufourc, E. J., Raymond, P., Rolin, D. 2002. The metabolic architecture of plant cells. Stability of central metabolism and flexibility of anabolic pathway during the growth cycle of tomato cells. *J. Biol. Chem.*, 277, 43948-43960.

Roscher, A. 1997. Etude du métabolisme énergétique végétal par RMN d'échange chimique du ^{3}1P. Développement et application de la technique 2D EXSY. University thesis, Grenoble 1.

Roscher, A., Emsley, L., Raymond, P., Roby, C. 1998. Unidirectional steady state rates of central metabolism enzymes measured simultaneously in a living plant tissue. *J. Biol. Chem.*, 273, 25053-25061.

Rudin, M., Sauter, A. 1992. Measurement of reaction rates *in vivo* using magnetization transfer techniques. In: *NMR basic principles and progress*, Springer-Verlag, Berlin, 27, 257-293.

Salon, C., Raymond, P. 1988. Quantification of metabolic fluxes in cyclic pathways by labelling experiments and modelization. *Plant Physiol. Biochem.*, 26, 187-195.

Salon, C., Raymond, P., Pradet, A. 1988. Quantification of carbon fluxes through the tricarboxylic acid cycle in early germinating lettuce embryos. *J. Biol. Chem.*, 263, 12278-12287.

Szyperski, T. 1995. Biosynthetically directed fractional 13C-labelling of proteinogenic amino acids. An efficient analytical tool to investigate intermediary metabolism. *Eur. J. Biochem.*, 232, 433-448.

Tran-Dinh, S., Hoerter, J.A., Mateo, P., Gyppaz, F., Herve, M. 1998. C-13-NMR spectroscopic evaluation of the citric acid cycle flux in conditions of high aspartate transaminase activity in glucose-perfused rat hearts. *Biochimie*, 80, 1013-1024.

Trethewey, R.N., Riesmeier, J.W., Willmitzer, L., Stitt, M., Geigenberger, P. 1999. Tuber-specific expression of a yeast invertase and a bacterial glucokinase in potato leads to an activation of sucrose phosphate synthase and the creation of a sucrose futile cycle. *Planta*, 208, 227-238.

18

CHAPTER

The Structure of Metabolic Networks or (Almost) All that is Known about the Simplest Metabolic Network

Jean-Pierre Mazat

INTRODUCTION: FROM GENOME TO PROTEOME AND TO METABOLOME

In principle, a knowledge of the genome of an organism should lead to an understanding of its metabolism. However, things are not so simple. First, genes have to be expressed and when expressed it is important to know to what extent; it is a crucial point in the modelling of metabolism and the study of fluxes. Second, a knowledge of the structure of a gene does not provide information concerning the properties of the corresponding encoded enzyme (*Km*, regulation, etc.). A first answer to these questions is given by transcriptome analysis using microarrays. It should be noted however that it is not exactly the answer to our questions: RNA expression is not necessarily proportional to protein expression even if this is generally true. A better answer, technically more difficult (2D gels for instance), is provided by proteome analysis at a given time under specific conditions, combined with knowledge of the kinetic properties of the enzymes.

Thus, after the rapid success of genomics, transcriptomics and to a less extent proteomics, there are limitations in metabolomics and fluxomics, i.e. the analysis of metabolism. It is a methodological bottleneck, which needs to be solved for different biological material (Sauer, 2004 ; Gombert and Nielsen, 2000; Westerhoff and Palsson 2004). It should be however noted, that, if it is difficult to characterise the kinetic properties of individual enzymes in different organisms or tissues, properties which are not provided by knowledge of the genome..., we have a good detailed knowledge of metabolic networks, accumulated in the 20th century by the experiments of biochemists, which has been reinforced by information on the sequence of genomes. In

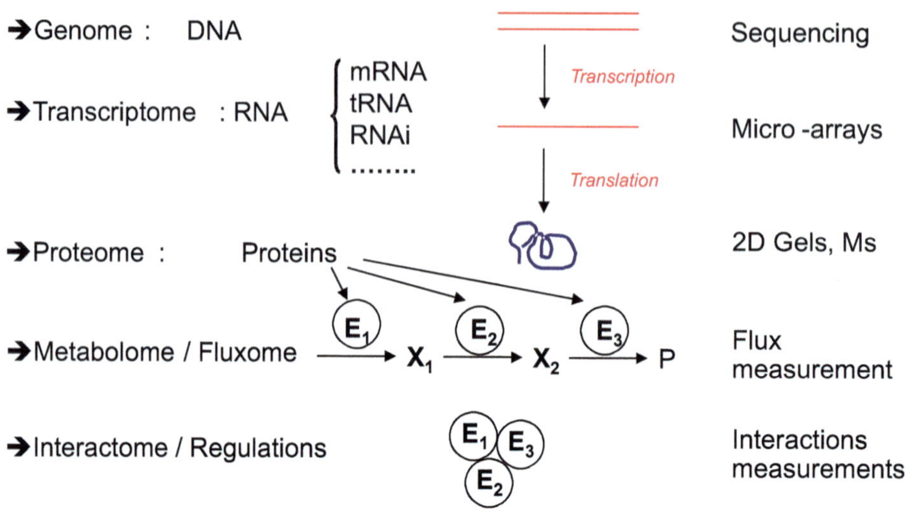

Fig. 18.1 From genome to proteome and to metabolome.

a first approach one can thus deal with the metabolic maps, i.e. with the *structure* of metabolic networks.

The aim of this chapter is to show that this study can be done independently of informations concerning the kinetic properties of the enzymes, which is of considerable interest. We will also provide and discuss some working hypotheses which are often used implicitly.

In order to illustrate our subject, we will use extensively the most simple of all metabolic networks represented in Fig. 2.

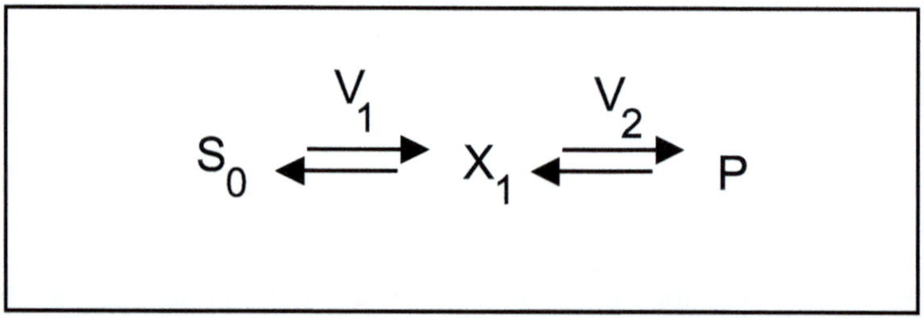

Fig. 18.2 The simplest metabolic network. S_0 and P are metabolites external to the network and could be omitted from the figure. X_1 is the only one intermediate metabolite of the metabolic network. V_1 and V_2 depict the reactions 1 and 2 supposed to be in general, reversible and catalysed by the enzymes E_1 and E_2.

WHAT IS METABOLIC NETWORK ?

In other words what is the difference between a network and the sum of its reactions?

The difference is shown in Fig. 18.3; it is also well illustrated in the common practice of enzymology laboratories. In enzyme studies, the substrate and product concentrations used are certainly very different from the cellular ones. Typically in the

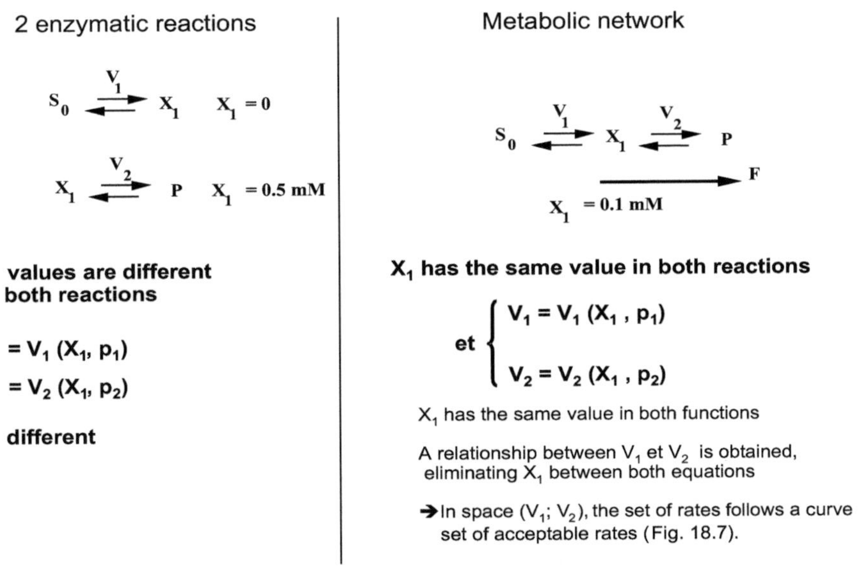

Fig. 18.3 What is metabolic network? In the left hand section, two independent reactions are shown, for which the concentration of X_1 is not necessarily the same. In the right hand section, a metabolic network is depicted, in which X_1, at the same concentration links both reactions. The elimination of X_1 provides a relationship between V_1 and V_2 which is the equation of the curve V followed, in the rate space, by the pairs (V_1 ; V_2) linked by X_1 (Fig. 18.7). (p_1 and p_2 are parameters, P is the product of reaction 2).

case of Fig. 18.3, the rate V_1 will be studied when the concentration of its product X_1 is zero. In contrast, the rate V_2 will be studied at variable (non zero) values of the X_1 concentration (X_1 is the substrate of the reaction). In cellular metabolism however, both reactions are linked by the metabolite X_1, which in the cell, has a single concentration (if we suppose that the cytosol is homogeneous; see the discussion below; see also Hofmeyr and Cornish-Bowden (1997) and Cornish-Bowden and Hofmeyr (2005) where there is a discussion on this point). As a matter of fact, the concentrations of the intermediate metabolites are determined by the structure of the network (i.e. the set of reactions which consume or produce a given intermediate metabolite) and the kinetics of the reactions (which depend upon the intermediate metabolite concentrations, substrates, products and possibly regulators). This point is depicted in the right hand part of Fig. 18.3. In a metabolic network, the intermediate metabolites are links between the various reactions of the network, and their concentrations, which are unique, are determined by the structure of the network and the kinetic properties of the various reactions.

First Principle: The Cellular Medium (Cytosol, Mitochondrial Matrix, etc.) is Homogeneous

This principle is indeed in general not verified. There are numerous examples of compartmentation.

It is however clear that most of the quantitative studies of metabolism, consider (without stating so, explicitly) that the homogeneity principle is satisfied. Even if this principle is probably seldom verified, it is useful to consider it true in a first quantitative approach of a metabolic network, for several reasons: the study of a heterogeneous metabolic network is much more difficult and the study of the same network in a homogeneous medium can give semi-quantitative indications on the actual metabolic network independently of the homogeneity of the medium. It means that a homogeneous medium is considered as the ideal case of a heterogeneous one. In addition, the actual nature of the heterogeneity of the cell medium is usually ignored and thus, it is difficult to take it into account. It must however be emphasised that the effects of a special type of heterogeneity, substrate channelling, has been studied in detail (Agius and Sherratt, 1997).

In all the following sections, it has been assumed that the principle of homogeneity has been verified.

Metabolic Networks as Open Systems

There are two ways to consider a metabolic network as shown in Fig. 18.4:

It can be viewed as a closed system which reaches an equilibrium with several zero concentrations, corresponding to (quasi)irreversible reactions. This situation does not fit exactly the cellular situation in which external metabolites are continuously taken up and products exported. On the contrary, it is more realistic to consider that external metabolites are found at constant concentrations (S_o in Fig. 18.4) or produced at a constant rate (P in Fig. 18.4). In these conditions, in general, the intermediate metabolite concentrations (inside the metabolic network) reach a steady-state, i.e. a constant non-zero value.

Second Principle: Steady-state

In order to calculate the fluxes in a metabolic network, one used to consider the intermediate metabolite concentrations as constant, i.e. their rate of consumption balances their production. This hypothesis has been readily verified in numerous cases. On time intervals of some minutes and sometimes more, one can actually consider that the concentrations of several metabolites are constant. If it were not the case, their assessment would be much more difficult. However, it is clear that this hypothesis does not hold true when the physiological conditions are changed: passing from rest to exercise for muscle; temperature variations or changes in light intensity for plants during the day. Glycolytic oscillations are also a well known example of metabolite concentrations not at steady-state.

In the following, we will accept the steady-state principle, knowing that it is a minimal hypothesis not always satisfied.

MATHEMATICAL REPRESENTATION OF A METABOLIC NETWORK: THE STOICHIOMETRY MATRIX

It is useful for a theoretical analysis of the structure of metabolic networks, to represent metabolic networks by a mathematical object which can be theoretically handled. This leads to the concept of stoichiometry matrix (Reder, 1988).

A metabolic network involves r reactions labeled by their enzymes $E_1, E_2, \ldots, E_r$ and their rates $V_1, V_2, \ldots, V_r$. These reactions in turn involve m metabolites $X_1, X_2, \ldots, X_m$ with

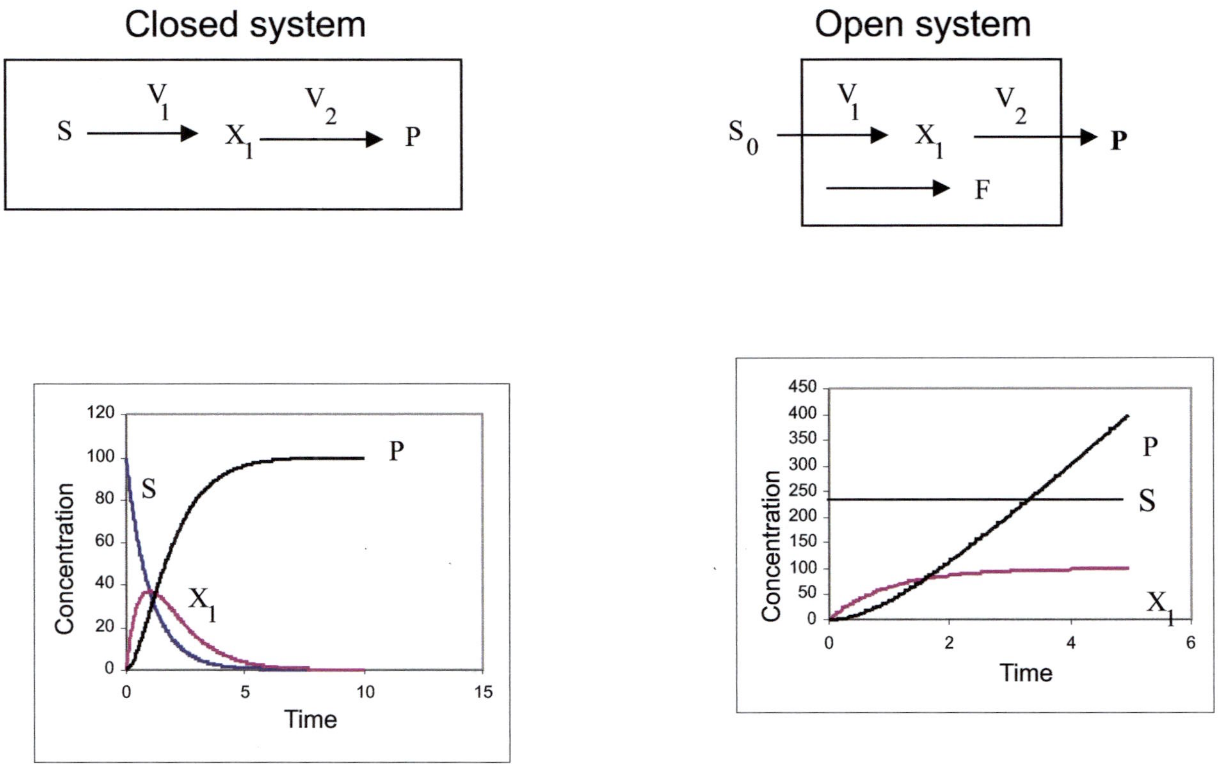

Fig. 18.4 A metabolic network is an open system.

The left hand section of the figure shows the metabolic network of Fig. 18.2 considered as a closed system. Curves show that if the reactions are irreversible, the substrate S_o and the intermediate metabolite X_1 are entirely converted after some time into the product P. The right hand section of the figure shows the same network considered as an open system in which S_0 concentration is supposed to be constant. Under these conditions, the concentration of X_1 reaches a stable value and P is produced at a constant rate.

stoichiometries $n_{i,j}$, i.e. $n_{i,j}$ molecules of the metabolite X_i react in the reaction V_j. The stoichiometries are counted positively for the products (synthesis of the metabolite) and negatively for the substrates (consumption of the metabolite). It means that a direction is defined for each reaction, particularly for the reversible ones; these directions are arbitrary (Fig. 18.5 – 18.6). We will give some examples. First, the example of Fig. 18.2 taken as example 1 in Fig. 18.5A: there is only one internal metabolite X_1, thus m = 1 and two reactions (r = 2) : V_1, $S_o \rightarrow X_1$ with the stoichiometry +1 et V_2, $X_i \rightarrow P$ with the stoichiometry –1 (S_o et P are missed out on Fig. 18.5A). The stoichiometry matrix is N = [+1 -1].

In example 2 (Fig. 18.5A) which represents the simplest form of a branched network, there is still only one intermediate metabolite X_1 produced by the first reaction (with stoichiometry +1), but with two consuming reactions (with stoichiometry –1) ; hence the stoichiometry matrix is : N = [+1 -1 -1].

Example 3 in Fig. 18.5B is more complex with three reactions and three metabolites. The stoichiometry matrix will be a square matrix with three rows and three columns (Fig. 18.6).

A more complex example is given in Fig. 18.7.

Three remarks concerning the concept of stoichiometry matrix:

1. The arrow representing an enzymatic reaction does not imply that the reaction is irreversible. It simply indicates an arbitrary orientation of the reaction which allows the definition of a metabolite as a substrate or a product and gives a sign to the rate function V.
2. There is a one-to-one correspondence between a metabolic network and its stoichiometry matrix. We have shown that the determination of the stoichiometry matrix was completely determined by the metabolic network. It is easy to verify that a unique metabolic network can be built from a given stoichiometry matrix.
3. It must be finally emphasized that the stoichiometric coefficients are not necessarily +1, –1 or 0. Integer coefficients (+2, -3, etc.) can be encountered depending on the reactions.

We will see in the following, the great interest in introducing the concept of stoichiometry matrix.

Dynamical Systems

In the simplest network of Fig. 18.1, it can be written without necessarily the steady-state hypothesis:

$$dX_1/dt = V_1\text{-}V_2,$$

In other words, the variation of the X_1 metabolite concentration dx_1/dt is the result of its production and of its consumption.

In example 4 of Fig. 18.6, there are three metabolites X_1, X_2 and X_3. Three equations can be written which depict the variations in the concentrations of the metabolites; they are represented in the dynamic system below :

$$\frac{dX_1}{dt} = V_1 - V_2 - V_4$$

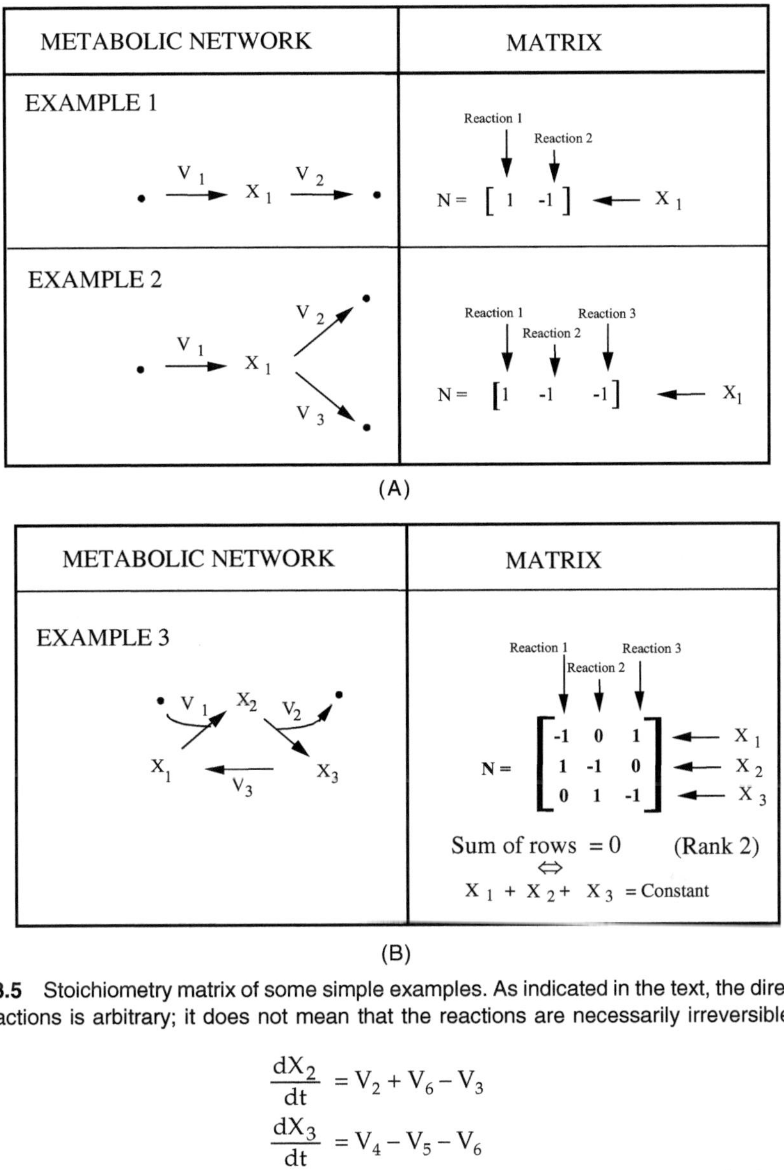

Fig. 18.5 Stoichiometry matrix of some simple examples. As indicated in the text, the direction of the reactions is arbitrary; it does not mean that the reactions are necessarily irreversible.

$$\frac{dX_2}{dt} = V_2 + V_6 - V_3$$

$$\frac{dX_3}{dt} = V_4 - V_5 - V_6$$

This dynamic system can be written under a matrix form (it means with the matrix multiplication law): $d[X]/dt = N.V$ (1) with:

$$\frac{d[X]}{dt} = \begin{bmatrix} \frac{dX_1}{dt} \\ \frac{dX_2}{dt} \\ \frac{dX_3}{dt} \end{bmatrix} \quad \text{and} \quad V = \begin{bmatrix} V_1 \\ V_2 \\ V_3 \\ V_4 \\ V_5 \\ V_6 \end{bmatrix} \quad \text{and} \quad N = \begin{bmatrix} 1 & -1 & 0 & -1 & 0 & 0 \\ 0 & -1 & 1 & 0 & 0 & 1 \\ 0 & 0 & 0 & 1 & -1 & -1 \end{bmatrix}$$

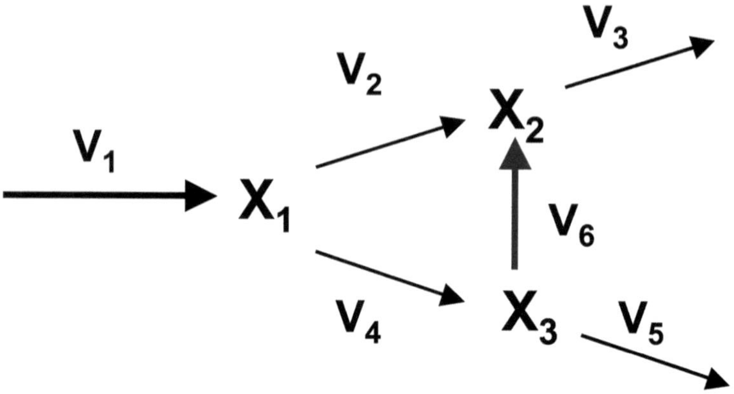

Fig. 18.6 Example 4: A metabolic network

One can now understand the advantage of the stoichiometry matrix, which allows us to reduce a system of numerous differential equations to a single matrix differential equation.

Third Principle: The Separation between the Structure and the Kinetics

The equation (1) d[X]/dt = N.V demonstrates that it is possible to separate the structure of a metabolic network, depicted by its stoichiometry matrix N (which tell us if there are loops, branches, linear parts etc.) from the rate equations which are collected in the rate column vector V.

The study of the stoichiometry matrix N will give the structural properties of the solutions of system (1).

The Steady-state Bis: Kernel of the Stoichiometry Matrix

The matrix equation (1) allows us to write the condition of steady-state in a simple expression: d[X]/dt = N.V = 0 (2) . In linear algebra, this equation means that the rate vector V (which is the column of all the rates of the network) belongs to what is called the Kernel of N (Ker[N]), the definition of which is just the equation (2) i.e. the set of vectors V, such that N.V = 0.

The concept of Kernel of matrix N depicts in a condensed way, the simple biochemical property, that, at steady-state, for all internal metabolites the rate of production equals consumption. The benefit of this representation is that we can take advantage of the general properties (which will not be detailed here) of the Kernel of linear applications depicted by matrices. We will use here one property, which is that the Kernel is a vectorial space.

What does it mean in the case of example 1, the simplest of all metabolic networks ?

In this network, N.V = 0 means V_1-V_2 = 0, i.e. $V_1 = V_2$. In the phase plane, i.e. the rate plane (V_1, V_2); it means that the stationary solutions are on the first bisecting line (Fig. 18.7).

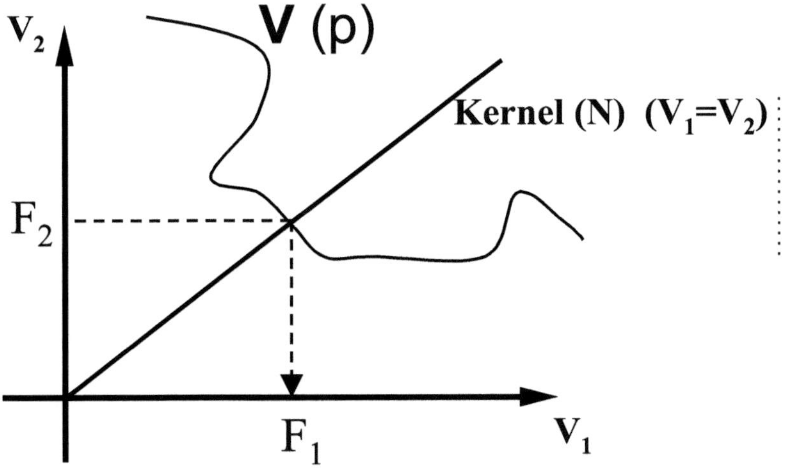

Fig. 18.7 Steady-state of example 1. The steady-state(s) are at the intersection of the Kernel , here the first bisecting line ($V_1 = V_2$), and of the set V(p) of the pair of acceptable rates (V_1; V_2) for which X_1 is the same.

Furthermore, $V_1 = V_1(X_1, p)$ were p depicts parameters (enzyme concentration, *Km*, temperature, pH, etc.); in the same way, $V_2 = V_2(X_1, p)$.V(p) designates the set of acceptable pairs of rates (V; V_2) when X_1 varies. The solutions, if they exit, are situated at the intersection of the Kernel of N (the first bisecting line $V_1 = V_2$) and of V(p).

Mathematically, $V_1 = V_2$, i.e. $V_1(X_1, p) = V_2(X_1, p)$ is an equation with the variable X_1, the solution(s) of which, if they exist are the concentrations $X_{1,0}$ at steady-state.

When we replace this $X_{1,0}$ value in $V_1(X_{1,0}, p) = F_1$ or $V_2(X_{1,0}, p) = F_2$, one gets the stationary flux already graphically obtained on Fig. 18.7 (with in this case $F_1 = F_2$).

In order to understand better the value of the matrix approach of metabolic networks, we will make the example 4, a little more complex. In this case the stoichiometry matrix is the 3 x 6 matrix (3 metabolites and 6 rates) (see above).

We will work, as above, in the rate space, the dimension of which is now six. N is a linear application from this space in the space of metabolites (or in the space of their derivatives), the dimension of which is less or equal to three; thus the dimension of the image of the rate space is of a dimension less or equal to three. It can be easily seen that in our case its dimension is exactly three (in N matrix, the column vectors 1, 2 and 4 for instance are independent). It can be demonstrated that:

Dimension of Image space by N (3 in our case) + Dimension of the Kernel of N = Dimension of the rate space (6 in our case). So that the dimension of Kernel (N) is 3. It means that all steady-state solutions can be expressed as a linear combination of 3 independent rate vectors. It is rather easy to find three independent vectors of the Kernel (solution of N.V = 0) (Fig. 18.8).

Each vector of a vector basis represents a particular metabolic pathway (Fig. 18.8). It is easy to understand in this figure that any set of fluxes at steady-state through this metabolic network (example 4) can be expressed as a linear combination of these

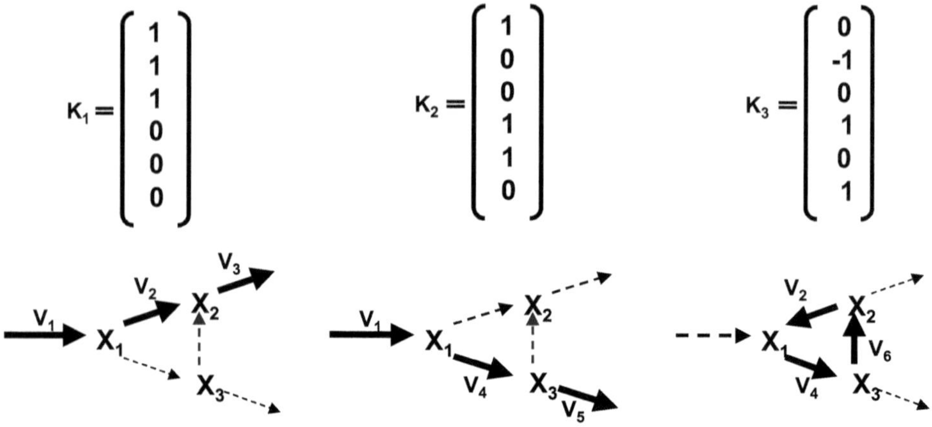

Fig. 18.8 Example 4. A vector basis of the Kernel of N. It should be noted that a vector basis is not unique. An infinity of other vector basis, i.e. of three independent vectors could be found. (Independent means that any of the three vectors cannot be generated by the two others).

particular metabolic pathways, i.e. all solutions can be written : $V = \lambda_1 K_1 + \lambda_2 K_2 + \lambda_3 K_3$. For instance the vector:

$$V = \begin{bmatrix} V_1 = 3 \\ V_2 = 2 \\ V_3 = 1 \\ V_4 = 1 \\ V_5 = 2 \\ V_6 = -1 \end{bmatrix} = 1 \bullet K_1 + 2 \bullet K_2 - 1 \bullet K_3$$

is a solution, which can be checked on the right hand part of Fig. 18.9, on which different rate values are indicated.

Apparently, the problem of the description of the stationary fluxes in a metabolic network can be solved using a basis of the Kernel of the stoichiometry matrix N: all the stationary fluxes belong to this Kernel and can thus be expressed as a linear combination of basis vectors of the Kernel. Algorithms can be used to find such a basis and to express the fluxes (Letellier *et al.*, 1991).

I said "apparently" because two problems remain.

The first problem is that biochemists would like to know all the possible pathways in a metabolic network instead of their theoretical description.

The second problem is that it is clear that some theoretical pathways will actually be impossible, because some reactions are irreversible or quasi-irreversible. For instance if reaction V_6 is irreversible in the direction indicated by the arrow, the solution above: $V = 1.K_1 + 2.K_2 - 1.K_3$ is impossible because V_6 will be operating in the wrong direction.

Both problems lead to the concept of **elementary flux modes (efm)**.

$$V = \begin{pmatrix} 3 \\ 2 \\ 1 \\ 1 \\ 2 \\ -1 \end{pmatrix} = 1K_1 + 2K_2 - 1K_3$$

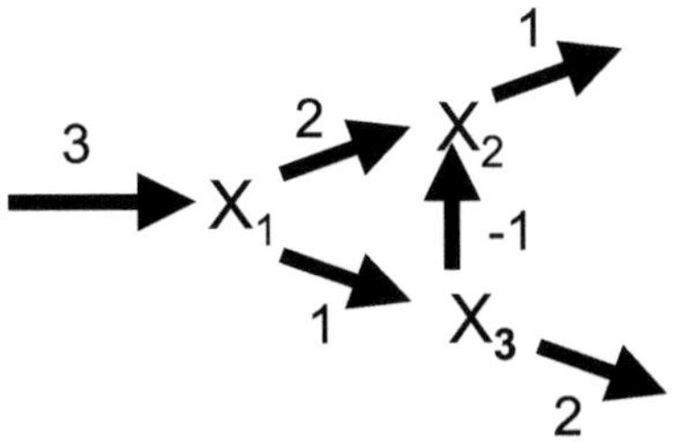

Fig. 18.9 Example of fluxes through the network of example 4. Figures on the scheme indicate the corresponding rates (which are proportional to the indicated rates). It can be checked that for each metabolite, a steady-state is reached (what enters equals what goes out). The rate V_6 = -1 indicates that the flux is in the direction opposite to that shown by the arrow.

ELEMENTARY FLUX MODE (EFM)

An elementary flux mode (efm) is a minimal set of reactions which can operate at steady-state using irreversible reactions in the right direction (Schuster and Hilgetag, 1994; Schuster, 2000; Stelling, 2002). "Minimal" means that the suppression of any reaction of the set will prevent the reaching of a steady-state. In other words, an elementary flux mode cannot be included in another bigger elementary flux mode.

In Fig. 18.8 (example 4), if we consider that all reactions are irreversible in the direction of the arrows, the K_1 and K_2 vectors are elementary modes (with all reactions in the right direction); they are minimal sets because if one reaction is suppressed, the steady-state property is lost. On the contrary, K_3 is not an elementary mode, because reaction 2 is operating in the wrong direction. The only other elementary mode is $K_4 = K_1+K_3$ depicted in Fig. 18.10. It can be shown that K_1, K_2 and K_4 are the only elementary modes of this metabolic network (example 4 with all reactions irreversible). If it is accepted that some of the reactions are reversible, new elementary modes are found.

For a metabolic network, the set of elementary modes corresponds to all possible (minimal) pathways leading from one (or several) entry metabolite(s) to one (or several) output metabolite(s) or to cycles.

It appears that for some pairs of entry and output metabolites, several pathway (efm) are possible. This is probably one of the reasons for the well-known robustness of metabolic networks and of the observed results that gene inactivation often has little effect on cell survival.

USEFULNESS OF THE ANALYSIS OF A METABOLIC NETWORK IN ELEMENTARY MODES

- Elementary modes are the only possible pathways taken in a metabolic network. For this reason it is useful to know them. The question as to whether a given efm is actually utilised will depend on the rate functions. For instance in the case of example 4, if V_4 or V_5 are null or quasi-null, the K_2 efm (pathway) will not be operating.

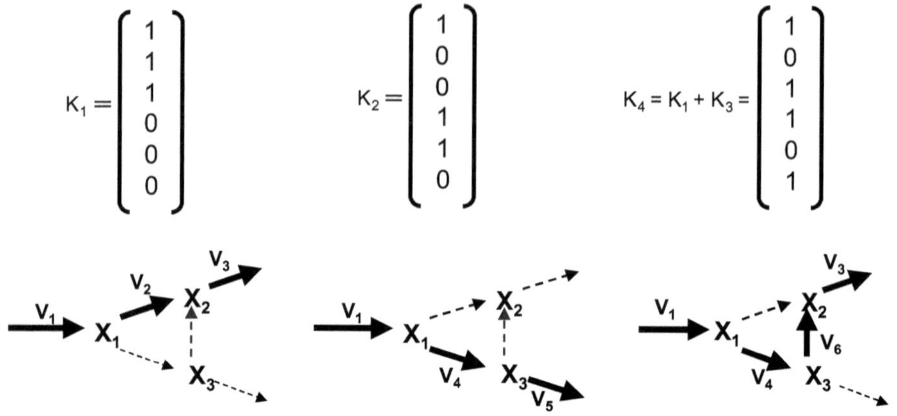

Fig. 18.10 The elementary modes of example 4. Metabolic pathways K_1, K_2, K_4, are the only three elementary modes of example 4 with all the reactions irreversible in the direction of the arrows. It is easy to check that there are elementary modes and that there is no other elementary mode.

- It has been established that tissue differences, are, (at least partly) based on metabolic differences, i.e. preferential use of different efm according to the tissue.
- For the reason evoked above, redundancy leading to metabolic network robustness, the knowledge of elementary modes can be useful to predict those mutations that can be tolerated and those that will be lethal.
- In contrast, in biotechnology, one can be lead to favour some particular elementary mode, the yield of which (for a given production) is high (Carlsson, 2002; Poolman, 2004a, b).

From a general point of view, all discussion on metabolism functioning has to be done in term of elementary modes. Examples of such use can be found in Fell (1998), Poolman, (2001), and Poolman, (2004a and b).

Finally, it should be noted that a quasi equivalent analysis called "extreme pathways" has been proposed by B.O. Palsson and colleagues (Schilling *et al.*, 2000 ; Covert *et al.*, 2001; Papin *et al.* 2004a, b).

THE PROBLEM OF ELEMENTARY FLUX MODE ANALYSIS

The problem of this approach is that the number of elementary modes increases rapidly with the number of reactions. For example, the metabolic map of mitochondrial energy metabolism–involves, according to the tissue or organism, about forty reactions and about thirty metabolites. The number of efm in this case varies around 8,000 (Pérès *et al.* 2007). Even if one can think that some of them are not possible due to kinetic reasons, a great number of efm can however remain. Because it is not possible to look at all the 8,000 efm, their classification appears as a pertinent answer to this problem. Several kinds of classification can be envisaged, based on reactions, balance equations etc. One can also consider efm as a sequence of reactions and treat them like protein or nucleic acid sequences using sequence comparison software. The traditional methods of classification do not suit very well in this case for essentially two reasons. The first reason is that some efm (i.e. some sequences of reactions) are atypical and disturb the classification. The second reason is that most of the classification methods are exclusive and do not allow a given efm to belong to several classes. We would like indeed to put an efm involving glycolysis and TCA cycle in a "glycolysis" class and in a "TCA-cycle" class. This lead Pérès *et al.* (2007) to develop a new classification algorithm which grouped together the efm through common motifs and permitted an efm to belong to several classes. Another advantage of the method is that the result does not depend upon the starting point (starting efm), used to perform the classification. Applied to mitochondrial energetic metabolism, this software allowed the grouping of about 8,000 efm (according to the tissues or organisms) into 15 – 35 classes to which a physiological meaning could be associated most of the time.

CONCLUSION

The analysis of the structure of metabolic networks in terms of elementary modes is essential to understand their function, as they depict all the pathways that can be taken. The problem is that their number increases rapidly with the number of metabolites and the reactions involved. For instance the determination of *E. coli* metabolism with more than 700 reactions and 500 metabolites is far greater than calculation capacity; even if the modes could be enumerated, their huge number would probably prevent any serious analysis.

Nevertheless, one question remains: are all the elementary modes of a network actually utilised, or only some of them for kinetic reasons; if so which ones?

This is a question that biochemists (I am one of them) do not even envisage, because they think that the only metabolic pathways are those of the textbooks. Of course, we know that from time to time a new pathway is described, or a new mutant is isolated for which a new pathway has to be proposed. In the case of mitochondrial energetic metabolism, for instance, it can be assumed that most of the 8,000 efm do not actually exist for kinetic reasons; there is presumably a small number among them that are used with a reasonable frequency (greater than 1%, 5%?). However, what is "a small number"? 2, 3, about 10, about 100? I believe that even about ten would appear as incredible for most biochemists.

I believe that this question is actually very important, because the correct answer will provide a different approach to considering metabolism and thus a different way to manage it (biotechnology, inborn errors of metabolism, etc.). To answer this question, one can imagine several approaches: firstly a theoretical approach, which can eliminate some efm; indeed some of them involve reaction stoichiometries with a high integer (about ten or about several hundreds) imposed by what B.O. Palsson calls "currency metabolites" (NAD-NADH; ATP-ADP; CoA; etc.) (Papin *et al.* 2004a, b; Pérès *et al.* 2007). It can be imagined that these great combinations of metabolites would be difficult to realize at the micromolecular level (by actual combination of the elementary molecules), but it remains to be proven. The second approach would be experimental; ideally it would be necessary to follow the fate of each molecule entering the network, i.e. to know which reaction it takes; experiments with this aim seem difficult to design.

Perhaps normal biochemists are right in thinking that the textbook metabolic pathways are the only ones. After all, their conviction comes from a long experimental practice, which never needed to envisage a multitude of pathways.

REFERENCES

Agius, L., Sherratt, H.S.A. ed. 1997. Channelling in Intermediary Metabolism. Portland Press Research Monograph London and Miami.

Carlsson, R., Fell, D., Srienc, F. 2002. Metabolic Pathway Analysis of a Recombinant Yeast for a Rational Strain Development. *Biotechnol. Bioeng.* 79; 121-134.

Cornish-Bowden, A., Hofmeyr, J.-H. S. 2005: Enzymes in context. Kinetic characterization of enzymes for systems biology. The Biochemist, 11-14.

Covert, M.W., Schilling, C.H., Famili, I., Edwards, J.S., Goryanin, I.I., Selkov, I., Palsson, B. Ø. 2001 : Metabolic modelling of microbial strains in silico. TIBS 26: 179-186

Fell, D.A. 1998. Increasing the flux in metabolic pathways: A metabolic control analysis perspective. *Biotechnol Bioeng.* 58: 121-124.

Gombert, A.K., Nielsen, J. 2000: Mathematical modelling of metabolism, *Current Opinion in Biotechnology* 11: 180-186.

Hofmeyr, J.-H. S., Cornish-Bowden, A. 1997: The reversible Hill equation: how to incorporate cooperative enzymes into metabolic models? Comput. Appl. Biosci (Bioinformatics). 13, 377–385.

Letellier, T., Reder, C., Mazat J.-P. 1991. 'CONTROL': A software for the Analysis of the Control of Metabolic Networks. Comp. Appl. in the Biosci. (Bioinformatics) 7, 383-390.

Mazat, J.-P., Reder, C., Letellier, T. 1990. *The structural approach to metabolic control theory : Geometrical aspects.* in Control of Metabolic Processes. Cornish-Bowden and Cardenas (Ed.) Plenum Press (N-Y and London) pp. 129-138.

Papin, J.A., Stelling, J., Price, N.D., Klamt, S., Schuster, S., Palsson B.Ø. 2004a. Comparison of network-based pathways analysis methods. TIBS, 22, 400-4005.

Papin, J.A., Reed, J.L. and Palsson, B.O. 2004b. Hierarchical thinking in network biology: the unbiased modularisation of biochemical networks. TIBS 29: 641-647.

Pérès, S., Beurton-Aimar, M., Mazat, J.-P. 2007. Classification des modes élémentaires : application au métabolisme énergétique mitochondrial Techinque et Science Informatique, in press.

Poolman, M.G., Olcer, H., Lloyd, J.C., Raines, C.A., Fell, D.A. 2001. Computer modelling and experimental evidence for two steady-states in the photosynthetic Calvin cycle. *Eur. J. Biochem.* 268: 2810-2816.

Poolman, M.G., Venkatesh, K.V., Pidcock, M.K., Fell, D.A. 2004a. A Method for the Determination of Flux in Elementary Modes, and its Application to Lactobacillus rhamnosus. *Biotechnol. Bioeng.* 88: 601-612.

Poolman, M.G., Assmus, H.E., Fell D.A. 2004b. Applications of metabolic modelling to plant metabolism. *J. Exp. Bot.* 55: 1177-1186.

Reder, C. 1988. Metabolic control theory: a structural approach, *J. Theor. Biol.* 135: 175-201.

Sauer, U. 2004. High-throughput phenomics: experimental methods for mapping fluxomes Current Opinion in Biotechnology 15: 58-63.

Schilling, C.H., Letscher, D., Palsson, B.Ø. 2000. Theory for the systemic definition of metabolic pathways and their use in interpreting metabolic function from a pathway-oriented perspective. *J. Theor. Biol.* 203: 229-248.

Schuster, S., Hilgetag, C. 1994: On elementary flux modes in biochemical reaction systems at steady-state. *J. Biol. Systems* 2: 165-182.

Schuster, S., Fell, D.A., Dandekar, T. 2000. A general definition of metabolic pathways useful for systematic organization and analysis of complex metabolic networks. *Nature Biotech.* 18: 326-332.

Stelling, J., Klamt, S., Bettenbrock, K., Schuster, S., Gilles, E.D. 2002. Metabolic network structure determines key aspects of functionality and regulation. *Nature*, 420: 190-193

Westerhoff, H.V., Palsson, B.O. 2004. The evolution of molecular biology into systems biology. *Nature Biotechnol.* 22: 1249-52.

Plant Models in Genomics

19

CHAPTER

The Model Species, *Arabidopsis thaliana*

X. Johnson, D. Bouchez

INTRODUCTION

To attain the status of a biological model, a living organism must have certain important criteria to validate its usage by the scientific community. The objective here is to accelerate acquisition of scientific knowledge by concentrating on a simple experimental system. In the first place, the latter must be generally valuable; its biological characteristics must therefore be globally representative of the taxonomic group under consideration, so that the information and tools obtained can easily be applied to other cases. Obviously, generalising that a set of results will hold true in another species cannot be done without caution, and most often requires experimental validation. In this case, the further taxonomically one is from the model species, the more caution is needed. Furthermore, the notion of a model underlines a relative facility for experimental manipulation. The model is therefore a simplified representation of a biological reality that is by essence varied, which has in its way a value of generality and presents some practical advantages for experimentation.

Arabidopsis thaliana, a small herbaceous species belonging to the Crucifer (Brassicaceae) family (Fig 19.1), has become in the last 20 years the plant species most used in plant biology. This species, described by Johannes Thal in the 16th century, has interested geneticists since the end of the 19th century, with the first mutant being described in 1873. The work by F. Laibach and his students in the years 1910-1950 gave place to the first collection of natural isolates (traditionally called ecotypes) and of artificially induced mutants. Following this, their successors (notably G. Rédei) have contributed greatly to popularizing this species in the scientific community. The grand scale adoption of the plant as a model began at the start of the 1980s, which saw a veritable explosion in the work on *A. thaliana*, from the level of biochemistry and physiology to developmental biology. Shortly after, genomic approaches were applied to the species

Fig. 19.1 *Arabidopsis thaliana*. Left, the plant in its vegetative state, before the appearance of the floral stem (lower left). Centre, the adult plant in a flowering/fruiting state. Right, flower, floral stem and seeds. The white bars represent 1 cm, except in the case of the flower and the seeds where they represent 1 mm. © D. Bouchez.

wherein it has shown itself to have incontestable advantages for studying the structure and function of the plant genome.

SPECIES BOTANY AND BIOLOGY

A. thaliana brings together many biological and genetic characteristics which make this a particularly interesting model species for experimentation in plant biology. It is a plant of small size, with a rapid life cycle (a few weeks from sow to sow) and is extremely prolific. On the whole the species is autogamous but can be easily crossed to produce hybrids and populations. These practical advantages for genetic experiments make it a veritable "plant drosophila" and initially it was these botanical and biological characteristics of the *A. thaliana* species which took the interest of geneticists like G. Rédei (Rédei, 1970).

Botany

Botanically speaking, *A. thaliana* is a part of the Brassicaceae family, like its close cousins radish, cabbage and canola. An abbreviated lineage can be shown as: Eukaryotes, Viridiplantae or Chlorobiotes; Streptophytes; Embryophytes; Tracheophytes; Spermatophytes; Magnoliophytes or Angiosperms; Eudicotyledonous; Core eudicots or Dicotyledonous with tetracyclic flowers; Rosidae; Eurosids II; Brassicales; Brassicaceae. No less than eleven synonyms have been used through the years to specify the species, and the boundaries of the *Arabidopsis* genus have had multiple revisions. The most recent study recognizes nine distinct species in the genus (Al-Shehbaz and O'Kane, 2002):

- *Arabidopsis thaliana* (2n = 10);
- *Arabidopsis arenosa* (2 species; 2n = 16 or 32);
- *Arabidopsis neglecta* (2n = 16 or 32);
- *Arabidopsis croatica;*
- *Arabidopsis cebennensis* (endemic species from French Cévennes);
- *Arabidopsis pedemontana* (endemic species from Piedmont);
- *Arabidopsis lyrata* (3 species; 2n = 16 or 32);
- *Arabidopsis halleri* (2 species; 2n = 16 or 32);
- *Arabidopsis suecica* (2n = 26) which is amphidiploid, the issue of a cross between *thaliana* (2n = 10) and diploid *arenosa* (2n = 16).

Region of Distribution

The natural geographic distribution of *A. thaliana* probably covers Europe and central Asia, but the species has been introduced into almost every country in the world, thus complicating the work of ecologists and botanists. The natural limit of the plant extends from the north of Scandinavia to North Africa and the south of the Middle East, a particularly large area of distribution for the relative simplicity of this species.

The species is described in the entirety of Europe[1], the greater part of Asia, Australia and New Zealand[2], Africa (Algeria, Ethiopia, Kenya, Libya, Morocco, South Africa, Tanzania, Tunisia, Uganda, Zaire), North America (United States and Canada) and South America (Argentina, Uruguay) (Pigliucci, 2002).

The species seems to be quite demanding, sensitive to competition and often occupies open or perturbed habitats: river banks, abandoned areas, old walls, cultivated lands, rocky outcrops, on sandy or rocky soils, most often bare or with a feeble covering. It can be locally abundant.

[1]Except Ireland

[2]Except Bahrain, Bangladesh, Bhutan, Cambodia, Iraq, Israel, Jordan, Kuwait, Kirghizistan, Laos, Malaysia, Burma, Nepal, Papua New Guinea, Oman, Philippines, Qatar, Sri Lanka, Taiwan, Thailand, Turkmenistan, United Arab Emirates, Yemen

Ecology

Surprisingly little information is available on the ecological requirements and life cycle of this species in its natural condition and only recently have these aspects been significantly explored (Mitchell-Olds, 2001).

A. thaliana is a typical, annual, herbaceous species with a simple life cycle that can be separated into four principal stages of development (Fig. 19.1):

- Plantlet: after germination, elongation and opening of cotyledons, the first leaves appear.
- Rosette: formation of a rosette of leaves. The number of leaves formed is extremely variable depending on genotype and growth conditions, ranging from five to six to many hundreds.
- Flowering: appearance and elongation of a principal stem from the centre of the rosette and opening of the first flowers. Secondary floral branches can develop from the base, and can be numerous. As a consequence, the production of seeds can be very variable.
- Seedset and senescence: senescence of leaves, ripening and dehiscence of fruits. The very small seeds germinate immediately, or after a period of dormancy, the length of which depends on the genotype. A plant can produce tens or tens of thousands of seeds depending on the conditions. *A. thaliana* is classically considered to be strictly autogamous but studies seem to show that in natural conditions out-breeding can occur between certain populations.

This very simple scenario can vary considerably depending on environmental conditions. In particular, early flowering time, which has been well studied, is an essential parameter affecting both the cycle and selective value of the plant. In the laboratory, the practice is to distinguish late or early ecotypes depending on the length of their vegetative (rosette) phase before floral initiation. Under natural conditions, many types can be identified from which the early flowering ecotypes are derived, about one month after being sown, which do not respond to vernalisation and which constitute most of the accessions currently used in genetic laboratories. This type corresponds reasonably well to the perennial plants that start their life cycle at the beginning of spring in Western Europe. Many other types can be defined based on their annual or bi-annual character and by their sensitivity to vernalisation, but no clear correlation exists as such with their geographic localisation and/or their conditions of origin and many uncertainties arise as to the ecological pertinence of this classification (Pigliucci, 2002).

Natural Variation

Regarding the extended area of distribution of the plant and the diversity of the natural regions where the species is present, from the islands of Cape Verde to the mountains of Pamir, one can expect a large diversity within this species which express themselves as adaptations to varied environmental conditions (Alonso-Blanco and Korneef, 2000). Numerous programmes are in place to characterise this variability and to use it to dissect the genetic basis of adaptation to an environment. Examples of such approaches are given in Chapter 27.

Genetic Transformation

One advantage of the species, and probably one of the most important, concerns its facility for genetic transformation, which can be accomplished by simply putting the plant in contact with the bacteria *Agrobacterium tumefaciens*. The first acts of transformation "*in planta*" were carried out in 1987 by Feldman, who soaked the seeds in a suspension of Agrobacterium conferring resistance to kanamycin in the transformed plant cells. Afterwards, the technique was considerably simplified by employing vacuum infiltration of adult plants in a suspension of Agrobacterium (Bechtold *et al.*, 1993). In the descendants of selected infiltrated plants, there can be one or multiple copies of the T-DNA conferring a trait such as resistance to the herbicide Basta (Bouchez *et al.*, 1993). All steps are performed in a glasshouse under standard horticultural conditions without the need for *in vitro* manipulations. Every laboratory now uses this technique to produce transgenic plants in this species, thus constituting an irreplaceable tool in most integrated biology programs. Furthermore, this technique has been employed on a large scale to produce insertion mutant populations for functional studies (Chapter 9). Globally, one could estimate that many hundreds of thousands of plants have been produced in this fashion.

Despite all these strong points, it would be an illusion to think that *A. thaliana* possesses only advantages. To count amongst its inconveniences – the happiness of some can be the misfortune of others – the small size of the plant (and therefore of the organs) can constitute a practical limit for physiological and biochemical approaches; furthermore, certain biological characteristics like mycorrhizal symbiosis, and nitrogen fixation do not exist in the species. Cellular genetic techniques and *in vitro* culture are clearly less easy than in the case of a species like tobacco, which was very popular in plant biology laboratories not very long ago. Finally, *A. thaliana* is far from being a species of agronomic importance and the responses of the plant to agricultural treatments are not necessarily representative of a cultivated species, submitted to human selection for many thousands of years.

Genetic and Genomic

It is not only the biological characteristics of *A. thaliana* which present advantages in the laboratory: equally convenient, is the genome, which is one of the most simple known for a higher plant (Table 19.1).

The species carries five pairs of chromosomes (2n = 10) and the total physical size of the chromosomes together equals 125 megabases (million base pairs; Mb), which in fact is the smallest known genome of a higher plant (Table 19.1). As a comparison, the size of a typical bacterial genome is 2-5 Mb, that of a unicellular eukaryotic organism like the yeast *Saccharomyces cerevisiae* around 15 Mb, that of an insect like *Drosophila* in the vicinity of 165 Mb, whilst the genome of a mammal adds up to around 3,000 Mb. In plants this size is very variable and can reach more than 120,000 Mb in certain species, notably due to the richness of repeatable elements, like retrotransposons, in these genomes (Chapter 26).

The organization of the *A. thaliana* genome is also particularly simple and favourable to molecular genetic and genomic experimentation (Fig. 19.2). The content of high

Table 19.1 Comparative size of a selection of plant genomes (in Mb per haploid genome)

Species	*Size (Mb)*	*Ploidy level*
Arabidopsis thaliana	125	2n = 2x = 10
Oryza sativa	420	2n = 2x = 24
Lycopersicon esculentum	1,000	2n = 2x = 24
Brassica napus	1,200	2n = 4x = 38
Nicotiana plumbaginifolia	2,300	2n = 2x = 20
Zea mais	2,500	2n = 2x = 20
Nicotiana tabacum	4,500	2n = 4x = 48
Triticum aestivum	16,000	2n = 6x = 42
Tulipa sp	30,000	
Fritillaria assyriaca	123,000	

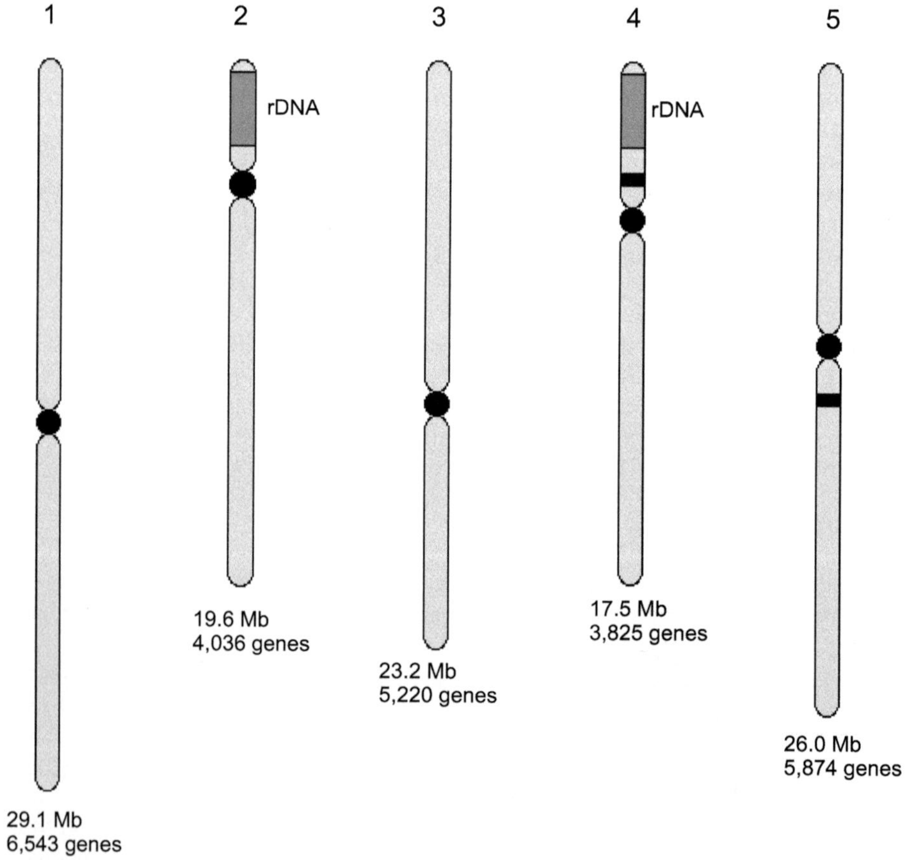

Fig. 19.2 The five chromosomes of *Arabidopsis thaliana*. The sequenced regions are represented in light grey. The size of these regions and their gene content is indicated beneath each chromosome. The regions of chromosome 2 and 4 coding for rRNA are represented in dark grey. The heterochromatic regions and centromeres are represented in black.

repeat sequences (10%) and medium repeat sequences (10%) is singularly low, as these repeatable elements can represent up to 95% of the DNA in the genomes of other plant species. Furthermore, these repeated sequences are not dispersed in the genome, but grouped in specialized regions of the genome: the first group essentially corresponds to repeated peri-centromeric sequences which surround the centromeres, the second to the two regions of the nucleolar organiser which codes for the ribosomal RNAs, and are localised at the extremity of chromosomes 2 and 4. Around 80% of the genome is therefore constituted of rarely repeated or unique sequences, essentially corresponding to the genic fraction. Thus, the density of information is exceptionally high, as the sequencing of the genome has confirmed.

Some very detailed genetic maps exist for *A. thaliana,* which includes a number of localised mutations (mutant.lse.okstate.edu/genepage/genepage.html), and a reference genetic map produced from recombinant inbred lines derived from a cross between the Landsberg *erecta* and Columbia ecotypes (nasc.nott.ac.uk/new-ri-map.html), based on 1,300 molecular markers. It is important to note that on top of the complete sequence for the ecotype Colombia we also have available, almost in its entirety, the sequence of the Landsberg parent, which should potentially give us access to the collection of genetic polymorphisms distinguishing the two accessions, for use in producing high resolution genetic maps.

The global size of the genetic map of A. *thaliana* is 500-600 cM, implying that on average a unit of genetic mapping corresponds to a physical size in the range of 200 kb, which is a size compatible with current techniques of cloning and sequencing. Note that this average equivalence can find some local disparity (from 20 to 2,000 kb/cM within certain regions).

Since the 1990s, much effort has been put into increasing our knowledge of the genome of this model species and developing high performance tools for genomics and functional genomics. In most cases, *A. thaliana* has been the first plant species to benefit from a new technology and methods have then been adapted for the study of more complex genomes, such as rice, *Medicago truncatula* or other species of agronomic interest.

More and more detailed physical maps have been constructed; at the start, using yeast artificial chromosome (YAC) clones (Schmidt *et al.*, 1995; Zachgo *et al.*, 1996; Camilleri *et al.*, 1998), then with the help of bacterial (BAC) clones (Marra *et al.*, 1999; Mozo *et al.*, 1999). The essential objective of these maps was to provide a genomic canvas to permit the integration of genetic and physical localizations, in particular to facilitate positional cloning of loci discovered by mutation. These maps have served as a basis for the genomic sequencing undertaken at the start of the 1990s (Chapter 2).

To go further with our knowledge of this model genome, some systematic sequencing approaches of transcript populations (expressed sequence tags or ESTs) were instigated at the start of the 1990s by French laboratory consortiums and then in the USA (Höfte, 1993). This permitted the first inventories of the most expressed genes in *A. thaliana* (Cooke *et al.*, 1996). More than 700,000 tags are available in the dbEST database (www.ncbi.nlm.nih.gov) and this approach has now been used in numerous other plant species.

To overcome a certain number of inherent limitations with the EST approach, the complete sequence of the nuclear genome of *A. thaliana* was put together by an international consortium of laboratories (Chapter 2). The final summary of this colossal work was published in December 2000 in *Nature* (AGI, 2000), and the results are available in their entirety in the diverse databases accessible on line; a good starting place to consult the data is the international database *The Arabidopsis Information Reserve* (TAIR: www.arabidopsis.org). The complete choroplast (154 kb, 79 genes) and mitochondrial (367 kb, 58 genes) genomes are also available.

The size of the genomic sequence is close to 115 million nucleotides covering all of the five chromosomes with the exception of zones of ribosomal DNA and the centromeres which represent around 10 Mb in total. This sequence, of high quality, has permitted the discovery of around 26,000 genes potentially coding for proteins. Close to 8% of these predicted proteins seem to be targeted to the chloroplasts compared to 1.5%, which are targeted to the mitochondria. As expected the genome is particularly dense with a gene occurring every 4.5 kb on average. The length of a gene is on average 1.9 kb and 79% of the genes are interrupted by one or more introns.

The complete catalogue of predicted genes has permitted functional classifications to be established and to give for the first time an idea of the global structure and of the content in genes, of the genome of a higher plant. At the same time, of the about 26,000 genes identified in the original publication, less than 10% have a biological function defined experimentally. Close to 30% have no sequence similarity with a gene of known function and for the remaining 60%, a function can be predicted, based on sequence comparison. For the most part, these predictions need to be treated with caution, before any experimental validation has been undertaken. This gives an idea of the enormity of the task waiting for biologists in the coming years.

One of the big surprises to come from this work is the quite high degree of redundancy identified within a genome that was expected to be particularly "compact". The some 26,000 predicted genes can be regrouped into 11,000 multigenic families and only 35% of these are "unique". Repeated and tandem genes are frequent (1,528 repetitions, counting upto 23 adjacent copies) and concerns 17% of the genes of *A. thaliana* . Furthermore, global analysis of the *A. thaliana* genome has brought to light a great number of duplications, which are more or less recent: 24 duplicated segments of more than 100 kb, representing close to 60% of the genome and presenting a divergence more or less pronounced (inversions, deletions, etc.) amongst the pairs of duplicated segments. The hypothesis put forward to explain these large scale duplications could be the existence of a tetraploid ancestor in the course of the evolutionary history of the species, which is now in the process of returning to a diploid state (Bevan *et al.*, 2001).

Functional Genomics

Now the complete sequence of the *A. thaliana* genome is available, and collections of ESTs also exist for many different plant species. Starting from this mass of information new perspectives are emerging, permitting "integrated approaches", from molecular genetics to population biology. However, the greater part of functional information derived from *in silico* analyses must be treated as hypothetical in the absence of any experimental confirmation. The way it works then is that we must reverse the traditional genetic approach (what are the genes behind this function?) and use a new "post-genomic" paradigm (what are the functions behind these genes?). How do we face the amplitude of this task? Out of more than 25,000 genes identified in *A. thaliana,* reliable functional information is only available for a few thousand amongst them.

Functional genomics proposes high throughput approaches permitting fast acquisition of knowledge on gene function (Bouchez and Höfte, 1998). It was an ambitious objective that the international community of *Arabidopsis* biologists fixed for themselves in the 10 coming years, that is to associate all the genes identified by sequencing and their products with a complete as possible biological function. This requires the gathering of all possible information for each gene and protein on structure, regulation, biochemical activity and sub-cellular localisation, interactions with other molecules, natural allelic variability, etc.

The last years have seen the spectacular development of these functional analysis high-throughput methodologies, which permit the simultaneous analysis of the activity of a complex network of genes and proteins. Again, most of the technologies are being developed on *A. thaliana.*

Different methodologies for the study of expression are being put into practice, visualising and quantifying the variations of gene products (transcripts and proteins) in time, in space, during the course of development, or in response to diverse environmental stimuli (Chapters 10, 12, 15). The biochemical and cytological approaches both can provide a wealth of information and despite the difficulty to make these disciplines high throughput, a number of techniques have been simplified, automated, and made reliable. Systematic cloning of sequences coding for interaction targets of protein-protein interactions, purification complexes, heterologous expression, construction of protein arrays, or the production of tagged proteins with a green or red fluorescent markers (GFP, RFP are good examples). Another example of a major breakthrough for functional analysis led by *A. thaliana,* concerns the availability of collections of mutants and the possibility of inactivation of genes by inverse genetics, Chapter 9). Quantitative genetic approaches (identification and characterisation of quantitative trait loci, association genetics, (Chapters 28-29), relying on the information and tools of the genome, should permit the fine dissection of complex characters and be used to place biological characters in an ecological and evolutionary perspective.

Beyond these biochemical and molecular aspects, there remains nevertheless a need to develop truly integrative approaches and for models that take into account all levels of biological complexity (metabolic network, organelle, cell, etc. to the level of organism and population) and their emergent properties.

A model system for plant biology, *A. thaliana* fulfils therefore two complementary roles: from the point of view of methodology, the plant is a platform for testing and for the development of new technologies for the functional analyses of genomes. It is also a formidable tool for the acquisition of knowledge on fundamental processes of a higher plant in relation to the environment. Some major advances in biology and development, the response to abiotic and biotic environments and in physiology have already been accomplished thanks to this model. In conclusion, the concentration of scientific effort on this species and the accumulation of results in a diverse field of subjects permit us to put into perspective various information contributing to the making of a truly integrative plant science.

Sequencing of the genome accomplished (*Nature*, pp. 408, 796 and following)

Some Chronological and Bibliographic Milestones	
1964	First publication of the Arabidopsis Information Service (AIS). Available on line at www.arabidopsis.org/ais/aisvols.html
1965	First international conference devoted to *A. thaliana*
1970	First major bibliographic review (Rédei, 1970).
1976	Second international conference.
1980	More and more work is consecrated to this model.
1983	First detailed genetic map (Korneef *et al.*, 1983)
1984	The size and complexity of the genome is determined (Leutwiler *et al.*, 1984).
1985	*Arabidopsis* is announced as the molecular genetic plant model (Meyerowitz and Pruitt, 1985)
1986	Transformation by Agrobacterium (Lloyd *et al.*, 1986). First sequences of *Arabidopsis* genes.
1987	First *in planta* transformations (Feldmann and Marks, 1987). Third international conference
1988	Launching of an "Arabidopsis" electronic newsgroup (www.arabidopsis.org/news/newsgroup.html) First map using molecular markers (Chang *et al.*, 1988)
1989	First cloning of a T-DNA tagged mutant (Feldmann *et al.*, 1989)
1990	Cloning of the *Agamous* gene by T-DNA mutagenesis (Yanofsky *et al.*, 1990). Launching of the international Arabidopsis Genome Project.
1991	Database (www.arabidopsis.org) and resource centre (www.arabidopsis.org/abrc/ and nasc.nott.ac.uk) operational.
1992	Chromosome walking reported (Arondel *et al.*, 1992; Giraudat *et al.*, 1992)
1993	Genetic transformation by infiltration (Bechtold *et al.*, 1993) First efforts to sequence ESTs (Höfte *et al.*, 1993)
1995	First complete YAC chromosome physical map (Schmidt *et al.*, 1995) First BAC and P1 banks (Choi *et al.*, 1995; Liu *et al.*, 1995)
1996	Launching of the genomic sequence project (Arabidopsis Genome Inititaive, AGI).
1997	Physical mapping of all 5 chromosomes is achieved.
1998	Publication of a large region of chromosome 4 (Bevan *et al.*, 1998).
1999	Complete sequence of chromosome 2 (Lin *et al.*, 1999) and 4 (Mayer *et al.*, 1999).
2000	Eleventh international *Arabidopsis* conference.

REFERENCES

Alonso-Blanco, C., Koornneef, M. 2000. Naturally occurring variation in *Arabidopsis*: an underexploited resource for plant genetics. *Trends Plant Science* 5, 22-29.

Al-Shehbaz, I., O'Kane, S. 2002. Taxonomy and phylogeny of *Arabidopsis* (Brassicaceae). In: The *Arabidopsis* book, American Society of Plant Biologists Ed. (http://aspb.org/publications/arabidopsis).

Arabidopsis Genome Initiative. 2000. Analysis of the genome sequence of the flowering plant *Arabidopsis thaliana*. *Nature* 408, 796-815.

Arondel, V., Lemieux, B., Hwang, I., Gibson, S., Goodman, H.M., Somerville, C.R. 1992. Map-based cloning of a gene controlling omega-3 fatty acid desaturation in *Arabidopsis*. *Science* 258, 1353-1355.

Bechtold, N., Ellis, J., Pelletier, G. 1993. *In planta Agrobacterium* mediated gene transfer by infiltration of adult *Arabidopsis thaliana* plants. *Comptes Rendus de l'Académie des Sciences. Série III, Sciences de la Vie* 316, 1194-1199.

Bevan *et al.* 1998. Analysis of 1.9 Mb of contiguous sequence from chromosome 4 of *Arabidopsis thaliana*. *Nature* 391, 485-488.

Bevan, M., Mayer, K., White, O., Eisen, J.A., Preuss, D., Bureau, T., Salzberg, S.L., Mewes, H.W. 2001. Sequence and analysis of the *Arabidopsis* genome. *Current Opinion in Plant Biology* 4, 105-110

Bouchez, D., Camilleri, C., Caboche, M. 1993. A binary vector based on Basta resistance for *in planta* transformation of *Arabidopsis thaliana*. *Comptes Rendus De l'Académie Des Sciences. Série III, Sciences De La Vie* 316, 1188-1193.

Bouchez, D., Höfte, H. 1998. Functional genomics in plants. *Plant Physiology* 118, 725-732.

Camilleri, C., Lafleuriel, J., Macadre, C., Varoquaux, F., Parmentier, Y., Picard, G., Caboche, M., Bouchez, D. 1998. A YAC contig map of Arabidopsis thaliana chromosome 3. *Plant Journal* 14, 633-642.

Chang, C., Bowman, J.L., DeJohn, A.W., Lander, E.S., Meyerowitz, E.M. 1988. Restriction fragment length polymorphism linkage map for *Arabidopsis thaliana*. *Proc. Natl. Acad. Sci. USA* 85, 6856-6860.

Choi, S.D., Creelman, R.A., Mullet, J.E., Wing, R.A. 1995. Construction and characterization of a bacterial artificial chromosome library of *Arabidopsis thaliana*. *Plant Molecular Biology Reporter* 13, 124-128.

Cooke, R. *et al.* 1996. Further progress towards a catalogue of all *Arabidopsis* genes: analysis of a set of 5000 non-redundant ESTs. *Plant Journal* 9, 101-124.

Feldmann, K.A., Marks, M.D. 1987. *Agrobacterium*-mediated transformation of germinating seeds of *Arabidopsis thaliana*: a non-tissue culture approach. *Molecular & General Genetics* 208, 1-9.

Feldmann, K.A., Marks, M.D., Christianson, M.L., Quatrano, R.S. 1989. A dwarf mutant of *Arabidopsis* generated by T-DNA insertion mutagenesis. *Science* 243, 1351-1354.

Giraudat, J., Hauge, B.M., Valon, C., Smalle, J., Parcy, F. 1992. Isolation of the *Arabidopsis* ABI3 gene by positional cloning. *Plant Cell* 4, 1251-1261.

Höfte, H. *et al.* 1993. An inventory of 1152 expressed sequence tags obtained by partial sequencing of cDNAs from *Arabidopsis thaliana*. *Plant Journal* 4, 1051-1061.

Koornneef, M., Van Eden, J., Hanhart, C.J., Stam, P., Braaksma, F.J., Feenstra, W.J. 1983. Linkage Map of *Arabidopsis thaliana*. *Journal of Heredity* 74, 265-272.

Leutwiler, L.S., Hough-Evans, B.R., Meyerowitz, E.M. 1984. The DNA of *Arabidopsis thaliana*. *Molecular & General Genetics* 194, 15-23.

Lin, X.Y., Kaul, S.S., *et al.* 1999. Sequence and analysis of chromosome 2 of the plant *Arabidopsis thaliana*. *Nature* 402, 761-768.

Liu, Y.G., Mitsukawa, N., Vazquez Tello, A., Whittier, R.F. 1995. Generation of a high-quality P1 library of *Arabidopsis* suitable for chromosome walking. *Plant Journal* 7, 351-358.

Lloyd, A.M., Barnason, A.R., Rogers, S.G., Byrne, M.C., Fraley, R.T., Horsch, R.B. 1986. Transformation of *Arabidopsis thaliana* with *Agrobacterium tumefaciens*. *Science* 234, 464-466.

Marra, M. *et al.* 1999. A map for sequence analysis of the *Arabidopsis thaliana* genome. *Nature Genetics* 22, 265-270.

Mayer, K., Schuller, C., *et al.* 1999. Sequence and analysis of chromosome 4 of the plant *Arabidopsis thaliana. Nature* 402, 769-777.

Meyerowitz, E.M., Pruitt, R.E. 1985. *Arabidopsis thaliana* and Plant Molecular Genetics. *Science* 229, 1214-1218.

Mitchell-Olds, T. 2001. *Arabidopsis thaliana* and its wild relatives: a model system for ecology and evolution. *Trends in Ecology & Evolution* 16, 693-700.

Mozo, T., Dewar, K., Dunn, P., Ecker, J.R., Fischer, S., Kloska, S., Lehrach, H., Marra, M., Martienssen, R., Meier-Ewert, S., Altmann, T. 1999. A complete BAC-based physical map of the *Arabidopsis thaliana* genome. *Nature Genetics* 22, 271-275.

Pigliucci, M. 2002. Ecology and evolutionary biology of *Arabidopsis thaliana.* In: The *Arabidopsis* book, American Society of Plant Biologists Ed. (http://aspb.org/publications/arabidopsis).

Redei, G.P. 1970. *Arabidopsis thaliana*: a Review of the Genetics and Biology. *Bibliographia Genetica* 20, 1-151.

Salanoubat, M. *et al.* 2000. Sequence and analysis of chromosome 3 of the plant *Arabidopsis thaliana. Nature* 408, 820-822.

Schmidt, R., West, J., Love, K., Lenehan, Z., Lister, C., Thompson, H., Bouchez, D., Dean, C. 1995. Physical Map and Organization of *Arabidopsis thaliana* Chromosome 4. *Science* 270, 480-483.

Tabata, S. *et al.* 2000. Sequence and analysis of chromosome 5 of the plant *Arabidopsis thaliana. Nature* 408, 823-826.

Theologis, A. *et al.* 2000. Sequence and analysis of chromosome 1 of the plant *Arabidopsis thaliana. Nature* 408, 816-820.

Yanofsky, M.F., Ma, H., Bowman, J.L., Drews, G.N., Feldmann, K.A., Meyerowitz, E.M. 1990. The protein encoded by the *Arabidopsis* homeotic gene agamous resembles transcription factors. *Nature* 346, 35-39.

Zachgo, E.A., Wang, M.L., Dewdney, J., Bouchez, D., Camilleri, C., Belmonte, S., Huang, L., Dolan, M., Coodman, H.M. 1996. A physical map of chromosome 2 of *Arabidopsis thaliana. Genome Research* 6, 19-25.

20

CHAPTER

Rice: A Model Plant for Cereal Genomics

M. Delseny

Rice is one of the most important plant species used by mankind, as it forms the staple diet of more than three billion people. Land surfaces cultivated with rice represented about 150 million hectares in 2000: they accounted for 11% of arable land in the world and about 10 times the available land in France. Rice grain production accounts for 29% of cereal production world wide. It is grown in all continents, but it is most important in Asia (Khush, 2005). It belongs to two related species, *Oryza sativa* and *Oryza glaberrima*, the latter being restricted to Africa. Within the *O. sativa* species, two subgroups which diverged two to three millions year ago on the two sides of the Himalaya mountain range can be distinguished: the *indica* type developed in tropical and subtropical areas, whereas the *japonica* type colonized more temperate regions Since the beginning of agriculture, rice has been a major target for plant selection and breeding, so that hundreds of thousands of accessions have been collected and stored in a few germplasm resource centers, the most important being at the IRRI (International Rice Research Institute, Philippines).

Cultivated rice behaves as a diploid species, and its genome, as analysed by cytogenetical methods, belongs to the AA group. Wild species occur besides cultivated ones. Some, like *Oryza rufipogon* are considered as the ancestor of present-day cultivated varieties. Others contain a genome belonging to a related but different cytogenetical group (BB, CC, EE, FF) and are often allotetraploids. Examples of wild rice species are *O. punctata* (BB and BBCC), *O. officinalis* (CC), *O. minuta* (BBCC), *O. australiensis* (EE), *O. latifolia* (CCDD) (Vaughan *et al.*, 2003). These wild species are an interesting reservoir of genes of interest, determining important agronomic traits such as disease resistance.

Independent of its economic value, rice was chosen as a model species in the early 1990's, on the basis of several criteria: the relatively small size of its genome compared to other cereals, its relatively well conserved collinearity with that of other cereals and the availability of a number of tools for functional genetics such as transformation (Sasaki and Moore, 1997; Delseny *et al.*, 2001). All the cereals (Poaceae family) are closely related species. Rice is separated from maize and wheat by approximately 60-70 millions years of evolution, as shown by the phylogenetic tree in Fig. 20.1.

GENERAL CHARACTERISTICS OF THE RICE GENOME

Rice has one of the smallest genomes of cereals. The estimated size of 400-440 Mbp is much smaller than that of maize (2,500 Mbp), or of wheat (16,000 Mbp). The rice genome is distributed into 12 chromosomes pairs. It was recognized very early on that the genome contained far less repeated sequences than that of other cereals. Such sequences represent approximately 50% of the rice genome, compared to more than 80% for wheat. This difference is more quantitative than qualitative, since all types of transposable elements are roughly equally represented amongst the different cereal genomes. These two properties rapidly made rice another attractive candidate for a whole genome sequencing project. Accordingly, the necessary resources were set up as early as 1990. Several genetic maps were constructed with high resolution from different crosses, usually between a *japonica* and an *indica* variety. These crosses were reasonably polymorphic and a large number of quantitative traits (QTL) have been shown to segregate in the progeny, allowing the identification of several major genes. Most of these maps have been constructed using RFLP markers, very often derived from at least partially sequenced cDNA (Sasaki and Moore, 1997; Delseny *et al.*, 2001).

The use of RFLP markers has demonstrated that the rice genome is largely collinear with that of other cereals. cDNA sequences are usually conserved within related species. This has allowed a comparison of RFLP maps constructed with this type of marker. This strategy rapidly revealed that the order of markers along a chromosome in rice was essentially conserved in the homologous chromosomal region of another cereal species, establishing that synteny order was conserved amongst cereals. A general model for the organisation of chromosome blocks in cereals was proposed (Moore *et al.*, 1995; Devos and Gale, 1997), whereby each genome is represented by concentric circles in which collinear regions between two genomes can be aligned. Thus, the 12 chromosomes of rice can be aligned with 10 maize chromosomes pairs, which are distributed along two circles because maize is an ancient tetraploïd; or with seven pairs of chromosomes of diploid wheat (Fig. 20.2). According to this model, wheat chromosome 7 corresponds to the fusion of rice chromosomes 6 and 8. In maize these two chromosomes are fused in chromosome 9, whereas rice chromosome 6 is collinear with a portion of maize chromosome 6 and rice chromosome 8 to a portion of maize chromosome 1. A major advantage of this model is that it allows one to predict the position of important genes controlling traits in different species, just by tracing a radius. Also represented in Fig. 20.2 are a few examples of a number of characters, which are common to maize, wheat and rice and are present at orthologous positions.

These observations and this model have considerably strengthened rice as a model species for genome sequencing. The rationale is that if one can map precisely an interesting character on the genetic map of a complex cereal genome, it should be possible to examine the sequence of the orthologous gene, or, if it is absent in rice, to find out several additional markers to saturate the fine genetic map and to restrict the

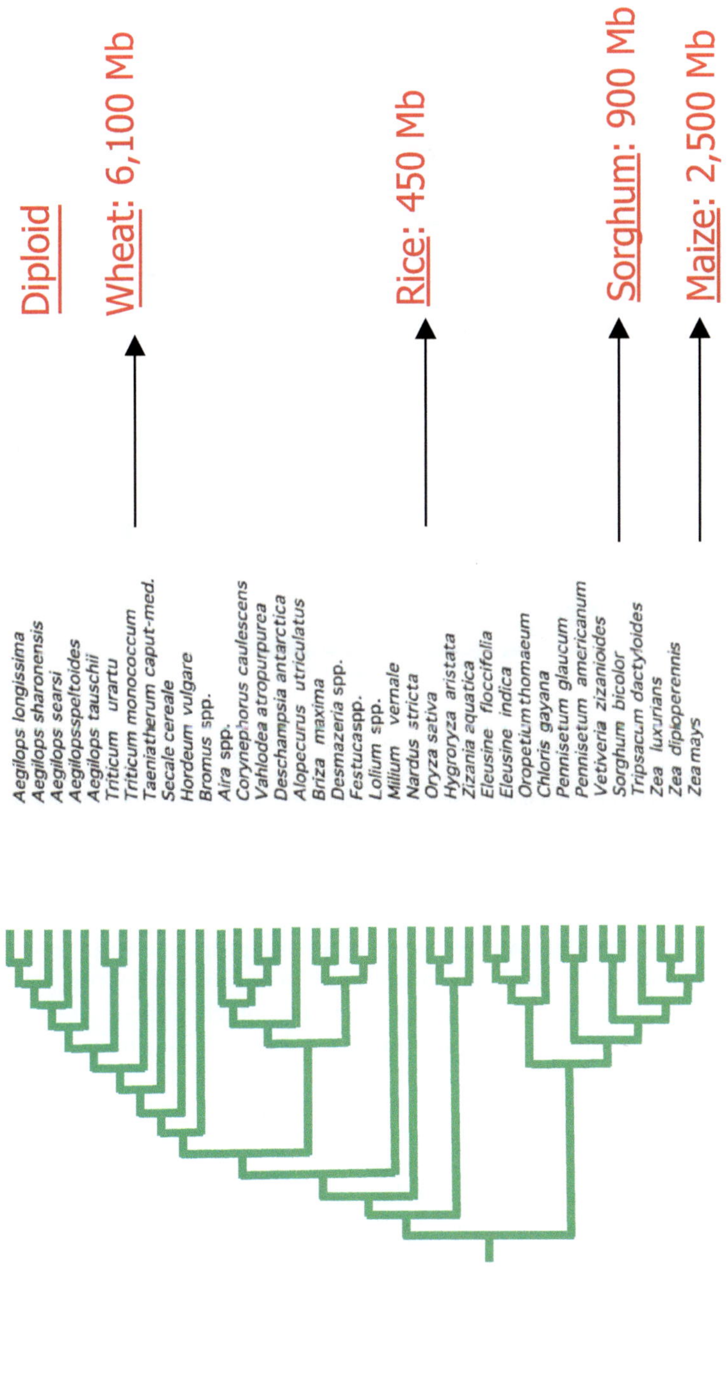

Fig. 20.1 Phylogenetic tree of the cereals

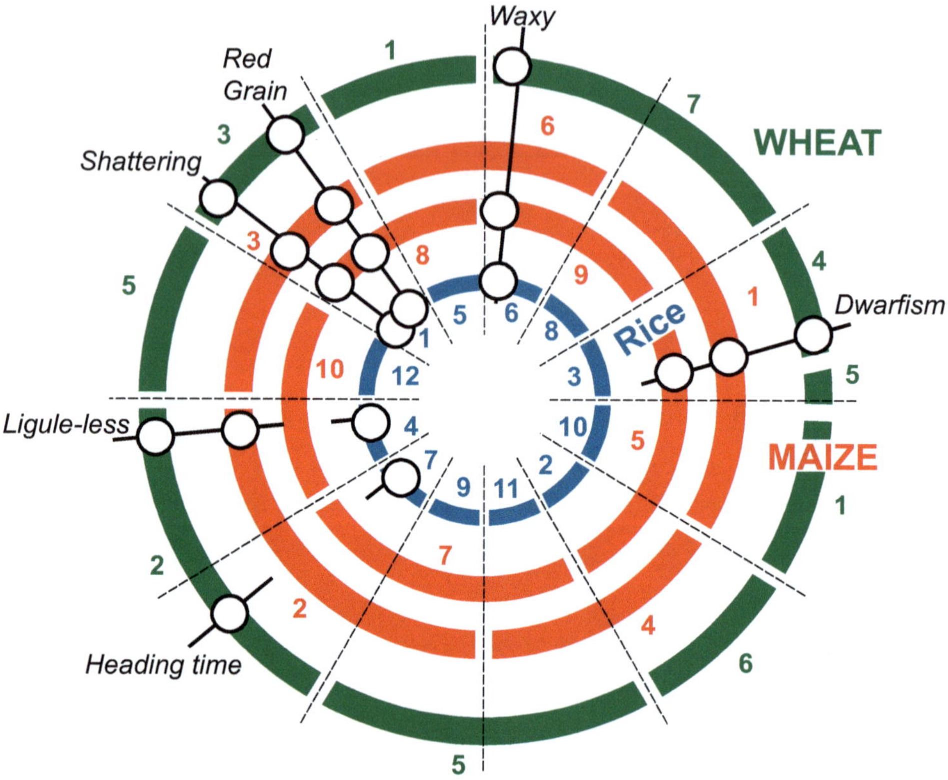

Fig. 20.2 Concentric circle model representing the rice, maize and wheat chromosomes. All maps are aligned relative to rice chromosomes. Numbers refer to chromosome number in each species nomenclature. Dashed radial lines represent the separation between rice chromosomes. A number of conserved traits have been positioned and are represented by open circles (O) along a radius (Devos and Gale, 1997).

candidate region in the complex genome. This type of shuttling strategy between model and target genomes has already proved successful in isolating genes from complex genomes and further justified a major effort in sequencing the rice genome.

However, now that rice has been sequenced and that much more detailed genetic maps are available for complex genomes, it is realized that synteny conservation is not as perfect as the model predicted and that many genes present in rice are not present at the orthologous position in maize or wheat and *vice versa* (Bennetzen and Ma, 2003). In addition, the situation is complicated by the fact that many genes have been duplicated as part of large segmental chromosome duplications, of ploidy events, or as part of localized tandem duplications (International Rice Genome Sequencing Project, 2005). In addition, cereal genomes have been invaded by retroelements which act as a driving force, both to increase genome size and to reduce it by allowing deletions via recombination between Long Terminal Repeat or LTR (Juretic *et al.*, 2004; Chapter 26). The many observed breakages in synteny conservation are indeed not unexpected, when one realizes that cereals started to differentiate as early as 70 millions years ago.

RICE GENOME SEQUENCING

Rice genome sequencing began with the accumulation of the necessary resources. Several dense saturated genetic maps had been produced in Japan, US, France, Korea, accumulating thousands of markers. A major initial effort was made by the Japanese Rice Genome Programme, who constructed numerous cDNA libraries, delivered more than 10,000 ESTs in the early 1990s, prepared YAC (Yeast Artificial Chromosome) and PAC (*P*1 phage derived *A*rtificial Chromosome) large insert libraries. Another key effort was made in the USA, supported by the Rockefeller and Novartis Foundations, to construct ordered BAC (Bacterial Artificial Chromosome) libraries (Sasaki and Moore, 1997; Delseny *et al.*, 2001). These BAC libraries, partially ordered by fingerprinting and contiging the clones and by BAC end sequencing, were available at the end of 1997 when the IRGSP (International Rice Genome Sequencing Project) was set up with the goal of obtaining the full sequence of the rice genome by 2007. This consortium decided to use the clone by clone strategy following a minimum tiling path for each chromosome (Chen *et al.* 2002; Wu *et al.*, 2002) and distributed specific chromosomes to the participating countries as for the Arabidopsis Genome Initiative. Japan received chromosomes 1, 2, 6, 7, 8 and 9, the US received chromosomes 3, 10 and 11, China received chromosome 4, Taiwan received chromosome 5 and France received chromosome 12. Several additional countries (Thailand, Korea, Brazil and India) sequenced limited portions of one chromosome (Delseny 2003; International Rice Genome Sequencing Project, 2005).

Sequencing indeed started efficiently by the end of 1999 and has been accelerated, due to three outsider competitors, who, independently from the consortium, decided to sequence the whole genome by different approaches. Two agrochemical companies, Monsanto and Syngenta have produced drafts of the sequence of the same *japonica* variety as IRGSP, Nipponbare, and the Beijing Genome Institute produced a draft shot gun sequencing of the two *indica* varieties which are (93-11 and Pei-Ai 645) the paternal and maternal parents of the Chinese super hybrid LYP 9. Monsanto and Syngenta respectively claimed in April 2000 and February 2001 that they had both achieved the sequencing of the rice genome but only Syngenta and the Beijing Group published their work in 2002. It turned out that their sequences were rather preliminary and incomplete, since they were distributed into 42,000 contigs for the Syngenta ones and 127,000 for the Beijing, Group. None of these groups were able to assemble the highly repetitive sequences (Barry, 2001; Goff *et al.*, 2002; Yu *et al.*, 2002). Monsanto and later on, Syngenta negotiated with the IRGSP a privileged access to their data, which accelerated the IRGSP and saved a substantial amount of money. The announcement of a web available high quality phase 2 (Chapter 2) draft was made by the IRGSP consortium at the end of 2002 and so far 5 chromosomes have been published (Sasaki *et al.*, 2002; Fang *et al.*, 2002; Rice chromosome 10 sequencing consortium 2003, Rice chromosome 11 and 12 sequencing consortia, 2005). The complete finished sequence was announced by the IRGSP in November 2004 (International Rice Genome Sequencing Project, 2005) and at the same time the Beijing Group reported a considerably elaborated draft of the Indica sequence (Yu *et al.*, 2005). All these sequences are available from Genbank (http://www.ncbi.nlm.nih.gov/Genbank).

Accordingly, the sequencing phase of the rice genome is essentially finished and the most difficult task, the physical and functional annotation, remains to be carried out.

SPECIFIC FEATURES OF RICE GENOME

A completely unexpected feature is the observation that the vast majority of rice genes contain a nucleotide composition gradient from the 5' to 3' end, the 5' end being more GC-rich than the 3'. This feature further complicates physical annotation.

With a fully sequenced genome, it should be possible to determinate the gene number. However, this issue is far from easy: first, annotation is being carried out independently by the various groups, with heterogeneous criteria and, secondly many predicted genes are likely to be remnants of transposable elements. The first reported numbers ranged between 32,000 and 62,000 genes for the shotgun *indica* genome (Goff *et al.*, 2002; Yu *et al.* 2002). Estimates deduced from the first two sequenced chromosomes were between 57,000 and 62,500 genes (Feng *et al.*, 2002; Sasaki *et al.*, 2002, Delseny 2003). After completion of the Nipponbare sequence it was reannotated homogeneously (Yuan *et al.*, 2005): 61,250 gene models were predicted, of which 53% corresponded to *bona fide* known genes, 6% corresponded to expressed sequences and 40% were completely hypothetical, without any experimental evidence. It is estimated that 22% of these gene models are in fact transposable elements. As a consequence, the gene number has been reduced and currently stands at 37,544 (International Rice Genome Sequencing Project, 2005) The most updated rice genome annotation can be found on the following websites:

- http://riceorgaas.dna.affrc.go.jp
- http://rad.dna.affrc.go.jp
- http://rice.tigr.org

Structural annotation was improved with the determination of more than one million ESTs, 28,000 full length cDNAs (Kikuchi *et al.*, 2003) and with the comparison of the sequence with *A. thaliana*. More than 85% of the *A. thaliana* genes have homologues in rice. However the gene order between the two species has been almost completely reshuffled. (Salse *et al.*, 2002). Approximately 71% of the predicted rice genes have homologues in *A. thaliana* and 2,859 genes seem to be unique to rice and other cereals (International Rice Genome Sequencing Project, 2005). About 95% of known and expressed genes of rice have an equivalent in other cereals, thus justifying the use of rice as a model.

Gene distribution along the chromosomes is not even. Gene density is usually higher in the distal regions of chromosome arms (18-19 genes/100 kb) than in the proximal ones (10-12 genes/100 kb). In contrast, repeated elements have a distinct distribution. The majority of retrotransposons are located in pericentromeric regions, whereas MITEs (Miniature Inverted-repeat Transposable Element) are preferentially associated with genes. Whether such a contrasting distribution is enough to account for a gene space in rice is not clear yet.

Amongst the identified genes, there are no major changes in the functional categories in comparison to *A. thaliana*. For instance the same transcription factor families are represented in both genomes in roughly equivalent proportions. Considering flowering time and morphology, some *A. thaliana* genes have no orthologue in rice, which is not so surprising given the biological differences between the two species. An important difference concerns disease resistance (R) genes. The TIR-NBS-LRR (Toll-Interleukin Receptor-like) Nucleotide Binding Site, Leucine Rich Repeat, class which accounts for 2/3 of the *A. thaliana* R genes is completely absent from the rice genome, and probably from other cereal genomes. The biological significance of this difference is not yet clear(Zhou *et al.*, 2004).

When the physical distances along the chromosomes are compared to genetical distances, a clear cut suppression of recombination is observed in centrometric regions. The ratio between physical and genetic distances thus varies between 210 kbp/cM in euchromatin, up to 636 kbp/cM in heterochromatin.

Sequencing of the rice genome revealed several additional important features. The first one is the presence of a large number of mitochondrial and chloroplastic DNA insertions along the chromosome. Chromosome 12 thus has 1% of its total DNA represented by such insertions (The Rice chromosomes 11 and 12 Sequencing Consortia, 2005). Present data indicates that several successive transfer events have ocurred occurred during the past 10 million years. A second feature, already outlined after sequencing of the *A. thaliana* genome, is the duplicated nature of this genome. Most of the genes are members of multigene families. Some result from tandem duplications and sometimes make large clusters, such as a group of 10 protein kinase genes on chromosome I. This organisation pattern is superimposed on to large segmental duplications representing several megabases. The rice genome can be organized into 18 pairs of duplicated segments covering nearly 70% of the genome. These duplications, with the exception of one between the short arms of chromosomes 11 and 12, are probably more ancient than the ones undergone by the *A. thaliana* genome, since there are on average only 17% of the genes which are present as pairs within the two segments of a duplication (Guyot and Keller, 2004; Salse *et al.*, 2004; Yu *et al.*, 2005; International Rice Genome Sequencing Project, 2005; The Rice Chromosomes 11 and 12 Sequencing Consortia, 2005). Distribution of heterochromatin is also quite uneven among chromosomes. This was revealed both by cytogenetic methods and by sequencing (Cheng *et al.*, 2002). The chromosome with the largest heterochromatin is chromosome 10. An intriguing issue is the organisation of the centromere. The centromere in rice contains many copies of a 155 bp satellite DNA and copies of a centromere specific retrotransposon. The number of repeats of this satellite varies from 1,980 on chromosome 1 to 80 on chromosome 8. The centromeric regions can also be delimited by immunostaining with antibodies against the centromere specific histone H3 variant. Interestingly the regions defined by the presence of the satellite DNA and by variant H3 staining do not completely overlap.

COMPARISON OF THE RICE GENOME WITH OTHER CEREAL GENOMES

As indicated previously, a major argument for sequencing the rice genome was that it could be used as a model to unravel the organisation of the other cereal genomes. Comparisons with other genomes can be made in two ways, the first being at the whole genome scale. For instance, by selecting all the mapped and sequenced cDNAs from maize and aligning them against the rice genomic sequence, one can identified the best homologue genes (likely the orthologues) between the two species. Such an analysis has now been made between rice, wheat, corn and sorghum (Goff *et al.*, 2002; Sorell *et al.*, 2003; Klein *et al.*, 2003; Paterson *et al.*, 2004; Salse *et al.*, 2004). The comparison has revealed, as expected, extensive synteny conservation, but many more rearrangements than anticipated from the scheme shown in Fig. 20.2. An illustration is given for the rice-maize comparison in Fig. 20.3 (Salse *et al.*, 2004). The second approach consists of sequencing an orthologous region in several species and is illustrated in Fig. 20.4 (Bennetzen and Ma, 2003) for the *Shrunken 2* (*sh2*)/*Anthocyaninless* (*A1*) locus in rice, sorghum, maize and wheat. It has demonstrated that the locus is considerably expanded in wheat and maize due to the invasion of retrotransposons. In sorghum, the *A1* gene has been tandemly duplicated. This specific locus has been completely reshuffled in wheat on 4 distinct chromosome segments. In addition, in wheat, two copies of the *X1* and *X2* genes have become pseudogenes.

Because the genome of two varieties of rice has been sequenced, it is possible to compare them directly. However this comparison is difficult, as about 15% of the two genomes cannot be simply aligned due to the presence of many retrotransposons (Han and Xue 2003). Nevertheless, with the improvement of sequencing quality, a total of 80 127 polymorphic sites between the two species could be recognized. SNP (Single Nucleotide Polymorphism) frequency varies from 0.53 to 0.78 %, which is 20 times more than that which is observed between the *A. thaliana* Columbia and Landsberg ecotypes (International Rice Genome Sequencing Project, 2005). These retrotransposons are extremely helpful to distinguish the two rice types and to date their divergence (Vitte *et al.*, 2004).

RICE FUNCTIONAL GENOMICS

Several tools can be used to analyse the function of rice genes. The first step is to use sequence comparison to infer a putative function to a given sequence. However, at this time, the major limitation is that many of the gene models in rice are not yet completely reliable, making it hazardous to compare them with other genes in the database. The release of a set of full length cDNA sequences (Kikuchi *et al.*, 2003) considerably improved the situation, although a number of problems persist. In addition, these full length cDNAs should turn out to be extremely useful in experiments involving ectopic expression, site-directed mutagenesis and preparing antibodies, or analysing interactions with partners.

A second step to analyse rice gene function is to engineer and analyse mutant collections. Because rice, particularly the *japonica* type, is easy to transform (Tyagi and Mohanty, 2000; Delseny *et al.*, 2001), several labs have set up collections of T-DNA and maize transposable element (*Ac*/*Ds* and *En*/*Spm*) insertion mutants. An alternative to this strategy is to use the endogenous *Tos17* retrotransposon to knock out genes. *Tos17* is a retrotransposon which can be activated by *in vitro* culture, so that regenerated plants contain 3-10 additional copies of the element. The advantage of using *Tos17* for insertion mutagenesis is that the collection can be grown and evaluated in the field, without any restriction. This is more difficult with T-DNA or Ac/Ds insertion lines due to GMO regulations. Most of the groups who have engineered such collections have started to sequence the genomic regions flanking the insertions (FST or Flanking Sequence Tags). So far approximately 110,000 FST have been identified from nearly 400,000 lines. Most of them have been positioned along the chromosome sequence as illustrated in Fig. 20.5. The estimate is that at least 430,000 lines are necessary to saturate

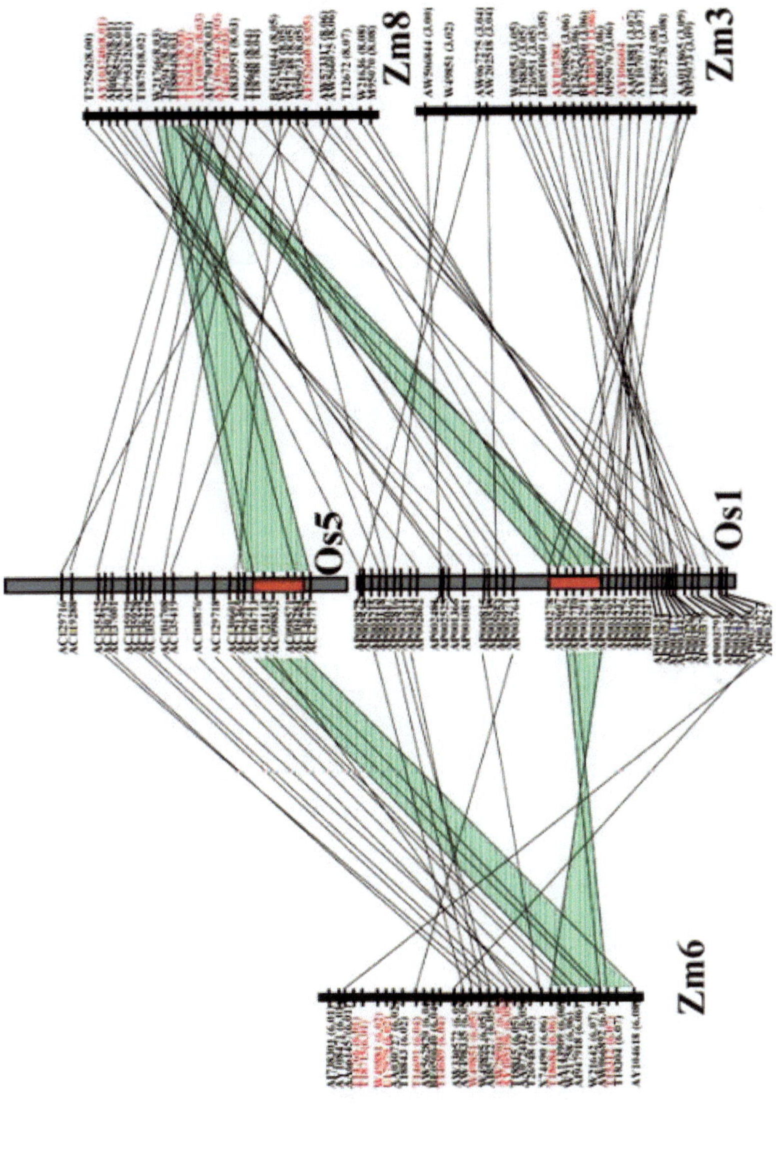

Fig. 20.3 Alignment between rice chromosomes Os 1 and Os 5 and maize chromosomes Zm 6, Zm 8 and Zm 3 (Salse *et al.*, 2004) Horizontal bars on rice chromosomes indicate BAC clones and their accession numbers. Each bar, on maize chromosomes correspond to a mapped and sequenced marker with an homologue in rice. Markers in common between maize and rice are linked by straight lines. Shading illustrates collinear regions between rice and maize chromosomes and identifiy a region in rice Os 5 and Os 1 which is duplicated and represented in a different colour.

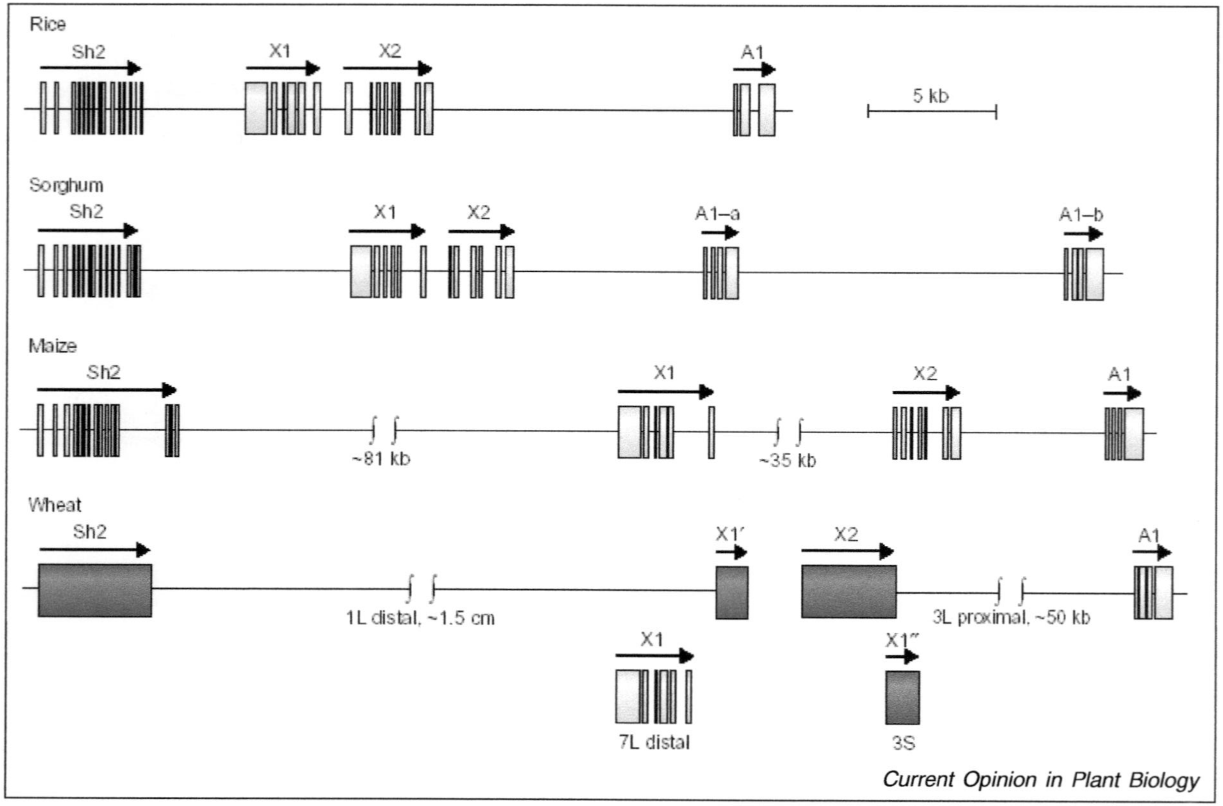

Fig. 20.4 Comparison of the gene organisation between rice, sorghum, maize and wheat around the *Sh 2* (*Shrunken 2*) locus–(Bennetzen and Ma, 2003). The exon /intron structure of the different genes is shown. Arrows show the positions, approximate sizes and transcriptional orientations of each gene. Note that the locus slightly expended in sorghum and much more in maize. The locus has been completely reshuffled in wheat and the genes are dispersed on 4 chromosome arms (1L, 3L, 3S and 7 S). In addition two copies of the *X1* genes are truncated on 1L and 3S and the *Sh2* and *X2* have lost their introns.

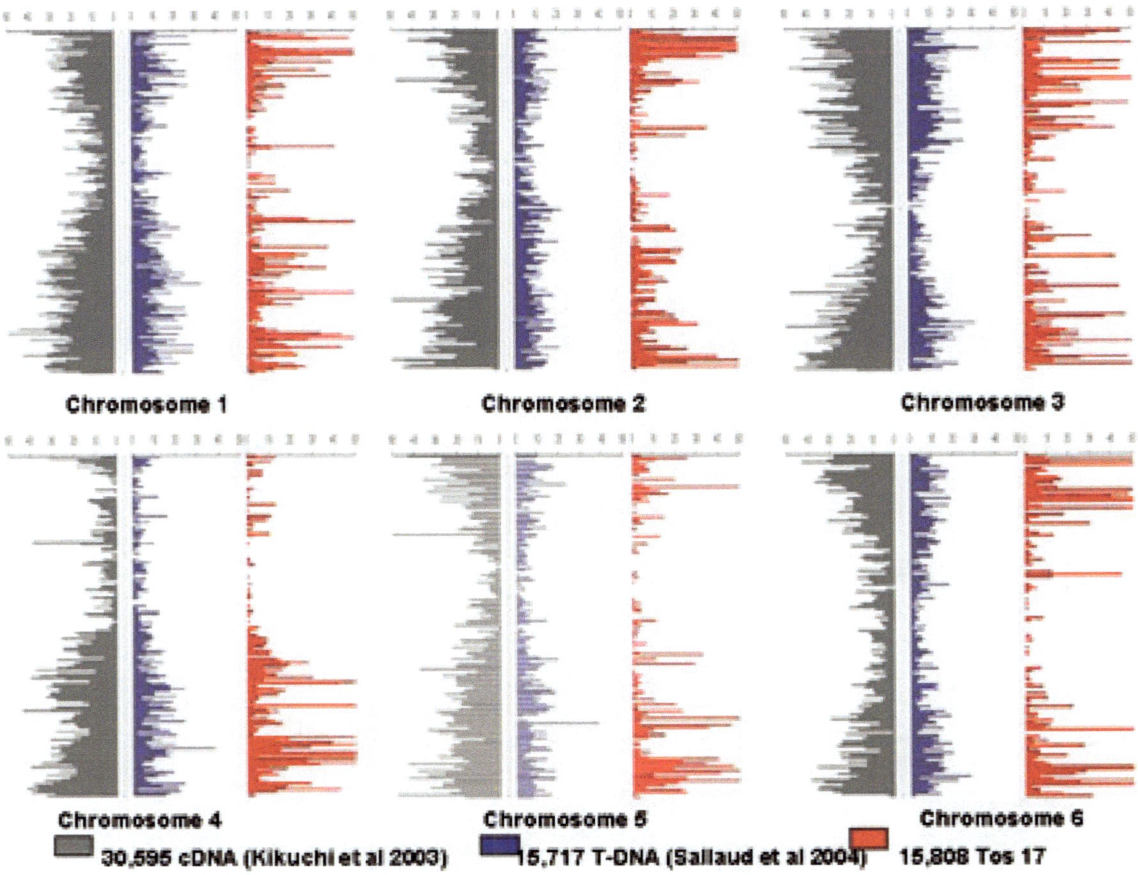

Fig. 20.5 Distribution of the full length cDNA, T-DNA insertions and *Tos 17* insertion along the rice chromosomes 1-6 (Sallaud *et al.,* 2004). Densities of T-DNA, and *Tos 17* FST and Full Length cDNA were plotted for each 200 kbp.

the rice genome (Miyao *et al.* 2003; Greco *et al.*, 2004; Hirochika *et al.*, 2004; Sallaud *et al.*, 2004; Enckevort *et al.*, 2005). Such collections can be screened according to two strategies. Forward genetic screening consists in observing putative mutants in the collection, establishing linkage between the phenotype and the insertion, characterizing the target gene and analysing its expression. The alternate strategy is based on reverse genetic screens, in which the FST database is interrogated for sequences matching the target gene, or its immediate environment. When a positive line is identified, *in silico*, it can be grown in the greenhouse or in the field and the phenotype can be analysed in detail. An interesting variation of the insertion mutants is the use of defective transposable elements from maize (*Ds*). Sets of lines with *Ds* insertion are engineered and the position of the insertion in the sequence is determined. These *Ds* elements can then be mobilized by crossing a specific *Ds* line with another line carrying the transposase gene. Because *Ds* usually jumps from its initial position to a nearby one, this strategy should prove useful to saturate a given chromosomal region with mutations. Sets of lines have also been recently engineered in order to tag specific cell lineages with a GFP (Green Fluorescent Protein) marker and to carry out transactivation experiments in these cells (Johnson *et al.*, 2005) The major limitation in using these strategies is to collect all the data in a unique searchable database and to collect enough seed material from each line in a single stock centre, which can distribute it on request to the scientific community.

Besides insertion lines, other classical mutant collections (EMS, irradiation...) have been set up (Wu *et al.*, 2005). They can be screened either by observing phenotypes and isolating corresponding genes by positional cloning, or by using the newly developed strategy of TILLING (Targeting Induced Local Lesions IN Genomes) which allows the identification of multiple alleles in known genes. However, several backcrosses are usually necessary to correlate the mutation to the phenotype, as there are many mutations in each line.

The availability of the rice sequence and of many ESTs has opened up the way to transcriptome analysis techniques (Dubouzet *et al.*, 2003). Several additional reports have already been published using either partial or pangenomic, oligonucleotides, or cDNA chips (Zhu *et al.*, 2003, Stolk *et al.*, 2005).

Other high throughput functional genomic methods, such as proteomic (Komatsu, 2005), metabolomic, and interactomic techniques are slowly emerging, the delays being due to the necessity to finish an accurate genome sequence, to re-annotate it and to mobilize money and manpower to develop and fully exploit these strategies.

CONCLUSION

Rice, with a small genome size, roughly conserved colinearity with other cereal genomes, a relative ease of transformation and a large mutant collection, has been confirmed as a good model. The complete DNA sequences of two genotypes are now available. A recent issue of *Plant Molecular Biology* (Wing, 2005) contains a series of updates on rice genomics. In addition, important efforts have been made to analyse related wild type genomes (Wing *et al.*, 2005) and to align their physical maps along side the cultivated rice maps. The numerous accessions and the vast germplasm of varieties adapted to multiple environments now provide the breeders and the biologists with an extensive allele reservoir, in which they can search both for improving the crop and understanding the molecular basis of its traits. Sequencing and analysing the rice genome has set up a platform for the analysis of the other more complex cereal genomes. The next challenge is to complete the maize sequence within the next five years.

REFERENCES

Barry, G., 2001. The use of the Monsanto draft rice genome sequence in research. *Plant Physiol,* 125, 1161-1165.

Bennetzen, J.L., Ma, J., 2003. The genetic colinearity of rice and other cereals on the basis of genomic sequence analysis. *Current Opinion Plant Biol.,* 6, 128-133.

Chen, M., Presting, G., Barbazuk, W.D., *et al.* 2002. An integrated physical and genetic map of the rice genome. *Plant Cell,* 14, 537-545.

Cheng, Z., Dong, F., Langdon,T., Ouyang, S., Buell, R., Gu, M., Blattner, F.R., Jiang, J. 2002. Functional rice centromeres are marked by a satellite repeat and a centromere-specific retrotransposon. *Plant Cell,* 14, 1691-1704.

Delseny, M., Salse, J., Cooke, R., Sallaud, C., Regad, F., Lagoda, P., Guiderdoni, E., Ventelon, M., Brugidou C., Ghesquière, A. 2001. Rice genomics: present and future. *Plant Physiol. Biochem.,* 39, 323-334.

Devos, K.M., Gale, M.D. 1997. Comparative genetics in the grasses. *Plant Mol. Biol.* 35, 3-15.

Dubouzet, J.G., Sakuma, Y., Ito, Y., Kasuga, M., Dubouzet, E.G., Miura, S., Seki, M., Shinozaki, K, Yamaguchi-Shinozaki, K. 2003. *OsDREB* genes in rice, *Oryza sativa* L., encode transcription activators that function in drought-, high salt- and cold-responsive gene expression. *Plant J.,* 33, 751-763.

Enckevort, L.J.G., Droc, G., Piffanelli, I.P. *et al.* 2005. EU-OSTID: A collection of transposon insertional mutants for functional genomics in rice. *Plant Mol. Biol.,* 59, 99-110.

Feng, Q., Zhang, Y., Hao, P. *et al.* 2002. Sequence and analysis of rice chromosome 4. *Nature,* 420, 316-320.

Goff, A.S., Ricke, D., Lan, T.H. *et al.* 2002. A draft sequence of the rice genome (*Oryza sativa* L. ssp *japonica*) *Science,* 2002, 296, 92-100.

Greco, R., Oowerkerk, P.B., Taal, A.J.C., Sallaud, C., Guiderdoni, E., Meijer, A.H., Hodge, J.H.C., Pereira, A. 2004. Transcription and somatic transposition of the maize *En/Spm* transposon system in rice. *Mol. Gen. Genomics,* 270, 514-523.

Guyot, R., Keller, B. 2004. Ancestral genome duplication in rice. *Genome,* 47, 610-614.

Han, B., Xue, Y. 2003. Genome wide intraspecific variations in rice. *Current Opin. Plant Biol.,* 6, 134-138.

Hirochika, H., Guiderdoni, E., An, G., Hsing, Y.I., Eun, M.Y., Upadhyaya, N., Ramachandran, S., Zhang, Q., Pereira, A., Sundaresan, V., Leung, H. 2004. Rice mutant resources for gene discovery. *Plant Mol. Biol.,* 54, 325-334.

International Rice Genome Sequencing Project. 2005. The map-based sequence of the rice genome. *Nature,* 436, 793-800.

Johnson, A.A.T., Hibberd, J.M., Gay, C., Essah, P.A., Haselhof, J., Tester, M., Guiderdoni, E. 2005. Spatial control of transgene expression in rice (*Oryza sativa* L.) using the GAL 4 enhancer trapping system. *Plant J.,* 41, 779-789.

Juretic, N., Bureau, T., Bruskiewich, R.M. 2004. Transposable element annotation of the rice genome. *Bioinformatics,* 20, 155-160.

Kikuchi, S., Satoh, K., Nagata, T. *et al.* 2003. Collection, mapping, and annotation of over 2,800 cDNA clones from *japonica* rice. *Science*, 301, 376-379.

Klein, P.E., Klein, R.R., Vrebalov, J., Mullet, J.E. 2003. Sequence-based alignment of sorghum chromosome 3 and rice chromosome 1 reveals extensive conservation of gene order and one major chromosomal rearrangement. *Plant J.*, 34, 605-621.

Khush, G.S. 2005. What it will take to feed 5.0 billion rice consumers in 2030? *Plant Mol. Biol.*, 59, 1-6.

Komatsu, S. 2005. Rice proteome database: a step towards of functional analysis of the rice genome. *Plant Mol. Biol.*, 59, 179-190.

Miyao, A., Tanaka, K., Murata, K. *et al.* 2003. Target site specificity of the *Tos 17* retrotransposon shows a preference for insertion within genes and against insertion in retrotransposon-rich regions of the genome. *Plant Cell*, 15, 1771-1780.

Moore, G., Devos, K.M., Wang, Z., Gale, M.D. 199. Cereal genome evolution. *Current Biol.*, 5, 737-739.

Paterson, A.H., Bowers, J.E., Chapman, B.A., 2004. Ancient polyploidization predating divergence of the cereals and its consequences for comparative genomics. *Proc. Natl. Acad. Sci.* USA, 101, 9903-9908.

Rice chromosome 10 sequencing consortium. 2003. In depth view of structure, activity and evolution of rice chromosome 10. *Science*, 300, 1566-1569.

Sallaud, C., Gay, C., Larmande, P. *et al.* 2004. High throughput T-DNA insertion mutagenesis in rice: A first step towards in silico reverse genetics. *Plant J.*, 39, 450-464.

Salse, J., Piegu, B., Cooke, R, Delseny, M. 2002. Synteny between *Arabidopsis thaliana* and rice at the genome level: a tool to identify conservation in the ongoing rice sequencing project. *Nucleic Acids Res.*, 30, 2316-2328.

Salse, J., Piegu, B., Cooke, R, Delseny, M., 2004. New *in silico* insight into the synteny between rice (*Oryza sativa* L.) and maize (*Zea mays* L.) highlights reshuffling and identifies new duplications in the rice genome. *Plant J.*, 38, 396-409.

Sasaki, T., Moore, G. (Eds). 1997. Oryza:from molecule to plant. *Plant Mol. Biol.*, 35, 1-254. (A rice special issue summarizing knowledge at that time).

Sasaki, T., Matsumoto, T., Yamamoto, K. *et al.* 2002. The genome sequence and structure of rice chromosome 1. *Nature*, 420, 312-316.

Sorrells, M.E., La Rota, M., Bermudez-Kandianis, C.E. *et al.* 2003. Comparative DNA sequence analysis of wheat and rice genomes. *Genome Res.*, 13, 1818-1827.

The Rice Chromosome 11 and 12 Sequencing Consortia. 2005. The sequence of rice chromosomes 11 and 12, rich in disease resistance genes and recent gene duplications. BMC Biology, 3, e20-38.

Stolk, V., Li, L., Wang, X. *et al.* 2005. A pilot study of transcription unit analysis in rice using oligonucleotide tiling path microarray. *Plant Mol. Biol.*, 59, 161-164.

Tyagi, A.K., Mohanty, A. 2000. Rice transformation for crop improvement and functional genomics. *Plant Sci.*, 158, 1-18.

Vaughan, D.A., Morishima, H., Kadowaki, K. 2003. Diversity in the *Oryza* genus. *Current Opin. Plant Biol.*, 6, 139-146.

Vitte, C., Lamy, F., Ishii, T., Brar, D., Panaud, O. 2004. Genomic paleontology provides evidence for two distinct origins of Asian rice (*Oryza sativa* L). *Mol. Gen. Genomics*, 272, 504-511.

Wing, R., 2005. Editorial. *Plant Mol. Biol.*, 59, vii.

Wing, R., Ammiraju, J.S., Luo, M., *et al.* 2005. The Oryza map alignment project: the golden path to unlocking the genetic potential of wild rice species. *Plant Mol. Biol.*, 59, 53-62.

Wu, J., Maehara, T., Shimokawa, T. *et al.* 2002. A comprehensive rice transcript map containing 6,591 expressed sequence tag sites. *Plant Cell,* 14, 525-535.

Wu, J.L., Wu, C., Lei, C. *et al.* 2005. Chemical- and irradiation-induced mutants of indica rice IR64 for forward and reverse genetics *Plant Mol. Biol.,* 59, 85-97.

Yuan, Q., Ouyang, S., Wang, A. *et al.* 2005. The institute for genomic research Osa1 rice genome annotation database. *Plant Physiol.,* 138, 18-26.

Yu, J., Hu, S., Wang, J. *et al.* 2002. A draft sequence of the rice genome (*Oryza sativa* L ssp *indica*). *Science,* 296, 79- 92.

Yu, J., Wang, J., Lin, W. *et al.* 2005. The genomes of *Oryza sativa*:a history of duplications. *PLoS Biol.,* 3(2) e38.

Zhou, T., Wang, Y., Chen, J.Q., Araki, H., Jing, Z., Jiag, K., Shen, J., Tian, D. 2004. Genome-wide identification of NBS genes in *japonica* rice reveals significant expansion of divergent nonTIR NBS-LRR genes. *Mol. Gen.* Genomics, 271, 402-415.

Zhu, T., Budworth, P., Chen, W., Provart, N., Chang, H.S., Guimil, S., Su, W., Estes, B., Zou, G., Wang, X. 2003. Transcriptional control of nutrient partitioning during rice grain filling. *Plant Biotech. J.,* 1, 59-70.

21

CHAPTER

The Model *Medicago truncatula*: from Plant-Microbe Symbioses to Crop Legume Breeding Boosted by Cross-Legume Genomics

E.P. Journet, P. Gamas

IMPORTANCE AND UNIQUE FEATURES OF LEGUMES

Legumes, otherwise known as the Fabaceae, constitute the third largest family of flowering plants, comprising approximately 650 genera and over 18,000 species (Polhill *et al.*, 1981). Legumes range from small herbaceous plants to giant trees, dominating many tropical rainforests. A major attribute of legumes is their almost unique capacity for symbiotic fixation of atmospheric dinitrogen. Under limiting conditions of mineral nitrogen–a frequent situation in soils–most legumes can establish a specific relationship with members of the *Rhizobium* family of soil bacteria. Following reciprocal partner recognition, root infection leads to the formation of specialized, symbiotic nitrogen-fixing organs, the root nodules, which host differentiated bacteroids. In this mutually beneficial interaction, the plant provides the endosymbiont bacteria with carbohydrate, in exchange for valuable fixed nitrogen, resulting from the conversion of N_2 into NH_3 by bacterial nitrogenase (Mylona *et al.*, 1995; Pawlowski and Bisseling, 1996; Gage, 2004; Patriarca *et al.*, 2004; Geurts *et al.*, 2005).

Thus, legumes play a crucial role in both natural ecosystems and traditional agrosystems, by providing an abundant supply of renewable fixed nitrogen. As major contributors to the global nitrogen cycle, legumes are key species in many ecosystems. The high protein content of legume vegetative parts and seeds is the reason for their high nutritional value for both humans and animals. Cultivated legumes, in addition to being nitrogen self-sufficient, are also a key source of organic soil fertilizer and an essential component of most agricultural rotations.

Since the emergence of human civilizations, legume species have been domesticated and cultivated for millennia in association with cereals. Pastures and forages worldwide consist of a mixture of legumes and grasses. Nowadays, grain and forage legumes are second only to cereal crops in agricultural importance, accounting for approximately 27% of the world's primary crop production, and grain legumes alone contribute about one-third of the dietary protein intake of humans. Legumes are also an important source of both fodder and forage for animals and of edible and industrial oils (Graham and Vance, 2003). Moreover, legumes have evolved a complex assortment of secondary metabolites, involved in many of the beneficial and pathogenic interactions with other organisms. These include various flavonoids, isoflavonoids and triterpene saponins, many of which have documented antimicrobial and pharmacological properties. A number of isoflavonoids and saponins may be of benefit to human health through anticancer and other health-promoting activities (Dixon and Sumner, 2003).

In terms of production and trade, soybean (*Glycine max*) is currently the most important legume in the world: most of the seed produced is used for oil extraction and as fodder for livestock. Common bean (*Phaseolus vulgaris*), pea (*Pisum sativum*), chickpea (*Cicer arietinum*), broad bean (*Vicia faba*), pigeon pea (*Cajanus cajan*), cowpea (*Vigna unguiculata*), and lentil (*Lens culinaris*) constitute the primary dietary grain legumes for populations in Asia, Africa, and South America. Important forage or pasture legume species in temperate climates include alfalfa (*Medicago sativa*), clovers (*Trifolium* spp.), trefoil (*Lotus corniculatus*), sweetclovers (*Melilotus* spp.) and vetches (*Vicia* spp.). Woody tree legumes represent an interesting potential for forestry, and for tree-fallow and alley-cropping systems for subsistence farming in Africa and Asia (Graham and Vance, 2003).

In addition to the rhizobial symbiosis, legume plants also have the capacity to associate with symbiotic soil fungi (the Glomeromycota) to form arbuscular endomycorrhiza (AM), where the fungus colonizes cortical root tissues and develops specialised exchange interfaces within the plant host cells. AM associations enhance the ability of plants to scavenge nutrients such as phosphate from the soil, partly due to the greater volume of soil exploited by the filamentous fungal partner (Gianinazzi-Pearson, 1996; Smith and Read, 1997; Harrison, 2005). This type of fungal-plant symbiosis is shared by more than 90% of land plants, a reflection of its ancient evolutionary origin and importance. Notable exceptions include the Crucifer (Brassicaeae) family, including the model-plant *Arabidopsis thaliana*.

Thus, legumes have the rare privilege of benefiting from both rhizobial and AM endosymbioses, which play crucial roles in their mineral nutrition, development patterns and population dynamics. The need to understand the molecular bases of these unique symbioses, which cannot be studied in parallel using either of the model plants *A. thaliana* or rice (*Oryza sativa*), have motivated basic research on legume molecular biology for the last two decades. Recent studies increasingly point to common or similar pathways for the two symbioses (Parniske, 2000; Oldroyd *et al.*, 2005). Legumes are also valuable models to study plant-microbe interactions in general, in order to directly compare in the same organism symbiotic, pathogenic and parasitic interactions, as well as the co-evolution of the host with its complex microbial environment (Parniske, 2000). Other particularities justify research on legume biology, such as plant architecture and development, protein-rich seed formation and the biosynthesis of specific secondary metabolites.

With respect to their agronomic uses, legumes are economic and environmentally friendly crop and pasture species. The manufacturing and application of nitrogen fertilisers requires large amounts of increasingly expensive fossil energy (two tons of fuel equivalents / ton N fertiliser). Thus, using legumes as a "free" N source to produce plant proteins and to fertilize soils results in reduced costs of farm energy inputs, lower emissions of CO_2 and other greenhouse gases, as well as a major reduction of environmental pollution from water run-off of nitrates. Legumes also contribute to crop diversity, which helps break disease cycles in cereal-rich rotations and should result in a reduced use of pesticides. Therefore, a better and larger use of legumes is one obvious solution for the necessary transition towards more sustainable agricultural systems (Jensen and Hauggaard-Nielsen, 2003). In recent decades, improvements in legume crop yields have not kept pace with those of cereals. To achieve this goal, more efforts will have to be dedicated to legume genetics, breeding, crop physiology and agronomy. Obvious targets include the quality of seeds as well as the predictability and performance of crops in response to biotic and abiotic stresses (diseases, pests, drought, as well as soil salinity, acidity and nutrient limitations; Graham and Vance, 2003; Gepts *et al.*, 2005).

NECESSITY OF MODEL LEGUMES FOR LEGUME BIOLOGY RESEARCH AND CROP LEGUME BREEDING

All of the above underline the need for carrying out both basic and applied research on legume biology, genetics and genomics. Although research on non-legume model species such as *A. thaliana* and rice provides insight into many fundamental features of plant biology, it cannot of course address many important aspects of legume biology nor be readily applied to legume breeding programmes. Equally, the large number of important legume species precludes their simultaneous in-depth characterization. The legume family is divided into three subfamilies: Mimosoideae, Caesalpinioideae, and the numerically dominant Papilionoideae (Doyle and Luckow, 2003). The Papilionoideae subfamily contains nearly all of the economically important crop legumes. With the notable exception of peanut, all of them fall into the Galegoid and Phaseoloid clades, which are often referred to as temperate ("cool season") and tropical legumes, respectively (Fig. 21.1). Despite their close phylogenetic affiliations, the genetic systems represented within this group are diverse, ranging from simple autogamous diploids to complex out-crossing polyploids. Genome size also varies widely among legumes, with pea having a genome size 10 times that of some related diploid genera (Table21.1). For those reasons, and with the goal of establishing a unified genetic system for legumes, the need was recognized early on for a model legume amenable to molecular genetics, and sharing sufficient synteny with crop legumes for efficient knowledge transfer in breeding programmes (Choi *et al.*, 2004b).

Medicago truncatula and *Lotus japonicus* as Primary Model Legumes

The two cultivated legume species, pea and soybean, were originally amongst the best genetically characterized crop plants, with well developed genetic maps based on

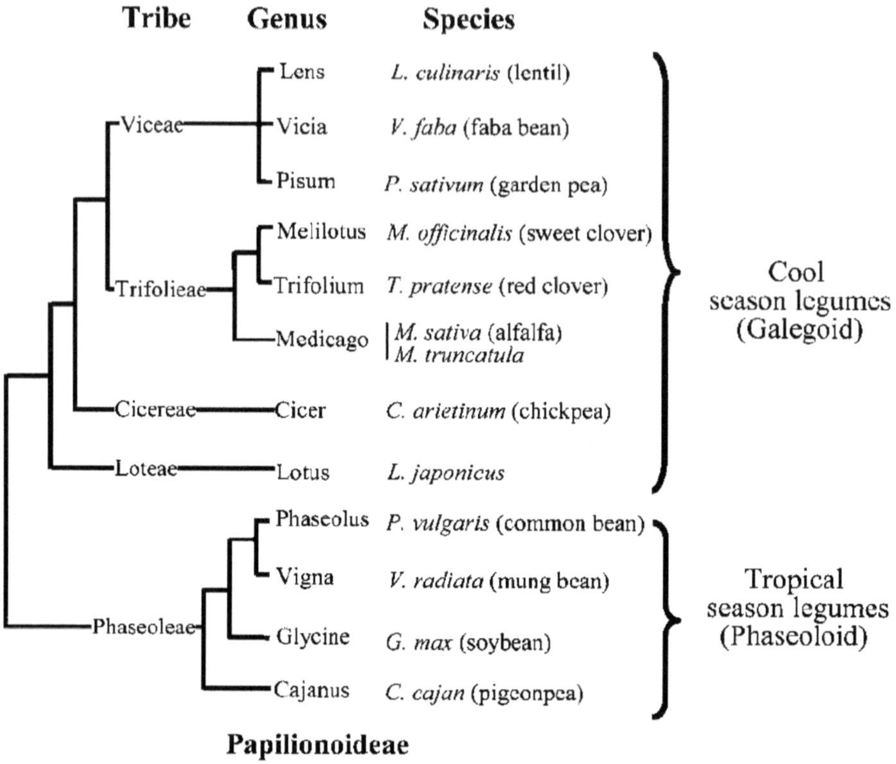

Fig. 21.1 Dendrogram depicting phylogenetic relationships of the legume subfamily *Papilionoideae*. Crop species belong mainly to the Galegoideae or Phaseoleae clades. Reproduced, with permission, from Zhu *et al.* (2005), based on Fig. 5 of Doyle and Luckow (2003).

numerous classical and molecular markers. However, as with other crop legumes, these species have one or more characteristics (e.g. medium... to large genomes, polyploid nature, recalcitrance to genetic transformation/regeneration; Somers *et al.*, 2003; Zhu *et al.*, 2005) that limit their utility as model systems. Moreover, most grain legumes (e.g. chickpea, *Vigna*, pea, beans and soybean) have large seeds, relatively few seeds per plant and large seedlings, which limits high-density culture.

In this context, the search to find a legume with more favourable attributes led, at the beginning of the 1990s, to the proposal of both *Medicago truncatula* (*Mt*) and *Lotus japonicus* (*Lj*) as model species. The choice of *Mt* initially resulted from a collaborative program run by the French public research institute INRA in 1985-86. The initial objective was to select a species within the genus *Medicago* with all the attributes of a model legume and that could be nodulated by *Sinorhizobium meliloti*, the well-studied N-fixing microsymbiont of alfalfa (*Medicago sativa*), in order to establish a model symbiotic couple (Barker *et al.*, 1990). On the other hand, *Lj* was chosen by laboratories previously studying the close relative *Lotus corniculatus* (Handberg and Stougaard, 1992). Both *Mt* and *Lj* are self-fertile diploids with modest genome sizes (around 500 Mb each), short

Table 21.1 Chromosome number, ploidy level and genome size of major crop and model legumes[a]

Tribe	*Genus*	*Species*	*Chromosome No. (2n) / ploidy level*	*Genome size (Mb / 1C)*	*Self-Compatibilty*
Viceae	*Lens*	*L. culinaris* (lentil)	14 / 2x	4,116	Selfing
	Vicia	*V. faba* (faba bean)	12 / 2x	13,059	Selfing
	Pisum	*P. sativum* (garden pea)	14 / 2x	4,778	Selfing
Trifolieae	*Melilotus*	*M. officinalis* (sweet clover)	16 / 2x	1,103	Outcrossing
	Trifolium	*T. pratense* (red clover)	14 / 2x	637	Outcrossing
		T. repens (white clover)	32 / 4x	956	Outcrossing
	Medicago	***M. truncatula*** (barrel medic)	16 / 2x	466	Selfing
		M. sativa (alfalfa)	32 / 4x	1,715	Outcrossing
Cicereae	*Cicer*	*C. arietinum* (chick pea)	16 / 2x	931	Selfing
Loteae	*Lotus*	***L. japonicus***	12 / 2x	466	Selfing
Phaseoleae	*Phaseolus*	*P. vulgaris* (common bean)	22 / 2x	588	Selfing
	Vigna	*V. radiata* (mung bean)	22 / 2x	515	Selfing
	Glycine	*G. max* (soybean)	40 / 2x	1,103	Selfing
	Cajanus	*C. cajan* (pigeon pea)	22 / 2x	858	Selfing
Aeschynomeneae	*Arachis*	*A. hypogaea* (peanut)	40 / 4x	2,813	Selfing

[a]Data were obtained from the Angiosperm C-value Database (http://www.rbgkew.org.uk/cval/homepage.html), Release 5.0 (Bennett and Leitch, 2004). Genome size values are "prime estimates" from the database.

seed-to-seed generation times, and both are tractable to genetic manipulation. In contrast with *Lj*, *Mt* is a member of the Galegoid clade which includes most temperate crop legumes (Trifolieae tribe: alfalfa, clover; Vicieae tribe: pea, vetch, lentil; Cicereae tribe: chickpea). The main biological differences are annual growth and compact spiral seed pods in *Mt* versus perennial growth and plant rejuvenation through side shoots, and straight seed pods with easily accessible seeds in *Lj*. One additional significant difference between the two species lies in the developmental type of nodules, indeterminate for *Mt* versus determinate for *Lj*. Thus, although the two models are complementary to some extent, they nevertheless encompass only a fraction of the evolutionary diversity of the entire legume family (Udvardi *et al.*, 2005). From the beginning, *Mt* and *Lj* have played central roles in symbiosis research (e.g. see Pichon *et al.*, 1992; Crespi *et al.*, 1994; Kapranov *et al.*, 1997; Schauser *et al.*, 1998). Both legumes then evolved into models for a broader range of research, as a base for developing ambitious functional genomics and genome sequencing programs, and both now possess strong international research communities (Cook, 1999; Stougaard, 2001; VandenBosch and Stacey, 2003; Udvardi *et al.*, 2005; Young *et al.*, 2005).

Finally, soybean has recently been proposed as a third reference legume species, since it is a representative and widely studied member of the Phaseoloid clade (warm-season legumes; Gepts *et al.*, 2005). For the rest of this review, we will mainly focus on *Mt*. For a recent review on *Lj*, see Udvardi *et al.* (2005).

GENERAL PROPERTIES OF *M. TRUNCATULA* AS A MODEL LEGUME

Commonly known as "barrel medic" because of the shape of its seed pods, *Mt* is an annual legume which originates from the Mediterranean basin and is a pasture species of economic importance in Australia (Crawford *et al.*, 1989). It is closely related to the tetraploid alfalfa, the major world forage legume. It is self-fertile, has a short generation time (~3 months from seed to seed) and abundant seed production within four months. It has a simple diploid genome divided across 8 chromosomes with a total size of ~500 Mbp (Blondon *et al.*, 1994), equivalent to four to five times the size of the *A. thaliana* genome and similar to that of rice.

Biodiversity within *Mt* is characterized by a high level of morphological and genetic variability within and between populations, and by a high level of homozygosity within each individual genome (Bonnin *et al.*, 1996). Genetic resources are abundant and freely accessible. *Mt* is nodulated by *S. meliloti*, the best characterized *Rhizobium* species and whose genome was the first to be completely sequenced (http://sequence.toulouse.inra.fr/S.meliloti; Galibert *et al.*, 2001). In particular, the many symbiotic mutants available for *S. meliloti* constitute excellent tools to dissect the mechanisms of the nitrogen-fixing symbiosis.

Several genotypes of *Mt* are now commonly used in research laboratories. In particular, the lines A17 (a single seed descent isolated from the cultivar Jemalong, virtually identical to another sibling line, J5) and R108-1 were initially selected for their symbiotic capacities and their tractability for *in vitro* regeneration. A17 is the genotype used for *Mt* genome sequencing, and the source of most mutants created so far. Handling of *Mt* in the laboratory is now routine practice: life cycle and amplification of genotypes, crossings, growth in a range of conditions, inoculation of root symbionts

(Journet *et al.*, 2001; Fig. 21.2). Protocols for genetic transformation mediated by *Agrobacterium tumefaciens* and subsequent *in vitro* regeneration have been progressively optimized (Thomas *et al.*, 1992; Trieu and Harrison, 1996; Trinh *et al.*, 1998; Chabaud *et al.*, 2003; Van Iantcheva *et al.*, 2005). A fast and efficient transformation procedure using *A. rhizogenes* (Boisson-Dernier *et al.*, 2001) is now widely used to obtain transgenic roots and has proved particularly useful for promoter and reverse genetics studies for genes that function in the root (e.g. during root symbioses; Limpens *et al.*, 2004; Boisson-Dernier, Andriankaja *et al.*, 2005). Amongst the numerous registered accessions, other *Mt* genotypes are now being studied as a source of natural variability in traits of interest, and also as crossing parents for genetic mapping (e.g. A20, DZA 315-16). *Mt* also offers promising prospects as a bioreactor for the production of functional recombinant proteins (Abranches *et al.*, 2005).

Mt has been the subject of a number of classical molecular biology studies, which have led to the identification and characterization of several dozens of genes involved in the rhizobial symbiotic interaction (nodulin genes; see Pawlowski, 1997; Schultze and Kondorosi, 1998) or in the endomycorrhizal symbiosis (Harrison, 1999; Parniske, 2004). The advent of genomics approaches and systematic mutant screens in recent years has strongly accelerated the rate of discoveries in legume biology (see below).

INTERNATIONAL INITIATIVES BOOSTING LEGUME GENOMICS

Following the lessons of the *A. thaliana* success story, an international network of research laboratories has grown within a few years around the model *Mt*, taking advantage of the increasingly favourable scientific and technical context for the development of plant genomics, and has become a strong international research community. Several national and international programs have merged the ongoing efforts of the community and made it possible to achieve major progress in the development of *Mt* genomic resources, thus paving the way for the future transfer to important crop legumes (VandenBosch and Stacey, 2003). In the US, the National Science Foundation (NSF) Plant Genome Program (2001-2005; www.medicago.org) and the Samuel Roberts Noble Foundation (Ardmore, Oklahoma; Center for Medicago Genomics Research; www.noble.org/medicago/index.htm) have played a key role in supporting and encouraging *Mt* genomics, proteomics, metabolomics and bioinformatics. In Europe, the 5th (2000-02) and 6th ("Grain Legumes Integrated Project" (GLIP), 2002-06; http://medicago.toulouse.inra.fr) EU Framework Programs have also made major contributions, in addition to several smaller scale national programmes (e.g. supported by INRA or the Genoscope in France; by the DFG in Germany). In Australia, the ARC Centre of Excellence for Integrative Legume Research (www.cilr.uq.edu.au) exploits resources from many different legumes, including Lotus, Medicago and soybean. All these initiatives have culminated in the current sequencing of the *Mt* genome, run by an international consortium (see below). In parallel, considerable progress has been made in *Lj* genomics (Udvardi *et al.*, 2005) including current genome sequencing (www.kazusa.or.jp/lotus). Although the sequencing of two model legumes might appear as a duplication of effort, it is now clear that this will permit valuable cross-comparisons and in-depth analyses of legume specificities.

Fig. 21.2 Aerial parts and endosymbiotic roots of *Medicago truncatula* (line Jemalong A17). a) Barrel-shaped, spiny pods; b) *Mt* seeds with truncated shape; c) *Mt* shoot bearing flowers / developing pods on each node; d) opening flower at the post-anthesis stage; e) 4 day-old seedling shoot bearing the first, monolobed leaf; f) young N-fixing, pink nodules on primary root 10 days post-inoculation with the microsymbiont *Sinorhizobium meliloti*; g) high-magnification view of nodule central zone showing infected cells packed with N-fixing bacteroids (photographed by Georges Truchet); h) root segment colonized by hyphae of the endosymbiotic mycorrhizal fungus *Glomus intraradices* (blue stain) that form arbuscules in numerous cells in the inner cortex (purple stain); i) magnified view of a fungal arbuscule (black stain) connected to the internal hyphae and located near xylem tissue; arbuscules are the major site of nutrient exchange between plant and fungal partners (h and i: courtesy of Mireille Chabaud).

Scale bars : a) 10 mm; b) 5 mm; c) 15 mm; d) 2 mm; e) 10 mm; f) 1 mm; g) 20 μm; h) 300 μm; i) 20 μm.

Finally, international meetings devoted to joint model and crop legume biology and genomics are organized every year. An EMBO course took place at Gif-sur-Yvette (Institut des Sciences du Végétal, CNRS, France) in 2001 and was backed up by the first edition of a practical *Medicago* handbook (www.isv.cnrs-gif.fr/embo01/index.html). A second edition with a more comprehensive coverage is pending for 2006.

FUNCTIONAL GENOMICS IN *MEDICAGO TRUNCATULA*

ESTs, cDNAs and *In Silico* Analyses

The first achievement in the rapid expansion of *Mt* functional genomics has been the generation of a large EST (Expressed Sequence Tags) collection. The first two sets of *Mt* ESTs derived from single-pass random sequencing of a root hair-enriched and young root nodule cDNA libraries (Covitz *et al.*, 1998; Gyorgyey *et al.*, 2000). Subsequently, systematic EST sequencing programs sponsored by the NSF *Mt* consortium and the Noble Foundation in the USA, and by a CNRS-INRA-Genoscope collaboration in France, led to over 180,000 new ESTs by the end of 2002. These ESTs were obtained from around 40 cDNA libraries representing various organs, developmental stages, and varied physiological contexts such as biotic interactions with symbionts, pathogens, pests, elicitors as well as abiotic stresses such as mineral deprivation, water stress or UV and gamma irradiation. Since then, random EST sequencing has been pursued at a slower pace, along with the targeted exploration of new cDNA libraries generated using suppressive subtractive hybridization (SSH) (Wulf *et al.*, 2003; Brechenmacher *et al.*, 2004; Frenzel *et al.*, 2005). As of January 2005, a collection of more than 220,000 *Mt* ESTs was publicly available.

The first step in EST data analysis consists in predicting the sequence of gene transcripts from assembled individual ESTs. Several databases propose such sequences assembled from Mt ESTs (and in some cases expressed transcripts, called ETs), following quality assessment and chimeric cDNA elimination. Different procedures have been used for each database. Thus, the MENS database (Laboratory of Plant Microbe Interactions CNRS-INRA, Toulouse; http://medicago.toulouse.inra.fr/MENS; Journet *et al.*, 2002) differs from others by the fact that ESTs determined from the 5' and 3' ends of a same cDNA clone have been deliberately grouped into the same EST core cluster, even when not overlapping. Nevertheless, a similar number of clusters were predicted in the three major *Mt* EST databases, namely the TIGR *Mt* Gene Index (MtGI) (The Institute for Genomic Research, Rockville, Maryland; www.tigr.org/tdb/tgi.shtml), MtDB 2.0 (Center for Computational Genomics and Bioinformatics, Univ. Minnesota; www.medicago.org/MtDB2; Lamblin *et al.*, 2003) and MENS. Thus, starting from approximately 190,000 ESTs, about 18,000 clusters (also designated TCs, for Tentative Consensus or Contigs) and as many non-clustered (singleton) ESTs were predicted.

The second step of sequence annotation has also been handled in different ways by these databases. The TIGR MtGI indicates the five best hits found in a non-redundant protein database, and uses the best scoring one to provide a tentative annotation. In addition, when a similarity is found with a non-automatic Gene Ontology (GO) assignment, the GO terms are transferred to the TC. On the other hand, MENS gives access to detailed results of blastn/x searches against DNA, protein databases and EST clusters (identification of gene and protein families), as well as to encoded proteins predicted with the FrameD software adapted to ESTs (Schiex *et al.*, 2003). Signatures and motifs found by searches with InterPro (Integrated Resource of Protein Domains and Functional Sites, which integrates the PROSITE, Pfam, ProDom and SMART databases; Apweiler *et al.*, 2001) are shown and used for an automatic description of the sequence (for approx. 14,000 clusters). Manual annotation is also provided for more than 6,000 clusters, and an interface is available for such annotation.

The potential of *in silico* analyses has been demonstrated for a number of plant species. The fact that available *M. truncatula* ESTs originate from a variety of libraries and physiological conditions makes it possible to infer patterns of gene expression from EST counts. Electronic northerns allowing the visualization of EST distributions in various cDNA libraries are provided by TIGR MtGI and MENS (which also gives a tool to compare several clusters, for example from the same gene family). Identification of clusters showing differential EST abundance between libraries or groups of libraries ("*in silico* screening") is also possible with MtDB 2.0, MtGI and MENS. Since it is not straightforward in such analyses to distinguish between random fluctuations due to cDNA sampling and those actually reflecting differential gene expression, a statistical index (likelihood ratio R) is provided in MtGI and MENS. *In silico* studies have been used to predict the differential regulation of hundreds of genes during the development of root endosymbiotic associations (Fedorova *et al.*, 2002; Journet *et al.*, 2002), and also of putative legume-specific genes (Graham *et al.*, 2004).

Unigene Sets, Microarrays and Transcriptome Analysis

One obvious application of EST sequencing is the production of macro- and micro-arrays for transcript profiling, using sets of appropriate EST cluster representatives (so-called "unigene sets"). At least six generations of *Mt* macro- and micro-arrays have been used over the past years. Macro-arrays made from over 2,000 clones were used to identify genes expressed during the arbuscular mycorrhizal symbiosis (Liu *et al.*, 2003). Macro- and micro-arrays containing approximately 6,000 probes generated exclusively from root, nodule and mycorrhiza cDNAs led to the identification of hundreds of genes either up- or down-regulated during both nodulation (El Yahyaoui *et al.*, 2004) and mycorrhization (Manthey *et al.*, 2004; Hohnjec *et al.*, 2005). Other 6 k cDNA micro-arrays representing a broader diversity of libraries were used to analyse early nodulation events (Lohar *et al.*, 2006). A new generation of micro-arrays based on a collection of more than 16 k 70-mer oligonucleotides representing TCs and putative transcription factor genes is now being used in different laboratories worldwide (notably in the framework of the European FP6 GLIP) to explore various biological topics (symbiotic or pathogenic interactions, seed formation...).

The Affymetrix technology has also been used, initially with DNA chips representing about 10 k *Mt* expressed genes used for gene profiling in *S. meliloti*-inoculated roots and in root nodules (Mitra *et al.*, 2004a). These Affymetrix chips have in addition allowed the cloning of a key symbiotic gene (*DMI3*) by a method based on transcript abundance (Mitra *et al.*, 2004b), and also assisted in the cloning of the symbiotic NSP2 putative transcriptional regulator (Kaló *et al.*, 2005). Recently, more comprehensive Affymetrix chips have been set up, with 61 k probe sets representing *Mt*, *M. sativa* and *S. meliloti* sequences (www.affymetrix.com/products/arrays/specific/Medicago.affx).

Another platform, based upon quantitative RT-PCR analysis and funded by TIGR and GLIP, has recently been developed in Golm (M. Udvardi, Max Planck Institute, Germany) for expression profiling of hundreds of predicted transcription factor genes, as previously carried out for *A. thaliana* (Czechowski *et al.*, 2004).

Collections of Natural Variants and Mutants

Sequence information and global gene expression analyses enable speculation about gene function, but definitive assignment of function for each gene requires a combination of molecular/genetic approaches: either analysis of existing natural variability, or analysis of the phenotypic repercussions of mutations generated by forward and reverse genetics.

Access to the wide natural biodiversity within *M. truncatula* is possible thanks to an extensive collection of several hundreds of natural geographic population samples ("ecotypes", renamed "accessions") and derived genotypic lines, created at INRA Montpellier, France (Prosperi, 2000). Many *Mt* accessions are also distributed by SARDI in Australia (Auricht *et al.*, 1999; www2.dpi.qld.gov.au/extra/asp/auspgris) and USDA (www.ars-grin.gov). Analysis of the natural genetic diversity in relation to various biological processes is currently being studied (e.g. for seed development; Djemel *et al.*, 2005). The screening of these genetic resources has uncovered natural variants in the *Medicago-Sinorhizobium* symbiotic recognition specificity (e.g. Tirichine *et al.*, 2000), as well as genotypes harbouring various levels of resistance or susceptibility to pathogenic bacteria (*Ralstonia solanacearum*; Vailleau *et al.*, S2P INP/ENSAT, Toulouse, France, pers. comm.), pathogenic fungi (*Colletotrichum trifolii, Phoma medicaginis, Mycosphaerella pinodes*), oomycetes (*Phytophtora medicaginis, Aphanomyces euteiches*) and nematodes (*Meloidogyne incognita, Ditylenchus dipsaci*) (Cook *et al.*, 2000; O'Neill and Bauchan, 2000; O'Neill *et al.*, 2003; Torregrosa *et al.*, 2004; Davis and Mitchum, 2005; Jacquet *et al.*, CNRS-UPS Toulouse, and Moussart, Pilet-Nayel *et al.*, INRA, Rennes, France, pers. comm.). A single dominant gene has been identified in two *Mt* accessions that confers resistance to the bluegreen aphid, an important pest of alfalfa (Klingler *et al.*, 2005). The detailed characterization of such pathosystems for *Mt* will open the way to positional cloning of the relevant resistance genes and to new opportunities for crop legume breeding.

A number of different approaches are either currently available or being developed to generate *Mt* mutants (reviewed in Tadege *et al.*, 2005). Several mutagenesis projects using classical physico-chemical agents have been implemented using γ-rays (Sagan *et al.*, 1995), ethyl methane sulfonate (EMS; Penmetsa and Cook, 2000), and fast neutrons (S. Long, Stanford University, USA; G. Oldroyd, John Innes Centre, Norwich, UK). These approaches have generated biological material for large scale screens selecting for various phenotypes, primarily for mutants affected in the rhizobial or mycorrhizal symbioses, but also in plant morphology and physiology (McCon and Nakata, 2002; Penmetsa and Cook, 2000; Ellis *et al.*, 2003; see also the Nodulation Mutant Database (NodMutDB), http://nodmutdb.vbi.vt.edu). A number of *Mt* nodulation mutants have thus been isolated and their analysis has proven to be extremely valuable for the genetic dissection of the mechanisms of plant-bacterium recognition, root infection and nodule development (Benaben *et al.*, 1995; Sagan *et al.*, 1995, 1998; Penmetsa and Cook, 1997; Catoira *et al.*, 2000, 2001; Wais *et al.*, 2000). Along with similar rapid progress in *Lj*, positional cloning of a growing number of symbiotic *Mt* genes has made a major contribution to our understanding of the molecular basis of legume symbiotic signalling (see for reviews Oldroyd and Downie, 2004; Geurts *et al.*, 2005; Mulder *et al.*, 2005; Oldroyd *et al.*, 2005). Positional cloning will of course be greatly facilitated in the near future by the availability of genome sequences for *Mt* and *Lj*. It is noteworthy that forward genetics approaches have made a crucial contribution in demonstrating that rhizobial and mycorrhizal symbioses share several key genes involved in early signal transduction (3 identified in *Mt* and pea, 6 in *Lj*).

Development of Insertional Mutagenesis and Reverse Genetics Tools

Insertional mutagenesis is a tractable genetic strategy in which a DNA sequence (T-DNA, transposon or retrotransposon) is used to mutate and tag the genome, with the convenience of enabling fast gene isolation from tagged loci. Collections of insertional mutagenized plant lines are also the material of choice for reverse genetics. Genome-wide insertional mutagenesis has proven to be highly successful in *A. thaliana* (Parinov and Sundaresan, 2000; Alonso *et al.*, 2003), and therefore would represent an important step forward for *Mt* functional genomics. At present, a standard transformation procedure via *in vitro* somatic embryogenesis has successfully generated about 2,000 independent T-DNA-tagged *Mt* lines (Scholte *et al.*, 2002). However, T-DNA tagging is not the method of choice for saturation mutagenesis in *Mt* because the estimated number of insertions needed to saturate the *Mt* genome is > 500,000, and, in contrast to *A. thaliana*, no simple and efficient *in planta* transformation system has been yet established for *Mt* to avoid tedious tissue-culture. Although Trieu *et al.* (2000) initially reported an *in planta* transformation system for *Mt*, this has not yet been successfully reproduced (Somers *et al.*, 2003). Finally, T-DNA tagging suffers from several drawbacks, such as somaclonal mutations occurring during tissue culture, and deletions or rearrangements at the target site sequences that complicate the isolation of tagged loci and reverse genetic screens (Tadege *et al.*, 2005).

The pioneering work of d'Erfurth and colleagues (2003) has shown that the tobacco retrotransposon *Tnt1* is a viable and attractive option for saturation mutagenesis of *Mt*. *Tnt1* mutagenesis has several significant advantages compared to T-DNA tagging. For example, in both *A. thaliana* and *Mt*, *Tnt1* insertions often occur within genes and therefore apparently favour gene-rich regions. In addition, deletions and rearrangements at *Tnt1* insertion sites are rare. Moreover, *Tnt1* transposition is highly efficient in *Mt*, with an average of 15 copies per line inserted into unlinked loci. This means that the number of mutant lines that need to be generated can be reduced by at least an order of magnitude when compared to T-DNA tagging (approx. 35,000 independent *Tnt1* lines needed for genome-wide mutagenesis). Finally, since *Tnt1* transposition in stable transformed lines can be re-activated by *in vitro* tissue culture, it is not necessary to generate new lines by repeated transformation. Thus, seeds from a single *Tnt1* "starter" line can be used to generate as many newly transposed lines as needed via rapid *in vitro* regeneration (Tadege *et al.*, 2005).

Collaborative efforts are under way in Europe (FP6 GLIP) and the USA (Noble Foundation) to generate ~20,000 *Tnt1*-tagged *Mt* lines by 2008. Initial *Tnt1* tagging studies were performed with the highly regenerating R-108-1-1c3 line, and this line is being used for a large scale program at the Noble Foundation. However, crosses between R108-1 and the lines most widely used for *Mt* genetic studies (A17 & J5) result in severe segregation distortion, most likely due to genomic incompatibility (Kulikova *et al.*, 2004). For this reason, efforts are under way in Europe to optimise a high-throughput protocol for *Tnt1*-tagging in the highly embryogenic A17 line 2HA (Rose *et al.*, 1999). Once they are available, *Tnt1*-tagged *Mt* lines will serve multiple purposes, including forward mutant screening, reverse genetic screening of DNA pools and possibly the establishment of a flanking sequence database for *in silico* gene cloning.

Two other valuable (transformation-independent) complementary strategies, TILLING and fast neutron mutagenesis, are being developed in *Mt* to generate large collections suitable for high-throughput reverse genetic screens. TILLING (Targeting-Induced Local Lesions In Genomes) is a PCR-based reverse-genetics strategy that can identify single nucleotide changes in a known sequence in EMS-mutagenized populations (McCallum *et al.*, 2000; Henikoff *et al.*, 2004). This makes it possible to identify a range of mutant alleles for a given gene and to carry out targeted studies on particular protein domains. TILLING has been developed both in *Lotus* (Perry *et al.*, 2003) and in *Medicago* (D. Cook's laboratory, University of California, Davis, CA, USA; R. Thompson's laboratory, INRA, Dijon, France, in collaboration with URGV, INRA-CNRS Evry, France, and John Innes Centre platform, UK).

Unlike EMS, fast neutron bombardment generates DNA deletions ranging in size from a few base pairs to more than 30 kb (Li and Zhang, 2002) as well as other chromosomal rearrangements. Thus, fast neutrons readily generate knockout mutants and are particularly desirable for tandemly repeated gene families. Detection of a single fast neutron deletion mutation within > 1000-fold excess of the wild-type sequence in pooled DNA preparations is now feasible, thus making this strategy much more sensitive and cheaper than TILLING. Large-scale fast neutron mutagenesis of *Mt* is currently in progress at the Noble Foundation (USA) and at the John Innes Centre (UK, financed by GLIP), with the objective of generating over 20,000 M_2 families available for screening at each site in the next few years.

RNA interference (RNAi) has proven to be a powerful tool to discover gene function. RNAi can be activated by generating transgenic lines that express RNAs capable of forming a double-stranded hairpin that guides sequence-specific mRNA degradation (Waterhouse and Helliwell, 2003). RNA-induced gene silencing can result in a range of phenotypes depending on the extent of RNA suppression, and thereby necessitates the analysis of sufficiently large numbers of transformants. Since the generation of transgenic lines is relatively time-consuming for *Mt*, the RNAi approach has been first applied using rapid *A. rhizogenes*-mediated root transformation to study gene function in roots (Limpens *et al.*, 2003; Limpens *et al.*, 2004; Ivashuta *et al.*, 2005).

The alternative virus-induced gene silencing (VIGS) strategy has been used successfully in pea using a newly developed vector based on the *Pea Early Browning Virus* (Constantin *et al.*, 2004). Using the same vector, 3 out of 22 *Mt* accessions tested were found to be hosts for the PEBV virus and showed efficient silencing of a phytoene desaturase reference gene (M. Grønlund, I.E. Johansen, Danish Institute of Agricultural Sciences; pers. comm.). Viral vectors devised for VIGS are also in development for *Mt* in other laboratories (K.S. Mysore, Noble Foundation; G. Dwyer, Murdoch Univ., Australia).

Proteome and Metabolome

Genomics approaches alone are insufficient to address complex biological questions. The direct evaluation of the encoded proteins and derived metabolites using integrated systems approaches is also essential for understanding the functional consequences of gene expression.

Proteomics is facilitated by high-throughput mass spectrometry, which provides information on the exact mass and/or sequence of proteins and derived peptides. This information, together with deduced protein sequence data, makes it possible to identify proteins and to follow changes in protein expression and post-translational modification. Successful applications of proteomics approaches have been reported in recent years for the legume species pea, soybean, *Lotus*, alfalfa and *Mt* (reviewed in Wan *et al.*, 2005). Targeted proteomes refer to individual organs and specific physiological situations such as single cell-types (root hairs, embryogenic cells) and subcellular compartments (peribacteroid membrane, symbiosome space, nodule mitochondria), as well as the proteome of associated symbiotic rhizobia (*Sinorhizobium meliloti* bacteroids in nodules; Djordjevic, 2004). Concerning *Mt*, a reference root proteome map was first published by Mathesius and coll. (2001). Bestel-Corre and coworkers (2002) used 2-DE and MALDI-TOF-MS to visualize and identify approximately 500 proteins associated with *Mt* root endosymbioses. Temporal changes in the *Mt* seed proteome during development have been described by Gallardo *et al.* (2003). A study of organ/tissue specific proteomes from *Mt* successfully identified 304 proteins in a survey of roots, leaves, stems, flowers, seed pods and suspension cultures (Watson *et al.*, 2003). Proteomic changes linked to somatic embryogenesis have been reported (Imin *et al.*, 2005) and over 1,300 proteins from a *Mt* cell suspension culture have been successfully identified (Lei *et al.*, 2005).

Recent developments or refinements of a variety of analytical platforms, including GC-mass spectrometry and liquid chromatography-MS, together with software developments, have enabled high-throughput, non-biased analysis of thousands of metabolites from plants and other organisms. These tools afford not only a much broader view of metabolites and metabolism but also the opportunity to discover novel metabolites and previously unknown aspects of metabolism (Fiehn *et al.*, 2000; Sumner *et al.*, 2003; Desbrosses *et al.*, 2005). The first large-scale metabolomic studies reported for *Mt* addressed the effects of biotic and abiotic elicitors on the metabolism of cell cultures (Broeckling *et al.*, 2005; Suzuki *et al.*, 2005).

STRUCTURAL GENOMICS AND GENOME SEQUENCING IN *MEDICAGO TRUNCATULA*

Integration of Genetic, Physical and Cytogenetic Maps

Two genetic maps with dense marker sets have been established for *Mt*. The first of these was constructed by T. Huguet and coworkers (CNRS-INRA, Toulouse), based on a cross between a homozygous line (J6) from the cultivar Jemalong (considered to be identical to A17) and a line from the Algerian natural population DZA315. This choice was determined by the high degree of molecular and phenotypic polymorphism between the two parental lines. An F_2 segregating population was used to construct a first version of the genetic map (Thoquet *et al.*, 2002) and F_6 recombinant inbred lines (or RILs, stabilized mapping populations) were obtained by single seed descent (INRA Toulouse and Montpellier, France). The genetic map obtained from RILs spans 1,090 cM (average 500 kb/cM) and comprises 613 markers including 294 microsatellites (SSR, identified within ESTs), RAPD and AFLP markers, known genes and isoenzymes, arranged in eight linkage groups (2n = 16). Markers are uniformly distributed throughout the map and segregation distortion is limited to only three linkage groups. Several RIL populations based on other crosses between polymorphic parental lines are currently under construction (INRA Toulouse and Montpellier), and will provide the community with valuable tools to map / clone individual genes or QTLs revealed by natural variation.

A second genetic map has been constructed by D. Cook (U.C. Davis, California) and colleagues, based on a cross between A17 and the A20 line (Choi *et al.*, 2004a). An F_2 population was used to analyse the segregation of about 300 sequence-characterized genetic markers (ESTs, BAC end sequence tags, and resistance gene analogues) covering 513 cM. In the case of EST-based markers, an intron-targeted marker strategy was used with primers designed to anneal in conserved exon regions and to amplify across intron regions. Polymorphisms were significantly more frequent in intron *vs.* exon regions, thus providing an efficient mechanism to map transcribed genes. Genetic and cytogenetic analysis produced eight well-resolved linkage groups, which had been previously correlated with the eight *Mt* chromosomes by means of fluorescent *in situ* hybridisation (FISH) of mapped BAC clones. Links between the two genetic maps have been established using SSR markers.

The availability of a physical genome map is a key requisite for advanced genomics. A physical map indicates the sequential order of overlapping cloned genomic fragments (e.g. in BAC libraries) needed to reconstitute the entire genome sequence. Several genomic BAC libraries (representing from 6 to 23x the *Mt* genome) have been constructed, initially at U.C. Davis (Nam *et al.*, 1999) and more recently at CNRS-INRA (Evry and Toulouse, France). The next steps for the construction of the *Mt* physical map involved an international collaboration including a number of laboratories (U.C. Davis; U. Minnesota, Saint Paul; U. Oklahoma; CNRS-INRA, Toulouse; Génoscope, Evry; John Innes Center, UK; Biological Research Center, Szeged, Hungary). Combined BAC "fingerprinting", which yields fingerprint-based contigs, and systematic BAC end sequencing, have established a physical map with approximately 11x coverage from over 80,000 BAC clones. Moreover, ~500 non-overlapping BAC clones, containing ESTs, were selected to be fully sequenced and serve as "seed" sequences to prepare the full-scale genespace sequencing (see below). In parallel, several hundreds of contigs were anchored on the genetic linkage map using SSR markers, creating contiguous or near-contiguous assemblies of genome sequence that comprise most of the genes. Details about physical and genetic map resources are accessible at http://mtgenome.ucdavis.edu.

It was also important to develop molecular cytogenetic tools in support of *Mt* genetic and physical mapping. Indeed, cytogenetic mapping is known to yield valuable information with respect to genome organization, distribution of euchromatin and heterochromatin regions, and location of centromeric regions. The advantage of such mapping based on DNA hybridization (FISH) is that it does not depend upon genetic recombination and does not require DNA sequence polymorphism. This approach therefore makes it possible to map genes within chromosomal regions with low recombination frequencies or little DNA polymorphism between parental genotypes used for linkage maps. An *Mt* karyotype was initially established for two genotypes using FISH on metaphasic plaques (Cerbah *et al.*, 1999). Studies carried out in Ton Bisseling's group (Wageningen Univ., Netherlands) have demonstrated that a detailed molecular cytogenetic map can be established by positioning and ordering BACs along pachytene chromosomes, with a much better resolution (~60 kb) than the much more compact metaphase chromosomes (Kulikova *et al.*, 2001; Kulikova *et al.*, 2004). The pachytene chromosomes of *Mt* are relatively easy to visualize, and all eight chromosomes can be identified by appearance.

The integration of the genetic, physical and cytogenetic maps, involving various molecular markers and BAC clones, is underway and has already contributed to the positional cloning of several *Mt* genes (see above).

Identification and Sequencing of the "Genespace" of *Medicago truncatula*

Rather surprisingly, phylogenetically related legume species can have genomes which are very dissimilar in size. For example, the genomes of pea and faba bean are respectively nine-fold (4,500 Mb) and 27-fold (13,500 Mb) larger than that of *Mt* (Table 21.1). Concerning pea, it seems that this size difference is not due to duplications within the genome, but mostly to the recent accumulation of retro-elements (Vershinin *et al.*, 2003) and of non-coding DNA, which have not disrupted macrosyntenic relationships with *Mt* (Choi *et al.*, 2004b).

The existence of gene-rich regions in *Mt*, as found for *A. thaliana*, has been revealed by cytogenetic studies which show a strikingly simple genome organization for *Mt*, by comparison with that of rice (Kulikova *et al.*, 2001; Choi *et al.*, 2004a; Kulikova *et al.*, 2004). Significantly, all chromosomes show distinct euchromatic arms and heterochromatic centromeres and pericentromeres, although chromosome 6 is substantially more heterochromatic than the rest. Kulikova *et al.* (2001) found that each of 20 gene-rich BAC clones hybridized exclusively to euchromatic arms, and subsequent FISH analysis confirmed and extended this initial observation, with 80 gene-rich BACs all localizing to euchromatin (Choi *et al.*, 2004a; R. Geurts, Wageningen Univ., Netherlands, pers. comm.). Thus, most *Mt* genes are likely to be located in euchromatic regions throughout chromosome arms, in the so-called "genespace", and therefore largely absent from the heterochromatin of centromeres and pericentromeres. A growing body of data suggests that other legume species, even those with much larger genomes, such as soybean (*Glycine max*), may actually comprise genespaces similar to *Mt* and *Lj* (Mudge *et al.*, 2005). In summary, cytogenetics has demonstrated that most of the *Mt* genespace could be covered by sequencing the collection of overlapping gene-rich BACs (Young *et al.*, 2005).

Following initial BAC survey sequencing at the University of Oklahoma, funded by the Noble Foundation (May and Dixon, 2004), the US NSF and EU GLIP programs initiated a large-scale genome sequencing project for *Mt* in 2003, coordinated by an international steering committee and performed at four centres worldwide (see www.medicago.org/genome). Annotation is coordinated by an international consortium known as the International *Medicago* Genome Annotation Group (IMGAG), and involves a multi-institution pipeline, relying on *Medicago*-trained FGENESH predictions, the EuGene predictor and gene combiner software, and the TIGR PASA procedure for aligning ESTs with genomic sequence (Cannon *et al.*, 2005; http://ensembl.ccgb.umn.edu/Medicago_truncatula).

By December 2005, approximately 170 Mb of the *Mt* genome sequence (~95 Mb finished and ~75 Mb in draft form) were publicly available. Analysis of these genome sequences fully validates the choice of the anchored, clone-by-clone strategy instead of a whole genome shotgun or methylated genome filtering method. Gene density is reasonably high in sequenced *Mt* clones: 149 genes/Mb (1 gene every 6.7 kb). Based on these estimates, gene density in *Mt* is slightly lower than in *A. thaliana* by a factor of approximately 1.5 (*Arabidopsis* Genome Initiative, 2000). Whilst there are repetitive elements on most sequenced clones, there are no cases of BACs from euchromatic regions devoid of genes, or consisting entirely of repeats. There is also no indication that *Mt* genes are found in islands separated by extensive stretches of retro-elements, as observed in maize. The *Mt* genespace is now estimated to encompass a total of 230 to 270 Mb, based upon the fact that 77 Mb of finished BACs contain approximately 11,500 genes out of the total of 35- 40,000 genes (estimate from the current *Mt* unigene set, see above). Hence, current genome sequence is ~60% complete and sequencing of the remaining ~60-100 Mb should cover most *Mt* genes (Young *et al.*, 2005).

M. truncatula, A Reference Species for Comparative Legume Genomics: towards a Composite Cross-legume Map.

One important goal of legume genomics is to transfer knowledge from model to crop legumes. The possibility of using the *A. thaliana* genome sequence as a support tool has been investigated. Although *A. thaliana* and legumes represent sister clades within the dicot subclass Rosidae, the conservation of their genome structure is limited. Approaches based on conserved genetic markers and BAC sequences revealed a lack of extensive macrosynteny between the *A. thaliana* and *Mt* genomes (Zhu *et al.*, 2003). Nevertheless, synteny is frequently maintained over small chromosomal segments (~500 kb of *A. thaliana*). In cases of localized synteny, genetically linked loci in *Mt* are often colinear with several segments of *A. thaliana*, consistent with the fact that the *A. thaliana* genome has experienced extensive segmental duplication and reshuffling accompanied by selective gene loss. Sequence analyses also revealed networks of microsynteny that are often highly degenerate. The erosion of microsynteny could be due either to selective gene loss from duplicated loci or the absence of close homologs of legume genes in *A. thaliana* (Zhu *et al.*, 2002; reviewed in Zhu *et al.*, 2005). These findings have confirmed that the *A. thaliana* genome is not an appropriate model for crop legumes. Consequently, an in-depth evaluation of genome structure conservation amongst legume species is required.

A growing number of studies have demonstrated macro- and microsynteny amongst legumes reviewed in (reviewed in Young *et al.*, 2005; Zhu *et al.*, 2005). In a study spanning multiple legume species, macrosynteny between *Mt*, *Lj*, soybean, alfalfa, pea and chickpea was examined in detail (Choi *et al.*, 2004b). The results indicate extensive genome-wide synteny between *Mt* and Galegoid legumes (alfalfa, pea, chickpea and *Lj*). Long tracts of macrosynteny were also observed between *Mt* and the Phaseoleae species soybean and mungbean (*Vigna radiata*), although at levels lower than with Galegoid legumes. These comparisons enabled the construction of a genome-wide picture of legume synteny in the form of concentric circles of chromosomes anchored by *Mt* (Choi *et al.*, 2004b; Fig. 21.3), similar to the model previously developed for rice and other grasses (Devos and Gale, 2000).

In addition, microsynteny studies describe similarities at the scale of individual BAC clones or clone contigs between legume genomes. Such microsynteny between either *Mt* or *Lj* and crop species like alfalfa and pea has already enabled the positional cloning of several symbiosis genes (Endre *et al.*, 2002; Stracke *et al.*, 2002; Limpens *et al.*, 2003). Interestingly, microsynteny with *Mt* also extends to the more evolutionarily distant soybean. Thus, the comparison of two sequenced soybean regions totalling 1 Mb in length revealed an impressive level of colinearity to the *Mt* genome (Choi *et al.*, 2004b; Mudge *et al.*, 2005). Altogether, the syntenic regions comprised nearly 500 predicted genes, with 75% of soybean genes colinear with their *Mt* homologs. Such microsynteny between *Mt* and soybean appears to be widespread throughout the two genomes. One study compared 50 pairs of *Mt* and soybean BAC contigs, comprising nearly 10 Mb in each species, and found that 35% of contigs compared *via* cross-hybridization and physical mapping exhibited substantial microsynteny, with another 20% showing more limited levels of microsynteny (Yan *et al.*, 2003).

All these results suggest that comparative mapping may have considerable utility for basic and applied research in legumes, although its predictive value is likely to be limited by phylogenetic distances and genome duplication. The construction of a legume composite map, by using conserved anchor genetic markers and genome sequence stretches, will greatly facilitate bridging the gap between model and crop legumes and open the way to map-based cloning of symbiotic or other agronomically important genes via their orthologs in the model legumes *Mt* and *Lj*.

M. TRUNCATULA BIOINFORMATICS AND WEB RESOURCES

All large-scale "omics" approaches require powerful bioinformatics tools, integrated databases and user-friendly web resources, in order to process and analyse the enormous amounts of data generated. This allows the integration of genomic, genetic and biological information from multiple legume species, and offers in-depth data mining facilities for the research community. In the case of *Mt*, bioinformatics efforts were initially devoted to EST databases (TIGR MtGeneIndex, www.tigr.org/tdb/tgi/mtgi, Lee *et al.*, 2005; "Medicago EST Navigation System" (MENS), http://medicago.toulouse.inra.fr/MENS, Journet *et al.*, 2002; "*Medicago truncatula* Consortium DataBase 2.0", http://www.medicago.org/MtDB2, Lamblin *et al.*, 2003) and have now been oriented towards the integration of accumulating genome sequence and cross-legume genomics data, which can now be accessed through a variety of websites. A primary *Medicago* portal at CCGB, Univ. of Minnesota (www.medicago.org), gives access to all local *Mt in silico* resources and relevant web links. Information related to the genome sequencing project (Cannon *et al.*, 2005) include an Ensembl (automatic annotation software system) implementation at CCGB (http://ensembl.ccgb.umn.edu/Medicago_truncatula), genetic map, finger-print-contig-based

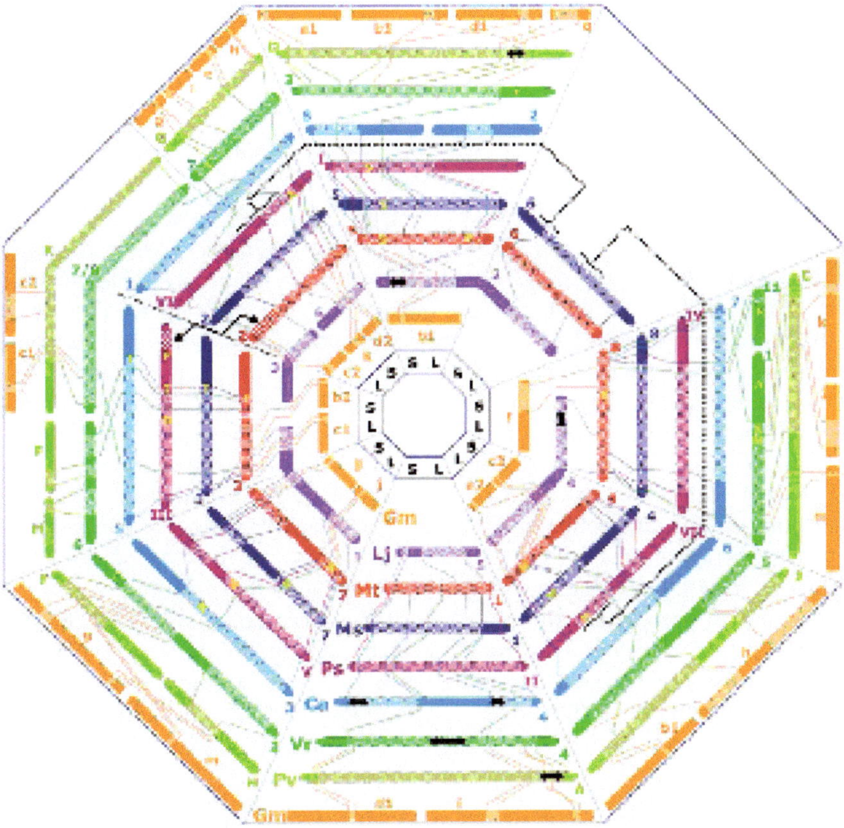

Figure 21.3 A simplified consensus map for eight legume species (reproduced, with permission, from Zhu *et al.*, 2005). Mt, *M. truncatula*; Ms, *Medicago sativa* (alfalfa); Lj, *L. japonicus*; Ps, *Pisum sativum* (pea); Ca, *Cicer arietinum* (chickpea); Vr, *Vigna radiata* (mungbean); Pv, *Phaseolus vulgaris* (common bean); Gm, *Glycine max* (soybean). S and L denote the short and long arms of each chromosome in *M. truncatula*. Syntenic blocks are drawn to scale based on genetic distance.

physical map and marker resources at UC Davis (http://mtgenome.ucdavis.edu) and genome viewers at the University of Oklahoma (www.genome.ou.edu), TIGR (www.tigr.org/tdb/e2k1/mta1) and the Munich Information for Protein Sequences Center (http://www.urmeldb.net, Schoof *et al.*, 2005). Thus, the *Mt* genome sites provide query and visualization tools for BAC-based sequence assemblies, marker-BAC associations, BAC-sequence browsers showing tentative gene identifications, and FTP downloads of large genome sequence datasets.

In addition, the National Center for Genome Resources (NCGR) in cooperation with the USDA has developed the Legume Information System (LIS; www.comparative-legumes.org), a comparative legume resource aimed at integrating genetic and molecular data from multiple legume species enabling cross-species (*Mt*, *Lj*, soybean and *A. thaliana*) genomic and transcript comparisons (Gonzales *et al.*, 2005). More generally, integrating the huge collection of information from different sources, e.g. from genome to expression data, biochemical pathways and regulation networks, is an objective shared by different bioinformatics groups worldwide to fully exploit the potential of genomics approaches.

GENERAL PROSPECTS

Thanks to the international efforts to develop the model *Medicago truncatula* by a growing scientific community, a large set of essential genomics tools is now available and will stimulate molecular and genetic studies on many aspects of legume biology: over 220,000 ESTs in public databases, microarrays, detailed genetic and cytogenetic maps, recombinant inbred lines taking advantage of a wide natural diversity and extensive genomic BAC libraries. Invaluable, complementary reverse genetics tools are accessible or in development, such as large collections of EMS / fast neutron mutants devised for TILLING and insertional mutants tagged with the *Tnt1* transposon. Genomic sequencing is advancing rapidly and has already uncovered about 60% of the *Mt* genespace. Several centres in Europe and the USA have started to monitor and distribute genetic and genomic *Mt* resources. Moreover, legume genomics is now supported by a strong bioinformatics community through multiple on-line databases.

Research on legume biology at the molecular level is spreading from symbiotic interactions to pathogenic and parasitic interactions (e.g. nematodes, *Striga*), plant architecture and development, protein content and quality of seeds, biosynthesis of specific secondary metabolites, response to abiotic stress (drought, soil salinity, acidity and nutrient limitations). Finally, beyond the ongoing and parallel integration of the *Mt* and *Lj* genetic, cytogenetic and physical maps, the construction of a legume composite map will greatly facilitate map-based cloning of symbiotic or other agronomically important genes in crop legumes via their orthologs in the model legumes *Mt* and *Lj*.

There are several very good reasons why the agricultural cultivation of legumes worldwide should increase in the relatively short-term. Firstly, the world demand for plant protein is rising, due to demographic expansion and changes in food habits. Secondly, the prices of fossil energy are expected to keep rising in the medium/long term due to structural constraints on the market (higher energy demand from emerging countries, decreasing oil reserves etc.), resulting in ever increasing costs of chemical N fertilization. Thirdly, there is an urgent need for limiting greenhouse gases emissions. In this context, cultivation of legume species offers alternative, sustainable solutions for efficient production of not only protein-rich food / feed, but also of non-food raw materials to substitute in part oil products for biomass, biofuels and substrates of "green chemistry". Whereas existing legume cultivars and tree species can already be used to address a number of these challenges, advances in legume genomics will provide essential tools for the improvement and diversification of legume agronomical and compositional traits, through traditional breeding or using the tools of biotechnology.

ACKNOWLEDGEMENTS

We are grateful to Rujin Chen (Noble Foundation, Ardmore, Oklahoma), R. Geurts (Wageningen Univ., Netherlands), Mette Grønlund and Elisabeth Johansen (Danish

Institute of Agricultural Sciences), Nevin Young (Univ. Minnesota, St Paul), Alain Baranger and Anne Moussart (INRA, Rennes, France), Pascal Ratet (ISV-CNRS, Gif / Yvette, France), Thierry Huguet and Fabienne Vailleau (S2P, INP/ENSAT, Toulouse, France), Christophe Jacquet (SCSV, CNRS-UPS, Toulouse), Mireille Chabaud, Frederic Debellé, Jean Dénarié and Jérôme Gouzy (LIPM, INRA-CNRS, Toulouse) for providing us with updated information and/or for helpful suggestions. We thank David Barker (LIPM) for critical reading of the manuscript. We also thank Hongyan Zhu (Univ. Kentucky, Lexington) for permitting the reproduction of published figures, and Mireille Chabaud (LIPM) for microscopical illustrations. The image of infected nodule cells was taken by our late colleague Georges Truchet.

REFERENCES

Abranches, R., Marcel, S., Arcalis, E., Altmann, F., Fevereiro, P., Stoger, E. 2005. Plants as bioreactors: a comparative study suggests that *Medicago truncatula* is a promising production system. *J. Biotechnol.*, 120, 121-134.

Alonso, J.M., Stepanova, A.N., Leisse, T.J., Kim, C.J., Chen, H., Shinn, P., Stevenson, D.K., Zimmerman, J., Barajas, P., Cheuk, R., Gadrinab, C., Heller, C., Jeske, A., Koesema, E., Meyers, C.C., Parker, H., Prednis, L., Ansari, Y., Choy, N., Deen, H., Geralt, M., Hazari, N., Hom, E., Karnes, M., Mulholland, C., Ndubaku, R., Schmidt, I., Guzman, P., Aguilar-Henonin, L., Schmid, M., Weigel, D., Carter, D.E., Marchand, T., Risseeuw, E., Brogden, D., Zeko, A., Crosby, W.L., Berry, C.C., Ecker, J.R. 2003. Genome-wide insertional mutagenesis of *Arabidopsis thaliana. Science*, 301, 653–657.

Apweiler, R., Attwood, T.K., Bairoch, A., Bateman, A., Birney, E., Biswas, M., Bucher, P., Cerutti, L., Corpet, F., Croning, M.D., Durbin, R., Falquet, L., Fleischmann, W., Gouzy, J., Hermjakob, H., Hulo, N., Jonassen, I., Kahn, D., Kanapin, A., Karavidopoulou, Y., Lopez, R., Marx, B., Mulder, N.J., Oinn, T.M., Pagni, M., Servant, F., Sigrist, C.J., Zdobnov, E.M. 2001. The InterPro database, an integrated documentation resource for protein families, domains and functional sites. *Nucleic Acids Res.*, 29, 37-40.

Arabidopsis Genome Initiative. 2000. Analysis of the genome sequence of the flowering plant *Arabidopsis thaliana. Nature*, 408, 796-815.

Auricht, G.C., Prosperi, J.M., Snowball, R., Hughes, S.J. 1999. *The characterisation and Preliminary Evaluation of Medicago and Trifolium Germplasm.* Eds. Bennett, S.J., and Cocks, P.S. Kluwer Academic Publishers, The Netherlands, 141-149.

Barker, D.G., Bianchi, S., Blondon, F., Dattee, Y., Duc, G., Essad, S., Flament, P., Gallusci, P, Genier, G., Guy, P., Muel, X., Tourneur, J., Dénarié, J., Huguet, T. 1990. *Medicago truncatula*, a model plant for studying the molecular genetics of the *Rhizobium*-legume symbiosis. *Plant Mol. Biol. Rep.*, 8, 40-49.

Bénaben, V., Duc, G., Lefebvre, V., Huguet, T. 1995. TE7- An inefficient symbiotic mutant of *Medicago truncatula* Gaertn cv. Jemalong. *Plant Physiol.*, 107, 53-62.

Bennett, M.D., Leitch, I.J. 2004. Angiosperm DNA C-values Database (release 5.0) www.rbgkew.org.uk/cval/homepage.html

Bestel-Corre, G., Dumas-Gaudot, E., Poinsot, V., Dieu, M., Dierick, J.F., van Tuinen, D., Remacle, J., Gianinazzi-Pearson, V., Gianinazzi, S. 2002. Proteome analysis and identification of symbiosis-related proteins from *Medicago truncatula* Gaertn. by two-dimensional electrophoresis and mass spectrometry. *Electrophoresis*, 23, 122-137.

Blondon, F., Marie, D., Brown, S., Kondorosi, A. 1994. Genome size and base composition in *Medicago sativa* and *M. truncatula* species. *Genome*, 37, 264-270.

Boisson-Dernier, A., Chabaud, M., Garcia, F., Bécard, G., Rosenberg, C., Barker, D.G. 2001. *Agrobacterium rhizogenes*-transformed roots of *Medicago truncatula* for the study of nitrogen-fixing and endomycorrhizal symbiotic associations. *Mol. Plant-Microbe Interact.*, 14, 693-700.

Boisson-Dernier, A., Andriankaja, A., Chabaud, M., Niebel, A., Journet, E.P., Barker, D.G., de Carvalho-Niebel, F. 2005. *MtENOD11* gene activation during rhizobial infection and mycorrhizal arbuscule development requires a common AT-rich-containing regulatory sequence. *Mol Plant Microbe Interact.*, 18, 1269-1276.

Bonnin, I., Huguet, T., Ghérardi, M., Prospéri, J.M., Olivиéri, I. 1996. High level of polymorphism and spatial structure in a selfing plant species *Medicago truncatula* Gaertn. using RAPDs markers. *Amer. J. Bot.*, 83, 843-855.

Brechenmacher, L., Weidmann, S., van Tuinen, D., Chatagnier, O., Gianinazzi, S., Franken, P., Gianinazzi-Pearson, V. 2004. Expression profiling of up-regulated plant and fungal genes in early and late stages of *Medicago truncatula-Glomus mosseae* interactions. *Mycorrhiza*, 14, 253-262.

Broeckling, C.D., Huhman, D.V., Farag, M.A., Smith, J.T., May, G.D., Mendes, P., Dixon, R.A., Sumner, L.W. 2005. Metabolic profiling of *Medicago truncatula* cell cultures reveals the effects of biotic and abiotic elicitors on metabolism. *J. Exp. Bot.*, 56, 323-336.

Cannon, S.B., Crow, J.A., Heuer, M.L., Wang, X., Cannon, E.K., Dwan, C., Lamblin, A.F., Vasdewani, J., Mudge, J., Cook, A., Gish, J., Cheung, F., Kenton, S., Kunau, T.M., Brown, D., May, G.D., Kim, D., Cook, D.R., Roe, B.A., Town, C.D., Young, N.D., Retzel, E.F. 2005. Databases and information integration for the *Medicago truncatula* genome and transcriptome. *Plant Physiol.*, 138, 38-46.

Catoira, R., Galera, C., de Billy, F., Penmetsa, R.V., Journet, E.P., Maillet, F., Rosenberg, C., Cook, D., Gough, C., Denarie J. 2000. Four genes of *Medicago truncatula* controlling components of a nod factor transduction pathway. *Plant Cell*, 12, 1647-1666.

Catoira, R., Timmers, A.C., Maillet, F., Galera, C., Penmetsa, R.V., Cook, D., Denarie, J., Gough, C. 2001. The *HCL* gene of *Medicago truncatula* controls *Rhizobium*-induced root hair curling. *Development*, 128, 1507-1518.

Cerbah, M., Kevei, Z., Siljak-Yakovlev, S., Kondorosi, E., Kondorosi, A., Trinh, T.H. 1999. FISH chromosome mapping allowing karyotype analysis in *Medicago truncatula* lines Jemalong J5 and R-108-1. *Mol. Plant-Microbe Interact.*, 12, 947-950.

Chabaud, M., de Carvalho-Niebel, F., Barker, D.G. 2003. Efficient transformation of *Medicago truncatula* cv. Jemalong using the hypervirulent *Agrobacterium* tumefaciens strain AGL1. *Plant Cell Rep.*, 22, 46-51.

Choi, H.K., Kim, D., Uhm, T., Limpens, E., Lim, H., Mun, J.H., Kalo, P., Penmetsa, R.V., Seres, A., Kulikova, O., Roe, B.A., Bisseling, T., Kiss, G.B., Cook, D.R. 2004a. A sequence-based genetic map of *Medicago truncatula* and comparison of marker colinearity with *M. sativa. Genetics*, 166, 1463-1502.

Choi, H.K., Mun, J.H., Kim, D.J., Zhu, H., Baek, J.M., Mudge, J., Roe, B., Ellis, N., Doyle, J., Kiss, G.B., Young, N.D., Cook, D.R. 2004b. Estimating genome conservation between crop and model legume species. *Proc. Natl. Acad. Sci.* USA, 101, 15289-15294.

Constantin, G.D., Krath, B.N., MacFarlane, S.A., Nicolaisen, M., Johansen, I.E., Lund, O.S. 2004. Virus-induced gene silencing as a tool for functional genomics in a legume species. *Plant J.*, 40, 622-631.

Cook, D.R. 1999. *Medicago truncatula*-a model in the making ! *Curr. Opin. Plant Biol.*, 2, 301-304.

Cook, D.R., Kim, D., Zhu, H., Uribe, P. 2000. Plant-pathogen interactions in *Medicago truncatula. Grain Legumes*, 28, 20.

Covitz, P.A., Smith, L.S., Long, S.R. 1998. Expressed sequence tags from a root-hair-enriched *Medicago truncatula* cDNA library. *Plant Physiol.*, 117, 1325-1332.

Crawford, E.J., Lake, A.W.H., Boyce, K.G. 1989. Breeding annual *Medicago* species for semiarid conditions in southern Australia. *Adv. Agronomy*, 42, 399-437.

Crespi, M.D., Jurkevitch, E., Poiret, M., d'Aubenton-Carafa, Y., Petrovics, G., Kondorosi, E., Kondorosi, A. 1994. *ENOD40*, a gene expressed during nodule organogenesis, codes for a non-translatable RNA involved in plant growth. *EMBO J.*, 13, 5099-5112.

Czechowski, T., Bari, R.P., Stitt, M., Scheible, W.R., Udvardi, M.K. 2004. Real-time RT-PCR profiling of over 1400 *Arabidopsis* transcription factors: unprecedented sensitivity reveals novel root- and shoot-specific genes. *Plant J.*, 38, 366-379.

Davis, E.L., Mitchum, M.G. 2005. Nematodes. Sophisticated parasites of legumes. *Plant Physiol.*, 137, 1182-1188.

d'Erfurth, I., Cosson, V., Eschstruth, A., Lucas, H., Kondorosi, A., Ratet, P. 2003. Efficient transposition of the *Tnt1* tobacco retrotransposon in the model legume *Medicago truncatula. Plant J.*, 33, 1-12.

Desbrosses, G.G., Kopka, J., Udvardi, M.K. 2005. *Lotus japonicus* metabolic profiling. Development of gas chromatography-mass spectrometry resources for the study of plant-microbe interactions. *Plant Physiol.*, 137, 1302-1318.

Devos, K.M., Gale, M.D. 2000. Genome relationships: the grass model in current research. *Plant Cell*, 12, 637–646.

Dixon, R.A., Sumner, L.W. 2003. Legume natural products: understanding and manipulating complex pathways for human and animal health. *Plant Physiol.*, 131, 878–885.

Djemel, N., Guedon, D., Lechevalier, A., Salon, C., Miquel, M., Prosperi, J.M., Rochat, C., Boutin, J.P. 2005. Development and composition of the seeds of nine genotypes of the *Medicago truncatula* species complex. *Plant Physiol. Biochem.*, 43, 557-566.

Djordjevic, M.A. 2004. *Sinorhizobium meliloti* metabolism in the root nodule: a proteomic perspective. *Proteomics*, 4, 1859-72.

Doyle, J.J., Luckow, M.A. 2003. The rest of the iceberg. Legume diversity and evolution in a phylogenetic context. *Plant Physiol.*, 131, 900-910.

El Yahyaoui, F., Kuster, H., Ben Amor, B., Hohnjec, N., Puhler, A., Becker, A., Gouzy, J., Vernie, T., Gough, C., Niebel, A., Godiard, L., Gamas, P. 2004. Expression profiling in *Medicago truncatula* identifies more than 750 genes differentially expressed during nodulation, including many potential regulators of the symbiotic program. *Plant Physiol.*, 136, 3159-3176.

Ellis, D.R., López-Millán, A.F., Grusak, M.A. 2003. Metal physiology and accumulation in a *Medicago truncatula* mutant exhibiting an elevated requirement for zinc. *New Phytol.*, 158, 207–218.

Endre, G., Kereszt, A., Kevei, Z., Mihacea, S., Kalo, P., Kiss, G.B. 2002. A receptor kinase gene regulating symbiotic nodule development. *Nature*, 417, 962-966.

Fedorova, M., van de Mortel, J., Matsumoto, P.A., Cho, J., Town, C.D., VandenBosch, K.A., Gantt, J.S., Vance, C.P. 2002. Genome-wide identification of nodule-specific transcripts in the model legume *Medicago truncatula*. *Plant Physiol.*, 130, 519-537.

Fiehn, O., Kopka, J., Dormann, P., Altmann, T., Trethewey, R.N., Willmitzer, L. 2000. Metabolite profiling for plant functional genomics. *Nat. Biotechnol.*, 18, 1157–1161.

Frenzel, A., Manthev, K., Perlick, A., Meyer, F., Puhler, A., Küster, H., Krajinski, F. 2005. Combined transcriptome profiling reveals a novel family of arbuscular mycorrhizal-specific *Medicago truncatula* lectin genes. *Mol. Plant Microbe Interact.*, 18, 771-782.

Gage, D.J. 2004. Infection and invasion of roots by symbiotic, nitrogen-fixing rhizobia during nodulation of temperate legumes. *Microbiol. Mol. Biol. Rev.*, 68, 280-300.

Galibert, F., Finan, T.M., Long, S.R., Puhler, A., Abola, P., Ampe, F., Barloy-Hubler, F., Barnett, M.J., Becker, A., Boistard, P., Bothe, G., Boutry, M., Bowser, L., Buhrmester, J., Cadieu, E.,

Capela, D., Chain, P., Cowie, A., Davis, R.W., Dreano, S., Federspiel, N.A., Fisher, R.F., Gloux, S., Godrie, T., Goffeau, A., Golding, B., Gouzy, J., Gurjal, M., Hernandez-Lucas, I., Hong, A., Huizar, L., Hyman, R.W., Jones, T., Kahn, D., Kahn, M.L., Kalman, S., Keating, D.H., Kiss, E., Komp, C., Lelaure, V., Masuy, D., Palm, C., Peck, M.C., Pohl, T.M., Portetelle, D., Purnelle, B., Ramsperger, U., Surzycki, R., Thebault, P., Vandenbol, M., Vorholter, F.J., Weidner, S., Wells, D.H., Wong, K., Yeh, K.C., Batut, J. 2001. The composite genome of the legume symbiont *Sinorhizobium meliloti*. *Science*, 293, 668-672.

Gallardo, K., Le Signor, C., Vandekerckhove, J., Thompson, R.D., Burstin, J. 2003. Proteomics of *Medicago truncatula* seed development establishes the time frame of diverse metabolic processes related to reserve accumulation. *Plant Physiol.*, 133, 664–682.

Gepts, P., Beavis, W.D., Brummer, E.C., Shoemaker, R.C., Stalker, H.T., Weeden, N.F., Young, N.D. 2005. Legumes as a model plant family. Genomics for food and feed report of the Cross-Legume Advances Through Genomics Conference. *Plant Physiol.*, 137, 1228-1235.

Geurts, R., Fedorova, E., Bisseling, T. 2005. Nod factor signaling genes and their function in the early stages of *Rhizobium* infection. *Curr. Opin. Plant Biol.*, 8, 346-352.

Gianinazzi-Pearson, V. 1996. Plant cell responses to arbuscular mycorrhizal fungi: getting to the roots of the symbiosis. *Plant Cell*, 8, 1871-1883.

Gonzales, M.D., Archuleta, E., Farmer, A., Gajendran, K., Grant, D., Shoemaker, R., Beavis, W.D., Waugh, M.E. 2005. The Legume Information System (LIS): an integrated information resource for comparative legume biology. *Nucleic Acids Res.*, 33, D660-665.

Graham, M.A., Silverstein, K.A., Cannon, S.B., VandenBosch, K.A. 2004. Computational identification and characterization of novel genes from legumes. *Plant Physiol.*, 135, 1179-1197.

Graham, P.H., Vance, C.P. 2003. Legumes: importance and constraints to greater use. *Plant Physiol.*, 131, 872-877.

Gyorgyey, J., Vaubert, D., Jimenez-Zurdo, J.I., Charon, C., Troussard, L., Kondorosi, A., Kondorosi, E. 2000. Analysis of *Medicago truncatula* nodule Expressed Sequence Tags. *Mol. Plant Microbe Interact.*, 13, 62-71.

Handberg, K., Stougaard, J. 1992. *Lotus japonicus*, an autogamous, diploid legume species for classical and molecular-genetics. *Plant J.*, 2, 487–496.

Harrison, M.J. 1999. Molecular and cellular aspects of the arbuscular mycorrhizal symbiosis. *Annu. Rev. Plant Physiol. Plant Mol. Biol.*, 50, 361-389.

Harrison, M.J. 2005. Signaling in the arbuscular mycorrhizal symbiosis. *Annu. Rev. Microbiol.*, 59, 19-42

Henikoff, S., Till, B.J., Comai, L. 2004. TILLING. Traditional mutagenesis meets functional genomics. *Plant Physiol.*, 135, 630–636.

Hohnjec, N., Vieweg, M.F., Puhler, A., Becker, A., Kuster, H. 2005. Overlaps in the transcriptional profiles of *Medicago truncatula* roots inoculated with two different *Glomus* fungi provide insights into the genetic program activated during arbuscular mycorrhiza. *Plant Physiol.*, 137, 1283-1301.

Imin, N., Nizamidin, M., Daniher, D., Nolan, K.E., R.J., Rose, Rolfe, B.J. 2005. Proteomic analysis of somatic embryogenesis in *Medicago truncatula*. explant cultures grown under 6-benzylaminopurine and 1-naphthaleneacetic acid treatments. *Plant Physiol.*, 137, 1250–1260.

Ivashuta, S., Liu, J., Liu, J., Lohar, D.P., Haridas, S., Bucciarelli, B., VandenBosch, K.A., Vance, C.P., Harrison, M.J., Gantt, J.S. 2005. RNA interference identifies a calcium-dependent protein kinase involved in *Medicago truncatula* root development. *Plant Cell*, 17, 2911-21.

Jensen, E.S., Hauggaard-Nielsen, H. 2003. How can increased use of biological N2 fixation in agriculture benefit the environment? *Plant Soil*, 252, 177–186.

Journet, E.-P., Barker, D.G., Harrison, M.J., Kondorosi, E. 2001. *M. truncatula* as biological material. *in* protocol manual of EMBO course "The new plant model system *Medicago truncatula*" Gif/Yvette, November 2001 (http://www.isv.cnrs-gif.fr/embo01/manuels/index.html).

Journet, E.P., van Tuinen, D., Gouzy, J., Crespeau, H., Carreau, V., Farmer, M.J., Niebel, A., Schiex, T., Jaillon, O., Chatagnier, O., Godiard, L., Micheli, F., Kahn, D., Gianinazzi-Pearson, V., Gamas, P. 2002. Exploring root symbiotic programs in the model legume *Medicago truncatula* using EST analysis. *Nucleic Acids Res.*, 30, 5579-5592.

Kaló, P., Gleason, C., Edwards, A., Marsh, J., Mitra, R.M., Hirsch, S., Jakab, J., Sims, S., Long, S.R., Rogers, J., Kiss, G.B., Downie, J.A., Oldroyd, G.E. 2005. Nodulation signaling in legumes requires NSP2, a member of the GRAS family of transcriptional regulators. *Science*, 308, 1786-1789.

Kapranov, P., de Bruijn, F.J., Szczyglowski, K. 1997. Novel, highly expressed late nodulin gene (*LjNOD16*) from *Lotus japonicus*. *Plant Physiol.*, 113, 1081-1090.

Klingler, J., Creasy, R., Gao, L., Nair, R.M., Calix, A.S., Jacob, H.S., Edwards, O.R., Singh, K.B. 2005. Aphid resistance in *Medicago truncatula* involves antixenosis and phloem-specific, inducible antibiosis, and maps to a single locus flanked by NBS-LRR resistance gene analogs. *Plant Physiol.*, 137, 1445-1455.

Kulikova, O., Gualtieri, G., Geurts, R., Kim, D.-J., Cook, D., Huguet, T., de Jong, H., Fransz, P., Bisseling, T. 2001. Integration of the FISH pachytene and genetic maps of *Medicago truncatula*. *Plant J.*, 27, 49-58.

Kulikova, O., Geurts, R., Lamine, M., Kim, D.J., Cook, D.R., Leunissen, J., de Jong, H., Roe, B.A., Bisseling, T. 2004. Satellite repeats in the functional centromere and pericentromeric heterochromatin of *Medicago truncatula*. *Chromosoma*, 113, 276-283.

Lamblin, A.F., Crow, J.J.A., Johnson, J.E., Silverstein, K.A.T., Kunau, T.M., Kilian, A., Benz, D., Stromvik, M., Endré, G., VandenBosch, K.A., Cook, D.R., Young, N.D., Retzel, E.F. 2003. MtDB: A database for personalized data mining of the model legume *Medicago truncatula* transcriptome. *Nucl. Acids Res.*, 31, 196-201.

Lee, Y., Tsai, J., Sunkara, S., Karamycheva, S., Pertea, G., Sultana, R., Antonescu, V., Chan, A., Cheung, F., Quackenbush, J. 2005. The TIGR Gene Indices: clustering and assembling EST and known genes and integration with eukaryotic genomes. *Nucl. Acids Res.*, 33, D71–D74.

Lei, Z., Elmer, A.M., Watson, B.S., Dixon, R.A., Mendes, P.J., Sumner, L.W. 2005. A 2-DE proteomics reference map and systematic identification of 1367 proteins from a cell suspension culture of the model legume *Medicago truncatula*. *Mol Cell Proteomics*, 4, 1812-25.

Li, X., Zhang, Y. 2002. Reverse genetics by fast neutron mutagenesis in higher plants. *Funct. Integr. Genomics*, 2, 254–258.

Limpens, E., Franken, C., Smit, P., Willemse, J., Bisseling, T., Geurts, R. 2003. LysM domain receptor kinases regulating rhizobial Nod factor-induced infection. *Science*, 302, 630–633.

Limpens, E., Ramos, J., Franken, C., Raz, V., Compaan, B., Franssen, H., Bisseling, T., Geurts, R. 2004. RNA interference in *Agrobacterium rhizogenes*-transformed roots of *Arabidopsis* and *Medicago truncatula*. *J. Exp. Bot.*, 55, 983-992.

Liu, J., Blaylock, L.A., Endre, G., Cho, J., Town, C.D., VandenBosch, K.A., Harrison, M.J. 2003. Transcript profiling coupled with spatial expression analyses reveals genes involved in distinct developmental stages of an arbuscular mycorrhizal symbiosis. *Plant Cell*, 15, 2106-2123.

Lohar, D.P., Sharopova, N., Endre, G., Penueal, S., Samas, D., Town, C., Silverstein, K., VandenBosch, K. 2006. Transcript analysis of early nodulation events in *Medicago truncatula*. *Plant Physiol.*, 140, 221-234.

Manthey, K., Krajinski, F., Hohnjec, N., Firnhaber, C., Puhler, A., Perlick, A.M., Kuster, H. 2004. Transcriptome profiling in root nodules and arbuscular mycorrhiza identifies a collection of novel genes induced during *Medicago truncatula* root endosymbioses. *Mol. Plant Microbe Interact.*, 17, 1063-1077.

Mathesius, U., Keijzers, G., Natera, S.H.A., Weinman, J.J., Djordjevic, M.A., Rolfe, B.G. 2001. Establishment of a root proteome reference map for the model legume *Medicago truncatula* using the expressed sequence tag database for peptide mass fingerprinting. *Proteomics*, 1, 1424-1440.

May, G.D., Dixon, R.A. 2004. *Medicago truncatula*. *Curr. Biol.*, 14, R180-181.

McCallum, C.M., Comai, L., Greene, E.A., Henikoff, S. 2000. Targeting induced local lesions in genomes (TILLING) for plant functional genomics. *Plant Physiol.*, 123, 439–442.

McConn, M.M., Nakata, P.A. 2002. Calcium oxalate crystal morphology mutants from *Medicago truncatula*. *Planta*, 215, 380–386.

Mitra, R.M., Gleason, C.A., Edwards, A., Hadfield, J., Downie, J.A., Oldroyd, G.E., Long, S.R. 2004b. A Ca2+/calmodulin-dependent protein kinase required for symbiotic nodule development: Gene identification by transcript-based cloning. *Proc. Natl. Acad. Sci.* USA, 101, 4701-4705.

Mitra, R.M., Shaw, S.L., Long, S.R. 2004a. Six nonnodulating plant mutants defective for Nod factor-induced transcriptional changes associated with the legume-rhizobia symbiosis. *Proc. Natl. Acad. Sci. U S A*, 101, 10217-10222.

Mudge, J., Cannon, S.B., Kalo, P., Oldroyd, G.E., Roe, B.A., Town, C.D., Young, N.D. 2005. Highly syntenic regions in the genomes of soybean, *Medicago truncatula*, and *Arabidopsis thaliana*. *BMC Plant Biol.*, 5, 15.

Mulder, L., Hogg, B., Bersoult, A., Cullimore, J.V. 2005. Integration of signalling pathways in the establishment of the legume-rhizobia symbiosis. *Physiol Plant.*, 123, 207-218

Mylona, P., Pawlowski, K., Bisseling, T. 1995. Symbiotic nitrogen fixation. *Plant Cell*, 7, 869-885.

Nam, Y.W., Penmetsa, R.V., Endre, G., Uribe, P., Kim, D.J., Cook, D.R. 1999. Construction of a bacterial artificial chromosome library of *Medicago truncatula* and identification of clones containing ethylene-response genes. *Theor.Appl. Genet.*, 98, 638-646.

Oldroyd, G.E., Downie, J.A. 2004. Calcium, kinases and nodulation signalling in legumes. *Nat. Rev. Mol. Cell Biol.*, 5, 566-576.

Oldroyd, G.E., Harrison, M.J., Udvardi, M. 2005. Peace talks and trade deals. Keys to long-term harmony in legume-microbe symbioses. *Plant Physiol.*, 137, 1205-1210.

O'Neill, N.R., Bauchan, G.R., 2000. Sources of resistance to anthracnose in the annual *Medicago* core collection. *Plant Disease*, 84, 261-267.

O'Neill, N.R., Bauchan, G.R., Samac, D.A. 2003. Reactions in the annual *Medicago* spp. core germ plasm collection to *Phoma medicaginis*. *Plant Disease*, 87, 557-562.

Parinov, S., Sundaresan, V. 2000. Functional genomics in *Arabidopsis* : large-scale insertional mutagenesis complements the genome sequencing project. *Curr. Opin. Biotechnol.*, 11, 157-161.

Parniske, M. 2000. Intracellular accommodation of microbes by plants: a common developmental program for symbiosis and disease? *Curr. Opin. Plant Biol.*, 3, 320-328.

Parniske, M. 2004. Molecular genetics of the arbuscular mycorrhizal symbiosis. *Curr. Opin. Plant Biol.*, 7, 414-421.

Patriarca, E.J., Tate, R., Ferraioli, S., Iaccarino, M. 2004. Organogenesis of legume root nodules. *Int. Rev. Cytol.*, 234, 201-62.

Pawlowski, K. 1997. Nodule-specific gene expression. *Physiol. Plant.*, 99, 617-631.

Pawlowski, K., Bisseling, T. 1996. Rhizobial and actinorhizal symbioses: What are the shared features? *Plant Cell*, 8, 1899-1913.

Penmetsa, R.V., Cook, D.R. 1997. A legume ethylene-insensitive mutant hyperinfected by its rhizobial symbiont. *Science*, 275, 527-530.

Penmetsa, R.V., Cook, D.R. 2000. Production and characterization of diverse developmental mutants of *Medicago truncatula*. *Plant Physiol.*, 123, 1387-1397.

Perry, J.A., Wang, T.L., Welham, T.J., Gardner, S., Pike, J.M., Yoshida, S., Parniske, M. 2003. A TILLING reverse genetics tool and a web accessible collection of mutants of the legume *Lotus japonicus*. *Plant Physiol.*, 131, 866–871.

Pichon, M., Journet, E.P., Dedieu, A., de Billy, F., Truchet, G., Barker, D.G. 1992. *Rhizobium meliloti* elicits transient expression of the early nodulin gene *ENOD12* in the differentiating root epidermis of transgenic alfalfa. *Plant Cell*, 4, 1199-1211.

Polhill, R.M., Raven, P.H., Stirton, C.H. 1981. Evolution and systematics of the Leguminosae. *In* R.M. Polhill, P.H. Raven, eds, Advances in Legume Systematics Part 1. Royal Botanic Gardens, Kew, UK, pp. 1–26.

Prosperi, J.M. 2000. The genus *Medicago* and the genetic resources of *Medicago truncatula*. *Grain Legumes*, 28, 14-15.

Rose, R.J., Nolan, K.E., Bicego, L. 1999. The development of the highly regenerative seed line Jemalong 2 HA for transformation of *Medicago truncatula*: implications for regenerability *via* somatic embryogenesis. *J. Plant Physiol.* 155, 788-791.

Sagan, M., de Larambergue, H., Morandi, D. 1998. Genetic analysis of symbiosis mutants in *Medicago truncatula*. *In* Biological nitrogen fixation for the 21st Century. C. Elmerich, A. Kondorosi, W.E. Newton, eds. (Kluwer Academic Publishers), pp. 317-318.

Sagan, M., Morandi, D., Tarenghi, E., Duc, G. 1995. Selection of nodulation and mycorrhizal mutants in the model plant *Medicago truncatula* (Gaertn.) after g-ray mutagenesis. *Plant Sci.*, 111, 63-71.

Schauser, L., Handberg, K., Sandal, N., Stiller, J., Thykjaer, T., Pajuelo, E., Nielsen, A., Stougaard, J. 1998. Symbiotic mutants deficient in nodule establishment identified after T-DNA transformation of *Lotus japonicus*. *Mol. Gen. Genet.*, 259, 414-423.

Schiex, T., Gouzy, J., Moisan, A., de Oliveira, Y. 2003. FrameD: A flexible program for quality check and gene prediction in prokaryotic genomes and noisy matured eukaryotic sequences. *Nucl. Acids Res.*, 31, 3738-41.

Scholte, M., d' Erfurth, I., Rippa, S., Mondy, S., Cosson, V., Durand, P., Breda, C., Trinh, H., Rodriguez Llorente, I., Kondorosi, E., Schultze, M., Kondorosi, A., Ratet, P. 2002. T-DNA tagging in the model legume *Medicago truncatula* allows efficient gene discovery. *Mol. Breeding*, 10, 203-215.

Schoof, H., Spannagl, M., Yang, L., Ernst, R., Gundlach, H., Haase, D., Haberer, G., Mayer, K.F. 2005. Munich information center for protein sequences plant genome resources: a framework for integrative and comparative analyses. *Plant Physiol.*, 138, 1301-1309.

Schultze, M., Kondorosi, A. 1998. Regulation of symbiotic root nodule development. *Annu. Rev. Genet.*, 32, 33-57.

Smith, S.E., Read, D.J., eds. 1997. *Mycorrhizal Symbiosis*. San Diego, CA: Academic Press

Somers, D.A., Samac, D.A., Olhoft, P.M. 2003. Recent advances in legume transformation. *Plant Physiol.*, 131, 892–899.

Stougaard, J. 2001. Genetics and genomics of root symbiosis. *Curr. Op. Plant Biol.*, 4, 328-335.

Stracke, S., Kistner, C., Yoshida, S., Mulder, L., Sato, S., Kaneko, T., Tabata, S., Sandal, N., Stougaard, J., Szczyglowski, K., Parniske, M. 2002. A plant receptor-like kinase required for both bacterial and fungal symbiosis. *Nature*, 417, 959-62.

Sumner, L., Mendes, P., Dixon, R. 2003. Plant metabolomics: large scale phytochemistry in the functional genomics era. *Phytochemistry*, 62, 817–836.

Suzuki, H., Reddy, M.S., Naoumkina, M., Aziz, N., May, G.D., Huhman, D.V., Sumner, L.W., Blount, J.W., Mendes, P., Dixon, R.A. 2005. Methyl jasmonate and yeast elicitor induce differential transcriptional and metabolic re-programming in cell suspension cultures of the model legume *Medicago truncatula. Planta*, 220, 696-707.

Tadege, M., Ratet, P., Mysore, K.S. 2005. Insertional mutagenesis: a Swiss Army knife for functional genomics of *Medicago truncatula. Trends Plant Sci.*, 10, 229-35.

Thomas, M.R., Rose, R.J., Nolan, K.E. 1992. Genetic transformation of *Medicago truncatula* using *Agrobacterium* with genetically modified Ri and disarmed Ti plasmids. *Plant. Cell Rep.*, 11, 113-117.

Thoquet, P., Gherardi, M., Journet, E.P., Kereszt, A., Ane, J.M., Prosperi, J.M., Huguet, T. 2002. The molecular genetic linkage map of the model legume *Medicago truncatula*: an essential tool for comparative legume genomics and the isolation of agronomically important genes. *BMC Plant Biol.*, 2, 1.

Tirichine, L., de Billy, F., Huguet, T. 2000. *Mtsym6*, a gene conditioning *Sinorhizobium* strain-specific nitrogen fixation in *Medicago truncatula. Plant Physiol.*, 123, 845-851.

Torregrosa, C., Cluzet, S., Fournier, J., Huguet, T., Gamas, P., Prosperi, J.M., Esquerre-Tugaye, M.T., Dumas, B., Jacquet, C. 2004. Cytological, genetic, and molecular analysis to characterize compatible and incompatible interactions between *Medicago truncatula* and *Colletotrichum trifolii. Mol. Plant Microbe Interact.*, 17, 909-920.

Trieu, A.T., Burleigh, S.H., Kardailsky, I.V., Maldonado-Mendoza, I.E., Versaw, W.K., Blaylock, L.A., Shin, H., Chiou, T.J., Katagi, H., Dewbre, G.R., Weigel, D., Harrison, M.J. 2000. Transformation of *Medicago truncatula* via infiltration of seedlings or flowering plants with *Agrobacterium. Plant J.*, 22, 531-541.

Trieu, A.T., Harrison, M.J. 1996. Rapid transformation of *Medicago truncatula*: regeneration via shoot organogenesis. *Plant Cell Rep.*, 16, 6-11.

Trinh, T.H., Ratet, P., Kondorosi, E., Durand, P., Kamate, K., Bauer, P., Kondorosi, A. 1998. Rapid and efficient transformation of diploid *Medicago truncatula* and *Medicago sativa* ssp. *falcata* lines improved in somatic embryogenesis. *Plant Cell Rep.*, 17, 345–355.

Udvardi, M.K., Tabata, S., Parniske, M., Stougaard, J. 2005. *Lotus japonicus*: legume research in the fast lane. *Trends Plant Sci.*, 10, 222-228.

Van Iantcheva H., Slavov, S., Prinsen, E., Vlahova, M., van Onckelen, H., Atanassov, A. 2005. Embryo induction and regeneration from root explants of *Medicago truncatula* after osmotic pre-treatment. *Plant Cell Tiss. Org. Cult.*, 81, 37–43.

VandenBosch, K.A., Stacey, G. 2003. Summaries of legume genomics projects from around the globe. Community resources for crops and models. *Plant Physiol.*, 131, 840–865.

Vershinin, A.V., Allnutt, T.R., Knox, M.R., Ambrose, M.J., Ellis, T.H. 2003. Transposable elements reveal the impact of introgression, rather than transposition, in *Pisum* diversity, evolution, and domestication. *Mol. Biol. Evol.*, 20, 2067-2075.

Wais, R.J., Galera, C., Oldroyd, G., Catoira, R., Penmetsa, R.V., Cook, D., Gough, C., Denarie, J., Long, S.R. 2000. Genetic analysis of calcium spiking responses in nodulation mutants of *Medicago truncatula. Proc. Natl. Acad. Sci.* USA, 97, 13407-13412.

Wan, J., Torres, M., Ganapathy, A., Thelen, J., DaGue, B.B., Mooney, B., Xu, D., Stacey, G. 2005. Proteomic analysis of soybean root hairs after infection by *Bradyrhizobium japonicum. Mol. Plant Microbe Interact.*, 18, 458–467.

Waterhouse, P.M., Helliwell, C.A. 2003. Exploring plant genomes by RNA-induced gene silencing. *Nat. Rev. Genet.*, 4, 29–38.

Watson, B.S., Asirvatham, V.S., Wang, L., Sumner, L.W. 2003. Mapping the proteome of barrel medic (*Medicago truncatula*). *Plant Physiol.*, 131, 1104–1123.

Wulf, A., Manthey, K., Doll, J., Perlick, A.M., Linke, B., Bekel, T., Meyer, F., Franken, P., Kuster, H., Krajinski, F. 2003. Transcriptional changes in response to arbuscular mycorrhiza development in the model plant *Medicago truncatula. Mol. Plant Microbe Interact.*, 16, 306-314.

Yan, H.H., Mudge, J., Kim, D.J., Larsen, D., Shoemaker, R.C., Cook, D.R., Young, N.D. 2003. Estimates of conserved microsynteny among the genomes of *Glycine max, Medicago truncatula* and *Arabidopsis thaliana. Theor. Appl. Genet.*, 106, 1256-1265.

Young, N.D., Cannon, S.B., Sato, S., Kim, D., Cook, D.R., Town, C.D., Roe, B.A., Tabata, S. 2005. Sequencing the genespaces of *Medicago truncatula* and *Lotus japonicus. Plant Physiol.*, 137, 1174-1181.

Zhu, H.Y., Cannon, S.B., Young, N.D., Cook, D.R. 2002. Phylogeny and genomic organization of the TIR and non-TIR NBS-LRR resistance gene family in *Medicago truncatula. Mol. Plant-Microbe Interact.*, 15, 529-539.

Zhu, H.Y., Choi, H.K., Cook, D.R., Shoemaker, R.C. 2005. Bridging model and crop legumes through comparative genomics. *Plant Physiol.*, 137, 1189-1196.

Zhu, H.Y., Kim, D.J., Baek, J.M., Choi, H.K., Ellis, L.C., Küster, H., McCombie, W.R., Peng, H.M., Cook, D.R. 2003. Syntenic relationships between *Medicago truncatula* and *Arabidopsis* reveal extensive divergence of genome organization. *Plant Physiol.*, 131, 1018-1026.

22

CHAPTER

Tomato: A Model Plant for Solanaceae Genomics

R. Stevens, M. Bouzayen, M. Causse, C. Etienne, C. Rothan

TOMATO, A MODEL SPECIES IN SEVERAL WAYS

The tomato (*Solanum lycopersicum*, previously *Lycopersicon esculentum*) belongs to the Solanaceae family which contains a number of economically important crops such as potato, pepper, aubergine and tobacco. Because of its favourable biological and genetic features (short life cycle, easy crossing, routine genetic transformation via *Agrobacterium*), the tomato has emerged as a bridging species for the whole Solanaceae family as well as the reference model system for fleshy fruit research.

Being adapted to varied and sometimes contrasting growth conditions, the tomato has become in recent times the most widely consumed vegetable in the world after the potato, either as a fresh or processed product. Since the 1940s, selection has been aimed at creating varieties specifically adapted to diverse conditions by simultaneously improving fruit quality and pathogen resistance. To reach these goals, genetic resources have been thoroughly explored. However, while the cultivated tomato presents a wide morphological diversity, the molecular diversity, as studied by RFLP mapping, is rather limited (Miller and Tanksley, 1990). In contrast, the wild species represent a real source of genetic variability though the variability potential differs among different species. Several thousand ancient cultivars have been collected and are now available in genetic resource conservation centres where representatives of the eight species most closely related to tomato are also found. These wild species have been frequently used in the search for pathogen resistance genes (http://tgrc.ucdavis.edu).

Though traditionally classified as a vegetable by consumers, the tomato is a fruit in the botanical sense of the term. Fleshy fruits undergo very similar developmental processes, from fertilization to ripening, despite their diversity and regardless of the tissue from which they originate: the 'fruit' corresponds to a swollen ovary in tomato and a swollen floral receptacle in strawberry, for example. Two processes are highly specific to fleshy fruits: (i) the large accumulation of water and solutes (sugars, acids, secondary metabolites) in large cells during the period of rapid fruit growth, which gives the fleshy characteristic to the fruit; and (ii) the ripening phase during which traits underlying the sensory quality of the fruit are acquired (colour, texture, aroma). Model plants for which a wealth of genetic and genomic resources have been developed, like *Arabidopsis*, rice or, more recently, *Medicago*, are useful for studying some aspects of fruit development such as ovary development and fruit set but they are not suitable for studying the later stages of fruit enlargement and ripening as these are specific to fleshy fruits. For this reason and because of the biological advantages mentioned above, the tomato has become the model system for fleshy fruits which include a number of agronomically interesting species such as grape, apple, peach, strawberry and melon.

GENETIC RESOURCES: RELATED SPECIES, MUTANTS AND GENETIC STOCKS

The CM Rick Tomato Genetic Resource Centre (TGRC) at the University of Davis in California (http://tgrc.ucdavis.edu) currently holds 1,160 wild accessions from *Solanum lycopersicum* and related species, collected mainly in their zone of origin, in South America. Also contributing to these genetic resources are 1,535 miscellaneous genetic stocks and two collections of introgression lines obtained by crossing related species (*S. pennellii,* previously *L. pennellii,* or *S. habrochaites,* previously *L. hirsutum*) with a cultivated tomato (*S. lycopersicum*). Each introgression line contains one or several chromosomal regions of the parent species introgressed into the *S. lycopersicum* genetic background and identified by genetic markers. An initial collection, developed by D Zamir (Hebrew University, Israel) using *S. pennellii* includes 75 introgression lines (Zamir, 2001). A second collection, developed by S. Tanksley (Cornell University, USA) from *S. habrochaites* comprises 57 introgression lines covering 85% of the genome of the wild species with more than 31 lines containing additional fragments allowing increased coverage of the genome. These collections are a powerful tool for gene mapping in cultivated tomato which generally shows few polymorphisms. Finally, stocks of tomato mutants displaying a variety of visual phenotypes and arising from either spontaneous mutations or mutations induced through chemical mutagenesis (EMS, ethyl methanesulfonate), X rays or fast neutron bombardment have been collected over recent decades and are available at TGRC (994 monogenic stocks for 614 loci). More recently, a collection of over 3,000 phenotypically characterised tomato mutants (M82 variety) has been created by screening a total of 13,000 M2 families obtained after EMS mutagenesis and fast neutron mutagenesis (Menda *et al.*, 2004). Individual mutants from this collection, characterised in the M2 families are available and described on the web site http://www.sgn.cornell.edu/mutants. Several other collections of tomato genetic resources exist in the world (Ukraine, Netherlands, Germany, France).

GENETIC MAPS FOR MAP-BASED CLONING OF GENES AND QTL

The tomato has one of the highest number of exploitable monogenic characteristics amongst cultivated varieties of vegetable. Several of these characteristics result from mutations that appeared spontaneously in cultivars or were discovered in species related to cultivated tomato. These traits concern pathogen resistance, plant architecture, colour, fruit composition or shelf life. So far, all resistances used in cultivars come from wild species and the majority are dominant monogenic traits (Tables 22.1 and 22.2).

The first molecular markers allowed the construction of a reference map for the tomato genome (Tanksley *et al.*, 1992). This map, composed of more than 1,000 RFLP markers spread over the 12 chromosomes, has allowed the localisation of a large number of mutations or genes of interest (Table 22.2). More recently, other markers (AFLP, Amplified Fragment Length Polymorphism; SSR, simple sequence repeat) have completed the map. A large part of the tomato genome mapping data are consultable on the web site http://www.sgn.cornell.edu.

CLONING GENES OF INTEREST

The number of mapped genes in tomato is growing rapidly and research is now focusing on their functional characterisation as an essential step towards the understanding of their regulation mechanism. Because of the small size of the tomato genome (950 Mb) and the considerable accumulation of data on major disease resistance genes, the tomato is one of the main species for cloning resistance genes and constitutes a model system for molecular characterisation of plant-pathogen interactions which illustrate the gene for gene model. In this regard, it is important to mention that the first resistance gene cloned was *Pto* which was shown to encode for resistance to *Pseudomonas tomato* (Martin *et al.*, 1993). The signal cascade set off by the interaction between *Pto* and *AvrPto* (pathogen avirulence gene) which leads to a resistance phenotype is currently the model where research on the 'gene for gene' relationship is the most advanced. Subsequently, a number of resistance genes were cloned in tomato such as *Cf-2*, *Cf-4*, *Cf-5* and *Cf-9* (*Cladosporium fulvum* resistance), *I-2* (*Fusarium* o.f.sp *lycopersici* resistance), *Mi* (nematode resistance) and *Sw-5* (TSWV resistance) (Brommonschenkel *et al.*, 1997; Ori *et al.*, 1997; Milligan *et al.*, 1998). It is of note that all these genes show numerous conserved domains (Hammond Kosack *et al.*, 1997) and their molecular dissection revealed that *Mi* and *Meu-1* are the same gene with a pleiotropic effect on pathogens as different as nematodes and aphids (Vos *et al.*, 1998). By taking advantage of the construction of YAC (yeast artificial chromosome) and BAC (bacterial artificial chromosome) libraries (http://www.genome.clemson.edu), several genes linked to fruit quality or plant architecture have also been cloned (Table 22.2). The sequencing of the BAC ends currently being performed will offer an anchorage point between the tomato genetic and physical maps (Budiman *et al.*, 2000) and hence will pave the way towards the genome sequencing effort.

THE PRINCIPAL PHYSIOLOGICAL FRUIT TRAITS

Tomato has for a long time been the favoured model for studying fruit development and ripening. An important part of these studies has been motivated by research into the molecular basis of sensorial quality. Fruit development, from anthesis to maturation, corresponds to a succession of genetically programmed processes taking place mainly under the control of endogenous hormones but influenced by external environmental factors (temperature, light). Many of the metabolic pathways contributing to the fruit sensory qualities are activated during the ripening process and

Table 22.1 Characteristics of resistances to major tomato pathogens

Pathogen	*Genes*	*Cloned*	*Chromosome*	*Origin*
Virus				
Tobacco Mosaic Virus (TMV)	*Tm-1*		2	*S. habrochaites* (previously *L. hirsutum*)
	Tm-2, Tm-22	*	9	*S. peruvianum* (previously *L. peruvianum*)
Potato Virus Y (PVY)	2 genes	*	3	*S. habrochaites*
Tobacco Etch potyVirus (TEV)	1 gene		-	*S. habrochaites*
Cucumber Mosaic Virus (CMV)	polygenic		-	Several species
Alfalfa Mosaic Virus (AMV)	*Am*		6	*S. habrochaites*
Tomato Spotted Wilt Virus (TSWV)	*Sw-5*	*	9	*S. peruvianum*
Tomato Yellow Leaf Curl Virus (TYLCV)	*Ty-1* + QTL		6	*S. chilense* (previously *L. chilense*)
	polygenic		+ others	and several other species
Fungi				
Cladosporium fulvum	Cf-4	*	1	*S. habrochaites*
	Cf-9, Cf-2	*	1, 6	*S. pimpinellifolium*
	Cf-5		6	(previously *L. pimpinellifolium*)
	+ other genes		-	*S. pimpinellifolium*
Verticillium daliae	Ve		*9*	*S. pimpinellifolium*
Stemphylium spp.	*Sm*		11	*S. pimpinellifolium*
Pyrenochaeta lycopersici / corky root of tomato	*py-1*		3	*S. peruvianum*
Leveillula taurica / tomato powdery mildew	*Lv*		12	*S. chilense*
Oidium lycopersicum /tomato powdery mildew	*Ol-1* + QTL polygenic		6 + others	*S. habrochaites*
Phytophthora infestans / Mildew	*Ph-1, Ph-2,* *Ph-3*		7, 10, 9 + others	*S. pimpinellifolium*
Fusarium oxysporum f.sp. *lycopersici* / Fusarium wilt	*I, I-2*		11	*S. pimpinellifolium*
	I-1	*	*7*	*S. pennellii* (previously *L. pennellii*)

Contd.

Table 22.1 Contd.

Pathogen	*Genes*	*Cloned*	*Chromosome*	*Origin*
Fusarium oxysporum f.sp *radicis lycopersici* / Crown and root rot	*Frl*		9	*S. peruvianum*
Botrytis cinerea	polygenic		-	*S. habrochaites*
Bacteria				
Pseudomonas tomato	*Pto*	*	5	*S. pimpinellifolium*
Clavibacter michiganensis / bacterial canker	polygenic		Location of several QTL	Several species
Xanthomonas campestris pv. vesicatoria / bacterial leaf spot	*Bs3-2* *+ other*			*S. lycopersicum* (previously *L. esculentum*)
	polygenic genes		-	*S. pimpinellifolium*
Ralstonia solanacearum / bacterial wilt	polygenic		Location of several QTL including chrom. 6	*S. pimpinellifolium* and *S.lycopersicum* var *cerasiforme*
Parasites and insects				
Nematode *Meloidogyne spp.*	*Mi, Mi-3* + other genes	*	6, 12	*S. peruvianum*
Nematode *Globodera rostochiensis*	*Hero*		4	*S. pimpinellifolium*
Aphid *Macrosiphum euphorbiae*	*Meu-1*	*	6	*S. peruvianum*
Leaf mining flies *Liriomyza spp.*	*polygenic*		-	*S. habrochaites* *S. cheesmaniae* (previously *L. cheesmanii*)
Glasshouse whitefly *Trialeurodes vaporariorum;* Insects in general	polygenic		Location of several QTL	*S. habrochaites f. glabratum*

By convention, recessive alleles are noted in lowe- case in contrast to dominant alleles which start with a capital letter

Table 22.2 The principal major genes used for the improvement of fruit quality and plant growth

Gene	*Phenotype*	*Origin*	*Cloned*	*Chromosome*
alc (alcobaca)	Inhibited maturation (semi-dominant)	*S. lycopersicum* (previously *L. esculentum)*		10
at (apricot)	Orange fruits	*S. lycopersicum*		5
B (Beta)	Yellow fruits	*S. habrochaites* (previously *L. hirsutum)*	*	6
Del (Delta)	Orange fruits	*S. habrochaites*	*	12
hp (high pigment)	Increased lycopene content	*S. lycopersicum*		12
j (jointless)	Absence of abscission zone from peduncle	*S. lycopersicum*	*	11
j-2 (jointless)	Absence of abscission zone from peduncle	*S. cheesmaniae* (previously *L. cheesmanii)*		11
nor (non ripening)	Inhibited maturation (semi-dominant)	*S. lycopersicum*	*	10
ogc (old gold-crimson)	Increased lycopene content	*S. lycopersicum*	*	6
r (yellow flesh)	Yellow fruits	*S. lycopersicum*	*	3
rin (ripening inhibitor)	Inhibited maturation (semi-dominant)	*S. lycopersicum*	*	5
sp (self pruning)	Determinate growth	*S. lycopersicum*	*	6
t (tangerine)	Orange fruits	*S. lycopersicum*	*	10
u (uniform colour)	Absence of green shoulder	*S. lycopersicum*		10
y	Uncoloured epidermis	*S. lycopersicum*		1

By convention, recessive alleles are noted in lower case in contrast to dominant alleles which start with a capital letter the organoleptic traits acquired at this stage are essential in making the fruit attractive to the consumer. It has now been clearly demonstrated that the ripening-associated modifications of gene expression are behind the metabolic changes affecting fruit colour, texture, flavour and aroma. Therefore, far from being a destructive process, ripening is now considered as an example of a highly regulated developmental program specific to higher plants. Fruits are usually divided into two classes: climacteric and non climacteric, according to whether the onset of ripening is accompanied by a sharp increase in respiration. In the case of climacteric fruits, such as tomato, ethylene plays a crucial role in the activation and progression of ripening. For this reason, ethylene biosynthesis and perception have been the initial target of a number of studies aimed at deciphering the genetic control of pre- and postharvest mechanisms of fruit physiology. The genes involved in the ethylene biosynthesis pathway were first isolated in tomato and then in a number of other species. Subsequently, the modulation of their expression via transgenesis, has allowed efficient control of the ripening process (Hamilton *et al.*, 1990; Ayub *et al.*, 1996). The generation and characterisation of transgenic lines with decreased ethylene production revealed the existence of ethylene-dependent and independent phenomena within the ripening process of climacteric fruits (Lelievre *et al.*, 1997). In the early nineties, the unravelling of the molecular basis of the ethylene transduction pathway in the model plant *Arabidopsis* opened new opportunities for the control of fruit ripening via the manipulation of ethylene perception. These advances identified the Nr (*never ripe*) natural mutation in tomato as an alteration of a gene encoding an ethylene receptor (Wilkinson *et al.*, 1995). More recently, the molecular cloning of the *rin* and *nor* ripening mutations using a map-based cloning strategy (Vrebalov *et al.*, 2002) represented another major step forward by revealing that the *rin* and *nor* genes act upstream of the ethylene regulatory cascade and may therefore control early and global events of the ripening process. Therefore, as an ethylene-independent phenomenon, the *RIN*-related regulation is likely to be shared among climacteric and non-climacteric fruit and may represent a common component involved in triggering the ripening process. While these findings open a new era towards understanding the global developmental process leading to the transition from unripe to ripe fruit, the exact mechanism underlying the attainment of the capacity to ripen remains to be elucidated.

The colour change from green to red which characterises ripening in tomato results from chlorophyll degradation and carotenoid pigment accumulation (β-carotene and lycopene). The numerous tomato mutants affected in pigmentation represent a valuable genetic resource which has been exploited to facilitate the identification of the genes involved in the carotenoid biosynthetic pathways and the understanding of the complex mechanisms regulating pigment accumulation (Bramley, 2002). In several cases, the role of light in the regulation of fruit pigmentation has been shown (Giovannoni, 2001).

The loss of firmness corresponds to a major phenomenon accompanying the maturation process. This phenomenon is genetically regulated and is thought to be tightly linked to the modifications that take place within the cell wall. A number of genes potentially involved in cell wall degradation, rearrangement and structure have been isolated and most of them were studied in the tomato model. Unexpectedly, it has been shown that the suppression of certain of these candidate genes, such as those encoding polygalacturonase, pectin-methyl-esterase and β-glucanase did not have a major impact on the evolution of fruit firmness (Giovannoni *et al.*, 1989; Tieman *et al.*, 1992; Brummell *et al.*, 2000). In contrast, the same transgenesis approaches showed that fruit softening is strongly affected by the deregulation of genes coding expansins, a cell wall protein which accumulates during fruit maturation (Brummell *et al.*, 1999). These results underline the importance of reverse genetic approaches to validate the physiological role of candidate genes.

QTL MAPPING IN TOMATO

Since 1980, with the arrival of molecular markers, quantitative trait loci (QTL) can be localised on genetic maps and their individual effects quantified. Studies in tomato have illustrated the power of this approach. Several studies have focused on identification of QTLs for quality and adaptation. Taking into account the low molecular polymorphism observed in cultivated tomato, the majority of these studies rely on interspecific progeny (Tanksley *et al.*, 1997). In all cases QTLs are detected often with strong effects: for a particular trait, it is normal to detect a major QTL responsible for 20% to 50% of the phenotypic variation which is modulated by other weaker QTLs. The effect of most QTLs is additive, but dominant QTLs, even overdominant QTLs, are often detected. We can distinguish "*generalist*" QTLs which remain stable across the environment, the year or the type of progeny, and QTLs more specific for a given condition (Paterson *et al.*, 1991). Certain regions involved in trait variation in progeny resulting from crosses with different accessions from the same species, or even different species, are also detected (Bernacchi *et al.*, 1998). The wild species, in spite of their characteristics inferior to those of cultivated species, can possess alleles useful for improving agronomic traits. Certain improvements are difficult to predict, like for example QTLs improving fruit colour, detected in a BC3 involving *S. habrochaites* (*L. hirsutum*), a green-fruited species. Tanksley and collaborators have proposed the advanced backcross technique to exploit the potential offered by wild species. By studying progeny from five species, they have revealed a considerable number of unexpected transgressions (Fulton *et al.*, 2002a). The molecular markers localised close to QTLs are used in marker assisted selection to create lines with increased potential, or, in contrast, to avoid certain unfavourable traits.

THE FIRST QTLS CLONED BY CHROMOSOME WALKING

Fine mapping allows precise localization of QTLs and the verification of the existence of several linked QTLs in the same region (Paterson *et al.*, 1991). By reducing the size of an introgressed fragment from *S. pennellii*, it has been possible to identify three linked QTLs controlling fruit weight (Eshed and Zamir, 1995). Fine mapping also constitutes an essential step towards QTL cloning. For example, a fruit weight QTL, common to several studies, has been precisely localised and then cloned (Frary *et al.*, 2000). Another QTL controlling sugar concentration in fruit has also been cloned following a study involving 7,000 F_2 plants (Fridman *et al.*, 2000). The cloned gene encodes a cell wall invertase and the functional polymorphism has been delimited to an amino acid near the catalytic site which affects enzyme kinetics and fruit sink strength (Fridman *et al.*, 2004).

COLLECTIONS OF EST AND MICROARRAYS

The NSF project on tomato has allowed the assembly of a high density collection of ESTs including more than 160,000 ESTs (5′ sequencing) from 26 libraries covering a variety of organs and normal and abnormal developmental stages (Moore *et al.*, 2002; Van der Hoeven *et al.*, 2002). Special attention has been paid to fruit maturation (24,000 sequences), plant defence reactions (10,000 sequences) and ovary development (10,000 sequences). The analysis of these ESTs has allowed the reconstruction of more than 27,000 distinct tomato cDNA sequences (unigenes) which could correspond to 17,500 genes and represent about half the total number of tomato genes estimated at 35,000 (Van der Hoeven *et al.*, 2002). The collection of non-redundant tomato ESTs has been deposited in the TIGR (The Institute for Genomic Research) database, the Tomato Gene Index (http://compbio.dfci.harvard.edu/tgi/cgi-bin/tgi/gimain.pl?gudb-tomato)[1] as well as in the Solanaceae Genomics Network database (http://www.sgn.cornell.edu), dedicated to genetic and genomic mapping data from the Solanaceae. The SGN EST database gathers 213,000 tomato ESTs assembled into 41,425 annotated unigenes (June, 2006). Taking into account the diversity of libraries studied and the number of cDNAs sequenced, the tomato EST databases provide an excellent overview of the identity, relative abundance and transcript localisation of a given gene. Thirteen thousand non redundant coding sequences have been deposited on microarrays (Cornell, http://bti.cornell.edu/CGEP/CGEP.html), which are publicly available for the analysis of the tomato transcriptome.

COMPARATIVE GENOMICS AND SYNTENY: CONSERVED ORTHOLOG SET (COS) MARKERS

Tomato and *Arabidopsis* belong to two botanic families: Solanaceae and Brassicaceae respectively, which diverged 112 to 156 million years ago, relatively early during the spread of the Dicotyledons. This taxonomic position has made tomato a model species for determining regions of synteny with *Arabidopsis*, possibly common to numerous dicotyledonous species of the tomato/*Arabidopsis* clade (Fig. 22.1). Within the framework of the American programme, the identification and mapping of orthologous tomato/*Arabidopsis* genes has been undertaken (Fig. 22.2). More than one thousand Conserved Ortholog Set (COS) markers have thus been defined by large scale systematic comparison of the high density EST collection from tomato and the genomic sequence from *Arabidopsis*. Five hundred and thirty four COS markers have been mapped onto a high density genetic map for tomato (about one marker per centimorgan), constructed from an interspecific cross between *S. lycopersicum* and *S. pennellii* showing interesting phenotypic differences for fruit developmental traits (Fulton *et al.*, 2002b; http://www.sgn.cornell.edu/markers/cos_markers. pl)

The transfer of COS markers within the tomato/*Arabidopsis* dicotyledonous clade is made possible by direct hybridisation with the heterospecific probe from *Arabidopsis* or

[1]This regularly updated database regroups all known tomato coding sequences: ESTs from the American programme of high throughput sequencing as well as coding sequences from general banks, both public and international (DDBJ/EMBL/Genbank).

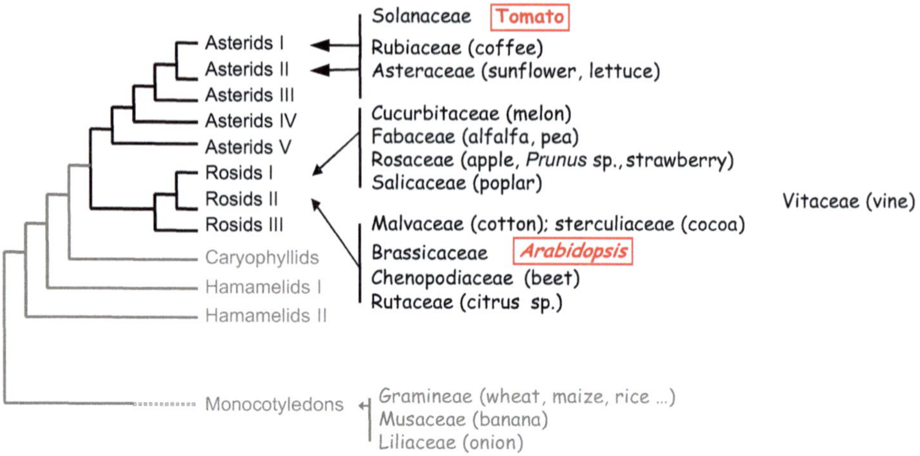

Fig. 22.1 Dendrogram indicating the phylogenetic relationships between higher plants (from Ku *et al.*, 2000). The monophyletic clade tomato/*Arabidopsis* is shown in bold while the taxonomic groups outside the clade are shown in normal type/grey. The clade tomato/Arabidopsis corresponds to the clade more commonly known as "Core Eudicots" (dicotyledonous plants with tetracyclic flowers) which breaks down into two main groups: the Asterids and the Rosids. The Vitaceae family (for example the vine) belongs to the Core Eudicots clade and is related to the Rosids but does not belong to them (J.J. Doyle, pers. comm.).

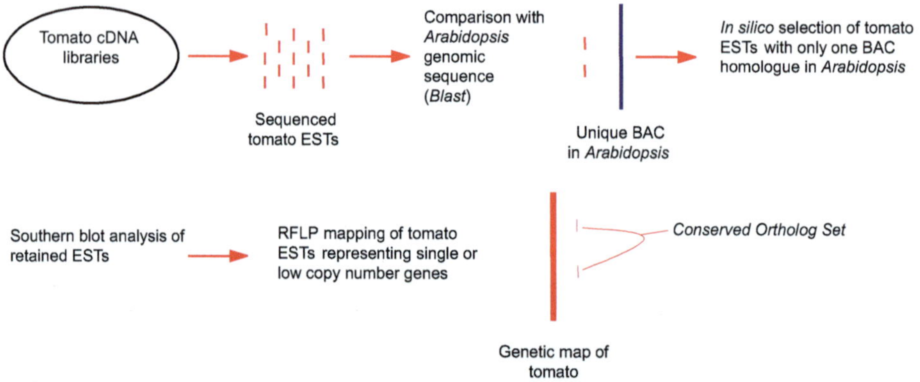

Fig. 22.2 From tomato ESTs to COS markers (from Tanksley *et al.*, http://plantgenome.sdsc.edu/AwardeesMeeting). In order to design conserved ortholog set (COS) markers, each tomato EST was compared to the *Arabidopsis* genome represented in a preliminary database: the *Arabidopsis* minimum Tiling path, corresponding to a super-assembly of unique BAC sequences ordered on each *Arabidopsis* chromosome (comparison of translated sequences tomato/*Arabidopsis* by the programme Tblastx). The tomato ESTs presenting a significant similarity with a unique BAC of the Arabidopsis genome (E-value < e-15) were selected. One thousand and twenty five COS markers were thus designed, of which 85% represent, after Southern blot analyses, single or low copy tomato genes. Over 1,000 COS markers have been mapped onto an interspecific tomato genetic map.

tomato, according to the genetic distance from the species considered. COS markers, defined from half the total tomato genes have been proposed as a way of establishing the macrosynteny in a large spectrum of plants (Fulton *et al.*, 2002b). In tomato, COS markers have revealed blocks of synteny between *Arabidopsis* and tomato which can reach 10 centimorgans. However, at the microsynteny scale, complex orthology relationships link coding sequences from tomato and Arabidopsis, suggesting a paleopolyploid origin for the *Arabidopsis* genome, a hypothesis confirmed by evolutive approaches applied to the complete genome sequence of *Arabidopsis* (Vision *et al.*, 2000; Lynch and Conery, 2000; Fig. 22.3). The elucidation of networks of microsynteny between tomato and Arabidopsis, possibly common to dicotyledonous species, should allow improved understanding of evolutionary processes which have shaped plant genomes and determination of a deductive strategy for transfer of genomic information between plants. Exploitation of tomato/*Arabidopsis* macrosynteny (Ku *et al.*, 2001), or indeed networks of microsynteny between tomato and *Arabidopsis* (Oh *et al.*, 2002), has allowed the identification of positional candidate genes for the characterisation of loci involved in early fruit development in tomato (for example *ovate* and *diageotropica*).

FUNCTIONAL ANALYSIS OF GENES BY REVERSE GENETICS

Several genes of interest have been cloned over the last few years in tomato by direct genetic approaches, due to the variability encountered in wild relatives (resistance or quality genes) or in natural mutants (maturation genes) and the use of recently developed tools (introgression lines, BAC libraries). However, the current fallout from genomic programmes (EST, transcriptome) makes the development of high throughput reverse genetic tools necessary for the functional analysis of genes in tomato.

INSERTIONAL MUTAGENESIS

Transformation by *in vitro* somatic embryogenesis has been used for more than two decades in tomato. One of the first genetically modified plant species to be commercialised for a quality trait such as firmness, was the appearance on the market of the modified tomato Flavr Savr in 1994 in the USA. Although in common use in laboratories, integrating more recent approaches such as gene inactivation by RNAi, has been difficult to implement. Transformation under vacuum or by dipping, as used in *Arabidopsis,* has not been conclusive in tomato. Mutagenesis by T-DNA insertion represents a huge task, which explains the existence of a unique collection of about 10,000 tomato T-DNA transformants (gain of function mutants from the company Exelixis, USA; Mathews *et al.*, 2003). Another insertional mutagenesis approach has relied on the use of transposons from heterologous species such as the non autonomous mobile elements *Activator(Ac)/Dissociation(Ds)* of maize. The advantage of mobile elements is their capacity to excise and reinsert in the plant genome, therefore a lower initial number of transformants are required than for a T-DNA mutagenesis approach. A first pilot collection of 3,000 tomato lines, based on the system *Ac/Ds,* has been developed in the miniature tomato, MicroTom, which is easily cultivated at high density in the greenhouse and has a short life cycle (four months, seed to seed)

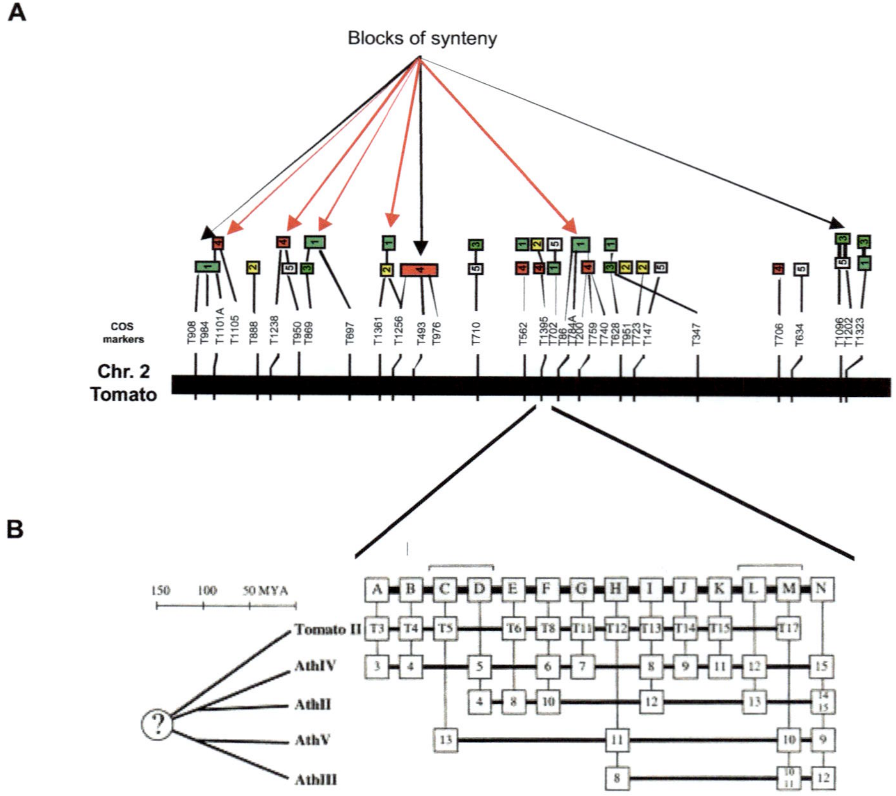

Fig. 22.3 Description of macrosynteny and microsynteny relationships between *Arabidopsis* and tomato revealed by COS markers. Above: tomato chromosome 2 (horizontal thick line) anchored to the Arabidopsis genome via the COS markers (Fulton *et al.*, 2002). The regions of synteny with *Arabidopsis* are indicated above the tomato chromosome by rectangles containing the *Arabidopsis* chromosome number. The width of the rectangles is proportional to the physical distance in *Arabidopsis*, which varies between 100 kb and 1 Mb. Certain regions correspond to genomic segment duplications in *Arabidopsis*. Below: a BAC clone of 105 kb from tomato chromosome 2 anchored to the *Arabidopsis* genome by a COS marker (Ku *et al.*, 2000). The tomato BAC fragment contains 17 coding sequences of which 14 show strong similarity with 4 distinct BACs in *Arabidopsis* (duplication of chromosomal segments). The most commonly used explanation is that a tetraploidisation has taken place during the evolutionary history of the *Arabidopsis* genome, of which the date is estimated, depending on the Arabidopsis genome evolution study referred to, at about 112 or 65 million years ago (Lynch and Conery, 2000; Vision *et al.*,, 2000) and that only certain duplicated chromosomal fragments were conserved.

(Meissner *et al.*, 1997, 2000). A second collection consists of a more restricted number of lines in the variety Moneymaker (405) but presents interesting technical advances (Gidoni *et al.*, 2003): 140 of the Ds elements are currently mapped, allowing an inactivation of nearby genes by excision and reinsertion of the Ds element. Besides the insertional inactivation of target genes (knockout mutants), the different vectors used in these collections offer other functions e.g. promoter trap, induction of chromosomal rearrangements in the insertion region. However, considering the tomato genome includes 30 to 40,000 genes (Van der Hoeven, 2002), the size of a collection of transposon tagged insertion mutants necessary to obtain 95% saturation of the tomato genome is estimated at near to 200,000 to 300,000 lines (Emmanuel and Levy, 2002), which remains considerable!

CHEMICAL MUTAGENESIS BY EMS/TILLING

Mutagenesis by EMS induces point mutations (base-base transitions) and thus a large number of alleles are obtained for a relatively small population (a few thousand plants), from the null allele (about 5% of mutations in a gene) to alleles where gene function is positively or negatively affected. The recent development of detection methods for point mutations allows us to envisage large scale screening of collections of tomato EMS mutants by *Tilling* (*Targeted Induced Local Lesions IN Genomes*) to detect mutated alleles for a given candidate gene. The candidate gene is amplified from pooled DNA extracted from mutagenized plants. For this, the PCR product is denatured and then hybridised with the equivalent PCR product from a wild type plant. In the presence of a mutant in the mutagenized plant pool, the hybridization of DNA from the two sources (mutant and wild type) leads to the formation of a heteroduplex with a DNA base pair mismatch at the mutation site. The high throughput detection of mismatches due to unknown point mutations is currently carried out using two major techniques: dHPLC (denaturing high pressure liquid chromatography) (McCallum *et al.*, 2000) and a technique based on the separation by gel electrophoresis of DNA fragments resulting from the cleavage of the heteroduplex formed at the mismatch by *Cel* I type endonucleases (Fig 22.4).

To be efficient, these strategies require the generation of a population of plants in which the genome is highly saturated in mutations. In *Arabidopsis* the screening of a collection of about 3,000 plants leads to about 14 allelic variants of which 5% will be null mutations and half should affect the encoded protein (Colbert *et al.*, 2001; Till *et al.*, 2003). These statistics should be the same in tomato (Emmanuel and Levy 2002). In this species, several EMS mutant collections are already available. In the variety M82 (field variety, industrial tomato, determinate growth), a moderate mutagenesis with EMS (0.5% EMS; LD15 carried out by the team of D. Zamir (Hebrew University, Israel) has led to the generation of 6,000 M2 families for which the mutations have been characterized at the phenotypic level. Using this population of tomato mutants, a *Tilling* project is already being successfully carried out (A. Bendhamane, INRA Evry, France; pers. comm.). Another tomato EMS mutant collection has been created in the dwarf variety, MicroTom, which is well adapted for high throughput screening of mutants by individual laboratories (Meissner *et al.*, 1997), through a collaborative study (C Rothan, INRA Bordeaux and A. Levy, Weizmann Inst., Israel). This collection has been obtained

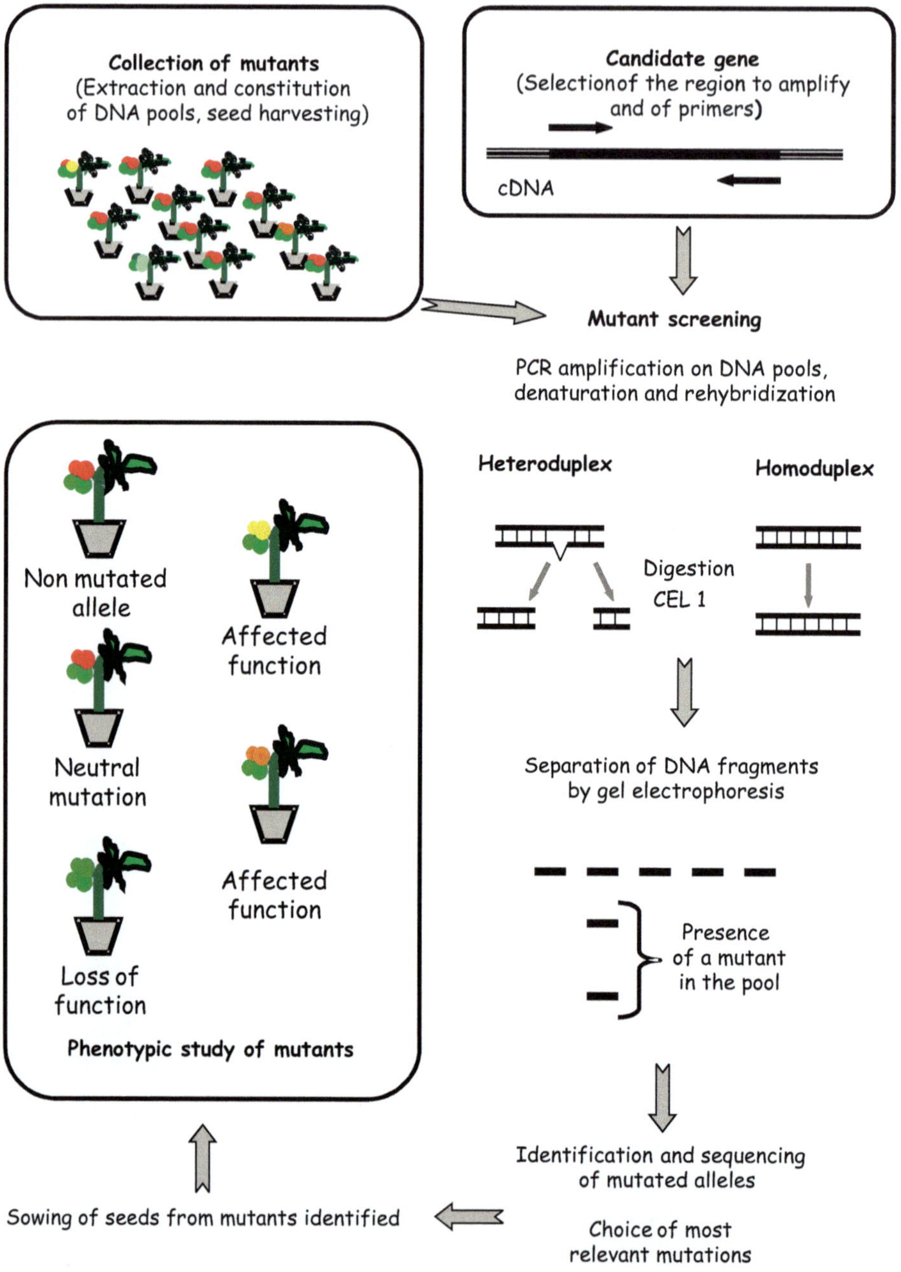

Fig. 22.4 Search for tomato mutants by TILLING (Targeted Induced Local Lesions in Genomes)

by several rounds of mutagenesis, in order to saturate the genome with mutations (seeds from EMS mutants were remutagenised with a dose of 1% EMS; LD80 or LD90). It currently holds over 3,000 M2 families, the number estimated necessary to obtain around 15-20 alleles per 1 kb of candidate gene, and has been thoroughly characterized at the phenotypic level. Phenotypic data from Microtom mutants will soon be available, together with M82 mutant data, through the Tomato Mutant database at SGN (http://zamir.sgn.cornell.edu/mutants). The MicroTom mutant collection has been successfully tested for detection of point mutations in candidate genes using the *Tilling* technology developed at INRA, Evry, (France) (http://www.evry.inra.fr/public/projects/tilling/tilling.html) and is currently available for *Tilling* projects in tomato (INRA, Bordeaux, France).

MUTAGENESIS BY FAST NEUTRONS

A collection of 7,000 M2 families mutagenised by fast neutrons has been obtained by D. Zamir (Hebrew University, Israel; http://zamir.sgn.cornell.edu/mutants). For this type of mutant, a direct genetic approach is currently the best adapted approach to clone the corresponding tomato genes (Isaacson *et al.*, 2002). However, in the near future, the sequencing of the tomato genome will allow the necessary information to be obtained for PCR screening with candidate genes following the method developed by Li *et al.* (2001).

VIRUS INDUCED GENE SILENCING (VIGS)

The high throughput functional analysis of genes by VIGS is now possible in tomato thanks to the use of a viral vector based on TRV (Tobacco Rattle Virus) (Liu *et al.*, 2002). The recombinant TRV virus infects plants and efficiently induces gene silencing, as has been shown for genes of the ethylene signalling pathway and carotenoid biosynthesis genes. The addition of Gateway recombination system sites to the vector facilitates the cloning of candidate genes to be tested, allowing high throughput screening of these genes.

PERSPECTIVES: INTERNATIONAL TOMATO GENOMIC PROGRAMMES, TOWARDS THE SEQUENCE OF THE TOMATO GENOME

As for many species of agronomic interest, an international consortium has recently been set up with the aim of coordinating tomato genomic initiatives and to reach a critical mass sufficient to tackle the genome sequencing of this species. The consortium's activities have at first backed up American initiatives developed in the framework of a wide ranging programme of the NSF (National Science Foundation) entitled Plant Genome Research Program.

The first phase of this American programme (1998-2001) on tomato genomics has been dedicated to three priority actions: the generation of a wide ranging EST collection (160,000 clones); the establishment of the first generation of DNA chips/*microarrays*; and the study of synteny relationships between tomato and Arabidopsis. The second phase (2001-2005) mainly focused on actions that aim at preparing the tools necessary for the genome sequencing project (BAC contig assembly, anchoring of the physical and genetic maps). One important decision made by the international consortium was to make all tomato genomic resources, past and future, available to all its members, without restriction.

The coordinated initiative for sequencing of the tomato genome was launched by the International Solanaceae Genome Project (SOL) in November 2003 to address two major questions:

- How does the small variation among the Solanaceae in terms of gene content, account for the huge variety of phenotypic differences observed within this large family of plants?
- How can the diversity be used to improve agricultural products?

The tomato was chosen for genome sequencing not only with regard to its well-established position as a model plant for fleshy fruit and for the Solanaceae family but also because it represents the smallest diploid genome for which homozygous inbreds are available. In addition, an advanced BAC-based physical map is now available for use in the genome sequencing project. It is expected that with the help of existing comparative maps among the Solanaceae and the availability of EST sequences from different Solanaceae species, the genome sequence information will reinforce the tomato as a high quality reference genome for comparison with the other Solanaceae. The information to be acquired on genome organisation and sequence diversity will give important clues to the genetic basis underlying the phenotypic and physiological variation within Solanaceae crops.

The size of the tomato genome is 950 Mb and the gene-rich regions (euchromatin) represent one quarter of the total size (220 Mb) but contain more than 90% of the tomato genes. Because the gene-rich regions are well clustered and the peri-centromeric heterochromatin is believed to be gene poor and difficult to sequence due to the presence of many repetitive sequences, it was decided that the sequencing would be selectively targeted to the euchromatin of the tomato genome. This strategy is expected to provide the sequence of nearly all tomato genes whilst limiting the sequencing effort to a fraction of the genome. *Solanum lycopersicum* var Heinz 1706 was chosen for sequencing and the approach to be followed will consist in sequencing ordered BACs derived from a well-characterised *HindIII* library (Budiman *et al.*, 2000). The US project will generate two additional BAC libraries and perform BAC-end sequencing as a resource for the entire project. Currently, 10 countries are taking part in the tomato genome sequencing effort and the 12 chromosomes have been distributed between the following partners: China, France, India, Italy, Japan, Korea, Netherlands, Spain, the UK and the USA. In addition, Argentina is committed to sequencing the mitochondrial genome and Italy the chloroplast genome. The sequencing effort will be coordinated by a multinational committee who have agreed that all data will be submitted to Genbank and the Solanaceae Genomics Network, which will serve as a repository and a web-portal for the data. Progress updates can be found at http://www.sgn.cornell.edu/help/about/tomato_sequencing.html.

The data from the tomato genome sequencing project will boost research on Solanaceae species and will open new perspectives for:

- establishing evolutionary history through comparison with other sequenced genomes such as *Arabidopsis* and rice.

- providing a better understanding of the basis of phenotypic diversity in the Solanaceae.
- rapid gene identification and cloning.
- uncovering the mechanisms of gene regulation, signalling, disease resistance and defence, fruit development and quality.
- contributing to agronomic improvement.

REFERENCES

Ayub, R., Ben Amor, M., Gillot, L., Roustan, J.P., Latché, A., Bouzayen, M., Pech, J.C. 1996. Expression of ACC oxidase antisense gene inhibits ripening of melon fruit. *Nature Biotech.*, 14, 862-866.

Bernacchi, D., Beck-Bunn, T., Emmatty, D., Eshed, Y., Inai, S., Lopez, J., Petiard, V., Sayama, H., Uhlig, J., Zamir, D., Tanksley, S. 1998. Advanced backcross QTL analysis in tomato. II. Evaluation of near-isogenic lines carrying single-donor introgressions for desirable wild QTL-alleles derived from *Lycopersicon hirsutum* and *L. pimpinnellifolium*. *Theor. Appl. Genet.*, 97 (1/2), 170-180.

Bramley, P.M. 2002. Regulation of carotenoid formation during tomato fruit ripening and development. *J Exp Bot.*, 53 (377), 2107-2113.

Brommonschenkel, S.H., Tanksley, S. 1997. Map-based cloning of the tomato genomic region that spans the *Sw-5* tospovirus resistance gene in tomato. *Mol. Gen. Genet.*, 256, 121-126.

Brummell, D.A., Harpster, M.H., Civello, P.M., Palys, J.M., Bennett, A.B., Dunsmuir, P. 1999. Modification of expansin protein abundance in tomato fruit alters softening and cell wall polymer metabolism during ripening. *Plant Cell*, 11, 2203-2216.

Brummell, D.A., Hall, B.D., Bennett, A.B. 2000. Antisense repression of tomato endo-1,4-beta-glucanase Cel2 mRNA accumulation increases the force required to break abscission zones but does not affect fruit softening. *Plant Mol. Biol.*, 40, 615-622.

Budiman, M.A., Mao, L., Wood, T.C., Wing, R.A. 2000. A deep-coverage tomato BAC library and prospects toward development of an STC framework for genome sequencing. *Genome Res.*, 129-136.

Colbert, T., Till, B.J., Tompa, R., Reynolds, S., Steine, M.N., Yeung, A.T., McCallum, C.M., Comai, L., Henikoff, S. 2001. High-throughput screening for induced point mutations. *Plant Physiol.*, 126, 480-484.

Emmanuel, E., Levy, A.A. 2002. Tomato mutants as tools for functional genomics. *Curr Opin Plant Biol.*, 5(2), 112-117.

Eshed, Y., Zamir, D. 1995. An introgression line population of Lycopersicon pennellii in the cultivated tomato enables the identification and fine mapping of yield-associated QTLs. *Genetics*, 141, 1147-1162.

Fitch, W.M. 1970. Distinguishing homologous from analogous proteins. *Sys. Zoo.* 19: 99-113.

Frary, A., Nesbitt, T.C., Grandillo, S., Knaap, E., Cong, B., Liu, J., Meller, J., Elber, R., Alpert, K.B., Tanksley, S.D. 2000. fw2.2: a quantitative trait locus key to the evolution of tomato fruit size. *Science*, 289, 85-88.

Fridman, E., Pleban, T., Zamir, D. 2000. A recombination hotspot delimits a wild-species quantitative trait locus for tomato sugar content to 484 bp within an invertase gene. *Proc Natl Acad Sci.*, 97(9), 4718-4723.

Fridman, E., Carrari, F., Liu, Y.-S., Fernie, A.R., Zamir, D. 2004. Zooming in on a quantitative trait for tomato yield using interspecific introgressions. *Science*, 305, 1786-89.

Fulton, T.M., Bucheli, P., Voirol, E., Lopez, J., Pétiard, V., Tanksley, S.D. 2002a. Quantitative trait loci (QTL) affecting sugars, organic acids and other biochemical properties possibly contributing to flavor, identified in four advanced backcross populations of tomato. *Euphytica*, 127, 163-177.

Fulton, T.M., Van der Hoeven, R., Eanetta, N.T., Tanksley, S.D. 2002b. Identification, analysis and utilization of conserved ortholog set markers for comparative genomics in higher plants. *Plant Cell,* 14, 1457-67.

Gidoni, D., Fuss, E., Burbidge, A., Speckmann, G.J., James, S., Nijkamp, D., Mett, A., Feiler, J., Smoker, M., de Vroomen, M.J., Leader, D., Liharska, T., Groenendijk, J., Coppoolse, E., Smit, J.J., Levin, I., de Both, M., Schuch, W., Jones, J.D., Taylor, I.B., Theres, K., van Haaren, M.J. 2003. Multi-functional T-DNA/Ds tomato lines designed for gene cloning and molecular and physical dissection of the tomato genome. *Plant Mol. Biol.,* 51, 83-98.

Giovannoni, J., DellaPenna, D., Bennett, A.B., Fischer, R.L. 1989. Expression of a chimeric polygalacturonase gene in transgenic rin (ripening inhibitor) tomato fruit results in polyuronide degradation but not fruit softening. *Plant Cell* 1(1), 53-63.

Giovannoni, J. 2001. Molecular biology of fruit maturation and ripening. *Annu Rev Plant Physiol Plant Mol Biol.* 52, 725-749.

Hamilton, A.J., Lycett, G.F, Grierson, D. 1990. Antisense gene that inhibits synthesis of the hormone ethylene in transgenic plants. *Nature,* 346, 284-287.

Hammond-Kosack, K.E., Jones, J.D.G. 1997. Plant disease resistance genes. *Ann Rev Plant Physiol Plant Mol Biol.,* 48, 575-607.

Isaacson, T., Ronen, G., Zamir, D., Hirschberg, J. 2002. Cloning of tangerine from tomato reveals a carotenoid isomerase essential for the production of beta-carotene and xanthophylls in plants. *Plant Cell,* 14(2), 333-342.

Ku, H.M., Liu, J., Doganlar, S., Tanksley, S.D. 2001. Exploitation of Arabidopsis-tomato synteny to construct a high-resolution map of the ovate-containing region in tomato chromosome 2. *Genome,* 44, 470-475.

Ku, H.M., Vision, T., Liu, J., Tanksley, S.D. 2000. Comparing sequenced segments of the tomato and Arabidopsis genomes : large-scale duplication followed by selective gene loss creates a network of synteny. *Proc. Natl. Acad Sci.,* USA, 97, 9121-9126.

Lelievre, J.M., Latche, A., Jones, B., Bouzayen, M., Pech, J.C. 1997. Ethylene and fruit ripening. *Physiol Plant.,* 101, 727-739.

Li, X., Song, Y., Century, K., Straight, S., Ronald, P., Dong, X., Lassner, M., Zhang, Y. 2001. A fast neutron deletion mutagenesis-based reverse genetics system for plants. *Plant J.,* 27, 235-242.

Liu, Y., Schiff, M., Dinesh-Kumar, S.P. 2002. Virus-induced gene silencing in tomato. *Plant J.,* 31, 777-786.

Lynch, M., Conery, J.S. 2000. The evolutionary fate and consequences of duplicate genes. *Science,* 290, 1151-1155.

Martin, G.B., Brommonschenkel, S.H., Chunwongse, J., Frary, A., Ganal, M.W., Spivey, R., Wu, R., Earle, E.D., Tanksley, S.D. 1993. Map-based cloning of a protein kinase gene conferring disease resistance in tomato. *Science,* 262, 1432-1436.

Mathews, H., Clendennen, S.K., Caldwell, C.G., Connors, K., Matheis, N., Schuster, D.K., Menasco, D.J., Wagoner, W., Lightner, J., Wagner, D.R. 2003. Activation tagging in tomato identifies a transcriptional regulator of anthocyanin biosynthesis, modification, and transport. *Plant Cell,* 15, 1689-1703.

McCallum, C.M., Comai, L., Greene, E.A., Henikoff, S. 2000. Targeted screening for induced mutations. *Nature Biotech.,* 18, 455-457.

Meissner, R., Jacobson Y., Melamed, S., Levyatuv, S., Shalev, G., Ashri, A., Elkind, Y., Levy, A.A. 1997. A new model system for tomato genetics. *Plant J.,* 12, 1465-1472.

Meissner, R., Chague, V., Zhu, Q., Emmanuel, E., Elkind, Y., Levy, A.A. 2000. A high throughput system for transposon tagging and promoter trapping in tomato. *Plant J.,* 22, 265-274.

Menda, N., Semel, Y., Peled, D., Eshed, Y., Zamir, D. 2004. In silico screening of a saturated mutation library of tomato. *Plant J.,* 38, 861-72.

Miller, J.C., Tanksley, S.D. 1990. RFLP analysis of phylogenetic relationships and genetic variation in the genus Lycopersicon. *Theor. Appl. Genet.*, 80, 437-448.

Milligan, S.B., Bodeau, J., Yaghoobi, J., Kaloshian, I., Zabel, P., Williamson, V. 1998. The root knot nematode resistance gene Mi from tomato is a member of the leucine zipper, nucleotide binding, leucine-rich repeat family of plant genes. *Plant Cell*, 10, 1307-1319.

Moore, S., Vrebalov, J., Payton, P., Giovannonni, J. 2002. Use of genomic tools to isolate key ripening genes and analyse fruit maturation in tomato. *J. Exp. Bot.*, 53, 2023-2030.

Oh, K., Hardeman, K., Ivanchenko, M.G., Ivey, M.E., Nebenführ, A., White, T.J., Lomax, T.L. 2002. Fine mapping in tomato using microsynteny with the *Arabidopsis* genome: the Diageotropica (Dgt) locus. *Genome Biol.*, 3 (9), research 0049.

Ori, N., Eshed, Y., Paran, I., Presting, G., Aviv, D., Tanksley, S., Zamir, D., Fluhr, R. 1997. The IC2 family from the wilt disease resistance locus I2 belongs to the nucleotide binding, leucine-rich repeat superfamily of plant resistance genes. *Plant Cell*, 9(4), 521-532.

Paterson, A.H., Damon, S., Hewitt, J.D., Zamir, D., Rabinowitch, H.D., Lincoln, S.E., Lander, E.S., Tanksley, S.D. 1991. Mendelian factors underlying quantitative traits in tomato: comparison across species, generations, and environments. *Genetics*, 127, 181-197.

Tanksley, S.D., Ganal, M.W., Prince, J.P., de Vicente, M.C., Bonierbale, M.W., Broun, P., Fulton, T.M., Giovannoni, J.J., Grandillo, S., Martin, G.B., Messeguer, R., Miller, J.C., Miller, L., Paterson, A.H., Pinedo, O., Roder, M.S., Wing, R.A., Wu, W., Young, N.D. 1992. High density molecular linkage maps of the tomato and potato genomes. *Genetics*, 132, 1141-1160.

Tanksley, S.D., McCouch, S.R. 1997. Seed banks and molecular maps: Unlocking genetic potential from the wild. *Science*, 277, 1063-1066.

Tieman, D.M., Harriman, R.W., Ramamohan, G., Handa, A.K., 1992. An antisense pectin methyleesterase gene alters pectin chemistry and soluble solids in tomato fruit. *Plant Cell*, 4, 667-679.

Till, B.J., Reynolds, S.H., Greene, E.A., Codomo, C.A., Enns, L.C., Johnson, J.E., Burtner, C., Odden, A.R., Young, K., Taylor, N.E., Henikoff, J.G., Comai, L., Henikoff, S. 2003. Large-Scale Discovery of Induced Point Mutations With High-Throughput TILLING. *Genome Res.*, 13(3), 524-30.

Van der Hoeven, R., Ronning, C., Giovannoni, J., Martin, G., Tanksley, S.D. 2002. Deductions about the number, organization and evolution of genes in the tomato genome based on analysis of a large expressed sequence tag collection and selective genomic sequencing. *Plant Cell*, 14, 1441-56.

Vrebalov, J., Ruezinski, D., Padmanbhan, V., White, R., Medrano, D., Drake, R., Such, W., Giovannoni, J. 2002. A MADS-Box gene necessary for fruit ripening at the tomato *ripeining-inhibitor* (*rin*) locus. *Science*, 296, 343-346.

Vision, T.J., Brown, D.G., Tanksley, S.D. 2000. The origins of genomic duplications in *Arabidopsis*. *Science*, 290, 2114-2117.

Vos, P., Simons, G., Jesse, T., Wijbrandi, J., Heinen, L., Hogers, R., Frijters, A., Groenendijk, J., Diergaarde, P., Reijans, M., Fierens-Onstenk, J., de Both, M., Peleman, J., Liharska, T., Hontelez, J., Zabeau, M. 1998. The tomato Mi-1 gene confers resistance to both root-knot nematodes and potato aphids. *Nature Biotech.*, 16, 1365-1369.

Wikinson, J., Lanahan, M., Yen, H., Giovannoni, J., Klee, H.J. 1995. An ethylene-inducible component of signal transduction encoded by *Never-ripe*. *Science*, 270, 1807-1809.

Zamir, D. 2001. Improving plant breeding with exotic genetic libraries. *Nature Rev Genet.*, 2(12), 983-989.

23

CHAPTER

Genomics as a Road for Grapevine Improvement

A.F. Adam-Blondon, C. Romieu, A, Bouquet

The cultivated grapevine, *Vitis vinifera* L., is a crop of high economic value. Grape berries are eaten fresh, dried, canned, as juice or jam, but their main use is in the winemaking. Total world acreage and grape production were estimated in 2002 to be 7.87 millions hectares and 6.15 million tonnes, from which 68% was used for wine production. Cultivation of the grapevine began during the neolithic era along the eastern shores of the Black Sea, but archaelogical findings of grape seeds indicate that wild grapevines were distributed throughout Europe. It is likely that prehistoric people ate the berries from the vines that climbed on the forest trees, but domestication of the plant was the result of the discovery of wine. Based on a recent finding of pottery fragments containing wine residues in Iran, Mc Govern *et al.* (1996) suggested 5400-5000 BC as the probable period of domestication of the grape. Grape growing and winemaking extended from Transcaucasia to the Middle East, all around the Mediterranean Sea and throughout Europe with the establishment of the Roman Empire (Johnson, 1989). As grapevine was originally dioecious, the first step in the domestication of the wild populations, consisted probably in the selection of plants showing hermaphroditic mutations of the flower type. A second step consisted in the selection of genotypes adapted to different environments (especially to cold stress) and their propagation through cuttings. With the fall of the Roman Empire, the wine trade was disrupted and grape cultivation declined but the association of wine with Christian religious practices was responsible for a second wave of dissemination in Europe during the Middle Ages. That led to the establishment of a new international trade of wine that soon outgrew its religious association (Johnson, 1989). A third wave of dissemination into other continents developed after the 15th century, following the roads of either missionaries or settlers escaping religious wars in Europe.

PROGRESS AND PROSPECTS FOR GRAPEVINE IMPROVEMENT

During the second half of the 19th century, European viticulture had to face several threatening diseases introduced by botanists together with wild species of *Vitis* from northern America: e.g. *Dactylosphaeria vitifoliae* (Phylloxera), *Uncinula necator* (powdery mildew), and *Plasmopara viticola* (downy mildew). This led to the use of rootstocks bred from American wild species resistant to the Phylloxera aphid and the need to protect the vines against mildews with chemical treatments. Another important consequence was a dramatic reduction of the diversity of the varieties cultivated (Bouquet and Boursiquot, 1999) but also the disappearance of most wild *V. vinifera* populations (Arnold *et al.*, 1998).

In viticulture, the main response to biological constraints or economic changes has been to manipulate the existing traditional cultivars by applying progressively higher inputs of husbandry. With regards to breeding, grapevines are highly heterozygous outcrossers and do not breed true from seeds. Although there has been progress in the development of new rootstocks during the past century, the impact of plant breeding at the level of the scion is controversial. During the first half of the 20th century, the selection of wine grape cultivars resistant to pathogens was considerably developed in France by hybridization between *V. vinifera* and American wild species. However, these hybrid varieties, that occupied one third of the grapevine acreage in 1958, were rapidly and completely discarded because of the poor quality of their wine.

Strong pressures for changes in methods of production are now being applied by economic and social forces. In particular, chemical protection is not only expensive but is also a concern regarding health and environment. Thus, the breeding of new varieties showing genetic resistance to pathogens, combined to a good quality in new or ancient (via biotechnological approaches) cultivars is certainly the challenge to be met in the next century by the scientists and the breeders (Bisson *et al.*, 2002, Ferreira *et al.*, 2004).

The recent advances into the functioning of genomes of model species together with the fact that the grape genome is small (around 475 Mbp; Lodhi and Reisch, 1995) and diploid, lead–the international community to set up an international consortium, the International Grape Genome Consortium (IGGP, http://www.vitaceae.org) aimed at the development of genomic resources for grapevine, to enhance grapevine improvement and knowledge transfer from model species. One of the priorities of the IGGP is to obtain the complete sequence of the grapevine genome.

THE VITACEAE FAMILY IS CLOSE TO THE ROSIDS AND SHOWS UNIQUE DEVELOPMENTAL FEATURES

Vitis vinifera (2n = 38) is part of the *Vitaceae* family, which comprises 17 genera, mostly woody or herbaceous lianas primarily intertropical in their distribution. Only one genus, *Vitis*, produces edible berries. It contains about 60 species that are distributed almost equally between America and Asia and represent a considerable resource for resistance to pathogens (Mullins *et al.*, 1992). *V. vinifera* L. is the only species originating from Eurasia and it has been spread throughout the world by human cultivation.

The number of chromosome per haploid genome across the *Vitaceae* family varies from 11 to 20 (Mullins *et al.*, 1992). Cytological observations of F_1 hybrids between *V.*

vinifera (2n = 38) and *Muscadinia rotundifolia* (2n = 40), suggested an allopolyploid origin of the *Vitis* genome (Mullins *et al.*, 1992). However, these observations were not supported by the results of *in situ* hybridization of rDNA probes on the *V. vinifera* chromosomes that showed only one rDNA locus (Haas *et al.*, 1994), nor by the construction of genetic maps. Thus, gene duplication if confirmed, is likely to be ancient and to have been followed by many chromosomes rearrangements.

The position of the *Vitaceae* in the phylogeny of flowering plants as a basal family of the Rosid (Soltis *et al.*, 2000) is however still not supported by a high jackknife parameter. Another highly probable hypothesis is that the *Vitaceae* family is a basal family of the eudicots (Goertzen, 2005; Fig. 23.1). The age of the split between the *Vitaceae* family and the Rosid to which Arabidopsis also belongs, was estimated as 108-117 Myr (Soltis *et al.*, 2000, Ingrouille *et al.*, 2002).

Perennial liana like *V. vinifera* have many developmental aspects different from Arabidopsis, poplar, Medicago and tomato, which include a unique shoot architecture with petiolated leaves opposite to either inflorescences or tendrils–(Mullins *et al.*, 1992). Inflorescences have a pyramidal branched structure with an asynchronous development of flowers. The five petals of the flowers are jointed into a cap structure which falls at anthesis, liberating the stamens and the pistil (Mullins *et al.*, 1992). Particularly adapted to avian dissemination, the berry pericarp undergoes multifaceted metabolic changes common to most fleshy fruits species as they shift from a repulsive to an attractive status. The transition between these two status is abrupt and the trigger remains unknown (Mullins *et al.*, 1992). Unlike in tomato, berry ripening occurs without an increase in respiration, or a significant emission of ethylene. Moreover, neither ethylene nor its antagonists, triggered or precluded berry ripening, even if anthocyanin accumulation and malate breakdown, showed some dependence on this hormone (El Kereamy *et al.*, 2003, Chervin *et al.*, 2004).

THE GERMPLASM COLLECTIONS ARE RICH, WELL CHARACTERISED AND WILL BE THE BASIS OF LINKAGE DISEQUILIBRIUM-BASED STUDIES

There are about 140 main germplasm collections around the world in 36 countries, mainly in Europe (Alleweldt and Dettweiler, 1994). The most important part of these collections is made up of *V. vinifera* varieties, that are a resource for quality traits and adaptation to an abiotic environment and to a lesser extent, of genotypes of other crossable species that are a resource for resistance to pathogens. At the European level, most of the collections have homogenised their descriptors and try to coordinate their actions through an ECP/GR network (http://www.genres.de/eccdb/vitis and http://www.ecpgr.cgiar.org/contacts/ecpgr_contact.htm). The INRA Domaine de Vassal (Hérault, France) contains the largest collection, supported by the French Board of Genetic Resources, with around 2,600 identified varieties of *V. vinifera*, 1,000 interspecific hybrids and rootstock varieties and 300 accessions of other species (Bouquet and Boursiquot, 1999). During the last 10 years, there has been an increasing interest in Europe and around the Mediterranean sea into the preservation of the few remaining wild populations of *V. vinifera* (Arnold *et al.*, 1998). Their study will increase our understanding of the domestication process of grape across the ages (Arroyo-Garcia, 2002; Aradhya *et al.*, 2003; Grassi *et al.*, 2003).

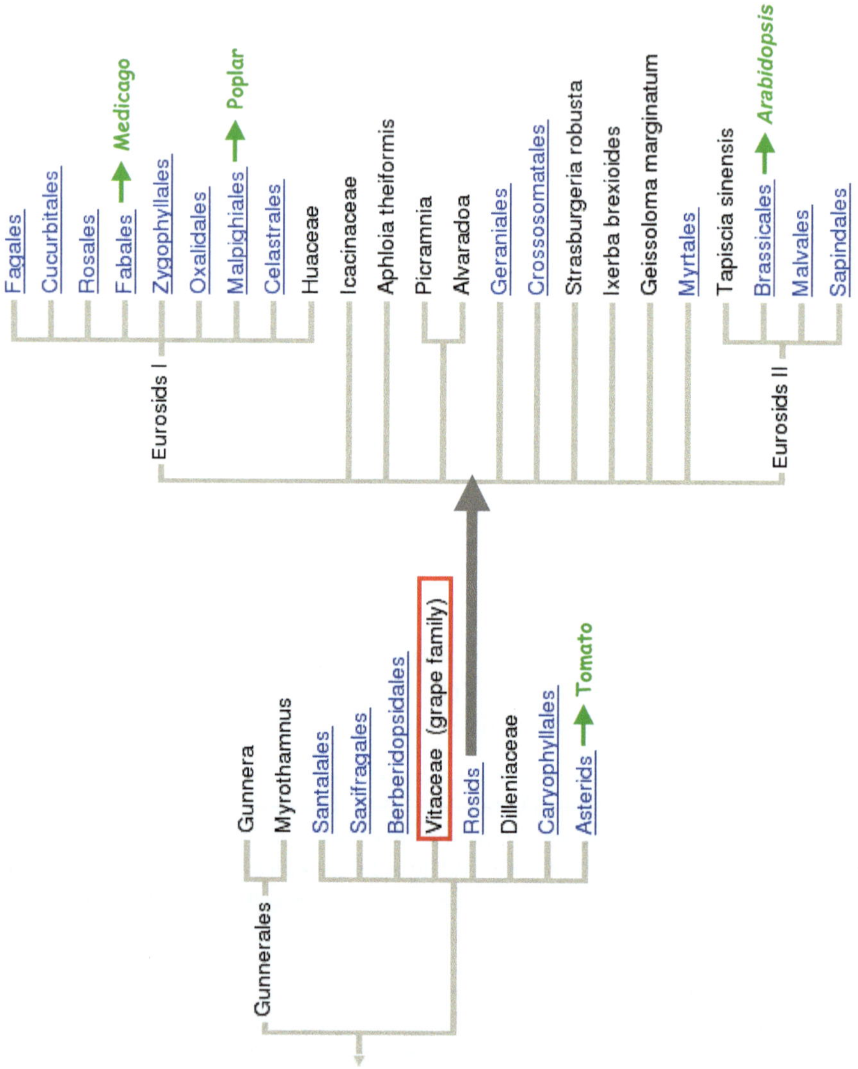

Fig. 23.1 The *Vitaceae* family in the flowering plants phylogeny according to the Tree of Life web project (http://tolweb.org) with regards to the position of the dicotyledonous species being sequenced (tomato, Medicago) or already sequenced (*Arabidopsis*, Medicago).

Two SSR markers have been enough to identify the 44 most important French cultivars, showing an important genetic diversity between cultivars (Tessier *et al.*, 1999; Siret, 2001). The gene pool of cultivated grapevine shows a weak genetic differentiation, which is consistent with the fact that this germplasm has been moved around all grape growing regions across the ages (Bowers *et al.*, 1999, Sefc *et al.*, 2000, Aradhya *et al.*, 2003). Grapevine cultivars are highly heterozygous with an average PIC value of SNP-based haplotypes of 0.61 (Salmaso *et al.*, 2004) and an average heterozygosity of SSR markers of 0.77 (Aradhya *et al.*, 2003) to 0.82 (Siret, 2001). The grapevine genome thus consists of two very different haplotypes. Core collections are currently being defined at the European level in order to facilitate (i) the assessment of the usefulness of the markers for SAM and (ii) linkage disequilibrium-based studies which is an interesting road for associating gene sequence and trait variations.

The majority of cultivars have been propagated vegetatively for sometimes more than 2000 years. In most cases, somatic mutations occurred along the generations of vegetative multiplication, leading to the formation of families of "clones" showing sometimes chimeric structures, among which is the famous Pinot family (Franks *et al.*, 2002, Crespan, 2004, Hocquigny *et al.*, 2004). The molecular SSR-based patterns of the clones of a same family are highly conserved (Franks *et al.*, 2002, Hocquigny *et al.*, 2004). Several transposable elements have been isolated (Verries *et al.*, 2000; Pelsy and Merdinoglu 2002; Kobayashi *et al.*, 2004) and their role in varietal and clonal variation was recently demonstrated (Kobayashi *et al.*, 2004, Pereira *et al.*, 2005). The natural mutants (more than 150) conserved in the Vassal collection, are affected in many different aspects of plant development and can be a very good material for the understanding of the molecular events involved in the variation of these traits (Boss and Thomas, 2002; Kobayashi *et al.*, 2004).

EFFICIENT TOOLS FOR GENETIC MAPPING HAVE BEEN DEVELOPED DURING THE LAST 10 YEARS AND ARE NOW USED FOR THE STUDY OF THE GENETIC DETERMINISM OF IMPORTANT TRAITS

The first genetic map of the grapevine genome is fairly recent, as it was published in 1995 by Lodhi *et al.*,. This map was however constructed mainly from RAPD markers which are dominant and not transferable from one cross to another. Moreover, this map and the following one published by the same team (Dalbó *et al.*, 2000), were based on progenies obtained by complex interspecific hybridization. The second genetic map, constructed in France and published in 2002 by Doligez *et al.*,, was based on a *V. vinifera* progeny, using AFLP and SSR markers.

In 1998, the international scientific community decided to join its effort to develop a large set of microsatellite markers (SSR) through a consortium, the Vitis Microsatellite Consortium, coordinated by the private company Agrogène. Such markers are indispensable for the genetic study of heterozygous species because they are co-dominant and highly transferable from one population to another one. They have also been shown to be useful across the *Vitaceae* family (Lin and Walker, 1998; Sefc *et al.*, 1999; Di Gaspero *et al.*, 2000; Decroocq *et al.*, 2003). Around 350 SSR markers were developed but until recently very few were publicly available (Riaz *et al.*, 2004; Adam-Blondon *et al.*,

2004a). More recently, two important sets were deposited into the UniSTS database at NCBI; a set of 107 markers developed at the University of Udine by Di Gaspero *et al.*, (2005) and a set of 175 markers developed in France through a Génoplante project (Merdinoglu *et al.*, 2005). There are currently 366 markers available in dbSTS (www.ncbi.nlm.nih.gov/dbSTS). The grapevine scientific community is now coping with is the development of SNP markers (Salmaso *et al.*, 2004).

The available SSR markers were quickly used for the construction of several reference maps (Riaz *et al.*, 2004, Adam-Blondon *et al.*, 2004a) and a consensus map which comprises more than 500 SSR loci is now under construction in the frame of the IGGP from the raw data of five maps (Doligez A., pers. comm.). The estimation of the genetic grape genome size lies between 2000 cM and 2300 cM (Adam-Blondon *et al.*, 2004a; Riaz *et al.*, 2004). These SSR marker-based maps, represent a major improvement for the study of the genetic determinism of traits of interest in grape through the detection of QTLs (Table 23.1), allowing a comparison between different studies. Marker assisted selection projects for seedless table grapes are beginning, using markers close to the main QTLs detected (This *et al.*, 2000). They follow the identification, by Bulk Segregant Analysis (Lahogue *et al* 1998) of a SCAR marker linked to the *sdI* (*seed development inhibitor*) gene, and the detection of QTLs for berry traits (Doligez *et al.*, 2002). First results have underlined the importance of a parental selection and the requirement for further theoretical modelling of the effect of different strategies in heterozygous plant species (Doligez A., pers. comm.).

Map-based cloning projects are still rare, mainly due to the length of the biological cycle and the space needed for large segregating populations. The most advanced program concerns the cloning of the *Run*1 gene that confers complete resistance to powdery mildew (Barker *et al.*, 2005). Preliminary work was the establishment of local integrated genetic and physical maps (Pauquet *et al.*, 2001; Donald *et al.*, 2002, Barker *et al.*, 2005) of this gene, which originates from *Muscadinia rotundifolia* and was previously introduced into *V. vinifera* by several back-cross cycles (Bouquet, 1986). Another project is the mapping and cloning of the *Rpv1* gene conferring partial resistance to downy mildew, which is linked to the *Run1* gene (Merdinoglu *et al.*, 2003). The development of genomic tools, enhancing the transfer of knowledge from model species (*Arabidopsis*, tomato…) to grape, will probably help the diversification of such projects and speed up their finalisation through chromosome landing approaches.

PHYSICAL MAPS ARE UNDER CONSTRUCTION AND WILL GREATLY FACILITATE THE MAPPING OF GENES AND CANDIDATE GENE APPROACHES THROUGH THEIR INTEGRATION WITH GENETIC MAPS

Several BAC libraries are now available in grape, opening the roads for structural genomics in this species (Table 23.2). The first grape BAC library was produced from the Syrah cultivar by Tomkins *et al.*, (2001). Five others were subsequently developed from important cultivars (Table 23.2). The two Cabernet Sauvignon BAC libraries were chosen as the reference libraries by the International Grape Genome Program, www.vitaceae.org. Such resources will allow direct access to genes and alleles for the analysis of sequence variation between the six different haplotypes harboured by these

Table 23.1 Genetic maps published in grape

Population type	*Population size*	*Map description*						
		Categories of markers	*Map$*	*Number of markers*	*Number of linkage groups*	*Total size (cM Kosambi)*	*mapped QTL/ major genes*	*Reference*
2 complex interspecific hybrids	60	422 RAPD 16 RFLP/ isozymes	F M	214 225	20 22	1,196 1,477		Lodhi *et al.*, 1995
2 complex interspecific hybrids	58	277 RAPD 25 SSR 4 CAPS, 12 AFLP	F M	153 179	20 20	1,199 1,470	Sex	Dalbó *et al.*, 2000
V. vinifera x *V. vinifera*	139	250 AFLP 44 SSR 6 isozymes, SCAR, RAPD	F M C	157 144 301	22 23 20	767 816 1,002	Berry color Seedlessness Berry size	Doligez *et al.*, 2002
V. riparia x *V. vinifera*	81	741 AFLP 47 SSR 3 SSCP	F M	388 429	20 19	1,639 1,518		Grando *et al.*, 2003
Complex interspecific hybrid x *V. vinifera*	153	185 AFLP 137 RAPD 85 SSR 22 SCAR/CAP	F M	265 164	20 26	1,277 1,158	Resistance to powdery and downy mildew; Veraison; Berry size and colour, Axillary shoot growth	Fischer *et al.*, 2003
Cross between two *V. rupestris* x *V. arizonica* interspecific	116	410 AFLP 32 RAPD 24 ISSR 9 SSR	F M	160 144	17 19	756 1,082		Doucleff *et al.*, 2004

Contd.

Table 23.1 Genetic maps published in grape

Population type	*Population size*	*Map description*						
		Categories of markers	*Map*$	*Number of markers*	*Number of linkage groups*	*Total size (cM Kosambi)*	*mapped QTL/ major genes*	*Reference*
hybrids								
V. vinifera x *V. vinifera*	153	158 SSR	F	120	19	1,397		Riaz *et al.*, 2004
			M	122	21	1,579		
			C	158	19	1,882		
V. vinifera x *V. vinifera*	99	220 SSR	F	177	19	1,172		Adam-Blondon *et al.*, 2004a
			M	178	18	1,361		
			C	220	19	1,406		
Selfing of a *V. vinifera*	99	110 SSR	110	19	1192			Adam-Blondon *et al.*, 2004a
V. vinifera x *V. vinifera*	184	110 SSR 203 AFLP	F M	Data not shown	Data not shown	Data not shown	Fruit yield components	Fanizza *et al.*, 2005

$ 'F' is for female map, 'M' is for male map and 'C' for the consensus map between the female and the male maps

Table 23.2 Grape BAC libraries published in grape

Genotype	*Number of clones*	*Average insert size (kb)*	*Genome coverage*	*Reference*
Syrah	55,296	144	16.5	Tomkins *et al.*, 2001
Syrah cl.ENTAV 173	23,040	158	7.7	Adam-Blondon *et al.*, 2005
Cabernet-Sauvignon cl.412	44,544	142	13.2	Adam-Blondon *et al.*, 2005
Cabernet-Sauvignon	56,448	139	13.9	www.vitaceae.org
Pinot noir cl.777	48,384	146	14.8	Adam-Blondon *et al.*, 2005
Pinot noir cl.115	23,424	92.7	4.5	Adam-Blondon *et al.*, 2005
VHR3294-R23	55,295	93	9.4	Barker *et al.*, 2005

three cultivars. The next BAC libraries should be constructed now from different species in order to develop comparative genomics between the *Vitaceae,* in particular with the objective to improve knowledge on disease resistances and their introgression into *V. vinifera.*

Two physical maps of the *Vitis vinifera* genome are under construction: one using a Pinot Noir BAC library (Moroldo *et al.*, 2004) and one using a Cabernet Sauvignon BAC library (Tual *et al.,* 2004). In both cases, several approaches have been developed: (1) fingerprinting a set of BAC clones covering 8 to 10 times the grape genome; (2) obtaining the sequences of the extremities of the inserts in BAC clones (BAC end sequences); (3) anchoring markers on the same BAC clones; and (4) mapping on a genetic map as many of the anchored markers as possible.

The fingerprinting method that is used for the construction of both maps is based on the fluorescent labelling of four different restriction sites and was adapted by Moroldo *et al.,* (2004) from the original paper of Luo *et al.,* (2003). Up to now, 44,983 BAC clones from the Pinot Noir library representing 10 times the grape genome and 30,151 clones from the Cabernet Sauvignon BAC library representing 8.7 times the grape genome have been fingerprinted (Moroldo *et al.,* 2004; Tual *et al.,* 2004; Paillard, S. unpubl. results). The high level of sequence divergence between grape haplotypes makes the construction of fingerprint-based maps more difficult than in a homozygous species (Tual *et al.,* 2004) and the first draft of the Pinot map is made up of more than 1,000 contigs (Moroldo *et al.,* 2004). An improvement of the map has thus been attempted through the anchoring of markers (SSR, derived from EST sequences or from BAC ends). Around 1,400 markers (including mapped SSR and ESTs) have been anchored on the Cabernet Sauvignon BAC clones (Adam-Blondon *et al.,* 2004b) and 800-1000 genetically mapped STS have been anchored in the Pinot BAC library (Velasco *et al.,* 2004). 77237 BAC end sequences have been obtained from the Cabernet Sauvignon BAC library (Adam-Blondon *et al.,* 2004b) and 36,000 from the Pinot BAC library (Velasco *et al.,* 2004). Good quality physical maps, integrated to genetic maps and on which many genes are located, should thus soon be available for the grape community and greatly accelerate candidate gene approaches.

Another output of these approaches will be a better knowledge of the grapevine genome composition and structure. A rough analysis of large sets of BAC end sequences shows that the percentage of repeated sequences is in same order of magnitude as that found in rice and their mapping on the *Arabidopsis* genome gave similar results to the mapping of Medicago BAC ends sequences (Guichard, C. ; Aubourg, S. pers. comm.). We thus expect the same order of conservation between the grapevine genome and the *Arabidopsis* genome, as there is between Medicago and *Arabidopsis*. Finally four BACs representing the two haplotypes of two regions of the Cabernet Sauvignon genome have been recently sequenced, confirming the high percentage of sequence divergence between these two haplotypes: 3-8%, mainly due to short insertion deletion events (Adam-Blondon, A.F., Wincker, P., unpub. results).

IMPORTANT EST SEQUENCING PROJECTS HAVE PROVIDED A BASIC RESOURCE FOR THE STUDY OF THE MOLECULAR BASIS OF TRAITS, BUT ALSO FOR AN EFFICIENT TRANSFER OF KNOWLEDGE FROM MODEL SPECIES TO GRAPEVINE

ESTs now represent a basic tool for the study of the molecular mechanisms involved in traits of interest for any species. They give a quick overview of the genes that are expressed and a fast access to the sequences of candidates genes. For instance, numerous candidate gene approaches have been focussed on berries, mostly on sugar, water and organic acid transporters (Boss and Davis, 2001; Delrot *et al.*, 2001; Terrier and Romieu, 2001), energy and secondary metabolism- (Bavaresco and Fregoni, 2001). The developmental patterns of these genes were confirmed in a first high throughput study of berry development using 15 K 50-mer microarrays (Terrier *et al.*, 2005). This study pinpointed the relative expression of the different members of gene families, the fact that most of the induced genes belonged to unknown functional categories and the fact that the majority of gene expression changes occured within a critical 24 hours period, just before any change in colouration, sugar and acid contents could be detected. A new hypothesis on hormonal regulation and transcription factors has also been proposed (Terrier *et al.*, 2005). Many global analyses of the grape transcriptome are now underway, addressing biotic and abiotic stress, — organ development and cultivar difference (Thomas *et al.*, 2004, Cushman *et al.*, 2005, Goes da Silva *et al.*, 2005, Terrier *et al.*, 2005).

EST collections also represent a necessary bridge between model species and others. For instance, based on knowledge of the molecular mechanisms involved in *Arabidopsis* flower development and flowering transition, some progress has been made in the deciphering of these mechanisms in grape and has shown that the grape orthologs of *Arabidopsis* genes involved in the flowering process are involved in the same process but do not have the same function, or are not regulated in the same manner (Carmona *et al.*, 2002, Boss *et al.*, 2003, Joly *et al.*, 2004). As large transcriptome analyses are developed on other fruits such as tomatoes, prunus and citrus, the comparison of the ripening processes in these species and grape will help us to understand which genes are involved in common processes like sugar accumulation and cell wall softening (Fei *et al.*, 2004). It has also been useful to identify the function of genes that are absent in *Arabidopsis* as for example the *ASR* genes that were recently shown to be involved in the cross talk between sugar and hormone signalling (Çakir *et al.*, 2003). Finally transcriptome analysis can enhance the identification of specifically induced promoters.

Large EST collections have been developed since 1999 from many different tissues and in many different laboratories leading to the deposition of 164,946 *Vitis* ESTs in the public databases (http://www.vitaceae.org/GrapecDNA.html; Table 23.3). These sequences have been used for the definition of two sets of unique genes (unigene sets) (www.tigr.org/tgi, www.ncbi.nlm.nih.gov/Unigene) which have been utilized for the development of an international Affymetrix Gene Chip® (14K *V. vinifera* and 1.7 K other *Vitis* species transcripts, Cramer *et al.*, 2004) and a 13 K 70-mer set from Qiagen SA that are commercially available. Efforts are now directed towards the development of full length cDNA libraries in order to gain knowledge on the structure of grape genes (intron-exon length, codon preferences, etc.), which will be useful for genome annotation and also to enhance studies aiming at the discovery of unknown gene functions. Proteomic approaches are also starting (Cramer *et al.*, 2004, Sarry *et al.*, 2004). Since 80% conservation is needed to get a possibility of protein identification, EST sequences allow- significant improvement in the identification of proteins. A platform for proteome and metabolome analysis in grape have been recently set up at the University of Nevada (Cramer *et al.*, 2004).

Table 23.3 dbEST release, 26th November 2004

Species	*Number of deposited ESTs*
Vitis vinifera	143,587
Vitis shuttleworthii	10,704
Vitis hybrid cultivar	*6,533*
Vitis aestivalis	2,101
Vitis riparia	1,910
Vitis cinerea x *Vitis rupestris*	61
Vitis cinerea x *Vitis riparia*	32
Vitis pseudoreticulata	13
Vitis labrusca	5
Total	164,946

THE DEVELOPMENT OF REVERSE GENETIC APPROACHES IN FUNCTIONAL GENOMICS NEEDS HIGHLY EFFICIENT TECHNIQUES OF GENETIC TRANSFORMATION.

The lack of high throughput system for reverse genetic approaches is a real bottleneck to be overcome in the future. A lot of work has been carried out to generate mutants in fruit crops and thus in grape, mostly using combinations of radiation and tissue culture techniques or by triggering somaclonal variations (Predieri, 2001). However, the expected mutations are usually recessive and are revealed in the F_2 generation which takes 6-10 years in grape. To overcome this difficulty, an interesting road is to generate mutations that enhance gene expression (Topping and Lindsey, 1995). An Ac/Ds transposon enhancer mutagenesis system has been tentatively applied to grapevine for this purpose (Perl *et al.*, 2000) but is far from being in routine use. Thus, the development of reverse genetic approaches in grapevine genomics has mainly resorted to genetic transformation of single genes.

The first report of *Agrobacterium*-mediated transformation of grapevine was made by Mullins *et al.*, (1990) and a few years later, biolistic-mediated transformation was successfully achieved (Kikkert *et al.*, 1996). The historical development, applications, limitations and constraints of grapevine genetic transformation have been reviewed

(Kikkert *et al.*, 2001; Bouquet *et al.*, 2003). Now, the *Agrobacterium*-mediated technology is currently used worldwide for improvement of rootstock cultivars (Xue *et al.*, 1999), winegrape cultivars (Iocco *et al.*, 2001) or table grape cultivars (Franks *et al.*, 1998) and its methodology has been well-defined (Perl *et al.*, 2004). The biolistic-mediated technology is more adapted to study transient expression of genes, than for stable transformation but its efficiency in this use has been proven (Vidal *et al.*, 2003).

Gene transfer technology is especially attractive in the wine industry, as it allows the improvement of the traditional cultivars with the minimum disturbance of their genome. This allows the maintenance of the varietal characteristics, which is not possible with breeding methods based on hybridization. For the most part, targeted traits included resistance or tolerance to diseases and pests and the genes introduced were mainly genes encoding viral coat-proteins, antifungal proteins, lytic peptides, detoxifying enzymes or lectins (Colova-Tsolova *et al.*, 2001; Bouquet *et al.*, 2003). Genes that influence fruit quality are starting to receive attention: the invertase gene for modifying the sweetness of the berry (Perl *et al.*, 1994), the polyphenoloxidase gene used in the antisense direction for reducing the browning of raisins (Thomas *et al.*, 2000) and the barnase gene to obtain seedless grapes (Perl *et al.*, 2003). Genetic transformation is also an invaluable tool in grapevine functional genomics (Bouquet *et al.*, 2004). We may now study, by promoter/reporter gene fusion and stable *Agrobacterium*- or transient biolistic-mediated transformation, the regulation of gene expression in specific organs or tissues (Gollop *et al.*, 2002; Pratelli *et al.*, 2002) during the various developmental stages of the plant, or in response to external stimuli such as anaerobiosis (Torregrosa *et al.*, 2002), light or sugar (Gollop *et al.*, 2002).

In a next future it will be possible to silence, by "RNA interference" (RNAi) technology, numerous genes in order to study their role in plant growth and development, primary or secondary metabolism and response to biotic and abiotic stresses. Until now, RNAi has been used in grapevines only for inducing resistance to viruses (Jardak-Jamoussi *et al.*, 2004).

Another limitation for reverse genetic approaches in grapevine genomics is the long time needed if the genes studied are involved in the development of the reproductive apparatus, and in particular the formation of the berry. However, the tool box of the grapevine scientific community could be considerably improved by the discovery of a dwarf mutant showing precocious and generalized floral induction (Boss and Thomas, 2002).

PERSPECTIVES: TOWARDS THE SEQUENCING OF THE GRAPE GENOME

The growing emergence of genomic resources in grape and several preliminary studies showing the high level of haplotypic sequence divergence, have allowed the production of alternative strategies that are under discussion for the sequencing of the grape genome.

The first one consists in the construction of a detailed physical map of the Cabernet Sauvignon genome. The minimum tilling path would be a patchwork of the two haplotypes defining the Cabernet Sauvignon genome.

The second strategy was initially proposed by a partnership of INRA and Genoscope teams. It relies on the existence of near homozygous lines that were created at the INRA of Colmar (Bronner and Oliveira, 1990). One of these lines will be sequenced with a whole genome shotgun strategy and the sequence contigs will be ordered and orientated by mapping on a genetic map.

This latter project is also associated with grape pathogen- genome sequence projects. The availability of the sequence of a pathogen and of its host would offer many opportunities to understand the molecular mechanisms underlying their interaction and the dynamic of these pathogens in the grape eco-system using complete set of genes on both sides.

REFERENCES

Adam-Blondon, A.F., Roux, C., Claux, D., Butterlin, G., Merdinoglu, D., This, P. 2004a. Mapping 245 SSR markers on the *Vitis vinifera* genome: a tool for grape genetics. *Theor. Appl. Genet.*, 109: 1017-1027.

Adam-Blondon, A.F. 2004b. Progress of the "BAC libraries, physical maps and genome sequencing" working group. 7th international symposium on grapevine physiology and biotechnology, Davis, CA, USA, June 21-25.

Adam-Blondon, A.F., Bernole, A., Faes, G., Lamoureux, D., Pateyron, S., Grando, M.S., Caboche, M., Velasco, R., Chalhoub, B. 2005. Construction and characterization of BAC libraries from major grapevine cultivars. *Theor. Appl. Genet.*, 110: 1363-1371.

Alleweldt, G., Dettweiler, E. 1994. The genetic resources of *Vitis*: world list of grapevine collections. 2nd edition.

Aradhya, M.K., Dangl, G.S., Prins, B.H., Boursiquot, J.M., Walker, M.A., Meredith, C.P., Simon, C.J. 2003. Genetic structure and differentiation in cultivated grape, *Vitis vinifera* L. *Genet. Res. Camb.* 81: 179-192.

Arnold, C., Gillet, F., Gobat, J.M. 1998. Situation de la vigne sauvage *Vitis vinifera* ssp *silvestris* en Europe. *Vitis* 37: 159-170.

Arroyo-Garcia, R., Lefort, F., de Andres, M.T., Ibanaez, J., Borrego, J., Jouve, N., Cabello, F., Martinez-Zapater, J.M. 2002. Chloroplast microsatellite polymorphisms in Vitis species. Genome, 45: 1142-1149.

Barker, C.L., Donald, T., Pauquet, J., Ratnaparkhe, A., Bouquet, A., Adam-Blondon, A.F., Thomas, M.R., Dry, I. 2005. Genetic and physical mapping of the grapevine powdery mildew resistance gene, *Run1*, using a bacterial artificial chromosome library. *Theor. Appl. Genet.*, 111: 370-377.

Bavaresco, L., Fregoni, C. 2001. Physiological role and molecular aspects of grapevine stilibenic compounds. *In*: KA Roubelakis-Angelakis (ed*) Molecular Biology and Biotechnology of the Grapevine*. Kluwer Academic Publishers, Dordrecht, The Netherlands, pp. 153-182.

Bisson, L.F., Waterhouse, A.L., Ebeler, S.E., Walker, M.A., Lapsley, J.T. 2002. The present and future of the international wine industry. *Nature,* 418: 696-699.

Bohlman, M.C., Ergul, A. Tattersall, E.A.R., Tillett, R.L., Figueroa-Paredes, R., Kabuloglu, E., Cushman, M.A., Spreeman, K.L., Schlauch, K., Mendes, P., Cramer, G.R., Cushman, J.C. 2004. EST database and gene expression studies of abiotically stressed grapevine *Vitis vinifera* L. 7th international symposium on grapevine physiology and biotechnology, Davis, CA, USA, June 21-25.

Boss, P.K., Buckeridge, R.J., Poole, A., Thomas, M.R. 2003. New insights into grapevine flowering. *Functional Plant Biology,* 30: 593-606.

Boss, P.K., Davies, C. 2001. Molecular Biology of sugar and anthocyanin accumulation in grape berries. *In*: KA Roubelakis-Angelakis (ed*) Molecular Biology and Biotechnology of the Grapevine*. Kluwer Academic Publishers, Dordrecht, The Netherlands, pp. 1-33.

Boss, P.K., Thomas, M.R. 2002. Association of dwarfism and floral induction with a grape 'green revolution' mutation. *Nature,* 416: 847-849.

Bouquet, A. 1986. Introduction dans l'espèce *Vitis vinifera* L. d'un caractère de résistance à l'oïdium (*Uncinula necator* Schw. Burr.) issu de l'espèce *Muscadinia rotundifolia* (Michnx) Small. *Vignevini,* 12 (suppl): 141-146.

Bouquet, A., Boursiquot, J.M. 1999. Safeguarding old varieties and developing recently obtained new cultivars : a joint step to conciliate tradition and innovation in France. *Bull. OIV* 825-826: 753-761.

Bouquet, A., Torregrosa, L., Chatelet, P. 2003. Grapevine genetic engineering: Tool for genome analysis or plant breeding method ? Which future for transgenic vines? AgBiotechNet, Vol 5, October ABN 116, p. 10.

Bowers, J., Boursiquot, J.M., This, P., Chu, K., Johansson, H., Meredith, C. 1999. Historical Genetics: the parentage of Chardonnay, Gamay and other wine grapes of northeastern France. *Science,* 285: 1562-1565.

Bronner, A., Oliveira, J. 1990. Creation and study of the Pinot Noir variety lineage. In: Proc. 5th Intern. Symp. Grape Breeding St Martin/Pflaz, Germany, Sept 1989, Vitis, special issue, pp. 69-80.

Çakir, B., Agasse, A., Gaillard, C., Saumonneau, A., Delrot, S., Atanassova, R. 2003. A grape ASR protein involved in sugar and abscisic acid signaling. *Plant Cell,* 15: 2165-2180.

Carmona, M.J., Cubas, P., Martinez-Zapater, J.M. 2002. VFL, the grapevine FLORICAULA/ LEAFY ortholog is expressed in meristematic regions independently of their fate. *Plant Physiol,* 130: 68-77.

Chervin, C., El-Kereamy, A., Roustan, J.P., Latche, A., Lamon, J., Bouzayen, M. 2004. Ethylene seems required for the berry development and ripening in grape, a non climacteric fruit. *Plant Science,* 167: 1301-1305.

Colova-Tsolova, V., Perl, A., Krastanova, S., Tsvetkov, I., Atanassov, A. 2001. Genetically engineered grape for disease and stress tolerance. *In*: KA Roubelakis-Angelakis (ed) *Molecular Biology and Biotechnology of the Grapevine.* Kluwer Academic Publishers, Dordrecht, The Netherlands, pp. 411-432.

Cramer, G.R., Cushman, J.C., Shooley, D.A., Quilici, D., Vincent, D., Bohlman, M.C., Schlauch, K., Mendes, P. 2004. Progress in bioinformatics-the challenge of integrating transcriptomic, proteomic and metabolomic information. 7th international symposium on grapevine physiology and biotechnology, Davis, CA, USA, June 21-25.

Crespan, M. 2004. Evidence on the evolution of polymorphism of microsatellite markers in varieties of *Vitis vinifera* L. *Theor. Appl. Genet.,* 108: 231-237.

Cushman, J.C., Bohlman, M.C., Ergul, A., Tillett, R.L., Sonderegger, J., Vincent, D., Tattersall, E.A.R., Spreeman, K.L., Schlauch, K., Mendes, P., Cramer, G.R. 2005. Expression Profiling Of Water Deficit And Salinity Stress Responses In *Vitis vinifera.* Plant Animal and Microbe Genome Conference XIII, SanDiego, CA, January 15-19, 2005.

Dalbó, M.A., Ye, G.N., Weeden, N.F., Steinkellner, H., Sefc, K.M., Reisch, B.I. 2000. A gene controlling sex in grapevines placed on a molecular marker-based genetic map. *Genome* 43: 333-340.

Decroocq, V., Favé, M.G., Hagen, L., Bordenave, L., Decroocq, S. 2003. Development and transferability of apricot and grape EST microsatellite markers across taxa. *Theor. Appl, Genet.,* 106: 912-922.

Delrot, S., Picaud, S., Gaudillère, J.P. 2001. Water transport and aquaporins in grapevine. *In*: KA Roubelakis-Angelakis (ed) *Molecular Biology and Biotechnology of the Grapevine.* Kluwer Academic Publishers, Dordrecht, The Netherlands, pp. 241-262.

Di Gaspero, G., Peterlunger, E., Testolin, R., Edwards, K.J., Cipriani, G. 2000. Conservation of microsatellite loci within the genus *Vitis. Theor. Appl. Genet.* 101:301-308.

Doligez, A., Bouquet, A., Danglot, Y., Lahogue, F., Riaz, S., Meredith, C.P., Edwards, K.J., This, P. 2002. Genetic mapping of grapevine (*Vitis vinifera* L.) applied to the detection of QTLs for seedlessness and berry weight. *Theor. Appl. Genet.*, 105: 780-795.

Donald, T.M., Pellerone, F., Adam-Blondon, A.F., Bouquet, A., Thomas, M.R., Dry, I.B. 2002. Identification of resistance gene analogs linked to a powdery mildew resistance locus in grapevine. *Theor. Appl. Genet.* 104: 610-618.

Doucleff, M., Lin, Y., Gao, F., Riaz, S., Krivanek, A.F., Walker, M.A. 2004. A genetic linkage map of grape, utilizing Vitis rupestris and Vitis arizonica. *Theor. Appl. Genet.* 109: 1178-1187.

El-Kereamy, A., Chervin, C., Roustan, J.P., Cheynier, V., Souquet, J.M., Moutounet, M., Raynal, J., Ford, C., Latché, A., Pech, J.P., Bouzayen, M. 2003. Exogeneous ethylene stimulates the long-term expression of genes related to anthocyanin biosynthesis in grape berries. *Physiologia Plantarum,* 119: 175-182.

Fanizza, G., Lamaj, F., Costantini, L., Chaabane, R., Grando, M.S. 2005. QTL analysis for fruit yield components in table grapes (Vitis vinifera). *Theor. Appl. Genet.,* 111: 658-664.

Fei, Z., Tang, X., Alba, R.M., White, J.A., Ronning, C.M., Martin, G.B., Tanksley, S.D., Giovannoni, J.J. 2004. Comprehensive EST analysis of tomato and comparative genomics of fruit ripening. *The Plant Journal,* 40: 47-59.

Ferreira, R.B., Monteiro, S.S., Piçarra-Pereira, M.A., Teixeira, A.R. 2004. Engineering grapevine for increased resistance to fungal pathogen without compromising wine stability. *Trends in Biotechnology,* 22: 168-173.

Fischer, B., Salakhutdinov, I., Akkurt, M., Eibach, R., Edwards, K.J., Töpfer, R., Zyprian, E.M. 2004. Quantitative trait locus analysis of fungal disease resistance factor on a molecular map of grapevine. *Theor. Appl. Genet.,* 108: 501-515.

Franks, T., Botta, R., Thomas, M.R., Franks, J. 2002. Chimerism in grapevines: implications for cultivar identity, ancestry and genetic improvement. *Theor. Appl. Genet.,* 104: 192-199.

Franks, T., He, D.G., Thomas, M.R. 1998. Regeneration of transgenic *Vitis vinifera* L. Sultana plants : Genotypic and phenotypic analysis. *Mol. Breed.* 4: 321-333.

Goertzen, L. 2005. The application of grapevine genomic resources to evolutionary studies, with an example concerning the higher-level classification of the Vitaceae. International Grape Genomics Symposium, July 12-14, Saint Louis, MO, USA.

Goes da Silva, F.D., Blandolino, A., Choi, H.K., Baek, J.M., Leslie, A.D., Xu, J., Cook, D.R. 2005. Characterizing the transcriptional response of grapes to the pierce's disease pathogen *Xylella fastidiosa.* Plant Animal and Microbe Genome Conference XIII, SanDiego, CA, January 15-19, 2005.

Gollop, R., Even, S., Colova-Tsolova, V., Perl, A. 2002. Expression of the grape dihydroflavonol reductase gene and analysis of its promoter region. *Journal of Experimental Botany,* 53: 1397-1409.

Grando, M.S., Bellin, D., Edwards, K.J., Pozzi, C., Stefanini, M., Velasco, R. 2003. Molecular linkage maps of *Vitis vinifera* L. and *Vitis riparia* Mchx. *Theor. Appl. Genet.,* 106: 1213-1224.

Grassi, F., Labra, M., Imazio, S., Spada, A., Sgorbati, S., Scienza, A., Sala, F. 2003. Evidence of a secondary grapevine domestication centre detected by SSR analysis. *Theor. Appl. Genet.,* 107: 1315-1320.

Haas, H.U., Budahn, H., Alleweldt, G. 1994. In situ hybridization in *Vitis vinifera* L. *Vitis* 33: 251-252.

Hocquigny, S., Pelsy, F., Dumas, V., Kindt, S., Heloir, M.C., Merdinoglu, D. 2004. Diversification within grapevine cultivars goes through chimeric states. *Genome,* 47: 579-589.

Ingrouille, M.J., Chase, M.W., Fay, M.F., Bowman, D., Bank, M.V.D., Bruijn, A.D.E. 2002. Systematics of Vitaceae from the viewpoint of plastid rbcl DNA sequence data. *Botanical Journal of the Linnean Society,* 138: 421-432.

Iocco, P., Franks, T., Thomas, M.R. 2001. Genetic transformation of major wine grape cultivars of *Vitis vinifera* L. *Transgenic Res.* 10: 105-112.

Jardak-Jamoussi, R., Ebel, R., Dubois, C., Manthley, T., Bassler, A., Mliki, A., Bouamama, B., Wetzel, T., Ghorbel, A., Krczal, G., Reutsle, G.M. 2004. Induction of silencing in transgenic grapevine (*Vitis* sp) plants. 7th international symposium on grapevine physiology and biotechnology, Davis, CA, USA, June 21-25.

Johnson, H. 1989. The Story of Wine. Mitchell Beazley Publishers London. p. 478.

Joly, D., Perrin, M., Gertz, C., Kronenberger, J., Demangeat, G., Masson, J.E. 2004. Expression analysis of flowering genes from seedling-stage to vineyard life of grapevine cv Riesling. *Plant Science,* 166: 1427-1436.

Kikkert, J.R., Hébert-Soulé, D., Wallace, P.G., Striem, M.J., Reisch, B.I. 1996. Transgenic plantlets of " Chancellor " grapevine (*Vitis* sp.) from biolistic transformation of embryogenic cell suspensions. *Plant Cell Rep.* 15, 311-316.

Kikkert, J.R., Thomas, M.R., Reisch, B.I. 2001. Grapevine genetic engineering. *In*: KA Roubelakis-Angelakis, (ed) *Molecular Biology and Biotechnology of the Grapevine.* Kluwer Academic Publishers, Dordrecht, The Netherlands, pp. 393-410.

Kobayashi, S., Goto-Yamamoto, N., Hirochika, H. 2004. Retrotransposon-induced mutations in grape skin color. *Science,* 304: 982.

Lahogue, F., This, P., Bouquet, A. 1998. Identification of a codominant SCAR marker linked to seedless character in grapevine. *Theor. Appl. Genet.* 97: 950-951.

Lin, H., Walker, A. 1998. Identifying grape rootstocks with Simple Sequence Repeat (SSR) DNA markers. *Am. J. Enol. Vitic.* 49: 403-407.

Lodhi, M.A., Daly, M.J., Ye, G.N., Weeden, N.F., Reisch, B.I. 1995. A molecular marker based linkage map of Vitis. *Genome* 38: 786-794.

Lodhi, M.A., Reisch, B.I. 1995. Nuclear DNA content of *Vitis* species, cultivars and other genera of the *Vitaceae. Theor. Appl. Genet.,* 90: 11-16.

Luo, M.C., Thomas, C., You, F.M., Hsiao, J., Ouyang, S., Buell, C.R., Malandro, M., McGuire, P.E., Anderson, O.D., Dvorak, J. 2003. High-throughput fingerprinting of bacterial artificial chromosomes using the SnaPshot labelling kit and sizing of restriction fragments by capillary electrophoresis. *Genomics* 82: 378-389.

Mac Govern, P.E., Glusker, D.L., Exner, L.J., Voigt, M.M. 1996. Neolithic resinated wine. *Nature,* 381: 480-481.

Merdinoglu, D., Butterlin, G., Bevilaqua, L., Chiquet, V., Adam-Blondon, A.F., Decroocq, S. 2005. Development and characterization of a large set of microsatellite markers in grapevine (*Vitis vinifera L.*) suitable for multiplex PCR. *Molecular Breeding,* 15: 349-366.

Merdinoglu, D., Wiedemann-Merdinoglu, S., Coste, P., Dumas, V., Haetty, S., Butterlin, G., Greif, C. 2003. Genetic analysis of downy mildew resistance derived from *Muscadinia rotundifolia. Acta. Hort.* 603: 451-456.

Moroldo, M., Scalabrin, S., Prete, G., Felice, N., Marconi, R., Faes, G., Velasco, R., Morgante, M. 2004. Construction of a contig-based physical map of grape using fluorescent fingerprinting technology, 7th International Symposium on Grapevine Physiology and Biotechnology, Davis, CA, USA, June 21-25.

Mullins, M.G., Bouquet, A., Williams, L.E. 1992. Biology of grapevine, Cambridge University Press, Cambridge.

Mullins, M.G., Tang, F.C.A., Facciotti, D. 1990. *Agrobacterium*-mediated genetic transformation of grapevines. Transgenic plants of *Vitis rupestris* Scheele and buds of *Vitis vinifera* L. *Bio/Technology* 8: 1041-1045.

Pauquet, J., Bouquet, A., This, P., Adam-Blondon, A.F. 2001. Establishment of a local map of AFLP markers around the powdery mildew resistance gene *Run1* in grapevine and assessment of their usefulness for marker assisted selection. *Theor. Appl. Genet.*, 103: 1201-1210.

Pelsy, F., Merdinoglu, D. 2002. Complete sequence of Tvv1, a family of Ty1 copia-like retrotransposons of Vitis vinifera L., reconstituted by chromosome walking. *Theor. Appl. Genet.*, 105: 614-621.

Pereira, H.S., Barão, A., Delgado, M., Morais-Cecìlio, Viegas, W. 2005. Genomic analysis of *Grapevine Retrotransposon* 1 (*Gret*1) in *Vitis vinifera. Theor. Appl. Genet.*, 111: 871-878.

Perl, A., Colova-Tsolova, V., Eshdat, Y. 2004. Agrobacterium-mediated transformation of grape embryogenic calli. In: IS Curtis (ed) *Transgenic crops of the world,* Kluwer Academic Publishers, The Netherlands, 17: 1-14.

Perl, A., Sahar, N., Eliassi, R., Baron, I., Spiegel-Roy, P., Bazak, H. 2003. Breeding of new seedless table grapes in Israel by conventional and biotechnological approaches. *Acta. Hort.* 603: 185-187.

Perl, A., Sahar, N., Spiegel-Roy, P., Gavish, S., Elyasi, R., Orr, E., Bazak, H. 2000. Conventional and biotechnological approaches in breeding seedless table grape. *In:* Bouquet, A. and Boursiquot, J.M. (eds) *Proceedings of the seventh symposium on Grapevine Genetics and Breeding,* ISHS publisher, Vol 2, pp. 607-61.

Pratelli, R., Lacombe, B., Torregrosa, L., Gaymard, F., Romieu, C., Thibaud, J.B., Sentenac, H. 2002. A grapevine gene encoding a guard cell K+ channel displays developmental regulation in the grapevine berry. *Plant Physiol.* 128: 564-577.

Predieri, S. 2001. Mutation induction and tissue culture in improving fruits. *Plant Cell, Tissue and Organ Culture,* 64: 185-210.

Riaz, S., Dangl, G.S., Edwards, K.J., Meredith, C.J. 2004. A microsatellite marker based framework linkage map of *Vitis vinifera* L. *Theor. Appl. Genet.*, 108: 864-872.

Salmaso, M., Faes, G., Segala, C., Stefanini, M., Salakhutdinov, I., Zyprian, E., Toepfer, R., Grando, M.S., Velasco, R. 2004. Genome diversity and gene haplotypes in the grapevine (*Vitis vinifera* L.), as revealed by single nucleotide polymorphisms. *Mol. Breed,* 14: 385-395.

Sarry, J.E., Sommerer, N., Sauvage, F.X., Bergoin, A., Rossignol, M., Albagnac, G., Romieu, C. 2004. Grape berry biochemistry revisited upon proteomic analysis of the mesocarp. *Proteomics,* 4: 201-215.

Sefc, K.M., Lopes, M.S., Lefort, F., Botta, R., Roubelakis-Angelakis, K.A., Ibàñez, J., Pejiæ, I., Wagner, H.W., Glössl, J., Steinkellner, H. 2000. Microsatellite variability in grapevine cultivars from different European regions and evaluation of assignment testing to assess the geographic origin of cultivars. *Theor. Appl. Genet.*, 100: 498-505.

Sefc, K.M., Regner, F., Tureschek, E., Glossl, J., Steinkellner, H. 1999. Identification of microsatellite sequences in *Vitis riparia* and their application for genotyping of different *Vitis* species. *Genome* 42: 367-373.

Siret, R. 2001. Etude du polymorphisme génétique de la vigne cultivée (*Vitis vinifera* L.) à l'aide de marqueurs microsatellites : application à la caractérisation des cépages dans les vins. Thèse de l'Université de Montpellier I, 11 janvier 2001, pp. 148.

Soltis, D.E., Soltis, P.S., Chase, M.W., Mort, M.E., Albach, T.D., Zanis, M., Savolaninen, V., Hahn, W.H., Hoot, S.B., Fay, M.F., Axtell, M., Swensen, S.M., Prince, L.M., Kress, W.J., Nixon, K.C., Farris, J.S. 2000. Angiosperm phylogeny inferred from 18S rDNA, rbcL and atpB sequences. *Botanical Journal of the Linnean Society,* 133: 381-461.

Terrier, N., Glissant, D., Grimplet, J., Barrieu, F., Abbal, P., Couture, C., Ageorges, A., Atanassova, R., Léon, C., Renaudin, J.P., Dedaldechamp, F., Romieu, C., Delrot, S., Hamdi, S. 2005. Isogene specific oligo arrays reveal multifaceted changes in gene expression during grape berry (*Vitis vinifera* L.) development. *Planta:* 6: 1-16.

Terrier, N., Romieu, C. 2001. Grape berry acidity. *In*: KA Roubelakis-Angelakis (ed*) Molecular Biology and Biotechnology of the Grapevine.* Kluwer Academic Publishers, Dordrecht, The Netherlands, pp. 35-57.

Tessier, C., David, J., This, P., Boursiquot, J.M., Charrier, A. 1999. Optimisation of the choice of molecular markers for varietal identification in *Vitis vinifera* L. *Theor. Appl. Genet.*: 98: 171-177.

This, P., Lahogue, F., Adam-Blondon, A.F., Doligez, A., Bouquet, A. 2000. Towards marker-assisted selection for seedlessness in grapevine. *Acta. Hort.* 528: 221-229.

Thomas, M.R., Hua, C.T., Iocco, P., Davies, C. 2004. Grapevine ESTs and their use in microarray analysis to study gene controlling grapevine berry ripening and development. 7th International Symposium on Grapevine Physiology and Biotechnology, Davis, CA, USA, June 21-25.

Thomas, M.R., Iocco, P., Franks, T. 2000. Transgenic grapevines : Status and future. *Acta. Hort.* 528: 279-284.

Tomkins, J.P., Peterson, D.G., Yang, T.J., Ablett, E.R., Henry, R.J., Lee, L.S., Holton, T.A., Waters, D., Wing, R.A. 2001. Grape (*Vitis vinifera* L;) BAC library construction, preliminary STC analysis, and identification of clones associated with flavonoid and stilbene biosynthesis. *Am. J. Enol. Vitic.*, 52: 287-291.

Topping, J.F., Lindsey, K. 1995. Insertional mutagenesis and promoter trapping in plants for the isolation of genes and the study of development. Transgenic Research, 4: 291-305.

Torregrosa, L., Verriés, C., Tesnière, C. 2002. Grapevine (*Vitis vinifera* L.) promoter analysis by biolistic-mediated transient transformation of cell suspensions. *Vitis*, 41: 27-32.

Tual, S., Paillard, S., Ratnaparkhe, M., Lamoureux, D., Clainche, I., Dossat, C., Cruaud, C., Wincker, P., Weissenbach, J., Adam-Blondon, A.F. 2004. Toward the construction of a fingerprint-based physical map of the grape genome. Plant GEMS, 22-25 September, Lyon, France.

Velasco, R., Grando, M.S., Troggio, M., Faes, G., Pindo, M., Malacarne, G., Segala, C., Coppola, G., Fontana, P., Zambanini, J., Jesse, T., Chalhoub, B., Adam-Blondon, A.F., Prete, G., Scalabrina, S., Felice, N., Marconia, R., Moroldo, M., Morgante, M. 2004. Integrating molecular and physical maps in grapevine (*Vitis vinifera* cv Pinot noir): powerful aids for gene isolation and breeding. 7th International Symposium on Grapevine Physiology and Biotechnology, Davis, CA, USA, June 21-25.

Verries, C., Bès, C., This, P., Tesnière, C. 2000. Cloning and characterization of Vine-1, a LTR retrotransposon-like element in *Vitis vinifera* L. and other species. *Genome*, 43: 366-376.

Vidal, J.R., Kikkert, J.R., Wallace, P.G., Reisch, B.I. 2003. High-efficiency biolistic co-transformation and regeneration of "Chardonnay" (*Vitis vinifera* L.) containing npt-II and antimicrobial peptide genes. *Plant Cell* Rep, 22: 252-260.

Xue, B., Ling, K.S., Reid, C.L., Krastanova, S., Sekiya, M., Momol, E.A., Sule, S., Mozsar, J., Gonsalves, D., Burr, T.J. 1999. Transformation of five grape rootstocks with plant virus genes and a vir E2 gene from *Agrobacterium tumefaciens. In vitro Cell Dev. Biol. Plant* 35: 226-231.

24

CHAPTER

Sugarcane: A Tropical Crop with a Highly Complex Genome

J.Y. Hoarau, G. Souza, A. D'Hont, M. Menossi, L. Rossini Pinto, A. Pereira de Souza, L. Grivet, C.F. Martins Menck, E. Cesar Ulian, M. Vincentz

INTRODUCTION

Sugarcane is a tall tropical perennial grass with its center of origin located in Southeast Asia, where it has probably been cultivated since prehistoric times. Sugar cane is able to convert the energy of the sun into carbohydrates more efficiently than any other plant and the unusual ability of storing sucrose in stem cell vacuoles, along with a high biomass production and ease of cultivation, makes it one of the most interesting and productive agricultural crops. Thriving in humid and warm climates, sugarcane is mainly cultivated in tropical and subtropical regions on more than 20 million hectares. The harvested biomass of sugar cane makes it the world's largest crop with 1,300 million metric tons produced in 2004 as reported by FAO (http://apps.fao.org). Mainly used to produce sugar, it accounts for about two-thirds of the total world sugar production, while beet sugar is responsible for one-third. By-products obtained from sugarcane include a wide range of derivatives (molasses, alcohol, fuel, livestock feed, paper, particleboard...) that can be used in the energy, food, chemical, pharmaceutical, biotechnology and other industries. Cultivated in more than 90 countries, this crop is one of the mainstays of many developing countries, principally in Latin America, the Caribbean, southern Africa and Asia, where it offers sustainable energy sources (ethanol and electricity) and new developmental alternatives.

Sugarcane is propagated vegetatively by stem cuttings. It re-sprouts (ratoons) annually from underground buds on basal portions of old stalks. Depending on the variety and growing conditions, stalks of 2 to 4 meters containing 15% sugar can be produced in about 12 months from an original planting or from regrowth. In favorable environments the plant can produce extremely high levels of biomass (200 tons /ha), that can be used for fiber or energy. Sugarcane belongs to the Poaceae family, along with rice and to the Andropogoneae tribe, like maize and sorghum. It belongs to the genus *Saccharum*, a complex genus characterized by high polyploidy and frequent aneuploidy. Sugarcane cultivars are wind pollinated outbreeders, highly heterozygous and intolerant to inbreeding. Substantial increases in yields have occurred over the past decades due to the cultivation of higher-yielding cultivars, obtained through successful breeding programs that bred genotypes resistant to diseases and tailored to modern industrial needs (Hogarth *et al.*, 1997).

The sugarcane industry must continually seek new approaches and adopt new technologies that might better guide research in cultivar development programs. For many years, sugarcane genetics had received little attention from plant scientists despite its economic importance. This may be due to the complexity of sugarcane genome, which appears difficult to deciphere. Recently, however, a vast array of genomic tools have become available and new opportunities have been opened to refine our understanding of the genetic architecture of sugarcane and to explore its functional systems. After a brief history of the plant's origin, this review focuses on recent advances in sugarcane genomics, including: genome structure and diversity, genetic mapping, gene tagging, transcriptome, functional genomics and lessons from these studies.

ORIGIN AND HISTORY

Sugarcanes belong to the genus *Saccharum* L. which encompasses only highly polyploid species (2n = 40 to about 200). The reference domesticated species is *S. officinarum* (2n = 80). It has a group of thick juicy canes, which were first cultivated for chewing only in SE Asia and in the Pacific islands prior to 2500 BC. Cultigens of this species are still used in traditional agriculture, especially in Melanesia (Daniels and Roach, 1987). Besides this species, sugarcanes indigenous to North India (2n = 81 - 124) and China (2n = 116 - 120) have been cultivated from prehistoric times in mainland Asia, where sugar-making industries were born. They are referred to as *S. barberi* and *S. sinense* respectively and are well adapted to sub-tropical environments. In the 16th century sugar manufacturing techniques and *S. officinarum* were spread worldwide over the inter-tropics. This latter species was the most cultivated until the end of the 19th century, before a decline due to disease epidemics. At that time, *S. spontaneum* (2n = 40 - 128), as well as a few *S. barberi* clones were used in introgression programs aimed at overcoming the disease outbreaks affecting *S. officinarum*. *S. spontaneum* is a very robust wild species, highly polymorphic, with thin stalks and no sugar, growing in tropics and subtropics, from Africa to Solomon Island. Restoration of the high-sugar-producing type was obtained by backcrossing of the hybrids to *S. officinarum*. Until now, sugarcane breeding has consisted of successive cycles of intercrossing and selection of those early interspecific derivatives. As a consequence, modern sugarcane cultivars possess a complex interspecific aneupolyploid genome encompassing between 100 and 130 chromosomes.

Sugarcane evolution and domestication has long been debated on botanical and historical terms. Recently Grivet *et al.* (2004) proposed a global review, thanks to new data obtained by molecular genetics. Sugarcane probably evolved from a specific lineage restricted to the current genus *Saccharum* and independently from lineages that are closely related to the genera *Miscanthus* and *Erianthus*, although inter-generic hybridization is possible (D'Hont *et al.*, 1995; Piperidis *et al.*, 2000). The cultigens of the sugar-producing species, *S. officinarum* (2n = 80) which has constituted the starting point of modern interspecific breeding, were domesticated in New Guinea from *S. robustum* (2n = 60 and 2n = 80), a wild species of long and thick stalks producing little sugar with a natural distribution from Borneo to the New Hebrides. North Indian (*S. barberi*) and Chinese (*S. sinensis*) cultivars formerly cultivated in mainland Asia, are natural hybrids between *S. officinarum* and *S. spontaneum* (D'Hont *et al.*, 2002). They represent horticultural groups rather than established species.

GENOME STRUCTURE

Genomic *in situ* hybridization (GISH) showed that 15 to 25% of the genome of modern cultivars are derived from *S. spontaneum* (D'Hont *et al.*, 1996; Piperidis and D'Hont, 2001) and together with genetic mapping (Grivet *et al.*, 1996; Hoarau *et al.*, 2001), demonstrated the occurrence of recombination between the chromosomes of the two parental species (Fig. 24.1).

Fluorescence *in situ* hydridization (*FISH*) of rDNA genes indicated that the 80 chromosomes of *S. officinarum* are organized in 8 homologous copies of a basic set of 10 different chromosomes (2n = 8x = 80), while the 40 to 128 chromosomes of *S. spontaneum* are organized in 5 to 12 homologous copies of a basic set of 8 different chromosomes 2n = 5x - 12x = 40 - 128 (D'Hont *et al.*, 1998; Ha *et al.*, 1999). A schematic representation of the genome of current cultivars as deduced from these FISH and GISH experiments, is presented in Fig. 24. 1.

This difference in the basic chromosome number implies that structural differences separate the two genomes. As a consequence, two distinct chromosome organizations coexist in present-day cultivars. This is illustrated by the mapping studies of the cultivar 'R570' (Grivet *et al.*, 1996, Hoarau *et al.*, 2001, Rossi *et al.*, 2003) that have shown that certain independent linkage groups inherited from *S. officinarum* correspond to a single linkage group derived from *S. spontaneum* (Fig. 24.2). These chromosomal patterns observed in 'R570', as well as a tentative alignment of the ancestral *Saccharum* species with the sorghum genome (Ming *et al.*, 1998), suggest that divergence between *S. officinarum* and *S. spontaneum* may have resulted from simple fusion or fission events. Similarly, GISH data of D'Hont *et al.* (1996) showing an interstitial location of rDNA sequences in *S. spontaneum* compared to a terminal position in *S. officinarum*, give credit to the hypothesis of some chromosome fusion in *S. spontaneum*.

The size of the monoploid genome (1x) was estimated to be 930 Mbp (0.96 pg) for *S. officinarum* (x = 10) and 750 Mpb (0.78 pg) for *S. spontaneum* (x = 8). This is similar the gametic genome size of sorghum and about twice that of rice. However, the size of the total genome of a sugarcane cultivar, 10,000 Mbp (R570, 2n = 115), is much greater than the one of maize, 5,500 Mbp (2n = 20), sorghum, 1,600 Mbp (2n = 20) or rice, 860 Mbp (2n = 24) reflecting the high polyploidy of sugarcane cultivars (D'Hont and Glaszmann, 2001).

Chromosome pairing was investigated in the modern cultivar, R570, taking advantage of a particular single copy RFLP probe (BNL 12.06), revealing 11 alleles in a locus belonging to the homologous group I. The result disclosed an atypical picture, with pairing frequencies ranging from 0 to 40% and differential affinities leading to the

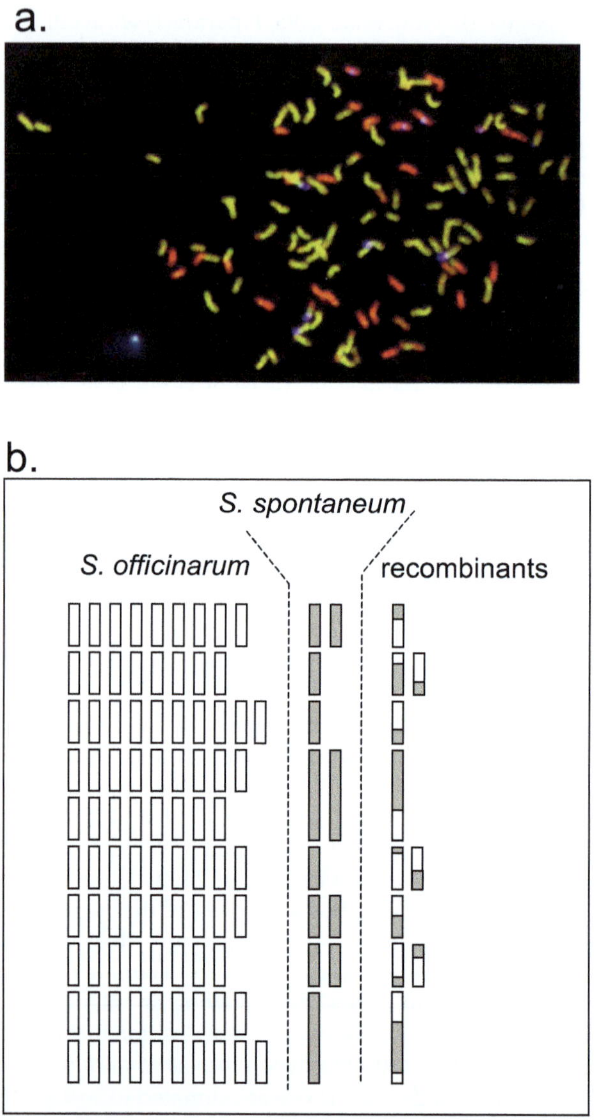

Fig. 24.1 **a.** Chromosome preparation of the sugarcane cultivar NCo 376 after GISH using *S. officinarum* genomic DNA (labeled with fluorescein isothiocyanate, green), *S. spontaneum* genomic DNA (labeled with Texas Red, red) and 18S-5.8S-25S rDNA (labeled with coumarin, blue). Around 70% of the chromosomes are inherited from *S. officinarum* (yellow), 20% from *S. spontaneum* (orange) and 10% are recombinant from both species. **b.** Schematic representation of the genome of current sugarcane cultivars as deduced from FISH and GISH experiments. Each bar represents a chromosome. White and grey bars correspond to *S. officinarum* and *S. spontaneum* chromosomes or chromosome segments, respectively. Chromosomes from the same row are homologous (or homeologous). The main characteristics of this genome are the high polyploidy, aneuploidy, the bi-specific origin of the chromosomes and the occurence of interspecific recombinant chromosomes. In addition, there are clear structural differences between the chromosomes of the two species.

identification of several chromosome subsets. This example illustrated the unsystematic meiotic behavior in sugarcane cultivars. It highlights a continuous range of pairing affinities between chromosomes and pinpoints a strong role of individual chromosome features, partly related to their ancestral origin, in the determination of these affinities (Jannoo *et al.*, 2004).

Genome organization of modern sugarcane cultivar is thus particularly intricate:

- the 'hybrid basic chromosome number' is likely to be greater than ten, and the proportion of duplicated regions in the 'monoploid hybrid genome' is potentially higher than in ancestral species;
- the frequent occurrence of unpaired chromosomes in several homology groups due to aneuploidy, implies a loss of a few chromosomes in each generation, which could result in a gradual erosion of the chromosome number after several rounds of recurrent breeding (Butterfield *et al.*, 2001);
- pairing schemes between chromosomes is thought to be more or less specific to each homo(eo)logous group and does not fit any pre-established clear-cut scheme, such as disomy or complete polysomy (Jannoo *et al.*, 2004; Grivet and Arruda, 2001).

MOLECULAR GENETIC DIVERSITY

A comprehensive understanding of genetic diversity within *Saccharum*, and knowledge of the contribution of founder species to the variability present in current working collections, could improve the effectiveness of breeding programs aimed at combining broadened complementary resources.

Cultivars representative of several international breeding programs have been studied with restriction fragment length polymorphism (RFLP), amplified fragment length polymorphism (AFLP) or single sequence repeats (SSR) markers in several investigations (Lu *et al.*, 1994; Jannoo *et al.*, 1999b; Lima *et al.*, 2002; Raboin, pers. comm.). As a general rule, a small differentiation between cultivars from different breeding stations has been observed, which is mainly explained by *S. spontaneum* specific markers. These results reflect; (i) the high ploidy and heterozygosity of modern cultivars, which has enabled them to keep individually a large proportion of the alleles contributed by the parents of initial interpsecific crosses; (ii) the tradition of a large exchange of parental material between sugarcane breeding stations.

Jannoo *et al.* (1999b) have shown that most of the *S. officinarum* variability is found in the modern cultivar pool exploited by the breeding programs. Only a small additional reservoir of *S. officinarum* alleles (15%) could be used to broaden the genetic base of modern cultivar pool. On the contrary, a large variability available within *S. spontaneum* germplasm has not been exploited yet.

Linkage disequilibrium (non random association between alleles from different loci) was investigated in modern cultivars by Jannoo *et al.* (1999a). Allele associations have been detected between loci separated by up to 10 cM in the whole genome. This global linkage disequilibrium is interpreted as being the result of the foundation bottleneck related to the very few interspecific combinations at the beginning of sugarcane breeding, coupled with the limited number of recombination events between the present-day cultivars and first interspecific hybrids (less than 10 meioses). The preferential allele associations thus created have been maintained through subsequent crosses when the loci were closely linked. This phenomenon is likely also to apply to genes of agricultural interest. A practical consequence is that markers can be used to track known QTLs in modern breeding materials without the necessity of repeatedly studying segregating progenies. This structure gives high value to the correlation between molecular markers and agricultural traits among cultivars.

GENETIC MAPPING

Constraints dictated by polyploidy to construct genetic maps have been analyzed by Wu *et al.* (1992). Regarding the high ploidy of modern sugarcane cultivars and the complex properties of their chromosome associations (Jannoo *et al.* 2004), only single dose alleles can be readily mapped, thanks to the use of standard methodologies developed for diploid organisms. Double dose markers are much less informative (Ripol *et al.* 1999) and appropriate mapping functions for their use are lacking.

Mapping strategies in sugarcane are based on pseudo-testcross approaches. In a biparental progeny, only single dose markers specific to each parent are available for mapping. However, in a selfed population all simplex markers present in the selfed genotype can be mapped. Therefore mapping out-crossed populations results in lower density maps than selfed populations (Raboin *et al.*, 2006).

Partial genetic maps have been produced for *S. spontaneum* (Ming *et al.*, 1998, 2002; Al Janabi *et al.*, 1993; da Silva *et al.*, 1995), *S. officinarum* (Mudge *et al.*, 1996; Guimaraes *et al.*, 1997, 1999; Ming *et al.*, 1998, 2002) and modern cultivars (D'Hont *et al.*, 1994; Grivet *et al.*, 1996; Hoarau *et al.*, 2001; Rossi *et al.*, 2003; Aiken *et al.*, 2005; Raboin *et al.*, submitted) using different molecular marker technologies (RAPD, AFLP, RFLP and SSR). The most advanced maps (Rossi *et al.*, 2003; Aiken *et al.*, 2005) were developed on the cultivar 'R570' ($2n \approx 115$) and Q165 respectively. These maps encompass around a thousand markers unevenly distributed along more than 100 linkage groups of heterogeneous size and are estimated to cover only around 50% of the genome of approximately 17,000 cM. Marker coverage appears uneven, with chromosomes inherited from *S. spontaneum* being covered more densely than *S. officinarum* chromosomes (Fig. 24.2). This discrepancy could be attributed to the general lower polymorphism observed in *S. officinarum* germplasm coupled with the higher ploidy level of this species in the constitution of present-day hybrid genomes. To overcome this latter problem and to obtain higher density maps of the *S. officinarum* component of the genome, specific material could be designed by backcrossing modern cultivars with *S. spontaneum* to reduce the number of *S. officinarum* chromosomes (Butterfield *et al.*, 2001).

The availability of genetics maps gave opportunities to undertake linkage analyses with agronomic traits of interest, such as sugar yield or disease resistance. The first mapping studies reported the segregation of a major rust resistance gene in a hybrid cultivar (Daugrois *et al.*, 1996), and loci linked to eyespot resistance (Mudge *et al.*, 1996), or to short day-flowering (Guimaraes *et al.*, 1997) in ancestral species. More recently Ming *et al.* (1998) detected QTLs controlling sugar content in interspecific (*S. officinarum* x *S. spontaneum*) F_1 populations. Attempts have also been made to detect QTLs related to yield components in commercial germplasm for use in breeding (Hoarau *et al.*, 2002). In such experiments the effect of individual QTLs are small, ranging from 3% to 7% of

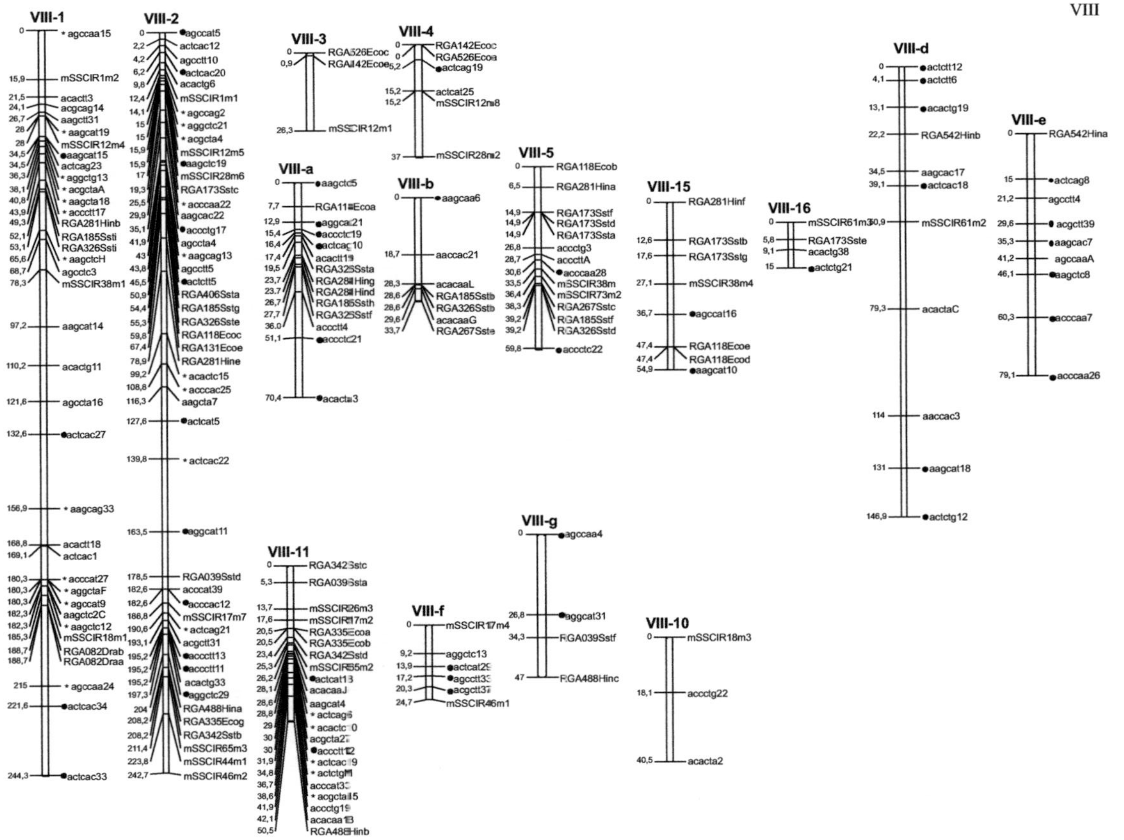

Fig. 24.2 Representation of one homology group (HG VIII) of the genetic map of the sugarcane cultivar R570 (from Rossi *et al.*, 2003; Hoarau *et al.*, 2001). This HG encompasses 15 cosegregation groups (CG) and includes AFLP, SSR (mSSCIR) and Resistance Gene Analogs (RGA) markers. Where the species origin of the markers has been identified (Hoarau *et al.*, 2001), it is indicated with a star for a putative *S. spontaneum* origin and with a black dot for a putative *S. officinarum* origin. In this HG, two large CGs with a putative *S. spontaneum* origin correspond to two separate groups of small CGs of putative *S. officinarum* origin, illustrating the difference in structure between the two genomes involved in the formation of modern cultivars.

the phenotypic variance. QTL effects fluctuate slightly across different environments (crop cycles and/or harvest seasons), on the verge to being perceptible given the experimental resolution. Multiple regression analyses were performed with all putative genetic factors revealed (simplex or mutiplex QTLs and di-genic linear x linear interactions). Altogether these different types of QTLs explained from 30 to 55% of the total phenotypic variance depending on the trait. These percentages are in agreement with the map coverage so far obtained (Hoarau *et al.*, 2002).

Polyploidy makes QTL mapping particularly challenging in sugarcane. When in a diploid plant, two alternative alleles per locus are compared, in sugarcane, between 8-12 segregating factors per locus have to be dealt with. In such a polyploid background with predominantly polysomic inheritance, one intuitively expects to be able to tag the most (or the least) favorable allele, amongst the whole allelic series. This confounding effect of the high ploidy level is probably reinforced in a modern genetic background in which favorable alleles have been concentrated by recurrent breeding, thus diminishing the contrast between alleles that determines the trait segregation and the magnitude of the allele effects.

The construction of theoretical models specifically adapted to analyze multiple-dose QTLs in a polyploid context has been attempted by Doerge and Craig (2000) despite challenging issues to be solved. However, the application of any theoretical model will inevitably rely heavily on simplistic assumptions to be made on sugarcane chromosome and gene functioning, in order to sort out practical QTL estimates.

Genetic maps also need to be massively refined to achieve coverage of the genome more suitable for QTL detection. To this end, the Diversity Array Technology (DArT) technology (http://www.cambia.org/DArT.html) may be worth testing as it may provide rapid and massive coverage of the genome. Microsatellites in particular, derived from ESTs could also be advantageously developed.

Since their discovery, microsatellites or simple sequence repeats (SSRs) have become one of the most suitable markers for genome analysis. To detect such polymorphisms, unique primers for PCR amplification reactions are designed from the conserved sequences flanking the SSR. The requirement of information on these conserved sequences is one of the drawbacks of SSR isolation from genomic libraries, which allied with the several distinct steps involved (construction of genomic library, enriched library process, screening of positive clones), may represent a relatively time-consuming and expensive process. Alternatively, electronic searches in genome databases, where information on the conserved sequence already exists, may represent a simple, fast and economically way to obtain SSRs. They can provide direct mapping of genes with known function allowing candidate genes involved in important pathways to be directly evaluated for associations with important agronomic traits. EST derived markers are ideal for marker-assisted selection as they themselves may be the gene responsible for the trait, thus, no recombination is expected between the marker and the trait (Cato *et al.*, 2001; Ma *et al.*, 2004).

The availability of microsatellites derived from ESTs (EST-SSRs) overcomes the difficulties normally associated with RFLP assays (a laborious and time consuming technique) and they display a high levels of polymorphism compared to those obtained using ESTs as probes. However, SSRs in ESTs are present in low frequencies with reported values of 2.5% in grape (Scott *et al.*, 2000), 2.9% for cotton, 3% for *Arabidopsis thaliana*, 4.4% for tomato, 5.2% for soybean (Cardle *et al.*, 2000) and 2.88% for sugarcane (Cordeiro *et al.*, 2001).

The Sugarcane Expressed Sequence Tag Project (SUCEST) produced 237,954 ESTs assembled in 43,000 clusters (Vettore *et al.*, 2001; http://sucest.lad.ic.unicamp.br/en). The SUCEST database facilitates candidate gene strategies in sugarcane breeding, and this large collection of clusters shows considerable potential to generate EST-SSRs. Taking into account a minimum number of repeat of $n \geq 7$ for dinucleotide, $n \geq 5$ for trinucleotide and $n \geq 3$ for tetranucleotide, 47 SSR motif types were used as an entry to search against the SUCEST Cluster sequences using BLASTn program. This simple procedure adopted for mining EST-SSRs identified 2005 clusters containing SSRs, that in turns represents around 5% of the entire cluster population of the database (Pinto *et al.*, 2004). Pre-selected clusters containing SSRs and displaying similarity to genes of interest were preferentially used to develop a working primer set of EST-SSRs. Until know a subset of 258 EST-SSRs were designed from the SUCEST database at the University of Campinas (UNICAMP, São Paulo, Brasil). These 258 EST-SSRs complement the 259 genomic SSRs developed by the International Consortium of Sugarcane Biotechnology (ICSB) of which 183 were developed by Cordeiro *et al.* (2000) and 76 from CIRAD (Centre de Cooperation Internationale en Recherché Agronomique pour le Développement, Montpellier, France).

The functional nature of EST-SSRs counterbalances their relatively low level of polymorphism. Polymorphism analysis of a subset of EST-SSRs screened against a set of sugarcane cultivars revealed a valuable level of polymorphism, corresponding to numerous segregating markers that could potentially be used for additional mapping efforts (Pinto *et al.*, 2004). Because EST-SSRs are more transferable among related species, they may also be used for comparative mapping (Yu *et al.*, 2004).

COMPARATIVE MAPPING

Conservation of chromosome organization has proved to be particularly important in the grass familly (Freeling *et al.*, 2001). Any knowledge about the genetic determinants of a given trait in a particular species could thus potentially be transferred to other grass species through map comparison (Gale and Devos, 2001). Given the high polyploidy of sugarcane and the absence of a *Saccharum* diploid, the advantage of investigating synteny with other Poaceae was realized at an early stage. The first comparisons were made with maize that had, at that time, more advanced maps. Synteny was found to be conserved (D'Hont *et al.*, 1994; Grivet *et al.*, 1994; Dufour *et al.*, 1996), although quite perturbed by the segmental allopolyploidy, followed by diploidization, and the increase in genome size caused by active transposition that affected the maize lineage (Gaut *et al.*, 2000). Rice also shows relatively global simple synteny relationships with sugarcane, with however many rearrangements explained by the large distance between the two species (Glaszmann *et al.*, 1997). Sorghum appears to be the species with the simplest synteny relationship with sugarcane (Grivet *et al.*, 1994; Dufour *et al.*, 1997; Glaszmann *et al.*, 1997, Guimaraes *et al.*, 1997, Ming *et al.*, 1998) and thus the most adapted to help sugarcane mapping.

MAPS-BASED CLONING

The major gene (*Bru1*) conferring resistance to brown rust identified in the sugarcane cultivar R570 (Daugrois *et al.*, 1996) and located at the edge of a co-segregation group belonging to homology group VII (Asnaghi *et al.*, 2000), is the target of a map-based cloning project in the particular complex genome of sugarcane with the financial support of ICSB. This project required the development of a Bacterial Artificial Chromosome (BAC) library of cultivar R570 (Tomkins *et al.*, 1999) and a fine map focused on the region surrounding the gene. One of the main challenges in developing a fine map around a target gene in such high polyploidy, is that the target region for fine mapping, around 2 cM, represents around 1/10,000 of the estimated size of the 17,000-cM genetic map (Hoarau *et al.*, 2001), as compared to around 1/1,000 for a typical diploid species, such as sorghum, with a genetic map of 1,500 cM.

To saturate the target region with markers, bulk segregant analysis was first used. More than 400 amplified fragment length polymorphism (AFLP) primer pairs were analyzed on two resistant and susceptible, bulks (each composed of DNA from five progenies) leading to the identification of four markers surrounding the resistance gene (at 2.2 and 1.9 cM) (Asnaghi *et al.*, 2004).

To further saturate the target area with markers, the good syntenic relationships between sugarcane, sorghum and rice (Dufour, 1997; Glaszmann, 1997; Ming *et al.*, 1998) were exploited (Asnaghi *et al.*, 2000; D'Hont, pers. comm.). The AFLP markers surrounding the gene were cloned and mapped on sorghum. The sorghum and rice physical regions orthologous to the sugarcane area were identified and exploited to saturate the genetic map and build up a physical map of the target sugarcane region.

SUGARCANE TRANSCRIPTOME, A REPRESENTATIVE SET OF GRASS SPECIFIC SEQUENCES

When considering large and complex genomes such as sugarcane, the sequencing of cDNA to produce "Expressed Sequence Tags" (EST) is an alternative to reveal the expressed portion of a genome (Rudd, 2003). ESTs can be assembled to generate a non redundant set of transcripts which represent an important source of genetic information for functional and evolutionary genomic analysis (i.e., gene discovery, annotation and analysis, genetic mapping, candidate genes for quantitative characteristics, evolution of gene content and architecture, among others).

To provide basic information for a better understanding of the biology of sugarcane and to improve breeding programs of this important crop, a large collection of new candidate sugarcane genes were identified by the Brazilian **su**garcane **EST** project (SUCEST; http://sucest.lbi.ic.unicamp.br/public), that has involved a network of laboratories. In order to obtain an exhaustive view of the sugarcane transcriptome, several cDNA libraries representing a set of organs, developmental stages, response to temperature stress and endophytic nitrogen fixing bacteria were constructed. From these libraries, a total of 237,954 high quality ESTs were obtained and assembled into 26,803 contigs and 16,338 singletons. The resulting 43,141 putative unique sugarcane transcripts were referred to as Sugarcane Assembled Sequences (SAS). Based on an estimation of the internal redundancy, it was concluded that the collection of SASs possibly identified around 33,000 sugarcane genes (Grivet *et al.*, 2001, 2003; Vettore *et al.*, 2003).

Functional annotation of the SASs was performed based on the similarity with known sequences in GenBank (Vettore *et al.*, 2003), which allowed a detailed categorization of sugarcane genes into signal transduction components, regulation of gene expression, development, biotic and abiotic stresses, transposable elements, metabolism and so on (for a complete description see the special issue of Genetics and Molecular Biology; http://sucest.lbi.ic.unicamp.br/public/year2001.html). This effort constitutes an invaluable contribution to improve sugarcane breeding strategies. Another important aspect of the annotation of SASs, concerns our understanding of angiosperm genome evolution. To contribute towards defining the genetic basis of the diversity found among flowering plants, we compared the SASs with DNA and protein sequences from the model eudicotyledon plant *Arabidopsis thaliana*, the model monocotyledon rice and from a set of other angiosperms (mainly ESTs). This approach resulted in the identification of a core set of eudicot-monocot sequences (75% of the SASs), which are likely to fulfill similar angiosperm functions. Two other classes representing monocot-specific sequences (14% of the SASs) and sequences restricted to sugarcane (13.5% of the SASs), were also defined. These classes could theoretically correspond to evolutionary novelties, or have been caused by differential gene loss or even horizontal gene transfer events. Of particular interest was the finding that a significant proportion of the monocot-specific sequences represent fast evolving sequences found in members of conserved angiosperm gene families, such as transcriptional regulatory factors and resistance genes, among others. A high rate of evolution may be related to functional diversification that could result in new protein functions and could therefore be involved in the differentiation of specific evolutionary lineages. Additionally, monocot-specific motives or domains that are integrated into conserved eudicot-monocot proteins, were also identified and may represent new protein architectures. Finally, a few monocot-specific proteins, most of which are rice putative or hypothetical proteins, were also found (Vincentz *et al.*, 2004). It appears therefore that the evolutionary events that underlie the differentiation of eudicots and monocots relies at least partly, on functional diversification (generating new protein functions) from duplicated copies of conserved gene families. The extent to which the sugarcane-specific SASs represent novelties restricted to this organism remains unclear (Ma *et al.*, 2004).

A small proportion of the SASs (2%) that were found to be conserved among angiosperms were missing in the *A. thaliana* genome, suggesting that the corresponding genes were present in the ancestor of monocots and eudicots and were subsequently lost in *A. thaliana* (Vincentz *et al.*, 2004). An example of a nuclear gene for a chloroplastic protein which has been lost in *A. thaliana* is shown in Fig. 24.3. Genes that are lost in specific lineages are likely to be unessential to angiosperms but may confer some selective advantage in the species in which they have been retained. We also detected eudicot-specific gene loss indicating that such events can occur at any stages of evolution. It appears therefore that differential gene loss is an active process in the shaping of angiosperm genomes. As a byproduct of this comparative genomic studies, some of the SASs were found to have potential orthologs of unanotated *A. thaliana* genomic sequences. These sequences may correspond to potentially new and highly

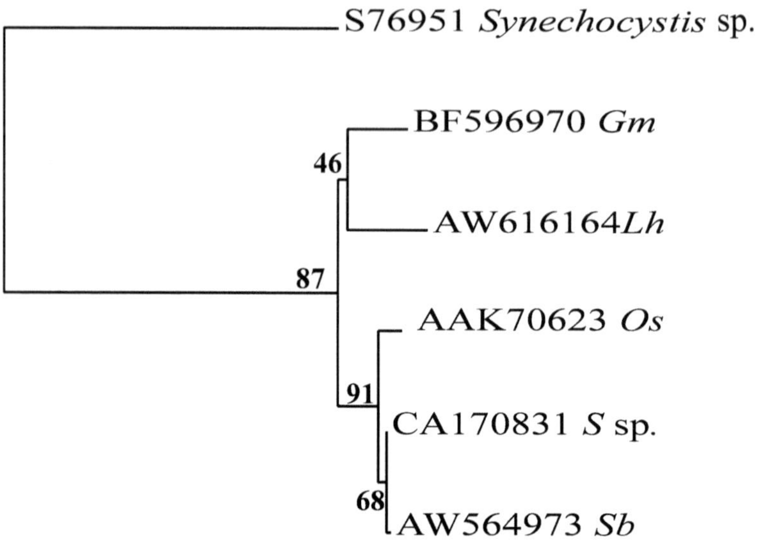

Fig. 24.3 Differential gene loss. Phylogeny illustrating a gene loss event in the *Arabidopsis thaliana* lineage. A gene that encode a putative chloroplast protein acquired from the ancestral cyanobacterial symbiont (*Synechocystis* sp.) was detected in several eudicot and monocot species but not in *Arabidopsis thaliana*. The protein is related to a ubiquinone metaquinone biosynthesis methyltransferase enzyme. The tree was inferred by the neighbor-joining method using PAM distances of a conserved motif of 105 amino acids. Bootstrap values for 1,000 replicates are indicated as percentages along the branches. Sequences are identified by their accession number. Species abbreviations are as follows: *Gm*, soybean (*Glycine max*); *Lh*, wild tomato (*Lycopersicon hirsutum*); *Os*, rice (*Oryza sativa*); *S* sp, sugarcane (*Saccharum officinarum*) and *Sb*, sorghum (*Sorghum bicolor*).

conserved *Arabidopsis thaliana* genes, improving our knowledge of the transcriptome of this thale cress.

In conclusion, our comparative genomic analysis confirmed that lineage-specific gene loss, high evolutionary rate of specific sequences and exon shuffling are important processes involved in the divergence of angiosperms. The challenge is now to define the function of monocot-specific sequences.

SUGARCANE FUNCTIONAL GENOMICS

The Sugarcane EST sequencing project (SUCEST) produced 43,000 transcripts which are believed to represent around 90% of the sugarcane genes and over 30% of the transcripts correspond to new genes (Vettore *et al.*, 2003; http://sucest.lbi.ic.unicamp.br/public).

The unraveling of all gene functions in such a complex organism is a formidable task and a necessary one if we are to understand the key features of this important grass. The unique capacity of sugarcane to accumulate large quantities of sucrose in its culm is indicative that novel gene functions are bound to be discovered in this organism. Preliminary information on gene function can be derived from sequence analysis alone (i.e., functional annotation; Vettore *et al.*, 2003) but additional efforts are required to confirm putative functions. When dealing with novel genes in particular, functional genomic approaches are essential to associate function to the new sequences and to identify markers of desired traits. These approaches include gene expression profiling of tissues from different varieties, as well as mapping efforts to associate genes to phenotypes. It is desirable to eventually manipulate the expression levels of these genes and test dosage effects using transgenic technology.

To increase our understanding of the complex instructions contained in the sequences obtained and how they are related to the different characteristics found in different cane cultivars a functional genomics project, SUCEST-FUN, is underway. The project aims to associate function to sugarcane genes and to identify putative targets for variety improvement (http://sucest-fun.org). A new database containing 57,339 ESTs representing 4,594 contigs selected from SUCEST was constructed to meet the new needs of the functional genomics project and is open to those involved in this new enterprise.

Plants are prone to diverse environmental challenges that cause huge annual losses in agriculture. Defining signal transduction pathways is of strategic importance to the understanding of fundamental processes such as growth and development, as well as cellular responses triggered by biotic and abiotic stresses. The SUCEST collection of 237,954 sugarcane ESTs was examined in search of signal transduction genes. Over 3,500 components involved in several aspects of signal transduction, transcription, development, cell cycle, stress responses and pathogen interaction were compiled into the Sugarcane Signal Transduction (SUCAST) Catalogue (Papini-Terzi *et al.*, 2005). Sequence comparisons and protein domain analysis revealed 477 receptors, 510 protein kinases, 107 protein phosphatases, 75 small GTPases, 17 G-proteins, 114 calcium and inositol metabolism proteins and over 600 transcription factors. The elements were distributed into 29 main categories subdivided in 409 sub-categories. Genes with no matches in the public databases and of unknown function were also catalogued. A cDNA microarray was constructed to profile individual variation of plants cultivated in the field and transcript abundance in six plant organs (flowers, roots, leaves, lateral buds, 1st and 4th internodes). From 1,280 distinct elements analyzed, 217 (17%) presented differential expression in two biological samples of at least one of the tissues tested. A total of 153 genes (12%) presented highly similar expression levels in all tissues.

A gene expression database was created using the open source BioArray Software Environment (BASE) software, which allows the storage, management, and analysis of cDNA array data using web-browsers (Saal *et al.*, 2002). A fully integrated tool is available that normalizes the data, identifies up- and down-regulated genes using the Student's t-test. In addition a new intensity-dependent algorithm, generates scatter plots, MA-plots, MS-plots, hierarchical clustering and self-organizing maps. These tools can be used either for data from experiments using radioactive or fluorescent dyes. Another tool allows the identification of SAGE (Serial analysis of gene expression) tags in the SUCEST database, and the expression profiles from these experiments can be integrated to those from cDNA arrays. Finally, a gene validation tool allows users to enter information associated with each gene (*in situ* hybridization, northern blots, RT-PCR, subcellular localization, etc.). The expression data obtained can aid in assigning function and will also be useful for promoter characterization of this and other economically important grasses. The identification of tissue-specific and of "housekeeping" genes is of great interest in expression studies, since they are valuable experimental controls and indicate promoters useful for plant biotechnology. In monocots the most frequently used promoters are from the maize ubiquitin *Ubi-1* and the rice actin *act1* genes. Even though they maintain a relatively constant expression pattern, they have distinct expression patterns in different species, cell types and cultivation conditions (Benfey and Chua, 1989; Neuteboom *et al.*, 2002). The identification of sugarcane tissue specific promoters is an important step that will allow controlled gene expression so that novel products can be accumulated in the desired part of the plant, such as the chloroplast in the leaves, or the vacuoles in stem parenchyma cells.

An additional 5,000 genes have been selected for gene expression profiling using cDNA microarrays from all categories in the SUCEST database. The SUCEST-FUN project is targeting efforts to identify differentially expressed genes in varieties with contrasting responses to insect attack, endophytic bacteria interaction, drought stress, phosphate defficiency and sucrose accumulation. Once differentially expressed DNA sequences are identified, there are a few approaches that can be taken to study sugarcane gene function, one step further. Genetic transformation offers the potential to either silence or overexpress these genes, creating the opportunity to produce new phenotypes not possible through conventional breeding. Well-known examples of this are the introduction of genes conferring herbicide and insect resistance. For many years now, genetic transformation of sugarcane has been a reality in different laboratories around the world. Varieties with herbicide (Falco *et al.*, 2000) and insect (Braga *et al.*, 2001) resistance have been produced, as well as with resistance to viral diseases (Joyce *et al.*, 1998; Ingelbrecht *et al.*, 1999). None of these have been commercialized, due either to intellectual property considerations, or concerns over public perception. These constraints on commercial release are likely to continue in the near future.

The ease with which many sugarcane genotypes can now be transformed, together with the identification of the sequences of thousands of genes that this plant expresses, raises the possibility of altering the expression of specific genes in the plant and identifying the effects this modification has on the phenotype. The development of RNAi and anti-sense technology allows for gene down-regulation. In commercial sugarcane fields, flowering is an undesirable characteristic. At the Cane Technology Center (CTC) in Piracicaba, Brazil, the anti-sense expression of a single gene found in the SUCEST database to be involved in flower development, has produced non-flowering plants of a sugarcane variety that, otherwise, flowers heavily every year (Figueiredo, 2003). Down-regulation and over-expression of genes involved in carbohydrate metabolism is also the subject of many studies at CTC and at other research institutions, with the aim of increasing the content of sucrose and other metabolites. An initial data mining survey of the large sugarcane EST database for genes involved in carbohydrate metabolism, allowed the identification of more than 150 expressed sequences representing these genes. Six of them were selected for individual down regulation in transgenic sugarcane plants, in an attempt to shunt sugar in the direction of sucrose accumulation. Preliminary results indicate that for at least three of these genes, down regulation increased by up to 30%, the sugar accumulated by transgenic sugarcane plants with no interference with their phenotype. The results of the expression studies mentioned above will certainly identify other target genes that are required to control sugar metabolism, as well as biotic and abiotic stresses.

In the last few years, sugarcane has also turned into a target for the production of novel products such as proteins with pharmaceutical properties and as a producer of biopolymers (Brumbley *et al.*, 2004). The plant is well suited for these approaches due to some of its characteristics, such as vegetative propagation, the absence of flowering in most commercial varieties, the production of a large biomass, the considerable amount of carbon partitioned into sucrose (up to 42% of the stalk dry weight), and the mobile pool of hexose sugars throughout most of its life.

ACKNOWLEDGMENTS

Part of this work was sponsored by Fundação de Amparo à Pesquisa do Estado de São Paulo (FAPESP), Conselho Nacional de Desenvolvimento Científico e Tecnológico (CNPq) and Copersucar.

REFERENCES

Al Janabi, S.M., Honeycutt, R.J., McClelland, M., Sobral, B.W. 1993. A genetic linkage map of *Saccharum spontaneum* L. 'SES 208'. *Genetics*, 134, 1249-1260.

Asnaghi, C., Paulet, F., Kaye, C., Grivet, L., Glaszmann, J.C., D'Hont, A. 2000. Application of synteny across the Poaceae to determine the map location of a rust resistance gene of sugarcane. *Theor. Appl. Genet.*, 10, 962-969.

Asnaghi, C., Roques, D., Ruffel, S., Kaye, C., Hoarau, J.Y., Télismart, H., Girard, J.C., Raboin, L.M., Risterucci, A.M., Grivet, L., D'Hont, A. 2004. Targeted mapping of a sugarcane rust resistance gene (*Bru1*) using bulked segregant analysis and AFLP markers. *Theor. Appl. Genet.*, 108, 759-764.

Aitken, K.S., Jackson, P.A., McIntyre, C.L. 2005. A combination of AFLP and SSR markers provides extensive map coverage and identification of homo(eo)logous linkage groups in a sugarcane cultivar. *Theor. Appl. Genet.*, 110, 789-801.

Benfey, P.N., Chua, N.-H. 1989. Regulated genes in transgenic plants. *Science* 244, 174-81.

Braga, D.P.V., Arrigoni, E.D.B., Burnquist, W.L., Silva-Filho, M.C., Ulian, E.C. 2001. A new approach for control of Diatraea saccharalis (Lepidoptera: Crambidae) through the expression of an insecticidal CryIa(b) protein in transgenic sugarcane. *Proc. Int. Soc. Sugar Cane Technol.*, 24(2), 331-336.

Brumbley, S.M., Purnell, M.P., Petrasovits, L.A., O'Shea, M.G., Nielsen L.K. 2004. Development of sugarcane as a biofactory for biopolymers. In: *Plant & Animal Genomes XII Conference*, 2004 San Diego, California, USA. W117. http://63.141.253.172/pag/12/abstracts/W27_PAG12_117.html

Butterfield M.K., D'Hont A., Berding, N. 2001. The sugarcane genome: a synthesis of current understanding, and lessons for breeding and biotechnology. *Proc. Soc. Afr. Sugarcane Technol. Ass.* 75, 1-5.

Cardle, L., Ramasay, L., Milbourne, D., Macaulay, M., Marshall, D., Waugh, R. 2000. Computational and Experimental Characterization of Physically Clustered Simple Sequence Repeats in Plants. *Genetics* 156, 847-854.

Cato, S.A., Gardner, R.C., Kent, J., Richardson, T.E. 2001. A rapid PCR-based method for genetically mapping ESTs. *Theor. Appl. Genet.*, 102, 196-306.

Cordeiro G.M., Taylor G.O., Henry R.J. 2000. Characterization of microsatellite markers from sugarcane (*Saccharum* sp.) a highly polyploid species. *Plant Science* 155, 161-168.

Cordeiro, G.M., Casu, R., McIntyre, C.L., Manners, J.M., Henry, R.J. 2001. Microsatellite markers from sugarcane (*Saccharum* spp.) ESTs cross transferable to erianthus and sorghum. *Plant Science* 160, 1115-1123.

Cho, Y.G., Ishii, T., Temnykh, S., Chen, X., Lipovich, L., McCouch, S.R., Park, W.D., Ayres, N., Cartinhour, S. 2000. Diversity of microsatellites derived from genomic libraries and GenBank sequences in rice (*Oryza sativa* L.) *Theor. Appl. Genet.*, 100, 713-722.

Daniels, J., Roach, B.T. 1987. Taxonomy and evolution. Pages 7-84 in *Sugarcane Improvement through Breeding*. Edited by D.J. Heinz, Elsevier press, Amsterdam.

Daugrois, J.H., Grivet, L., Roques, D., Hoarau, J.Y., Lombard, H., Glaszmann, J.C., D'Hont, A. 1996. A putative major gene for rust resistance linked with a RFLP marker in sugarcane cultivar 'R570'. *Theor Appl Genet* 92, 1059-1064.

Da Silva, J., Honeycutt, R.J., Burnquist, L., Al-Janabi, S.M., Sorrells, M.E., Tanksley, S.D., Sobral, W.S. 1995. *Saccharum spontaneum* L., 'SES 208' gentic linkage map combining RFLP- and PCR-based markers. *Molecular Breeding* 1, 165-179.

D'Hont, A., Lu, Y., Gonzalez de Leon, D., Grivet, L., Feldmann, P., Lanaud, C., Glaszmann, J.C. 1994. A molecular approach to unraveling the genetics of sugarcane, a complex polyploid of the Andropogoneae tribe. *Genome* 37, 222-230.

D'Hont, A., Rao, P.S., Feldmann, P., Grivet, L., Islam-Faradi, N., Taylor, P., Glaszmann, J.C. 1995. Identification and characterisation of sugarcane intergeneric hybrids, *Saccharum officinarum* x *Erianthus arundinaceus*, with molecular markers and DNA in situ hybridisation. *Theor. Appl. Genet.* 91, 320-326.

D'Hont, A., Grivet, L., Feldmann, P., Rao, P., Berding, N., Glaszmann, J.C. 1996. Characterisation of the double genome structure of modern sugarcane cultivars (*Saccharum* spp.) by molecular cytogenetics. *Mol. Gen. Genet.* 250, 405-413.

D'Hont, A., Ison, D., Alix, K., Roux, C., Glaszmann, J.C. 1998. Determination of basic chromosome numbers in the genus *Saccharum* by physical mapping of ribosomal RNA genes. Genome 41, 221-225.

D'Hont, A., Glaszmann, J.C. 2001. Sugarcane genome analysis with molecular markers : a first decade of research. *Proc. Int. Soc. Technol* 24, 556-559.

D'Hont, A., Paulet, F., Glaszmann, J.C. 2002. Oligoclonal interspecific origin of 'North Indian' and 'Chinese' sugarcanes. *Chromosome Research* 10, 253-262.

Doerge, R.W. Craig, B.A. 2000. Model selection for quantitative trait locus analysis in polyploids. *Proc. Natl. Acad. Sci.* USA 97, 7951-7956.

Dufour, P., Deu, M., Grivet, L., D'Hont, A., Paulet, F., Bouet, A., Lanaud, C., Glaszmann, J.C., Hamon, P. 1997. Construction of a composite sorghum genome map and comparison with sugarcane, a related complex polyploid. *Theor. Appl. Genet.*, 94, 409-418.

Dufour, P., Grivet, L., D'Hont, A., Deu, M., Trouche, G., Glaszmann, J.C., Hamon, P. 1996. Comparative genetic mapping between duplicated segments on maize chromosomes 3 and 8 and homoeologous regions in sorghum and sugarcane. *Theor. Appl. Genet.*, 92, 1024-1030.

Falco, M.C., Neto, A.T., Ulian, E.C. 2000. Transformation and expression of a gene for herbicide resistance in a Brazilian sugarcane. *Plant Cell Rep.*, 19, 1188-1194.

Figueiredo, L.H.M. 2003. Caracterização do gene LEAFY de Saccharum ssp e análise filogenética entre diferentes espécies vegetais utilizando as seqüências da família LEAFY/ Floricaula. Master´s thesis 137 p. University of São Paulo, Ribeirão Preto, Brazil

Freeling, M. 2001. Grasses as a single genetic system. Reassessment. *Plant Physiol.* 125, 1191-1197.

Gale, M.D. Devos, K.M. 1998. Plant comparative genetics after 10 years. *Science*, 282, 656-659.

Glaszmann, J.C, Dufour, P., Grivet, L., D'Hont, A., Deu, M., Paulet, F., Hamon, P. 1997. Comparative genome analysis between several tropical grasses. *Euphytica*. 96, 13-21.

Grivet, L., Daniels, C., Glaszmann, J.C., D'Hont, A. 2004. A review of recent molecular genetics evidence for sugarcane evolution and domestication. *Ethnobotany Research and Applications* 2, 9-17.

Grivet, L., Arruda, P. 2002. Sugarcane genomics: depicting the complex genome of an important tropical plant. *Current Opinion in Plant Biology* 5, 122-127.

Grivet, L., D'Hont, A., Roques, D., Feldmann, P., Lanaud, C., Glaszmann, J.C. 1996. RFLP mapping in cultivated sugarcane (*Saccharum* spp.) : genome organization in a highly polyploid and aneuploid interspecific hybrid. *Genetics* 142, 987-1000.

Grivet, L., D'Hont, A., Dufour, P., Hamon, P., Roques, D., Glaszmann, J.C. 1994. Comparative genome mapping of sugar cane with other species within the Andropogoneae tribe. *Heredity* 73, 500-508.

Grivet, L., Glaszmann, J.C., Arruda, P. 2001. Sequence polymorphism from EST data in sugarcane: a fine analysis of 6-phosphogluconate dehydrogenase genes. *Genet. Mol. Biol.*, 24, 161-167.

Grivet, L., Glaszmann, J.C., Vincentz, M., daSilva, F.R., Arruda, P. 2003. ESTs as a source for sequence polymorphism discovery in sugarcane: example of the ADH genes. *Theor. Appl. Genet.* 106, 190-197.

Guimaraes, C.T., Sills, G.R., Sobral, B.W.S. 1997. Comparative mapping of Andropogonae : *Saccharum* L. (sugarcane) and its relation to sorghum and maize. *Proc. Natl. Acad. Sci.* USA 94, 14261-14266.

Guimaraes, C.T., Honeycutt, R.J., Sills, G.R., Sobral, B.W.S. 1999. Genetic maps of *Saccharum officinarum* L. and *Saccharum robustum* Brandes & Jew. Ex Grassl. *Genet. Mol. Biol.* 22, 125-132.

Ha, S., Moore, P., Heinz, D., Kato, S., Ohmido, N., Fukui, K. 1999. Quantitative chromosome map of the polyploid *Saccharum spontaneum* by multicolor fluorescent in situ hybridisation and imaging methods. *Plant Molecular Biology* 39, 1165-1173.

Hoarau, J.Y., Offmann, B., D'Hont, A., Risterucci, A.M., Roques, D., Glaszmann, J.C., Grivet, L. 2001. Genetic dissection of a modern sugarcane cultivar (*Saccharum* spp.). I. Genome mapping with AFLP markers. *Theor. Appl. Genet.*, 103, 84-97.

Hoarau, J.Y., Grivet, L., Offmann, B., Raboin, L.M., Diorflar, J.P., Payet, J., Hellmann, M., D'Hont, A., Glaszmann, J.C. 2002. Genetic dissection of a modern sugarcane cultivar (*Saccharum* spp.). II. Detection of QTLs for yield components. *Theor. Appl. Genet.*, 105, 1027-1037.

Hogarth, D.M., Cox, M., Bull, J.K. 1997. Sugarcane improvement: past achievements and future prospects. In : *Crop improvement for the 21 st century,* Chapter 3: 29-56.

Jannoo, N., Grivet, L., Dookun, A., D'Hont, A., Glaszmann, J.C. 1999a. Linkage disequilibrium among sugarcane cultivars. *Theor. Appl. Genet.*, 99, 1053-1060.

Jannoo, N., Grivet, L., Seguin, M., Paulet, F., Domaingue, R., Rao, P.S., Dookun, A., D'Hont, A., Glaszmann, J.C. 1999b. Molecular investigation of the genetic base of sugarcane cultivars. *Theor. Appl. Genet.*, 99, 171-184.

Jannoo, N., Grivet, L., David, J., D'Hont, A. Glaszmann, J.C. 2004. Differential chromosome pairing affinities at meiosis in polyploidy sugarcane revealed by molecular markers. *Heredity* 93, 460-467.

Jarne, P., Lagoda, P.J.L. 1996. Microsatellites, from molecules to populations and back. *Trends in Molecular Ecology* 11, 424-429.

Joyce, P.A., McQualter, R.B., Handley, J.A., Dale, J.L., Harding, R.M., Smith, G.R. 1998. Transgenic sugarcane resistant to sugarcane mosaic virus. *Proc. Aust. Soc. Sugar Cane Technol.* 20, 204-210.

Ingelbrecht, I.L., Irvine, J.E., Mirkov, T.E. 1999. Posttranscriptional gene silencing in transgenic sugarcane. Dissection of homology-dependent virus resistance in a monocot that has a complex polyploid genome. *Plant Physiol.* 119, 1187-1197.

Lima, M.L., Garcia, A.A., Oliveira, K.M., Matsuoka, S., Arizono, H., De Souza, C.L. Jr., De Souza, A.P. 2002. Analysis of genetic similarity detected by AFLP and coefficient of parentage among genotypes of sugar cane (Saccharum spp.) *Theor. Appl. Genet.* 104, 30-38.

Lu, Y.H., D'Hont, A., Paulet, F., Grivet, L., Arnaud, M., Glaszmann, J.C. 1994. Molecular diversity and genome structure in modern sugarcane varieties. *Euphytica* 78, 217-226.

Ma, H.M., Schulze, S., Lee, S., Yang, M., Mirkov, E., Irvine, J., Moore, P., Paterson, A. 2004. An EST survey of the sugarcane transcriptone. *Theor. Appl. Genet.*, 108, 851-863.

Ming, R., Liu, S.C., da Silva, J., Wilson, W., Braga, D., van Deynze, A., Wenslaff, T.F., Wu, K.K., Moore, P., Burnquist, W., Sorrells, M.E., Irvine, J., Paterson, A.H. 1998. Detailed alignment of *saccharum* and *sorghum* chromosomes: comparative organization of closely related diploid and polyploid genomes. *Genetics* 150, 1663-1682.

Ming, R., Liu, S.C., Moore, P.H., Irvine, J., Paterson, A.H. 2001. QTL analysis in a complex autopolyploid: genetic control of sugar content in sugarcane. *Genome Research* 11, 2075-2084.

Ming, R. Liu, S.C., Bowers, J.E., Moore, P.H., Irvine, J.E., Paterson, A.H. 2002. Construction of a *Saccharum* consensus genetic map from two interspecific crosses. *Crop Sci.* 42, 570-583.

Mudge, J., Andersen, W.R., Kehrer, R.L., Fairbanks, D.J. 1996. A RAPD gentic map of *Saccharum officinarum*. *Crop Sci.* 36, 1362-1366.

Neuteboom, L.W., Kunimitsu, W.Y., Webb, D., Christopher, D.A. 2002. Characterization and tissue-regulated expression of genes involved in pineapple (*Ananas comosus* L.) root development. *Plant Science* 163, 1021-1035.

Papini-Terzi, F.S., Rocha, F.R., Vêncio, R.Z.N., Oliveira, K.C., Felix, J.M., Vicentini, R., Rocha, C.S., Simões, A.C.Q., Ulian, E.C., di Mauro, S.M.Z., Pereira, C.A.B., da Silva, A.M., Menossi, M., Souza, G.M. 2005. Transcription profiling of signal transduction-related genes in sugarcane tissues. *DNA Res.* 12, 43-54.

Piperidis, G., Christopher, M.J., Carroll, B.J., Berding, N., D'Hont, A. 2000. Molecular contribution to selection of intergeneric hybrids between sugarcane and the wild species *Erianthus arundinaceus*. *Genome* 43, 1033-1037.

Piperidis, G., D'Hont, A. 2001. Chromosome composition analysis of various *Saccharum* interspecific hybrids by genomic in situ hybridization (GISH). *Proc. Int. Soc. Technol* 24, 565-566.

Pinto, L.R., Oliveira, K.M., Ulian, E.C., Garcia, A.A., de Souza, A.P. 2004. Survey in the sugarcane expressed sequence tag database (SUCEST) for simple sequence repeats. *Genome* 47, 795-804.

Raboin, L.M., Oliveira, K.M., Lecunff, L., Telismart, H., Roques, D., Butterfield, M., Hoarau, J.Y., D'Hont, A. 2005. Genetic mapping in the high polyploid sugarcane using a biparental progeny; identification of a gene controlling stalk colour and a new rust resistance gene (*Theor Appl Genet.* 112: 1382-1391).

Ripol, M.I., Churchill, G.A., da Silva, J.A.G., Sorrells, M. 1999. Statistical aspects of genetic mapping in autopolyploids. *Gene* 235:31-45.

Rossi, M., Araujo, P.G., Paulet, F., Garsmeur, O., Dias, V.M., Chen, H., Van Sluys, M.A., D'Hont, A. 2003. Genomic distribution and characterization of EST-derived resistance gene analogs (RGAs) in sugarcane. *Mol. Gen. Genomics* 269, 406-419.

Rudd, S. 2003. Expressed sequencetags: alternative or complement to whole genome sequences? *Trends Plant Science* 8, 321-329.

Saal, L.H., Troein, C., Vallon-Christersson, J., Gruvberger, S., Borg, A., Peterson, C. 2002. Bioarray software environment: A platform for comprehensive management and analysis of microarray data. *Genome Biology*, 3:software0003.1-software0003.6

Scott, K.D., Eggler, P., Seaton, G., Rossetto, M., Ablett, E.M., Lee, L.S., Henry, R.J. 2000. Analysis of SSRs derived from grape ESTs. *Theoretical and Applied Genetics* 100, 723-726.

Thiel, T., Michalek, W., Varshney, R.K., Granner, A. 2003. Exploiting EST databases for the development aqnd characterization of gene-devired SSR-markers in barley (*Hordeum vulgare* L.) *Theor. Appl. Genet.*, 106, 441-422.

Tomkins, J.P., Yu, Y., Miller-Smith, H., Frisch, D.A., Woo, S.S., Wing, R. 1999. A bacterial artificial chromosome library for sugarcane. *Theor. Appl. Genet.*, 99, 419-424.

Vettore, A.L., Da Silva, F.R., Kemper, E.L., Arruda, P. 2001. The libraries that made SUCEST. *Genetics and Molecular Biology* 24, 1-7.

Vettore, A.L., da Silva, F.R., Kemper, E.L., Souza, G.M., da Silva, A.M., Ferro, M.I., Henrique-Silva, F., Giglioti, E.A., Lemos, M.V., Coutinho, L.L., Nobrega, M.P., Carrer, H., Franca, S.C., Bacci-Junior, M., Goldman, M.H., Gomes, S.L., Nunes, L.R., Camargo, L.E., Siqueira, W.J., Van Sluys, M.A., Thiemann, O.H., Kuramae, E.E., Santelli, R.V., Marino, C.L., Targon, M.L., Ferro, J.A., Silveira, H.C., Marini, D.C., Lemos, E.G., Monteiro-Vitorello, C.B., Tambor, J.H., Carraro, D.M., Roberto, P.G., Martins, V.G., Goldman, G.H., de Oliveira, R.C., Truffi, D., Colombo, C.A., Rossi, M., de Araujo, P.G., Sculaccio, S.A., Angella, A., Lima, M.M., de Rosa Junior, V.E., Siviero, F., Coscrato, V.E., Machado, M.A., Grivet, L., Di Mauro, S.M., Nobrega, F.G., Menck, C.F., Braga, M.D., Telles, G.P., Cara, F.A., Pedrosa, G., Meidanis, J., Arruda P. 2003. Analysis and functional annotation of an expressed sequence tag collection for tropical crop sugarcane. *Genome Res.* 13, 2725-35.

Vincentz, M., Cara, F.A.A., Okura, V.K., da Silva, F.R., Pedrosa, G.L., Hemerly, A.S., Capella, A.N., Marins, M., Ferreira, P.C., França, S.C., Grivet, L., Vettore, A.L., Kemper, E.L., Burnquist, W., Targon, M.L.P., Siqueira, W.J., Kuramae, E.E., Marino, C.L., Camargo, L.E.A., Carrer, H., Coutinho, L.L., Furlan, L., Lemos, M.V.F., Nunes, L.R., Gomes, S.L., Santelli, R.V., Goldman, M.H., Bacci Jr, M., Giglioti, E.A., Thiemann, O.H., Silva, F.H., Van Sluys, M.A., Nobrega, F.G., Arruda, P., Menck, C.F.M. 2004. Evaluation of monocot and eudicot divergence using the sugarcane transcriptome. *Plant Physiol.* 134, 951-959.

Wu, K.K., Burnquist, W., Sorrells, M.E., Tew, T.L., Moore, P.H., Tanksley, S.D. 1992. The detection and estimation of linkage in polyploids using single-dose restriction fragments. *Theor. Appl. Genet.* 83, 294-300.

Yu, J.K., La Rota, M., Kantety, R.V., Sorrells, M.E. 2004. EST derived SSR markers for comparative mapping in wheat and rice. *Mol. Gen. Genomics* 271, 742-751.

Genomics, Genetic Variability and Plant Breeding

25

CHAPTER

Molecular Markers and High-throughput Genotype Analysis

M. Falque, S. Santoni

DNA polymorphism is generated over the course of evolution by various molecular mechanisms producing either nucleotide substitutions or insertions or deletions of one or several bases. The frequency of these phenomena may vary greatly depending on the mechanisms involved.

In genomics studies, it is often necessary to visualize such DNA polymorphisms between different individuals of a single species, or even of different species.

This is the case, for example, each time a genetic mapping approach is used to locate a gene or another sequence on chromosomes, or to detect QTL (Chapter 27).

It is also the case in functional genomics in searching for polymorphisms between different alleles of a single gene, which will be responsible for the character variation that is hypothesized to be controlled by this gene. Statistical associations are then sought between the allele at each of the polymorphic sites detected and the phenotypic value measured for the character on the corresponding individual.

Finally, it is the case in comparative genomics, for example, when the focus of interest is the evolution and comparison of sequences constituting the genome of different species, to understand their dynamics.

It is of course possible to visualize sequence polymorphism between two DNA fragments by sequencing the fragments. However, despite advances in automation, this remains quite costly and time consuming on a large scale. At present, sequencing justifies the effort only when searching exhaustively for existing sequence polymorphisms for a given DNA fragment (for example a gene) between different genotypes. This chapter will not cover sequencing techniques properly speaking.

When it is not necessary to visualize all the sequence polymorphisms, some indirect methods much faster and cheaper than sequencing can be used: these are molecular markers.

Molecular markers are numerous and rapidly improving with the appearance of new concepts and analytical processes. The abbreviations used to name these markers are numerous, sometimes redundant or imprecise, and often incomprehensible to non-specialists. The various types of molecular markers can be classified by technical criteria based on the analytical methods used or genetic criteria based on the quality of information they generate. The two aspects are important in choosing the technique best suited to a given genomics programme, taking into account the information required and the technical or budgetary constraints of the laboratory.

The overall organization of this chapter is based primarily on genetic and molecular criteria and the various techniques are then detailed for each group of markers.

In terms of genetics, in the discussion that follows the term "marker" is synonymous with "marker locus", i.e., what indicates a unique position of the genome. Moreover, the term "dominant marker" is used when a heterozygous diploid individual cannot be distinguished from one of the two homozygotes (for example, in the case of a polymorphism of presence/absence of a single band). The term "codominant marker" is used when the three genotypes can be distinguished without ambiguity (for example, when each allele is manifested by the appearance of a distinct band).

Thus, two groups of techniques will be considered: those that allow individual visualization of markers (monolocus markers), which are generally codominant, and those that simultaneously reveal several markers, generally dominant, in the form of multilocus complex profiles. These are also called "genetic fingerprints". This classification is practical but necessarily imperfect because certain techniques considered in this chapter as producing markers detected individually can, in some cases, generate dominant markers, or may detect several markers simultaneously (e.g., RFLP).

Practically speaking, because of the large numbers of samples generally analysed in genomic programmes, the analysis throughput (number of analyses realized per unit of time) is a critical factor. High performance genotyping techniques are therefore increasingly sought, for which the speed of analysis, the potential for automation, and the cost of consumables are critical elements. In this chapter, the emphasis will thus be on techniques that produce a high yield, while others (such as RFLP) that are certainly important but have a lower yield are presented more briefly.

CODOMINANT MARKERS DETECTED INDIVIDUALLY

These molecular markers are based on the combination of various basic principles, including in particular:

— measurement of length of DNA fragments;
— detection of presence or absence of restriction sites;
— hybridization or non-hybridization between two DNA strands;
— detection of differences in DNA conformation.

Most of the time, in the context of markers detected individually, these sequence polymorphisms are detected within a DNA fragment amplified by polymerase chain reaction (PCR) from the total genomic DNA of the individual to be analysed, using a pair of specific primers of the ends of the region to be studied.

PCR is considered to be a well-known technique and will not be explained as such in this work. This technique, based on the use of a thermo-resistant DNA polymerase, can be used to synthesize a very large number of copies of a DNA fragment (up to some thousands of nucleotides in length), as long as there are single-stranded DNA fragments of around 20 bases available (called primers) complementary to the two ends of the sequence that is to be copied. The fragment is thus considered "amplified" and the important point is that this amplification is exponential. In fact, the reaction is made cyclical because of temperature variations, and each newly formed DNA strand can be used as a template for synthesis of the next cycle. PCR has virtually revolutionized techniques of genome analysis, because it requires very little DNA to start with and makes it possible to specifically "target" a single region of the genome flanked by the two primers.

The specificity of marker locus based on PCR thus relies on the primer pair used. These markers are called "sequence tagged sites" (STS).

With the exception of RFLP, all the markers described in this section are STS.

Relationships between Physical and Genetic Mapping

For plant species for which there is a physical map, composed of large cloned and ordered DNA fragments representing the entire genome, it is possible to physically map genes or markers (Wu *et al.,* 1997). It is not necessary then to search for a polymorphism, but it must be determined on which large fragments the sequence that is to be mapped is to be found. The STS, which make it possible to visualize the polymorphism, also allow PCR screening of libraries of large DNA fragments in order to determine on which clones the sequence corresponding to the marker is found. This is a way of establishing the indispensable bridges between a genetic map and a physical map. With RFLP markers, locus specificity is obtained not by PCR, but by hybridization of a DNA fragment of a target locus, or probe, on the total fragmented genomic DNA (see below). This same probe can also thus be used to screen a library of large fragments by hybridization on membranes on which small drops of DNA of each of the clones have earlier been deposited in a grid pattern. Hybridization of the labelled probe makes it possible to determine on which clone the marker is found, and so to link genetic and physical map.

Restriction Fragment Length Polymorphism (RFLP)

The restriction fragment length polymorphism (RFLP) technique involves cleavage of the DNA by restriction enzymes, then electrophoresis and molecular hybridization (Fig. 25.1).

Restriction enzymes cleave the DNA at specific sites generally comprising an even number of bases (4, 6, or 8). If the bases are random in the genome, an enzyme having a recognition site with six bases will cleave the DNA at every 4096 bases on average (4^6). A genome of 10^9 bases could thus produce around 250,000 restriction fragments of variable length. The specificity is often such that replacement of a single base in a site suffices to prevent cleavage of the DNA. It is this specificity that is exploited to detect polymorphism: a presence/absence of a restriction site leads to a fragment length polymorphism. Since this phenomenon is not rare, the digestion of the genomic DNA from any two individuals in a given species produces a very large number of differences in fragment lengths. But although it is easy to obtain polymorphic fragments, it is

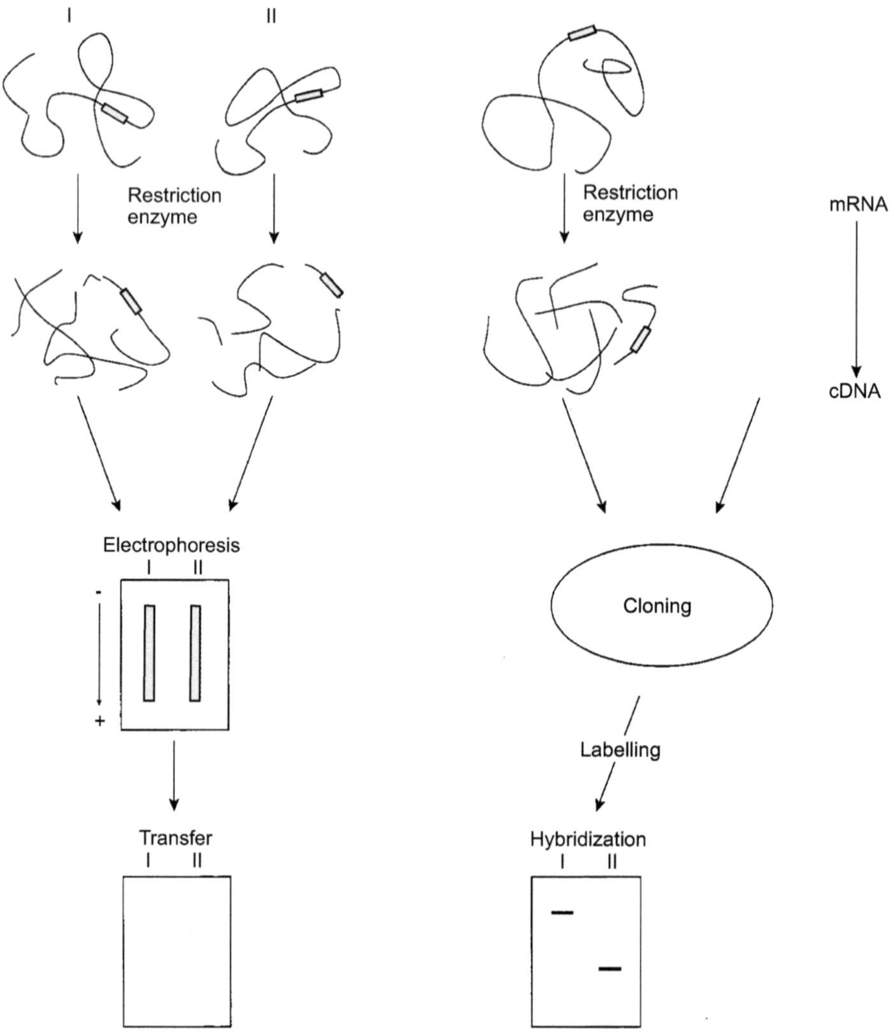

Fig. 25.1 Principle of RFLP technique. I and II represent two individuals. The DNA molecule is drawn as a filament. The grey rectangle on the DNA represents a particular region recognized by a probe. After digestion by a restriction enzyme, the fragment carrying this region is longer in individual I than in individual II. This length polymorphism will be revealed by hybridization of the probe labelled on the DNA after migration (de Vienne, 2003).

difficult to visualize them. In fact, gel electrophoresis on a large number of genomic DNA digestion products, followed by coloration, produces a continuous smear image. Particular fragments that are homologous between several individuals, and possibly allelic, can be identified only by means of molecular probes using the Southern technique (Southern, 1975).

The use of RFLP thus involves the following manipulations: Total genomic DNA extracted and purified in sufficiently large quantity from each genotype is subjected to digestion by a restriction enzyme. The fragments generated are separated according to size by agarose gel electrophoresis. The DNA is then transferred from the gel on to a membrane, conserving the relative positions of fragments. The last step is molecular hybridization with a DNA probe previously labelled either chemically or by radioactivity. The probe then hybridizes with the DNA fragment or fragments with which it presents sufficient sequence similarity. The membrane is a solid support that can be subjected to several hybridization/dehybridization cycles. It can thus be used with several probes successively (up to about 10).

Molecular probes are DNA fragments isolated and individualized by cloning or by PCR amplification. They may come from fragmented total genomic DNA (gDNA) and thus contain coding or non-coding sequences, unique or repeated, of nuclear or cytoplasmic origin. They may also be complementary DNA (cDNA) that are DNA copies of messenger RNA and correspond always to the transcribed sequences.

A probe is called homospecific (or homologous) when it corresponds to a DNA fragment isolated in the species analysed, and heterospecific (or heterologous) when it comes from the genome of a different, more or less similar species. The percentage of usable heterospecific probes decreases as the genetic distance between the two species increases, more so when the probe sequence is not subjected to selection pressure during evolution (e.g., non-coding sequence). Probes corresponding to genes (e.g., cDNA) are generally more effective for interspecific uses (Table 25.1). The heterospecific probes allow comparisons on the positions and order of loci between different species and constitute the basis of comparative gene mapping studies (Devos and Gale, 1997).

Table 25.1 Efficiency (%) of heterospecific probes in Poaceae (de Vienne, 1998)

	Oat	*Rice*	*Wheat*	*Barley*	*Sugarcane*
cDNA (oat)	100	83	65	71	60
gDNA (rice)	2	100	0	0	0

Probes of oat and rice have been hybridized on the DNA of a series of Poaceae species. With a high success rate, cDNA probes of oat hybridize on the DNA of species belonging to fairly distant tribes (rice and sugarcane). They can thus be used routinely, for example in comparative mapping studies. In contrast, the gDNA of rice recognize the DNA of oat at a low rate, and not at all with other species. This is due to the fact that the cDNA probes generally correspond to genes, the sequences of which cannot diverge too far over evolution without a loss of efficiency of their function.

RFLP-detected polymorphism results from mutation events in restriction sites and/or insertion/deletion in the sequence targeted by the probe or near it (Fig. 25.2). Since a probe reveals all the sequences that are homologous to it, a nuclear DNA probe may, in heterozygous plants, allow visualization of the two alleles simultaneously and thus produce codominant markers.

Systematic cDNA sequencing programmes (EST programmes) are used on certain plant species but generally only a small number of these cDNA have been mapped. In plants on which no physical map is available, genetic mapping is necessary in order to know the position of these genes on chromosomes, for example in studying their co-localization with QTL. The RFLP technique, which is used to visualize polymorphism outside the probe sequence, remains practically the only easily used tool when the cloned cDNA is monomorphic between the parents of mapping populations.

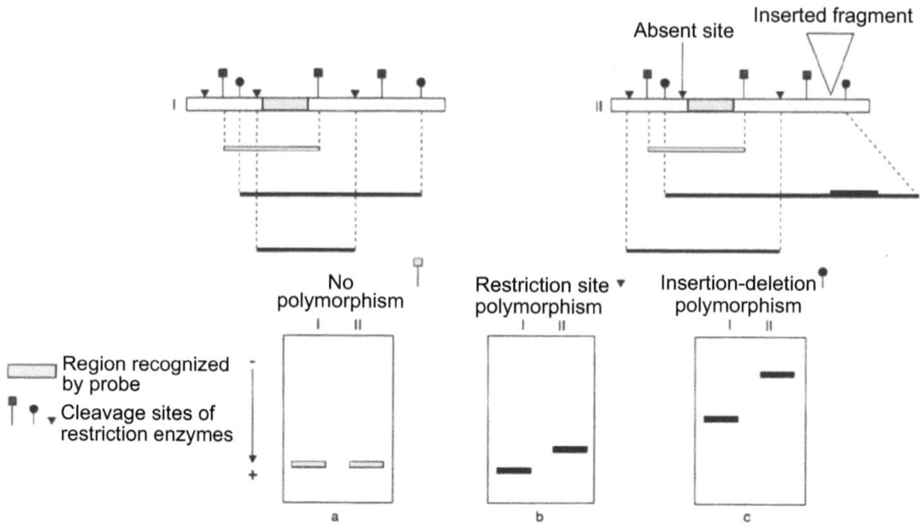

Fig. 25.2 Elementary situations of polymorphism revealed by the RFLP technique between two individuals I and II: absence of polymorphism, insertion-deletion polymorphism, and restriction site polymorphism. The cleavage sites of three restriction enzymes near a region recognized by a probe are represented by different symbols (de Vienne, 2003).

Otherwise, it is better to consider SNP mapping, which is easier and more suitable to high throughput analysis. In man and in maize, a very large number of cNDA loci have been mapped by RFLP (Botstein et al., 1980; Coe *et al.*, 2002).

The use of RFLP requires competence in molecular biology as well as a well-equipped laboratory, especially in the case of radioactively labelled probes. It is difficult to automate the steps of this technique. The quantity of DNA required per genotype to be analysed is high (a few micrograms) and the time taken to detect the result of an RFLP analysis is often several days.

Special Case of RFLP: CAPS

The cleaved amplified polymorphic sequence technique (CAPS) or PCR-RFLP consists of digesting a PCR-amplified fragment with one or several restriction enzymes, and detecting the polymorphism of presence/absence of restriction sites by measuring the resulting fragment length by electrophoresis. The choice of enzymes with 4 bases, which cleave the DNA on average every 256 bases (4^4), is imposed by the average length of PCR products, between 0.5 and 3 kb, and makes it possible to have a high cleavage probability. Like conventional RFLP, this technique provides codominant and locus-specific markers. The polymorphism, contained only with the amplified fragment, is sometimes difficult to detect (numerous enzymes must be tried). In grasses, for example, this technique has been used to convert certain RFLP probes into STS (Williams *et al.*, 1991). The PCR-RFLP technique is mostly suitable when a few markers must be analysed on a large number of plants.

SNP Genotyping

The term SNP (single nucleotide polymorphism) refers to a nucleotide site at which a substitution polymorphism has been observed with significant frequency between different individuals. The sequence information initially necessary to discover SNP can be produced specially for the purpose by sequencing PCR products obtained with the same primers from DNA of different individuals. It can also be obtained from sequences present in the databases, for example when the cDNA of a single gene have been cloned from different genotypes and sequenced (certain massive sequencing programmes of EST). In this case, a certain number of SNPs can be discovered at low cost by *in silico* studies. Moreover, such SNPs are located in transcribed sequences, which could be useful.

Like CAPS, SNP genotyping techniques described below are part of STS markers. But unlike CAPS, they allow in principle the visualization of any nucleotide substitution. Moreover, these techniques are generally developed to be used in high-throughput genotyping programmes, where the possibilities of automation as well as the rapidity of analyses are critical. For example, the long steps of electrophoresis or multiple transfers of liquids from tube to tube are preferably limited, in favour of approaches based on direct detection of fluorescence or luminescence, ideally in a single reaction tube.

The techniques described below rely on the following basic mechanisms:

- allele-specific oligonucleotide hybridization (ASOH): allele-specific PCR, 5′ nuclease test, DNA chips, microbeads;
- elongation of a DNA strand by a DNA polymerase: primer extension, pyrosequencing;
- single-strand conformation polymorphism (SSCP) or high performance liquid chromatography of double-stranded DNA (DHPLC).

The experience of large-scale use of these techniques in plants is still relatively modest. The different approaches mentioned above are described in this chapter, and each can be adapted to certain types of SNP genotypinge projects, depending on the objectives and on the equipment and finances available.

This list is far from being exhaustive, and numerous methodological studies on the subject are ongoing, motivated by significant economic implications. For example, there are techniques using DNA ligase, which can be combined with DNA chips or microbeads techniques. For a more complete and detailed survey of different methods of high throughput SNP genotyping, the synthetic article of Gut (2001) is recommended.

Allele-specific PCR

The basic idea of allele-specific PCR is that the PCR amplification is greatly reduced if the base at the very end of the 3′ end of the primers is not complementary to the template DNA. An allele-specific test can then be designed by choosing one of the two PCR primers with the 3′ end on the SNP (Fig. 25.3). This very simple test generates information on the presence or absence of the PCR product (which can be visualized by electrophoresis or direct fluorescence in the tube). In practice, things are not so simple

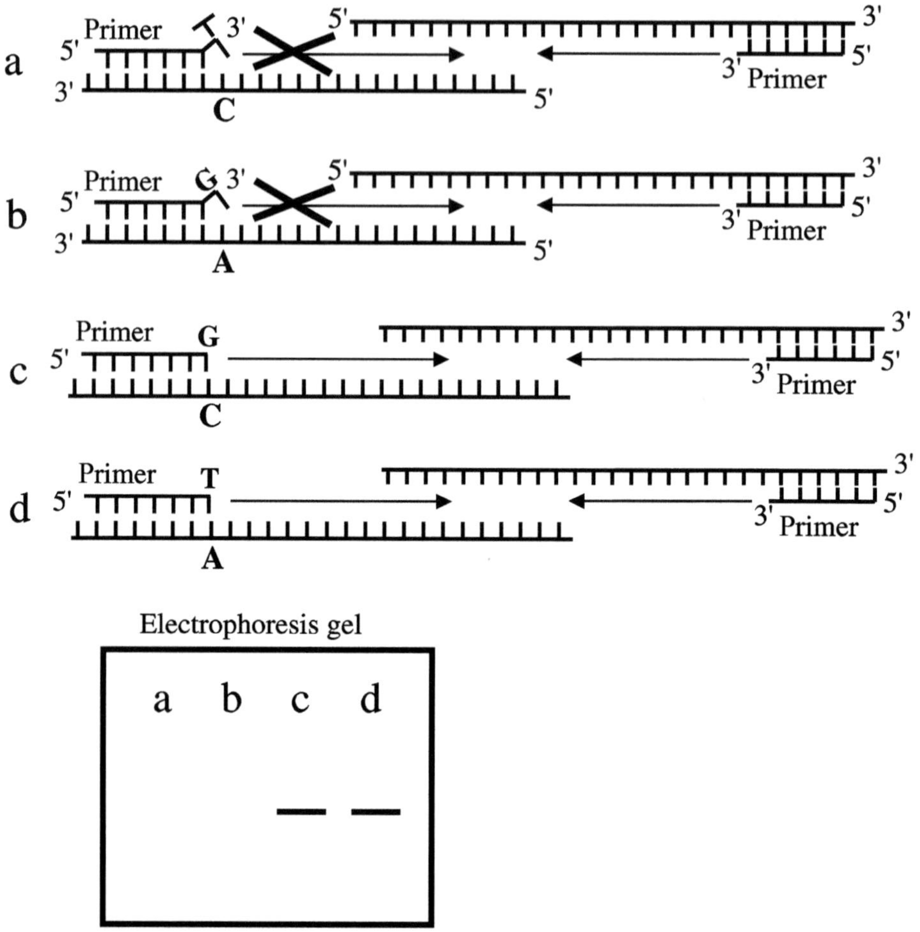

Fig. 25.3 Principle of allele-specific PCR. a, b: When the base at the 3' end of the primer is not complementary to the template DNA, the amplification does not occur and no band is obtained. c, d: When the base at the 3' end of the primer is complementary to the template DNA, the amplification occurs and a band is observed on the gel.

for two reasons: a mismatch of the 3′ end base of the primer does not always suffice to prevent the PCR from functioning; the absence of the product may be due to a certain allele of SNP, but also to a failure of the PCR for other reasons, and it may not be possible to distinguish between the two situations. The first point may be improved by optimizing the PCR conditions or by adding other mismatches to the primer to reduce the stability of its 3′ end, but such optimization must be repeated for each new SNP. The second problem can be resolved by carrying out double reactions, with for each a specific primer of one of the two alleles that is to be distinguished. Data will then be valid only if one of the primers gives a product and not the other. In case of failure of the PCR, neither of the two reactions will give a product. If the two PCRs give a product, the test lacks specificity, and the data is no longer valid. The four primers necessary for the double PCR can be mixed in a single reaction tube, as long as the sizes of products of each of the reactions is chosen carefully to be sufficiently different so that they can be distinguished by electrophoresis, and so that the absence of undesirable reactions between the primers can be verified. Such a test is simple to carry out and requires minimal molecular biology equipment, but most of the SNPs require a fairly long specific development, and the reliability is not always sufficient. The overall throughput remains low, and this technique is better suited for projects involving only a few different SNPs.

5' nuclease test assay (TaqMan™ test)

The 5′ nuclease assay uses a quantitative PCR method based on an "annex" property of certain Taq polymerases (used for PCR amplifications): the 5′exonuclease activity (Gibson *et al.*, 1996). The method is used to carry out PCR amplification and allele-specific hybridization of an oligonucleotide at the same time and in a single tube. The analysis is thus quite fast and can be easily automated. The PCR amplification is carried out in the presence of an oligonucleotide probe chosen to hybridize on the PCR product in the region containing the SNP. The probe is paired at its 5′ end with a fluorochrome "reporter" and at its 3′ end with another fluorochrome that acts as a "quencher" of fluorescence. This signifies that when the two fluorochromes are physically very close to one another, the complex is not fluorescent and when they are separated, the fluorochrome reporter appears fluorescent. During the extension of the PCR amplification, the probe hybridizes on the template strand, and when the DNA polymerase reaches the probe, its 5′ exonuclease activity cleaves it, releasing a fluorochrome reporter separated from its quencher, and thus fluorescent. The increase in fluorescence intensity is measured in a special thermocycler. When the SNP allele on the PCR product does not correspond to the probe sequence, the latter hybridizes poorly, and the Taq polymerase will not be able to degrade the probe to liberate the fluorochrome. It is possible to define a probe labelled with a different colour for each SNP allele, so that the ratio of fluorescence intensities between the two colours indicates the allele present (Fig. 25.4). This method is simple to use and its yield is high, but apart from the high cost of the equipment, the double fluorescent labelling of probes is also costly, which could be a limiting factor, particularly for projects demanding numerous SNPs on a few individuals.

DNA chips

DNA chips are small metal or glass plates on which different types of single-stranded DNA molecules can be fixed using a chemical covalent bond. Each type of DNA molecule is fixed at a precise position on the chip, on such a miniaturized scale that is possible to deposit close to a million different sequences on a surface of 1 cm^2. The DNA molecules deposited on the chip are oligonucleotides of 25 bases corresponding to the sequences flanking the SNP, with the variable base close to the medium. For each allele of each SNP, the corresponding oligonucleotide is fixed at a particular position on the chip. The various fragments (e.g., the various genes) containing the SNP to be analysed are amplified by PCR (with fluorescent markers) for a given individual, then the

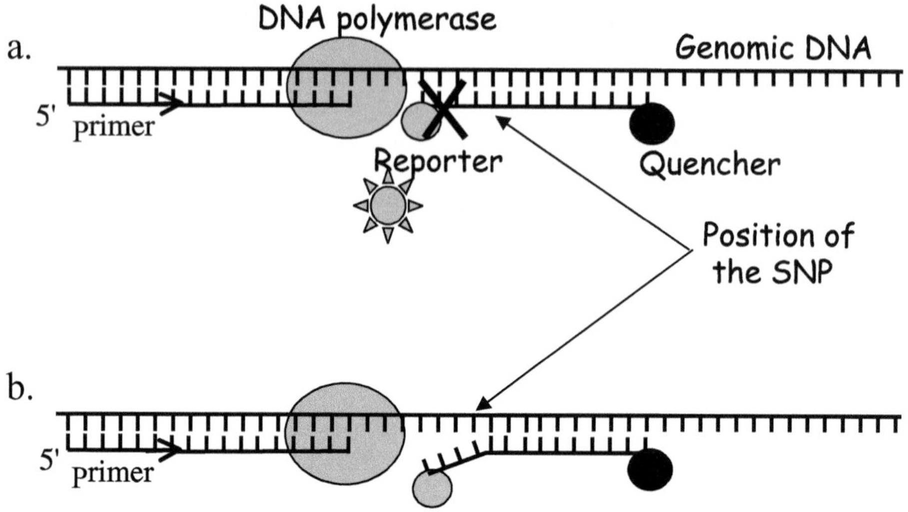

Fig. 25.4 Principle of 5' nuclease assay (TaqMan™ test): the reaction is directly effected with the genomic DNA because the PCR is carried out at the same time as the SNP allele detection. a: When the SNP allele corresponds well to the probe sequence, the sequence hybridizes completely and could thus be cleaved by DNA polymerase. This liberates the fluorochrome reporter, which then becomes fluorescent. b: When the SNP allele does not correspond to the probe sequence, it does not hybridize completely and cannot be cleaved by the DNA polymerase. The fluorochrome reporter remains paired with the quencher and does not fluoresce.

different PCR products are mixed and hybridized on the chip. After washing of non-hybridized molecules, the fluorescence intensity is measured at each position on the chip using an image analysis software. In practice, to obtain reliable results, five slightly different oligonucleotides are used for each allele of each SNP, and it each oligonucleotide must be repeated at two or three different positions of the chip.

The use of DNA chips for SNP genotyping has been described by Hu *et al.*, (2005) for humans. Since it is necessary to carry out numerous hybridizations simultaneously and in the same experimental conditions, the choice of sequences of oligonucleotides is sometimes delicate and requires suitable expertise and software. Moreover, one factor makes this technique difficult to use beyond the human domain: the very high cost of manufacturing a "custom" chip. In fact, these chips are manufactured like electronic circuits by photolithography, using masks that are extremely expensive to fabricate. Once the mask is available, however, a large number of identical chips can be manufactured at a moderate price. The process is consequently useful only when it is necessary to produce many identical chips, as could be the case with medical diagnostics. Processes designed to use a marketable "universal" chip for genotyping any SNP are being studied but to our knowledge have not yet been applied on a large scale.

Techniques using microbeads

With DNA chips, allele-specific hybridization (ASOH) is carried out on different types of oligonucleotides identified by their position on the chip. Microbead techniques also involve hybridization of a combination of different oligonucleotides, but each of them is identified by being fixed on fluorescent microbeads that emit light of a particular wavelength and intensity when they are excited by a laser. By combining 10 levels of intensity and two wavelengths, it is thus possible to distinguish between 100 types of different beads. Beads of each type are fixed on oligonucleotides in which the sequence corresponds to a given allele of a given SNP, then all the beads are mixed. With 100 different types of beads, it is thus possible to analyse simultaneously up to 50 SNPs (with two alleles per SNP) in a single tube.

The PCR fragments to be analysed can be labelled by fluorescence, then hybridized with the mixture of beads recovered from oligonucleotides, but there are variants of the technique using primer extension or oligonucleotide ligation in place of hybridization. The beads are then charged in a flow cytometer, where they pass one by one in a capillary before a combination of laser beams and fluorescence detectors. These detectors measure the fluorescence level at two wavelengths that identify the type of bead and thus indicate which allele of which SNP is being analysed, as well as the fluorescence level at the wavelength, which indicates whether the sample presents this allele or not. When numerous beads are then passed in front of the detectors (thousands in a few seconds), the average fluorescence intensity for each type of bead can be used to determine which allele is present for which SNP. The potential throughput of this technique is high, but the fine-tuning is sometimes delicate for some SNPs.

Primer extension

Primer extension is a term used to describe mini-sequencing, single-base extension, or GOOD assay (Sauer *et al.*, 2002). These techniques are often combined with mass spectrometry (Fig. 25.5) or with electrophoresis with an automated DNA sequencer. In these techniques, the fragment to be analysed is first amplified by PCR (STS approach), then used as a template for the synthesis of a mini-strand of DNA from a primer that hybridizes immediately before the SNP. This reaction is carried out by a DNA polymerase in the presence of normal nucleotides (dNTP) and/or modified nucleotides (ddNTP). The latter stop the elongation because they cannot fix themselves on the following nucleotide, while the dNTP allow indefinite elongation. In mini-sequencing, one nucleotide in the form of ddNTP and a combination of three others in their normal form (dNTP) are added to the reaction. A nucleotide is incorporated if it is complementary to the base located on the template just after the 3′ end of the primer. If it is a ddNTP, no other nucleotide can polymerize after it, and the elongation of the primer is limited to a single base. On the other hand, if the nucleotide incorporated is a dNTP, the polymerization reaction can continue with the addition of other nucleotides, until the ddNTP is incorporated. The SNP allele is manifested thus by a difference of size of the elongation product. This difference is then visualized by mass spectrometry or electrophoresis.

A variant of this method is single-base extension. In this case, only ddNTP are added to the elongation reaction, so that no matter what the SNP allele, the primer will be elongated by only one base. But it is possible to distinguish subsequently which ddNTP (among the four possible) was incorporated. For this purpose, ddNTP labelled with fluorochromes of different colours are used, or the mass of the extension product is

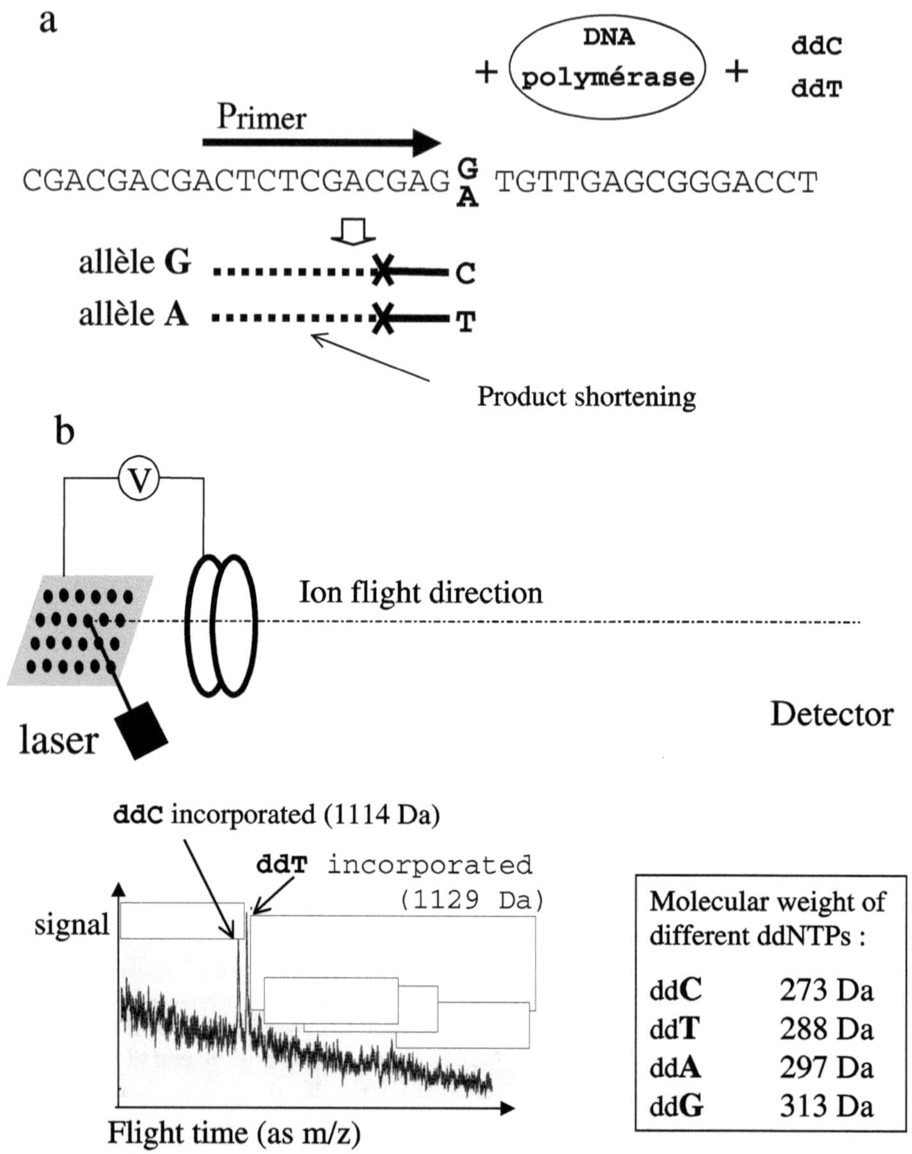

Fig. 25.5 SNP genotyping by extension of primers and MALDI-TOF mass spectrometry. a: The primer, the 3' end of which is found just before the SNP, is lengthened by a single base by incorporation of a dideoxynucleotide that thus depends on the SNP allele. The product of elongation is then cut short by the action of a phosphodiesterase, so that only three or four bases are conserved at the 3' end, which are protected by phosphorothioate bonds. b: The molecules obtained are volatilized and ionized by a laser, then the ions are accelerated by an electric field and cross a flight tube until they reach a detector. The equipment very precisely measures the flight time of ions, which is proportionate to the mass/charge ratio (m/z) of the ion. In this way, we can recognize which ddNTP has been incorporated (de Vienne, 2003).

measured very precisely (after shortening of the 5′ end of the product) using a mass spectrometry that makes it possible to distinguish the differences in mass of different ddNTP. This last technique was used routinely at the Centre National de Genotypage (France) with an extremely high throughput. It gives good results but can be used only with large structures to justify the investment involved.

Pyrosequencing

With pyrosequencing, a DNA polymerase synthesizes a DNA strand on which the SNP is to be analysed. The reaction begins with an oligonucleotide primer that hybridizes slightly upstream of the SNP, and the peculiarity of the system is that the four nucleotides (dNTP) are provided one by one sequentially to the reaction. If the nucleotide provided is complementary to the template strand, it will be incorporated at the newly formed strand by liberating a pyrophosphate that will then be detected via a cascade of enzymatic reactions (Fig. 25.6a) ending in a luminous emission. If the nucleotide provided is not complementary to the template, it will not be incorporated and there will be no light emission. The four dNTP are thus provided cyclically until 20 to 30 nucleotides have been incorporated. Before each new addition of dNTP, the remaining molecules of the preceding dNTP are destroyed by the action of an enzyme. The intensity of luminescence emitted after the addition of a dNTP is proportionate to the number of times this base has been incorporated, that is, the number of identical repeated bases on the template, complementary to the dNTP provided. The reactions occur in an automated apparatus that adds the different reagents sequentially and records the light emission over time, as well as the time each nucleotide has been added. It is thus easy to reconstitute the entire nucleotide sequence of a newly synthesized strand (Fig. 25.6b). This is nearly a complete sequencing method, but the length of the sequence read cannot exceed 20 to 30 bases, against 500 to 800 for conventional sequencing. The technique offers a high yield, is very easy to implement, and provides reliable information, but apart from the cost of the apparatus, the consumables required for the analysis are costly, which could limit the usefulness of the technique for large-scale projects.

Single strand conformation polymorphism (SSCP) or stability polymorphism (S/TGGE)

The single strand conformation polymorphism (SSCP) technique (Orita *et al.* 1989) is also included among STS markers because it applies to the analysis of PCR amplification products. When such a fragment of double-stranded DNA is heat-denatured at 95°C, then rapidly cooled, the single-stranded molecules have no time to reconnect with each other but form a stable secondary structure by intramolecular reassociations at complementary sequence zones. The sequence differences can lead to conformation polymorphism, which is detected by migration in non-denaturing conditions in an acrylamide gel. In small DNA fragments, most of the sequence polymorphisms are detectable, but this proportion diminishes when the length of the fragment increases. In plants, SSCP has been used to map ESTs in beetroot (Schneider *et al.*, 1999) or to analyse the gene flows between oak populations (Bodenes *et al.*, 1996). The SSCP process is interesting because it does not require prior knowledge of the position of the SNP or of

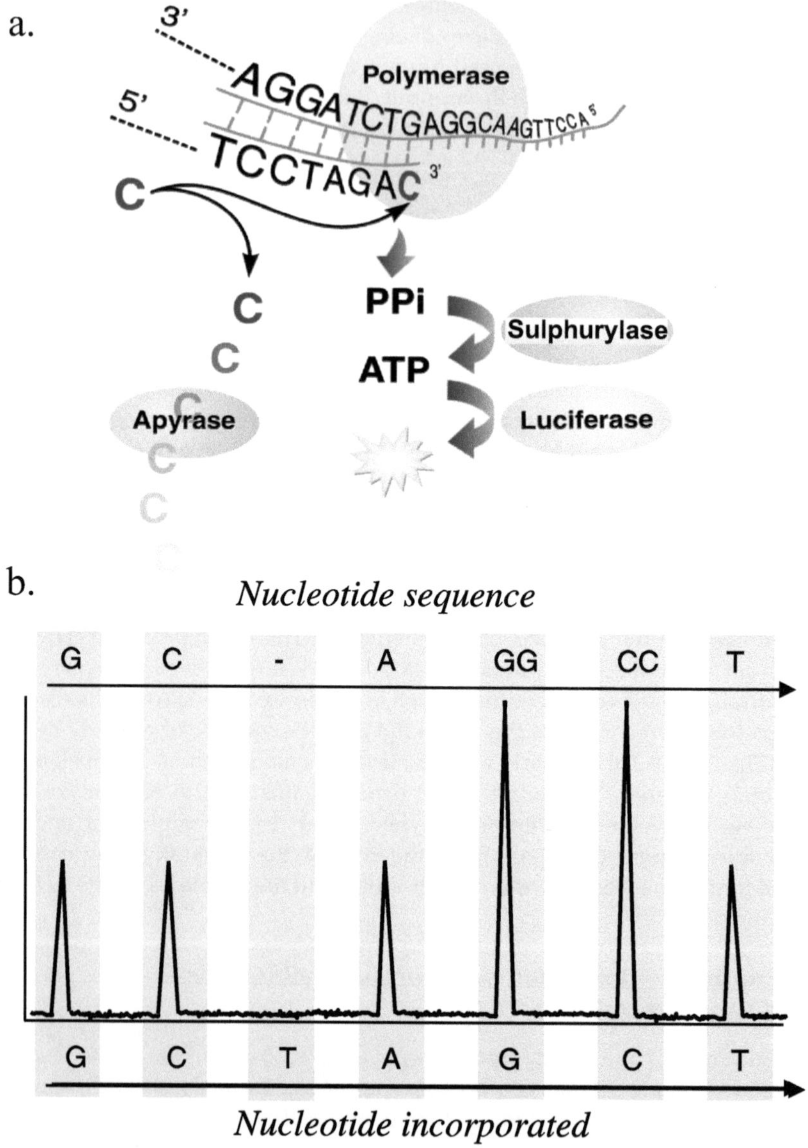

Fig. 25.6 SNP genotyping by pyrosequencing (http:www.pyrosequencing.com). a: The primer is elongated by a DNA polymerase that copies the template strand. Each type of dNTP is introduced sequentially. If the dNTP is incorporated, there is liberation of pyrophosphate, which is converted into ATP and then detected in the form of a peak of light emission. If the dNTP is not complementary to the template, there is no light emission. All the remaining dNTP are then destroyed by an apyrase, then the cycle begins with another dNTP. b: When the order of introduction of each dNTP is known, it is easy to reconstitute the sequence of the template according to the luminescence profile recorded by the apparatus. When the dNTP is incorporated several times at consecutive positions, the peak is so much higher.

the sequence that flanks it, but it is rarely used mostly because of difficulties in controlling the conditions of electrophoresis. There are also techniques based on temperature gradients (TGGE) or denaturing agent (DGGE), but these are longer and difficult to carry out.

Denaturing HPLC (DHPLC)

Denaturing HPLC uses reverse phase HPLC at a precisely controlled temperature to separate the perfect duplexes of DNA from heteroduplexes containing one or several mismatches. To detect a mutation with a SNP, the PCR product to be analysed is mixed with the PCR product of the reference individual. If the sample contains exactly the sequence of the reference individual, all the double-stranded molecules will be perfect homoduplexes, and there will be only one elution peak. If the sample includes at least one substitution as compared to the sequence of the reference individual, half of the molecules will be homoduplexes, and the other half will be heteroduplexes, which gives two distinct elution peaks. To obtain a clear separation between the peaks, the temperature of the column as well as the conditions of the gradient must be defined precisely for each sequence. Software and specially adapted HPLC equipment are available for such applications. The analysis throughput is generally better than with SSCP, and the primary interest of this approach is that it requires no prior knowledge of the position of SNP. This method can be used in functional genomics to find mutations on a given gene from a collection of plants treated by a mutagenic agent.

Capillary electrophoresis in temperature gradient

Recently, a technique of capillary electrophoresis has been developed to maximize the differences of rates of electrophoretic migration between two sequences of the same size, one of which has a mismatch. As with DHPLC, the sample to be analysed is mixed with a reference sample and the mixture is analysed. This technique makes it possible to distinguish all the SNP polymorphisms on fragments of several hundreds of bases with no prior knowledge of the sequence, which constitutes a certain advance over SSCP or DHPLC. Moreover, the analysis throughput from this type of equipment is very high because of the use of 96, or even 384 capillaries in parallel.

Simple Sequence Repeats (SSR)

Microsatellite sequences, or simple sequence repeats (SSRs), are made up of tandem repeats of mono-, di-, tri-, or tetranucleotide. The most common are (A)n, (AT)n, (GA)n, (GT)n, (TAT)n, (GATA)n, etc., the value of n ranging from a few units to several tens. Such motifs are abundant in the genome of eukaryotic organisms. In higher plants, the mass sequencing programmes make it possible to evaluate the abundance of SSRs in coding and non-coding portions of the nuclear genomes of different species (Morgante *et al.,* 2002). SSR have also been identified in the chloroplast genome of various angiosperm and gymnosperm species (Vendramin *et al.,* 1996).

Apart from their distribution throughout the genome, the advantage of SSRs in genetics lies in their extremely high polymorphism (Nakamura *et al.,* 1987). The polymorphism pertains to the number of repeat units that constitute the microsatellite sequence (Tautz, 1989). This is subject to mechanisms of rapid evolution of the tandem repeat sequences, such as asymmetrical recombinations and, probably and most likely, errors due to "slippage" of the polymerase during chromosome replication. The rates of mutation of SSR are about 4.10^{-4} to 5.10^{-6} per allele and per generation (Primmer *et al.*, 1996). This high rate, compared to other genomic DNA sequences, explains why SSRs are very rarely present in coding sequences, since they are too unstable and greatly counter-selected (Jarne and Lagoda, 1998).

A given SSR is not locus-specific, but the flanking sequences are. By means of PCR, a pair of specific oligonucleotide primers of these flanking regions can be used to amplify this single SSR locus and thus make available a sequence tagged microsatellite site (STMS Bechman & Soller, 1990). The length polymorphism is revealed by acrylamide or agarose gel electrophoresis at high resolution. Figure 25.7 is a diagram of the principle of SSR genotyping.

The development of SSR, that is, the definition of specific flanking primers, is a painstaking task. Conventionally, a genomic library must be put together and screened in order to find clones that carry SSR. These clones are sequenced, primers are designed from their sequence, and the conditions of amplification and detection of polymorphism are perfected. Protocols for elaboration of libraries rich in SSR sequences are used to facilitate this task of cloning and screening (Kijas *et al.*, 1994). The larger genetics programmes based on microsatellites (human, animal, and plant genomes) show that, to obtain one or two usable SSRs, five must be sequenced. It is, however, possible to detect sequences that carry SSR motifs simply by querying the databases of published DNA sequences. For each plant, 1% to 5% on average of available gene sequences or cDNA sequences contain SSRs (Kantety *et al.*, 2002). These are nearly exclusively located in non-coding regions: introns, sequences upstream of the start codon, or non-translated 3′ region (Milbourne *et al.*, 1998).

SSRs have been transferred to similar species of the same genus in annual plants (Peakall *et al.*, 1998) and forest trees. Within a very similar group of species, the zones flanking the SSR are sufficiently conserved, which allows transfer of amplification primers. In species of other genera of the same family, the obtaining of amplification of

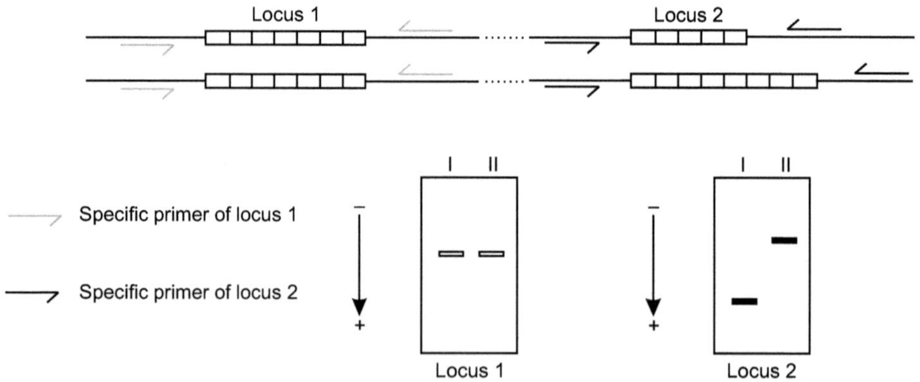

Fig. 25.7 Principle of microsatellite genotyping. Using targeted PCR primers, the polymorphism of different microsatellite loci can be revealed individually. The differences of size between alleles will then be visualized by electrophoresis (de Vienne, 2003).

SSRs and their polymorphism is less successful. Conservation of these sequences rapidly disappears with increasing genetic distance of taxa. In *Eucalyptus* species, the transfer of amplification and polymorphism of some microsatellite loci tested was 100% in the same section, and only 50% or 0% in species of other sections (Byrne *et al.*, 1996). However, rather generalist primers are used to amplify chloroplastic microsatellite loci in various species of the Pinaceae family (Vendramin *et al.*, 1996).

SSRs are among the most efficient markers to detect polymorphism. They are increasingly often used for numerous studies. They are excellent genetic markers, locus-specific, codominant, and highly polymorphic. They essentially label the non-coding regions, but they are quite uniformly distributed on the genome. They can be manipulated by PCR and their visualization can be automated by means of automated sequencers. However, their low portability between species, even similar ones, necessitates sometimes considerable development effort for each species.

MULTIPLE ARBITRARY AMPLICON PROFILING: GENETIC FINGERPRINTING

Unlike the techniques described above, which all attempt to detect polymorphism in or near a particular sequence (locus-specific technique), the techniques described under the term MAAP (multiple arbitrary amplicon profiling), all based on PCR, do not "target" a particular region of the genome fixed in advance. Nearly always, they simultaneously detect several loci (up to 100), which often makes it possible, in a few experiments, to characterize a genotype without ambiguity. They are naturally used to realize rapid genotyping (or fingerprinting), but also serve for genetic mapping and each time a particular region of the genome must be saturated with markers.

Randomly Amplified Polymorphic DNA (RAPD)

The principle of randomly amplified polymorphic DNA (RAPD) consists of a PCR reaction on the DNA of the individual studied using a short primer, of generally 10 nucleotides, of arbitrary sequence. The primer will hybridize each time a sequence is found in the DNA that is complementary to it (or includes a limited number of mismatches). If two hybridization sites are similar to one another (at least 3000 bp) and in opposite directions, that is, in a configuration allowing the PCR, the amplification will take place. If one of these two sites is absent in another individual, there will be no amplification and a presence/absence polymorphism will be observed. RAPD amplification products, which rarely exceed about 10, are generally separated by agarose gel electrophoresis.

In most cases, the polymorphism detected by RAPD is of the presence/absence kind: for a given locus, there is or is not fragment amplification. In these conditions, the homozygotes for the "absence of band" allele are detected without ambiguity, while the presence of the band does not allow us to distinguish between the heterozygote and the homozygote for the "presence of band" allele. These markers are therefore dominant.

When a RAPD marker has been detected near an interesting gene, for example a gene that is useful to follow over the course of generations in a breeding programme, it can be converted into an individually detected marker specific for the locus considered. For this, the RAPD amplification band of interest is extracted from the gel. The DNA is then cloned and sequenced. Primers of about 20 bases are then defined in order to amplify the locus marker specifically. It then becomes a SCAR marker (sequence characterized amplified region). The various techniques of polymorphism detection within the STS amplification products could be used.

RAPD and similar techniques, arbitrarily primed PCR (AP-PCR) or DNA amplification fingerprinting (DAF) using arbitrary primers, often present significant problems of reproducibility because of the low temperature of hybridization used in PCR.

Amplified Fragment Length Polymorphism (AFLP)

The amplified fragment length polymorphism (AFLP) technique is based on the joint indication of polymorphism of restriction sites and hybridization of arbitrary primers (Vos. *et al.*, 1995). It was perfected by the Netherlands company Keygene for initial use in plant improvement and is covered by a patent. Figure 25.8 summarizes the principle of AFLP. The purified genomic DNA is first cleaved by two restriction enzymes (generally EcoR1 and MseI) having recognition sites respectively of six and four bases. Known sequence adaptors of 10 to 15 bases, adapted at the sticky ends of the restriction sites, are added to the ends of DNA fragments by a ligation reaction. A primary amplification, called pre-selective, amplifies the DNA fragments using primers corresponding to the sequence of adaptors and the restriction site with a 3′ end of an arbitrarily defined base added to it. A mismatch of this terminal 3′ base of the primer prevents amplification. In this reaction, only 1/16 of the initial restriction fragments are amplified. The second amplification, called selective, uses primers identical to the first, with 1, 2, or 3 supplementary bases at their 3′ end. A subset of fragments is then amplified. The final amplification products are separated by electrophoresis in denaturing acrylamide gel (sequence gel). They can then be visualized by silver nitrate staining or detected by a radioactive or fluorescent marker during the second amplification. The profiles are generally complex (50 to 100 bands). The number of arbitrary bases of primers of the selective amplification determine the number of bands produced as a function of the genome size of the organisms studied.

It is theoretically possible to use several tens of combinations of restriction enzymes, at sites of four to six bases and a large number of combinations of selective bases on the amplification primers. The restriction-amplification combinations are nearly infinite. The AFLP technique allows detection of an important polymorphism in all plants, essentially of the presence/absence type, and can produce markers derived from all the regions of the genome. It constitutes a technique of choice for rapidly labelling a large number of loci in a genome. This technique thus allows us to carry out highly informative genetic fingerprinting in order to compare genetically very similar genotypes, to easily chart genetic maps, or to saturate a particular region of the genome.

As with RAPD, SCAR can be derived from AFLP markers.

The AFLP technique requires more technical know-how than RAPD but generates more markers and more polymorphs, and with greater reliability. Numerous studies of diversity or mapping use the AFLP procedure.

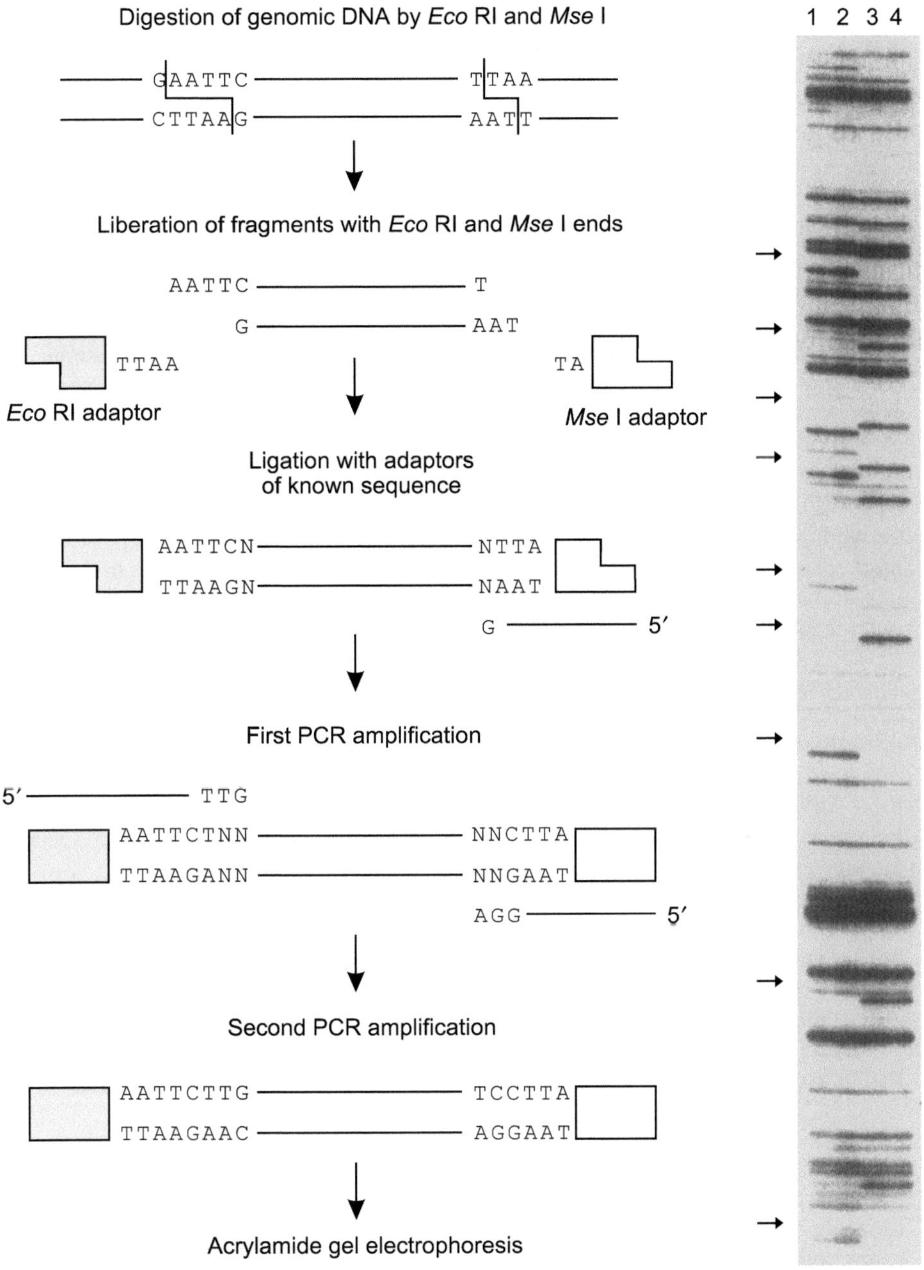

b

Fig. 25.8 Principle of AFLP. Profiles of two genotypes of the ecotype Columbia (a, tracks at left), and of two genotypes of the ecotype Landsberg (b, tracks at right) of *Arabidopsis thaliana* (de Vienne, 2003).

Several genotyping techniques are derived from the AFLP technique. Two of them, cDNA-AFLP and sequence-specific amplification polymorphism (SSAP Waugh *et al.*, 1997) focus on polymorphism at the gene level. The first, cDNA-AFLP, involves an AFLP analysis from cDNA populations (Bachem et al., 1996; Money *et al.*, 1996). Therefore, it is not genetic markers but rather physiological markers that are sought by studying the expression of genes. The latter is based on conventional AFLP, and the first steps are identical. However, the selective amplification is carried out using an AFLP primer with two or three flanking bases at the 3′ end coupled with a primer defined in a known sequence. This could be a retrotransposon sequence or a conserved motif of a multigenic family. In these cases, all the amplification fragments correspond to flanking parts of the sequences that are targeted. The SSAP with retrotransposons is used to generate numerous markers from DNA sequences considered relatively unstable. The SSAP profiles are used to visualize modifications of insertion sites of retrotransposons. They constitute an interesting tool to search for polymorphism between very similar plants, possibly for intra-clonal characterizations. SSAP with primers defined in the repeat gene sequences (resistance genes, protein kinases, transcription factors, etc.) allow us to generate, as easily as for a conventional AFLP analysis, markers that are all derived from potentially coding sequences. The elaboration of genetic maps with these markers will permit, for example, testing of "candidate gene" hypotheses in the course of searches for QTL. In the same manner, the selective amplification of microsatellite polymorphic loci (SAMPL) associates microsatellite sequences with an AFLP primer.

The ISSR Technique

A more simple exploitation of SSR is mass detection, following the RAPD principle. For this, a primer made partly of a microsatellite sequence and partly of arbitrary bases is used. Two types of primers can be used, depending on the relative positions of these two parts (Fig. 24.9). The PCR will amplify flanking fragments of SSR. In specially adapted amplification conditions, the ISSR technique (Inter SSR-PCR) is used to produce several tens of products that are visualized either on agarose gels or on acrylamide gels. The polymorphism detected is essentially of the presence/absence type, as with RAPD, but sometimes corresponds to fragment length polymorphism, as with SSR. The ISSR markers prove highly polymorphic. They make it possible to distinguish genetically very similar varieties (Prevost and Wilkinson, 1999).

PRINCIPAL APPLICATIONS OF DIFFERENT TYPES OF MARKERS

The two types of criteria to be taken into account in selecting a molecular marker involve partly the kind of genetic information that can produce different markers (polymorphism, dominance, number of loci revealed simultaneously, etc.) and partly the technical constraints linked to their use. These two questions must of course be examined in light of the particular objectives of the project and resources of the laboratory (equipment, available plant material, cost, etc.). In this summary, as before, we will distinguish successively the case of markers detected individually and that of profiles detecting multiple loci.

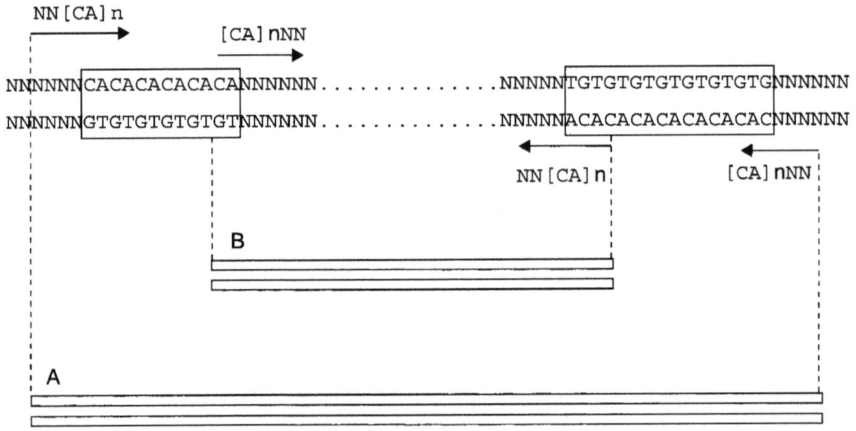

Fig. 25.9 ISSR markers. If the arbitrary bases of the primer are on the 5' side, one obtains the amplification product A. If they are on the 3' side, one obtains product B.

Techniques Visualizing One Locus at a Time

For gene mapping, these markers have the advantage of being specific to a given locus. They can thus easily be exchanged between laboratories as markers of a particular position of the genome. Moreover, it is often useful to map a particular sequence, for example a gene. For this, techniques of the STS type, based on PCR, are suitable, particularly different techniques of SNP genotyping. Any kind of polymorphism present in the gene (nucleotide substitution, etc.) can be visualized using these techniques, without it being necessary for the gene to contain a particular type of sequence (e.g., microsatellite type). The RFLP technique, even though time-consuming, low in yield, and requiring a great deal of plant material, is the only one that allows visualization of the polymorphism outside the gene that is to be mapped (but very close to it), when the gene itself is monomorphic. Moreover, these locus-specific markers are generally codominant, and thus allow validation of the genetic information contained in any type of mapping population. In the particular case of SSR, the very high level of polymorphism is also a critical asset for mapping.

For marker-assisted selection, the objective is to analyse a large number of individuals for some markers that are previously identified, for example because they flank a QTL that needs to be investigated. It is thus generally preferable to use techniques visualizing a single locus, since there is a great chance that other loci that are associated with it in a multilocus profile will not be of interest. Markers of the STS type (e.g., microsatellites or SNP), easy to genotype on numerous individuals and working with rudimentary DNA extractions, are the best suited for the purpose.

When researchers are looking for statistical associations between a SNP and a phenotypic character, they use this type of marker, since it targets a particular gene. In light of the number of samples to be genotyped to obtain significant results, high-yield SNP genotyping techniques will be most suitable because they are particularly effective in analysing SNP on a large number of samples. Techniques using a MALDI-TOF mass spectrometer are presently one of the best technical solutions for this type of programme, but they involve a very high investment and are generally justified only in service organizations such as the Centre National de Genotypage at Evry (France). In smaller laboratories, the primer extension techniques, 5′ nuclease assay or pyrosequencing, give satisfactory results. One important factor in choosing between these solutions is the cost of consumables in the various approaches, for which different limits are fixed depending on the objectives in number of individuals and of SNP to be analysed.

In population genetics, when the allelic richness of a population must be evaluated for certain given loci, to study the structuration of the populations, etc., "mono-locus" techniques generating codominant markers are also recommended. The microsatellites here will have the advantage of being extremely variable, and it is not rare to find more than 50 different alleles at a single microsatellite locus, as long as the population sample is sufficiently large. In certain very particular cases, this could be a disadvantage because two alleles observed at the same gel position are sometimes not identical by descent, which could falsify the interpretations.

To identify individuals or study the genetic diversity of a species, multi-locus techniques are inherently better adapted, as will be seen later, but two aspects may still bring about a preference for microsatellite markers: the very high level of polymorphism of microsatellites sometimes renders a single microsatellite locus more informative than a large number of AFLP bands, and the implementation of microsatellites is easy (once the difficult phase of marker development has been carried out), particularly when very little plant material is available and a quick and rudimentary method of DNA extraction must be used.

Techniques for Simultaneous Visualization of Several Loci

For reasons of reproducibility and efficiency, multilocus markers used at present are mostly AFLP, and rarely RAPD.

These markers are particularly indicated for identifying individuals, when the position on the genome of different loci visualized is not of concern. Thus, with a minimum of effort and cost, information is obtained about the whole genome.

On the other hand, these markers are generally dominant, which limits their informativity about heterozygous genotypes. Moreover, the level of polymorphism of each of the loci detected is generally low, which could be compensated for by the number of loci. An interesting exception: the ISSRs combine the advantage of a multi-locus technique with the high polymorphism of this type of microsatellite. This technique is thus interesting on species with low polymorphism but is still not widespread, and there is little information on its reproducibility.

Another important field of application of multi-locus markers is genetic mapping, when a large number of markers are sought for quick construction of a map. AFLP markers are nearly indispensable when a genetic map is to be made more dense (fine mapping) in a particular region, for example "bulk segregant analysis". Nevertheless, these markers being dominant, they will give more information on fixed populations (recombinant lines) than on very large populations used in fine mapping, which are generally not fixed.

Additional Technical Considerations

Generally, all the STS techniques based directly on PCR require little DNA and support rudimentary methods of extraction producing DNA of poor quality. AFLP, on the other hand, requires sufficiently clean and abundant DNA to ensure good digestion by restriction enzymes and good ligation of adaptors. This necessarily involves more constraints on the mode of extraction of the plant material and more complicated manipulations to extract the DNA. The case of RFLP is extreme, since there is no PCR, and a large amount of DNA of very good quality is required to carry out the analyses.

Moreover, the RFLP technique is difficult to use on numerous samples (membrane production) and requires a well-equipped laboratory well versed in all the basic techniques of molecular biology.

In genomics programmes, the samples to be analysed are often large in number, and the idea of analysis throughput (number of analyses carried out per unit of time) is of primary importance). When a large rate of analyses must be carried out, the possibilities of automation offered by different techniques is an important criterion because, apart from the gain in handling time, automation considerably limits the risks of error. In practice, simple robots are capable only of transferring liquids from one tube into another, or rather from one plate to another: in fact, in high throughput techniques, isolated tubes are not handled, but rather microtitre plates composed of 96 or even 384 wells, each containing one sample. These plates can be identified by bar codes that the robot can recognize. All the steps that require anything other than transfer of liquid from plate to plate may thus pose problems of automation. This is the case for example of loading electrophoresis gels. Therefore, when possible, it is better to use capillary electrophoresis, for which the apparatus will directly sample the samples in the plate, or even better, homogeneous assays that are carried out integrally in a tube, with for example a reading of the fluorescence level directly in the tube.

REFERENCES

Bachem, C.W., van der Hoeven, R.S., de Bruijn, S.M., Vreugdenhil, D., Zabeau, M., Visser, R.G. 1996. Visualization of differential gene expression using a novel method of RNA fingerprinting based on AFLP: analysis of gene expression during potato tuber development. *Plant J.*, 9(5), 743-753.

Beckman, J.S., Soller, M. 1990. Toward a unified approach to gene mapping of eukaryotes based on sequence tagged microsatellite sites. *Bio/Technology*, 8, 930-932.

Bodenes, C., Laigret, F., Kremer, A. 1996. Inheritance and molecular variations of PCR-SSCP fragments in pedunculate oak (*Quercus robur* L.). *Theor. Appl. Genet.*, 93, 348-354.

Bostein, D., White, R.L., Skolnic, M., Davis, R.W. 1980. Construction of a genetic linkage map in man using restriction fragment length polymorphism. *Am. J. Hum. Genet.*, 32, 314-331.

Byrne, M., Marquez-Garcia, M.I., Uren, T., Smith, D.S., Moran, G.F. 1996. Conservation and genetic diversity of microsatellite loci in the genus *Eucalyptus*. *Aus. J. Botany*, 44, 331-341.

Coe, E., Cone, K., McMullen, M., Chen, S.S., Davis, G., Gardiner, J., Liscum, E., Polacco, M., Paterson, A., Sanchez-Villeda, H., Soderlund, C., Wing, R. 2002. Access to the maize genome: an integrated physical and genetic map. *Plant Physiol.*, 128(1), 9-12.

De Vienne, D., Santoni, S. & Falque, M. 2003. Principal Sources of Molecular Markers. *In* Molecular Markers in Plant Genetics and Biotechnology. D. de Vienne Ed. Science Publishers, USA

Devos, K.M., Gale, M.D. 1997. Comparative genetics in the grasses. *Plant Mol. Biol.*, 35, 3-15.

Gibson, U.E., Heid, C.A., Williams, P.M. 1996. A novel method for real time quantitative RT-PCR.. *Genome Res* 6(10), 995-1001.

Gut, I.G. 2001. Automation in genotyping of single nucleotide polymorphisms. *Hum Mutat.* 17(6), 475-92.

Hu, N., Wang, C., Hu, Y., Yang, H.H., Giffen, C., Tang, Z.Z., Han, X.Y., Goldstein, A.M., Emmert-Buck, M.R., Buetow, K.H., Taylor, P.R., Lee, M.P. 2005. Genome-wide association study in esophageal cancer using GeneChip mapping 10K array. *Cancer Res.* 65(7), 2542-6.

Jarne, P., Lagoda, P. 1996. Microsatellites, from molecules to populations and back. *Trends Ecol. Evol.*, 11, 424-429.

Kantety, R.V., La Rota, M., Matthews, D.E., Sorrells, M.E. 2002. Data mining for simple sequence repeats in expressed sequence tags from barley, maize, rice, sorghum and wheat. *Plant Mol. Biol.*, 48(5-6), 501-510.

Kijas, J.M.H., Fowler, J.C.S., Garbett, C.A., Thomas, M.R. 1994. Enrichment of microsatellites from citrus genome using biotinylated oligonucleotide sequences bound to streptavidin-coated magnetic particles. *BioTechniques*, 16, 656-662.

Milbourne, D., Meyer, R.C., Collins, A.J., Ramsay, L.D., Gebhardt, C., Waugh, R. 1998. Isolation, characterization and mapping of simple sequence repeat loci in potato. *Mol. Gen. Genet.*, 259, 233-245.

Money, T., Reader, S., Qu, L.J., Dunford, R.P., Moore, G., 1996. AFLP-based mRNA fingerprinting. *Nucleic Acids Res.* 24(13), 2616-7.

Morgante, M., Hanafey, M., Powell, W. 2002. Microsatellites are preferentially associated with nonrepetitive DNA in plant genomes. *Nat. Genet.*, 30(2), 194-200.

Nakamura, Y., Leppert, M., O'Connell, P. *et al.* 1987. Variable number of tandem repeat (VNTR) markers for human gene mapping. *Science*, 235(4796), 1616-1622.

Orita, M., Iwahana, H., Kanazawa, H., Hayashi, K., Sekiya, T. 1989. Detection of polymorphism of human DNA by gel electrophoresis as single-strand conformation polymorphisms. *Proc. Natl. Acad. Sci.* USA, 86, 2766-2770.

Peakall, R., Gilmore, S., Keys, W., Morgante, M., Rafalski, A. 1998. Cross-species amplification of soybean (*Glycine max*) simple sequence repeats (SSR) within the genus and other legume genera: implications for the transferability of SSRs in plants. *Mol. Biol. Evol.*, 15(10), 1275-1287.

Prevost, A., Wilkinson, M.J. 1999. A new system of comparing PCR primers applied to ISSR fingerprinting of potato cultivars. *Theor. Appl. Genet.*, 98, 107-112.

Primmer, C.R., Ellengren, H., Saino, N., Moller, A.P. 1996. Directional evolution in germline microsatellite mutations. *Nat. Genet.*, 13, 391-393.

Sauer, S., Gelfand, D.H., Boussicault, F., Bauer, K., Reichert, F., Gut, I.G., 2002. Facile method for automated genotyping of single nucleotide polymorphisms by mass spectrometry. *Nucleic Acids Res.* 30(5), e22.

Schneider, K., Borchardt, D.C., Schafer-Pregl, R., Nagl, N., Glass, C., Jeppsson, A., Gebhardt, C., Salamini, F. 1999. PCR-based cloning and segregation analysis of functional gene homologues in *Beta vulgaris*. *Mol. Gen. Genet.*, 262(3), 515-524.

Southern, E.M. 1975. Detection of specific sequences among DNA fragments separated by gel electrophoresis. *J. Mol. Biol.*, 98, 503-517.

Tautz, D. 1989. Hypervariability of simple sequence as a general source for polymorphic DNA markers. *Nucl. Acids Res.*, 17, 6463-6471.

Vendramin, G.G., Lelli, L., Rossi, P., Morgante, M. 1996. A set of primers for the amplification of 20 chloroplast microsatellites in Pinaceae. *Mol. Ecol.*, 5, 595-598.

Vos, P., Hogers, R., Bleeker, M., *et al.* 1995. AFLP: a new technique for DNA fingerprinting. *Nucl. Acids Res.*, 23, 4407-4414.

Waugh, R., McLean, K., Flavell, A.J., Pearce, S.R., Kumar, A., Thomas, B.B.T., Powell, W. 1997. Genetic distribution of Bare-1-like retrotransposable elements in the barley genome revealed by sequence-specific amplification polymorphisms (S-SAP). *Mol. Gen. Genet.*, 253, 687-694.

Williams, M.N.V., Pande, N., Nair, S., Mohan, M., Bennett, J. 1991. Restriction fragment length polymorphism analysis of polymerase chain reaction products amplified from mapped loci of rice (*Oryza sativa* L.) genomic DNA. *Theor. Appl. Genet.*, 82, 489-498.

Wu, J., Shimokawa, T., Maehara,T., Koike, K., Umehara, Y., Matsumoto, T. 1997. Mapping rice EST's onto YAC clones. *Rice Genome*, 6, 2-3.

Zietkiewicz, E., Rafalski, A., Labuda, D. 1994. Genome fingerprinting by simple-sequence repeat (SSR)-anchored polymerase chain reaction amplification. *Genomics*, 20, 176-183.

26

CHAPTER

Transposable Elements and the Analysis of Plant Biodiversity

S.M. Tam, C. Mhiri, M.-A. Grandbastien

INTRODUCTION

The evolution of mankind has always been intricately linked to the exploitation of natural plant resources. Following plant domestication, the genetic diversity of a crop species would not only be influenced by nature but also subjected to pressures from human interventions. In the last decades, human action has been responsible for intensive reductions in the genetic diversities of crop species, thus highlighting the issues of over-exploitation of natural resources and consequently, the importance of conservation. Conservation not only allows us to confront issues threatening biodiversity but also to develop priorities and programs to better manage our natural resources. This is critical for ensuring the long-term maintenance of evolutionary and ecological equilibria of the living world, necessary for continued development of economic crops and ultimately safe guarding our own future.

An integral part of conservation focuses on constant research to develop novel and improve existing methods for more efficient evaluation and monitoring of genetic diversity. Traditional approaches for the measurement of diversity rely upon the ability to resolve differences in morphological characters, which then expanded to include anatomical, biochemical and phytochemical assays (Karp *et al.*, 1996). The past two decades have seen molecular markers emerge as the most used indicators of genetic diversity, with the tidal wave of data emerging from DNA based techniques creating a powerful impact in the study of plant diversity and evolution. The many advantages of employing molecular markers include the direct use of genetic material to provide fundamental characters that reflect genealogical relationships among individuals. In addition, the ability to access nearly unlimited pools of genetic variability throughout the entire biological world, provide common yardsticks for measuring divergence and facilitate mechanistic appraisals of evolution (Avise, 2004). Molecular techniques which directly reveal modifications in genetic inheritance (DNA), have the ability to provide numerous and independent characters with lower levels of homoplasy (though this assumption is debatable) and are considered less subjective in handling types of character analysis (Sytsma, 1990; Smith and Sytsma, 1994).

In the exploration of genetic diversity, conventional molecular techniques either do or do not employ direct DNA sequencing. Examples of the former include detecting SNPs (Single Nucleotide Polymorphism) or sequence variations, while the latter consist of a wide range of methods that includes detecting physical and chemical differences of proteins (isozymes, allozymes), DNA hybridization, restriction analyses such as RFLP (Restriction Fragment Length Polymorphism) or AFLP (Amplified Fragment Length Polymorphism), small sequence variations (RAPD, Randomly Amplified Polymorphic DNA), minisatellites and SSRs (simple sequence repeats)/microsatellites (genomic or EST-derived). However, an additional level of modification in genetic inheritance can be found in the presence and activity of mobile DNA sequences, also known as transposable elements (TEs). Transposable elements are capable of mobility and inserting themselves at different locations in the genome. For a long time, they were misunderstood as "parasite" DNA, but have recently been shown to be major contributors of genetic variability and phenotypic plasticity in eukaryotic genomes, especially in plants. Therefore, it is not surprising that new strategies are actively being developed in order to detect and exploit element insertion sites as molecular markers (Schulman *et al.*, 2004).

PRINCIPAL TYPES OF TRANSPOSABLE ELEMENTS

Transposable elements are present in both prokaryotic and eukaryotic organisms, and were originally discovered in plants because of mutagenic activity associated with the mobility of the elements. Mutation events causing phenotypic modifications are easily spotted on a plant, such as changes in the colour of grains or leaves. One of the major impacts of transposable element activity is generating genomic variability due to insertions and deletions (indels). Two genotypes may differ by an insertion at a particular location (insertion site), which exists in only one of two genotypes, creating an insertion polymorphism for this site between them. Insertions of transposable elements can also generate mutations that may abolish or modify gene function, for example, by inserting into or in proximity to a gene (Bennetzen, 2000; Jordan *et al.*, 2003) that may consequently alter phenotype. In addition, the repetitive nature and mobility of transposable elements may also entail chromosome reshuffling on a much larger scale (Lönnig and Saedler, 2002), qualifying them as important sources of genetic diversity.

A common characteristic of transposable elements is the ability to utilise the cellular machinery of the host for their expression, where well-defined genetic entities generally code for the specific functions necessary for clean transposition. Families of transposable elements are grouped according to structural characteristics and classified into two main classes, where their transposition mechanism differs radically. Class I elements (Fig. 26.1a) or retroelements, transpose via a RNA intermediate that is reverse transcribed into a daughter copy DNA and then re-inserted into the genome. Class II elements (Fig. 26.1b) or DNA transposons, move directly from DNA to DNA, generally by excision followed by reinsertion in the case of eukaryotes. Therefore, there is an appreciable difference in the level of amplification of elements between the two classes, which also determines their quantitative and qualitative importance. Class I retroelements are specific to eukaryotic genomes, while class II DNA transposons are found both in the prokaryotes as well as the eukaryotes.

Retroelements

Transposition of retroelements involves a RNA intermediate resulting from the transcription of the inserted mother copy. This RNA is subsequently retro-transcribed into a daughter copy of double stranded DNA that will then be reinserted elsewhere in the genome. The immobility of the mother copy and creation of new daughter copies show that the transposition cycle of a retroelement is actually also an amplification process, more easily envisioned as a "copy-and-paste" mechanism. Two major consequences of this kind of transposition mechanism are that, theoretically, the insertions are irreversible and the retroelements are capable of amplifying to very high copy numbers. Indeed, retroelements are present from many thousands up to hundreds of thousands of copies in the genomes of complex eukaryotes. Recent data from genomic sequencing projects have reported that retroelements form 42% of the human genome (International Human Genome Sequencing Consortium, 2001) and are estimated to represent more than 60% of the genomes of certain plant species (Bennetzen, 2002).

There are two principal categories of retroelements (Fig. 26.1a): retroelements with LTR (Long Terminal Repeat), which comprises retrotransposons and retroviruses, and retroelements without LTR, or retroposons. There are also other viral elements such as pararetrovirus, which utilize reverse transcriptase in the course of their replication cycle, but will not be discussed here. Retroviruses and retrotransposons are structurally, functionally and evolutionarily very close, the major difference being that retroviruses possess an additional *env* function, a gene that encodes for proteins involved in interacting with cellular receptors and mediate fusion of the host and viral membranes, and is implicated in viral infection by permitting an extracellular stage, whilst the multiplication cycle of retrotransposons is, in theory, restricted to the intracellular level. However, the delimitation between retroviruses and retrotransposons is sometimes vague, as certain retroviruses have lost their *env* activity, and are thus restricted to an intracellular cycle of amplification (endogenous retrovirus, ERV), while certain retrotransposons carry additional open reading frames (ORF) of the type *env*, though their infectious capacity remains to be proven. New categories of retrotransposons are continuously being discovered, such as TRIMs (Terminal-repeat Retrotransposons In Miniature) and LARDs (LArge Retrotransposon Derivatives). TRIMs consist of small LTRs framing a short central domain of non-coding sequences (Witte *et al.*, 2001), and LARDs are larger defective derivatives that contain large internal domains without significant similarities to retrotransposon coding domains (Kalendar *et al.*, 2004). Both TRIMs and LARDs share many of the structural characteristics of LTR retrotransposons, such as reverse transcriptase priming sites, and are highly amplified, suggesting efficient trans-activation by other retroelements. The existence of TRIMS and LARDS indicates that eukaryotic genomes contain a large range of retrotransposons, which structural features are much more diverse and complex

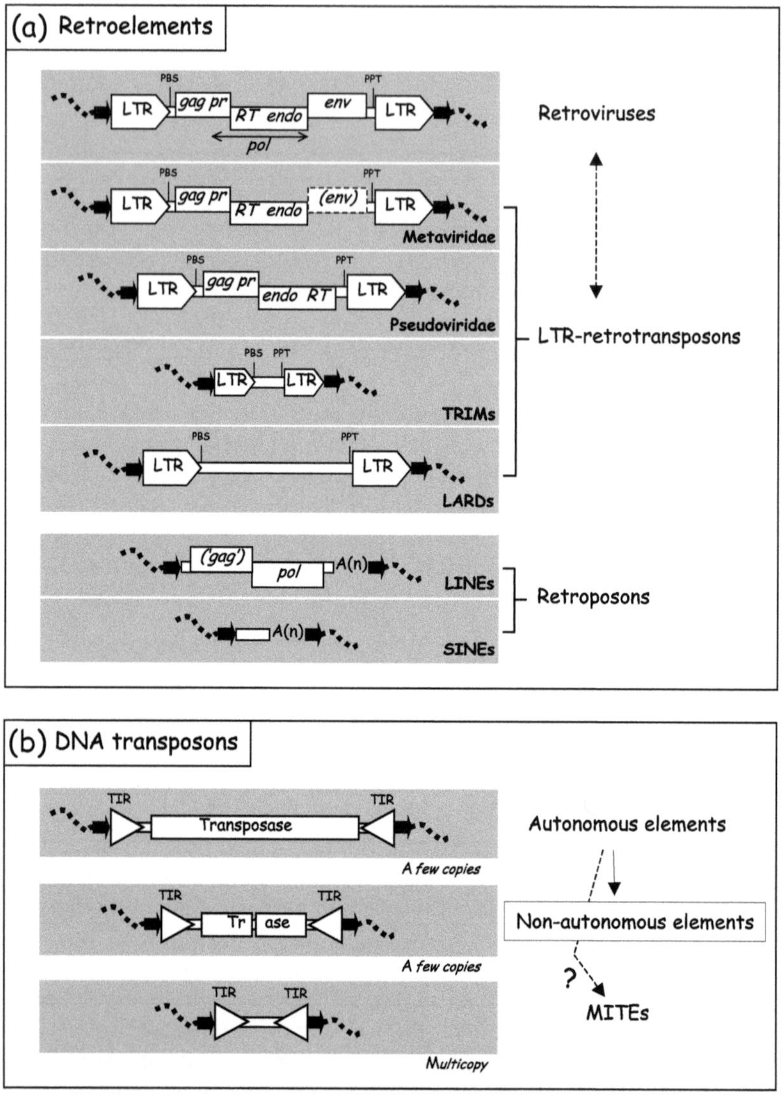

Fig. 26.1 Principal types of eukaryotic transposable elements

(a) Retroelements. Retrovirus inserts in the genome of an infected host and replicates via a RNA intermediate. The integrated form of a retrovirus is bordered by two long terminal repeat sequences (LTR), where transcription begins (5' LTR) and ends (3' LTR) and contains Open Reading Frames coding for proteins required for viral replication. The RNA matrix is encapsulated in a viral particle that is essentially formed by GAG proteins and surrounded by host cellular membranes associated to the ENV protein. The protein ENV permits the virion to infect a new cell and the template RNA is reverse transcribed using the RT (reverse transcriptase) into a daughter copy DNA that will enter into the nucleus and be inserted into the genome using the ENDO (endonuclease) protein. The PBS (primer binding site) and PPT (polypurine tract) sites are implicated in priming the synthesis of the two DNA strands. Retrotransposons are structurally and functionally closely related to retroviruses. However, *bona fide* retrotransposons are not

Contd... on next page

than previously thought. Retroposons are devoid of LTRs and include LINEs (Long Interspersed Nuclear Elements), which are elements with a few kb of coding sequences, and SINEs (Short Interspersed Nuclear Elements), small non-coding elements of a few hundred bp that exploit the transposition machinery of LINEs to ensure their amplification.

LTR retrotransposons form the largest majority of retroelements found in plant genomes (Kumar and Bennetzen, 1999), and all known categories of retroelements have been discovered in plants except for actual infectious retroviruses. Although plant retrotransposons with additional ORFs of the type *env* (Laten *et al.*, 1998; Vicient *et al.*, 2001b; Wright and Voytas, 2001) have been detected, the potential role of such a function to ensure entry of a particular virus by fusion with a target cell membrane, remains for the moment, very hypothetical in plants. The replication cycle of retroelements involves a cytoplasmic stage, and in the case of retrotransposons, requires the formation of a VLP (Virus-Like-Particle), which facilitates the reverse transcription of the intermediate RNA that produces the daughter copy DNA outside of the nucleus. This amplification cycle engages multiple steps of control, with transcription notably being the most important. Transcription controls the synthesis of the messenger RNA utilized for the production of proteins needed for the transposition cycle and the synthesis of the RNA matrix, which will be reverse transcribed into a copy of double stranded DNA. Transcription is absolutely crucial for retrotransposition, as a consequence, the mode of amplification of a particular element will be conditioned by its transcriptional regulation.

DNA Transposons

Class II elements move in a more direct DNA to DNA manner. With these elements, two distinct transposition mechanisms are known – one involving direct interaction between the donor transposon and the target site, resulting in the copying of the donor element (replicative transposition) and the second consisting of the excision of the element from its original site and reintegration at a new site (conservative transposition) (Brown, 1999). In the eukaryotes, transposition of DNA transposons is

Fig. 26.1 Contd...

infectious and their amplification cycle involving a cytoplasmic VLP (Virus-Like Particle), is exclusively intracellular. However, the delimitation between retrovirus and retrotransposons is sometimes vague (see text). LTR-retrotransposons are classified as several types, such as Metaviridae (also known as Ty3-gypsy type) and Pseudoviridae (also known as Ty1-copia type), depending on the organisation of the RT/ENDO functions. The TRIM and LARD elements are devoid of clear coding domains, but display many similar characteristics to LTR-retrotransposons (LTR, PBS, PPT). The LINE retroposons are devoid of LTRs but also amplify via a RNA intermediate and contain gag- and pol-like coding domains. The small non-coding SINEs use the coding functions provided in *trans* by LINEs.

(b) DNA transposons. These are bounded by TIRs (Terminal Inverted Repeats), unique for each family and comprise completely autonomous elements that have an active transposase (as well as several other functions for certain complex elements), and non-autonomous defective elements that can be trans-activated by an autonomous copy of the same family. MITEs are devoid of coding sequences, and represent particular forms of non-autonomous, derived, highly amplified elements. Most transposable elements are flanked by short target site duplications (black arrows), generated upon insertions.

generally conservative, resembling the "cut and paste" mechanism. Such a mechanism consequently limits the number of DNA transposons, making them less common in a genome and often resulting in unstable mutations. It is precisely this characteristic of unstable insertion that facilitated the discovery of DNA transposons in maize by Barbara McClintock, some 50 years ago.

There are many families of DNA transposons, composed individually of a series of elements mobilised by the same transposase (Fig. 26.1b). Each family contains a very small number of autonomous copies capable of coding for a functional transposase that assures clean mobility. Other members are defective elements, derived from autonomous copies following diverse forms of mutations and deletions, but still conserving the capacity to be mobilized by a transposase produced by corresponding autonomous elements. Contrary to retroelements, the mobilization of a DNA transposon does not require transcription, only the presence of terminal or sub-terminal sequences (particularly short terminal inverted repeats or TIRs) that are specifically recognised by a related transposase produced *in trans*. A specific category of class II elements consists of MITEs (Miniature Inverted-repeat Transposable Elements) that are disturbingly peculiar (Feschotte *et al.*, 2002a, b). MITEs are very small elements (few hundred bp) without coding sequences, but possessing TIRs, which in certain cases, permit associations with a bigger, potentially autonomous element and probably represent a particular non-autonomous form derived from a complete element. The excision of a MITE has not been documented until recently (Kikuchi *et al.*, 2003; Nakazaki *et al.*, 2003). Intriguingly, MITEs are generally distributed in families of elements with very conserved homogenous sizes, and are present in very high copy numbers (thousands to tens of thousands), suggesting that MITEs are able to amplify to high levels using a yet unknown mechanism.

Why Apply Transposable Elements to Evaluate Genetic Diversity?

Transposable elements can be exploited to evaluate genetic diversity, because the transposition activity of elements generates genetic variation in the form of heritable insertion polymorphisms. These insertion polymorphisms can be useful molecular markers that contribute to the better understanding of processes that influence biodiversity. Basically, this requires the detection of the two possible alternative allelic states for each insertion site: an empty site (element not inserted) or a full site (element inserted) (Fig. 26.2). Depending on the integration timing of a transposable element copy, the insertion polymorphism will enable the differentiation of lines, groups or species of organisms. In higher eukaryotes such as plants and vertebrates, there is generally a very low probability of convergence of insertions: an insertion found at the same position in two genotypes was inherited from a common ancestor and can be considered as a synapomorphy (shared derived character).

The most practical elements for evaluating genetic diversity are those with high copy numbers that provide a larger number of useful characters, and those that generate stable insertions, important for ensuring transmission to descendants. Therefore, retroelements such as LTR retrotransposons and SINEs are preferentially used. However, MITEs, even though susceptible to excision, have also recently been employed as molecular markers, as they exist in high copy numbers and possess certain interesting insertion characteristics (see next page). In addition, the genomic

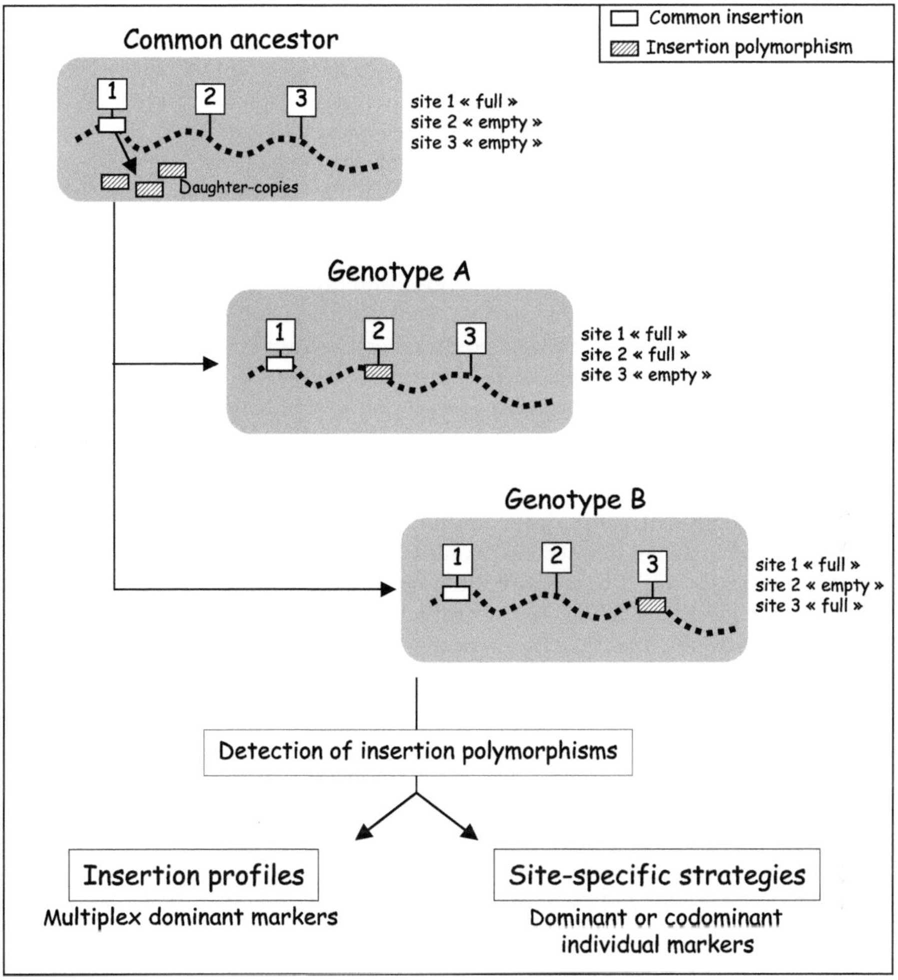

Fig. 26.2 Application of transposable elements as molecular markers
The basis of the utilization of transposable elements, is the detection of insertion polymorphisms. A particular copy is either present at a given site (full site) or absent (empty site), depending on the timing of the transposition event. Insertion polymorphisms are useful for differentiating genotypes: insertions of sites 2 and 3 have appeared after the divergence between genotypes A and B, hence can be used for distinguishing between these genotypes (e.g. site 2 is full in A and empty in B, and site 3 is empty in A and full in B). Older insertions serve to trace similarities between genotypes: insertion at site 1, that is present in the common ancestor is also found in the two derived genotypes, A and B. The robustness of this information is much greater when the insertions are stable (e.g. retrotransposons), or relatively stable (e.g. MITEs). The combination of many insertion sites present in each genotype, permits the estimation of genetic diversity.

distribution of elements is an important factor, particularly for element insertions near to genic regions, which will definitely increase the chances of generating useful markers for tracking genes of agronomic interest.

APPLIED STRATEGIES

In recent years, several molecular marker methods based mainly on retrotransposons and MITEs, have been developed. All methods rely on the ability to detect the link formed between genomic DNA and the element during integration. This is done by PCR amplification between primers anchored on the element and matching a nearby motif in the genome (Schulman *et al.*, 2004), which may be the sequence of the host or other transposable elements. Detection of insertion polymorphisms can be performed either individually, or using a multiplex profile exhibiting a significant fraction of the total number of insertions in the genome (insertion profiles). An insertion profile may show the sum of amplified fragments and generates a representation "map", reflecting the different insertion sites of the transposable element found in the sampled genotype. Individual detection of the allelic states of insertion sites serves as site-specific markers that are extremely useful for screening samples on a large scale.

Insertion Profiles

SSAP (Sequence Specific Amplification Polymorphism; Waugh *et al.*, 1997) is the most utilised method to generate insertion profiles, and was derived from AFLP (Vos *et al.*, 1995). The primary aim is to amplify the fragment bordering an insertion of a given transposable element, up to the closest restriction site located on the genomic flanking region (Figs. 26.3 and 26.4). The efficacy of this strategy is based on the assumption that each inserted copy is found statistically at a different distance from the restriction site, and on the use of high-resolution systems to clearly detect fragments of different sizes. The general implementation of SSAP is as follows: genomic DNA is digested by restriction enzyme(s), followed by ligation of adaptor(s) at the extremities of the fragments (Fig. 26.4a). PCR is subsequently carried out with primers corresponding to the extremity of the transposable element and to the adaptor. The TE primer is normally tagged (radioactive, fluorescent) in order to reveal only element-anchored fragments - each fragment will therefore consist partly of the element's terminal sequences and the host flanking sequences (between element and restriction site). Amplified products are then resolved according to size using different high-resolution techniques (PAGE/sequencing gels, capillary sequencers) in order to produce insertion profiles (Fig. 26.4b). Comparison of the insertion profiles from different genotypes permits the identification of insertion polymorphisms, scored as presence of a fragment corresponding to an insertion being present, versus an absence of this fragment in certain genotypes, indicating in theory that the site is devoid of the same insertion. SSAP markers are dominant where the heterozygous state (empty site/ full site) cannot be discriminated from the homozygous state (full site/full site), and the scored binary matrices from insertion profiles allow analyses of genetic variation and relationships.

SSAP, being the first retrotransposon-based method to be described and now the most applied, was originally used in a study of barley using retrotransposon BARE-1 (Waugh *et al.*, 1997). This method is also known as Transposon Display, described from the study of transposon dTph of petunia (Van der Broek *et al.*, 1998). Together, SSAP and Transposon Display are very similar technically to other methods used for the analysis of T-DNA flanking sequences in collections of insertion mutants (Devic *et al.*, 1997; Frey *et al.*, 1998). All these techniques differ essentially in their adoption or not, of

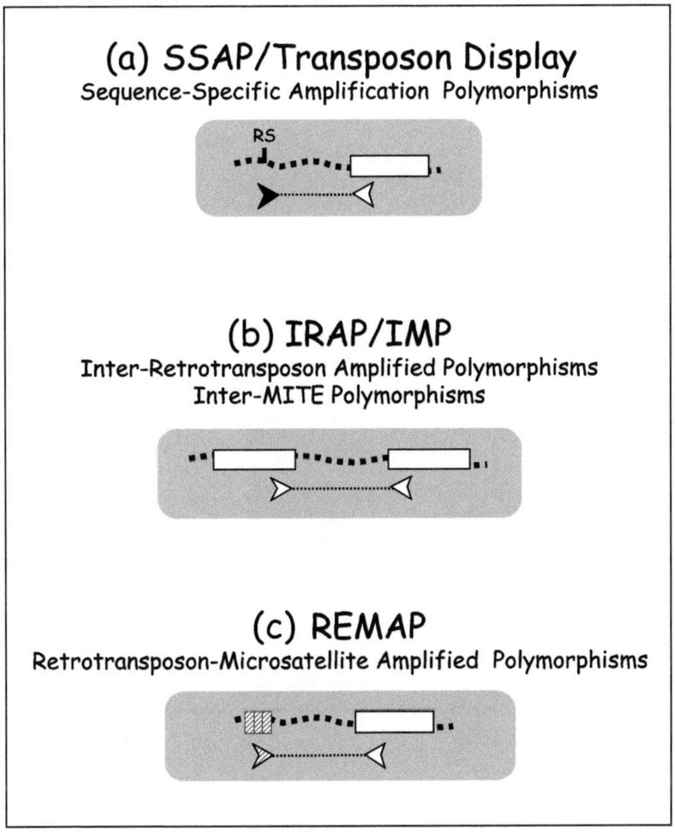

Fig. 26.3 Principal strategies for generating insertion profiles
Based on simultaneous amplification of flanking regions of many insertions present within the same genotype, SSAP reveals amplified fragments between the element and adjacent restriction sites (RS). It is the most popular technique whereby success is less dependent on copy numbers. IRAP and IMP, which reveal insertion polymorphisms between closely situated elements, are efficient only for very high copy numbers. REMAP reveals insertion polymorphisms between elements and adjacent microsatellites.

procedures designed to limit the amplification of parasite AFLP fragments from the link formed between adaptor primers - for example, using specific asymmetric primers or a biochemical selection of fragments anchored in the element. Frequently as well, selective nucleotides at the 3′ end of the adaptor primer are added in order to fractionate the pools of amplified fragments, notably when working with very high copy number elements. All these techniques share the same advantages (multiplex system) and inconveniences (eg. requiring digestion and ligation of DNA) as AFLP due to their conceptual similarities.

Other multiplex methods, such as IRAP (Inter-Retrotransposon Amplified Polymorphisms) or IMP (Inter-MITE Polymorphisms), are essentially based on the detection of insertions by direct PCR between elements (Kalendar *et al.*, 1999; Chang *et al.*, 2001) (Fig. 26.3b). These methods are easier to apply than SSAP since it is

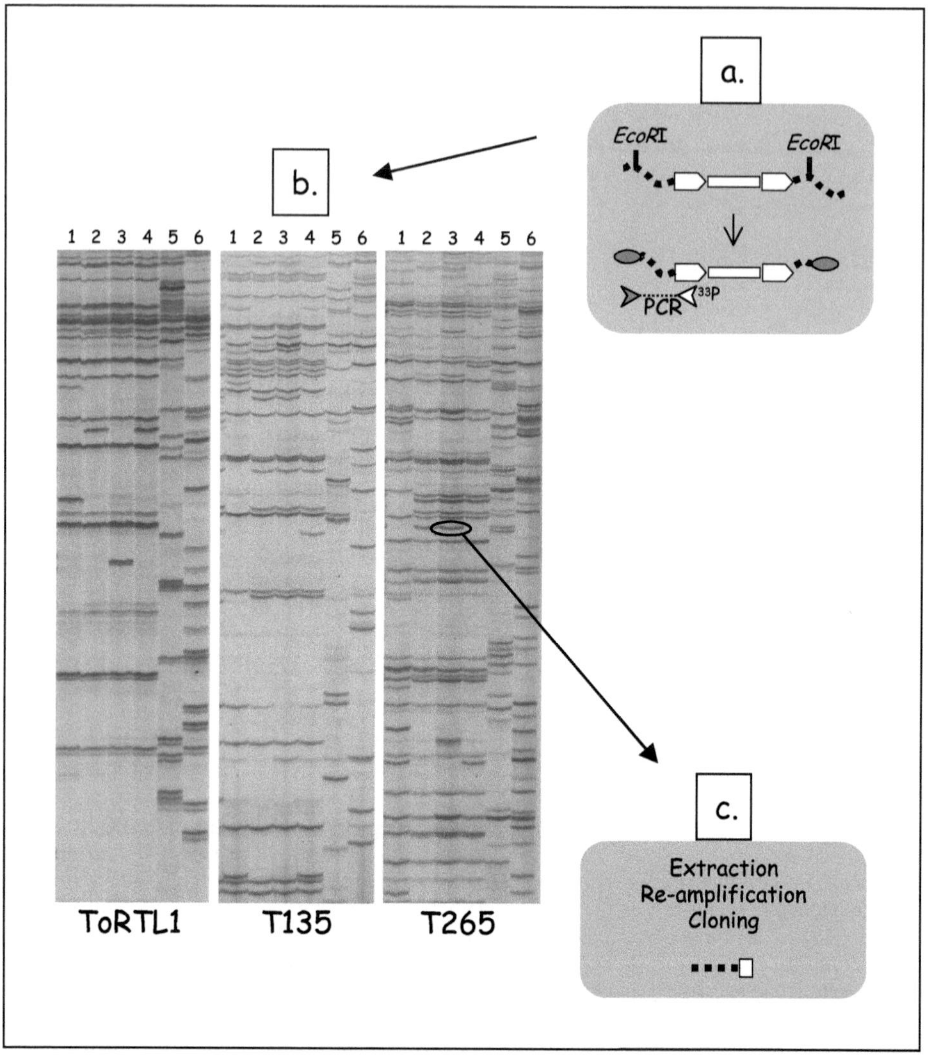

Fig. 26.4 Examples of insertion profiles

SSAP profiles were obtained with three different tomato retrotransposon families (ToRTL1, T135, T265, Tam *et al.*, 2005) from six accessions of *Solanum* subsection *Lycopersicon*.

(a) Genomic DNAs were digested using *EcoR*I, ligated with adaptors and amplified using primers corresponding to the adaptor (without selective bases) and anchored in the beginning of the LTR of each three elements, using ^{33}P labelled retro-primers.

(b) Amplified fragments are subsequently separated using high-resolution polyacrylamide gels in order to identify polymorphic SSAP bands, and scored for presence/absence along the length of the insertion profile.

(c) SSAP bands can be isolated from the gel and sequenced in order to analyse flanking sequences and develop site-specific markers.

unnecessary to digest genomic DNA but mostly work efficiently with high copy number elements (e.g. thousands of copies). Another promising method is REMAP (Retrotransposon-Microsatellite Amplified Polymorphisms) or *Copia*-SSR, based on direct PCR between an element and microsatellites (Kalendar *et al.*, 1999; Provan *et al.*, 1999) (Fig. 26.3C). There seems to be an association between retrotransposons and SSRs, at least in barley (Ramsay *et al.*, 1999) and REMAP can generate both dominant and co-dominant markers (SSR length variation).

Site-specific Strategies

An insertion polymorphism corresponds to a precise biological event that is the presence or absence of an element at a specific location and may be analysed as a single locus, if the specific location in question had been identified beforehand. Site-specific strategies are essentially simple PCRs designed to detect the two alternative states (empty or full) of a given insertion site. A reaction using primers from both the flanking DNA alone (PCR external/external) detects the unoccupied empty site while a reaction using a primer from the flanking DNA together with a primer in the transposable element (PCR external/internal) detects the occupied full site (Fig. 26.5). The major advantage of these strategies is that they are based on simple PCRs and can be developed into high throughput analyses of a large number of samples.

For insertion polymorphisms of small elements (e.g. SINEs, MITEs), the alternative states (empty or full) are visualised as a size difference of the PCR product amplified by the two primers flanking the insertion site (Fig. 26.5a), thus producing co-dominant markers. Additionally, the more straightforward PCR external/internal strategy, that produces dominant markers (Fig. 26.5b), will amplify a product only when the site is occupied, rendering it unnecessary even to know the size of the expected fragment because the results are simply based on the presence and absence of the PCR product. Samples can be analysed using simple, rapid techniques for detecting DNA such as dot-blots with a probe of the insertion site, or using fluorescent primers that actually permits the combination of multiple polymorphic sites.

For larger sized retrotransposons, the difficulty of amplifying between two flanking primers spanning the long element insertion, actually permitted the development of an interesting method known as RBIP (Retrotransposon-Based Insertion Polymorphisms) that generates single locus co-dominant markers (Flavell *et al.*, 1998). RBIP combines PCR external/external with PCR external/internal (Fig. 26.5b′), whereby the former PCR will only succeed in amplifying a product if the site is empty, while the latter PCR will only be successful when there is an insertion (full site). Reversed results produced from each PCR reaction for the full and empty sites act as control reactions for one another, which prevents the formation of false positives or false negatives. RBIPs generate co-dominant markers because the heterozygous state of a locus is known when the two PCRs succeed in amplifying products.

The major inconvenience of these site-specific strategies is that sequences flanking at least one side of the insertion must be available. In order for this, polymorphic SSAP bands can be extracted from gels, cloned and sequenced to provide the flanking sequence of one side of the insertion (Fig. 26.4c). For RBIP, the flanking sequences of the other side of the insertion must also be acquired, using techniques such as adding the

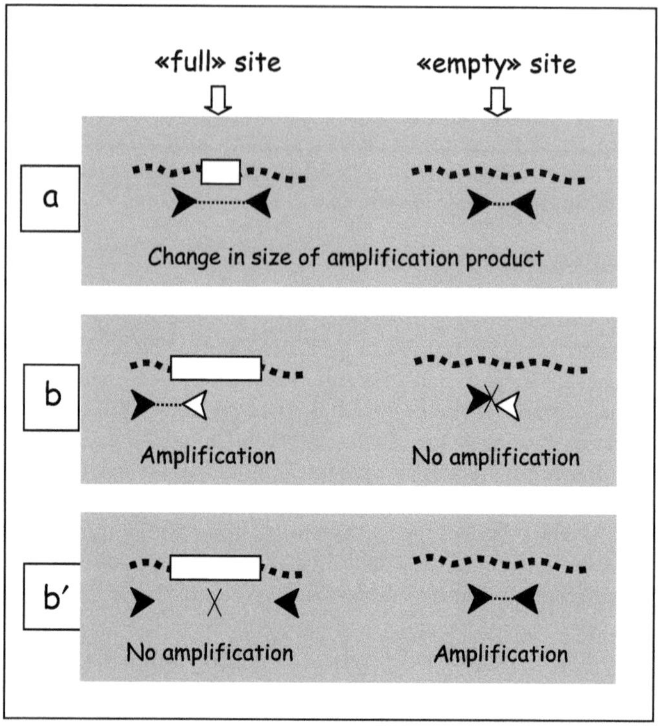

Fig. 26.5 Principal site-specific strategies

Two allelic states (empty or full) of an individual insertion site are detected. (a) In the case of small sized elements (MITEs, SINEs), two primers designed for the flanking regions of each side of the insertion distinguishes the two alleles because of the size difference in amplified products (co-dominant markers). (b) It is equally possible to distinguish the two alleles by combining primers designed for a flanking region and anchored on the element: an amplified product will be produced only when the site is occupied. Not dependent on element size, this strategy is particularly useful when flanking sequences of only one side of the element insertion are known, though it generates dominant markers. (b') For large sized elements (retrotransposons, LINEs), two primers designed for each flanking side of the insertion will not produce an amplified product except when there is an empty site, giving the opposite results to that of (b). The combination of (b + b') with each PCR controlling each other, generates co-dominant markers, and this strategy is known as RBIP. All these strategies that are based on the presence and absence of amplified products have the advantage of possibly being automated for high-throughput screening of large number of samples.

known sequences of target site duplication as selective bases at the 3′ end of a primer specific for the un-sequenced end of the insertion, or alternatively, using commercial kits. This procedure involves a specific primer derived from the host flanking DNA orientated for synthesis towards the insertion site used on accessions lacking the insertion to reveal sequences on the other side (Schulman *et al.*, 2004). For species that are the object of concentrated genomic sequencing efforts, it is easier to obtain flanking sequence data, and even to identify insertion sites *in silico* by mining relevant databases. Once the sequence data is available, these site-specific strategies are easy to manipulate and adapt. The methods may be automated to obtain high-performance data by adopting, for example, a dot-blot assay, or robotic spotting and data capturing. The results can be processed using technology developed for microarrays in order to rapidly analyse in a reproducible manner very large numbers of samples for the presence and/or absence of an insertion at a given location, to increase the performance of the method (Flavell *et al.*, 2003; www.biocenter.helsinki.fi/bi/tegerm).

POSSIBLE UTILIZATIONS

Molecular markers based on transposable elements fundamentally share similar applications as other conventional molecular markers such as RFLP, RAPD, AFLP or microsatellites. Additionally, retroelements which generate stable insertions, may be particularly useful for the study of molecular phylogenetics because they provide polarized characters. The ancestral state of each insertion site (marker) is empty and the occupied site is considered as the derived character state, while for other marker systems, the ancestral or anterior state is unknown or ambiguous. Molecular markers based on transposable elements have been applied in various studies, from genetic mapping and cloning genes of interests to the analyses of genetic diversity and interspecific relationships.

Mapping and Cloning of Genes of Interest

Different studies have integrated transposable element-based markers into genetic maps, such as in:

- barley, using SSAP, IRAP and REMAP (Waugh *et al.*, 1997; Provan *et al.*, 1999; Manninen *et al.*, 2000), and IMP (Chang *et al.*, 2001);
- oat, using SSAP (Yu and Wise, 2000);
- wheat, using IRAP and REMAP (Boyko *et al.*, 2002), and SSAP (Queen *et al.*, 2004);
- pea, using SSAP (Ellis *et al.*, 1998);
- maize, using MITE Display (Casa *et al.*, 2002);
- citrus and poncirus, using IRAP (Bernet and Asins, 2003)
- tomato and pepper, using SSAP (Tam *et al.*, in press Tam *et al.*, still in preparation)

With the exception of certain accumulation hot-spots, particularly in the cereals and tomato, retrotransposon markers generally showed widespread distribution along all linkage groups of different species, and were located precisely enough to permit the cloning of genes of interests. IRAP and REMAP markers based on BARE-1 allowed more accurate mapping of net blotch resistance genes in barley (Manninen *et al.*, 2000). In addition, IMP markers are envisioned to play an important role in future mapping studies because of the reported preferential associations of MITEs with coding sequences (Jiang and Wessler, 2001; Santiago *et al.*, 2002). Transposable element-based molecular markers are also being used in the localisation of qualitative characters or the positioning of QTLs with agronomic importance (Kumar and Hirochika, 2001).

Phylogenetic Applications

Different studies have employed retroelement-based molecular markers for constructing phylogenies. Insertion polymorphisms of SINEs, for example, have been used for inferring phylogenetic relationships of animal species, such as the phylogenies of fish (Murata *et al.*, 1996; Hamada *et al.*, 1998; Takahashi *et al.*, 1998) and mammals (Shimamura *et al.*, 1997; Nikaido *et al.*, 1999; Nikaido *et al.*, 2003) and to study the similarity and migration of human populations (Shedlock and Okada, 2000). In plants, the first example was from a study of *Brassica* using the SINE element S1Bn, where 21 insertion sites allowed phylogenetic inference with a much higher resolution than a previous study based on more than a hundred RFLP markers (Tatout *et al.*, 1999). In the genus *Pisum*, SSAP insertion polymorphisms of the retrotransposon PDR1 (Ellis *et al.*, 1998; Vershinin and Ellis, 1999) and the combination of insertion polymorphisms of PDR1 and three other retrotransposons (Pearce *et al.*, 2000) and a DNA transposon (Vershinin *et al.*, 2003) permitted detailed analyses of intra and interspecific relationships. It was also reported that SSAP based trees derived from the *Grande* retrotransposon reflected established species phylogeny and were useful in differentiating maize lines (Garcia-Martinez and Martinez-Izquierdo, 2003), and that MITE-SSAP was amenable for studying the genetic variation and species relationship in rice (Park *et al.*, 2003). More recently, PCR as well as *in silico* studies of retrotransposon insertion polymorphisms provided evidence that *Indica* and *Japonica* rice subspecies arose from independent domestication events (Vitte *et al.*, 2004). The combined SSAP analysis of insertions from retrotransposons ToRTL1, T135 and Tnt1 also allowed the inference of interspecific relationships of tomato and eight related wild species of the subsection *Lycopersicon.* In addition, clear species delimitation and separation of self compatible, self incompatible, red fruited and green fruited groups were observed (Tam *et al.*, in press), comparable to the ITS based phylogeny (Marshall *et al.*, 2001). Similarly, the combined insertions of retrotransposons T135 and Tnt1 also revealed interspecific relationships of pepper and six related species of *Capsicum* (Tam *et al.*, still in preparation), which generally agreed with previous studies using *atpB-rbcL* and *waxy* (Walsh and Hoot, 2001) and RAPD markers (Rodriguez *et al.*, 1999).

Analysing Genetic Resources and Genotyping

Insertion polymorphisms have proven to be equally useful as molecular markers for differentiating genotypes within species, especially for species with limited genetic diversity and when only a few useful conventional molecular markers were available. In addition to studies mentioned in the previous chapter, SSAP markers were also used to differentiate types of grapevine (Pelsy *et al.*, 2003; Labra *et al.*, 2004) and in the analysis of the genetic diversity of sweet potato varieties (Berenyi *et al.*, 2002). In addition, SSAPs proved to be useful for evaluating genetic diversities of commercial collections of tomato and pepper lines (Tam *et al.*, 2005), and public germplasm collections of tomato, pepper and their respective wild relatives (Tam *et al.*, in press; Tam *et al.*, still in preparation). IMP markers (*Stowaway* and *Barfly*) were successfully applied to genotyping barley cultivars (Chang *et al.*, 2001). Additionally, MITE *Heartbreaker (Hbr)* derived markers were useful for the genetic characterisation of a set of maize inbred lines belonging to stiff stalk and non-stiff stalk heterotic groups. Results obtained with *Hbr* markers agreed well with pedigree, RFLP and SSR data (Casa *et al.*, 2002). Automated site-specific strategies, in particular, have immense potential for use in breeding programs such as structuring large collections of genetic resources. They can be used for variety protection and quality control by evaluating the degree of purity of seed batches and determining the genetic origin of dried or transformed plant material among mixed or pure samples. Therefore, one of the main objectives of the studies carried out in the frame of a CEE project (QLRT-1999-31502 TEGERM) was to develop useful RBIP markers for structuring large-scale collections of genetic resources of peas, barley, tomato and pepper (www.biocenter.helsinki.fi/bi/tegerm).

Comparison with Other Molecular Markers

Previous studies that employed transposon-based molecular markers since their development, showed that the power of these techniques in analysing biodiversity lies in the higher levels of polymorphisms revealed by element based markers when compared with conventional marker systems. Several studies have been carried out to compare the performance of element based marker systems with others such as RAPD, AFLP and SSR. The percentage of BARE-1 polymorphisms observed by SSAP was 25% higher than AFLP in barley (Waugh *et al.*, 1997) and 30% higher in oat (Yu and Wise, 2000). In pea, SSAPs using PDR1 recorded a three-fold improvement compared to AFLP in discriminating accessions (Ellis *et al.*, 1998) while for *Medicago sativa*, SSAP showed the highest assay efficiency index and marker index compared to SAMPL (Selective Amplification of Microsatellite Polymorphic Loci) and AFLP (Porceddu *et al.*, 2002). SSAP also recorded the highest polymorphism level when compared with AFLP and RAPD in an analysis of sweet potato varieties (Berenyi *et al.*, 2002). A recent study comparing the usefulness of SSAP, AFLP and SSR for analysing genetic diversities of tomato and pepper collections also identified SSAP as the most informative system because it showed the highest ratio of number of polymorphic bands per assay and not surprisingly, the highest marker index (Tam *et al.*, 2005).

One explanation behind the higher levels of polymorphisms shown by SSAP is its complexity. In fact, insertion profiles not only simply reveal the transposition dynamics of the element used, they also reflect possible sequence polymorphisms at the flanking restriction site (similar to AFLP) and sequence variations at the extremities of the transposable elements where the primer is anchored. Hence, insertion profiles represent a combination of diverse sources of variability, where only site-specific strategies such as RBIP reflect insertion polymorphisms *sensu stricto*.

For Which Species?

Transposable elements, particularly retrotransposons, are present in all plant species in which they have been studied and therefore present no theoretical limit to the field of applications as molecular markers. Sequences of retrotransposons and MITEs isolated from many diverse species have been deposited in databases, and can be readily used for defining primers at the first stage of marker development (TIGR: http://www.tigr.org/tdb/e2k1/plant.repeats/index.shtml; TREP: http://wheat.pw.usda.gov/ITMI/Repeats/index.shtml; Repbase: http://www.girinst.org). In addition, bioinformatics programs are also available to help search and identify new transposable elements from full or partial genomic sequence data that are rapidly accumulating for many species (Le *et al.*, 2000; McCarthy and McDonald, 2003).

For species of interest where no element or sequence data is currently available, there is a strategy to isolate LTR sequences of retrotransposons, which was originally developed for isolating Ty1/*copia* type retrotransposons (Pearce *et al.*, 1999). This technique utilizes the presence at the end of the coding sequence (RNAseH region of the RT), of protein motifs that are very conserved and allow the design of universal degenerate primers in order to non-specifically amplify all the elements belonging to the Ty1/*copia* group in the genome (Fig. 26.6). Amplified products are internal fragments containing the end of the ORF and the beginning of the 3′ LTR, the latter being useful for defining specific primers in order to obtain polymorphic insertions. New retrotransposons were successfully characterised using this technique in *Pisum, Vicia fabia* and *Picea abies* (Pearce *et al.*, 1999), tomato (Tam *et al.*, 2005), as well as tobacco, pepper and aubergine (unpublished data). This technique can provide new families of retrotransposons relatively quickly, with varying degrees of insertion polymorphisms for use either individually or in combination, according to the objectives of study.

CHARACTERISTICS OF MOLECULAR MARKERS BASED ON TRANSPOSABLE ELEMENTS

Useful insertion polymorphisms were readily obtained from elements with very different levels of amplification, for example, BARE-1 is present in more than 70,000 copies in the barley genome, while less than 200 copies of PDR1 was estimated for the pea genome. Element-based techniques do not require that the activity rate of the retrotransposons be known, since essentially, these techniques describe insertions that are a consequence of the past activities (recent or distant) of the element. The efficiency of these strategies will depend on the genomic distribution of the transposable elements, which may not always be homogenous in the genome, depending on their amplification dynamics. It is important to realize that transposable elements are not inert sequences. They amplify following distinctive modes that vary according to the elements and respective expression conditions, and are possibly influenced by the evolutionary history and environment of the host species. In addition, even when normally stable insertions are generated, notably by retrotransposons, there is always the possibility of control mechanisms aimed at removing the insertions in the course of evolution, where hosts strive to control deleterious mutations that may arise from transposable element activity.

Genomic Distribution

Many cytogenetic studies of the distribution of retroelements in plant chromosomes (Kumar and Bennetzen, 1999) reported that most elements were well dispersed along the chromosomes, with the notable exception of the model plant *Arabidopsis thaliana* and of tomato where elements occupied centromeric and pericentromeric regions (The Arabidopsis Genome Initiative, 2000; Tam *et al.*, in press). Genetic mapping of insertions such as BARE-1 in barley (Waugh *et al.*, 1997) and *Aegilops* (Boyko *et al.*, 2002), PDR1 in pea (Ellis *et al.*, 1998) and different MITEs in barley (Chang *et al.*, 2001), in fact, showed good distributions of insertions along different chromosome arms.

However, the fine distribution of elements is not so homogenous, varying according to host species and element. One of the most well studied species is maize, where an important fraction of retrotransposons was found in the form of nested insertions into

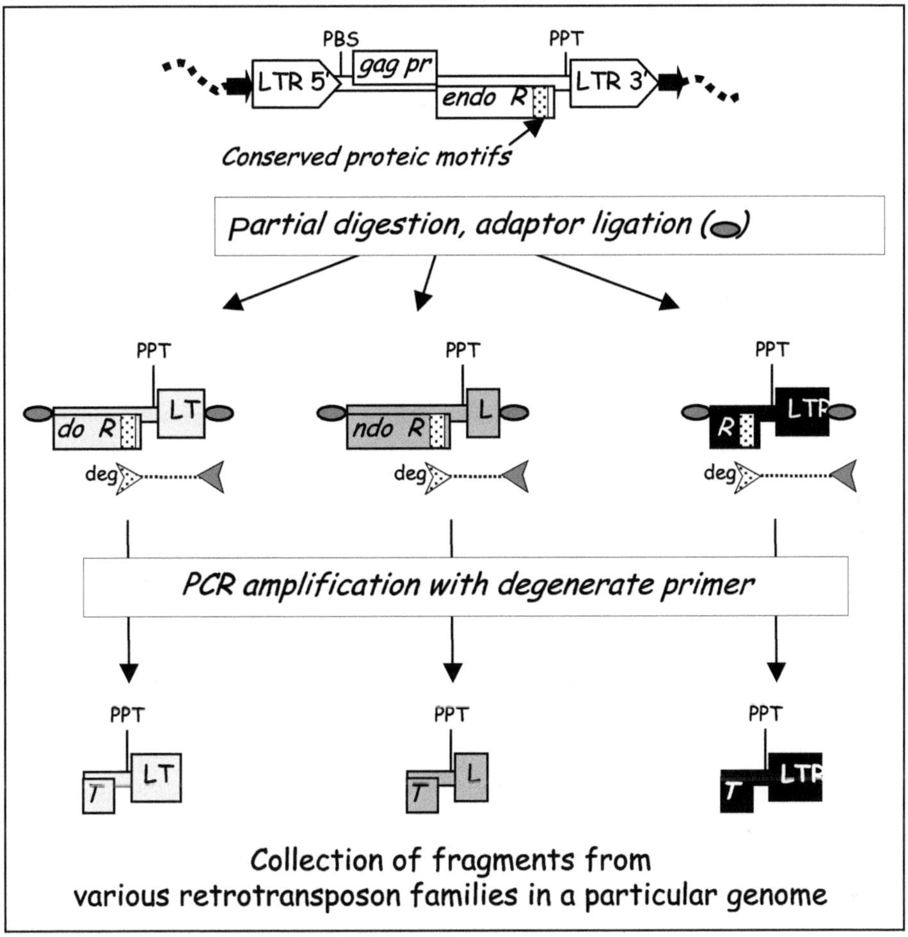

Fig. 26.6 Isolation of LTR regions of Ty1/*copia* retrotransposons

The method (Pearce *et al.*, 1999) is based on the existence of conserved protein motifs at the end of the RT coding domain. These motifs are useful for defining degenerate primers that can non-specifically recognize retrotransposons of the type Ty1/copia (Pseudoviridae) present in a genome. Firstly, genomic DNA is digested to obtain fragments containing part of the terminal ORF and the beginning or complete 3' LTR. Fragments are then ligated with adaptors, and a PCR is performed using adaptor and degenerate RT primers. The latter is biotinylated in order to select PCR products. After cloning, a collection of sequences belonging to many families of Ty1/copia retrotransposons are obtained – sequences that each consists of the end of the RT and various lengths of downstream element sequences. These sequences potentially contain the beginning of the 3' LTR of the element. The 3' LTR is by definition homologous to the 5' LTR, therefore it will be possible to design primers from each new family of retrotransposons for SSAP assays, in order to observe insertion polymorphisms. Abbreviations are as shown in Fig. 26.1.

each other in intergenic regions (San Miguel *et al.*, 1996). The inflation of intergenic regions due to retrotransposon blocks contributed to the difference of genome size between maize and sorghum (Tikhonov *et al.*, 1999). Generally, it appears that large sized retroelement insertions were relatively excluded from strictly genic regions in many plant or animal eukaryotes (Medstrand *et al.*, 2002). Differences in the distribution of *Arabidopsis thaliana* LTR-retrotransposons were also recently observed, depending on the type of element (Pereira, 2004; Peterson-Burch *et al.*, 2004). On the other hand, small elements, such as MITEs or SINEs, display different distributions, being frequently found in proximity to coding sequences (Jiang and Wessler, 2001; Tikhonov *et al.*, 2001; Santiago *et al.*, 2002), which might cause these insertions, particularly the MITEs, to be considered as more polymorphic than retrotransposons, though precise comparison remains to be done.

In addition, the distribution of transposable elements in the genome is dynamic, and trends seem pronounced, correlated with the age of the insertions. The association frequencies of small elements (MITEs, SINEs) with genes increased with age (Medstrand *et al.*, 2002; Santiago *et al.*, 2002), while the opposite was noted for large sized elements, where older elements were more excluded from genic regions, as was reported for the human genome (Medstrand *et al.*, 2002). Similarly, results from a study of tobacco showed that newly transposed Tnt1 copies associate preferentially with genes, while older copies were more often associated with non-genic and repetitive DNA (Le *et al.*, still submitted).

Progressive exclusions of elements from genic regions can be attributed to the negative impact caused by an insertion into or near to a gene, constituting a mutation that could be selected against. The remarkable observation of progressive strengthening in association between MITEs/SINEs and genes, suggests that these insertions may be either neutral or have a positive effect. A recent review of more than 2000 human genes found that close to a quarter of the 5′ promoter regions contained transposable elements, the majority being SINE elements, and the *cis*-regulatory regions of a significant proportion of these genes were derived from transposable elements (Jordan *et al.*, 2003). Regardless of the precise reasons responsible for the association between small elements and genic regions, this result suggests immense potential for the use of these small transposable elements as molecular markers for gene-related studies.

Transposition Control and the Dynamics of Amplification

The distribution of a family of transposable elements in a range of hosts is also not certain to be homogeneous. Some elements are restricted to one of two host species, while others are distributed in many genera of a plant family. Moreover, the number of copies and polymorphism levels of the same element may differ from one host to another. Even though reasons accounting for differences are not definitely known, without doubt, it reflects the differences in transposition regulation according to genomic contexts. These differences can be directly linked to conditions of expression, particularly in the case of retrotransposons. The Tnt1 family of retrotransposons in the Solanaceae, for example, shows considerable variations in transcription regulatory regions within and between host species (Vernhettes *et al.*, 1998; Araujo *et al.*, 2001). In tobacco, this variability is strongly associated with diverse transcriptional regulation (Beguiristain *et al.*, 2001). This observation suggests that the Tnt1 family of retrotransposons evolved by modifying their conditions for expression, which resulted in the establishment of different modes of amplification that could be specific to each host and/or its environmental history. It thus follows that a family of elements in one species will not necessarily behave in similar manner in another species of the same plant family.

With only a handful of active retrotransposons and MITEs characterised today, little is known about the modes of activation of these elements. The development of molecular markers based on insertion polymorphisms should promote a better understanding of element behaviour. Amplification of the PDR1 element, for example, appeared relatively constant among different *Pisum* species, except in *Pisum abyssinicum*, where PDR1 had ceased amplifying very early (Ellis *et al.*, 1998). In contrast, BARE-1 insertions are polymorphic in many cereal species such as wheat, rye and oat (Gribbon *et al.*, 1999) and even in the grasses (Vicient *et al.*, 2001a), indicating that the transposition activity of this family of elements had been maintained across a very large spectrum of hosts. Similarly, Tnt1 appeared to have maintained its transposition activity across many genera of Solanaceae, though the levels of amplification differed within each species (Grandbastien *et al.*, 2005).

Indeed, because transposable elements may adopt different transposition strategies, there can be strong selection for conservation of function in low copy number elements that have the capability of long-term persistence or are subjected to host control mechanisms. Therefore, it is incorrect to generally assume that higher transposition activity is automatically associated with high copy number elements (Vieira and Biemont, 1997; Navarro-Quezada and Schoen, 2002; Miura *et al.*, 2004). Recent studies of species in *Solanum* subsection *Lycopersicon* and in *Capsicum*, using similar elements, showed different results. After discounting for phylogeny, no significant linear correlation was found between the average number of insertion sites per accession (estimate of copy number) and polymorphism values (estimate of transposition activity), in species of subsection *Lycopersicon* (Tam *et al.*, in press). However, in *Capsicum*, significant positive correlations were detected between copy number and transposition activity, again showing differences in transposition strategy of similar element families in different hosts (Tam *et al.*, still in preparation).

Certain plant retrotransposons are activated by stress or environmental modifications (Grandbastien, 1998; Melayah *et al.*, 2001; Takeda *et al.*, 2001). An interesting study demonstrated that the dynamics of BARE-1 retrotransposons in the wild barley *Hordeum spontaneum*, differed in response to microclimatic divergence (Kalendar *et al.*, 2000). The environmental history of the host will thus play a crucial role in the amplification dynamics of a particular element and in generating potentially useful insertion polymorphisms. Another important factor influencing element dynamics is the evolutionary history of the host, especially important events such as hybridization, introgression and allopolyploidy. Interspecific hybridisation activated amplification of retroelements in certain animals (O'Neill *et al.*, 1998). In plants, the situation remains controversial, as many examples showed that interspecific hybridisation was able to activate the expression of retrotransposons, but not necessarily their amplification (Baumel *et al.*, 2002; Kashkush *et al.*, 2003).

Neutrality, Linkage Disequilibrium and Life-history

The neutral theory (Kimura, 1968; King and Jukes, 1969) suggested that most polymorphisms are selectively neutral, with strongly deleterious mutations being quickly eliminated and contributing little to variation within and between species. Indeed, many diversity studies using molecular marker systems today, adopt this underlying assumption of neutrality for analysing and interpreting results (Ford, 2002). However, because the nature of transposable elements make them susceptible to natural selection and hitch-hiking events such as selective sweeps and background selection, it is also important to examine the neutrality of element based molecular markers. Ultimately, researchers have the opportunity to choose molecular markers (neutral or not), ideally suited to the particular questions being addressed (Avise, 2004).

Vershinin *et al.* (2003) observed high polymorphisms but constant copy numbers for PDR1 in *Pisum* and suggested that segregation and genetic drift were effective mechanisms determining the loss of elements. Studies of retrotransposons in tomato, pepper and related species also generally showed neutral distributions of insertions and thus identified genetic drift as one of the primary forces influencing the maintenance of element insertions (Tam *et al.*, 2005; Tam *et al.*, in press). Recombination rate is another important well-known factor that influences genetic variation, and is generally linked to the life-history of the hosts such as mating system (outcrossers, inbreeders), effective population size and geographical and historical factors such as population structure, genetic bottleneck and domestication. More specifically, many recent studies have focussed on obtaining a better understanding of the roles of recombination rates and breeding systems in driving evolutionary changes in transposable element dynamics. Various population genetics models with different theoretical predictions exist, that discuss the regulation of abundance and frequencies of element insertions (Wright and Schoen, 1999; Morgan, 2001; Wright *et al.*, 2001; Petrov *et al.*, 2003). In our recent study of species in *Solanum* subsection *Lycopersicon*, insertion site frequencies showed significant positive correlations with mating systems where self incompatible species had higher levels of polymorphisms.

Closely related to the neutrality issue is linkage disequilibrium, or the non-random association of alleles. For example, if hitch-hiking is a significant factor, then even if much observed variation is neutral, the dynamics of the neutral variation could be governed more by linkage to selected sites than by genetic drift (Ford, 2002). Linkage disequilibrium observed in a population is a consequence of interplay of many similar factors determining genetic variation such as recombination rate, genetic isolation, population subdivision, natural and artificial selection, population size, frequency of new mutations and others (Rafalski and Morgante, 2004). Linkage disequilibria may be specifically important for studies of genetic mapping (no segregation distortions) and its influence on genetic relationships of breeding material (due to high artificial selection for certain phenotypes). The MITEs *Stowaway* and *Barfly*, with tendencies for inserting into genic regions (higher probabilities of selection and linkage disequilibria), actually performed well as molecular markers in a mapping study of barley (Chang *et al.*, 2001). In tomato, genetic relationships depicted by SSAP were more similar to AFLP but differed slightly from SSR, which showed clearer delimitation of tomato lines based on fruit morphology (Tam *et al.*, 2005). This suggested that SSAP and AFLP essentially reflected the same underlying divergences and were useful for inference of overall genetic variation and relationships, while SSR was found to be more sensitive to linkage disequilibria caused probably by the complex inheritance of fruit characteristics and was more useful for distinguishing breeding material.

The Informative TE Markers

For all the reasons listed above, a family of elements may not be informative at all taxonomic levels and their usage will depend on the dynamics of amplification. Insertion polymorphisms of PDR1 in pea, replicated well the genetic relationships of most of the *Pisum* species, including the eco-geographic distribution of the accessions except, for *Pisum abyssinicum* (Ellis *et al.*, 1998). We also observed that retrotransposon insertions were useful for inferring the phylogeny of *Solanum* subsection *Lycopersicon* and *Capsicum* (Tam *et al.*, in press, Tam *et al.*, still in preparation). Similarly, insertion polymorphisms of SINE S1Bn were useful for inferring the phylogeny of the genus *Brassica* (Tatout *et al.*, 1999). In contrast, Ty1-*copia* elements showed mostly unique insertions in sweet potato and were not useful for studying genetic relationships among selected eight geographical genotypes (Berenyi *et al.*, 2002). Inter-variety polymorphisms of BARE-1 reflected well the evolution of barley varieties but were too polymorphic at the interspecific level to reveal species relationships in *Hordeum* (Kalendar *et al.*, 1999). A similar observation was made for Tnt1 insertions between species of *Nicotiana* (Melayah *et al.*, 2004), suggesting that in *Hordeum* and *Nicotiana*, post-speciation amplifications generated many new insertions, masking phylogenetic signals from the older insertions.

Therefore, the suitability of transposable elements for use at a particular taxonomic level has to be evaluated according to individual cases (Leigh *et al.*, 2003). The ideal element should have a constant and relatively low rate of transposition over the course of time, such as seems to be the cases of PDR1 in pea (Ellis *et al.*, 1998) and of Tnt1 and T135 in tomato and pepper (Tam *et al.*, 2005). This limitation, however, can be overcome either by using another family of more suitable retrotransposon, or a combination of elements possessing different dynamics of expansion. Each genome comprises many different families of elements, for example, maize and rice were estimated to have many dozens of retrotransposon families each (Bennetzen *et al.*, 1998; McCarthy *et al.*, 2002). Retrotransposon sequences isolated from one species can also be used in other genera of the same family (Baumel *et al.*, 2002; Tam *et al.*, 2005). Each family of elements present within a given genome can be activated *a priori* by different stimuli, hence resulting in different levels of amplification and polymorphism. The combination of insertion polymorphisms from different elements in the same hosts should permit a more refined study of relationships between groups of species.

Are Genomes Undergoing Permanent Expansion?

Retroelement insertions are considered stable because no specific mechanism exists for the total excision of a retroelement from its insertion locus. Nonetheless, plant genomes are neither undergoing permanent nor irreversible expansion, so in fact, there must be mechanisms that decelerate and even reverse genome expansion. Recent information has shown that plants were capable of repeatedly increasing and reducing the size of their genomes in the course of evolution (Bennetzen, 2002; Wendel *et al.*, 2002; Bennetzen *et al.*, 2005).

Element copies can be lost in a non-specific manner caused by deletion and/or recombination events (e.g. ectopic recombination), further generated by interactions between repetitive elements. However, the primary control specifically limiting the increment of genome size is the very strict control of transposition. For many plants, retrotransposon expression is restricted to particular conditions or tissues, less likely to transmit new insertions to the descendants. This regulation is often mediated by element-specific regulatory sequences and involves specific host transcriptional factors (Grandbastien *et al.*, 1998; Sugimoto *et al.*, 2000). Epigenetic mechanisms are also implicated in repressing expression of plant retrotransposons (Okamoto and Hirochika, 2001). In addition, transcriptional activation of retroelements is not necessarily followed by transposition (Kashkush *et al.*, 2003), which indicates that control might happen at different times during the transposition cycle. A second specific mechanism responsible for reducing the retrotransposon load, is recombination between LTRs within the same element. This mechanism eliminates the majority of the element, leaving only a unique LTR, called a solo LTR. For example, there are twenty times more BARE-1 solo LTRs than complete elements in the genomes of certain species of *Hordeum*, with their relative proportions varying according to the host (Vicient *et al.*, 1999). Therefore, this is an extremely efficient, and probably host-specific control on the number of active copies in the genome. The presence of a solo LTR at a specific site will not affect the SSAP band generated on an insertion profile, if the retrotransposon primer was designed for the LTR. However, it will complicate the utilisation of the site for RBIP, since the PCR external/external will also result in a product.

A third control mechanism applies at the level of maintenance of a particular insertion. Insertions of transposable elements that have an effect on the phenotype will consequently influence allelic frequencies of both the full and empty sites until there is eventual fixation or elimination in a specific population. Mechanisms of this sort are probably responsible for the interesting discrepancies seen in the distributions of elements in function of their age. Finally, it was recently proposed that the rate of illegitimate recombination might be fundamental in controlling the number of retrotransposons, especially in small genomes. The content of retrotransposons in *Arabidopsis* is very low (2%) and the majority of the elements are inactive. Nevertheless, the *Arabidopsis* genome retains partial traces of the presence of a large number of retrotransposons, which suggests that the size of its genome had been reduced by fragmentation of old endogenous elements perpetuated by many illegitimate recombination events (Bennetzen, 2002; Devos *et al.*, 2002). Similarily, recent analysis of the rice genome show that a large number of retrotransposon insertions are found as truncated remains, and that a major part of the retrotransposon content of rice has been deleted in the last few millions years (Gao *et al.*, 2004; Ma *et al.*, 2004; Vitte *et al.*, in preparation). All these observations indicate the existence of efficient mechanisms that limit and reverse genomic expansion driven by the activity of repetitive elements that transpose by amplification. Therefore, the assumption of stable insertions may be violated and should be treated with caution notably when used for phylogenetic studies. Generally, this necessitates using many polymorphic insertions in order to confidently validate observed evolutionary information.

CONCLUSION

The discovery of transposable elements by Barbara McClintock had been a major step in our progressive understanding of the concepts of heredity. Considered for a long time as parasite DNA, transposable elements now actually appear to be one of the principal agents of genome fluidity and dynamics. Their ability to rapidly modify their positions and copy numbers in the genome, makes transposable elements probably one of the major factors determining variations in genome size. Transposable elements have also recently been shown to acquire and fuse fragments of cellular genes, creating the raw material for the evolution of new genetic functions (Jiang *et al.*, 2004; Bennetzen, 2005; Lai *et al.*, 2005; Morgante *et al.*, 2005). Transposable elements also generate fine-scale mutations and are capable of modifying the expression of cellular genes. They are implicated in the origin of a significant proportion of regulatory regions of genes (Jordan *et al.*, 2003) and their response to genomic shocks or developmental signals can in turn mediate the modulation of multiple adjacent host genes (Kashkush *et al.*, 2003; Peaston *et al.*, 2004), leading to reprogramming of batteries of genes as part of the organism's response to specific stimuli. The repetitive and widespread nature of transposable elements, combined with their ability to respond to specific signals, are thus expected to be fundamental in the fine-tuning of host gene expression and function, increasingly consolidating the original concept of "Controlling Elements" described by Barbara McClintock.

Transposition activity depends on subtle and complex interactions with host genomes, and this gives rise to the somewhat perturbing concept in classical genetics, the acknowledgement of the possibility of direct environmental influence on genome structure. Transposable elements also have the ability of changing their sequences to adapt their behaviour to the individual host. All these particularities also make transposable elements fascinating study models for understanding processes of gene regulation and evolution, and are sometimes considered as important factors in macroevolution and speciation (McDonald, 1990; Hamdi *et al.*, 2000).

It is the essential role that transposable elements play in generating genetic diversity that plainly justifies their utilisation as molecular markers. They can be applied in many domains, from analysing genetic diversity, molecular phylogenetics to genotyping using exciting high throughput automated strategies. For species whose genomes are being sequenced, the development of element based molecular marker strategies is generally correlated with active searches for insertions from databases, sometimes also providing *in silico* evidence of insertion polymorphisms. For the majority of other species, it is possible to isolate retrotransposons using certain conserved motifs of their coding regions, making it possible to develop transposable elements as molecular markers for the species of interest.

Transposable element-based molecular markers more than fulfil the important objectives of conservation genetics by providing the tools to assess genetic diversity and to better understand the evolutionary processes that promote and sustain biodiversity (Avise, 2004). The development and general application of more transposable elements as molecular markers in diverse species will ultimately provide clearer and more precise visions of the behaviour of elements, enabling the further understanding of their biological impact and importance as drivers of genome evolution.

ACKNOWLEDGEMENTS

We thank Quang Hien Le for his comments. The data presented in Fig. 26.4 was obtained in collaboration with Steven Pearce (University of Brighton, UK) and the Unité de Génétique et Amélioration des Fruits et Légumes (UGAFL) at INRA-Montfavet, in the frame of the CEE TEGERM project (QLRT-1999-31502).

REFERENCES

Araujo, P.G., Casacuberta, J.M., Costa, A.P.P., Hashimoto, R.Y., Grandbastien, M.-A., Van Sluys, M.-A. 2001. Retrolyc1 subfamilies defined by different U3 regulatory regions in the *Lycopersicon* genus. *Mol. Gen. Genom.*, 266, 35-41.

Avise, J.C., 2004. Molecular markers in conservation genetics. In: Molecular markers, natural history and evolution". Sinauer Associates, USA (second edition), pp: 475-541.

Baumel, A., Ainouche, M., Kalendar, R., Schulman, A.H. 2002. Retrotransposons and genomic stability in populations of the young allopolyploid species *Spartina anglica* Hubbard, C.E. (Poaceae). *Mol. Biol. Evol.*, 19, 1218-27.

Beguiristain, T., Grandbastien, M.-A., Puigdomenech, P., Casacuberta, P. 2001. Three Tnt1 subfamilies show different stress-associated patterns of expression in tobacco. Consequences for retrotransposon control and evolution in plants. *Plant Physiol.*, 127, 212-221.

Bennetzen, J.L., SanMiguel, P., Chen, M., Tikhonov, A., Francki, M., Avramova, Z. 1998. Grass genomes. *Proc. Natl. Acad. Sci.* USA, 95, 1975–1978.

Bennetzen, J.L. 2000. Transposable element contributions to plant gene and genome evolution. *Plant Mol. Biol.* 42, 251-269.

Bennetzen, J.L. 2002. Mechanisms and rates of genome expansion and contraction in flowering plants. Genetica, 115, 29-36.

Bennetzen, J.L. 2005 Transposable elements, gene creation and genome rearrangement in flowering plants. *Curr. Op. Genet. Dev.*, 15, 621-7.

Bennetzen, J.L., Ma, J., Devos, K.M. 2005. Mechanisms of recent genome size variation in flowering plants. Ann. Bot. (Lond.), 95, 127-32.

Berenyi, M., Gichuki, T., Schmidt, J., Burg, K. 2002. Ty1-copia retrotransposon-based S-SAP (sequence-specific amplified polymorphism) for genetic analysis of sweet potato. *Theor. Appl. Genet.*, 105, 862-869.

Bernet, G.P., Asins, M.J. 2003. Identification and genomic distribution of gypsy like retrotransposons in *Citrus* and *Poncirus*. *Theor. Appl. Genet.*, 108, 121-130.

Boyko, E., Kalendar, R., Korzun, V., Fellers, J., Korol, A., Schulman, A.H., Gill, B.S. 2002. A high-density cytogenetic map of the *Aegilops tauschii* genome incorporating retrotransposons and defense-related genes: insights into cereal chromosome structure and function. *Plant Mol. Biol.*, 48, 767-790.

Brown, T.A. 1999. Genome Anatomies. In "Genomes". BIOS Scientific Publishers, USA, pp: 116-142.

Casa, A.M., Brouwer, C., Nagel, A., Wang, L., Zhang, Q., Kresovich, S., Wessler, S.R. 2000. The MITE family *Heartbreaker* (*Hbr*): Molecular markers in maize. *Proc. Natl. Acad. Sci.* USA, 97, 10083-10089.

Casa, A.M., Mitchell, S.E., Smith, O.S., Register, J.C. 3rd, Wessler, S.R., Kresovich, S. 2002. Evaluation of Hbr (MITE) markers for assessment of genetic relationships among maize (*Zea mays* L.) inbred lines. *Theor. Appl. Genet.*, 104, 104-10.

Chang, R.Y., O'Donoughue, L.S., Bureau, T.E. 2001. Inter-MITE polymorphisms (IMP): a high throughput transposon-based genome mapping and fingerprinting approach. *Theor. Appl. Genet.*, 102, 773-781.

Devic, M., Albert, S., Delseny, M., Roscoe, T.J. 1997. Efficient PCR walking on plant genomic DNA. *Plant Physiol. Biochem.*, 35, 331-339.

Devos, K.M., Brown, J.K.M., Bennetzen, J.L. 2002. Genome size reduction through illegitimate recombination counteracts genome expansion in Arabidopsis. *Genome Res.*, 12, 1075-1079.

Ellis, T.H.N., Poyser, S.J., Knox, M.R., Vershinin, A.V., Ambrose, M.J. 1998. Polymorphism of insertion sites of *Ty1-copia* class retrotransposons and its use for linkage and diversity in pea. *Mol. Gen. Genet.*, 260, 9-19.

Feschotte, C., Jiang, N., Wessler, S.R. 2002a. Plant transposable elements: where genetics meets genomics. *Nat. Rev. Genet.*, 3, 329-41.

Feschotte, C., Zhang, X., Wessler, S.R. 2002b. Minitature Inverted-repeat Transposable Elements and their relationship to established DNA transposons. In "Mobile DNA II", Eds Craig N.L. *et al.*, Chapter 50, 1147-1158.

Flavell, A.J., Knox, M.R., Pearce, S.R., Ellis, T.H. 1998. Retrotransposon-based insertion polymorphisms (RBIP) for high throughput marker analysis. *Plant J.*, 16, 643-650.

Flavell, A.J., Bolshakov, V.N., Booth, A., Jing, R., Russell, J., Ellis, T.H., Isaac, P. 2003. A microarray-based high throughput molecular marker genotyping method: the tagged microarray marker (TAM) approach. Nucleic Acids Res., 31: e115.

Ford, M.J. 2002. Applications of selective neutrality tests to molecular ecology. *Mol. Ecol.*, 11, 1245-1262.

Frey, M., Stettner, C., Gierl, A. 1998. A general method for gene isolation in tagging approaches: amplification of insertion mutagenised sites (AIMS). *Plant J.*, 13, 717-721.

Gao, L., McCarthy, E.M., Ganko, E.W., McDonald, J.F. 2004. Evolutionary history of *Oryza sativa* LTR retrotransposons: a preliminary survey of the rice genome sequences. BMC Genomics, 5, 18.

Garcia-Martinez, J., Martinez-Izquierdo, J.A. 2003. Study on the evolution of the grande retrotransposon in the *Zea* genus. *Mol. Biol. Evol.*, 20, 831-41.

Grandbastien, M.-A. 1998. Activation of plant retrotransposons under stress conditions. Trends Plant Sci., 3, 181-187.

Grandbastien, M.-A., Audeon, C., Bonnivard, E., Casacuberta, J.M., Chalhoub, B., Costa, A.-P.P., Le, Q.H., Melayah, D., Petit, M., Poncet, C., Tam, S.M., Van Sluys, M.-A., Mhiri, C. 2005. Stress activation and genomic impact of Tnt1 retrotransposons in Solanaceae. Cytogenet. Genome Res., 110, 229-241.

Gribbon, B.M., Pearce, S.R., Kalendar, R., Schulman, A.H., Paulin, L., Jack, P., Kumar, A., Flavell, A.J. 1999. Phylogeny and transpositional activity of Ty1-*copia* group retrotransposons in cereal genomes. *Mol. Gen. Genet.* 261, 883-891.

Hamada, H., Takasaki, N., Reist, J.D., DeCicco, A.L., Goto, A., Okada, N. 1998. Detection of the ongoing sorting of ancestrally polymorphic SINEs toward fixation or loss in populations of two species of charr during speciation. *Genetics,* 150, 301-311.

Hamdi, H.K., Nishio, H., Tavis, J., Zielinski, R., Dugaiczyk, A. 2000. *Alu*-mediated phylogenetic novelties in gene regulation and development. *J. Mol. Biol.*, 299, 931-9.

International Human Genome Sequencing Consortium. 2001. Initial sequencing and analysis of the human genome. *Nature,* 409, 860-921.

Jiang, N., Wessler, S.R. 2001. Insertion preference of maize and rice miniature inverted repeat transposable elements as revealed by the analysis of nested elements. *Plant Cell,* 13, 2553-2564.

Jiang, N., Bao, Z., Zhang, X., Eddy, S.R., Wessler, S.R. 2004. Pack-MULE transposable elements mediate gene evolution in plants. *Nature,* 431, 569-73.

Jordan, I.K., Rogozin, I.B., Glazko, G.V., Koonin, E.V. 2003. Origin of a substantial fraction of human regulatory sequences from transposable elements. *Trends Genet.*, 19, 68-72.

Kalendar, R., Grob, T., Regina, M., Suomeni, A., Schulman, A. 1999. IRAP and REMAP: two new retrotransposon-based DNA fingerprinting techniques. *Theor. Appl. Genet.*, 98, 704-711.

Kalendar, R., Tanskanen, J., Immonen, S., Nevo, E., Schulman, A. 2000. Genome evolution of wild barley (*Hordeum spontaneum*) by BARE-1 retrotransposon dynamics in response to sharp microclimatic divergence. *Proc. Natl. Acad. Sci.* USA, 97, 6603-6607.

Kalendar, R., Vicient, C.M., Peleg, O., Anamthawat-Jonsson, K., Bolshoy, A., Schulman, A. 2004. Large retrotransposon derivatives: abundant, conserved but non-autonomous retroelements of barley and related genomes. *Genetics,* 166, 1437-50.

Karp, A., Seberg, O., Buiatti, M. 1996. Molecular Techniques in the assessment of Botanical Diversity. *Ann. Bot.,* 78, 143-149.

Kashkush, K., Feldman, M., Levy, A.A. 2003. Transcriptional activation of retrotransposons alters the expression of adjacent genes in wheat. *Nature Genet.,* 33, 102-6.

Kikuchi, K., Terauchi, K., Wada, M., Hirano, H.Y. 2003. The plant MITE mPING is mobilized in anther culture. *Nature,* 421, 167-170.

Kimura, M. 1968. Evolutionary rate at the molecular level. *Nature,* 217, 624-626.

King, J.L., Jukes, T.H. 1969. Non-Darwinian evolution. Science, 164, 788-798.

Kumar, A., Bennetzen, J.L. 1999. Plant retrotransposons. *Annu. Rev. Genet.,* 33, 479-532.

Kumar, A., Hirochika, H. 2001. Applications of retrotransposons as genetic tools in plant biology. Trends Plant Sci., 6, 127-134.

Labra, M., Imazio, S., Grassi, F., Rossoni, M., Sala, F. 2004. Vine-1 retrotransposon-based sequence-specific amplified polymorphism for *Vitis vinifera* L. genotyping. Plant breeding, 123, 180-185.

Lai, J., Li, Y., Messing, J., Dooner, H.K. 2005. Gene movement by Helitron transposons contributes to the haplotype variability of maize. *Proc. Natl. Acad. Sci.* USA, 102, 9068-73.

Laten, H.M., Majumdar, A., Gaucher, E.A. 1998. *SIRE*-1, a *copia*/Ty1-like retroelement from soybean, encodes a retroviral envelope-like protein. *Proc. Natl. Acad. Sci.* USA, 95, 6897–6902.

Le, Q.H., Wright, S., Yu, Z., Bureau, T. 2000. Transposon diversity in *Arabidopsis thaliana*. *Proc. Natl. Acad. Sci.* USA, 97, 7376-7381.

Le, Q.H., Melayah, D., Bonnivard, E., Grandbastien, M.-A. still submitted Distribution of the Tnt1 retrotransposon in tobacco varies with age of insertion.

Leigh, F., Kalendar, R., Lea, V., Lee, D., Donini, P., Schulman, A.H. 2003. Comparison of the utility of barley retrotransposon families for genetic analysis by molecular marker techniques. *Mol. Genet. Genomics,* 269, 464-474.

Lönnig, W.-E., Saedler, H. 2002. Chromosome rearrangements and transposable elements. *Annu. Rev. Genet.,* 36, 389-410.

Ma, J., Devos, K.M., Bennetzen, J.L. 2004. Analyses of LTR-retrotransposon structures reveal recent and rapid genomic DNA loss in rice. *Genome Res.,* 14, 860-9.

Manninen, O., Kalendar, R., Robinson, J., Shulman, A.H. 2000. Application of BARE-1 retrotransposon markers to map a major resistance gene for net blotch in barley. *Mol. Gen. Genet.,* 264, 325-334.

Marshall, J.A., Knapp, S., Davey, M.R., Power, J.B., Cocking, E.C., Bennett, M.D., Cox, A.V. 2001. Molecular systematics of *Solanum* section *Lycopersicum* (*Lycopersicon*) using nuclear ITS rDNA region. *Theor. Appl. Genet.,* 103, 1216-1222.

McCarthy, E.M., Liu, J., Lizhi, G., McDonald, J.F. 2002. Long terminal repeat retrotransposons of *Oryza sativa. Genome Biol.,* 3, RESEARCH0053.

McCarthy, E.M., McDonald, J.F. 2003. LTR_STRUC: a novel search and identification program for LTR retrotransposons. *Bioinformatics,* 19, 362-367.

McDonald, J.F. 1990. Macroevolution and retroviral elements. *BioScience,* 40, 183-191.

Medstrand, P., Van de Lagemaat, L.N., Mager, D.L. 2002. Retroelement distributions in the human genome: variations associated with age and proximity to genes. *Genome Res.,* 12, 1483-1495.

Melayah, D., Bonnivard, E., Chalhoub, B., Audéon, C., Grandbastien, M.-A. 2001. The mobility of the tobacco Tnt1 retrotransposon correlates with its transcriptional activation by fungal factors. *Plant J.*, 28, 159-168.

Melayah, D., Lim, K.Y., Bonnivard, E., Chalhoub, B., Dorlhac de Borne, F., Mhiri, C., Leitch, A.R., Grandbastien, M.-A. 2004. Distribution of the Tnt1 retrotransposon family in the amphidiploid tobacco (*Nicotiana tabacum*) and its wild relatives. *Biol. J. Linnean Soc.*, 82, 639-649.

Miura, A., Kato, M., Watanabe, K., Kawabe, A., Kotani, H., Kakutani, T. 2004. Genomic localization of endogenous mobile CACTA family transposons in natural variants of *Arabidopsis thaliana. Mol. Genet. Genomics*, 270, 524-32.

Morgan, M.T. 2001. Transposable element number in mixed mating populations. *Genet. Res.*, 77, 261-75.

Morgante, M., Brunner, S., Pea, G., Fengler, K., Zuccolo, A., Rafalski, A. 2005. Gene duplication and exon shuffling by helitron-like transposons generate intraspecies diversity in maize. *Nat. Genet.*, 37, 997-1002.

Murata, S., Takasaki, N., Saitoh, M., Tachida, H., Okada, N. 1996. Details of retropositional genome dynamics that provide a rationale for a generic division: the distinct branching of all the pacific salmon and trout (*Oncorhynchus*) from the Atlantic salmon and trout (*Salmo*). *Genetics*, 142, 915-26.

Nakazaki, T., Okumoto, Y., Horibata, A., Yamahira, S., Teraishi, M., Nishida, H., Inoue, H., Tanisaka, T. 2003. Mobilization of a transposon in the rice genome. *Nature*, 421, 170-172.

Navarro-Quezada, A., Schoen, D.J. 2002. Sequence evolution and copy number of Ty1-copia retrotransposons in diverse plant genomes. *Proc. Natl. Acad. Sci.* USA, 99, 268-73.

Nikaido, M., Rooney, A.P., Okada, N. 1999. Phylogenetic relationships among cetartiodactyls based on insertions of short and long interspersed elements: Hippopotamuses are the closest extant relatives of whales. *Proc. Natl. Acad. Sci.* USA, 96, 10261-10266.

Nikaido, M., Nishihara, H., Hukumoto, Y., Okada, N. 2003. Ancient SINEs from African endemic mammals. *Mol. Biol. Evol.*, 20, 522 527.

O'Neill, R.J., O'Neill, M.J., Graves, J.A. 1998. Undermethylation associated with retroelement activation and chromosome remodelling in an interspecific mammalian hybrid. *Nature*, 393, 68-72.

Okamoto, H., Hirochika, H. 2001. Silencing of transposable elements in plants. *Trends Plant Sci.*, 6, 527-34.

Pearce, S.R., Stuart-Rogers, C., Knox, M.R., Kumar, A., Ellis, T.H.N., Flavell, A.J. 1999. Rapid isolation of plant Ty1-*copia* group retrotransposon LTR sequences for molecular marker studies. *Plant J.*, 19, 711-717.

Park, K.C., Lee, J.K., Kim, N.H., Shin, Y.B., Lee, J.H., Kim, N.S. 2003. Genetic variation in *Oryza* species detected by MITE-AFLP. *Genes Genet Syst.* 78:235-43.

Pearce, S.R., Knox, M.R., Ellis, T.H.N., Flavell, A.J., Kumar, A. 2000. Pea Ty1-*copia* group retrotransposons: transpositional activity and use as markers to study genetic diversity in *Pisum. Mol. Gen. Genet.*, 263, 898-907.

Peaston, A.E., Evsikov, A.V., Graber, J.H., de Vries, W.N., Holbrook, A.E., Solter D. 2004. Retrotransposons regulate host genes in mouse oocytes and preimplantation embryos. *Developmental Cell*, 7, 597–606.

Pelsy, F., Schehrer, L., Merdinoglu, D. 2003. Development of grapevine retrotransposon-based molecular markers. Proceedings of the Eighth International Conference on Grape Genetics and Breeding. *Acta. Horticulturae*, 603, 83-87.

Pereira, V. 2004. Insertion bias and purifying selection of retrotransposons in the *Arabidopsis thaliana* genome. *Genome Biol.*, 5, R79.

Peterson-Burch, B.D., Nettleton, D., Voytas, D.F. 2004. Genomic neighborhoods for *Arabidopsis* retrotransposons: a role for targeted integration in the distribution of the Metaviridae. *Genome Biol.*, 5, R78.

Petrov, D.A., Aminetzach, Y.T., Davis, J.C., Bensasson, D., Hirsh, A.E. 2003. Size matters: non-LTR retrotransposable elements and ectopic recombination in *Drosophila. Mol. Biol. Evol.*, 20, 880-92.

Porceddu, A., Albertini, E., Barcaccia, G., Marconi, G., Bertoli, F.B., Veronesi, F. 2002. Development of SSAP markers based on an LTR-like sequence from *Medicago sativa* L. *Mol. Genet. Genomics*, 267, 107-114.

Provan, J., Thomas, W.T.B., Forster, B.P., Powell, W. 1999. *Copia*-SSR: A simple marker technique which can be used on total genomic DNA. *Genome*, 42, 363-366.

Queen, R.A., Gribbon, B.M., James, C., Jack, P., Flavell, A.J. 2004. Retrotransposon-based molecular markers for linkage and genetic diversity analysis in wheat. *Mol. Genet. Genomics*, 271, 91-7.

Rafalski, A., Morgante, M. 2005. Corn and humans: recombination and linkage disequilibrium in two genomes of similar size. *Trends Genet.*, 20,103-11.

Ramsay, L., Macaulay, M., Cardle, L., Morgante, M., Degli Ivanissevich, S., Maestri, E., Powell, W., Waugh, R. 1999. Intimate association of microsatellite repeats with retrotransposons and other dispersed repetitive elements in barley. *Plant J.*, 17, 415-425.

Rodriguez, J.M., Berke, T., Engle, L., Nienhuis, J. 1999. Variation among and within *Capsicum* species revealed by RAPD markers. *Appl. Genet.*, 99, 147-156.

SanMiguel, P., Tikhonov, A., Jin, Y.K., Motchoulskaia, N., Zakharov, D., Melake-Berhan, A., Springer, P.S., Edwards, K.J., Lee, M., Avramova, Z., Bennetzen, J.L. 1996. Nested retrotransposons in the intergenic regions of the maize genome. *Science*, 274, 765-768.

Santiago, N., Herraiz, C., Goni, J.R., Messeguer, X., Casacuberta, J.M. 2002. Genome-wide analysis of the Emigrant family of MITEs of *Arabidopsis thaliana. Mol. Biol. Evol.*, 19, 2285-2293.

Schulman, A., Flavell, A.J., Ellis, T.H.N. 2004. The application of LTR Retrotransposons as molecular markers in plants. In: "Mobile Genetic Elements protocols and genomic applications". Eds: Miller, W.J., Capy, P. Humana Press, Totawa, NJ, USA, pp. 145-174.

Shedlock, A.M., Okada, N. 2000. SINE insertions: powerful tools for molecular systematics. Bioessays. 2000, 22, 148-160.

Shimamura, M., Yasue, H., Ohshima, K., Abe, H., Kato, H., Kishiro, T., Goto, M., Munechikak, I., Okada, N. 1997. Molecular evidence from retroposons that whales form a clade within even-toed ungulates. *Nature*, 388, 666-670.

Smith, J.F., Sytsma, K.J. 1994. Molecules and morphology: congruence of data in *Columnea* (Gesneriaceae). *Plant Syst. Evol.*, 193, 37-52.

Sugimoto, K., Takeda, S., Hirochika, H. 2000. MYB-Related transcription factor NtMYB2 induced by wounding and elicitors is a regulator of the tobacco retrotransposon *Tto1* and defense-related genes. *Plant Cell*, 12, 2511–2527.

Sytsma, K.J. 1990. DNA and morphology: Inference of plant phylogeny. Trends Ecol. Evol. 5, 104-110.

Takahashi, K., Terai, Y., Nishida, M., Okada, N. 1998. A novel family of short interspersed repetitive elements (SINEs) from cichlids: the patterns of insertion of SINEs at orthologous loci support the proposed monophyly of four major groups of cichlids fishes in Lake Tanganyika. *Mol. Biol. Evol.*, 15, 391-407.

Takeda, S., Sugimoto, K., Kakutani, T., Hirochika, H. 2001. Linear DNA intermediates of the Tto1 retrotransposon in Gag particles accumulated in stressed tobacco and *Arabidopsis thaliana. Plant J.*, 28, 307-317.

Tam, S.M., Mhiri, C., Vogelaar, A., Kerkveld, M., Pearce, S.R., Grandbastien, M.-A. 2005. Comparative analyses of genetic diversities within tomato and pepper collections detected by retrotransposon-based SSAP, AFLP and SSR. *Theor. Appl. Genet.* 110, 819-83.

Tam, S.M., Causse, M., Garchery, C., Burck, H., Mhiri, C., Grandbastien, M.-A., (2007) Understanding the distribution dynamics of *copia*-type retrotransposons in tomato and related wild species. J. Evolution. Biol. (in press).

Tam, S.M., Lefebvre, V., Palloix, A., Grandbastien, M.-A. (still in preparation). Retrotransposon-based genetic diversity and interspecific relationships of *Capsicum.*

Tatout, C., Warwick, S., Lenoir, A., Deragon, J.M. 1999. SINE insertions as clade markers for wild crucifer species. *Mol. Biol.Evol.*, 16, 1614-1621.

The Arabidopsis Genome Initiative. 2000. Analysis of the genome sequence of the flowering plant Arabidopsis thaliana. *Nature,* 408, 796-815.

Tikhonov, A.P., Lavie, L., Tatout, C., Bennetzen, J.L., Avramova, Z., Deragon, J.M. 2001. Target sites for SINE integration in *Brassica* genomes display nuclear matrix binding activity. *Chromosome Res.,* 9, 325-37.

Tikhonov, A.P., SanMiguel, P.J., Nakajima, Y., Gorenstein, N.M., Bennetzen, J.L., Avramova, Z. 1999. Colinearity and its exceptions in orthologous adh regions of maize and sorghum. *Proc. Natl. Acad. Sci.* USA, 96, 7409-14.

Van den Broeck, D., Maes, T., Sauer, M., Zethof, J., De Keukeleire, P., D'Hauw, M., Van Montagu, M., Gerats, T. 1998. Transposon Display identifies individual transposable elements in high copy number lines. *Plant J.,* 13,121-129.

Vernhettes, S., Grandbastien, M.-A., Casacuberta, J.M. 1998. The evolutionary analysis of the Tnt1 retrotransposon in Nicotiana species reveals the high plasticity of its regulatory sequences. *Mol. Biol. Evol.,* 15, 827-836.

Vershinin, A.V., Ellis, T.H.N. 1999. Heterogeneity of the internal structure of *PDR1*, a family of *Ty1/copia*-like retrotransposon in pea. *Mol. Gen. Genet.,* 262, 703-713.

Vershinin, A.V., Allnutt, T.R., Knox, M.R., Ambrose, M.J., Ellis, T.H. 2003. Transposable elements reveal the impact of introgression, rather than transposition, in *Pisum* diversity, evolution, and domestication. *Mol. Biol. Evol.,* 20, 2067-75.

Vicient, C.M., Suoniemi, A., Anamthawat-Jonsson, K., Tanskanen, J., Beharav, A., Nevo, E., Schulman, A.H. 1999. Retrotransposon BARE-1 and its role in genome evolution in the genus *Hordeum. Plant Cell,* 11, 1769-1784.

Vicient, C.M., Jääskeläinen, M., Kalendar, R., Schulman, A.H. 2001a. Active retrotransposons are a common feature of grass genomes. *Plant Physiol.,* 125, 1283-1292.

Vicient, C.M., Kalendar, R., Schulman, A.H. 2001b. Envelope-class retrovirus-like elements are widespread, transcribed and spliced, and insertionally polymorphic in plants. *Genome Res.,* 11, 2041-2049.

Vieira, C., Biemont, C. 1997. Transposition rate of the 412 retrotransposable element is independent of copy number in natural populations of *Drosophila simulans. Mol. Biol. Evol.,* 14, 185-8.

Vitte, C., Ishii, T., Lamy, F., Brar, D., Panaud, O. 2004. Genomic paleontology provides evidence for two distinct origins of Asian rice (*Oryza sativa L.*). *Mol. Genet. Genomics,* 272, 504-11.

Vitte, C., Panaud, O., Quesneville, H. (in preparation). LTR retrotransposons in rice: recent amplification followed by rapid DNA loss.

Vos, P., Hogers, R., Bleeker, M., Reijans, M., van de Lee, T., Hornes, M., Fritjers, A., Pot, J., Peleman, J., Kuiper, M., Zabeau, M. 1995. AFLP: a new technique for DNA fingerprinting. *Nucleic Acids Res.,* 23, 4407-4414.

Walsh, B.M., Hoot, S.B. 2001. Phylogenetic relationships of *Capsicum* (Solanaceae) using DNA sequences from two noncoding regions: the chloroplast *atpB-rbcL* spacer region and nuclear *waxy* introns. Int. *J. Plant Sci.,* 162, 1409-1418.

Waugh, R., McLean, K., Flavell, A.J., Pearce, S.R., Kumar, A., Thomas, B.B.T., Powel, W. 1997. Genetic distribution of Bare-1-like retrotransposable elements in the barley genome revealed by sequence-specific amplification polymorphisms (S-SAP). *Mol. Gen. Genet.,* 253, 687-694.

Wendel, J.F., Cronn, R.C., Johnston, J.S., Price, H.J. 2002. Feast and famine in plant genomes. *Genetica,* 115, 37-47.

Witte, C.P., Le, Q.H., Bureau, T., Kumar, A. 2001. Terminal-repeat retrotransposons in miniature (TRIM) are involved in restructuring plant genomes. *Proc. Natl. Acad. Sci.* USA, 98, 13778-13783.

Wright, S.I., Schoen, D.J. 1999. Transposon dynamics and the breeding system. *Genetica,* 107, 139-48.

Wright, S.I., Le, Q.H., Schoen, D.J., Bureau, T.E. 2001. Population dynamics of an Ac-like transposable element in self- and cross-pollinating arabidopsis. *Genetics,* 158,1279-88.

Wright, D.A, Voytas, D.F. 2001. *Athila4* of *Arabidopsis* and *Calypso* of soybean define a lineage of endogenous plant retroviruses. *Genome Res.,* 12, 122–131.

Yu, G.X., Wise, R.P. 2000. An anchored AFLP- and retrotransposon-based map of diploid *Avena*. *Genome,* 43, 736-749.

27

CHAPTER

Analysis of Complex Characters into Elementary Components through the Use of Molecular Quantitative Genetics in *Arabidopsis thaliana*

O. Loudet, F. Daniel-Vedele

Traditional genetic screens have mostly relied on the existence of null (or close to null), or constitutively active variants of genes, usually obtained through chemical or insertional mutagenesis and the production of transgenic plants. This has allowed the function of an impressive number of genes to be assigned in model plants such as *Arabidopsis thaliana*. However, there are still a large number of genes that have no putative function, as they have no strong homology with known genes already identified in the same, or other organisms, and when mutated, they show no obvious phenotype.

An alternative source of variation that has only recently started to be exploited in *A. thaliana* (Koornneef *et al.*, 2004) is contained in the accessions (lines derived from natural populations collected in the wild). The type of molecular diversity observed in these lines at the DNA level (mostly point mutations in the form of Single Nucleotide Polymorphisms -SNPs-, Insertion/Deletions -Indels- and small rearrangements), would be expected to give rise to more subtle and complex variation. The rationale behind this belief is that special alleles that have been selected for during evolution might represent variants that are best characterized as change-of-function as opposed to loss- or gain-of-function alleles, and which are therefore difficult or impossible to find by mutagenesis in the laboratory (Koornneef *et al.*, 2004). In addition, natural alleles are particularly likely to affect specific traits in a quantitative rather than qualitative manner, and to be effective in response to specific environmental conditions. Hence, most of the variation observed between these accessions is continuous, revealing multigenic control, where each locus accounts for only a fraction of the total phenotypic variation. A locus controlling the variation of a quantitative trait is then called a QTL (Quantitative Trait Loci) and represents a polymorphic region segregating between lines.

In model plants at least, especially when a complete sequence is available, it is now becoming feasible to identify the individual gene(s) and polymorphism(s) responsible for the QTL effect, and then study the molecular origin of the quantitative phenotype (molecular quantitative genetics). This explains the recent interest in QTL mapping in *A. thaliana* and the increasing amount of material being developed to perform these studies. Illustrated here using the example of the Bay-0 × Shahdara Recombinant Inbred Lines (RILs), a population developed at INRA Versailles, we will explain the principle of the first part of this work, i.e. the generation of the RIL population, its genotyping and phenotyping, the QTL mapping itself and its interpretation.

QUANTITATIVE TRAITS AND SEGREGATING MATERIAL

Traits can vary in a qualitative mode with the occurrence of discrete phenotypes that can be distinguished into classes; this is usually due to the segregation of a single gene (monogenic trait). On the other hand, continuous variation is the definition of quantitative traits that are evaluated as variable numbers. Mendelian genetics can be reconciliated with this continuous variation by accepting the fact that several loci are together controlling the trait, each one accounting for only a fraction of the total variance. These loci segregate more or less independently in a given polymorphic population. One way to identify them is to make a link (through the genetic markers and map) between the phenotype and the genotype of individuals in a segregating population.

In contrast to animal species, the segregating populations of plants are often derived from homozygous lines, and these are the most efficient for mapping. In the case of *A. thaliana*, lines taken from wild populations quickly become fully homozygous, due to the very low outcrossing rate in the species. Two parent populations exist of a diverse type, all derive from the F1 individual obtained after crossing the two parents. Backcrosses and F2 (selfing of the F1) have the disadvantage of not being permanent, thus limiting the amount of genotyping and phenotyping that can be done on the lines, as each individual is unique and cannot be reproduced identically because it is still partly heterozygous. Recombinant Inbred Lines (RIL) populations are immortal populations because they are made of (almost) completely fixed material. After the F2 seeds are obtained, lines are self fertilized individually until the F6 or F7 stage by several steps of Single Seed Descent -SSD (a cycle during which a single randomly-chosen plant is selfed to obtain the next generation of seeds). After the F6 stage, the residual heterozygosity is very low (1.6% at the F7), so that lines can be multiplied and remain genetically identical. It can be considered that the summed effects of recombination that occurred between the F1 plant and the RILs is equivalent to about two meiotic divisions (one at the F1- > F2 step and less than one in total between the F2 stage and the fixed material, because the decrease in residual heterozygosity reduces the effect of recombination). The result is that each RIL in a given population represents a unique mosaic of the parental genomes.

The Versailles *Arabidopsis thaliana* Resource Centre for Genomics is developing many RIL populations, for which the Bay-0 x Shahdara population is a good example (Fig. 27.1). It was developed from a cross between two distant accessions: Bay-0 is a European accession collected from open-fields in Germany, whereas Shahdara was collected in Central-Asia from the mountains in Tajikistan. One would expect them to be adapted to rather different environments and they have been shown to be genetically distant, based on sequence divergence studies of European and Asian accessions. This population is the largest available in *A. thaliana* (as of 2005), with 420 RILs genotyped at the F6 stage. Many new and even more extended sets of RILs are expected to be available soon, thanks to multiple efforts from laboratories in the Arabidopsis community.

GENOTYPING A RIL POPULATION

Genotyping is where quantitative genetics has made most progress in the last decade, allowing real genome-wide scans for QTL and precise targeted fine-mapping. The first genetic maps were originally made of visual or isozyme markers, however, these are no longer of any interest. They have been replaced by molecular markers that reveal a polymorphism in the DNA sequence. Molecular markers have many advantages: (1) they are neutral (they do not necessarily reveal a polymorphism with any phenotypic or selective impact); (2) they are usually codominant (heterozygotes can be

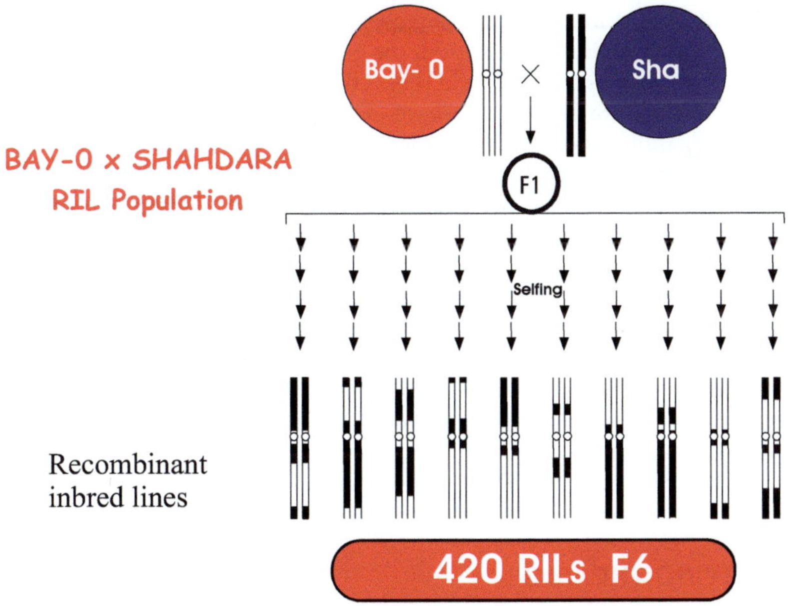

Fig. 27.1 Recombinant Inbred Lines (RIL) derived from a cross between two *Arabidopsis thaliana* accessions, Bay-0 and Shahdara. Each chromosome pair at the bottom of the figure represents the structure of a RIL genome, with white and black portions being inherited from the parents (Bay-0 and Shahdara). 420 RIL were conducted until the F7 generation by Single Seed Descent after genotyping at the F6 stage.

distinguished from both homozygous types); and (3) they are easily observable at any stage in any conditions, using a small amount of tissue.

Microsatellite markers (Tautz, 1989) are a good example of markers that are very useful for mapping, as they are also robust and easy to use routinely in any laboratory. The polymorphism, a variation in the number of repeats of a one to four nucleotide-long motif, is simply revealed on an agarose gel after a PCR reaction using primers specific for the border of the microsatellite locus. Insertion/Deletions markers (InDels) are technically identified in the same way. One additional advantage of microsatellites is that they are highly polymorphic, so that a single locus can be used to distinguish many different genotypes. In *A. thaliana* , microsatellite markers can be defined and targeted directly from the complete sequence so that the markers are anchored physically on the chromosomes.

For the Bay-0 x Shahdara RIL population, we have chosen 38 microsatellite markers polymorphic between Bay-0 and Shahdara and regularly distributed along the chromosomes. These markers were then used to genotype the 420 RILs to determine the allele present at each locus for each line. Using these genotypes, the recombination estimates between each pair of markers can be applied to calculate their position on a genetic map relative to each other (Fig. 27.2), employing such tools as Mapmaker. When carrying out QTL mapping, markers need to be unambiguously assigned a position on the map and regularly distributed. Figure 27.2 shows that for the Bay-0 x Shahdara genetic map, the average distance between markers is 10.8 cM, ranging from 2 to 17.2 cM (respectively MSAT4.39-MSAT4.8 interval on chromosome 4 and MSAT3.19-MSAT3.32 interval on chromosome 3). Other important information for QTL mapping that can be obtained from the genotypic data, is the allelic segregation. Theoretically, each locus tested should reveal the same number of lines with the Bay-0 allele and the Shahdara allele. In practice, all RIL populations show some extent of localised segregation distortion. In the case of the Bay-0 x Shahdara RIL population, a segregation distortion affects two regions on portions of chromosome 4 (36% of Bay-0 versus 64% of Shahdara) and chromosome 2 (63% of Bay-0 versus 37% of Shahdara). The extent of the distortion is in this case limited to small regions, with limited bias from the equilibrium. Moreover, the impact of these distortions is attenuated by the size of the population. Even for the most affected locus, the Bay-0 allele (36%) is still represented in 151 RIL, with is more than enough for a good estimate of QTL parameters in this region.

Other ways of genotyping RIL populations now often include more high-thoughput methods for SNP genotyping. SNP markers are extremely frequent in the *A. thaliana* genome and, although they are usually only bi-allelic, one can find rare SNP alleles that can be used to genotype several different populations derived with one common parent and create bridges between genetic maps (El-Lithy *et al.*, 2006).

PHENOTYPING THE RILS

Whatever quantitative trait is chosen, the main difficulty is that many lines need to be phenotyped for this individual trait (usually between 150 and 400). Quantitative traits of interest can be very integrative (growth...), or closer to a specific function (for example a metabolic function estimated through some enzyme activity or compound

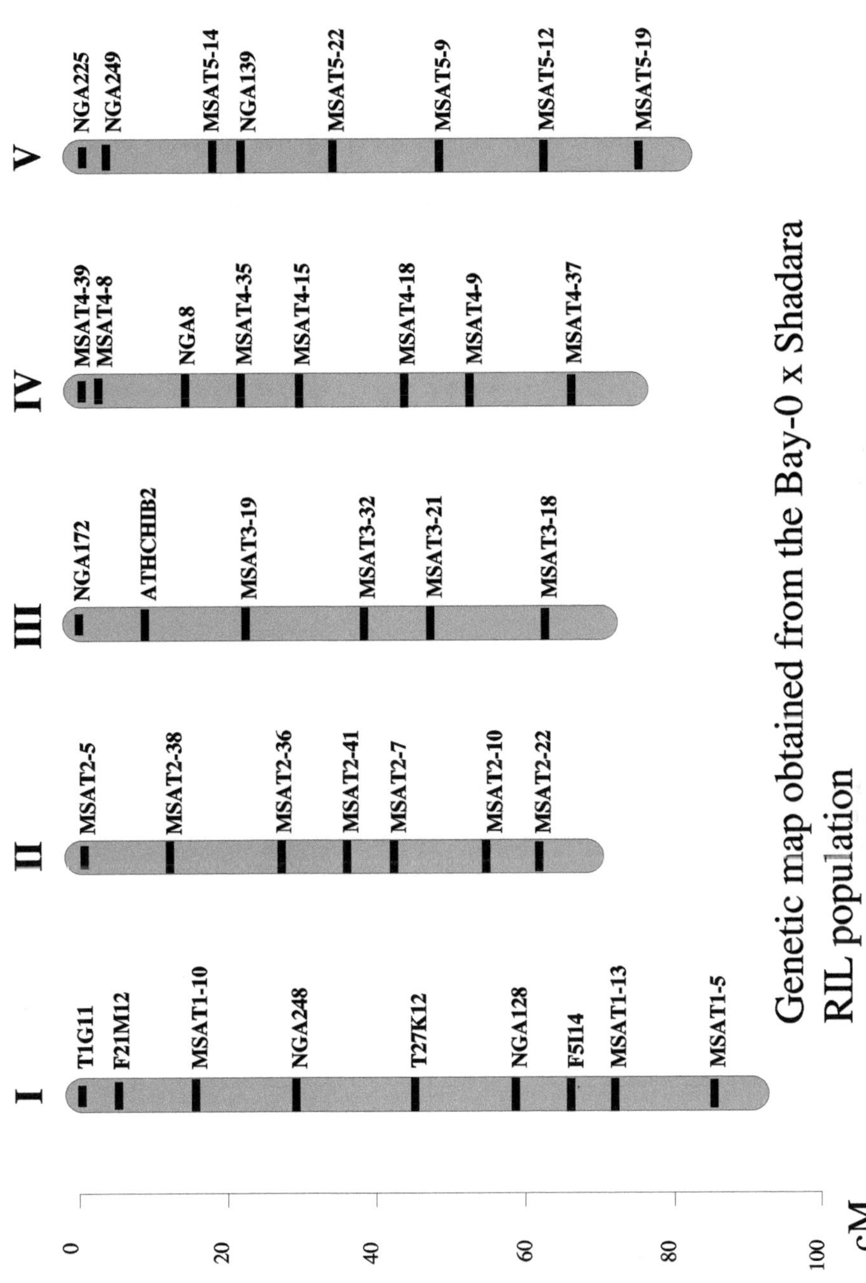

Fig. 27.2 Genetic map drawn from the Bay-0 x Shahdara RIL population and 38 microsatellite markers

content...). We usually expect that the more integrative traits could be controlled by more loci than traits representing a more precise function. All this of course relies on the structure of the variation in the population, because one can only detect a locus that is controlling a trait when it is polymorphic in the population studied. Another important factor is how much of the observed phenotypic variation is due to genetic effects, as compared to environmental effects. This is measured through the heritability of a given trait and decides how much the phenotyping experiment needs to be repeated. Phenotypes from highly heritable traits (h^2 above 0.60) will be easier to repeat than phenotypes from lower heritability traits. One needs to keep in mind however that for a fixed-size experiment, it is always more interesting to study more RILs than more repetitions, because more RILs gives more information from new genomic combinations and RILs are somehow partial repetitions of each other (through shared genomic regions).

Flowering time is among the most well investigated of biological functions in *A. thaliana* and is now systematically used as a starting trait in most RIL populations (Koornneef *et al.*, 2004). Growing the 420 Bay-0 x Shahdara RILs in long-days (16 hours of daylight) in the greenhouse, we have been able to measure flowering time as the number of days between germination and bolting (when the floral stem emerges from the rosette). The distribution of phenotypes in the population is depicted in Fig. 27.3. Strikingly, there is a huge variation among RILs, although parental accessions flower only a few days apart, many RILs flower well before or well after the parents. This phenomenon, known as transgression, is classically explained by the fact that a similar phenotype can be the result of different genotypes, i.e. different combinations of alleles at loci with positive or negative effects on the trait. With the independent segregation of these loci in the RIL population, combinations with many late alleles or many early alleles will be formed and show very different phenotypes. In the case of flowering time variation represented in Fig. 27.3, it is very likely that the unbalanced shape of the transgression toward early and late phenotypes is due to other constraints on the minimum duration of the vegetative cycle.

At this point, each of the RIL has been assigned a genotype at many markers along the chromosomes and a quantitative phenotypic value. Now is the time to reveal and estimate the regions of the genome that are responsible for the variations in the trait, the so-called QTL.

QTL MAPPING

Mapping a QTL relies on the existence of a genetic linkage between this locus and some of the molecular markers placed on the genetic map. In a given region, if there is a locus affecting the trait of interest, then the allele present in that region should have an effect on the observed phenotype. *A. contrario*, if a marker is not linked to any QTL, then the genotype should not matter for the expressed phenotype. One can either analyse the genome by examining markers individually, or by trying to get more information from several markers at a time.

In the first strategy, a simple Anova or Student test answers the following question: does the genotype at this marker (Bay-0 or Shahdara) have an effect on the trait value. This method is very simple and robust but it does not use all of the marker information and can represent a certain lack of precision as to the QTL position, when the marker

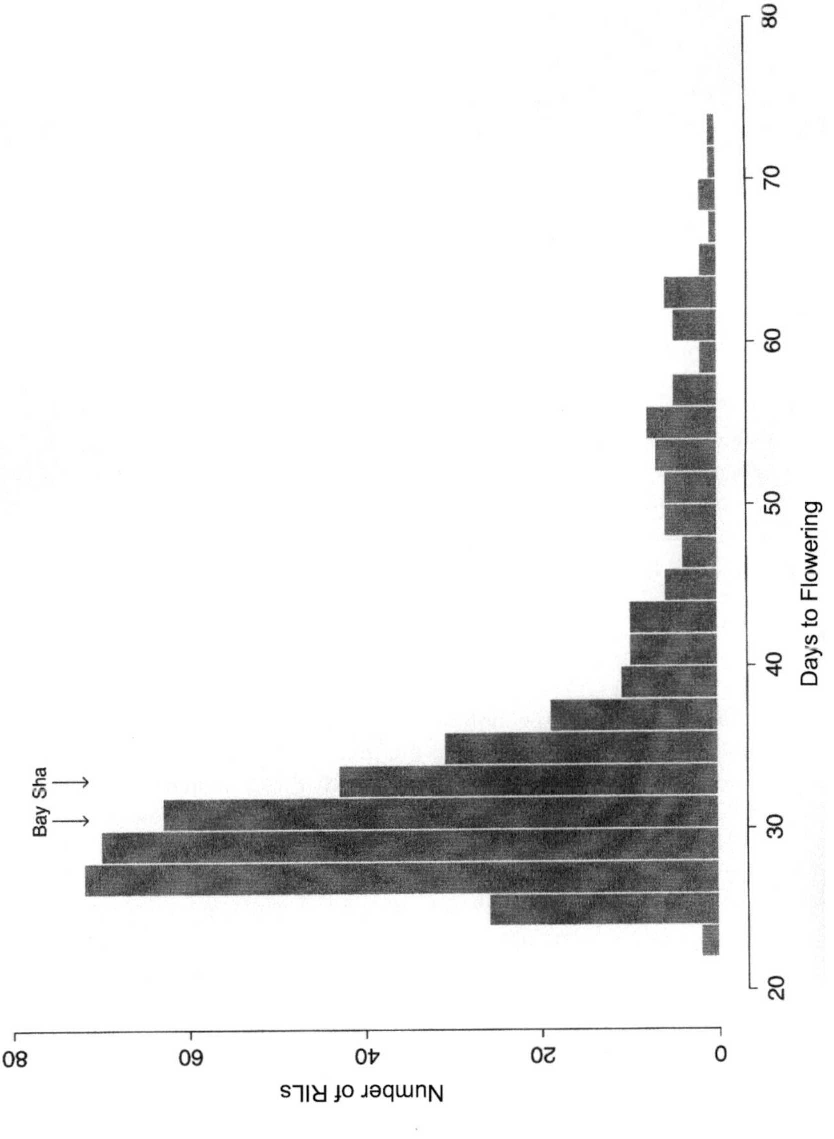

Fig. 27.3 Distribution of flowering time phenotypes across the 420 Bay-0 x Shahdara RILs. Flowering time was estimated as the number of days between germination and bolting.

density is not high. Another strategy known as Interval Mapping (Lander and Botstein, 1989), tests for the presence of a QTL at any position in a marker interval by using the genotypes of the flanking markers to estimate the probability of each allele at the tested position. Then the presence of a QTL is estimated by the LOD Score, which represents the chance that there is a significant QTL at this site. The LOD Score follows a log scale so that a LOD Score of three means that it is 1,000 times more likely that there is a QTL than there is not. Each peak of the LOD Score curve above a certain threshold (usually between two and three) is the best predicted position of a QTL. A further extension of this method is called Composite Interval Mapping and uses some specific markers as cofactors to take into account (and remove) the effect of the main QTLs, before testing again for smaller effect loci.

For each significant QTL detected, one can estimate the effect of the QTL, by two parameters. First, the '2a' value (the additive effect) is calculated from the difference between the mean phenotypic value of the lines with Bay-0 alleles and the mean phenotypic value of the lines with Shahdara alleles (2a = 'BB' - 'SS'). This parameter is expressed in the trait units and represents the expected effect on the trait value of the replacement of Shahdara alleles by Bay-0 alleles (at the homozygous state). Second, the R^2 parameter defines the QTL effect relative to the global phenotypic variation observed in the population. This is calculated from the Anova results as the ratio between the sum of squares for the marker effect and the total sum of squares (expressed as a percentage). Arguably (depending on the extent of variation in a specific population), it is usually considered that R^2 classes distinguish between different types of locus: major effect loci ($R^2 > 25\%$), medium effect loci (R^2 between 7 and 25%) and minor effect loci ($R^2 < 7\%$).

Applied to the flowering time variation described earlier in the Bay-0 x Shahdara population, composite interval mapping reveals that three different loci explain most of the variation for this trait. LD1 is located on chromosome 5 and is very highly significant with a LOD Score of 60. The segregation of LD1 in the population explains as much as 35 % of the total phenotypic variation and the presence of Bay-0 alleles at this locus decreases the time to flowering by 14 days, relative to Shahdara alleles. The two other loci, LD2 and LD3, respectively on chromosomes 5 and 2, exert a less strong effect, and the effect of LD2 is strongly dependent on the allele present at LD1 (QTL x QTL interaction or epistatis).

Although QTL mapping was initially introduced for crop species like maize, rice or tomato, *A. thaliana* has many advantages for the analysis of these results, in terms of identifying a molecular basis for this variation. Indeed, for the flowering time pathway many candidate genes can be localised on chromosomes and their position and effect compared to the detected QTL. Using such an approach, and combining it with sequencing information and mutant phenotypes, elements indicate that LD1 corresponds to *FRIGIDA* (*FRI*; Johanson *et al.*, 2000) and LD2 to *FLOWERING LOCUS C* (*FLC*; Sheldon *et al.*, 1999).

OTHER EXAMPLES OF THE USE OF *A. THALIANA* AS A QUANTITATIVE GENETICS MODEL

Flowering time is only one example of the many quantitative traits that can be dissected through quantitative genetics in *A. thaliana*, although it has been very widely used (because of obvious evolution impact and usually high heritability) and is now almost always systematically mapped in any new material. A limited number of major-effect loci for flowering time (often directly segregating as Mendelian factors) had been shown early on to explain most of the natural variation of the species. Among these, the above-mentioned *FRI* and *FLC* are key-players. But the variation for many other traits is expected to rely on a much more complicated genetic basis. In any case, the QTL mapping approach remains a "no *a priori*" approach, since only the variation observed for a chosen phenotype guides the analysis. Using the above-described genetic material or similar resources developed from other crosses, authors have described phenotypic variation for a large diversity of traits.

Other developmental factors such as circadian rhythms have been studied using *A. thaliana* RIL populations. For example, Swarup *et al.* (1999) observed leaf movements to estimate the period of the circadian clock and confirmed an interaction with the flowering time pathway (with QTL at genes like *FLC*...), but also mapped new loci controlling this trait. Morphological development of the flower and inflorescence has also been studied in detail (Juenger *et al.*, 2000; Ungerer *et al.*, 2002). Seed size and dormancy show a very dramatic and interesting variation in some specific RIL populations, like Ler x Cvi (Alonso-Blanco *et al.*, 2003).

In the field of plant-pathogen interactions, many studies have analysed plant resistance and tolerance to fungi, bacteria or viruses (Buell and Somerville, 1997; Wilson *et al.*, 2001; Denby *et al.*, 2004). In addition, Kliebenstein *et al.* (2001) and Kroymann *et al.* (2003) have used QTL mapping in *A. thaliana* to establish and confirm a link between insect resistance and secondary metabolites (glucosinolates).

The small size of *A. thaliana* plants and seedlings can be considered as an advantage for the geneticist, in that it allows the growth of thousands of plants under carefully controlled environmental conditions and/or to compare many genotypes in different environments. Thus, genotype x environment interaction analysis is now a major theme in the quantitative genetics study of *A. thaliana.*

Borevitz *et al.* (2002) have measured hypocotyl elongation in a series of different light environments (white, blue, red, far-red light and dark). These environmental conditions provide a better understanding of the variation in response to different stimuli and different light signalling pathways. A strong QTL x environment interaction was observed, illustrating the specific response of hypocotyl etiolation to some light conditions. For example, QTL were obtained specifically in one light condition (either in blue, or in red light). On the other hand, some loci were detected in all environments with comparable effects. The authors then attempted to organize these loci along a light pathway according to their specific or general pattern of effects. Many of these loci did not correspond to already known genes in the pathway.

Using the very large Bay-0 x Shahdara RIL population, Loudet *et al.* (2003a, b) have studied growth and nitrogen (N) metabolism in two contrasting N environments, one of which was limiting plant growth. The chosen traits (all measured at the vegetative stage) were either very integrative (e.g. shoot growth and water content), related to N metabolism (total-N, nitrate, free amino-acids contents...), or established links with other pathways (phosphate and chloride contents...). Most QTL were found to be interacting with the N environment and their effect was modified (or cancelled) depending on the amount of N in the nutrient solution. As expected, this was especially true for traits linked to N physiology. Such loci are of course good candidates for revealing regulatory factors directly involved in N metabolism. On the other hand, some stable loci were detected with the same effect whatever the N environment. Two examples of this stability concerned shoot water contents and it seems that these loci corresponded to the main flowering time QTLs segregating in the population. Epistatic interactions between QTLs have been found to explain a large part of the variation for some traits, such as the amino acid contents. Finally, several medium-effect loci may be particularly of value in increasing our understanding of the relationship between biomass accumulation and N use efficiency and are candidates for cloning.

Deciphering quantitative variation at the gene level is the reason behind the development of these approaches in a model plant like *A. thaliana*. The identification of the genes and ultimately the polymorphisms underlying the QTL is made possible by the availability of numerous genomic tools and resources developed by the scientific community (sequence diversity analysis, T-DNA and tilling mutants, transcriptomics, transgenic etc....). However, the elucidation of the complete sequence of the *A. thaliana* genome has probably been our greatest help in this endeavour (Jander *et al.*, 2002; Koornneef *et al.*, 2004).

Despite the success in gene mapping, cloning a QTL is not straightforward. A complication is that many traits have an inherent environmental variation, which introduces a variable error when trying to measure the genetic contribution. The point is that when focusing on a specific QTL, the phenotypic effect is reduced to that of the individual locus, while the environmental variation is still as strong. Understandably, the major limiting factor related to QTL cloning is the phenotype. Actually, the '2a' phenotypic effect of the QTL and the heritability of the trait are the two factors that decide whether a QTL can be Mendelized (confirmed individually in Near-Isogenic Lines -NILs) and followed up or not. As of 2005, only a small number of large-effect loci originally detected as QTL have been characterized and cloned in *A. thaliana* (Koornneef *et al.*, 2004). For example, El-Assal *et al.* (2001) cloned a novel natural allele of *CRY2* which regulates flowering time variation. However more recently, medium- or small-effect loci have been cloned (Kroymann and Mitchell-Olds, 2005). Finally, a debate has arisen as to what proves that a QTL has been cloned (Weigel and Nordborg, 2005) and some say that fine-mapping in NILs can be conclusive and might be considered as important as other more classically accepted proofs of cloning (transgenics, complementation ...).

REFERENCES

Alonso-Blanco, C., Bentsink, L., Hanhart, C.J., Blankestijn-De Vries, H., Koornneef, M. 2003. Analysis of natural allelic variation at seed dormancy loci of *Arabidopsis thaliana*. *Genetics* 164, 711-729.

Borevitz, J.O., Maloof, J.N., Lutes, J., Dabi, T., Redfern, J.L., Trainer, G.T., Werner, J.D., Asami, T., Berry, C.C., Weigel, D., Chory, J. 2002. Quantitative Trait Loci controlling light and hormone response in two accessions of *Arabidopsis thaliana*. *Genetics* 160, 683-696.

Buell, C.R., Somerville, S.C. 1997. Use of *Arabidopsis* recombinant inbred lines reveals a monogenic and a novel digenic resistance mechanism to *Xanthomonas campestris* pv campestris. *Plant Journal* 12, 21-29.

Denby, K.J., Kumar, P., Kliebenstein, D.J. 2004. Identification of Botrytis cinerea susceptibility loci in *Arabidopsis thaliana*. *Plant Journal* 38, 473-86.

El-Din El-Assal, S., Alonso-Blanco, C., Peeters, A. J., Raz, V., Koornneef, M. 2001. A QTL for flowering time in *Arabidopsis* reveals a novel allele of CRY2. *Nature Genetics* 29, 435-440.

El-Lithy, M.E., Bentsink, L., Hanhart, C.J., Ruys, G., Rovito, D., Broekhof, J.L., Van Der Poel, H.J., Van Eijk, M.J., Vreugdenhil, D., Koornneef, M. 2006. New *Arabidopsis* recombinant inbred line populations genotyped using SNPWaveTM and their use for mapping flowering time QTLs. *Genetics*, 172, 1867-1876.

Jander, G., Norris, S.R., Rounsley, S.D., Bush, D.F., Levin, I.M., Last, R.L. 2002. *Arabidopsis* map-based cloning in the post-genome era. *Plant Physiology* 129, 440-450.

Johanson, U., West, J., Lister, C., Michaels, S.D., Amasino, R.M., Dean, C. 2000. Molecular analysis of *Frigida*, a major determinant of natural variation in *Arabidopsis* flowering time. *Science* 290, 344-347.

Juenger, T., Purugganan, M.D., Mackay, T.F.C. 2000. Quantitative trait loci for floral morphology in *Arabidopsis thaliana. Genetics* 156, 1379-1392.

Kliebenstein, D.J., Kroymann, J., Brown, P., Figuth, A., Pedersen, D., Gershenzon, J., Mitchell-Olds, T. 2001. Genetic control of natural variation in *Arabidopsis thaliana* glucosinolate accumulation. *Plant Physiology* 126, 811–825

Koornneef, M., Alonso-Blanco, C., Vreugdenhil, D. 2004. Naturally Occurring Genetic Variation in *Arabidopsis thaliana. Annual Review of Plant Biology* 55, 141-172.

Kroymann, J., Mitchell-Olds, T. 2005. Epistasis and balanced polymorphism influencing complex trait variation. *Nature* 435, 95-8.

Kroymann, J., Donnerhacke, S., Schnabelrauch, D., Mitchell-Olds, T. 2003. Evolutionary dynamics of an *Arabidopsis* insect resistance quantitative trait locus. *Proc Natl Acad Sci* USA. 100, 14587-92.

Lander, E.S., Botstein, D. 1989. Mapping mendelian factors underlying quantitative traits using RFLP linkage maps. *Genetics* 121, 185-199.

Loudet, O., Chaillou, S., Merigout, P., Talbotec, J., Daniel-Vedele, F. 2003a. Quantitative trait loci analysis of nitrogen use efficiency in *Arabidopsis. Plant Physiology* 131, 345-358.

Loudet, O., Chaillou, S., Krapp, A., Daniel-Vedele, F. 2003b. Quantitative trait loci analysis of water and anion contents in interaction with nitrogen availability in *Arabidopsis thaliana. Genetics* 163, 711-722.

Sheldon, C.C., Burn, J.E., Perez, P.P., Metzger, J., Edwards, J.A., Peacock, W.J., Dennis, E.S. 1999. The FLF MADS box gene: A repressor of flowering in *Arabidopsis* regulated by vernalization and methylation. *Plant Cell* 11, 445-458.

Swarup, K., Alonso-Blanco, C., Lynn, J.R., Michaels, S.D., Amasino, R.M., Koornneef, M., Millar, A.J. 1999. Natural allelic variation identifies new genes in the *Arabidopsis* circadian system. *Plant Journal* 20, 67-77.

Tautz, D. 1989. Hypervariability of simple sequences as a general source for polymorphic DNA markers. *Nucleic Acids Research* 17, 6463-71.

Ungerer, M.C., Halldorsdottir, S.S., Modliszewski, J.L., Mackay, T.F., Purugganan, M.D. 2002. Quantitative trait loci for inflorescence development in *Arabidopsis thaliana. Genetics* 160, 1133-1151.

Weigel, D., Nordborg, M. 2005. Natural variation in Arabidopsis. How do we find the causal genes? *Plant Physiology* 138, 567-8.

Wilson, I.W., Schiff, C.L., Hughes, D.E., Somerville, S.C. 2001. Quantitative trait loci analysis of powdery mildew disease resistance in the *Arabidopsis thaliana* accession Kashmir-1. *Genetics* 158, 1301-1309.

28

CHAPTER

Characterization of Major Genes and QTLs by a Candidate Gene Approach

R. Stevens, S. Muños, M. Causse, A. Bendahmane, V. Lefebvre

Genetically controlled traits such as pathogen resistance, stress adaptation or quality can be controlled by a single locus (major gene, Mendelian inheritance), or by several loci (Quantitative Trait Loci, QTLs) with a partial effect on the trait variation. The location of a QTL can correspond to the segregation of one or several genetically linked genes. Major genes and QTLs can be easily mapped onto the genome of the appropriate species with the help of linked molecular markers (Chapters 26, 30). The characterization of a major gene or QTL involves isolating the gene responsible for the variation studied and identifying the function that it controls. The understanding of the nature of the genes controlling the simple or complex phenotypic trait variations has both a fundamental and practical interest. Different approaches can be used to isolate the gene of interest, which all include the identification of candidate genes for the QTL or major gene concerned. This chapter is not concerned with all the possible methods for identifying candidate genes (including transcriptome and proteomic approaches, see Chapters 10 and 12) but deals with the selection and validation of candidate genes likely to control a trait, or to correspond to a QTL.

PRINCIPLES OF THE CANDIDATE GENE APPROACH

This approach consists of proposing known nucleotidic sequences (candidate genes) which are likely to correspond to the major gene or QTL of interest. Different types of candidate genes can thus be identified based on either their position on the genome, their function or the phenotype they induce when mutated.

- *Positional candidate genes* are often used and are indispensable in the absence of *a priori* knowledge about the gene potentially involved in the trait. They are commonly identified by using a map-based cloning strategy (Rommens *et al.*, 1989, Tanksley *et al.*, 1995). Comparative genomic strategies and gene maps are also useful for their identification (Marra *et al.*, 1998).

- *Functional candidate genes* are interesting if all or some of the functions involved in the expression of the trait are known. In this case, candidate genes are selected as they might explain all or part of the trait variation (Pflieger *et al.*, 2001). Functional candidate genes can correspond to differentially expressed genes between two relevant conditions for the trait, identified by either transcriptomic (Chapter 10) or proteomic (Chapter 12) approaches.
- If the phenotype is suitable, a collection of mutants can be used, if available, to search for the gene(s) involved in a certain trait or to validate candidate genes (Bouchez and Hofte, 1998; Chapter 9).

In all three cases, the underlying hypothesis is that a polymorphism exists in the candidate gene which is responsible for the phenotypic variation observed. For quantitative traits, resulting from the effect of several genes with partial effects, often influenced by the environment, where the genetic map location remains relatively imprecise, the *a priori* candidate gene approach can be more promising than map-based cloning. Once the candidate genes have been isolated, a final stage of validation is necessary to confirm the role of the cloned gene in the original phenotype.

The approach includes three chronological stages. Firstly, the candidate genes are proposed on the basis of physiological or molecular studies (functional candidate genes), or on the basis of genetic linkage data from mapping and fine mapping (all genes linked to a locus are positional candidate genes). In the latter case, gene maps represent sources of candidates for the characterisation of a locus. Systematic EST (expressed sequence tag) sequencing programmes in plants should help accelerate the development of such maps. A molecular polymorphism must then be identified between the two parents in order to locate the candidate gene on the relevant genetic map, or to look for statistical associations between the polymorphisms of the candidate gene and the trait variation in a series of independent individuals. These two strategies are complementary and parallel. Candidate genes not linked to the locus of interest are rejected. The final validation of the role of a candidate gene relies on different approaches depending on whether it concerns a major locus or a QTL.

IDENTIFICATION OF CANDIDATE GENES

Map-based Cloning and Positional Candidate Genes

In the absence of hypotheses concerning the gene potentially responsible for a particular trait, the surest way to identify the gene governing the trait is to initiate map-based cloning. As a starting point, the localization of the trait on the genome will allow map-based cloning or chromosome walking (Fig. 28.1). The procedure will often generate several candidate open reading frames (ORFs) and has been used successfully to clone several genes. The first key examples in plant science are shown in Table 28.1. For the model plant *Arabidopsis thaliana*, the time required to identify a gene corresponding to a trait was estimated at around four person-years in 1995 and was only one person-year in 2002 (Jander *et al.*, 2002). To date, with the increase in molecular tools available for several plant species, it is becoming easier to use this strategy and about fifteen major genes have been isolated in the first part of 2005 (Chen *et al.*, Diener *et al.*, Green *et al.*, Hemsley *et al.*, Koizumi *et al.*, Krusell *et al.*, Larsen *et al.*, Li *et al.*,

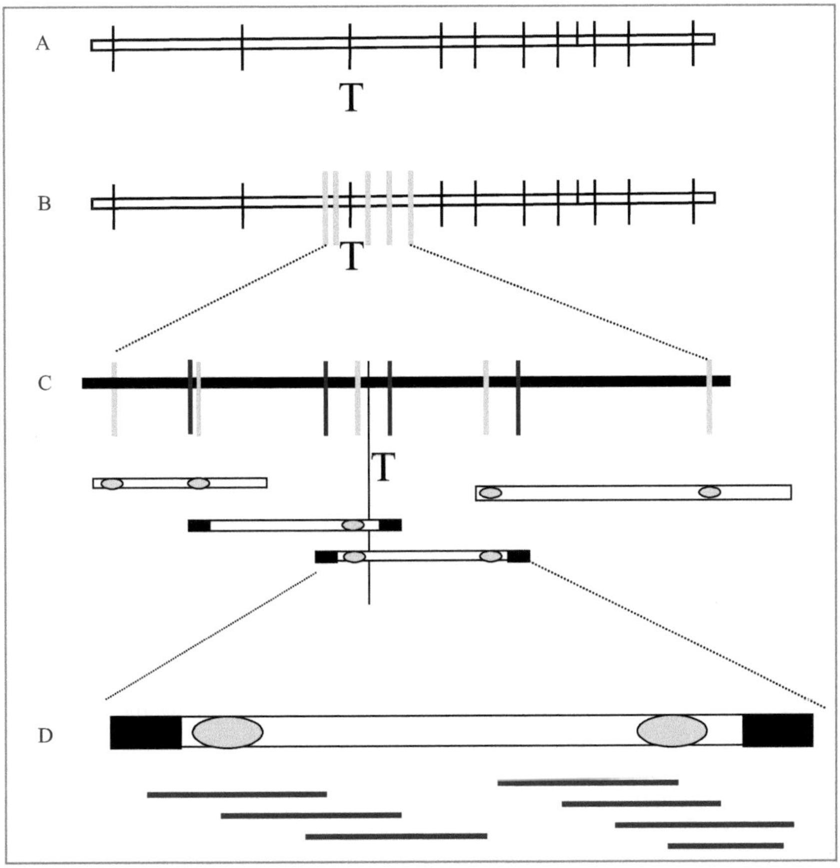

Fig. 28.1 Main steps for map-based cloning (A) Genetic mapping of target gene T with molecular markers; (B) Fine mapping by adding new markers (grey) and increasing the number of individuals studied; (C) Screening of a large fragment library (YAC, BAC, ...); addition of new markers corresponding to BAC ends (black) and estimation of the ratio kb/cM; (D) Search for ORFs in the fragment carrying two markers either side of T; screening of a cosmid library to select the clones carrying the target gene and validation of ORF candidates by plant transformation.

Table 28.1 Examples of the first genes cloned by chromosome walking

Gene	*Species*	*Trait*	*Reference*
ms14	Tomato	Male sterility	Gorman *et al.*, 1996
I2	Tomato	Resistance to *Fusarium oxysporum*	Ori *et al.*, 1997
Pto	Tomato	Resistance to *Pseudomonas tomato*	Martin *et al.*, 1993
Rx	Potato	Resistance to *potato virus X*	Bendahmane *et al.*, 1999
mlo	Barley	Resistance to *Erysiphe graminis*	Buschges *et al.*, 1997
Rps2	*Arabidopsis*	Resistance to *Pseudomonas syringae*	Mindrinos *et al.*, 1994
Omega-3	*Arabidopsis*	Desaturation of fatty acids	Arondel *et al.*, 1992
Hs1	Beet	Resistance to beet cyst nematode (*Heterodera schachtii* Schmidt)	Cai *et al.*, 1997
Xa21	Rice	Resistance to *Xanthomonas oryzae*	Song *et al.*, 1995

Noutoshi *et al.*, Stein *et al.*, Suetsugu *et al.*, Tamura *et al.*, Tanabe *et al.*, van der Vossen *et al.*, Zhang *et al.*, Zhang and Li) and two QTLs (Nishimura *et al.*, Ren *et al.*,)

Successful map-based cloning is carried out in several stages. The different stages are linked and a new stage can be started only if the previous one has been completed. It requires:

- the identification of molecular markers and recombination events in the target locus region;
- *fine mapping*: markers are positioned precisely according to the recombination events to develop a high resolution genetic map;
- *physical mapping*: the physical distance in the target region can be estimated by pulsed-field gel electrophoresis and the closest markers are used to screen a genomic library;
- *candidate gene validation*: when a clone that covers the interval between the markers closest to the target gene has been isolated and sequenced, the gene of interest can be identified by comparing its sequence to those of sequence databases, or of mutant alleles, or by complementation after integration of the gene into plants lacking the trait concerned.

Fine Mapping

Fine mapping involves finding and ordering markers located in a small interval around the locus of interest. The search for new markers is carried out using near-isogenic lines or, more generally, by analysing mixtures of individuals which will finally only differ for the trait of interest (Bulked Segregant Analysis, (BSA) Michelmore *et al.*, 1991). The techniques that allow screening of a large number of markers simultaneously, like AFLP (amplified fragment length polymorphism) markers, are particularly useful at this stage (Vos *et al.*, 1995; Chapter 24). The AFLP markers identified by BSA will not all be tightly linked to the target locus. They cover in general a genetic interval of 10 to 20 centimorgans.

To precisely map the new markers closest to the gene of interest, it is necessary to screen a large segregating population. Generally, two PCR markers spanning a 2 cM window around the locus are used on more than 1,000 segregating plants. Plants for which a recombination event has occurred between the two PCR markers are selected and used to precisely map the markers in the region. Combining genotyping and phenotyping data from recombinant plants allows the high resolution map of the locus to be obtained which is necessary for the next steps. To improve these technical steps, automation of the stages from seed germination to DNA extraction and genotyping is necessary. High throughput genotyping is often based on SNP detection (Syvänen, 2001). When the high resolution map of the locus has been obtained, it is possible to link the genetic region to the physical map.

Physical Mapping

Construction of a large-fragment library

When markers are judged to be sufficiently close to the locus of interest (maximum of 100 to 300kb from the gene), the physical mapping stage can start. Table 28.2 shows the

Table 28.2 Average relationship between map size in cM and genome size in kb

Species	*Size of genetic map*	*Size of physical map*	*kb/cM*	*Size of BAC library*[1]
Arabidopsis	1,000 cM	145 Mb	145	7,250 BAC
Rice	1,500 cM	500 Mb	350	25,000 BAC
Potato	1,200 cM	900 Mb	750	45,000 BAC
Pepper	1,500 cM	3,000 Mb	2,000	100,000 BAC
Barley	1,500 cM	5,300 Mb	3,500	265,000 BAC

[1] The estimations are carried out for a BAC library of five genome equivalents, with an average insert size of 100 kb.

average relationship between the map size in cM (genetic distance) and the genome size in kb (physical distance) for different species. This relationship is not linear and the ratio varies from one species to another. Furthermore, the relationship between genetic distance (cM) and physical distance (bp number), fluctuates considerably depending on the region of the genome considered (Table 28.3). This relationship is evaluated for each target gene, in order to estimate the work that is left to be carried out.

Physical mapping consists, initially, of creating a library of large genomic fragments. Yeast Artificial Chromosome (YAC, Burke *et al.*, 1987) libraries, made of clones of up to several hundred kilobases, but which are difficult to multiply and are often chimeric (Monaco and Larin, 1994), were used but Bacterial Artificial Chromosome (BAC, Shizuya *et al.*, 1992) libraries, where clones reach on average 100 to 200 kb, are now preferred.

The construction of a BAC library is carried out by partially digesting nuclear DNA with at least two restriction enzymes to avoid the library being biased for DNA sequences specific for one restriction site. The organization of the library into 384-well plates is carried out using robots. The size of the BAC library is often expressed in number of genome copies. The number of clones that a BAC library should contain to have a good chance of the gene of interest being present depends on the physical size of the genome (Table 28.2). Statistically, for a 99% probability of the library containing the gene of interest, the size of the library must be at least 4.6 times the size of the genome.

Table 28.3 Relationship between genetic distance (cM) and local physical distance (kb) for genes cloned from tomato

Gene	*Relationship kb/cM*	*Reference*
ms14 (male sterility)	24	Gorman *et al.*, 1996
I2 (resistance)	43	Ori *et al.*, 1997
Sw-5 (resistance)	100	Brommonschenkel and Tanksley, 1997
Rin (maturation)	200	Vrebalov *et al.*, 2002
j (jointless)	200	Mao *et al.*, 2000
Pto (resistance)	1,090	Martin *et al.*, 1993
Tm-2a (resistance)	4,000	Yang *et al.*, 1988
Genome average	**750**	

Library screening

The BAC library is often screened with PCR markers that are closest to the gene of interest. In other words, we look for the BAC clones hybridising with these markers. Screening is performed by PCR and this system is based on the organisation of the library into pools and superpools. Each DNA pool often corresponds to 384 colonies, then each superpool is generally made up of 10 pools. The identification of a positive colony in a library of 100,000 clones requires 76 PCR reactions (Fig. 28.2). A local map of clones that hybridise with the available markers is then established consisting of a series of enzymatic digestions which allow identification of the fragments common to several clones; the clones can then be ordered. A set of ordered clones constitutes a contig. The BAC ends are sequenced and new PCR markers are designed (Chapter 24) and used to anchor the genetic map to the physical map and to identify the BAC clone carrying the gene of interest, or to increase the contig size. Repeated sequences are frequent in plants and cause problems for chromosome walking. For example, the re-screening of the library by a BAC end can lead to the identification of BAC clones that do not belong to the same contig but can be assembled into this contig by error. It is therefore preferable to screen the BAC library with markers that are tightly linked to the target gene so that the first clone identified carries this gene. A BAC clone should carry the target gene when it positively hybridises with two markers either side of the target locus, so called chromosome landing. As an alternative to PCR-based screening, library screening can be carried out by Southern-type hybridization.

Sub-cloning for validation by complementation

When a clone carrying the gene of interest is identified, it is either sub-cloned and entirely sequenced to look for open reading frames (ORFs) or, alternatively, a cosmid library is screened so a smaller fragment can be sequenced. The polymorphism present in the identified ORFs is then studied and possible relationships between the polymorphism and the trait variation are looked for. Plasmids carrying these ORFs are produced and used to transform plants to validate their role in the trait variation by complementation (see section on genetic transformation). Tables 28.1 and 28.3 show the first examples of major genes cloned in this way.

Numerous genes involved in pathogen recognition, signal transduction or defence mechanisms have been isolated (Lamb *et al.*, 1989; Bent, 1996; Hammond-Kosack and Jones, 1996; Dangl and Jones; 2001). In five years, between 1993 and 1998, more than 20 resistance genes (R-genes) have been isolated from several species by map-based cloning or insertional mutagenesis (Staskawicz *et al.*, 1995; Hammond-Kosack and Jones, 1997; Gebhardt, 1997). These genes act against a large set of pathogens (bacteria, viruses, nematodes, fungi, oomycetes, insects). Isolation of R genes has shown that their structure is highly conserved which has allowed them to be classified into four main classes of sequences whose products appear to activate a similar range of defence mechanisms (Bent, 1996). The elucidation of R gene structure has facilitated the identification of resistance gene analogues (RGA) in plant different species.

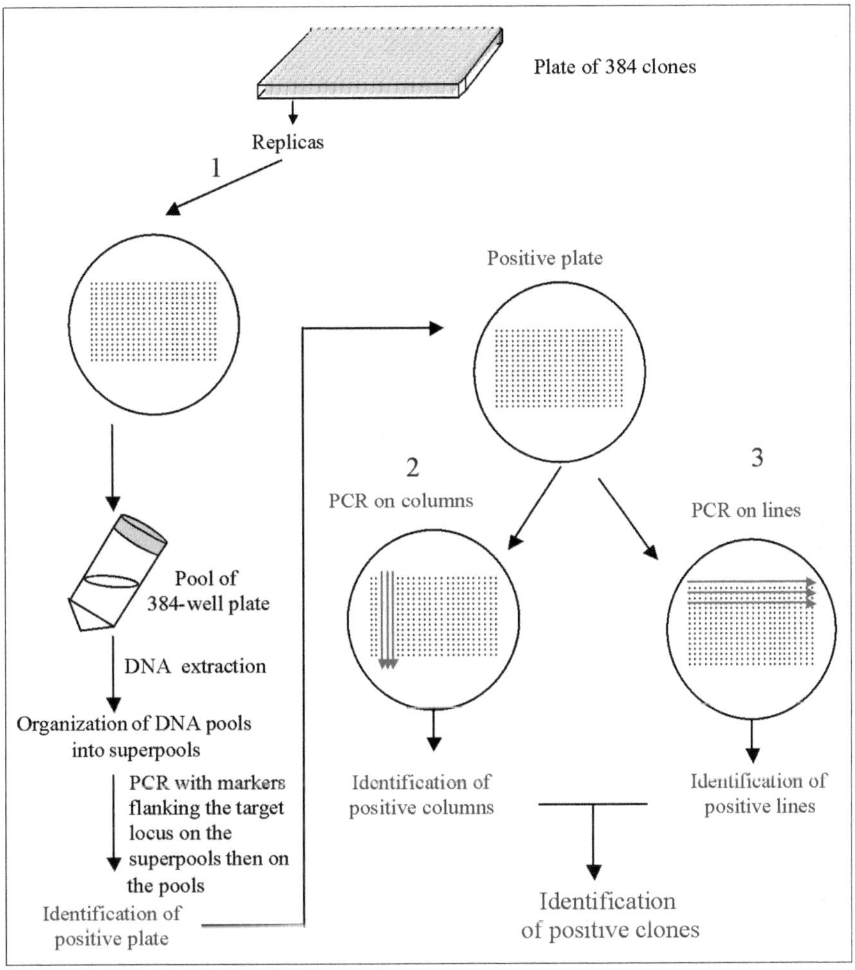

Fig. 28.2 Screening of a genomic library using the pooling method: screening of superpools (1) then DNA pools in columns (2) and lines (3)

Map-based QTL cloning, a special case

Direct cloning of a QTL is more difficult than cloning a major gene, because a QTL only partially influences trait variation and its effect can only be appreciated by statistical methods. For this reason, the resources required are more considerable. For the creation of nearly-isogenic lines or for BSA, flanking markers either side of the QTL are required to sort the genotypes, as sorting by phenotype is generally less easy. Therefore it is necessary that the location of the QTL to be cloned is precisely known and this requires the use of fine mapping (Paterson *et al.*, 1990, Salvi and Tuberosa, 2005). Currently, map-based cloning can only realistically be envisaged if the QTL is precisely localised within a marker interval of less than 1 cM. To confirm that the isolated gene corresponds to the QTL of interest, certain problems arise from validation by transformation. If the candidate QTL is, for example, a gene for dwarfism, we have the classic situation where a deficient gene must be complemented by an active gene. Generally, however, we aim to modify the value of a trait by introducing a favourable allele, no easy task when the effect of the environment, the genetic background, as well as of the transformation (dose effects, gene silencing) are taken into account. Constructions to overexpress the gene can be used but carry a risk of seeing artefactual positive effects on the trait. Therefore, other methods for proving the association between gene and QTL are required. High-resolution mapping of the candidate ORFs can allow identification of the clone which co-segregates with a given proportion of the trait. Sequencing can also provide an indication of gene function, which will aid in choosing the most likely candidate gene. Sequencing of alleles conferring distinct phenotypes can also identify the origin of the observed variation. Finally, for certain traits, a study of gene expression helps determine which gene is behind the QTL. For example, for a QTL involved in the response to stress, the expression of the gene could be induced or controlled by the stress in question.

The first QTLs cloned by map-based cloning correspond to QTLs with strong effects that are independent of the environment. A fruit weight QTL in tomato responsible for about 30% of the variation of this trait has been isolated using the classical strategy of high resolution mapping by screening 3,472 F_2 plants with two markers surrounding the QTL, which allowed identification of 55 recombinants, and screening a YAC library (Alpert et Tanksley, 1996). From a YAC likely to contain the required gene, a cosmid library was screened and three clones used to transform a tomato variety. The cosmid leading to differences in fruit size after transformation was sequenced and the two sequences corresponding to ORFs used in a second round of transformation. This allowed definitive identification of the clone corresponding to the QTL for fruit weight: *fw2.2* (Frary *et al.*, 2000). The gene controls carpel cell number and may be structurally similar to a human *ras* oncogene. Gene dosage experiments have shown that *fw2.2* transcripts are highly correlated (negatively) with fruit mass (Liu *et al.*, 2003). The cloning of a QTL controlling the variation in sugar content of tomato fruit followed the same initial stages (Fridman *et al.*, 2000) but benefited from the existence of recombinations within the gene and the QTL was localised to a region of 484 bp covering the sequence of a cell-wall invertase. The functional polymorphism has been delimited to a conserved aspartate near the catalytic site which affects enzyme kinetics and fruit sink strength. This result has shown that a QTL can be restricted to a single nucleotide polymorphism, leading to the definition of Quantitative Trait Nucleotide (Fridman *et al.*, 2004). A summary of cloned QTLs from different plants is presented in two reviews by Paran and Zamir (2003) and by Salvi and Tuberosa (2005), showing that the use of NILs (near isogenic lines), which isolate a single QTL region, has facilitated the task of QTL cloning. The powerful genetic tools now available enable the cloning of major QTLs but cloning of minor ones, under environmental control, remains a challenge.

COMPARATIVE MAPPING FOR THE CHARACTERIZATION OF GENES OF INTEREST

With the increasing number of genetic maps and then with comparative mapping, the concept of colinearity between plant genomes has emerged. Closely related species often have a conserved genome organization (Lagercrantz *et al.*, 1996; Gale and Devos 1998; Livingstone *et al.*, 1999). The colinearity, also named synteny, makes model species with relatively small genomes (rice and *Arabidopsis* for example) a source of markers and genes for species with larger or less well characterised genomes (Barnes, 2002; Schmidt, 2000). This observation is more helpful when considering species with completely sequenced genomes. However, it is often more restricted genomic regions that are colinear, the term of microsynteny is then employed.

As a direct application, genes isolated in model species could be candidate genes for species of agronomic interest. The information available in related species can be used to deduce consensus motifs and to define degenerate PCR primers to amplify the orthologous gene in the species in question. With the growing number of ESTs or full-length cDNA sequences in databases for many species, orthologous genes can be obtained by *in-silico* screening using sequence alignment software.

Several genes have been identified in *Arabidopsis* by random insertion of T-DNA (Bouchez and Hofte, 1998; Chapter 9). Certain mutants have been physiologically characterised and the genes involved in the expression of the mutated traits have been isolated. By sequencing and/or mapping these genes in other species, possible homologies between *Arabidopsis* sequences and important loci in these species have been revealed. For example, the gene *CO* (*Constans*) isolated in *Arabidopsis* and involved in the onset of flowering time (Putterill *et al.*, 1995), is a candidate for two QTLs controlling this trait in black mustard, *Brassica nigra* (Lagercrantz *et al.*, 1996). Similarly, analysis of the sequence and expression of the *Sp* gene (*Self pruning*) controlling indeterminate growth in tomato has shown that this gene is orthologous to the *Terminal Flower 1* gene from *Arabidopsis* which maintains the indeterminate state of the floral meristem (Pnueli *et al.*, 1998).

It has also been shown that even between distant genomes such as tomato and *Arabidopsis*, gene order can be conserved over a few hundred kilobases, even several megabases (the phenomenon of microsynteny; Ku *et al.*, 2000). Comparative mapping of orthologous genes, single copy and highly conserved between these two species, serves as an anchor on the *Arabidopsis* genome which could be useful to a large number of dicotyledonous plants (Fulton *et al.*, 2002).

It has been reported that R-gene clusters frequently lack synteny between related species of cereals and crucifers, but that they appear to be positionally well conserved among the species of the Solanaceae family. Huang *et al.*, (2005) developed a comparative analysis between the *I2* complex locus in the model Solanaceae, tomato, and the *R3* complex locus in potato. This comparative analysis has accelerated cloning of *R3a*, a potato gene that confers race-specific resistance to the late blight pathogen *Phytophthora infestans*. Comparative analyses of the *R3* complex locus with the corresponding *I2* complex locus in tomato suggest that it derives from an ancient locus involved in plant innate immunity against at least oomycete and fungal pathogens together.

FUNCTIONAL CANDIDATE GENES

As shown above for comparative mapping, knowledge about the physiology of a plant could provide information on the possible functions involved in the trait. For others, mutants with extreme phenotypes are available. If the corresponding genes are available, the question of whether these genes are responsible for the trait variation studied becomes instead, whether the genes are polymorphic and whether the polymorphism affects the variation of the trait considered.

When the metabolic pathways linked to a trait are known, candidate genes can be chosen amongst structural or regulational genes involved in these pathways. The number of genes sequenced is the limiting factor for this step. Nevertheless, the number of identified genes or EST sequences is growing rapidly for several species. The construction of cDNA libraries, specific for different organs, developmental stages or stress responses, coupled to differential screening, will allow isolation of functional candidate genes. The analysis of the transcriptome by high-throughput approaches (DNA chips) while simultaneously studying a large number of differentially expressed mRNAs (Wang *et al.*, 1999; Chapter 10) is another source of functional candidate genes, as has already been shown in strawberry (Aharoni *et al.*, 2000). Proteomic approaches, by revealing polymorphism within genes, as well as differences in protein expression, are complementary to DNA marker and mapping approaches.

Following the characterization of the first resistance genes, the candidate gene approach has been used to characterise new resistance genes. Degenerate primers, designed from consensus amino acid motifs from NBS (nucleotide-binding sites), LRR (leucine rich repeat) and kinase domains (corresponding to domains commonly found in resistance genes) have been used to amplify resistance gene analogues (RGA or R-gene analogues) in several species (Leister *et al.*, 1996; Aarts *et al.*, 1998). Certain RGAs have been mapped to regions where loci involved in resistance are found, thus constituting potential candidate genes. Several examples can be mentioned in agronomically important species, such as pepper (Pflieger *et al.*, 1999), soybean (Kanazin *et al.*, 1996, Yu *et al.*, 1996), lettuce (Shen *et al.*, 1998), bean (Geffroy *et al.*, 1999) or tomato (Pan *et al.*, 2000). Cloning of genes in other species has allowed cloning of *RPP8* from *Arabidopsis* which is homologous to a resistance gene of type *NBS-LRR* (McDowell *et al.*, 1998) or with the recessive resistance gene *pvr2* from pepper which corresponds to a eukaryotic translation factor (Ruffel *et al.*, 2002).

Cloning by Mutant Analysis

Occasionally mutants can be used to clone genes of interest based on the presence of a clearly defined phenotype. If a mutant strategy is envisaged, it is important to employ an efficient phenotype test for screening the population in order to identify a mutant for the trait. Mutants are very useful as they can accelerate gene isolation by avoiding or accelerating map-based localization. In this case, the mutants contain one or several well identified elements inserted in the genome. The inserted elements are either T-DNA (Azpiroz *et al.*, 1997), or mobile elements such as *Ac/Ds*, *En/Spm*, *Mu* in maize, *Tnt-1* in Solanaceae (Chapter 25). The insertion of such elements modifies the expression or activity of the affected gene, and as a consequence, the phenotype controlled by the gene. In species where such libraries are available, the identification of genes associated with characterised phenotypes in a mutant collection is relatively direct and requires the identification, for the mutants showing a modified phenotype, of the flanking sequences surrounding the inserted element. Insertion mutant collections are especially used in reverse genetic approaches, to look for phenotypes associated with a given gene. These collections are however reserved to a limited number of species. In *Arabidopsis,* several collections of many thousand lines have been developed (Bouchez and Hofte, 1998). In this species, as in maize, the aim is to generate a mutant in each gene (AGI, 2000; Gai *et al.,* 2000). With FST (Flanking Sequence Tag), it is possible to identify a mutant in a gene using a simple Blast on *SIGnAL "T-DNA Express" Arabidopsis Gene Mapping Tool* (http://signal.salk.edu/cgi-bin/tdnaexpress). In rice, several libraries are in the process of construction (Izawa *et al.,* 1997; Agrawal *et al.,* 2001), while in tomato several tens of lines having integrated Ac/Ds elements from maize have been produced (Meissner *et al.,* 2000), allowing cloning of several genes of agronomic interest (Jones *et al.,* 1994, Bishop *et al.,* 1996). Screening for deletion mutants by a PCR based strategy is also possible for reverse genetics. Deletion mutants, generated by fast neutron mutagenesis, can be obtained, for a gene of interest, and characterised (Li *et al.,* 2001).

SELECTION OF THE MOST PROBABLE CANDIDATE GENES

When one or several candidate genes have been chosen, experiments must be carried out to select the most likely, or best, candidates. In plants, two strategies are used (Fig. 28.3). The genetic location of the candidate gene and the locus of interest are compared, necessitating fine mapping of the locus of interest (Wei *et al.,* 1999) for functional candidate genes that have not been isolated by map-based cloning methods. Simultaneously, associations between the phenotypic variation of the trait and molecular polymorphisms in the candidate gene, are looked for in a series of non-related individuals.

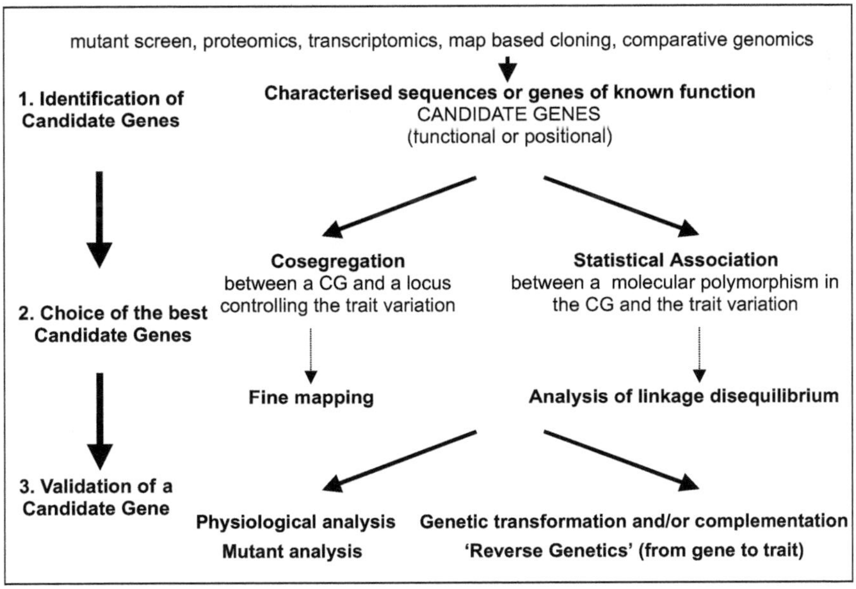

Fig. 28.3 Principles of the candidate gene (CG) approach

Cosegregation of a Candidate Gene and a Trait Controlled by a Single Locus

Mapping is essentially useful for rejecting all candidate genes that do not co-segregate with the locus of interest. In the case of major genes, the association between a candidate gene and the locus must be complete. As soon as a recombinant individual is obtained, the candidate gene is rejected. Thus fine mapping in the region of interest, by studying several hundred individuals leads to the rejection of genes previously judged interesting when smaller numbers of plants were involved. In parallel the polymorphism in the candidate gene can be studied to look for possible mutations explaining the variation in expression.

For example, in oilseed rape, the gene encoding the enzyme *FAE1* (*Fatty Acid Elongation 1*), involved in the metabolism of fatty acids, cosegregates with the locus E1 controlling the concentration of erucic acid in a double haploid population of 152 individuals (Barret *et al.*, 1998; Fourmann *et al.*, 1998). Several deletions have been detected in the alleles of the gene *FAE1* linked to low levels of erucic acid. These deletions lead to frame shifts or stop codons and are thought to be responsible for the low acid content phenotype (Roscoe *et al.*, 2001).

Cosegregation of a Candidate Gene and a QTL

The association between a candidate gene and a QTL is less easily shown. In contrast to a major gene, the identification of the position of a QTL is imprecise, the confidence interval generally covers several megabases (Dupuis and Siegmund, 1999), which can represent several hundred or thousand genes. Fine mapping is therefore necessary to reduce the confidence interval around the QTL. In the same way, the creation of nearly isogenic lines (differing for the region of the QTL) shows the effect of the QTL in a homogeneous genetic background. Subsequently, other QTLs influencing the trait are no longer segregating and this allows the position of the QTL to be defined more precisely (Paterson *et al.*, 1990, Touzet *et al.*, 1995).

Several examples of QTLs co-segregating with genes encoding enzymes involved in carbon metabolism have been published. The gene *Shrunken-2* which encodes a subunit of ADP-glucose pyrophosphorylase (ADPGppase), a key enzyme of starch synthesis in seeds, cosegregates with a QTL affecting starch content (Goldman *et al.*, 1993) in a population of a hundred or so maize F_3 families. In another cross, a QTL controlling seed amylose content has been observed near to the same locus (Prioul *et al.*, 1999).

For species with full sequenced genome, such as *Arabidopsis*, finding co-segregating candidate genes is easier. However, this species is lacking agronomic genetic data and there is an increasing number of scientific programmes aimed at mapping QTLs. For example, nitrate content or nitrate-use efficiency is on the way to being well described (Loudet *et al.*, 2003; Harada *et al.*, 2004) and these data will contribute to the integration of genetic and molecular data at a physiological level.

Statistical Associations between a Candidate Gene Polymorphism and the Phenotype

The molecular variation of a candidate gene can be studied in a series of unrelated individuals to search for statistical associations between the candidate gene polymorphism and the phenotype variation. Alternatively the variability can be studied in a population of EMS mutants by TILLING (targeted induced local lesions in genomes). A statistical correlation or tight genetic linkage between a candidate gene polymorphism and the trait variation is not sufficient to demonstrate irrefutably a cause and effect relationship between the gene and the phenotype. These relationships are necessary but not sufficient: the candidate gene polymorphism can simply show linkage disequilibrium with the gene responsible for the phenotype in question.

Association with simple traits

A good example of a simple trait is the physiological study of the wrinkled seed phenotype in pea mutants which suggests that the mutated gene is involved in starch metabolism (Bhattacharyya *et al.*, 1993). The wrinkled seed phenotype is determined by the *r* (rugosus) locus and is recessive while the smooth seed phenotype is dominant. The enzymatic activity of *SBE1* (*Starch-Branching Enzyme*) is reduced in wrinkled seeds. The gene *SBE1* has therefore been proposed as a candidate gene for the *r* locus. *rr* mutants have an 800 bp transposable element in the coding sequence of the gene *SBE1* (Bhattacharyya *et al.*, 1990). The allelic polymorphism of the *SBE1* gene observed in nearly isogenic lines except at the *r* locus has been associated with the seed phenotype variation.

In pepper, fruit colour is determined at the *y* locus. The dominant allele *y+* confers a red colour while the recessive allele *y* confers a yellow colour. At the biochemical level, the last step in the metabolic pathway to carotenoid synthesis is catalysed by the enzyme CCS (Capsanthin-Capsorubin Synthase; Bouvier *et al.*, 1994), whose activity is detected only in red fruits. A polymorphism for the *CCS* gene has been found to be completely linked to the fruit colour variation in an F_2 population of 145 individuals. Yellow fruited peppers are characterised by the absence of the *CCS* gene, likely to result from a deletion. This presence-absence polymorphism has been statistically associated with fruit colour in 31 unrelated individuals, which reinforces the hypothesis that the gene *CCS* corresponds to the *y* locus (Lefebvre *et al.*, 1998).

Associations with complex traits

The research for associations has been mainly used in human genetics, and applications are starting to be envisaged for QTL characterisation. Significant statistical associations have been detected for maize growth parameters and polymorphisms in three candidate genes (two genes encoding sucrose synthase *Sh1* and *Sus1*, and the gene *Sh2* encoding ADPGppase; Prioul *et al.*, 1999). The linkage disequilibrium amongst 46 genetically unrelated individuals has been calculated at two levels: intragenically, i.e. between polymorphic sites within the same gene and intergenically, i.e. between a candidate gene polymorphic site and a polymorphic site of another unlinked gene. On average, the intragenic linkage disequilibrium is greater than the intergenic linkage disequilibrium, suggesting that polymorphic sites within a candidate gene could be the cause of the phenotypic variation observed. The population structure can be an obstacle to this procedure but special statistical tests can take the structure into account (Remington *et al.*, 2001, Vigouroux *et al.*, 2002). In maize a significant association between *Dwarf8* gene polymorphism and variation in earliness has been shown (Thornsberry *et al.*, 2001). The usefulness of investigating genetic diversity to identify favourable alleles to be used in molecular breeding has been clearly demonstrated in more recent publications: for example in maize, where polymorphisms in three candidate genes for lignin biosynthesis were correlated with cell wall digestibility (Guillet-Claude *et al.*, 2004). In potato, late blight resistance is correlated with late maturity, an undesirable characteristic, but the use of PCR markers allowed correction of the maturity effect in resistant lines (Bormann *et al.*, 2004). Again in potato, starch and sugar content of tubers are important quantitative traits for the quality of processed products such as potato chips and french fries. Genes encoding invertases were shown to be involved in tuber sugar content. The duplicated invertase genes invGE and invGF colocalise with the cold-sweetening QTL Sug9. The two invertase alleles invGE-f and invGF-d were associated with better chip quality in three breeding populations. Allele invGF-b was associated with lower tuber starch content in 188 tetraploid potato cultivars (Li *et al.*, 2005).

VALIDATION OF THE ROLE OF A CANDIDATE GENE

The confirmation of the role of a candidate gene in the variation of a trait is not direct and includes:

- Statistical correlation of gene polymorphisms and the trait variation in populations where the linkage disequilibrium is at a minimum (in such populations, composed of unrelated individuals, only a direct cause and effect relationship ensures the correlation is maintained over generations);
- Analysis of the variation at a biochemical and metabolic level (without being sufficient, a condition necessary for an enzyme to be a QTL is that the enzyme activity is variable);
- Molecular analysis of alleles (restriction maps, sequencing) in order to find a molecular basis for the variation;
- Transformation, which in plants can be the ultimate proof, even if it is more difficult in the case of a QTL.

Analysis of Candidate Gene Expression

The relationship between transcript level, protein level or enzyme activity and trait value can be determined and can participate in the validation of candidate genes. Classic methods for expression analysis such as northern and western blots are often not sufficiently specific for a given member of a gene family. In this case, for transcript levels the 3′ UTR of the gene can be used or quantitative RT-PCR is employed, and for proteins, monoclonal antibodies will discriminate between the different members.

At least five QTLs controlling phenotypic variations between teosinte (maize ancestor) and maize are supposedly linked to stages in the domestication of maize (Doebley and Stec, 1993). One of them is located near the *tb1* locus (*teosinte branched 1*) determining apical dominance. The gene *tb1* has been cloned and encodes a protein with homology to the cycloidea gene of snapdragon. *tb1* seems to act both to repress the growth of axillary organs and to enable the formation of female inflorescences and exhibits an expression level two-fold higher in maize than in teosinte, confirming its suspected role in the control of apical dominance and its involvement in the evolution of teosinte into maize. The proof that *tb1* controls branching has been shown by genetic complementation experiments (Doebley *et al.*, 1997).

In tomato, the gene coding for an acid invertase, an enzyme that cleaves sucrose into glucose and fructose in fruits, is a good candidate gene for the sucrose content of fruit. A statistical association exists between a polymorphism in the acid invertase gene and the sucrose content (Chetelat *et al.*, 1993) and also between the expression level of the invertase and the sucrose content (Klann *et al.*, 1993), corroborating the presumed role for the gene in sucrose content.

Genetic Transformation

Genetic transformation is the ultimate way of validating the involvement of a candidate gene in the variation of a monogenic trait. In the case of complex traits, the validation by transformation is not immediate because of the partial effect of the genes involved. Transformations with antisense or RNAi constructions are often used, notably in the case of genes belonging to multigenic families. In this case, the transgene inhibits the expression of the functional gene and the proof is provided if the transformant shows a deficient phenotype directly linked to the trait. The inactivation of a gene by T-DNA insertion or by a transposon, when mutants are available, constitutes an alternative validation strategy. However, the technique is reserved to very few species because a collection of insertion mutants is required as well as a way of carrying out systematic screening for the phenotype studied. Transient loss-of-function transformation techniques, such as VIGS (virus-induced gene silencing) have also been shown to be useful to assess the function of candidate genes (Brigneti *et al.*, 2004).

The validation by transformation has limits. It is known that the expression level of a transgene depends on the insertion site and the number of copies inserted. Gene silencing phenomena have also frequently been described. The inhibition of a gene by an antisense strategy shows the involvement of the family of genes in question in the control of the trait (Ori *et al.*, 1997) but does not give information on the activity of a particular member of the gene family. The main limit is linked to the partial effects of QTLs. Several independent transformants must be obtained and tested to detect a statistically significant effect due to the transgene. In certain cases, we can imagine that genetic transformation generates artefactual phenotypes. This could be due to epistatic interactions between the transgene and the genetic background of the transformed plant, or due to allele dose effects, which prevent the effect of the transgene being measured. In order to overcome such effects, ideally the transgene should replace the wild type allele by homologous transformation. Currently strategies for gene replacement using homologous recombination only work in bacteria, yeast and a few rare plants such as moss or otherwise occur at frequencies too low to be useful (Mengiste and Paszkowski, 1999; Schuermann *et al.*, 2005).

CONCLUSION

Different approaches leading to the characterisation of genes controlling a phenotype are all based on the identification of candidate genes and include different strategies (genetic and map-based cloning, functional analysis or screening of mutants). These strategies are not mutually exclusive but instead very complementary. The fine mapping phase can be used for map-based cloning but is more efficient when coupled to a functional candidate gene approach. These two approaches have been shown to be good ways of characterizing loci of agronomic interest to isolate the underlying genes, in particular those involved in determining simple traits. Efforts are necessary for the application of these approaches to complex traits.

More and more tools contribute to the development of the candidate gene approach as the number of nucleic acid sequences and markers available increases following genome sequencing projects for model plants (*Arabidopsis*, rice, *Medicago truncatula*, tomato, wheat, *Populus* ...) and the systematic sequencing of EST libraries. Due to these new tools, high density gene maps are today being developed for several species. Similarly the arrival of transcriptomics, notably DNA chip technology, (Ramsay, 1998) opens new horizons via the identification of genes expressed specifically in one organ, at a precise developmental stage, or in response to biotic or abiotic stress. Together these new technologies contribute to a judicious choice of candidate genes. Moreover, the alignment of maps integrating anonymous markers, genes of known function and loci of interest, with physical maps of a species reinforces the usefulness of the approach. Finally, comparative genome mapping and the identification of regions where microsynteny is conserved will contribute to the success of this approach. The application of the candidate gene approach has, until now, been limited by insufficient knowledge about the physiology of the traits and the still limited number of genes cloned from species of agronomic interest. However, it is evident that as the number of genes mapped grows, more cases of cosegregation between a locus and a candidate gene will be found purely by chance, thus imposing the need for high resolution mapping on progeny from large numbers of individuals. The development of robots and the generalization of high-throughput approaches mean map-based cloning is becoming an envisageable alternative for numerous species.

The characterization of genes of agronomic interest can have several both practical and fundamental applications. Cloned and validated genes have a major interest in marker-assisted selection as they eliminate the risk of recombination between marker and locus and consequently ensure perfect control over the gene introgressed. Moreover, the cloning of a gene will often allow the cloning of new genes in other species. Finally QTL cloning should contribute to a better understanding of the molecular and physiological basis of polygenic traits and their organization in the genome.

REFERENCES

Aarts, M., Hekkert, B., Holub, E., Beynon, J., Stiekema, W., Pereira, A. 1998. Identification of R-gene homologous DNA fragments genetically linked to disease resistance loci in *Arabidopsis thaliana*. *Mol. Plant Microbe Interact.*, 11, 251-258.

AGI. 2000. The *Arabidopsis* Genome Initiative: analysis of the genome sequence of the flowering plant *Arabidopsis thaliana*. *Nature*, 408, 796-815.

Agrawal, G.K., Yamazaki, M., Kobayashi, M., Hirochika, R., Miyao, A., Hirochika, H. 2001. Screening of the rice viviparous mutants generated by endogenous retrotransposon Tos17 insertion. Tagging of a zeaxanthin epoxidase gene and a novel ostate gene. *Plant Physiol.*, 125, 1248-1257.

Aharoni, A., Keizer, L.C.P., Bouwmeester, H.J., Sun, Z., Alvarez-Huerta, M., Verhoeven, H.A., Blaas, J., van Houwelingen, A.M.M.L., De Vos, R.C.H., van der Voet. H., Jansen; R.C., Guis,

M., Mol, J., Davis, R.W., Schena, M., van Tunen, A.J., O'Connell, A.P. 2000. Identification of the SAAT gene involved in strawberry flavor biogenesis by use of DNA microarrays. *Plant Cell,* 12, 647-661.

Alpert, K.B., Tanksley, S.D. 1996. High-resolution mapping and isolation of a yeast artificial chromosome contig containing fw2.2: a major fruit weight quantitative trait locus in tomato. *Proc Natl Acad Sci* USA, 93, 15503-15507.

Arondel, V., Lemieux, B., Hwang, I., Gibson, S., Goodman, H.M., Somerville, C. R. 1992. Map-based cloning of a gene controlling Omega-3 fatty acid desaturation in *Arabidopsis. Science,* 258, 1353-1355.

Azpiroz-Leehan, R., Feldmann, K.A. 1997. T-DNA insertion mutagenesis in Arabidopsis: going back and forth. *Trends Genetics,* 13, 152-156.

Barnes, S. 2002. Comparing Arabidopsis to other flowering plants. *Curr. Opin. Plant. Biol.,* 5, 128-134.

Barret, P., Delourme, R., Renard, M., Domergue, F., Lessire, R., Delseny, M., Roscoe, T.J. 1998. A rapeseed *FAE1* gene is linked to the *E1* locus associated with variation in the content of erucic acid. *Theor. Appl. Genet.,* 96, 177-186.

Bendahmane, A., Kanyuka, K., Baulcombe, D.C. 1999. The *Rx* gene in potato controls separate virus resistance and cell death responses. *Plant Cell,* 11, 197-206.

Bent, A.F. 1996. Plant disease resistance genes: function meets structure. *Plant Cell,* 8, 1757-1771.

Bhattacharyya, M., Martin, C., Smith, A. 1993. The importance of starch biosynthesis in the wrinkled seed shape character of peas studied by Mendel. *Plant Mol. Biol.,* 22, 525-531.

Bhattacharyya, M., Smith, A.M., Ellis, T.H., Hedley, C., Martin, C. 1990. The wrinkled-seed character of pea described by Mendel is caused by a transposon-like insertion in a gene encoding starch-branching enzyme. *Cell,* 60, 115-122.

Bishop, G.J., Harrison, K., Jones, J.D.G. 1996. The tomato *Dwarf* gene isolated by heterologous transposon tagging encodes the first member of a new cytochrome P450 family. *Plant Cell,* 8, 959-969.

Bormann, C.A., Rickert, A.M., Ruiz, R.A., Paal, J., Lubeck, J., Strahwald, J., Buhr, K., Gebhardt, C. 2004. Tagging quantitative trait loci for maturity-corrected late blight resistance in tetraploid potato with PCR-based candidate gene markers. *Mol. Plant-Microbe Interact.,* 17, 1126-38.

Bouvier, F., Hugueney, P., d'Harlingue, A., Kuntz, M., Camara, B. 1994. Xanthophyll biosynthesis in chromoplasts: isolation and molecular cloning of an enzyme catalyzing the conversion of 5,6-epoxycarotenoid. *Plant J.,* 6, 45-54.

Bouchez, D., Hofte, H. 1998. Functional genomics in plants. *Plant Physiol.,* 118, 725-732.

Brigneti, G., Martin-Hernandez, A.M., Jin, H., Chen, J., Baulcombe, D.C., Baker, B., Jones, J.D. 2004. Virus-induced gene silencing in Solanum species. *Plant J.,* 39, 264-72.

Brommonschenkel, S.H., Tanksley, S. 1997. Map-based cloning of the tomato genomic region that spans the Sw-5 tospovirus resistance gene in tomato. *Mol. Gen. Genet.,* 256, 121-126.

Buschges, R., Hollricher, K., Panstruga, R., Simons, G., Wolter, M., Frijters, A., van Daelen, R., van der Lee, T., Diergaarde, P., Groenendijk, J., Topsch, S., Vos, P., Salamini, F., Schulze-Lefert, P. 1997. The barley *Mlo* gene: a novel control element of plant pathogen resistance. *Cell,* 88, 695-705.

Burke, D.T., Carle, G.F., Olson, M.V. 1987. Cloning of large segments of exogenous DNA into yeast by means of artificial chromosome vectors. *Science,* 236, 805-811.

Cai, D., Kleine, M., Kifle, S., Harloff, H.J., Sandal, N.N., Marcker, K.A., Klein-Lankhorst, R.M., Salentijn, E.M., Lange, W., Stiekema, W.J., Wyss, U., Grundler, F.M., Jung, C. 1997. Positional cloning of a gene for nematode resistance in sugar beet. *Science,* 275, 832-834.

Chen, G., Bi, Y.R., Li, N. 2005. EGY1 encodes a membrane-associated and ATP-independent metalloprotease that is required for chloroplast development. *Plant J.*, 41, 364-375.

Chetelat, R.T., Klann, E., Deverna, J.W., Yelle, S., Bennett, A.B. 1993. Inheritance and genetic mapping of fruit sucrose accumulation in *Lycopersicon chmielewskii*. *Plant J.*, 4, 643-650.

Dangl, J. L., Jones, J.D.G. 2001. Plant pathogens and integrated defence responses to infection. *Nature*, 411, 826-833.

Diener, A.C., Ausubel, F.M. 2005. Resistance to Fusarium oxysporum 1, a dominant Arabidopsis disease-resistance gene, is not race specific. *Genetics*, 171, 305-321.

Doebley, J., Stec, A. 1993. Inheritance of the morphological differences between maize and teosinte: comparison of results for two F2 populations. *Genetics*, 134, 559-570.

Doebley, J., Stec, A., Hubbard, L. 1997. The evolution of apical dominance in maize. *Nature*, 386, 485-488.

Dupuis, J., Siegmund, D. 1999. Statistical methods for mapping quantitative trait loci from a dense set of markers. *Genetics*, 151, 373-386.

Fourmann, M., Barret, P., Renard, M., Pelletier, G., Delourme, R., Brunel, D. 1998. The two genes homologous to *Arabidopsis FAE1* co-segregate with the two loci governing erucic acid content in *Brassica napus*. *Theor. Appl. Genet.*, 96, 852-858.

Frary, A., Nesbitt, T.C., Grandillo, S., Knaap, E., Cong, B., Liu, J., Meller, J., Elber, R., Alpert, K.B., Tanksley, S.D. 2000. fw2.2: a quantitative trait locus key to the evolution of tomato fruit size. *Science*, 289, 85-88.

Fridman, E., Pleban, T., Zamir, D. 2000. A recombination hotspot delimits a wild-species quantitative trait locus for tomato sugar content to 484 bp within an invertase gene. *Proc. Natl. Acad. Sci.*, 97, 4718-4723.

Fridman, E., Carrari, F., Liu, Y.-S., Fernie, A.R., Zamir, D. 2004. Zooming in on a quantitative trait for tomato yield using interspecific introgressions. *Science*, 305, 1786-1789.

Fulton, T.M., Van der Hoeven, R., Eannetta, N.T., Tanksley, S.D. 2002. Identification, analysis, and utilization of Conserved Ortholog Set markers for comparative genomics in higher plants. *Plant Cell*, 14, 1457-1467.

Gai, X., Lal, S., Xing, L., Brendel, V., Walbot, V. 2000. Gene discovery using the maize genome database ZmDB. *Nucl. Acids Res.*, 28, 94-96.

Gale, M.D., Devos, K.M. 1998. Comparative genetics in the grasses. *Proc. Natl. Acad. Sci.*, USA, 95, 1971-1974.

Gebhardt, C. 1997. Plant genes for pathogen resistance–variation on a theme. *Trends Plant Sci.*, 2, 243-244.

Geffroy, V., Sicard, D., de Oliveira, J., Sévignac, M., Cohen, S., Gepts, P., Neema, C., Langin, T., Dron, M. 1999. Identification of an ancestral gene cluster involved in the coevolution process between *Phaseolus vulgaris* and its fungal pathogen *Colletotrichum lindemuthianum*. *Mol. Plant Microbe Interact.*, 12, 774-784.

Goldman, I.L., Rocheford, T.R., Dudley, J.W. 1993. Quantitative trait loci influencing protein and starch concentration in the Illinois Long Term Selection maize strains. *Theor. Appl. Genet.*, 87, 217-224.

Gorman, S.W., Banasiak, D., Fairley, C., McCormick, S. 1996. A 610 kb YAC clone harbors 7 cM of tomato (Lycopersicon esculentum) DNA that includes the male sterile 14 gene and a hotspot for recombination. *Mol. Gen. Genetics*, 251, 52-59.

Green, K.A., Prigge, M.J., Katzman, R.B., Clark, S.E. 2005. CORONA, a member of the class III homeodomain leucine zipper gene family in *Arabidopsis*, regulates stem cell specification and organogenesis. *Plant Cell*, 17, 691-704.

Guillet-Claude, C., Birolleau-Touchard, C., Manicacci, D., Fourmann, M., Barraud, S., Carret, V., Martinant, J.P., Barriere, Y. 2004. Genetic diversity associated with variation in silage corn digestibility for three O-methyltransferase genes involved in lignin biosynthesis. *Theor. Appl. Genet.*, 110, 126-35.

Hammond-Kosack, K.E., Jones, J.D.G. 1996. Resistant gene-dependent plant defence response. *Plant Cell*, 8, 1773-1791.

Hammond-Kosack, K.E., Jones, J.D.G. 1997. Plant disease resistance genes. *Ann. Rev. Plant Physiol. Plant Mol. Biol.*, 48, 575-607.

Harada, H., Kuromori, T., Hirayama, T., Shinozaki, K., Leigh, R.A. 2004. Quantitative trait loci analysis of nitrate storage in *Arabidopsis* leading to an investigation of the contribution of the anion channel gene, AtCLC-c, to variation in nitrate levels. *J Exp Bot.*, 55, 2005-2014.

Hemsley, P.A., Kemp, A.C., Grierson, C.S. 2005. The TIP GROWTH DEFECTIVE1 S-acyl transferase regulates plant cell growth in *Arabidopsis. Plant Cell*, 17, 2554-2563.

Huang, S., van der Vossen, E.A., Kuang, H., Vleeshouwers, V.G., Zhang, N., Borm, T.J., van Eck, H.J., Baker, B., Jacobsen, E., Visser, R.G. 2005. Comparative genomics enabled the isolation of the R3a late blight resistance gene in potato. *Plant J.* 42, 251-261.

Izawa, T., Ohnishi, T., Nakano, T., Ishida, N., Enoki, H., Hashimoto, H., Itoh, K., Terada, R., Wu, C., Miyazaki, C., Endo, T., Iida, S., Shimamoto, K. 1997. Transposon tagging in rice *Plant Mol. Biol.*, 35, 219-229.

Jander, G., Norris, S.R., Rounsley, S.D., Bush, D.F., Levin, I.M., Last, R.L. 2002. *Arabidopsis* map-based cloning in the post-genome era. *Plant Physiol.*, 129, 440-450.

Jones, D.A., Thomas, C.M., Hammond-Kosack, K.E., Balint-Kurti, P.J., Jones, J.D.G. 1994. Isolation of the tomato Cf-9 gene for resistance to Cladosporium fulvum by transposon tagging. *Science*, 266, 789-792.

Kanazin, V., Marek, L., Shoemaker, R.C. 1996. Resistance gene analogs are conserved and clustered in soybean. *Proc. Natl. Acad. Sci.*, USA, 93, 11746-11750.

Klann, E.M., Chetelat, R.T., Bennett, A.B. 1993. Expression of acid invertase gene controls sugar composition in tomato (*Lycopersicon*) fruit. *Plant Physiol.*, 103, 863-870.

Koizumi, K., Naramoto, S., Sawa, S., Yahara, N., Ueda, T., Nakano, A., Sugiyama, M., Fukuda, H. 2005. VAN3 ARF-GAP-mediated vesicle transport is involved in leaf vascular network formation. *Development*, 132, 1699-1711.

Krusell, L., Krause, K., Ott, T., Desbrosses, G., Kramer, U., Sato, S., Nakamura, Y., Tabata, S., James, E.K., Sandal, N., Stougaard, J., Kawaguchi, M., Miyamoto, A., Suganuma, N., Udvardi, M.K. 2005. The sulfate transporter SST1 is crucial for symbiotic nitrogen fixation in Lotus japonicus root nodules. *Plant Cell*, 17, 1625-1636.

Ku, H.M., Vision, T., Liu, J., Tanksley, S.D. 2000. Comparing sequenced segments of the tomato and Arabidopsis genomes: Large-scale duplication followed by selective gene loss creates a network of synteny. *Proc. Natl. Acad. Sci.*, USA, 97, 9121-9126.

Kurata, N., Nagamura, Y., Yamamoto, K., Harushima, Y., Sue, N., Wu, J., Antonio, B.A., Shomura, A., Shimizu, T., Lin, S.-Y., Inoue, T., Fukuda, A., Shimano, T., Kuboki, Y., Toyama, T., Miyamoto, Y., Kirihara, T., Hayasaka, K., Miyao, A., Monna, L., Zhong, H.S., Tamura, Y., Wang, Z.-X., Momma, T., Umehara, Y., Yano, M., Sasaki, T., Minobe, Y. 1994. A 300 kilobase interval genetic map of rice including 883 expressed sequences. *Nat. Genet.*, 8, 365-372.

Lagercrantz, U., Putterill, J., Coupland, G., Lydiate, D. 1996. Comparative mapping in *Arabidopsis* and Brassica, fine-scale genome collinearity and congruence of genes controlling flowering time. *Plant J.*, 9, 13-20.

Lamb, C.J., Lawton, M.A., Dron, M., Dixon, R.A. 1989. Signals and transduction mechanisms for activation of plant defense against microbial attack. *Cell*, 56, 215-224.

Larsen, P.B., Geisler, M.J., Jones, C.A., Williams, K.M., Cancel, J.D. 2005. ALS3 encodes a phloem-localized ABC transporter-like protein that is required for aluminum tolerance in *Arabidopsis. Plant J.*, 41, 353-363.

Lefebvre, V., Kuntz, M., Camara, B., Palloix, A. 1998. The capsanthin-capsorubin synthase gene: a candidate gene for the *y* locus controlling the red fruit colour in pepper. *Plant Mol. Biol.,* 36, 785-789.

Leister, D., Ballvora, A., Salamini, F., Gebhardt, C. 1996. A PCR-based approach for isolating pathogen resistance genes from potato with potential for wide application in plants. *Nat. Genet.,* 14, 421-429.

Li, X., Song, Y., Century, K., Straight, S., Ronald, P., Dong, X., Lassner, M., Zhang, Y. 2001. A fast neutron deletion mutagenesis-based reverse genetics system for plants. *Plant J.,* 27, 235-242.

Li, C., Schilmiller, A.L., Liu, G., Lee, G.I., Jayanty, S., Sageman, C., Vrebalov, J., Giovannoni, J.J., Yagi, K., Kobayashi, Y., Howe, G.A. 2005. Role of beta-oxidation in jasmonate biosynthesis and systemic wound signaling in tomato. *Plant Cell,* 17, 971-986.

Li, L., Strahwald, J., Hofferbert, H.R., Lubeck, J., Tacke, E., Junghans, H., Wunder, J., Gebhardt, C. 2005. DNA variation at the invertase locus invGE/GF is associated with tuber quality traits in populations of potato breeding clones. *Genetics,* 170, 813-821.

Liu, J., Cong, B., Tanksley, S.D. 2003. Generation and analysis of an artificial gene dosage series in tomato to study the mechanisms by which the cloned quantitative trait locus *fw2.2* controls fruit size. *Plant Physiol.,* 132, 292-299.

Livingstone, K.D., Lackney, V.K., Blauth, J.R., van Wijk, R., Jahn M.K. 1999. Genome mapping in *Capsicum* and the evolution of genome structure in the *Solanacea. Genetics,* 152, 1183-1202.

Loudet, O., Chaillou, S., Merigout, P., Talbotec, J., Daniel-Vedele, F. 2003. Quantitative trait loci analysis of nitrogen use efficiency in *Arabidopsis. Plant Physiol.,* 131, 345-358.

Mao, L., Begum, D., Chuang, H.W., Budiman, M.A., Szymkowiak, E.J., Irish, E.E., Wing, R.A. 2000. JOINTLESS is a MADS-box gene controlling tomato flower abscission zone development. *Nature,* 406, 910-913.

Marra, M.A., Hillier, L., Waterson, R.H. 1998. Expressed-sequence tags – ESTablishing bridges between genomes. *Trends Genetics,* 14, 4-7.

Martin, G.B., Brommonschenkel, S.H., Chunwongse, J., Frary, A., Ganal, M.W., Spivey, R., Wu, T., Earle, E.D., Tanksley, S.D. 1993. Map-based cloning of a protein kinase gene conferring disease resistance in tomato. *Science,* 262, 1432-1436.

McDowell, J.M., Dhandaydham, M., Long, T.A., Aarts, M.G.M., Goff, S., Holub, E.B., Dangl, J.L. 1998. Intragenic recombination and diversifying selection contribute to the evolution of downy mildew resistance at the *RPP8* locus of *Arabidopsis. Plant Cell,* 10, 1861-1874.

Meissner, R., Chague, V., Zhu, Q., Emmanuel, E., Elkind, Y., Levy, A.A. 2000. A high throughput system for transposon tagging and promoter trapping in tomato. *Plant J.,* 22, 265-274.

Mengiste, T., Paszkowski, J. 1999. Prospects for the precise engineering of plant genomes by homologous recombination. *Biol. Chem.,* 380, 749-758.

Michelmore, R.W., Paran, I., Kesseli, R.V. 1991. Identification of markers linked to disease resistance genes by bulked-segregant analysis: a rapid method to detect markers in specific genomic regions by using segregating populations. *Proc. Natl. Acad. Sci.,* USA, 88, 9828-9832.

Mindrinos, M., Katagiri, F., Yu, G.L., Ausubel, F.M. 1994. The A. thaliana disease resistance gene RPS2 encodes a protein containing a nucleotide-binding site and leucine-rich repeats. *Cell,* 23, 1089-1099.

Monaco, A.P., Larin, Z. 1994. YACs, BACs, PACs and MACs: Artificial Chromosomes as Research Tools. *Trends Biotechnol.,* 12, 280-286.

Nishimura, A., Ashikari, M., Lin, S., Takashi, T., Angeles, E.R., Yamamoto, T., Matsuoka, M. 2005. Isolation of a rice regeneration quantitative trait loci gene and its application to transformation systems. *Proc. Natl. Acad. Sci.* USA, 102, 11940-11944.

Noutoshi, Y., Ito, T., Seki, M., Nakashita, H., Yoshida, S., Marco, Y., Shirasu, K., Shinozaki, K. 2005. A single amino acid insertion in the WRKY domain of the *Arabidopsis* TIR-NBS-LRR-WRKY-type disease resistance protein SLH1 (sensitive to low humidity 1) causes activation of defense responses and hypersensitive cell death. *Plant J.*, 43, 873-888.

Ori, N., Eshed, Y., Paran, I., Presting, G., Aviv, D., Tanksley, S., Zamir, D., Fluhr, R. 1997. The *I2C* family from the wilt disease resistance locus *I2* belongs to the nucleotide binding, leucine-rich repeat superfamily of plant resistance genes. *Plant Cell*, 9, 521-532.

Pan, Q., Liu, Y.-S., Budai-Hadrian, O., Sela, M., Carmel-Goren, L., Zamir, D., Fluhr, R. 2000. Comparative genetics of nucleotide binding site-leucine rich repeat resistance gene homologues in the genomes of two dicotyledons: tomato and *Arabidopsis*. *Genetics*, 155, 309-322.

Paran, I., Zamir, D. 2003. Quantitative traits in plants: beyond the QTL. *Trends Genetics*, 19, 303-306.

Paterson, A.H., de Verna, J.W., Lanini, B., Tanksley, S.D. 1990. Fine mapping of quantitative trait loci using selected overlapping recombinant chromosomes in an interspecific cross of tomato. *Genetics*, 124, 735-742.

Pflieger, S., Lefebvre, V., Causse, M. 2001. The candidate gene approach. A Review. *Molecular Breeding*, 7, 275-291.

Pflieger, S., Lefebvre, V., Caranta, C., Blattes, A., Goffinet, B., Palloix, A. 1999. Disease resistance gene analogs as candidates for QTLs involved in pepper/pathogen interactions. *Genome*, 42, 1100-1110.

Pnueli, L., Carmel-Goren, L., Hareven, D., Gutfinger, T., Alvarez, J., Ganal, M., Zamir, D., Lifschitz, E. 1998. The *self-pruning* gene of tomato regulates vegetative to reproductive switching of sympodial meristems and is the ortholog of *CEN* and *TFL1*. *Development*, 125, 1979-1989.

Prioul, J.-L., Pelleschi, S., Sene, M., Thévenot, C., Causse, M., de Vienne, D., Leonardi, A. 1999. From QTL for enzyme activity to candidate gene. *J. Exp. Bot.*, 50, 1281-1288.

Putterill, J., Robson, F., Lee, K., Simon, R., Coupland, G. 1995. The *constans* gene of *Arabidopsis* promotes flowering and encodes a protein showing similarities to zinc finger transcription factors. *Cell*, 80, 847-857.

Ramsay, G. 1998. DNA chips: state-of-the-art. *Nat Biotech.*, 16, 40-44.

Remington, D.L., Thornsberry, J.M., Matsuoka, Y., Wilson, L.M., Whitt, S.R., Doebley, J., Krezovich, S., Goodman, M.M., Buckler, E.S. 2001. Structure of linkage disequilibrium and phenotypic associations in the maize genome. *Proc. Nat. Acad. Sci.*, USA, 98, 11479-11484.

Ren, Z.H., Gao, J.P., Li, L.G., Cai, X.L., Huang, W., Chao, D.Y., Zhu, M.Z., Wang, Z.Y., Luan S., Lin H.X. 2005. A rice quantitative trait locus for salt tolerance encodes a sodium transporter. *Nat. Genet.*, 37, 1141-1146.

Rommens, J.M., Iannuzzi, M.C., Kerem, B.-S., Drumm, M.L., Melmer, G., Dean, M., Rozmahel, R., Cole, J.L., Kennedy, D., Hidaka, N., Zsiga, M., Buchwald, M., Riordan, J.R., Tsui, L.-C., Collins, F.S. 1989. Identification of the cystic fibrosis gene: chromosome walking and jumping. *Science*, 245, 1059-1065.

Roscoe, T.J., Lessire, R., Puyaubert, J., Renard, M., Delseny, M. 2001. Mutations in the fatty acid elongation 1 gene are associated with a loss of beta-ketoacyl-CoA synthase activity in low erucic acid rapeseed. *FEBS Lett.* 492, 107-111.

Ruffel, S., Dussault, M.H., Palloix, A., Moury, B., Bendahmane, A., Robaglia, C., Caranta, C. 2002. Natural recessive resistance gene against Potato virus Y in pepper corresponds to the eukaryotic initiation factor 4E (eIF4E). *Plant J.*, 32, 1067-1075.

Salvi, S., Tuberosa, R. 2005. To clone or not to clone plant QTLs: present and future challenges. *Trends Plant Sci.*, 6, 297-304.

Schmidt, R. 2000. Synteny: recent advances and future prospects. *Curr. Opin. Plant. Biol.*, 3, 97-102.

Schuermann, D., Molinier, J., Fritsch, O., Hohn, B. 2005. The dual nature of homologous recombination in plants. *Trends Genetics,* 21, 172-181.

Shen, K.A., Meyers, B.C., Islam-Faridi, M.N., Chin, D.B., Stelly, D.M., Michelmore, R.W. 1998. Resistance gene candidates identified by PCR with degenerate oligonucleotide primers map to clusters of resistance genes in lettuce. *Mol. Plant Microbe Interact.*, 11, 815-823.

Shizuya, H., Birren, B., Kim, U.J., Mancino, V., Slepak, T., Tachiiri, Y., Simon, M. 1992. Cloning and stable maintenance of 300-kilobase-pair fragments of human DNA in *Escherichia coli* using an F-factor-based vector. *Proc. Natl. Acad. Sci.*, 89, 8794-8797.

Song, W.Y., Wang, G.L., Chen, L.L., Kim, H.S., Pi, L.Y., Holsten, T., Gardner, J., Wang, B., Zhai, W.X., Zhu, L.H., Fauquet, C., Ronald, P. 1995. A receptor kinase-like protein encoded by the rice disease resistance gene, *Xa21*. *Science,* 270, 1804-1806.

Staskawicz, B.J., Ausubel, F.M., Baker, B.J., Ellis, J.G., Jones, J.D.G. 1995. Molecular genetics of plant disease resistance. *Science,* 268, 661-667.

Stein, N., Perovic, D., Kumlehn, J., Pellio, B., Stracke, S., Streng, S., Ordon, F., Graner, A. 2005. The eukaryotic translation initiation factor 4E confers multiallelic recessive Bymovirus resistance in Hordeum vulgare (L.). *Plant J.*, 42, 912-922.

Suetsugu, N., Kagawa, T., Wada, M. 2005. An auxilin-like J-domain protein, JAC1, regulates phototropin-mediated chloroplast movement in Arabidopsis. *Plant Physiol.*, 139, 151-162.

Syvänen, A.C. 2001 Accessing genetic variation: genotyping single nucleotide polymorphisms. *Nat Rev Genet.*, 12, 930-942.

Tamura, K., Shimada, T., Kondo, M., Nishimura, M., Hara-Nishimura, I. 2005. KATAMARI1/MURUS3 is a novel golgi membrane protein that is required for endomembrane organization in *Arabidopsis. Plant Cell,* 17, 1764-1776.

Tanabe, S., Ashikari, M., Fujioka, S., Takatsuto, S., Yoshida, S., Yano, M., Yoshimura, A., Kitano, H., Matsuoka, M., Fujisawa, Y., Kato, H., Iwasaki, Y. 2005. A novel cytochrome P450 is implicated in brassinosteroid biosynthesis via the characterization of a rice dwarf mutant, dwarf11, with reduced seed length. *Plant Cell,* 17, 776-790.

Tanksley, S.D., Ganal, M.W., Martin, G.B. 1995. Chromosomal landing: a paradigm for map-based cloning in plants with large genomes. *Trends Genetics,* 11, 63-68.

Thornsberry, J.M., Goodman, M.M., Doebley, J., Kresovich, S., Nielsen, D., Buckler, E.S. 2001. *Dwarf8* polymorphisms associate with variation in flowering time. *Nature Genetics,* 28, 286-289.

Touzet, P., Winkler, R.G., Helentjaris, T. 1995. Combined genetic and physiological analysis of a locus contributing to quantitative variation. *Theor. Appl. Genet.*, 91, 200-205.

van der Vossen, E.A., Gros, J., Sikkema, A., Muskens, M., Wouters, D., Wolters, P., Pereira, A., Allefs, S. 2005. The Rpi-blb2 gene from Solanum bulbocastanum is an Mi-1 gene homolog conferring broad-spectrum late blight resistance in potato. *Plant J.*, 44, 208-222.

Vigouroux, Y., McMullen, M., Hittinger, C.T., Houchins, K., Schulz, L., Kresovich, S., Matsuoka, Y., Doebley, J. 2002. Identifying genes of agronomic importance in maize by screening microsatellites for evidence of selection during domestication. *Proc. Nat. Acad. Sci.,* USA, 99, 9650-9655.

Vrebalov, J., Ruezinsky, D., Padmanabhan, V., White, R., Medrano, D., Drake, R., Schuch, W., Giovannoni, J. 2002. A MADS-box gene necessary for ripening at the tomato ripening-inhibitor (rin) locus. *Science,* 296, 343–346.

Vos, P., Hogers, R., Bleeker, M., Reijans, M., Van de Lee, T., Hornes, M., Fritjers, A., Pot, J., Peleman, J., Kuiper, M., Zabeau, M. 1995. AFLP: a new technique for DNA fingerprinting. *Nucl. Acids Res.,* 23, 4407-4414.

Wang, K., Gan, L., Jeffery, E., Gayle, M., Gown, A.M., Skelly, M., Nelson, P.S., Ng, W.V., Schummer, M., Hood, L., Mulligan, J. 1999. Monitoring gene expression profile changes in ovarian carcinomas using cDNA microarray. *Gene,* 229, 101-108.

Wei, F., Gobelman-Werner, K., Morroll, S.M., Kurth, J., Mao, R., Wing, D., Leister, D., Schulze-Lefert, P., Wise, R.P. 1999. The *Mla* (powdery mildew) resistance cluster is associated with three NBS-LRR gene families and suppressed recombination within a 240-kb DNA interval on chromosome 5S (1HS) of barley. *Genetics,* 153, 1929-1948.

Yang, N.D., Zamir, D., Ganal, M., Tanksley, S.D. 1988. Use of isogenic lines and simultaneous probing to identify DNA markers tightly linked to the Tm2-a gene in tomato. *Genetics,* 120, 579-585.

Yu, Y.G., Buss, G.R., Maroof, M.A. 1996. Isolation of a superfamily of candidate disease-resistance genes in soybean based on a conserved nucleotide-binding site. *Proc. Natl. Acad. Sci.,* USA, 93, 11751-11756.

Zhang, X., Dyachok, J., Krishnakumar, S., Smith, L.G., Oppenheimer, D.G. 2005. IRREGULAR TRICHOME BRANCH1 in *Arabidopsis* encodes a plant homolog of the actin-related protein2/3 complex activator Scar/WAVE that regulates actin and microtubule organization. *Plant Cell,* 17, 2314-2326.

Zhang, Y., Li, X. 2005. A putative nucleoporin 96 is required for both basal defense and constitutive resistance responses mediated by suppressor of npr1-1, constitutive 1. *Plant Cell,* 17, 1306-1316.

29

CHAPTER

Molecular Evolution and Validation of Candidate Gene

D. Manicacci, J.B. Veyrieras, A. Charcosset

Neutral molecular markers, such as isozymes, RFLP (restriction fragment length polymorphism) or SSR (simple sequence repeat), have brought about important advances in the understanding of the structure of genetic diversity in numerous species. For example, in maize, SSR has made it possible to determine precisely the geographic region and modalities of domestication, as well as the way maize expanded over the American continent (Matsuoka *et al.*, 2002). RFLP markers have resulted in a reconsideration of the modalities of the introduction of maize in Europe during the time of the discovery of America (Rebourg *et al.*, 2003).

Genomics offers great potential to go further in the study of biodiversity. It first led to a considerable increase in the number of markers available, and the development of more efficient genotyping techniques. For example, out of approximately 2,100 SSR markers publicly available for maize in late 2003, 1,770 resulted from the discovery of SSR during sequencing work on cDNA libraries (M. Falque, pers. comm.). In *Arabidopsis thaliana*, the complete sequence is an extremely important source of markers (Chapter 27). This increase in the number of markers facilitates diversity studies and improves their precision (Charcosset *et al.*, 1998 for the precision of estimation of genetic distance between individuals as a function of the number of markers used). Beyond this increase of the number of usable markers, we have seen in the preceding chapters that genomics makes it possible to identify on a large scale genes potentially involved in the variation of quantitative traits (or candidate genes). Major advances can therefore be expected in light of the diversity of these genes, rather than that of neutral markers without direct effect on the traits of interest.

The reader interested in the use of neutral markers to analyse biodiversity should refer to syntheses in recent studies (Kremer and Mariette, 2003). In the rest of this chapter, the focus is essentially on the analysis of the diversity of genes of interest and the study of this molecular diversity for the purpose of functional validation or fine mapping.

SEQUENCE POLYMORPHISM AND METHODS OF ANALYSIS

Vertical sequencing studies (or sequencing of alleles) of candidate genes have resulted in the identification of intraspecific polymorphisms and are increasing in number. They involve a wide range of species, including maize, wheat, *A. thaliana*, various species of the genus *Arabidopsis* or *Arabis*, various *Lycopersicon*, etc. The objectives of these analyses range from the description of molecular diversity to validation of candidate genes (agronomic or domestication traits) and include the identification of evolutionary processes (natural or human selection, genetic drift, population structure) that have shaped this diversity.

The first step of this approach is amplification or cloning of the region of interest for the different individuals concerned. This step requires the design of specific primers from sequence elements available for the region. It is followed by sequencing and then alignment of sequences. Unlike humans, in which nearly all polymorphisms are single nucleotide polymorphisms (SNP), which means the substitution of one nucleotide base for another, plants present numerous insertion-deletions. This sometimes makes it difficult to align sequences and thus define polymorphisms. This is true of some genes of species that are particularly polymorphic (e.g., *Opaque2* in maize, Henry *et al.*, 1997, 2005). Moreover, insertion-deletions are rarely taken into account as polymorphisms by statistical analyses described further in this chapter.

Various indices of diversity can be deduced from an alignment. Under the hypothesis of the infinite site model, that is, when the mutation rate is low enough to prevent recurrent mutations at the same site, the nucleotide diversity of a population (θ) depends on the effective size of the population (Ne) and the rate of mutation per base pair (μ): $\theta = 4.Ne.\ \mu$. The two most frequently used estimators of this diversity are π (Nei and Li, 1979) and θw (Watterson, 1975). The nucleotide diversity π is defined as the average number of nucleotide differences per base between two sequences taken at random from a sample (Nei and Li., 1979). The size of the sample studied (n) can be taken into account in estimating this divergence at the population level: $(n/(n-1))\ \pi$. The diversity can, moreover, be estimated from the number of segregated sites (K) per base, affected by a correction for the number of sequences described in the sample, and generally called the θ_w index of Watterson (1975):

$$\theta_w = K/\alpha_{n-1}, \text{ where } \alpha_{n-1} = \sum_{i=1}^{n-1} 1/i$$

These two estimators differ in that only π takes into account the allelic frequencies, while θw ascribes the same weight to all the polymorphic sites but practises a correction for the number of sequences sampled. The early published results show clearly a wide variation for these parameters between species, maize presenting a particularly high average level of diversity, as well as a wide variation between genes of a single species (Table 29.1).

From these two estimators of θ, Tajima (1983) developed an index (D_T) that made it possible to test the classical null hypothesis of population genetics that combines an absence of selection, or linkage disequilibrium with a selected locus, an absence of population structure, and a constant population size over the preceding generations:

$$D_T = \frac{\pi - \theta}{\sqrt{Var(\pi - \theta)}}$$

According to the null hypothesis, the distribution of allelic frequencies is such that $\pi = \theta$ and $D_T = 0$. Tajima (1989) proposed intervals of confidence of D_T, allowing identification of sequences presenting a value of D_T significantly different from 0. This test is often incorrectly called a neutrality test because, as has just been shown, the null hypothesis is complex and the test is difficult to interpret. For example, an excess of rare alleles ($D_T < 0$) may be interpreted as the result of directional selection or increasing population size. On the contrary, an excess of intermediate allele frequencies ($D_T > 0$) will come from balanced selection or a demographic decrease. Moreover, population structure, depending on the case, tends to generate positive or negative values of D_T.

Discriminating between some of these hypotheses is easier when several genes are sequenced in a single species. In fact, this comparison makes it possible to conclude on either general effects (demography or population structure) if the results are similar between genes, or more specific selective effects for genes presenting an original pattern. The presence of an external group will also help in discriminating between different neutral rates of evolution between loci and a difference in the selection intensity. The HKA test developed by Hudson, Kreitman and Aguade (Hudson *et al.*, 1987) assumes that in a situation of selective neutrality, the ratio between intraspecific variation and interspecific divergence will remain constant between loci. This test is conservative, among other reasons because it does not take into account possible recombinations. When the sequenced regions contain a coding part, it is possible to distinguish synonymous substitutions (which do not affect the amino acid sequence) or silent substitutions (including non-coding sites) from non-synonymous substitutions. Here also, according to the neutrality hypothesis, the ratio between these two types of substitutions is expected to be constant between loci. This is the principle of the test developed by McDonald and Kreitman (1991). If, as expected, the selective effects are specific to non-synonymous sites, this test will be more efficient than a global test on all the sites. Finally, the distinction between different effects of selection, such as selective sweep (reduction of diversity due to the invasion of a favourable allele into a population) or background selection (reduction of diversity due to the recurrent elimination of harmful mutations), requires a higher level of precision (larger sample size, supplementary information on evolutionary history of alleles and linkage disequilibrium) (Fay and Wu, 2000; Kim and Stephan, 2000).

Numerous studies have described the molecular diversity of genes, mostly in maize and *A. thaliana* (Table 29.1). Maize has a overall high molecular diversity that varies considerably among genes. In this species, the effects of evolutionary pressures can rarely be detected by means of significant values of D_T. This may be explained partly by the fact that this test is highly conservative (low statistical power) and the early studies were often carried out on small sequence samples. Nevertheless, some genes show extreme levels of diversity, which suggests a specific selection process rather than demographic change or population structure that should affect the whole genome. One of the most beautiful examples is the *Teosinte branched* (*Tb1*) gene, the promoter of

Table 29.1 Diversity of sequences for several genes of maize, *Arabidopsis thaliana*, and other species (n = sample size). The parameters π and θw (x10^3) are presented for the total sequence or silent sites only, depending on the publication.

Species	*Publication*	*Gene*	*Remarks*	*Size (bp)*	*Sample size*	π *tot*	π *sil*	θw *tot*	θ *w* sil
Zea mays	White and Doebley, 1999	*te1*	Cultivars	1,399	12	2.1	4.9	1.7	4.0
			Wild type	1,399	5	3.7	8.7	3.8	8.7
	Eyre-Walker *et al.*, 1998	*adh1*	Cultivars	~1,400	9	13.6	21.6	14.9	18.1
			Wild type	~1,400	8	19.5	23.4	17.9	24.5
	Goloubinof *et al.*, 1993	*adh2*	Cultivars	315		25.0	30.1	25.8	29.5
			Wild type	315		32.8	39.6	32.8	39.6
	Hanson *et al.*, 1996	*c1*	Cultivars	960	13	2.7	2.8	3.5	4.1
			Wild type	960	5	12.2	15.0	12.2	15.0
	Hilton and Gaut, 1998	*glb1*	Cultivars	~1,200	9	17.2	18.0	18.9	19.0
			Wild type	~1,200	8	25.8	27.3	31.7	23.2
	Wang *et al.*, 1999	*tb1-5'*	Cultivars	1,143	13	0.5		0.9	0.9
			Wild type	1,143	17	33.1		35.5	
	"	*tb1-TU*	Cultivars	1,729	13	1.7		2.4	
			Wild type	1,729	17	3.9		6.1	
	Whitt *et al.*, 2002	*ae1*	Cultivars	6,781	30		2.9		
	"	*bt2*	Cultivars	6,098	30		2.3		
	"	*sh1*	Cultivars	6,176	30		12.1		
	"	*sh2*	Cultivars	6,754	30		5.0		
	"	*su1*	Cultivars	9,378	30		2.7		
	"	*wx1*	Cultivars	2,978	30		11.5		
	Tenaillon *et al.*, 2001; 2004	*fus6*	Cultivars	248	23				8.7
		csu1171	Cultivars	468	25		6.7		6.2
			Wild type	467	9		11.7		14.2
		umc230	Cultivars	478	22				17.8
		asg26	Cultivars	464	24				9.2
		umc13	Cultivars	412	25				9.6
		ts2	Cultivars	946	24		5.5		4.8

(*Contd.*)

Table 29.1 Contd.

Species	*Publication*	*Gene*	*Remarks*	*Size (bp)*	*Sample size*	π *tot*	π *sil*	θw *tot*	θw sil
			Wild type	975	10		16.0		6.2
		asg75	Cultivars	342	24				7.1
		csu1138	Cultivars	324	25				15.7
		umc67	Cultivars	383	23				9.9
		asg11	Cultivars	375	24		8.9		8.6
			Wild type	492	6		7.3		8.9
		csu1150	Cultivars	397	25				2.8
		csu1132	Cultivars	544	25		11.8		14.8
			Wild type	531	8		25.1		22.5
		asg62	Cultivars	420	25				25.9
		umc128	Cultivars	526	23				15.5
		an1	Cultivars	661	25				6.6
		bz2	Cultivars	605	25		13.0		12.2
			Wild type	590	12		14.8		8.4
		glb1	Cultivars	968	23		29.7		35.8
			Wild type	991	8		35.8		29.6
		tb1	Cultivars	2,740	23		1.1		4.9
			Wild type	2,702	7		15.3		13.8
		d8	Cultivars	800	25		9.1		10.1
			Wild type	791	13		35.9		18.7
		adh1	Cultivars	1,406	25		18.8		17.6
			Wild type	1,257	8		25.2		17.5
		csu381	Cultivars	916	25		18.2		12.8
			Wild type	917	10		15.1		16.2
		zagl1	Cultivars	560	16		0.0		0.0
			Wild type	572	15		9.0		10.2
	Manicacci *et al.*, in press	*Sh2*	Cultivars	4,774	41				4.2
	Fontaine and Barriere, 2001	*OMT*	Cultivars	~2,500	6	10.05	14.7	9.4	13.5
	Zhang *et al.*, 2002	*hm1*	Cultivars	1,026	10		33.8		

(*Contd.*)

Table 29.1 Contd.

Species	*Publication*	*Gene*	*Remarks*	*Size (bp)*	*Sample size*	π *tot*	π *sil*	θw *tot*	θw sil
			Wild type	1,016	10		41.4		
		hm2	Cultivars	1,142	10		10.5		
			Wild type	1,109	11		12.6		
Arabidopsis thaliana	Kamiya *et al.*, 2002	*Atmyb2*		2,222	20	2.7			
	Innan *et al.*, 1996	*Adh*		2,412	17	8.0	11.5	9.3	
	Kawabe *et al.*, 1997	*ChiA*		1,799	17	10.9	14.9	20.2	
	Kawabe and Miyashita, 1999	*ChiB*		2,246	17	9.1	13.6	11.2	
	Kuittinnen and Aguade, 2000	*CHI*		1,829	24	4.0	5.4	5.3	
	Kawabe *et al.*, 2000	*PgiC*		~4,700	21	3.8		7.5	
	Purugganan and Suddith, 1998	*Cal*		~2,200	17	7.4	6.9	12.2	
	Purugganan and Suddith, 1999	*AP3*		1,680	19	6.4	7.5	13.2	
	"	*PI*		2,050	16	5.3	6.1	10.0	
	Olsen *et al.*, 2002	*API*		~4,700	15		4.7		
	"	*LFY*		~3,900	15		1.9		
	"	*TFL1*		~1,800	14		0.7		
	Aguade, 201	*FAH1*		~2,200	20	3.0	5.0		
	"	*F3H*		~1,300	20	7.0	17.5		
	Le Corre *et al.*, 2002	*FRI*		2,802	26	2.1		3.6	
	Caicedo *et al.*, 1999	*Rps2*		2,836	17	13.4		13.3	
	Tian *et al.*, 2002	*Rps5*		2,670	8	0.8	0.04		
Arabis gemmifera	Kawabe *et al.*, 2002	*ChiA*		~1,800	15	3.2			
Betula pendula	Jarvinen *et al.*, 2003	*BpMADS2*	(Region I)	770	40	2.3		1.6	
	"	*BpMADS2*	(Region II)	1,680	40	5.4		7.2	
Leavenworthia	Liu *et al.*, 1998	*Adh1*		264			45.6		47.1
	Filatov and Charlesworth, 1999	*PgiC*		~2,300	11	43.0	55.4	44.0	57.3
Dioscorea tokoro	Terauchi *et al.*, 1997	*Pgi*	(Introns)	2,049	15		2.9		4.5
	"	*Pgi*	(Exons)	1,698	20	1.1	1.5	1.3	1.5
Pinus sylvestris	Dvornyk *et al.*, 2002	*Pal1*		2,045	20	1.4	4.9	1.7	5.8
Oryza rufipogon	Yoshida *et al.*, 2004	*Adh1*		3,185	17	2.0	2.5	3.1	3.8

which exhibits almost no diversity in cultivated maize as compared to related wild types, the teosintes (Wang *et al.*, 1999). This gene can drastically modify the plant and ear architecture, leading to the morphological transformation of teosinte into maize. In validating *Tb1* as the major domestication gene, molecular study have made possible a giant advance in the understanding of mechanisms that allowed the rapid and effective transformation of teosinte into maize, as suggested by archaeological data but till then unexplained (Thornserry *et al.*, 2001).

In *A. thaliana*, the study of an increasingly large number of genes has made it possible to distinguish general effects on the genome from more specific effects of certain genes. An excess of rare alleles for several genes (*Cal, Ap3, P1,* and *Chi*), leading to often significant negative values of D_T, was interpreted as a record of the recent demographic expansion of the species (Price *et al.*, 1994). Nevertheless, the accumulated data on other genes showed that this situation was not general. In fact, the genes *Adh, ChiA, ChiB, RpmI, Fah1,* and *F3H* often present two or three major haplotypes and positive values of D_T. Integrating all these data, Aguade (2001) and Kuittinen *et al.* (2002) proposed a unifying hypothesis of metapopulation, including strong effects of drift because of the extinction and foundation of populations and the autogamous breeding system of the species. An analysis of intrapopulation diversity was proposed to validate this hypothesis (Kuittinen *et al.*, 2002).

The gene *Rpm1* has a peculiar evolutionary history because of its phenotypic role in resistance to *Pseudomonas* in *A. thaliana*. The genealogy of the *Rpm1* alleles reveals an ancient divergence between alleles of sensitivity and resistance, which suggests that the model of resistance allele turnover (according to which the resistance alleles derived repeatedly from sensitivity alleles) can be rejected in favour of the "gene for gene" hypothesis, which states that the two types of alleles have coexisted for a long time (Stahl *et al.*, 1999).

In this sense, detailed analysis of the sequence diversity of numerous genes with known functions allows certain deductions about the evolutionary history of species in general and of certain genes in particular. Several studies have carried out a comparative analysis of the molecular evolution of genes involved in a single metabolic pathway. For example, among six genes involved in the regulation of floral development in *A. thaliana*, two show a reduction of diversity that the authors attributed to a recent selective effect (Olsen *et al.*, 2002). Similarly, a comparison of six genes of the starch biosynthesis pathway in maize endosperm showed that two of them have reduced diversity in cultivated maize as compared to teosinte, a reduction greater than that generally observed among maize genes evolving under neutrality. These genes could thus play an important role in maize domestication (Whitt *et al.*, 2002).

It is of course tempting to compare species among themselves and to search for possible relationships between the patterns of diversity and the traits of life history. Some studies have tilted toward the relationship between molecular diversity and the breeding system. The expected relationship, that is a higher diversity in allogamous or self-incompatible populations or species, has been observed in the genera *Leavenworthia* (Liu *et al.*, 1999) and *Lycopersicon* (Baudry *et al.*, 2001). On the other hand, no difference in rate of protein evolution was found when the autogamous species *A. thaliana* was compared with the self-incompatible related *A. lyrata* (Wright *et al.*, 2002). These authors suggest that contrasted selective events between genes, or different demographic histories of the two species (such as a recent bottleneck in *A. lyrata*), could mask the effects of breeding system. These interspecific comparisons, which are still very rare today, offer rich information on the precise mechanisms that have shaped the diversity of genes and genomes.

LINKAGE DISEQUILIBRIUM AND RECOMBINATION

Linkage disequilibrium is defined within a population of individuals, as the statistical association between two polymorphisms located at different positions (different loci or positions within a gene). If i and j are possible forms of two polymorphic positions I and J, with respective frequencies f_i and f_j within the population, their linkage disequilibrium is $D_{ij} = f_{ij} - f_i f_j$, where f_{ij} is the frequency of gametes carrying i and j. In the case of positive disequilibrium between i and j, the probability of observing the allele j is greater in an individual that possesses the allele i than for an individual that does not possess it.

For a large isolated population evolving under panmixis (random union of gametes) and in the absence of selection, linkage disequilibrium diminishes at each generation under the effect of recombination : $D_{ij}(t+1) = D_{ijt}(1-\mathrm{r})$, where t is the initial generation and r is the rate of recombination between the two sites considered. The rate of recombination thus plays a critical role in the evolution of linkage disequilibrium after foundation of a population or migration. Linkage disequilibrium lasts over the long term in a panmictic population only for tightly linked polymorphisms. Deviation from the panmixis hypothesis, and notably population structure, can also play a very important role. If a population is in fact made up of a limited number of isolated subpopulations, linkage disequilibrium can be observed between the polymorphisms presenting contrasting frequencies in the different subpopulations (Otha, 1982), even if these are not linked physically.

Linkage disequilibrium is an essential parameter in understanding the structure of diversity and plays a critical role in interpretation of association studies (Flint-Garcia and Buckler, 2003; Gupta *et al.*, 2005). In fact, when a statistical association exists between a polymorphism and a phenotypic trait, the question arises whether the site involved is really responsible for the phenotypic variation or whether it is in linkage disequilibrium with the responsible site, which can be located in another non-sequenced part of the gene, a nearby gene, or even a genetically independent gene. In maize, a strongly allogamous species, linkage disequilibrium generally decreases some hundreds of base pairs within (Fig. 29.1; Remington *et al.*, 2001; Tenaillon *et al.*, 2002), even though this may be quite variable between genes and, in some cases, close associations may be maintained over several thousands of base pairs (*Sh2* gene; Manicacci *et al.*, 2007). It should be noted that maize genes behave like recombination "hotspots" and exhibit rates of recombination per unit of physical size up to 100 times those observed for the intergenic regions. In *A. thaliana*, an autogamous species, the linkage disequilibrium was shown to be maintained over several tens or hundreds of kilobases (Fig. 29.2; Hagenblad *et al.*, 2002), which confirms theoretical expectation. Within a species, the relationship between linkage disequilibrium and genetic distance depends, as has been seen previously, on the history of the material studied. For example, in comparison with studies cited earlier on maize, the study of Ching *et al.* (2002), conducted essentially on elite lines, indicates a slower decrease of linkage disequilibrium as a function of the physical distance between polymorphisms.

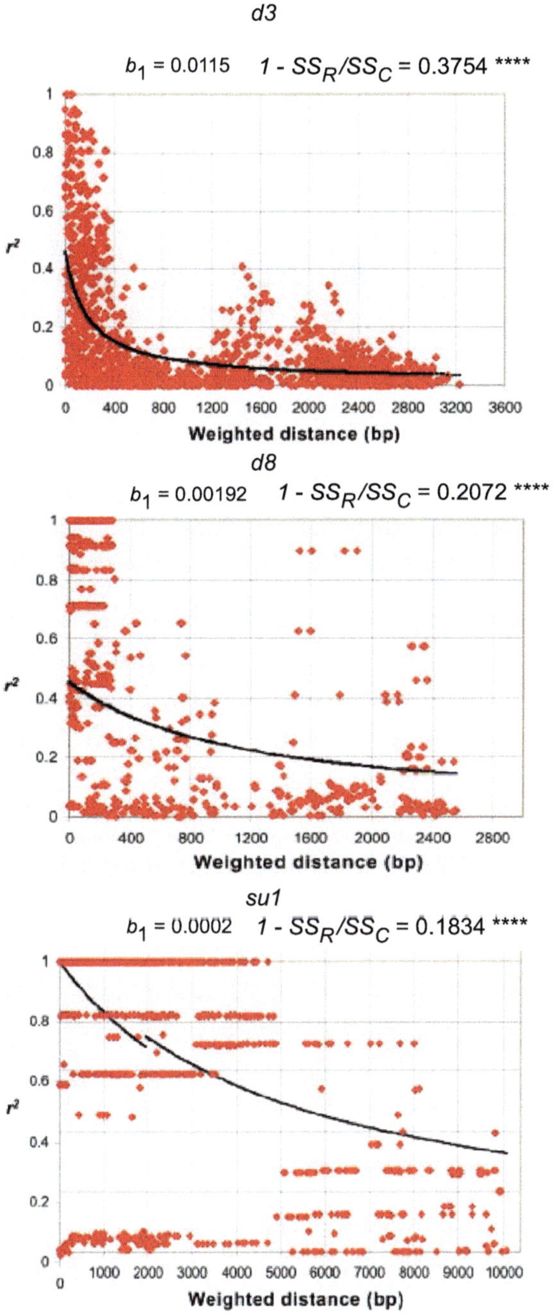

Fig. 29.1 Linkage disequilibrium (r^2 = correlation between polymorphic sites) as a function of physical distance for three genes of maize: *Dwarf-3* (*D3*), *Dwarf-8* (*D8*) and *Sugary-1* (*Su1*) (Remington *et al.*, 2001)

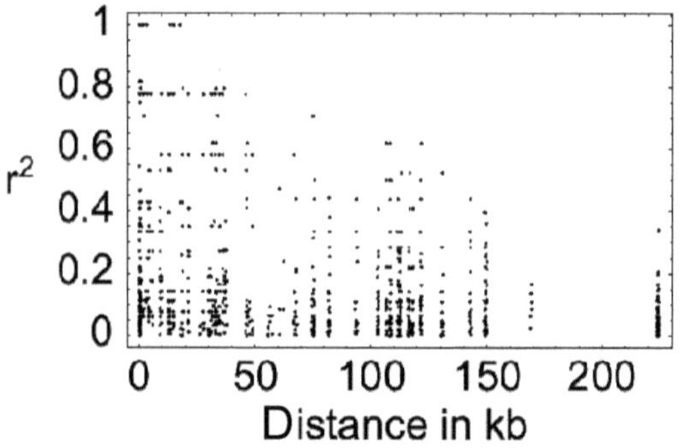

Fig. 29.2 Decrease in linkage disequilibrium (r^2 = square of correlation between polymorphic sites) as a function of physical distance in a region of 250 kb around the *Frigida* gene in *A. thaliana* (Nordborg *et al.*, 2002).

On the larger scale of the entire genome, there may nevertheless remain some traces of the history of populations, particularly in material as structured as American and European cultivated maize. This is why it is important to detect this structure, using highly variable neutral markers such as SSR, and to take it into account during studies of association.

FUNCTIONAL VALIDATION OF CANDIDATE GENES AND FINE MAPPING

The relationships between molecular diversity and the variation of quantitative traits can be used as a tool for functional validation of a candidate gene, or for QTL fine mapping. In this chapter, we will focus on the validation of candidate genes. These approaches, first used in human genetics to identify genetic factors for sensitivity to disease (Tabor *et al.*, 2002), are presently tested in the animal and plant kingdom (Buckler *et al.*, 2002). In plants, the pioneering studies of Thornsberry *et al.* (2001) are particularly notable. Their general principle is summarized in Fig. 29.3.

Constitution of the Population Studied and Phenotypic Evaluation

The choice of individuals studied is critically important and must reach two major objectives. The first is to represent, as well as possible, polymorphisms present within the species studied. The second is to reduce the impact of linkage disequilibrium by choosing material that has accumulated many generations of recombination. This last point is necessary to guarantee the precision of a fine mapping experiment, or to avoid confusing the effect of a candidate gene with that of a linked gene. The study of Thornsberry *et al.* (2001) focused on a collection of 100 maize lines representing different genetic origins (populations and regions of origin), as well as different cycles of selection: "first cycle" lines obtained by selfing of landraces, lines of the second cycle derived from crossing between two lines, up to the recent elite material representing 5 to 10 cycles of intercrossing of lines. It may be interesting to optimize the constitution of the collection by a "core collection" sampling. Such an approach aims to determine a subsample of individuals that best represents the total diversity of a collection, taking

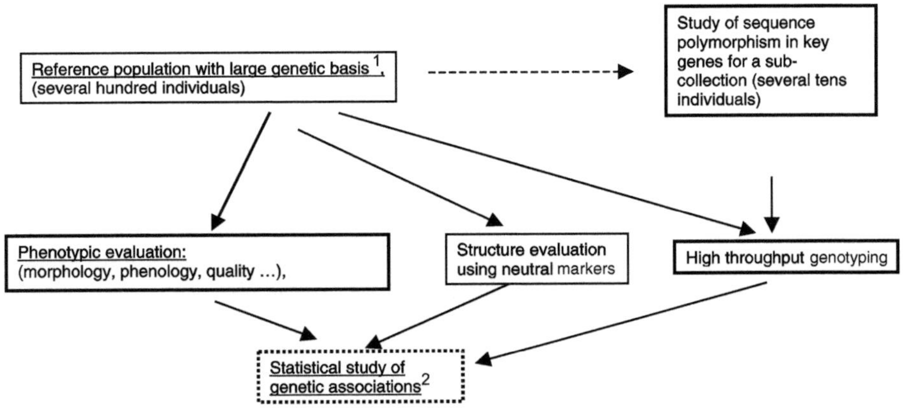

Fig. 29.3 Principle of search for associations between polymorphisms of candidate genes and variation of quantitative traits
[1]The choice of the plant material (fixed inbred lines if possible) should take the structure of the species diversity into account.
[2] see Box 29.1

into account the available information: neutral markers, first series of SNP(single nucleotide polymorphism)-type markers and even quantitative data. Several kinds of software are available to compile the samples, including MSTRAT (Gouesnard *et al.*, 2001), developed on the basis of studies by Bataillon *et al.* (1996). This was notably used to determine a representative collection of *A. thaliana* lines (McKhann *et al.*, 2003).

As in the search for QTL, the power of statistical tests will be determined by the collection size (Charcosset and Gallais, 1996). It would be preferable, for constant phenotypic evaluation cost, to increase the number of individuals studied, rather than the number of repetitions per individual. The ongoing programmes to our knowledge on maize and *A. thaliana* are run on 200 to 400 lines. These results will bring knowledge on the sample sizes to be considered for future programmes. Finally, approaches allowing high-flow phenotyping of "fine" traits can be particularly useful in this context: e.g., prediction of quality traits by near infrared spectroscopy, automation of ecophysiological measurements.

Study of the Collection Structure Using Neutral Markers

It is important to note that the collections used generally present an underlying structure into subpopulations. This is true, for example, for populations that diverged long ago in distant geographic regions, populations created from a limited number of founders and maintained in isolation. This structure may generate an association between a quantitative trait and a polymorphism, even if the latter is not linked physically to genes involved in the variation of the trait. In fact, a polymorphism exhibiting different frequencies between two subpopulations will generally be linked to traits for which these subpopulations differ. The structure of the collection must be analysed, therefore, lest an erroneous conclusion be formed on the effect of a gene (studies of associations), or on its presence in a given region (fine mapping). Various classical approaches of diversity analysis can be used for this purpose, such as factorial representations (principal component analysis, factorial analysis of correspondences), or analyses of classification based on distance criteria (e.g. unweighted pair group method with arithmetic mean–UPGMA) or minimal variance (e.g. Ward method). The classification approaches nevertheless do not help in managing the case of intermediate individuals, resulting for example from hybridization between subpopulations. Pritchard *et al.* (1999, 2000) developed a new approach to manage this "mixed" situation. By means of the Bayesian algorithm developed by these authors, using neutral marker data, the number of underlying subpopulations of the collection and the allelic frequencies that characterize them can be determined, and for each individual the proportion of the genome inherited from different subpopulations can be estimated. By construction, the subpopulations thus constituted have no linkage disequilibrium between loci that are not physically linked. These calculations can be made using Structure software, available on the site http://pritch.bsd.uchicago.edu. Applied to the collection of maize lines mentioned earlier, this approach made it possible to conclude the presence of three underlying subgroups (Fig. 29.4; Remington *et al.*, 2001).

Genotyping of Polymorphisms

Even with the fast evolution of sequencing techniques and costs, it may be valuable to conduct the process of genotyping in two stages: the sequencing of a subsample of the collection and then the genotyping of the entire collection with appropriate marker techniques. The subsample of individuals can be chosen according to the sampling approaches described below. It may also be relevant to integrate individuals presenting extreme phenotypes for the trait of interest. The question of the sequencing strategy, number of sequences per gene versus number of genes studied, remains still wide open. The knowledge available at present on the structure of the linkage disequilibrium in maize tends to recommend a sequence "run" (around 500 bp) for every 1 to 2 kb (E. Buckler, pers. comm.). For autogamous species, it may be supposed from linkage disequilibrium data that a single region per gene is sufficient, but this point remains to be confirmed.

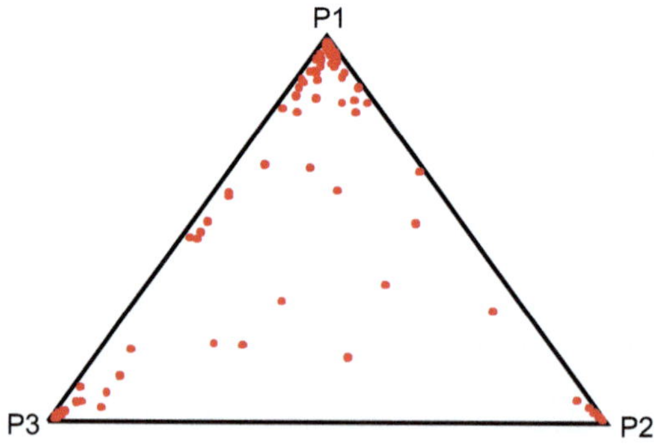

Fig. 29.4 Application of method of Pritchard *et al.* (2000) to a collection of 102 maize lines of diverse origins. The results lead to three underlying subpopulations called P1, P2 and P3 and allow estimation for each line of the percentage of the genome derived from each subpopulation (Remington *et al.*, 2001).

The sequence data obtained on the subsample can be used to determine the polymorphisms that will be analysed for the entire collection. The high throughput SNP genotyping techniques currently available (Chapter 25) allow rapid characterization of a very large number of individuals (up to several millions for ongoing projects on animal genetics; I. Gut, pers. comm.) but may require a fairly long development time. This raises the question of selecting the polymorphisms that will be analyzed for each gene. Tabor *et al.* (2002) set out the principles of selecting polymorphisms. These are likely to evolve but several points can be noted.

— Polymorphisms leading to a loss of functionality of the gene (stop codon, insertion or deletion in the coding part) must naturally be favoured. Observation of several independent events of this type may be particularly useful in concluding unambiguously about the effect of a gene. It is in this way that Johanson *et al.* (2000) and then Le Corre *et al.* (2002) indicated that mutations in the *Frigida* gene confer early flowering to *A. thaliana*.

— Following the same logic, polymorphisms leading to a substitution of amino acid may be particularly interesting, especially if this substitution may affect the protein structure.

— It would be preferable to choose from the polymorphisms in high linkage disequilibrium in order to avoid redundancy of information, even if it is not certain that linkage disequilibrium will be confirmed within a larger sample. From up to now, several methods have been devised to carry out such a selection. Most of these methods rely on the assumption that the linkage disequilibrium patterns are broken into a finite number of blocks. A block is characterized by a low genetic diversity and is separated from another block by shorter regions of higher diversity. First reported in human (Daly *et al.*, 2001), this LD block structure seems to reflect the historical effects of recombination, in particular block limits may line up with recombination hotspots. To date, direct connection between recombination hotspots and structure of LD produced through evolution has not been demonstrated in plants, nevertheless "it is likely that this connection does exist, as in humans" (Flint-Garcia *et al.*, 2003). Based on this empirical interpretation of linkage disequilibrium structure, Zhang *et al.* (2005) have developed a serie of bioinformatic tools to both determine the block partition and to select the minimal number of polymorphisms which best capture this partition. However, these algorithms are strongly dependent on the tuning parameters used to define the degree of diversity variation between and within blocks, so that block determination still remains subtle and must be taken with caution with regard to the underlying genetic factors which have shaped the observed linkage disequilibrium. Besides, most of these methods require that the phase (i.e the haplotypes) can be resolved, which is generally not obvious in diploid species, unless assuming simple hypotheses on the haplotype sampling scheme (for a review of haplotype reconstruction methods see Niu *et al.*, 2004). That's why other "block-free" methods have been proposed to select polymorphisms. These methods, which do not require the knowledge of the phase and do not assume a block partition, are based either on classical multivariate analysis frameworks (Meng *et al.*, 2003; Horne *et al.*, 2004) or on the optimization of some predictive functions (Halperin *et al.*, 2005). A related point is that it may be useful to optimize the choice of polymorphisms in order to characterize the haplotypes detected in the subsample. This point will be particularly important for autogamous species for which intragenic recombination is rare. Nevertheless, it is still unclear how these polymorphism selection procedures impact on the power of further association studies (Zhang *et al.*, 2002, 2003; De Bakker *et al.*, 2005).

— Finally, it is possible to take into account the results of preliminary association tests for individuals characterized by sequencing. These results must be considered with greater caution if the sample is small. This can be thought as a preliminary exclusion procedure where one aims at preserving enough power for association studies in larger population while removing polymorphisms for which there is a strong evidence of no association. In particular, Hoh *et al.* (2000) have proposed an interesting bootstrapping strategy to determine which polymorphisms must be retained.

Statistical Tests of Association

It has been recommended that the stucture of the collection be taken into account in studies of association for the candidate gene test (susceptibility to disease) in humans. This approach makes it possible to test whether a polymorphism exhibits frequency differences between sick individuals and healthy individuals, without it being due to an effect of structure into subpopulations of different susceptibility levels for that disease. In other words, it makes it possible to test whether there is an association within subpopulations. This approach was adapted to the study of quantitative traits by Thornsberry *et al.* (2001) (Box 29.1) and allowed the authors to conclude the probable effect of the candidate gene *Dwarf8* on flowering time in maize, and to identify an insertion of 6 bp associated with a eight days earlier flowering. Other models based on the same principles can be considered, such as multiple regression (Box 29.1), but the comparison of statistical properties of these different approaches remains to be studied. It is worth noting that multiple regression could offer a more flexible framework to evaluate complex genetic model including more than two alleles and/or several candidate polymorphisms. In particular, when haplotypes are available, there are generally more than two haplotypic classes. Instead of testing each haplotype versus all others using logistic regression, the haplotypes can then be tested jointly into a single linear regression model. However, when there is a high number of haplotypes, the risk is to reduce dramatically the power of the tests. "Tree scanning" strategies have been devised by Templeton *et al.* (1987) to address this issue. Assuming that recombination is negligible relative to mutation in the candidate gene region, the haplotypes can be classified by means of a cladogram constructed by standard clustering techniques. Then, the topology of the cladogram can be used to successively group haplotypes into a few number of clusters within which distance between haplotypes is minimal. A similar approach, which makes use of a sliding window to prevent from recombination effects, have been also proposed by Durrant *et al.* (2004).

Whatever the genetic model (single polymorphisms or haplotypes) used to carry out the association studies, multiple regression strategies seem to be more appropriate to integrate into the tests the uncertainty on the estimation of underlying population structure (Q, see Box 29.1). For example, based on a multiple regression framework, Hoggart *et al.* (2003) have proposed an hybrid Frequentist/Bayesian procedure to average the association tests over the most likely values of Q. Since this procedure integrates jointly the information on the genetic structure, the candidate gene polymorphisms and the trait variation, it could be particularly interesting when population structure is highly correlated with the trait variation (e.g. flowering time is often linked to the geographical origins of the sampled individuals).

Box 29.1 Test of Association between a Sequence Polymorphism and Trait Variation

For the collection studied (N lines):

Let C be the genotype observed for the candidate polymorphism (presence or absence in the simple case of homozygous lines and for two alleles), let T be the value (phenotype) of the trait studied (qualitative or quantitative), let Q be the proportion of genome issued from each of the subpopulations, determined by the approach of Pritchard *et al.* (2000).

According to the null hypothesis (H0), the genotype observed for the candidate polymorphism is independent of the phenotype observed for the trait; according to the alternative hypothesis (H1), it is associated with the phenotype. The principle of the test is to calculate the likelihood of the observed polymorphisms under each of these two hypotheses. According to the null hypothesis, the probability that an individual has a given genotype depends on $\hat{Q}$ and the polymorphism frequencies within the subpopulations (Pritchard *et al.*, 2000) and is written as $Pr0(C;\hat{Q})$. The probability under the hypothesis H1, $Pr1(C;T,\hat{Q})$, depends on the type of trait considered. Thornsberry *et al.* (2001) extended this approach to the case of quantitative traits, using regression software in which the presence or absence of a polymorphism in an individual is explained by Q and T. The likelihood of each of the two models were then compared conventionally by the following ratio:

$$\Lambda = \frac{Pr1(C;T,\hat{Q})}{Pr0(C;\hat{Q})}$$

The threshold of this test statistic is determined using simulations, according to the principle of Pritchard *et al.* (2000). For each simulation, a genotype for the polymorphism considered is randomly attributed to each individual, as a function of $\hat{Q}$ and polymorphism frequencies within subpopulations. Several hundreds of simulations are carried out to obtain the distribution of the test statistic according to the null hypothesis. It is thus possible to compare the value observed with this distribution, in order to deduce the type 1 risk of the test (probability of having obtained a value higher than or equal to that observed, under the null hypothesis of absence of effect of polymorphism). If several polymorphisms are tested for a single gene, it is necessary to consider the distribution of the maximum value of Ë for all the polymorphisms tested.

Other approaches based on the same principles but closer to models conventionally used to search for QTL can be considered, such as multiple regression of the type $T = \alpha\hat{Q} + C$, where α is the average of each of the subpopulations for the character considered. This model can be used to test whether a polymorphism has a significant effect on the variation of the trait, once the effect of structure has been taken into account. It may have the advantage of being easily adapted to several polymorphisms, within a single gene or different genes. Methodological studies are ongoing on the statistical properties (power, bias) of these different tests for the structures of populations observed in plants.

This section is based on the studies of Pritchard *et al.* (2000) and Thornsberry *et al.* (2001).

Notes on the Use of These Approaches in Autogamous Species and for Fine Mapping

It is important to note that the preceding approach is very useful in validation of functions of genes in allogamous species. The early results seem promising but require further validation studies. It is thus prudent at present to consider association genetics as one of the elements of a function validation strategy, in parallel with transgenesis and mutant search. In autogamous species, the elements available at present on the impact of linkage disequilibrium raise questions on the interest of this approach as a tool of functional validation. If several physically close genes show high linkage disequilibrium and are linked to a trait, it will be difficult to determine which of them is the cause of the association. This characteristic of linkage disequilibrium could nevertheless become an advantage if there are no pre-existing hypotheses on the genes causing the variation of a trait. The use of a very dense map of markers and the study of relationships between the polymorphism of these markers and the trait could contribute to fine mapping of the QTL. Finally, some methods are being developed to combine approaches of fine mapping within familial and to take into account pedigree relatedness in association tests (Yu et al., 2006) structures and approaches based on linkage disequilibrium (Meuwissen *et al.*, 2002).

REFERENCES

Aguade, M. 2001. Nucleotide sequence variation at two genes of the phenylpropanoid pathway, the *FAH1* and *F3H* genes, in *Arabidopsis thaliana. Mol. Biol. Evol.*, 18, 1-9.

Baudry, E., Kerdelhue, C., Innan, H., Stephan, W. 2001. Species and recombination effects on DNA variability in the tomato genus. *Genetics*, 158, 1725-1735.

Buckler, E.S., Thornsberry, J.M. 2002. Plant molecular diversity and applications to genomics. *Curr. Opin. Plant Biol.*, 5, 107-111.

Charcosset, A., Gallais, A. 1996. Estimation of the contribution of quantitative trait loci (QTL) to the variance of a quantitative trait by means of genetic markers. *Theor. Appl. Genet.*, 93, 1193-1201.

Ching, A., Caldwell, K.S., Jung, M., Dolan, M., Smith, O.S., Tingey, S., Morgante, M., Rafalski, J.A. 2002. SNP frequency, haplotype structure and linkage disequilibrium in elite maize inbred lines. *BMC Genet.*, 3(19), 1-14.

Daly, M.J., Rioux, J.D., Schaffner, S.F., Hudson, T.J., Lander, E.S. 2001. High-resolution haplotype structure in the human genome. *Nat Genet.*, 29, 229-232.

de Bakker, P.I.W., Yelensky, R., Pe'er, I., Gabriel, S.B., Daly, M.J., Altshuler, D.E 2005. Efficiency and power in genetic association studies. *Nat Genet.*, 37, 1217-1223.

Durrant, C., Zondervan, K.T., Cardon, L.R., Hunt, S., Deloukas, P., Morris, A.P. 2004. Linkage disequilibrium mapping via cladistic analysis of single-nucleotide polymorphism haplotypes. *Am. J. Hum. Genet.*, 75, 35-43.

Fay; J.C., Wu, C.I. 2000. Hitchhiking under positive Darwinian selection. *Genetics*, 155, 1405-1413.

Flint-Garcia, S.A., Thornsberry, J.M., Buckler, E.S. 2003. Structure of linkage disequilibrium in plants. *Annu. Rev. Plant. Biol.*, 54, 357-374.

Gouesnard, B., Bataillon, T.M., Decoux, G., Rozale, C., Schoen, D.J., David, J.L. 2001. MSTRAT: An algorithm for building germplasm core collections by maximizing allelic or phenotypic richness. *J. Heredity*, 92, 93-94.

Gupta, P.K., Rustigi, S., Kulwal, P.L. 2005. Linkage disequilibrium and association studies in higher plants: Present status and future prospects. *Plant Molecular Biology*, 57, 461-485.

Hagenblad, J., Nordborg, M. 2002. Sequence variation and haplotype structure surrounding the flowering time locus *FRI* in *Arabidopsis thaliana*. *Genetics*, 161, 289-298.

Halperin, E., Kimmel, G., Shamir, R. 2005. Tag SNP selection in genotype data for maximizing SNP prediction accuracy. Bioinformatics 21 Suppl. 1, 195-203.

Henry, A.M., Damerval, C. 1997. High rates of polymorphism and recombination at the Opaque-2 locus in cultivated maize. *Mol. Gen. Genet.*, 256, 147-157.

Henry, A.M., Manicacci, D., Falque, M., Damerval, C. 2005. Molecular evolution of the Opaque-2 gene in Zea mays L. *Journal of Molecular Evolution*, 61, 1-8.

Hoggart, C.J., Parra, E.J., Shriver, M.D., Bonilla, C., Kittles, R.A., Clayton, D.G., McKeigue, P.M. 2003. Control of confounding of genetic associations in strati ed populations. *Am. J. Hum. Genet.*, 72, 1492-1504.

Hoh, J., Wille, A., Zee, R., Cheng, S., Reynolds, R., Lindpaintner, K., Ott, J. 2000. Selecting SNPs in two-stage analysis of disease association data : a model-free approach. *Ann. Hum. Genet.*, 64, Pt 5: 413-417.

Horne, B.D., Camp, N.J. 2004. Principal component analysis for selection of optimal SNP-sets that capture intragenic genetic variation. *Genet Epidemiol.*, 26, 11-21.

Hudson, R.R., Kreitman, M., Aguade, M. 1987. A test of neutral molecular evolution based on nucleotide data. *Genetics*, 116, 153-159.

Johanson, U., West, J., Lister, C., Michaels, S., Amasino, R., Dean, C. 2000. Molecular analysis of *Frigida*, a major determinant of natural variation in *Arabidopsis* flowering time. *Science*, 290, 344-347.

Kim, Y., Stephan, W. 2000. Joint effects of genetic hitchhiking and background selection on neutral variation. *Genetics*, 155, 1415-1427.

Kremer, A., Mariette, S. 2003. Molecular markers in population genetics. In: *Molecular markers in plant genetics and biotechnology*. D. de Vienne, ed., Science Publishers Inc., Enfield, 119-138.

Kuittinen, H., Salguero, D., Aguade, M. 2002. Parallel patterns of sequence variation within and between populations at three loci of *Arabidopsis thaliana*. *Mol. Biol. Evol.*, 19, 2030-2034.

Le Corre, V., Roux, F., Reboud, X. 2002. DNA polymorphism at the *Frigida* gene in *Arabidopsis thaliana*. Extensive nonsynonymous variation is consistent with local selection for flowering time. *Mol. Biol. Evol.*, 19, 1261-1271.

Liu, F., Charlesworth, D., Kreitman, M. 1999. The effect of mating system differences on nucleotide diversity at the phosphoglucose isomerase locus in the plant genus *Leavenworthia*. *Genetics*, 151, 242-257.

Manicacci, D., Falque, M., Le Guillou, S., Piegu, B., Henry, A.M., Damerval, C., de Vienne, D. 2007. Maize *Sh2* gene is constrained by natural selection but escaped domestication. *Genetics*, in press.

Matsuoka, Y., Vigouroux, Y., Goodman, M.M., Sanchez, J.G., Buckler, E.S., Doebley, J. 2002. A single domestication for maize shown by multilocus microsatellite genotyping. *Proc. Natl. Acad. Sci.* USA, 00, 6080-6084.

McDonald, J.H., Kreitman, M. 1991. Adaptive protein evolution at the Adh locus in *Drosophila*. *Nature*, 351, 652-654.

McKhann, H.I.., Camilleri, C., Berard, A., Bataillon, T.M., David, J.L., Reboud, X., Le Corre, V., Caloustian, C., Gut, I.G., Brunel, D. 2004. Nested core collections maximizing genetic diversity in *Arabidopsis thaliana*. *Proc. Natl. Acad. Sci.* USA, *Plant J.*, 38(1), 193-202.

Meng, Z., Zaykin, D.V., Xu, C.-F., Wagner, M., Ehm, M.G. 2003. Selection of genetic markers for association analyses, using linkage disequilibrium and haplotypes. *Am. J. Hum. Genet.*, 73, 115-130.

Meuwissen, T.H.E., Karlsen, A., Lien, S., Olsaker, I., Goddard, M.E. 2002. Fine mapping of a quantitative trait locus for twinning rate using combined linkage and linkage disequilibrium mapping. *Genetics*, 161, 373-379.

Nei, M., Li, W.H. 1979. Mathematical model for studying genetic variation in terms of restriction endonucleases. *Proc. Natl. Acad. Sci.* USA, 76, 5269-5273.

Niu, T. 2004. Algorithms for inferring haplotypes. *Genet Epidemiol.,* 27, 334-347.

Nordborg, M., Borevitz, J.O., Bergelson, J., Berry, C.C., Chory, J., Hagenblad, J., Kreitman, M., Maloof, J.N., Noyes, T., Oefner, P.J., Stahl, E.A., Weigel, D. 2002. Contrasting evolutionary forces in the *Arabidopsis thaliana* floral developmental pathway. *Genetics*, 160, 1641-1650.

Otha, T. 1982. Linkage disequilibrium due to random drift in subdivided populations. *Proc. Natl. Acad. Sci.* USA, 79, 1940-1944.

Price, R.A., Palmer, J.D., Al-Shehbaz, I.A. 1994. Systematic relationships of *Arabidopsis*: a molecular and morphological perspective. In: *Arabidopsis*. Meyerowitz, E.M., Somerville, C., eds. Cold Spring Harbor Laboratory Press, New York, 7-19.

Prichard, J.K., Rosenberg, N.A., 1999. Use of unlinked markers to detect population stratification in association studies. *Am. J. Hum. Genet.*, 65, 220-228.

Pritchard, J.K., Stephens, M., Donnelly, P. 2000. Inference of population structure using multilocus genotype data. *Genetics*, 155, 945-959.

Pritchard, J.K., Stephens, M., Rosenberg, N.A., Donnelly, P. 2000. Association mapping in structured populations. *Am. J. Hum. Genet.*, 67, 170-181.

Rebourg, C., Chastanet, M., Gouesnard, B., Welcker, C., Dubreuil, P., Charcosset, A. 2003. Maize introduction into Europe: The history reviewed in the light of molecular data. *Theor. Appl. Genet.*, 106, 895-903.

Remington, D.L., Thornsberry, J.M., Matsuoka, Y.,Wilson, L.M., Whitt, S.R., Doebley, J., Kresovich, S., Goodman, M.M., Buckler, E.S. 2001. Structure of linkage disequilibrium and phenotypic associations in the maize genome. *Proc. Natl. Acad. Sci.* USA, 98, 11479-11484.

Stahl, E.A., Dwyer, G., Mauricio, R., Kreitman, M., Bergelson, J. 1999. Dynamics of disease resistance polymorphism at the Rpm1 locus of *Arabidopsis. Nature*, 400, 667-671.

Tabor, H.K., Risch, N.J., Myers, R.M. 2002. Candidate-gene approaches for studying complex genetic traits: practical considerations. *Nature Rev. Genet.*, 3, 107.

Tajima, F. 1983. Evolutionary relationships of DNA sequences in finite populations. *Genetics*, 105, 437-460.

Tajima, F. 1989. Statistical method for testing the neutral mutation hypothesis by DNA polymorphism. *Genetics*, 123, 585-595.

Templeton, A.R., Boerwinkle, E., Sing, C.F. 1987. A cladistic analysis of phenotypic associations with haplotypes inferred from restriction endonuclease mapping. I. Basic theory and an analysis of alcohol dehydrogenase activity in Drosophila. *Genetics*, 117, 343-351.

Tenaillon, M.I., Sawkins, M.C., Anderson, L.K., Stack, S.M., Doebley, J., Gaut, B.S., 2002. Patterns of diversity and recombination along chromosome 1 of maize (*Zea mays* ssp. *mays* L.). *Genetics*, 162, 1401-1413.

Tenaillon, M.I., U'Ren, J., Tenaillon, O., Gaut, B.S., 2004. Selection versus demography : a multilocus integration of the domestication process in maize. *Mol. Biol. Evol.*, 21, 1214-1225.

Thornsberry, J.M., Goodman, M.M., Doebley, J., Kresovich, S., Nielsen, D., Buckler, E.S. 2001. Dwarf8 polymorphisms associate with variation in flowering time. *Nat. Genet.*, 28, 286-289.

Tian, D., Araki, H., Stahl, E., Bergelson, J., Kreitman, M., 2002. Signature of balancing selection in *Arabidopsis. Proc. Natl. Acad. Sci.* USA, 99, 11525-11530.

Wang, R.L., Stec, A., Hey, J., Lukens, L., Doeley, J. 1999. The limits of selection during maize domestication. *Nature*, 398, 236-239.

Watterson, G.A. 1975. On the number of segregating sites in genetical models without recombination. *Theor. Popul. Biol.*, 7, 256-276.

Whitt, S.R., Wilson, L.M., Tenaillon, M.I., Gaut, B.S., Buckler, E.S.T. 2002. Genetic diversity and selection in the maize starch pathway. *Proc. Natl. Acad. Sci.* USA, 99, 12959-12962.

Wright, S., Lauga, B., Charlesworth, D. 2002. Rates and patterns of molecular evolution in inbred and outbred *Arabidopsis*. *Mol. Biol. Evol.*, 19, 1407-1420.

Yoshida, K., Miyashita, N.T. Ishii, T., 2004. Nucleotide polymorphism in the *Adh1* locus region of the wild rice *Oryza rufipogon*. *Theor. Appl. Genet.*, 109, 1406-1416.

Yu, J., G., Pressoir, W.H., Briggs Bi. I., Vroh, Doebley M.Y., McMullen, J.E., Gaut, M.D., Nielsen R.S., Holland, D.M., Kresovich J.B., E.S. SBuckler. 2006. A unified mixed-model method for association mapping that accounts for multiple levels of relatedness. Genetics 38(2), 203-208.

Zhang, K., Calabrese, P., Nordborg, M., Sun, F. 2002. Haplotype block structure and its applications to association studies : power and study designs. *Am. J. Hum. Genet.*, 71: 1386-1394.

Zhang, W., Collins, A., Morton, N.E. 2004. Does haplotype diversity predict power for association mapping of disease susceptibility? *Hum. Genet.*, 115, 157-164.

Zhang, K., Qin, Z., Chen, T., Liu, J.S., Waterman, M.S., Sun, F. 2005. HapBlock: haplotype block partitioning and tag SNP selection software using a set of dynamic programming algorithms. *Bioinformatics* 21, 131-134.

Zhang, L., Peek, A.S., Dunams, D., Gaut, B.S., 2002. Population genetics of duplicated disease-defense genes, *hm1* and *hm2*, in maize (*Zea mays* ssp. *mays* L.) and its wild ancestor (*Zea mays* ssp. *parviglumis*). *Genetics* 162, 851-860.

30

CHAPTER

Genomics and Plant Breeding: Maize and Wheat

B. Hirel, J. Le Gouis, M. Bernard, P. Perez, M. Falque, F. Quetier, J. Joets, P. Montalent, P. Rogowsky, A. Murigneux, A. Charcosset

WHEAT AND MAIZE, TWO LARGE-SCALE CROPS

Maize and wheat are, along with rice, the three major cereals cultivated in the world. The total cultivation area is 147 million (M) hectares (ha) for maize and 216 Mha for wheat and their production reached respectively 721 M and 627 M tonnes (t) in 2004 (http://apps.fao.org). In Europe, maize is cultivated over 4 Mha for a production of 39 Mt of grain; it ranks second behind wheat, which is cultivated over 14 Mha for a production of 100 Mt in 2004 (http://www.coceral.com). Including the production of silage, maize cultivation amounts to 7.7 m ha in the European Union

The situation is the same in France (www.agreste.agriculture.gouv; 2004), where wheat ranks first (5.2 Mha and 39.7 Mt) and maize second (1.8 Mha and 16.4 Mt for grain alone and 1.4 Mha for silage). However, maize ranks first in terms of yield per hectare, 89.8 q/ha against 75.8 q/ha for wheat in 2004. Half of the maize production consists of silage meant for animal feed. The other half is meant for production of grain used to propagate seeds, a wide variety of agrofood uses, and mostly animal feed, especially for monogastric livestock (pig, chicken). With respect to seed production, Europe ranks second after the USA.

Maize and wheat belong to the Poaceae family but differ in a large number of characters. Maize is an annual, monoecious, large-scale crop, possessing for most genotypes cultivated at present a life cycle of around 6 months. Maize was domesticated in Mexico from teosinte around 7,000 years ago (Matsuoka *et al.*, 2002). It is important to note that this domestication was not accompanied by speciation in the biological sense, so that gene flow from the original genetic pool remains possible. Maize was then distributed over the American continent and the rest of the world, reaching Europe around the time of the discovery of the New World (Rebourg *et al.*, 2003). During this expansion, maize acquired characters of climatic adaptation such as insensitivity to photoperiod and early flowering in limiting temperature conditions. Wheat is also an annual species planted mostly in the autumn in Europe, with a life cycle of 8 to 10 months. It is an autogamous species that was already cultivated more than 8,000 years ago in the Fertile Crescent of South West Asia and its genetic origin has been extensively studied. The complex genome of wheat is termed allohexaploid, because it is made up of three genomes derived from the cross of related wild species, close to the present species *Triticum monococcum*, *T. speltoides*, and *T. tauschii*. Over time, the various stages of domestication and selection led to an explosion in the genetic diversity of maize and wheat and their cultivation is now adapted to practically every kind of environment.

Even though our understanding of the genomics of these species is less advanced than with model plants (*Arabidopsis thaliana*, rice), mainly because of the size of their genomes, the technical advances in the field, as well as national and international programmes (such as Génoplante and programmes in the United States), suggest that this gap will be rapidly shortened (Ladugah *et al.*, 2001). The advances have been very rapid in maize; the genetics is particularly well known and major investment has been justified by the economic importance of the plant in the USA. A large amount of information is available at the Graingenes website (http:// wheat.pw.usda.gov) for wheat, barley, rye, and oats Carollo *et al.*, 2005) and for maize on the MaizeGDB database (http://www.maizeGDB.org). Apart from the fundamental interest, the information obtained on maize and wheat has important implications for breeding programmes. These are discussed at the end of the chapter.

MAPPING

Genetic mapping began very early in maize, with numerous mutations available (Emerson *et al.*, 1935). These studies were carried out using biochemical markers (isozymes). Since the emergence of molecular markers (RFLP, then SSR and SNP markers developed using PCR techniques), genetic mapping of maize has been the subject of major studies coordinated by a Missouri team (Coe, 2001). Efforts have recently paid off on a reference population, Intermated B73 x Mo17 (IBM), developed after four cycles of random crosses within the initial F_2 population. This programme involved accumulation of recombination events and increased the precision of the mapping by about a factor of three, in comparison with conventional mapping populations of equivalent size. The current public maps recently comprised around 2,000 SSR and 6,000 loci corresponding to genes or ESTs (see later). Many such studies have also been conducted by private groups or within the framework of private-public collaboration. The Genoplante program resulted in the mapping of around 2,500 additional loci corresponding to ESTs of candidate genes. This programme used the IBM population and a complementary population of the same type (developed from F2 and F252 lines), which allowed mapping of loci that are monomorphic on IBM (Falque *et al.*, 2005). Finally, AFLP markers can be used to make a region dense in markers, especially for the purpose of positional cloning.

At the molecular level, the first, often partial, wheat genome maps (Chao *et al.*, 1989) were established in the late 1980s; since then, the number of populations mapped has gone up despite the low polymorphism often found. The reference map (ITMImap), which has more than 1,500 RFLP and SSR markers (Gupta *et al.*, 2002; Song *et al.*, 2005) for a size of around 3,700 cM, corresponds to a very broad cross between a synthetic wheat (W7984) reconstituted from a durum wheat ("Altar 84") and a line of *T. tauschii* and the cultivar "Opata 85" (Nelson *et al.*, 1995). A microsatellite consensus map was recently constructed joining four independent genetic maps including the ITMI map (Somers *et al.*, 2004). Populations of intermediate agronomic value, such as the "Courtot" x "Chinese Spring" population (Cadalen *et al.*, 1997; Sourdille *et al.*, 2003), or of high agronomic value, such as "Renan" x "Recital" (Groos *et al.*, 2002), are now available and were mapped satisfactorily. The number of microsatellites described and mapped increases regularly (around 3,500 in 2005) and the process is now helped by the detection of SSR motifs in ESTs (Kantety *et al.*, 2002). Examining more than 170,000 ESTs from public and Génoplante databases previously clustered in 46,510 contigs, Nicot *et al.* (2004) identified 3,530 SSR which represented 7.5% of the total number of contigs.

Programmes have been set up to locate genes or EST on the wheat genome (Qi *et al.*, 2004; and associated papers). The strategy has been to use aneuploid or deletions lines of the "Chinese Spring" cultivar that are viable in wheat because of its hexaploid structure. Deletion lines were produced using a gametocidal chromosome of *Aegilops cylindrical, A. triuncialis*, or a chromosome fragment of *T. speltoides*: a series of 436 lines with terminal deletions (Endo and Gill, 1996) is available. A series of 101 of these lines was used (Qi *et al.*, 2003), it represented around 130 deletions; each deleted segment is called a 'bin' and corresponds to genetic distances from 10 to 40 cM. Qi *et al.* (2004) reported the mapping of 7104 EST with an average of 2.8 loci per EST. The assignation of microsatellite markers to the deletion bins (Sourdille *et al.*, 2004; Goyal *et al.*, 2005) enabled the establishment of a link between genetic and physical maps.

With the development of contiguous BAC libraries for the purpose of genome sequencing (see further), gene mapping methods presently tend toward assignment of genes to BAC, themselves positioned and oriented with respect to genetic maps. For maize, around 4,800 ESTs must serve as a bridge between the genetic and physical maps. Around 10,000 additional ESTs have been assigned to BAC in maize by hybridization of "overgo probes" defined on 10,000 sequences of consensus cDNA in the framework of the NSF Maize Mapping Project (http://www.maizemap.org). Some of them are being genetically mapped, in order to generate new bridges between the physical map and the genetic map. This last step constitutes a major bottleneck for the programme, because numerous duplications of maize gene sequences render the specific mapping of a particular copy of a gene difficult. The strategy is nevertheless to attempt as far as possible to limit the investigation to unique sequences in the genome: for this, ESTs that assign only a single contig of BAC are selected for genetic mapping. In wheat, several BAC libraries have also been produced and can be used for sequencing, physical mapping and gene cloning (Janda et al., 2004).

SYNTENY

Maize, wheat, and rice belong to the same family and share a common ancestor. On the basis of the rice genome sequence (International Rice Genome Sequencing Project, 2005), which is now nearly complete (Chapter 20), numerous studies of comparative genetics are being carried out. Information from comparative mapping of grasses can be found on the site http://www.gramene.org (Ware *et al.*, 2003). Overall, synteny (grouping of genes on a single chromosome) and colinearity (the order of genes on the chromosomes) are conserved between these different species (Devos and Gale, 1997, 2000) and genomes of 10 species of grasses can be described considering less than 30 blocks of the rice genome. Within cereals, therefore, a conservation of the order of genes on the scale of small regions (a few cM) can be observed. This similarity of genomes makes it possible to transpose part of the knowledge from one species to another and is particularly favourable for taking the rice sequence information into account in wheat or maize genomics. A fine mapping study of a quantitative trait locus (QTL) in wheat or maize can thus allow identification of candidate genes within a micro-syntenic region of rice. Nevertheless, it cannot be excluded that the desired gene may be absent in rice because of microdeletion or other chromosome rearrangements.

STRUCTURAL GENOMICS

The size of the maize genome is around 3,000 Mb (2,700 to 3,200 Mb depending on the lines). From early BAC sequencing studies, it was estimated that 80 to 85% of the genome is made up of repeat sequences, greatly implicating mobile elements (San Miguel *et al.*, 1996). Maize is certainly one of the largest "graveyards" of retrotransposons, which constitute 65 to 70% of its genome (Meyers *et al.*, 2001). Wheat is very close in this aspect (Bennetzen, 2000). In fact, most of these elements, when they are active, multiply a large number of times in the genome by inserting copies of their sequence. They may then lose their activity but all the inserted sequences survive, which increases the size of the genome. These two unfavourable characteristics of the maize genome militate against the sequencing of this cereal. They are nevertheless counteracted by the economic importance of the species in the USA and the potential advantages to be gained from understanding its genomics. Three BAC libraries have been compiled with three different restriction endonucleases, totalling 400,000 clones (coverage of 26x–http://www.genome.clemson.edu; http://www.chori.org). Two reduced representation shotgun (RRS) approaches have been evaluated to select gene-rich regions by selective cloning of hypomethylated regions or according to their kinetics of reassociation (Barbazuk *et al.*, 2003; Whitelaw *et al.*, 2003). These sequences have been clustered and assembled by two groups. TIGR constructed a database of 145,000 Assembled *Zea Mays* (AZM) sequences, with an average size of 1,500 kb, and 99,000 singletons in version 4.0. Recently AZM version 5 has been released (http://maize.tigr.org/release4.0/assembly.shtml) a point 163,390 Maize. Assembled Genomics Island (MAGI) and 536,377 singletons were computed by the Schnable group (Fu *et al.*, 2005) (http://www.plantgenomics.iastate.edu/maize). MAGI also includes additional genomic sequences from GenBank. In an effort to evaluate these two RSS methods 100, randomly selected BACs were sequenced (Haberer *et al.*, 2005). Three hundred and thirty genes were predicted, most of them supported by ESTs or protein sequences from maize or closely related species. While a large majority of these genes can be aligned to at least one sequence from the gene-enrichment GSS, only 30% are fully covered. If RSS methods are effective at sampling unique sequences, they will fail at providing a complete representation of the maize genes. This work also refined the estimate of genome repeat content to 66% at least. Repeats (mostly DNA transposons) were found in introns of 11% of genes. However, repeat expansion is not sufficient to fully explain the size difference between rice and maize genome. A three-fold increase in repeat-free sequence is observed in the maize genome compared to rice. It can be explained in part by the whole-genome duplication event that occurred 4.8 mya in the maize lineage. No clustering of genes was observed at the scale of BAC clones, questioning the widely accepted theory that the maize genome consists of gene islands separated by blocks of repeats.

Another approach for sequencing gene-rich regions was developed in the Maize Gene Discovery programme at Stanford University (Walbot, 2003), using knowledge of preferential insertion of the transposon Mutator in the genes and the sequencing of insertion borders (Walbot, 2003). A conventional approach of sequencing of BAC ends (154293 BES are available to date) and "contiguing" of BACs by automated analysis of "*Hind*III fingerprint" (FPC map) is ongoing (http://www.genome.arizona.edu/fpc/maize). The July 2005 release of the maize agarose FPC map contains 721 contigs that covers 2,149.6 Mb of the B73 genome. This information is at the moment a highly valuable resource for the development of markers at a high density in the regions of interest. A minimum tiling path of 17,704 clones provides contiguous sequences of 2.11 Gb or 90% of the genome. Complete sequence of these clones is expected by the end of 2009. The maize genome sequencing programme will probably be achieved exclusively by American teams, two consortia being present.

Finally, the knowledge available at present on certain targeted regions of the maize genome has provided new information on the importance of transposable elements (see above) or the gene duplication in certain regions (e.g., zeins). The early studies of comparative sequencing of genotypes showed differences in the number of copies of genes as well as differences in terms of the presence or absence of genes in a chromosomal region (Fu and Dooner, 2002, Brunner *et al.*, 2005, Morgante *et al.*, 2005). This last phenomenon leads notably to a new explanatory hypothesis of heterosis (hybrid vigour).

The size of the wheat genome is estimated at around 16,000 Mb with each of the 21 chromosomes being larger than the entire rice genome. Around 80% of the genome is made up of repeat sequences with approximately 70% of intergenic sequences represented by retroelements (Feuillet and Keller, 2002). Despite these very strong constraints, the first BAC libraries at different levels of ploidy of the genus *Triticum* appeared in 1999. Libraries of *T. tauschii* (Moullet *et al.*, 1999; Akhunov *et al.*, 2005), T. *speltoides* (Akhunov *et al.*, 2005), *T. monococcum* (Lijavetzky *et al.*, 1999) or *T. urartu* (Akhunov *et al.*, 2005), and *T. durum* (Cenci *et al.*, 2003) were produced. More recently, libraries of hexaploid wheat were obtained (Lagudah *et al.*, 2001, Shen *et al.*, 2005), notably at INRA (Evry), where a "7x library" of more than a million clones of the cultivar "Renan" was produced (Chalhoub *et al.*, 2003) and a library of the cultivar "Chinese Spring" has been developed in collaboration with John Innes Center. One strategy to reduce the complexity of the wheat genome approach was the creation of chromosome and chromosome-arm BAC libraries using flow cytometry (Janda *et al.*, 2005), such as the construction of the first chromosome-specific plant BAC library for the chromosome 3B (Safar *et al.*, 2004). This may open the possibility of an international effort to sequence large parts of the wheat genome (Gill *et al.*, 2004).

The first BAC sequences containing genes of interest have been published, such as the BAC sequence of *T. tauschii* containing the two paralogues of glutenins of high molecular weight (Anderson *et al.*, 2003) and, more recently, those of the BAC from diploid, tetraploid and hexaploid wheat species corresponding to the *Hardness* locus

(Chantret *et al.*, 2005). These first studies showed in wheat and barley that gene density in certain chromosomal regions was around one gene in every 5-20 kb, which is much higher than the density expected in the case of random distribution of genes, that is, one gene in every 250 kb (Devos and Gale, 2000; Feuillet and Keller, 2002). These numbers confirm the hypothesis of an organization of the genome into gene-rich regions. Moreover, this approach highlighted some sources of major differences between homoeologous genomes and opened the way to comparative genomics and evolution studies.

FUNCTIONAL GENOMICS AND STUDIES OF EXPRESSION

The term *functional genomics* includes approaches to determine precisely the function of genes. The use of different conditions of expression, in terms of organs, stages of development, and biotic or abiotic constraints, plays an important role in these approaches. The construction of cDNA libraries from contrasting tissues (organs, stages, culture, etc.), followed by their partial sequencing to obtain ESTs, has played a very important role in the genomics programmes of wheat and maize. This approach allows researchers in the first place to obtain sequence information on the expressed fraction of the genome, which is of major interest in the attempt to sequence the entire genome.

By December 2005, 582,988 ESTs of maize were available in public databases (GenBank/EMBL/DDBJ), emanating mostly from the NSF programme coordinated by V. Walbot (2003). These ESTs were clustered with cDNA and High Throughput cDNA sequences (HTC) by different groups into:

- 77,042 PlantGDB-assembled Unique Transcript (PUT) (4th November 2005 release, http://www.plantgdb.org/prj/ESTCluster/progress.php). This is the latest contribution to tentative unique transcript construction. It will be updated every three months following each new GenBank release.
- 36,625 Tentative Consensus (TC) and 35,422 singletons (2005/10 release) by TIGR (http://www.tigr.org/tigr-scripts/tgi/T_index.cgi?species=maize).
- 27,824 Unigene; 5th December 2005 by GenBank (http://www.ncbi.nlm.nih.gov/UniGene/UGOrg.cgi?TAXID=4577). The main goal of Unigene is to provide full length coverage. As a result there are fewer accessions.

Other very large databases have been developed by private companies (notably Dupont, Pioneer, Monsanto) and recently made available to the scientific community, provided engagements for intellectual property (http://www.maizeseq.org).

A considerable effort was made within public or private programmes to generate ESTs from different tissues of wheat. The International Triticeae EST Consortium (ITEC, http://www.ncbi.nlm.nih.gov/dbEST) generated 416,000, which placed wheat in the sixth rank for EST content, before barley with 315,000 ESTs. More than 100,000 supplementary ESTs from the Genoplante programmes were published in 2003 (Samson *et al.*, 2003). Programmes have now begun to localize these ESTs on the wheat genome (see above).

Today, the ESTs publicly available and those obtained within programmes such as Genoplante do not represent all the genes of maize or wheat. They nevertheless give us access to information on the sequence of numerous genes. This sequence information can especially be utilized by searching for homologies with genes, the function of which was established in other species. Corresponding EST can be mapped to search for candidate genes of QTL (see below).

By grouping ESTs in contigs and knowing their tissues of origin, it is possible to formulate a preliminary estimation of conditions of expression of a gene. The precision of this approach, generally called "Northern *in silico*", depends on the number of ESTs obtained for each library. It applies mostly to "standard" libraries that have not been standardized or subtracted. With libraries of the usual size (a few thousand ESTs per library), it may allow identification of genes that have very different profiles. It is nevertheless important to validate the results obtained in a second step by specific study of the expression of the gene, for example by reverse transcriptase PCR. The creation of DNA chips is now a major tool for studying the expression of genes more exhaustively and precisely (Chapter 10). The first public chip (Master Unigene-1) carrying 20,736 ESTs was created from cDNA. Even though this chip may be called Unigene, it must be noted that two to three ESTs may correspond to a single gene. Other projects are being developed to increase the number of genes represented and improve the specificity of hybridizations by synthesizing specific oligogenes. Certain groups have also developed micro-arrays that are smaller but chosen according to a given requirement, such as the chip developed by Lee *et al.* (2002) to study changes in expression during embryo development. Several projects on wheat are now well advanced and the first applications have been reported (Clarke and Rahman, 2005). An oligonucleotide array including 55,000 70 mers derived from ESTs, cDNA, genomic sequences (including AZM) is now available at cost recovery price from the Maize Oligoarray project (http://www.maizearray.org). An affymetrix maize GeneChip representing more than 13,000 genes is also available. Finally, expression at the protein level offers additional information, especially for the study of stress response phenomena (Thiellement *et al.*, 1999; Bahrman *et al.*, 2004).

FUNCTIONAL GENOMICS, MUTANTS AND GENETIC TRANSFORMATION

Numerous spontaneous mutations have been recorded in maize (Coe, 2001). Isolation of mutant genes began in the early 1980s, leading to characterization of the first genes implicated in the metabolism of starch, anthocyanins, and storage proteins (e.g., *Waxy, Factor R, Opaque* 2). These studies profited from the abundant information on transposable elements in this species. Either the phenotypic screening was carried out directly on a population mutagenized with a transposon (usually *Ac* or *Mu*), or an allele with an insertion was sought from a point mutation (targeted tagging). In the latter case, plants carrying the mutation studied were crossed with a line carrying an active transposon (*Mutator, Ac/Ds*or *En/Spm* systems). The mutant phenotypes observed within the progeny carried the original mutation and a new mutation created by the insertion of the transposon. In both cases, the causative gene was identified through cloning of the borders of insertion of the transposon, after verification of the co-segregation between the insertion and the phenotype. To date, some tens of mutations have been identified using transposons, including the *Hm1* gene, which confers resistance to the pathogenic fungus *Cochliobolus* (Johal and Briggs, 1992), the *An1* gene intervening in gibberellin synthesis and resulting in a dwarf phenotype (Bensen *et al.*,

1995), and the key domestication gene *Tb1* (Doebley *et al.*, 1997). More recently, transposition phenomena have been used to develop "gene machines" designed to identify plants carrying a mutation in a gene for which the function is to be defined (the reverse genetics approach). Many programmes of this type were set up for private research: the first was made accessible by Pioneer to public researchers in 1995 (Meeley and Briggs, 1995). These programmes use the *Mutator* element *Mu* (Hanley et al., 2000; May *et al.*, 2003), the *Ac-Ds* system for approaches more targeted in terms of positions, and finally genetically modified *Mutator* elements *Mu* to reduce the number of active copies and facilitate the cloning of sequences flanking the insertion (*Rescue-Mu* programme led by the team of V. Walbot). This reverse genetics approach has so far been used to identify insertions for several hundreds of genes, of which around 20% present interesting phenotypes.

In addition to these developments, the production of transgenic plants, which was achieved for the first time in maize in 1990 (Gordon-Kamm *et al.*, 1990), has been mastered by many research teams. Biolistics and *Agrobacterium tumefaciens* are also currently used methods for the production of transgenic maize. These techniques have been used to study the effect of constructions modifying the level or conditions of expression of a gene and also to suppress the effect of a gene by "antisense" (Jeanneau *et al.*, 2002) or, more recently, by "RNA interference" (RNAi) technology. These last approaches may offer advantages over the use of insertion mutants, particularly when genetic functions are duplicated (functional paralogues), or the complete inactivation of the function leads to a lethal phenotype. The genetic transformation also makes it possible to study the spatio-temporal, tissue, and developmental regulation of gene expression by considering the effect of fusions of promoter with reporter genes (GUS, GFP). Finally, this tool of functional analysis may prove valuable in maize to complement a mutant (Magnard *et al.*, 2003), particularly when a single allele is available. It may be used with confidence to demonstrate the cloning of a QTL by complementing the unfavourable allele of the candidate gene with the allele obtained from the parent containing the favourable effect. All these approaches thus play a key role in elucidating the effect of a gene. Some of them involve field assays, particularly to understand the effects on complex characters. In addition, they offer original perspectives of application.

For wheat, more than a thousand natural or artificial mutations have been recorded and characterized (McIntosh *et al.*, 1998), which makes it a high-performing tool of genetics. On the other hand, because of the large size of its genome, bread wheat remains a difficult species to transform. The average efficiency of transformation is around 0.1 to 5% (Lagudah *et al.*, 2001), with results obtained most often by biolistics and through *A. tumefaciens* (Cheng *et al.*, 1997; Shewry and Jones, 2005). Even though this success rate is low and often dependent on particular genotypes of low agronomic performance that are chosen for their aptitude to regenerate *in vitro*, it is sufficient for preliminary studies, as indicated by the increasing number of publications on wheat transformation. These publications particularly involve applications with respect to flour quality (Beecher *et al.*, 2002), *Fusarium* tolerance (Okubara *et al.*, 2002; Anand *et al.*, 2003), drought tolerance (Abede *et al.*, 2003), and resistance to herbicides (Melchiorre *et al.*, 2002).

For functional validation, the development of a gene machine for insertion mutations is presently difficult because of the low transformation success rate. The use of a heterologous transposition system is possible in principle (Lagudah *et al.*, 2001) but has not occurred in practice. Projects to create populations by chemical mutagenesis (EMS) or by radiation are being developed and have been successfully used to identify waxy starches (Slade *et al.*, 2004). Polyploidy and gene duplication limit these techniques, which could demand a time-consuming series of backcrosses to obtain a series of mutations in homeologous genes. The use of gene silencing systems may be a solution as was shown by either RNAi (Yan *et al.*, 2004) or virus-induced methods (Scofield *et al.*, 2005).

For maize and wheat, transversal functional genomics that is based, for example, on the study of rice mutants, through direct and reverse genetics, also holds promise of a good future (Chapter 20). Already, researchers at Dupont-Pioneer have isolated a gene controlling embryo size using a *Giant embryo* mutant of rice. They showed that at least one of the orthologous genes in maize co-localizes with a QTL of endosperm size (Sakai *et al.*, 2002).

MULTIFACTORIAL GENETICS: COMBINING PHYSIOLOGY AND AGRONOMY

The tools of multifactorial genetics, also called quantitative genetics, are at present best adapted to understanding in an integrated manner, the determinism of certain complex characters (organ development, phenotype, yield, metabolic efficiency) with continuous variation, whose expression is controlled by a large number of genes (QTL). This approach, even though it is time-consuming experimentally because it requires the study of a large number of genotypes, is presently one of the surest means of identifying regulator or structural genes controlling plant physiology and/or development (Wang *et al.*, 1999; Chapter 28).

The principle of this approach is to associate the physiological functions, characters linked to development, or agronomic characters, taken singly or all together, with molecular markers (Chapters 25 and 29). DNA mapping using RFLP, RAPD, SSR or SNP molecular markers is carried out from a population of segregating lines (recombinant or double haploid lines) generally derived from two homozygous parents. From the percentage of recombination of markers, in relation to one another, a genetic map can be charted to create a complete marking out of the genome. If there is a significant relationship between the character variation considered and the presence of one or another of the alleles of a marker, it signifies that at this point on the genome at least one gene partly controls the variability of the character (Chapters 20 and 21). Several QTL are generally identified, so that a complex character can be broken down into several determinants that can be studied individually in greater detail.

When the QTL of agronomic or phenotypic characters are localized on a genetic map, their genetic significance can then be discovered by studying their co-localization with QTL of physiological or biochemical characters and with the genes that are supposed to intervene in the determinism of the trait studied, candidate genes possessing a known function and already mapped on the genetic map. The candidate gene can then be validated by transgenesis, mutagenesis, or the study of the relationship between the polymorphisms of alleles, detected by sequencing, and the character can be studied (Chapters 28 and 29).

This type of approach also makes it possible to address the study of interactions between diverse regulatory elements and determine their adjustment as a function of environmental constraints and agronomic and economic needs (de Vienne and Causse, 1998). These tools can be used not only to map and characterize the QTL, but also to quantify their effects and search for reciprocal interactions as a function of different environmental constraints. This method is used to incorporate the results of the genetic manipulation/mutation approach by situating it within the contribution of other genetic factors controlling character variation. Existing or potential genetic variability can therefore be exploited in a more efficient and targeted manner, and a given character can be optimized qualitatively or quantitatively (Prioul *et al.*, 1997). The new combinations of genes or alleles and modifications of their interactions are the basis of selection of new genotypes, which earlier was carried out in a more or less empirical manner, because it was based on obtaining an improved phenotype for a given character.

GENETIC BASES OF NITROGEN USE EFFICIENCY IN MAIZE AND WHEAT

The search for physiological and agronomic bases of genetic variability of nitrogen use efficiency in wheat and maize is an example of the use of multifactorial genetics combined with integrated biological studies on the whole plant.

The evolution of economic constraints (cost, production quantity and quality) and environmental constraints (pollution, use of fossil fuels) gradually limits farming practices based on high inputs (fertilizers, herbicides, pesticides, and water in some cases). This means, particularly, a reduction of nitrogen supply to crops. Nitrogen, after carbon, is the second major element involved in the composition of biological molecules and the first mineral element of the soil necessary for plant growth. However, nutritional requirements, especially the need for nitrogen, differ greatly from one plant species to another. The genetic bases of the degree of nitrogen requirement are still very sketchily known. If they were understood, new cultivars could be bred that require less nitrogen than existing cultivars. Research has been undertaken to locate QTL associated with the quantitative character "degree of nitrogen need" or "nitrogen use efficiency" (absorption or assimilation of mineral nitrogen or cycling of protein nitrogen). Moreover, it is well known that nitrogen metabolism in plants changes over time. In fact, over its life cycle, the plant passes progressively from a metabolism assimilating mineral nitrogen, principally nitrate, to a remobilizing metabolism ensuring the cycling of protein nitrogen stored in vegetative organs.

The mobilization of nitrogen associated with plant senescence is a genetically determined process that develops naturally over the life of the plant and the evolution of which can be influenced significantly in favourable or unfavourable environmental conditions (climate, drought or nutrient stress, pathogen attack). The degradation and cycling of reserve organic nitrogen forms, chiefly stored as proteins in the leaves, stems, or even roots, along with primary assimilation of inorganic nitrogen, determine the plant life cycle. In fact, the turnover of nitrogen compounds plays a critical role in the redistribution of organic nitrogen within the plant and in the formation of reserve proteins. Thus, 60 to 70% of proteins of aerial organs are degraded during senescence and the nitrogen metabolized during degradation is exported towards the developing reserve organs (grain, fruits, tubers). This phenomenon is clearly illustrated in cereals such as maize and wheat, in which grain proteins contribute to nutritional value and the quality of products derived from them (Brouquisse *et al.*, 2001).

Since assimilation efficiency and nitrogen cycling are complex characters controlled by a large number of genes, a study of quantitative genetics was carried out in maize and wheat, associating the metabolic functions linked to nitrogen metabolism (levels of metabolites, enzymatic activities) with molecular markers having served to establish a genetic map (Causse *et al.*, 1996). The rest of the study consisted of researching whether QTL for these metabolic functions are located at the same place on the genetic map as QTL linked to yield (quantity and number of grains) and the quality of grains (protein content).

In a population of recombinant lines of maize (RIL) resulting from crosses of two parental lines Io and F2 field-tested for agronomic characteristics (Bertin and Gallais, 2001), researchers determined the physiological characters linked to nitrogen metabolism, such as nitrate level, nitrate reductase (NR) activity, and glutamine synthase (GS) activity. Within the population of these RIL, a positive correlation was observed between the nitrate level and the GS activity of leaves and the different components of grain yield. In contrast, a negative correlation was obtained with the leaf NR activity. These data suggest that yield increase in maize is partly linked to its capacity to accumulate nitrate in the leaves during the vegetative phase and its efficiency in mobilizing this mineral nitrogen during grain filling. The search for QTL was then undertaken to identify chromosome regions containing genes involved in the expression of the character(s) studied. Several co-localizations between QTL of grain yield, GS activity, and genes coding for this enzyme were determined. In particular, on chromosome 5, at the level of one of the GS genes (*gln1-3*), a set of co-localizations between the QTL of leaf nitrate level, NR activity, GS activity, and yield were observed. These results suggested the hypothesis that the expression of these three physiological parameters is co-regulated and partly determines maize grain yield (Hirel *et al.*, 2001) (Fig. 30.1).

Experiments have been carried out to verify whether the candidate gene *gln1-3* plays a major role in the metabolism and/or cycling of nitrogen required for grain filling and formation. To validate the GS/grain yield QTL, a programme was designed to attempt to produce transgenic plants modified in the expression of the *gln1-3* gene (overexpression of gene or introduction of a favourable allele to an unfavourable genetic basis) and to isolate a mutant deficient for this gene. The genetically modified plants had increased grain yield and yield was reduced in the mutant, thus validating the function of the candidate gene *gln1-3*. (Martin et al., 2006). The introgression of favourable alleles of this gene could then be used in marker-assisted breeding programmes.

In wheat, a population of 220 DH lines derived from two contrasting parents Arche (tolerant to N deficiency) and Récital (susceptible to N deficiency) was produced and characterized at two N levels (Laperche *et al.*, 2005). The population was shown to vary in the response to N deficiency. The QTL detection is currently being carried out to identify chromosomic regions implicated in the NUE at low N level and in the Genotype x N level interaction. In parallel, the localisation of the main genes involved in N metabolism has been initiated (Boisson *et al.*, 2005; Habash et al., 2006). This study will certainly take advantage of a comparison with maize regions already identified.

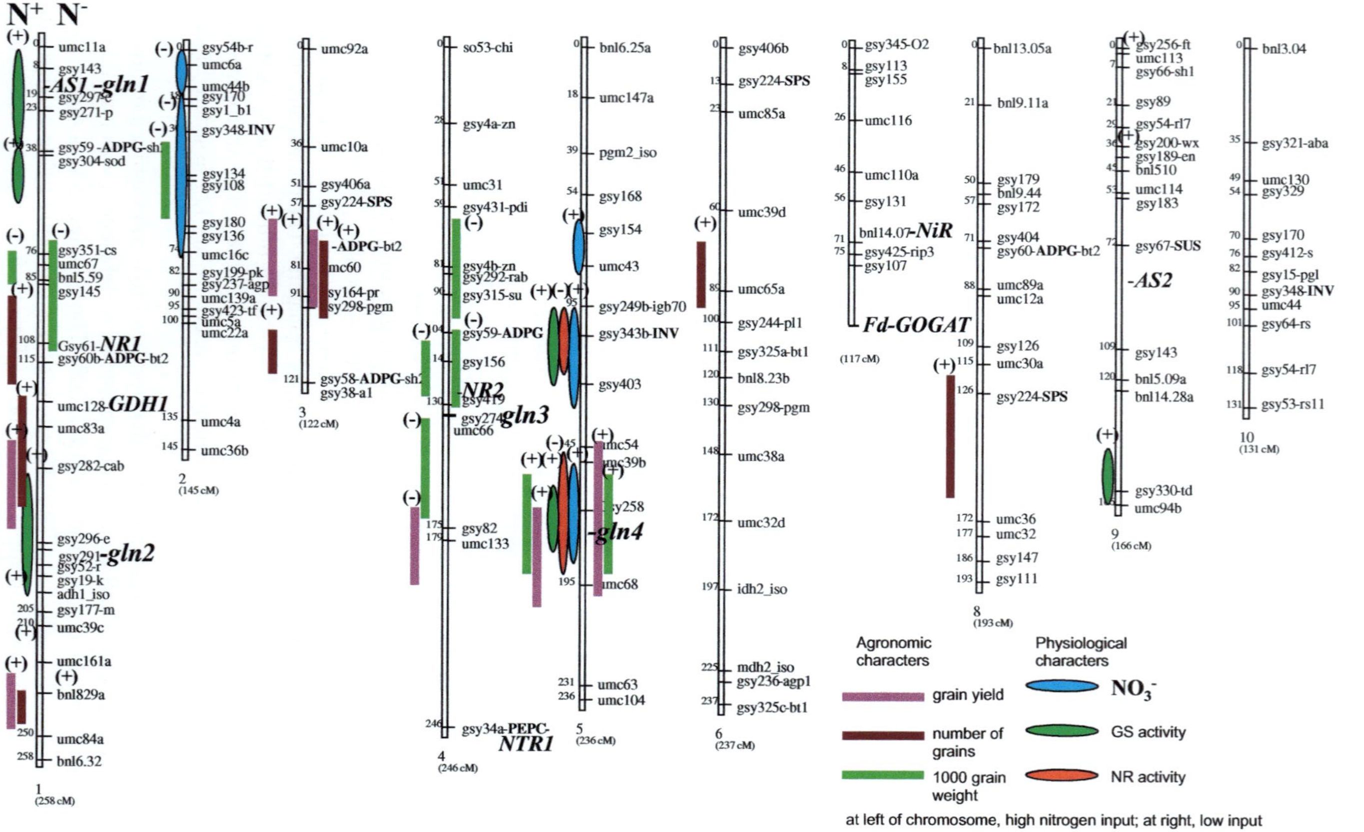

Fig. 30.1 Genetic map of maize (RFLP) and co-localization between the QTL for physiological characters and those linked to yield and to its components. Abbreviations of enzymes implicated in the nitrogen metabolism: ASN, asparagine synthetase; Fd-GOGAT, ferredoxin-dependent glutamate synthase; GS, glutamine synthetase; NR, nitrate reductase; NiR, nitrite reductase. Hirel *et al.,* 2001. © American Society of Plant Biologists.

CONTRIBUTION OF GENOMICS TO PLANT BREEDING

Advances in the understanding of genes and their biological effects can have a major impact on plant improvement. The identification of genes involved in the variation of characters of agronomic interest is a major advance with respect to the location of QTL, most often estimated with intervals of confidence of more than 10 cM. Beyond its interest as an approach of functional validation, the association genetics (Chapter 29) offers new perspectives to identify particularly interesting alleles. Finally, advances in the biological understanding of the effect of genes (see above the gene *gln* for GS) and their interactions must allow in the long term a better modelling of character variation (Martin et al., 2006). This new information tends to improve significantly the efficiency of marker-assisted selection and to extend its domains of application: "driving" the introgression of favourable alleles in a recipient genetic material, assembling favourable combinations of alleles, predicting performances of individuals. The perspectives of these applications of genomics are discussed in the following paragraphs. Readers interested in more precise information on marker-assisted selection can refer to the work of Charcosset and Gallais (1998), and Charcosset and Moreau (2004).

Introgression

The objective of introgression programmes is to modify existing genetic material in a targeted manner by replacing an initial chromosomal segment with a chromosomal segment carrying favourable characteristics contributed by another plant (donor parent). This approach involves crossing the initial genetic material (recipient parent) with the donor parent, then gradually eliminating the genome of the donor parent by successive backcrosses, while conserving the segment of interest in a targeted manner. It has been conducted successfully without genetic markers for monogenic characters, for which the presence of the interesting allele could be determined from phenotypic observations (such as a pathological test for resistance to a disease).

The primary advantage of the use of genetic markers is to accelerate considerably the return to the genome of the recipient parent, especially near the introduced segment. This approach was used to accelerate line conversion programmes from transformation events. Markers close to the gene of interest can be used to identify the presence of the favourable allele without phenotypic observation. This could be useful when phenotypic tests are time-consuming or difficult (recessive favourable alleles, test for resistance to a disease that is still absent in the cultivation zone). The use of markers has also made possible the introgression of favourable alleles of QTL, the presence of which cannot be detected reliably on the basis of phenotypic observations (Bouchez *et al.*, 2002). Nevertheless, it must be noted that the imprecision of the QTL position leads to the use of several markers in the region of a QTL to avoid the loss of the favourable allele because of recombination events (Hospital and Charcosset, 1997). The identification of a gene corresponding to a QTL and the development of specific markers makes it possible to eliminate the risk of loss of the target allele by recombination and to reduce the number of markers used as well as the number of individuals to be tested.

Finally, it is noted that through association studies described in Chapter 29 interesting alleles can be identified without the need to develop mapping populations.

These alleles can then be introgressed into the elite material, thus considerably facilitating the validation of genetic resources. This approach nevertheless remains to be experimentally validated.

Prediction of Genetic Values and Recurrent Marker-assisted Selection

On the basis of QTL location and estimation of their effects, new approaches have been proposed to aid prediction of genetic values of candidate individuals for selection (Lande and Thompson, 1990). Markers close to a QTL can be used to estimate, for each individual, the probability that it will carry a given genotype at this QTL. On the basis of these probabilities and effects associated with different genotypes, the value of the individual at the QTL can be predicted. This approach can be extended to all QTL detected and thus allows prediction of the genetic value of an individual. It will then become possible to predict the value of individuals without evaluating them phenotypically, or to combine this prediction with available phenotypic performances to obtain an optimal prediction of the genetic value.

Even though few experimental results on this approach are available to date (Johnson *et al.*, 2004; Moreau *et al.*, 2004), the conditions determining its efficiency (notably large populations) are well documented at the theoretical level. Genomics programmes allow much more precise identification (up to the scale of a gene) of the position of the loci causing the variation of quantitative traits. The development of markers very close to these loci will inherently offer a significant increase in efficiency of marker-assisted selection, especially by limiting the risks of recombination between markers and loci of interest during programmes conducted over several generations (Bernardo and Charcosset, 2006).

Prediction of Hybrid Value

The possibilities of predicting the performance of F_1 hybrids, or of heterosis by genetic distances estimated with neutral makers are now well known (Melchinger, 1999). Their efficiency depends on linkage disequilibrium (Chapter 29) between the neutral markers considered and the QTL implicated in the phenomenon of heterosis (presenting effects of dominance or superdominance). Because of this, it seems impossible to predict the heterosis of hybrids between non-related lines and those of different genetic origins (Charcosset and Essioux, 1994). These limits could be overcome if the genetic distances can be estimated directly at the level of genes implicated in the heterosis. It is nevertheless important to note that to be effective the distances must be calculated on a set of genes that explain a sufficiently large part of the phenomenon. Ongoing studies on maize yield emphasize a determinism involving numerous QTL, the apparent superdominance effects of which are probably linked in large part to genetic linkages between loci. The detection of genes implicated may be quite time-consuming, even though approaches based on their expression can guide the research. Another hypothesis is that of genic complementation, as it has been shown that a proportion of genes is absent from a particular genotype, although present in one another.

Finally, the marker-assisted prediction of hybrid performance has led to the adaptation of Best Linear Unbiased Prediction models conventionally used in animal breeding. Markers are used in this context to evaluate the "relationship" between genotypes and the weight that should be given to existing information (hybrids already evaluated), as well as to predict the value of a hybrid that has not yet been evaluated. This conceptual framework was used by Bernardo (2001) to measure the increase in efficiency of prediction that would be achieved if we knew all the genes for a quantitative trait in hybrid crops. The conclusions of this study (a relatively moderate increase) are specific to this use and do not call into question the vast potential of genomics mentioned earlier.

REFERENCES

Abebe, T., Guenzi, A.C., Martin, B., Cushman, J.C. 2003. Tolerance of mannitol-accumulating transgenic wheat to water stress and salinity. *Plant Physiol.*, 131, 1748-1755.

Akhunov, E.D., Akhunova, A.R., Dvorak, J. 2005. BAC libraries of *Triticum urartu*, *Aegilops speltoides* and *Ae. tauschii*, the diploid ancestors of polyploid wheat. *Theor. Appl. Genet.* DOI 10.1007/s00122-005-0093-1

Anand, A., Zhou, T., Trick, H.N., Gill, B.S., Bockus, W.W., Muthukrishnan, S. 2003. Greenhouse and field testing of transgenic wheat plants stably expressing genes for thaumatin-like protein, chitinase and glucanase against *Fusarium graminearum*. *J. Exp. Bot.*, 54, 1101-1111.

Anderson, O.D., Rausch, C., Moullet, O., Lagudah, E.S. 2003. The wheat D-genome HWM-glutenin locus: BAC sequencing, gene distribution, and retrotransposon clusters. *Funct. Integr. Genomics*, 3, 56-68.

Barbazuk, B., Whithelaw, C., Quackenbush, J., Bennetzen, J., Schubert, K. 2003. The maize genome sequencing project at the Donald Danforth Plant Science Center. 45th Annual Maize Genetics Conference, 13-16 March, Lake Geneva, Talk Abstract 5. Accessible at Maize DB http://www.agron.missouri.edu

Bahrman, N., Le Gouis, J., Negroni, L., Amilhat, L., Leroy, P., Lainé, A.-L., Jaminon, O. 2004. Differential protein expression assessed by two-dimensional gel electrophoresis for two wheat varieties grown at four nitrogen levels. *Proteomics*, 4, 709-719.

Beecher, B., Bettge, A., Smidansky, E., Giroux, J. 2002. Expression of wild-type pinB sequence in transgenic wheat complements a hard phenotype. *Theor. Appl. Genet.*, 105, 870-877.

Bennetzen, J.L. 2000. Transposable element contributions to plant gene and genome evolution *Plant Mol. Biol.*, 42 (1): 251-269

Bensen, R.J., Johal, G.S., Crane, V.C., Tossberg, J.T., Schnable, P.S., Meeley, R.B., Briggs, S.P. 1995. Cloning and characterization of the maize *An1* gene. *Plant Cell*, 7(1), 75-84.

Bernardo, R. 2001. What if we knew all the genes for a quantitative trait in hybrid crops? *Crop Sci.*, 41, 1-4.

Bernardo R., Charcosset A. 2006. Usefulness of Gene Information in Marker-assisted Recurrent Selection: A Simulation Appraisal. *Crop Science*, in press 46: 614-621.

Bertin, P., Gallais, A. 2001. Physiological and genetic basis of nitrogen use efficiency in maize. II. QTL detection and coincidences. *Maydica*, 46, 53-68.

Boisson, M., Mondon, K., Torney, V., Nicot, N., Lainé, A.-L., Bahrman, N., Gouy, A., Daniel-Vedele, F., Hirel, B., Sourdille, P., Dardevet, M., Ravel, C., Le Gouis, J. 2005 Partial sequences of nitrogen metabolism genes in hexaploid wheat. *Theor. Appl. Genet.*, 110, 932-940.

Bouchez, A., Hospital, F., Causse, M., Gallais, A., Charcosset, A. 2002. Marker-assisted introgression of favourable alleles at quantitative trait loci between maize elite lines. *Genetics*, 162, 1945-1959.

Brouquisse, R., Masclaux, C., Feller, U., Raymond, P. 2001. Protein hydrolysis and nitrogen remobilisation in plant life and senescence. In: *Plant Nitrogen.*, Lea, P., Morot-Gaudry, J.F., eds. INRA-Springer-Verlag, Berlin, Paris, 275-293.

Brunner, S., Fengler, K., Morgante, M., Tingey, S., Rafalski, A. 2005. Evolution of DNA sequence nonhomologies among maize inbreds. *Plant Cell,* 17: 343-360

Cadalen, T., Boeuf, C., Bernard, S., Bernard, M. 1997. An intervarietal molecular marker map in *Triticum aestivum* L. Em Thell. and comparison with a map from a wide cross. *Theor. App. Genet.,* 94, 367-377.

Carollo, V., Matthews, D.E., Lazo, G.R., Blake, T.K., Hummel, D.D., Lui, N., Hane, D.L. Anderson, O.D. 2005. GrainGenes 2.0. An improved resource for the small-grains community. *Plant Physiol.,* 139, 643-651.

Causse, M., Santoni, S., Damerval, C., Maurice, A., Charcosset, A., Deatrick, J., de Vienne, D. 1996. A composite map of expressed sequences in maize. *Genome,* 39, 418-432.

Cenci, A., Chantret, N., Kong, X., Gu, Y., Anderson, O.D., Fahima, T., Distelfeld, A., Dubcovsky, J. 2003. Construction and characterization of a half million clone BAC library of durum wheat (*Triticum turgidum ssp. durum*). *Theor. Appl. Genet.,* 107, 931-939.

Chalhoub, B., Safar, J., Lefevre, A., Roault, P., Pateyron, S., Bellec, A., Janda, J., Faivre-Rampant, P., Dolezel, J., Caboche, M. 2003. New approaches for genome characterization: genomic and chromosome-specific BAC libraries from hexaploid wheat. *Plant Anim. Genome,* XI: 48, www.intl-pag.org.

Chantret, N., Salse, J., Sabot, F., Rahman, S., Bellec, A., Laubin, B., Dubois, D., Dossat, C., Sourdille, P., Joudrier, P., Gautier, M.-F., Cattolico, L., Beckert, M., Aubourg, S., Weissenbach, J., Caboche, M., Bernard, M., Leroy, P., Chalhoub, B. 2005. Molecular basis of evolutionnary events that shaped the hardness locus in diploid and polyploid wheat species (*Triticum and Aegilops*). *Plant Cell,* 17, 1033-1045.

Chao, S., Sharp, P.J., Worland, A.J., Warham, E.J., Koebner, R.M.D., Gale, M.D. 1989. RFLP-based genetic maps of wheat homeologous group-7 chromosomes. *Theor. Appl. Genet.,* 78, 495-504.

Charcosset, A., Essioux, L. 1994. The effect of population structure on the relationship between heterosis and heterozygosity at marker loci. *Theor. Appl. Genet.,* 89, 336-343.

Charcosset, A., Gallais, A. 1998. Interet des marqueurs en selection. In: *Les Marqueurs Moleculaires en Genetique et Biotechnologies Vegetales.* de Vienne, D., ed. INRA, Paris, 139-159.

Charcosset A., Moreau L. 2004 Use of molecular markers for the analysis of genetic diversity and the development of new cultivars. *Euphytica* 137: 81-94

Cheng, M., Fry, J.E., Pang, S., Zhou, H., Hironaka, C.M., Duncan, D.R., Conner, T.W., Wan, Y. 1997. Genetic transformation of wheat mediated by *Agrobacterium tumefaciens. Plant Physiol.,* 115, 971-980.

Clarke, B., Rahman, S. 2005. A microarray analysis of wheat grain hardness. *Theor. Appl. Genet.,* 110, 1259-1267.

Coe, E.H. 2001. The origins of maize genetics. *Nat. Rev. Gen.,* 2(11), 898-905.

Devos, K.M., Gale, M.D. 2000. Genome relationships: the grass model in current research. *Plant Cell,* 12, 637-646.

Doebley, J.F., Stec, A., Hubbard, L. 1997. The evolution of apical dominance in maize. *Nature,* 386, 485-488.

Emerson, R.A., Beadle, G.W., Fraser, A.C. 1935. A summary of linkage studies in maize. *Cornell Univ. Agric. Exp. Stn. Mem.,* 180, 1-83.

Endo, T.R., Gill, B.S. 1996. The deletion stocks of common wheat. *J. Hered.,* 87, 295-307.

Falque, M., Décousset, L. Dervins, D. Jacob, A.M. Joets, J. Martinant, J.P. Raffoux, X. Ribière, N. Ridel, C. Samson, D. Charcosset A., Murigneux, A. 2005. Linkage mapping of 1454 new maize candidate gene loci. *Genetics,* 170: 1957-1966.

Feuillet, C., Keller, B. 2002. Comparative genomics in the grass family: molecular characterization of grass genome structure and evolution. *Ann. Bot.*, 89, 3-10.

Fu, H., Dooner, H.K. 2002. Intraspecific violation of genetic colinearity and its implications in maize. *Proc. Natl. Acad. Sci.* USA, 99, 9573-9578.

Fu, Y., Emrich, S.J., Guo, L., Wen, T.J., Ashlock, D.A., Aluru, S., Schnable, P.S. 2005. Quality assessment of maize assembled genomic islands (MAGIs) and large-scale experimental verification of predicted genes. *Proc. Natl. Acad. Sci.* USA, 102, 12282–12287

Gill, B.S., Appels, R., Botha-Oberholster, A.-M., Buell, C.R., Bennetzen, J.L., Chalhoub, B., Chumley, F., Dvorak, J., Iwamaga, M., Keller, B., Li, W., McCombie, W.R., Ogihara, Y., Quetier, F., Sasaki, T. 2004. A workshop report on wheat genome sequencing: International Genome Research on Wheat Consortium. *Genetics*, 168, 1087-1096.

Gordon-Kamm, W.J., Spencer, T.M., Mangano, M.L., Adams, T.R., Daines, R.J., Start, W.G., O'Brien, J.V., Chambers, S.A., Adams, W.R., Willetts, N.G., Rice, T.B., MacKey, C.J., Krueger, R.W., Kausch, A.P., Lemaux, P.G. 1990. Transformation of maize cells and regeneration of fertile transgenic plants. *Plant Cell*, 2(7), 603-618.

Goyal, A., Bandopadhyay, R., Sourdille, P., Endo, T.R., Balyan, H., Gupta, P. 2005. Physical molecular map of wheat chromosomes. *Funct. Integr. Genomics*, 5, 260-263.

Groos, C., Gay, G., Perretant, M.r., Gervais, L., Bernard, M., Dedryver, F., Charmet, G. 2002. Study of the relationship between pre-harvest sprouting and grain color by quantitative trait loci analysis in a white x red grain bread wheat cross. *Theor. Appl. Genet.*, 104, 39-47.

Gupta, P.K., Balyan, H.S., Edwards, K.J., Isaac, P., Korzun, V., Roder, M., Gautier, M.F., Joudrier, P., Schlatter, A.R., Dubcovsky, J., De la Pena, R.C., Khairallah, M., Penner, G., Hayden, M.J., Sharp, P., Keller, B., Wang, R.C.C., Hardouin, J.P., Jack, P., Leroy, P. 2002. Genetic mapping of 66 new microsatellite (SSR) loci in bread wheat. *Theor. Appl. Genet.*, 105, 413-422.

Habash, D.Z., Bernard, S., Shondelmaier, J., Weyen, J., Quarrie, S.A. (2006). The genetic of nitrogen use in hexaploid wheat: N utilization, development and yield. Theor Appl Genet (on line).

Haberer, G., Young, S., Bharti, A.K., Gundlach, H., Raymond, C., Fuks, G., Butler, E., Wing, R.A., Rounsley, S., Birren, B., Nusbaum, C., Mayer, K.F., Messing, J. 2005. Structure and architecture of the maize genome. *Plant. Physiol.* 139, 1612-24.

Hanley, S., Edwards, D., Stevenson, D., Haines, S., Hegarty, M., Schuch, W., and Edwards, K.J. (2000). Identification of transposon-tagged genes by the random sequencing of Mutator-tagged DNA fragments from *Zea mays*. Plant J. 23, 557-566.

Hirel, B., Bertin, P., Quillere, I., Bourdoncle, W., Attagnant, C., Dellay, C., Gouy, A., Cadiou, S., Retailliau, C., Falque, M., Gallais, A. 2001. Towards a better understanding of the genetic and physiological basis for nitrogen use efficiency in maize. *Plant Physiol.*, 125(3), 1258-1270.

Hospital, F., Charcosset, A. 1997. Marker-assisted introgression of quantitative trait loci. *Genetics*, 147, 1469-1485.

International Rice Genome Sequencing Project. 2005 The map-based sequence of the rice genome. *Nature*, 436, 793-800.

Janda, J., Bartos, J., Safar, J., Kubalakova, M., Valarik, M., Cihalikova, J., Simkova, H., Caboche, M., Sourdille, P., Bernard, M., Chalhoub, B., Dolezel, J. (2004). Construction of a subgenomic BAC library specific for chromosome 1D, 4D and 6D of hexaploid wheat. Theor Appl Genet 109: 1337-1345.

Janda, J., Bartos, J., Safar, J., Kubalakova, M., Valarik, M., Cihalikova, J., Simkova, H., Caboche, M., Sourdille, P., Bernard, M., Chalhoub, B., Dolezel, J. 2004 Construction of a subgenomic BAC library specific for chromosomes 1D, 4D and 6D of hexaploid wheat. *Theor. Appl. Genet.*, 109, 1337-1345.

Jeanneau, M., Gerentes, D., Foueillassar, X., Zivy, M., Vidal, J., Toppan, A., Perez, P. 2002. Improvement of drought tolerance in transgenic maize: towards the functional validation of the Zm-Asr1 gene and increase of water use efficiency by over-expressing the photosynthetic C4-PEPC. *Biochimie*, 84, 1127-1135.

Johal, G.S., Briggs, S.P. 1992. Reductase activity encoded by the HM1 disease resistance gene in maize. *Science*, 258, 985-987.

Johnson, R. 2004. Marker-assisted selection. *Plant Breed. Rev.* 24(1): 293-309.

Kantety, R.V., La Rota, M., Matthews, D.E., Sorrells, M.E. 2002. Data mining for simple sequence repeats in expressed sequence tags from barley, maize, rice, sorghum and wheat. *Plant Mol. Biol.*, 48, 501-510.

Lagudah, E., Dubcovsky, J., Powell, W. 2001. Wheat genomics. *Plant Physiol. Biochem.*, 39, 335-344.

Lande, R., Thompson, R. 1990. Efficiency of marker assisted selection in the improvement of quantitative traits. *Genetics*, 124, 743-756.

Laperche, A., Brancourt-Hulmel, M., Heumez, E., Gardet, O., Le Gouis, J. 2005 Estimation of genetic parameters of a DH wheat population grown at different N stress levels characterized with probe genotypes. *Theor. Appl. Genet.*

Lee, J.M., Williams, M.E., Tingey, S.V., Rafalski, J.A. 2002. DNA array profiling of gene expression changes during maize embryo development. *Funct. Integr. Genomics*, 2, 13-27.

Lijavetzky, D., Muzzi, G., Wicker, T., Keller, B., Wing, R., Dubcovsky, J. 1999. Construction and characterization of a bacterial artificial chromosome (BAC) library for the A genome of wheat. *Genome*, 42, 1176-1182.

Magnard, J.L., Heckel, T., Massonneau, A., Lassagne, H., Perez, P., Dumas, C., Rogowsky, P. 2003. ZmPRPL35-1 encodes a plastid ribosomal protein required for suspensor morphogenesis in maize embryos. 45th Annual Maize Genetics Conference, 13-16 March, Lake Geneva, Talk Abstract 9. Accessible at Maize DB (http://www.agron.missouri.edu).

Martin, A., Lee, J., Kichey, T., Gerentes, D., Zivy, M.,Tatou, C., Balliau, T., Valot, B., Davanture, M., Dubois, F., Tercé-Laforgue, T., Coque, M., Gallais, A., Gonzalez-Moro, M.B., Bethencourt, L., Quilleré, I., Habash, D.Z., Lea, P.J., Charcosset, A., Perez, P., Murigneux, A., Sakakibara, H., Edwards, K.J., Hirel, B. 2006 - Two cytosolic glutamine synthetase isoforms of maize (*Zea mays* L.) are specifically involved in the control of grain production. *The Plant Cell* (on line).

Matsuoka, Y., Vigouroux, Y., Goodman, M.M., Sanchez, G.J., Buckler, E., Doebley, J. 2002. A single domestication for maize shown by multilocus microsatellite genotyping. *Proc. Natl. Acad. Sci.* USA, 99(9), 6080-6084.

May, B.P., Liu, H., Vollbrecht, E., Senior, L, Rabinowicz, P.D., Roh, D., Pan, X., Stein, L., Freeling, M., Alexander, D., and Martienssen, R. (2003) Maize-targeted mutagenesis: A knockout resource for maize. Proc. Natl. Acad. Sci. 100, 11541–11546.

McIntosh, R.A., Hart, G.E., Devos, K.M., Gale, M.D., Rogers, W.J. 1998. A catalogue of gene symbols for wheat. *Proc. Int. Wheat Genet. Symp.*, Vol. 5. wheat.pw.usda.gov/ggpages/wgc/2003upd.html

Meeley, R.B., Briggs, S.P. 1995. Exploring gene function with reverse genetics technology. 37th Annual Maize Genetics Conference, Asilomar, California. Accessible at Maize DB http://www.agron.missouri.edu

Melchiorre, M.N., Lascano, H.R., Tripp, V.S. 2002. Transgenic wheat plants resistant to herbicide Basta obtained by microprojectile bombardment. *Biocell*, 26, 217-223.

Melchinger, A.E. 1999. Genetic diversity and heterosis. In: *The Genetics and Exploitation of Heterosis in Crops*. Coors, J.G., Pandey, S., eds. CJGPS, American Society of Agronomy, Madison, Wisconsin, 99-118.

Meyers, B.C., Tingey, S.V., Morgante, M. 2001. Abundance, distribution, and transcriptional activity of repetitive elements in the maize genome. *Genome Res*, 11, 1660-1676.

Moreau, L., Charcosset, A., Gallais, A. 2004. Experimental evaluation of several cycles of marker assisted selection in maize. *Euphytica*, 137:111-118.

Morgante, M., Brunner, S., Pea, G., Fengler, K., Zuccolo, A., Rafalski, A. 2005. Gene duplication and exon shuffling by helitron-like transposons generate intraspecies diversity in maize. *Nat. Genet.* 37, 997-1002.

Moullet, O., Zhang, H.B., Lagudah, E.S. 1999. Construction and characterization of a large DNA insert library from the D genome of wheat. *Theor. Appl. Genet.*, 99, 305-313.

Nelson, J.C., Van Deynze, A.E., Autrique, E., Sorrells, M.E., Lu, Y.H., Merlino, M., Atkinson, M., Leroy, P. 1995. Molecular mapping in bread wheat. Homeologous group-2. *Genome*, 38, 516-524.

Nelson, W.M., Bharti, A.K., Butler, E., Wei, F., Fuks, G., Kim, H., Wing, R.A., Messing, J., Soderlund, C. 2005. Whole-genome validation of high-information-content fingerprinting. *Plant. Physiol.* 139, 27-38.

Nicot, N., Chiquet, V., Gandon, B., Amilhat, L., Legeai, L., Leroy, P., Bernard M., Sourdille, P. 2004. Study of simple sequence repeat (SSR) markers from wheat expressed sequence tags (ESTs). *Theor. Appl. Genet.*, 109,800-805.

Okubara, P.A., Blechl, A.E., McCormick, S.P., Alexander, N.J., Dill-Macky, R., Hohn, T.M. 2002. Engineering deoxynivalenol metabolism in wheat through the expression of a fungal trichothecene acetyltransferase gene. *Theor. Appl. Genet.*, 106, 74-83.

Prioul, J.L., Quarrie, S., Causse, M., de Vienne, D. 1997. Dissecting complex physiological functions into elementary components through the use of molecular quantitative genetics. *J. Exp. Bot.*, 48, 1151-1163.

Qi, L.L., Echalier, B., Chao, S., Lazo, G.R. *et al.* 2004 A chromosome bin map of 16,000 expressed sequence tag loci and distribution of genes among the three genomes of polyploid wheat. *Genetics*, 168, 701-712.

Qi, L., Echalier, B., Friebe, B., Gill, B.S. 2003. Molecular characterization of a set of wheat deletion stocks for use in chromosome bin mapping of ESTs. *Funct. Integr. Genomics*, 3, 39-55.

Rebourg, C., Chastanet, M., Gouesnard, B., Welcker, C., Dubreuil, P., Charcosset, A. 2003. Maize introduction into Europe: the history reviewed in the light of molecular data. *Theor. Appl. Genet.*, 106, 895-903.

Safar, J., Bartos, J., Janda, J., Bellec, A., Kubalakova, M., Valarik, M., Pateyron, S., Weiserova, J., Tuskova, R., Cihalikova, J., Vrana, J., Simkova, H., Faivre-Rampant, P., Sourdille, P., Caboche, M., Bernard, M., Dolezel, J., Chalhoub, B. 2004. Dissecting large and complex genomes: flow sorting and BAC cloning of individual chromosomes from bread wheat. *Plant J.*, 39, 960-968.

Sakai, H., Nagasawa, N., Heppard, E., Maxwell, C., Lynch, J., Meeley, R., Williams, M., Cahoon, R. 2002. Giant embryo: a genomic approach to dissect genetic circuits controlling embryo/endosperm size. 44th Annual Maize Genetics Conference, Kissimee, Florida, Talk Abstract 9. Accessible at Maize DB http://www.agron.missouri.edu

Samson, D., Legeai, F., Karsenty, E., Reboux, S., Veyrieras, J.B., Just, J., Barillot, E. 2003. GenoPlant-Info (GPI): a collection of databases and bioinformatics resources for plant genomics. *Nucl. Acids Res.*, 31, 179-182.

San Miguel, P., Tikhonov, A., Jin, Y.K., Motchoulskaia, N. Zakharov, D., Melake-Berhan, A., Springer, P.S., Edwards, K.J., Lee, M., Avramova, Z., Bennetzen, J.L. 1996. Nested retrotransposons in the intergenic regions of the maize genome. *Science*, 274, 765-768.

Scofield, S.R., Huang, L., Brandt, A.S., Gill, B.S. 2005. Development of a virus-induced gene-silencing system for hexaploid wheat and its use in functionnal analysis of the Lr-21 mediated leaf rust resistance pathway. *Plant Physiol.*, 138, 2165-2173.

Sears, E.R. 1954. The aneuploids of common wheat. *Univ. Missouri Res. Bull.*, 572, 1-58.

Shen, B., Wang, D.M., McIntyre, C.L., Liu, C.J. 2005. A 'Chinese Spring' wheat (*Triticum aestivum* L.) bacterial artificial chromosome library and its use in the isolation of SSR markers for targeted genome regions. *Theor. Appl. Genet.* DOI 10.1007/s00122-005-0077-1.

Shewry, P.R. , Jones, H.D. 2005. Transgenic wheat: where do we stand after the first 12 years? *Ann. Appl. Biol.*, 147,1-14.

Slade, A.J., Fuerstenberg, S.I., Loeffler, D., Steine, M.N., Facciotti, D. 2004 A reverse genetic, nontransgenic approach to wheat crop improvement by TILLING. *Nature Biotechnology*, 23, 75-81.

Somers, D., Isaac, P., Edwards, K. 2004. A high-density microsatellite consensus map for bread wheat (Triticum aestivum L.). *Theor. Appl. Genet.*, 109, 1105-1114.

Song, Q.J., Shi, J.R., Singh, S., Fickus, E.W., Costa, J.M., Lewis, J., Gill, B.S., Ward, R., Cregan, P.B. 2005. Development and mapping of microsatellite (SSR) markers in wheat. *Theor. Appl. Genet.*, 110, 550-560.

Sourdille, P., Cadalen, T., Guyomarch, H., Snape, J.W., Peretant, M.R., Charmet, G., Boeuf, C., Bernard, S., Bernard, M. 2003. An update of the Courtot x Chinese Spring intervarietal marker linkage map for the QTL detection of agronomic traits in wheat. *Theor. Appl. Genet.*, 106, 530-538.

Sourdille, P., Singh, S., Cadalen, T., Brown-Guerida, G.L., Gay, G., Qi, L., Gill, B.S., Dufour, P., Murigneux, A., Bernard, M. 2004. Microsatellite-based deletion bin system for the establishment of genetic-physical map relationships in wheat (Triticum aestivum L.). *Funct Integr. Genomics*, 4, 12-25.

Thiellement, H., Bahrman, N., Damerval, C., Plomion, A., Rossignol, M., Santoni, V., de Vienne, D., Zivy, M. 1999. Proteomics for genetic and physiological studies in plants. *Electrophoresis*, 20, 2013-2026.

de Vienne, D., Causse, M. 1998. La cartographie et la caracterisation des locus controlant la variation des caracteres quantitatifs. In: *Les Marqueurs Moleculaires en Genetique et Biotechnologies Vegetales*. de Vienne, D., ed. INRA, Paris, pp. 204.

Ware, D.H., Jaiswal, P., Ni, J., Yap, I.V., Pan, X., Clark, K.Y., Teytelman, L., Schmidt, S.C., Zhao, W., Chang, K., Cartinhour, S., Stein, L.D., McCouch, S.R. 2003. Gramene, a tool for grass genomics. *Plant Physiol.*, 130, 1606-1613.

Walbot, V. 2003. Maize gene discovery project—update. 45th Annual Maize Genetics Conference, 13-16 March, Lake Geneva, Abstracts Poster 35. Accessible at Maize DB http: //www.agron.missouri.edu

Wan, R.L., Stec, A., Hey, J., Lukens, L., Doebley, J. 1999. The limits of selection during maize domestication. *Nature*, 398, 236-239.

Whithelaw, C., Quackenbush, J., Schubert, K., Beachy, R., Lakey, N., Bennetzen, J. 2003. Consortium for maize genomics—approach evaluation for targeted sequencing of maize. 45th Annual Maize Genetics Conference, 13-16 March, Lake Geneva, Talk Abstract 6. Accessible at Maize DB (http://www.agron.missouri.edu

Yan, L., Loukoianov, A., Blechl, A., Tranquilli, G., Ramakrishna, W., San Miguel, P., Bennetzen, J.L., Echenique, V., Dubcovsky, J. 2004. The wheat VRN2 gene is a flowering repressor down regulated by vernalization. *Science*, 12, 1640-1644.

Genomics and Technology

31

CHAPTER

From Genomics to Self-assembly

V. Planchot, S.G. Ring, A.J. MacDougall, M. Axelos, Y. Popineau, P. Colonna

INTRODUCTION

Tissues and plant organs result from the assembly of hundreds of molecules connected with low-energy bonds. These constructions lead to the formation of units which are heterogeneous on several scales from nm to cm and that have important biological functions. These units often persist through transformations that determine the nutrient and sensory characteristics of the final products (e.g., food) or the mechanical properties of fibres manufactured from the plant material.

The second characteristic of these biological units is that they are basically made up of biopolymers. The physics of polymers offers the theoretic framework through which the observed structures can be understood. Various non-covalent interactions (Table 31.1) can be implicated. Their characteristics, which differ considerably (Daoud and Williams, 1999), are the source of a wide variety of structures.

Thereby biopolymer structures *in planta* result simultaneously from a complex spatio-temporal genetic programming and from the application of physicochemical laws that determine the formation of plant cell walls, teguments, protein bodies, or starch granules, in grains and seeds.

Considering these plants organs as sinks able to accumulate molecules has been the dominant paradigm in biology for a long time.

At present, two experimental elements have radically modified this situation:

— the explosion of proteomics, which will lead to a complete inventory and chronological approach to cellular events;

— the development of methods (surface plasmonic resonance, calorimetry, double hybrids, proteins chips) that make it possible to visualize functional (and no longer solely catalytic) interactions between macromolecules.

Table 31.1 General characteristics of molecular interactions between two biopolymers similar in solution

Type	*Sign*	*Intensity*	*Distance of action*	*Influence of pH*	*Influence of ionic force*	*Influence of temperature*
Hydrophobic	Attractive	High	Long	No	Yes	Increase
Electrostatic	Repulsive or attractive. Energy decreasing exponentially with distance between objects.	Variable as function of pH and ionic force	Variable as function of pH and ionic force	Yes	Decrease	Increase
Hydrogen	Attractive	Low	Short	No	No	Decrease
Hydration	Repulsive	High	Short	No	No	Decrease
van der Waals	Strongly attractive at short length, then repulsive on contact (Born energy)	Low	Short	No	No	Without object
Interaction of excluded volume	Resulting between the repulsive potential between two links and the attractive or repulsive potential between two links and the solvent. If the link prefers to surround itself with solvent then the interaction between two links is repulsive at long distance; if the interaction between two links is attractive at long distance there is either precipitation–insolubility (very strong attraction) or disappearance of the excluded volume (low attraction)	High	Short	No	No	Without object
SS links	Attractive	Very high	Without object	Without object	Without object	No
Coordination of metallic ions	Attractive	High	Without object	No	No	No

There is thus a possibility of systematically comprehending the biological functions of macromolecules and then modelling them. The resulting product of either the direct expression of a gene (protein) or various anabolic pathways involving several enzymes must be compared to populations of molecules constructed according to the same structural pattern, but with significant variations in the detailed structure and the distribution of molecular mass.

Structural biology is used to understand and model the structure, assembly, and interactions of biological molecules and of macromolecules in solution, at interfaces, and in condensed systems, *in vivo* as well as *in vitro*. It aims at a better understanding of their biological functions as well as their spatial and temporal organizations over the course of development. It leads to a better understanding of the chronology of physicochemical events implicated in the interactions of these constituents *in situ*, as well as *in vitro*. The appearance of structures and functions over the course of development *in vivo* has become open to study in a mechanistic vision.

The general objective of this discipline is the establishment of structure-function relationships in the organizations on the supramolecular scale *in vivo*. The conceptual approach is based on the existence of biological functions optimized by the adaptive advantages conferred on organisms in which they are present. This supramolecular approach leads to a supposition of rational adaptation of molecular groups to their functions.

For the purposes of validation, the critical experiment is association of the different chemical partners to find *in vitro* the activity and function identified *in vivo*, or even to express a new function. This approach differs from that conventionally taken for the synthesis of molecules, where the research work relies on the purification of homogeneous chemical families: the properties can thus be studied, which then opens up access to the structural study of the mechanism.

This field is situated at the intersection of chemistry, physics and biology. These three disciplines concur in the elaboration of molecular models used to comprehend the chemical bases of major biological functions. Success depends on multidisciplinary groups. A continuum needs to be established between biosynthesis, catabolism and the comprehension of these complex molecular systems' behaviour.

THE DIFFERENT SCALES OF ORGANIZATION

The biological objects mentioned earlier are polymolecular constructions, the structure of which can be described at different scales.

From a monomer (0.5-1 nm amino acid or osidic unit), the formation of covalent bonds will lead to polymers, the mass of which may reach several hundreds of thousands for the largest proteins and some hundreds of millions for polysaccharides and lignin.

The dimensions in solution, though illustrative (R_g, radius of gyration : globular protein 2-10 nm; amylose, 20-50 nm; amylopectin, 300-400 nm), are to be considered with caution because the peculiarity of these constructions is that they are always based on organized, condensed, and even crystalline states. The dimensions of the biological object *in vivo* at the condensed state correspond best to the object of this chapter: protein bodies 1-2 μm, ribosome 30 nm, starch granule 1-100 μm, cellulose fibers up to 1 cm (Fig. 31.1).

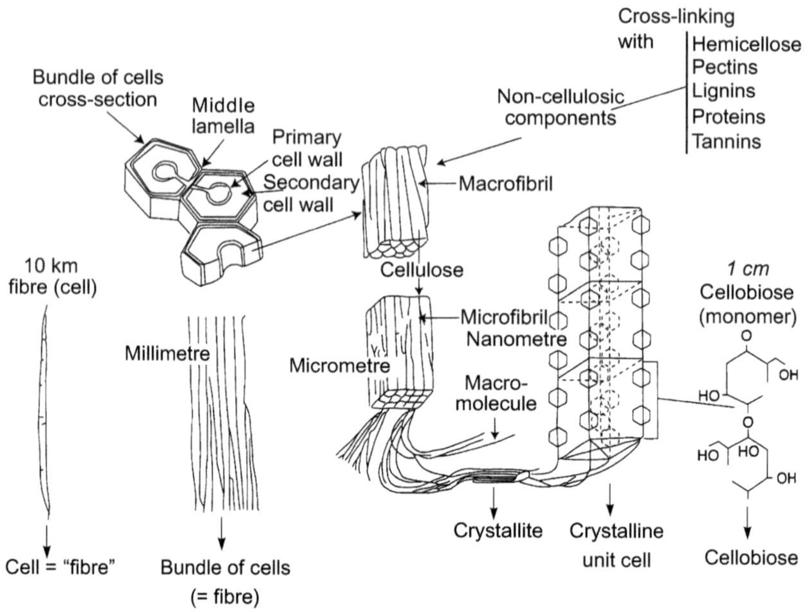

Fig. 31.1 Levels of increasing complexity, from glucosyl monomer (at right) to cellulose fibre (at left) (Monties, after Esau, 1977)

The size of the base element of the construction determines the size of simple constructions. Thus, in tobacco mosaic virus, the 2,130 globular proteins form a serrated helix of 17 proteins per turn, enclosing the RNA strand. The width of the virus is determined by the forces and shape of the elementary protein.

These objects belong to the class of colloids, that is to say, they are between 1 nm and 1 μm in size. One consequence is the considerable increase of the area-volume ratio of these colloids when their size diminishes.

Starch provides a complete example of a structure which can be studied at different scales. The starch granule is composed of two major polysaccharides, amylose and amylopectin, glucans linked in α-(1→4) and branched in α-(1→6), associated with proteins which are very sparsely represented but eminently specific. Amylose is an essentially linear polymer constituted of D-glucosyl units linked by bonds of type α-(1→4). A few bonds of type α-(1→6) are present to the extent that the molecular mass is high. Amylopectin, the major polysaccharide in the granule, forms the fraction that simultaneously is the most branched (4-5% of branches) and has the highest molecular mass (degree of polymerization or DP 10^7 to 10^8). It is also an organized macromolecule, with a tree structure from chains of DP 6 to 40 associated in semi-crystalline clusters. The clusters are separated from one another by longer chains (Fig. 31.2). In plants, the crystalline organization (Fig. 31.3) may correspond to two allomorphs (A and B).

This storage polymer may be compared with glycogen, with an equivalent structure in the animal world. However, the degree of ramification of the latter is very high, with one α-(1→6) bond for every 10 residues in α-(1→4). It is present in the form of low dense particles with 10 to 40 nm diameter. The molecular structure of glycogen allows linear links to adopt the coil-helix transition. However, the random positioning of α-

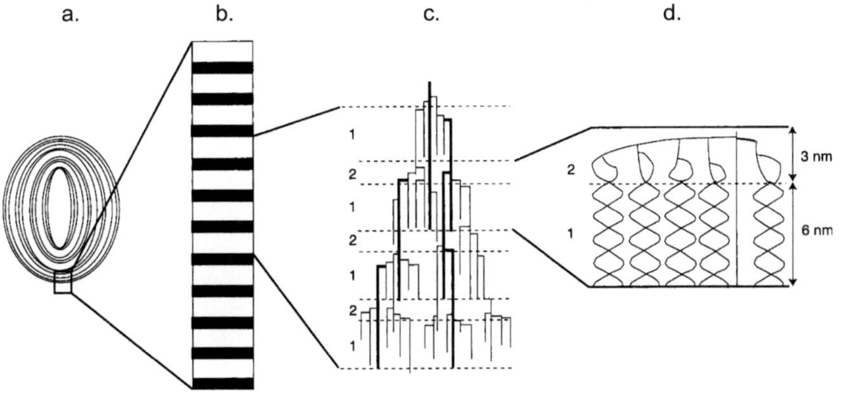

Fig. 31.2 Successive enlargements of cross-section of a starch granule. a: alternation of amorphous and semi-crystalline growth rings. b: enlargement of a semi-crystalline region in which amorphous lamellae (in white) and crystalline lamellae (in black) alternate regularly every 9 nm. c: succession of amorphous and crystalline lamellae related to the primary "cluster" structure proposed for amylopectin. d: crystalline lamellae with double helices. The position of amylose within this structure remains uncertain is thus not represented.

(1>6) bonds and the short length of these links do not allow them to form stable crystals. In consequence, the glycogen remains a soluble macromolecule, very rapidly metabolized, unlike starch, which undergoes a slow form of mobilization during germination.

The importance of the glycogen's molecular structure is illustrated in Andersen's disease. In this pathology, an enzyme which causes branching during biosynthesis is defective, leading to very long external ramifications capable of crystallizing. The consequence is hepatic cirrhosis, leading to death within two years.

FOLDING OF PROTEINS

The first level where the laws of physicochemistry can be used to predict tertiary structures of proteins is subject to the informatics difficulty of calculating conformations *ab initio* because of the weight of calculations. The rules that regulate the foldling of proteins are not known precisely, even though lateral polar chains and non-polar chains are certainly important in this domain.

To date, the 3D structure of each protein cannot be predicted simply from knowledge of the primary structure. Moreover, proteins having different primary structures such as the α and β chains of haemoglobin and of myoglobin have homologous tertiary structures. There is no unequivocal correspondence between the primary structure and the tertiary structure.

However, because of the rapid publication of nucleic acid sequences, databases have been specializing in the establishment of links between the primary amino acid sequence of a protein and its biological function. Thus, the database Interpro (http://www.ebi.ac.uk/Interpol) allows annotation of a sequence and attributes it automatically to a biological function or a particular structural domain. The more

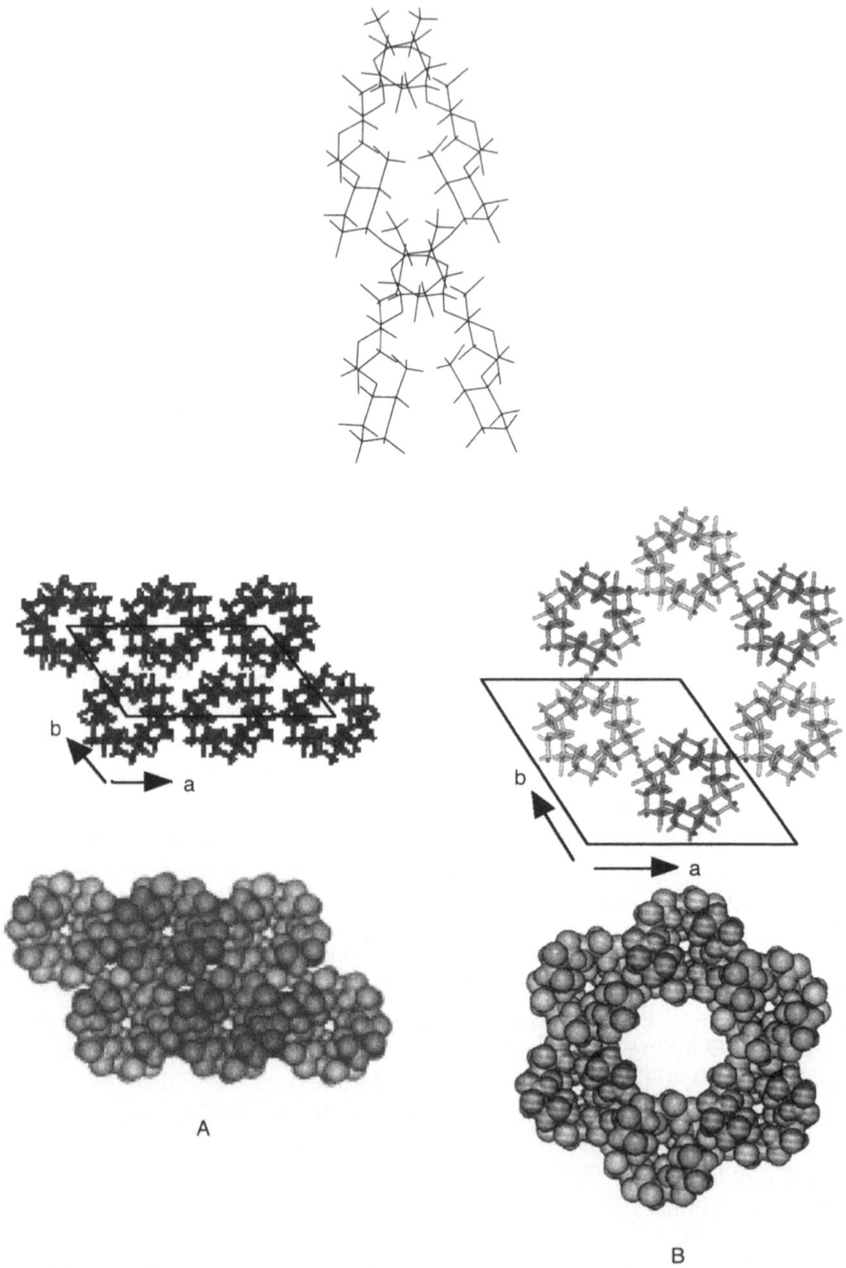

Fig. 31.3 Organization models of double helixes within crystalline structures of types A and B (a cross-section of Fig. 31.2d). Buleon *et al.*, 1998. © Elsevier.

rigorous Swiss-Prot database (http://www.expasy.org.sprot) proposes a high quality manual annotation of protein sequences.

Two approaches open up once the primary structure is determined:

- X-ray diffraction or high-field NMR to make use of experimental analysis.
- Comparison with related proteins, in which the similarity of amino acids can be used to determine the structural homology. The prediction of β sheets and α helixes has become reliable. The basic hypothesis was subsequently that the folding of proteins is better conserved than their sequences. Functional classifications about the enzymes involved in the formation or removal of glycosidic bonds (Cazy, http://afmb.cnrs.mrs.fr/CAZY) (Henrissat and Bairoch, 1996; Campbell *et al.*, 1997) have been proposed. The prediction of the 3D structure of a protein from a known protein, based on the hypothesis that the rolls and curves of the polypeptide chain are conserved, is a highly time-consuming calculation and the result is not reliable. Comparison of annotations with experimental results turns up serious inconsistencies: for example, 27 glucosyl transferases are annotated as ORF xylanases (Henrissat, 2003).

A set of molecules may interfere to accelerate the filling of polypeptides (disulfide isomerase for rearrangements of disulfur bridges, proline isomerase for peptide linkages with proline). Chaperone proteins prevent aggregation of newly synthesized polypeptides and reduce incorrect interactions within the molecule (Agashe and Hartl, 2000; Feldman and Frydman, 2000).

This question of protein folding is essential for the validation of studies form recombinant proteins.

TRANSITIONS INVOLVING POLYMERS

Molecules require thermal agitation in order to come into contact with each other. It must be emphasized that molecules that are not imbedded in molecular assemblies must be considered to be mobile and therefore subject to Brownian motion on large scales and to translational, rotational diffusion on a more local scale as well as to atomic vibrations. Thus, a globular protein rotates about 100,000 times a second around its center of gravity. This molecular mobility is essential to permit transitions involving polymers. Di Marzo (1999) reviewed the phase transitions involving polymers and leading to self-assembly phenomena.

Among the 10 transitions reported, eight are of major importance for biological structures.

Phase Separation in Polymer Mixtures

Unlike small solutes, polymers often are thermodynamically incompatible in solution, which leads to the appearance of a phase separation also called demixing. From the variations of average free energy of mixing as a function of the composition, three regions can be distinguished, depending on the quality of the solvent and the size of polymers. For a given temperature, upon increasing the polymer fraction in the mixture, one can thus obtain homogeneous solutions with a single stable phase (miscibility region), then a region in which two metastable phases coexist – in this case the separation is controlled kinetically by the nucleation and the growth of nuclei, leading to the dispersion of one phase in the other – and finally a zone where solutions are unstable and demix spontaneously into two phases. This last phase separation is known as spinodal decomposition. By varying the solvent conditions, the interaction parameters can be changed and influence in turns the characteristic structures of each type of phase separation. It must be emphasized that the nucleation and the growth or spinodal decomposition are two distinct phase separation mechanisms with very different kinetics, even though both phenomena are based on the diffusion of the constituents.

Coil-helix Transition

The coil form is a disordered structure present when several local conformations are permitted. In certain polymers the stabilization of the minimal energy structure is the helical structure and this is conserved in solution. The latter may be a single strand (α helix of proteins), double strand (DNA), or triple strand (collagen). In the last case, the polypeptide chain is outstandingly regular, with one proline for every three residues, which facilitates the formation of the triple helix of collagen. However, the same overall helix conformation may hide different molecular mechanisms of structuration. While α helixes of proteins are all stabilized by hydrogen bonds, each of the three helixes of collagen is stabilized by the steric repulsion of pyrrolidone cycles of proline and hydroxyproline residues. Then the three strands are linked by hydrogen bonds, between the NH of glycine residues and the CO groups of residues of other strands.

Adsorption of an Isolated Polymer

Solid-solid interfaces (plant cell wall), solid-liquid interfaces (lignified plant cell wall, cutin), and liquid-liquid interfaces (membrane, oleosome) cover a large range of examples of adsorption of an isolated polymer.

2D Crystallization

Proteins may assemble into flat sheets, in which each protein is integrated into a hexagonal organization. Closed structures such as rings, tubes, and spheres rely on cooperative interactions between these proteins.

3D Crystallization

3D crystallization is a major event for starch and cellulose: the formation of limited zones (some nm thick) with chains may be part of several crystalline chains.

Collapse

Collapse, phenomenon with no homology to the precipitation of small solutes, is the consequence of a positive assessment of potentials of polymer-polymer interactions and polymer-solvent interactions due to highly attractive interactions between the segments of chains. Unlike the transition of gelation, collapse makes it possible to obtain dense entities (starch granules).

Transition of Liquid Crystalline Phases

Transition of liquid crystalline phases operates from a nematic phase (no positional order, presence of an orientational order: discs or rods are parallel to one another) towards cholesteric phases (variation of the director of the orientational order across the sample) or smectic phases (double presence of a positional and orientational order: oriented liquid plane), which behave very much like a solid. This transition involves oriented liquid and is found mostly with small amphiphilic molecules and lipopolysaccharides.

Sol-gel Transition

The sol-gel transition is a critical phenomenon that corresponds to a connectivity phase transition. This transition may be analysed in terms of percolation theory. Starting from monomers or polymers, the connection of elements leads to a network that becomes macroscopic at the transition. At the sol-gel transition, the mechanical properties diverge in a singular fashion. This analysis has a universal character in the sense that it is material independent and has been confirmed for polysaccharide gels such as pectins (Axelos and Kolb, 1990). The polymer networks do not have long range order in contrast with crystalline networks. Beyond these remarkable mechanical properties, gels also have the capacity to trap solvent and other components. Plant cell wall viewed as a pectin gel, can be assimilated with a regulatory system for ions and various micro-constituents.

In biological mechanisms, these different transitions are sometimes coupled, making it a more complex task to dismantle them into elementary mechanisms. They are also affected by the temperature and presence of small solutes (e.g., ions, oses and oligosides, emulsifiers).

Vitreous Transition

Vitreous transition (at temperature T_g) deserves to be included in this list. Even though it cannot organize a mixture of molecules, it has the property of stabilizing organized systems. This transition (Roos, 1995) involves only variations of molecular mobility, as a function of the temperature and the medium (water level, small solutes). It is a second-order transition (no discontinuity of primary derivatives of G), which has been evoked to explain that the germination capacity of seeds is function of kinetic regimes of physical or chemical transformations in the smaller domains at T_g or larger domains at T_g + 100°C on the one hand and in the zone $T_g - T_g$ + 100°C on the other.

In the first case, temperature-dependence is described by the Arrhenius equation:

$$k = k_0 \exp(-E_a/RT)$$

where k is the rate constant at temperature T, E_a is the reaction activation energy, and R is the constant of perfect gases.

In the second case, temperature-dependence is described by the Williams-Landel-Ferry law (WLF):

$$\log a_T = -C_1 (T - T_{ref}) / (C_2 + (T - T_{ref})$$

where C_1 and D_2 are coefficients depending on the system considered and a_T is the ratio between the phenomenon of relaxation at T and the reference temperature T_{ref}. The variation of rate is increased by a factor of *10, comparative to the Arrhenius regime, in this critical zone for the conservation of living organisms at low water contents.

In biological systems, tolerance of desiccation is seen as an aptitude to form a glassy coating (Walters, 1998). Most plant organs (fruits, vegetables) can be considered as solution of small solutes, chiefly oses and oligosaccharides, minerals and organic acids, in a matrix of biopolymers gelled by the presence of pectins, hemicellulose, proteins, and sometimes lignins. The presence of oses (glucose, fructose, etc.) and oligosaccharides (sucrose, trehalose, etc.) is an argument in favour of a biological function of these small solutes. Bernard-Lugo and Leopold (1995) concluded a role of the raffinose-saccharose pair in the conservation of maize grains. Nevertheless, the importance of this factor has not been confirmed since. On the other hand, the LEA (late embryogenesis abundant) protein could play a similar role in a glassy coating formed of oses, oligosaccharides, and LEA. The relevance of vitreous transition has been reinforced by the absence of change in kinetics in molecular mobility, at a critical temperature Tc, where the collapse of glassy state sugars is generally observed (Wolkers, 1998). Henry *et al.* (2003) determined the mobility of water by dielectric analysis seed by seed. They emphasized the variations as a putative link with seed viability (Henry *et al.*, 2003). The molecular base remains unexplained, no doubt because the scale of observation chosen, the grain or seed as a whole, is not adapted *in vivo* to the identification of local phenomena, because of its multiphasic compartmentalization.

For micro-organisms, dehydration is most often addressed from the angle of osmotic shock. The stabilization of cellular structures is led by the synthesis of osmoprotective solutes, particularly trehalose.

For these two biological examples, preservation of structures needs to be linked to maintenance of the organization of phases of biopolymers. The positioning of each subcellular organization in the diagram of states is a way for researchers to identify the molecular mobility, and in that way deduce the molecules that could be critical (Fig. 31.4).

SELF-ASSEMBLY IN THE ESTABLISHMENT OF MOLECULAR CONSTRUCTIONS

These forms of organization result from spontaneous auto-assembly of molecules when they are present after biosynthesis in solution or in a solid-liquid interface.

Self-assembly, in the general sense of the term, can be defined as the spontaneous formation of complex hierarchical structures from simple "bricks". It implies by definition multiple levels of energy and multiple degrees of liberty.

The superstructures formed by assembled molecules are based on weak covalent forces, as opposed to strong covalent forces. These weak forces authorize the mobility.

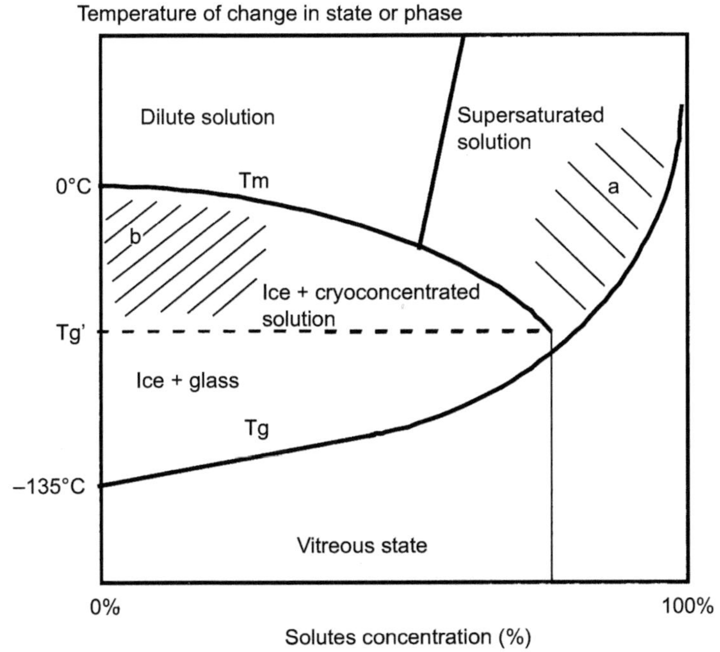

Fig. 31.4 States of an aqueous solution of small solutes. The vitreous transition curve (Tg) is added to the phase diagram of the water to obtain the diagram of states. Tm, melting temperature of pure water (0°C). Tm', melting temperature of water crystals in a condition of supersaturation, Tg' of the frozen solution concentrated at maximum. Zone *a* is the most susceptible of leading to phenomena of collapse, crystallization of solutes and of biopolymers. Zone *b* is most susceptible to growth of ice crystals.

Nevertheless, the cooperative nature of weak links leads to stable constructions (example of ionic forces) comparable to covalent bonds. The ionic bond takes up a particular place because of its larger shape than the other weaker bonds, and its modulation by pH and ionic force.

One of the best examples of simple supramolecular constructions occurs in the aqueous solutions of concentrated acetic acid: since the oxygen atom that is linked by a double bond to a carbon atom carries excessive electrons and since the hydrogen atom of the hydroxyl group (-OH) of a nearby molecule lacks electrons (the latter are attracted by the oxygen atom to which it is bound), the molecules of acetic acid pair by two hydrogen bonds.

The presence of chemical groups that could be of the electron donor or acceptor type, depending on the atoms that constitute these groups, leads to spontaneous associations of molecules, by the play of forces between donor groups and acceptor groups, which give rise to self-assembled structures.

The formation of organized molecular systems in living organisms, self-assembly, is thus based on the combination of molecules well suited for spontaneous assembly.

The solvent (or dispersion) medium largely determines the resulting form. The lipids, as well as the combinations of anionic or cationic surfactants, give rise to bilayers with an organization highly variable on the large scale, from vesicles to liquid crystals. Some bilayers, in the presence of anionic and cationic surfactants and in the absence of ionic force, may yield isometric regular forms. These constructions are stabilized by pores located at the tips of the isocahedron. Their dimensions are of about one μm, with a 10^{10} Da mass, which indicates that they are larger than protein constructions or viral capsids (Dubois *et al.*, 2001). It is important to know the micro-environment of assembly reactions.

The formation of amphiphilic helices leads to processes of self-assembly. An amphiphilic helix is determined by the regular presence of hydrophilic and hydrophobic amino acids along the sequence, when their periodicity corresponds exactly to the number of amino acids per turn of the helix. For an á helix, a typical amphiphilic sequence comprises three or four amino acids, in accordance with the 3.6 residues per turn of the α helix.

The polyproline II sequence has thus acquired a recognized importance in zones of interprotein contact with SH3 domains (Wu *et al.*, 1995) and WW domains (Staub and Rotin, 1996). The local conformation of PPII is an extended left helix, with a helix step of 0.93 nm and exactly three residues per turn. Because of this, one residue out of three is aligned along a generator along the helix. When the third residue is hydrophobic, the resulting PPII helix is amphiphilic.

The plant storage proteins are rich in proline. Zein-γ is a major protein of protein bodies, localized against the inner face of the membrane that delimits them. Zein-γ comprises three domains (Geli, 1994): a proline-rich N-terminal repeat sequence (γ-ZNPRD), a central repeat domain $(PX)_n$, and a cysteine-rich C-terminal domain. The γ-ZNPRD sequence contains a highly repetitive sequence VHLPPP. Herman and Larkins (1999) have suggested that the presence of this sequence leads to the formation of amphiphilic polyproline helixes.

In vivo observation is difficult. On the other hand, the self-assembly mechanism has been experimentally demonstrated *in vitro* by Kogan *et al.* (2001). From synthetic peptides mimicking the repeat sequence VHLPPP, they have shown that the octamer $(VHLPPP)_8$ adopts the conformation of an amphiphilic PPII helix, which self-assembles to give parallel fibrils above a graphite plate. Two pathways of self-assembly have been proposed: the first geometric by alignment of peptides above the graphite, the second by formation of cylindrical micelles that are deposited subsequently on the graphite plate. The major point of the second pathway is that it is initiated in solution independently of an adsorption surface. The latter is necessary only for visualization of the organization by microscopy of atomic force. The concentration needed for the micellization to appear is low, from 10^{-7} to 10^{-4}; it is perfectly compatible with biological mechanisms.

The formation of cylindrical micelles is made possible by the conformation of peptides in PPII helix $(VHLPPP)_n$ with $n = 1, 3$ and 8. The longitudinal growth of micelles occurs by superimposition of adjacent peptides.

The key structural element is the presence of the N-terminal sequence γ-ZNPRD, the self-assembly of which could result in a layer on the inner face of the maize protein bodies. This layer will then be stabilized by intramolecular SS covalent bonds from either the cysteine-rich C-terminal region or the cysteine residues adjacent to the γ-ZNPRD sequence (Kogan *et al.*, 2002).

CHARACTERISTICS OF SELF-ASSEMBLY

Self-assembly presents three characteristics that rank it in the category of nanotechnologies.

A Process Ascending in Hierarchical Steps

Its initial scale is the mesoscopic domain of 1-1,000 nm. The simplest examples are the formation of lipid bilayers, 3D folding of proteins, interactions between histones and DNA that play a critical role in compaction and regulation of the expression of genetic material, and formation of the triple helix of collagen.

Nevertheless, multi-protein structures or plant cell walls illustrate the capacity for self-assembly over larger scales.

Spontaneous Growth

The growth of construction is spontaneous because it is governed only by the equilibrium between the thermodynamics (the most stable structure) and kinetics (the fastest process). The competition between these two processes determines the final state. Any modification of the local physicochemical environment at the time of assembly leads to variations of structure. The reverse of this property is that the constructions can be destabilized by modifications of environmental conditions, having an impact on other constructions, in thermodynamic adequation with new environmental conditions.

Microfilaments illustrate this sensitivity of self-assemblages in micro-environment conditions. The intermediate filaments are one of the three constituents of the eukaryote cytoskeleton. They are made up of cores formed of two or three helical strands, rolled up, flanked by various sequences on either side of the core. The intermediate filaments are composed principally of proteins in á helixes that assemble in linear polymers of about 10 nm and several microns in length. These filaments of the cytoskeleton have a larger diameter than actin filaments, but they are more flexible with persistent lengths of a few hundreds of nm. In different classes of intermediate neurofilaments, the neurofilaments are distinguished by the presence of polypeptides, charged alternately positively and negatively, which are the surface of the filament and interact with other filaments and membranes. By diffusion of light and atomic force microscopy, and diffusion of neutrons, the effect of multivalent counter-ions has made it possible to detect the role of multivalent counter-ions in the formation and dissolution of complexes of neurofilaments (Shah *et al.*, 2000).

Stereospecificity

The established microstructures are determined by the fine structure of molecules: the bonds are based at the atomic level by the stereospecificity of each segment of molecules. Local modifications of the molecule structure induce modifications of assembling capacity. The preceding comparison between glycogen and starch is an illustration of stereochemistry.

Simultaneity between Synthesis and Assembly

The microstructures that form proceed without a component accumulation phase: there is a simultaneity of synthesis and assembly, and the limiting stage in terms of kinetics is always the synthetic phase.

For biomineral structures (mollusc shells, algal skeletons, vertebrate skeletons), a final characteristic has been recognized: the need for an assembly-initiating molecule, called the "template". These molecules, by associating with each other, create structures that support the development of inorganic structures with a base of calcium carbonate and phosphate.

Monod (1970) has already extended the concept of formation of stereospecific complex well beyond just enzymatic reactions. According to this author, the construction of a tissue or the differentiation of an organ, which are macroscopic phenomena, must be considered the integrated result of multiple microscopic interactions due to proteins and based on their properties of specific recognition (Monod, 1970).

The molecular forms determine the macroscopic forms. A set of molecular interactions corresponds only to a single macroscopic form. In fact, this approach, valid for lock-and-key mechanisms with enzymes, is more complex when the molecular relationships are not strictly exclusive or specific.

DEVELOPMENT OF NEW STRUCTURES *AB INITIO*

Once the laws of interaction between molecules and then between their assembled constructions are understood, new constructions can be considered. The amphiphilic models, their pairs of affinity, such as avidin-biotin or antigen-antibody, serve as molecular "bricks" for constructions which are self-assembled step by step.

Thus, the lipidic vesicles may serve as a base module for constructions. These vesicles are units that may interact, for selective adhesion, exchange of lipids, and membrane fusion between vesicles (Marchi-Artzner *et al.*, 2001).

Interaction between vesicles may be reinforced by the establishment of metallic complexes between lipidic ligands incorporated in the membrane and cations, which are linked to vesicles. In playing on the nature of the ion-ligand pair, the formation of multivesicular systems can be induced selectively.

We have described above biological objects, the construction of which comprises several steps. Nevertheless, these steps are simply described one after another without the proposal of a law of change in scale. There is still a requirement to have a technique such as the standardization group that will allow calculations to be carried out on an infinite number of scales simultaneously.

In the line of studies of Turing and Wiener, Wolfram (2002) proposed that all the steps are based on information patterns, the implementation of which achieves each object. Matter is put into form only by these information patterns. The author concludes that once these patterns are revealed, they will confirm that each detail of our universe follows rules that can be represented by a simple programme and that nothing can deflect this programme (Wolfram, 2002).

Derived from the theory of dynamic systems (particularly chaos theory), this approach of information patterns needs to be adapted at the biological level with cellular and molecular automatons. Their aim is to produce spontaneously, from simple rules, forms that can be observed in nature, some very complicated and highly organized and others appearing random.

If the correspondence between calculated and observed forms clearly illustrates the advantage of exploring the morphology of development, without explaining it, models and informatics simulations are certainly capable of contributing new hypotheses. The major advantage of this approach is that it helps to identify the appropriate scale and therefore to understand the formation of a structure, instead of relying entirely on molecular sequences that are very distant from the mechanisms of assembly.

This controversy revitalized by Wolfram has obliged us to distinguish clearly the description act from the comprehension act.

SYNERGY BETWEEN BIOLOGICAL AND PHYSICOCHEMICAL APPROACHES

Molecular interactions are described thoroughly by many theories. However, biological interactions differ from classic colloidal interactions in several ways: a biological interaction is generally a process involving a sequence of elementary interactions that occur in space and in time. Thus, a local interaction may have a consequence elsewhere (spatial dependence); some aspects of thermodynamic non-equilibrium, dependences on speeds or times of relaxation, may play an essential role (time-dependence).

Any biological construction is a combination of interactions over short and long distances, specific and non-specific, with interrelationships (Leckband *et al.*, 1994; Leckband and Israelachvili, 2003).

The examples of cell wall and starch granule illustrate simultaneously its limits and the characteristics of phenomena of self-assembly. They will be presented in the next two sections.

THE PECTIN NETWORK OF THE PLANT CELL WALL AND MIDDLE LAMELLA

Recent progress in understanding the structure and function of the plant cell wall provides another example of the value of taking a biophysical approach in studying biological systems.

The plant cell wall is a complex macromolecular assembly which performs a range of functions (Carpita and Gibeaut, 1993). Its behaviour is being examined from different perspectives. Plant biologists are examining the notion that the behaviour of the cell wall *in vivo* is controlled by direct enzyme action on the cell wall, or through the action of proteins such as the expansins (Cosgrove, 2000, 2001; Cosgrove *et al.*, 2002). An established approach is to examine the effect of mutation on the characteristics of the cell wall. Physical chemists are conscious that the dense structure of the cell wall could be an effective barrier preventing the access of enzymes and proteins. The environmental responsiveness of the cell wall to the apoplastic environment could play an important role in affecting the access of cell wall enzymes and proteins to their substrate.

During cell division, a cell plate is formed, which becomes the middle lamella, as the cell wall is deposited on either side. The middle lamella plays a major role in intercellular adhesion. In dicotyledenous plants it is formed from pectic polysaccharides which are complex polyelectrolytes. The primary cell wall consists of cellulose microfibrils dispersed within a hydrated matrix of cell wall polysaccharides which include the pectic polysaccharides, and the hemicellulose xyloglucan. Xyloglucan can bind to the cellulose microfibril and in some models of the cell wall can crosslink the cellulose fibrils to form a loose scaffold. The arrangement of the pectic polysaccharides within this scaffold is still the subject of debate.

Native pectins have an extremely complex substructure, which varies markedly with plant development and subcellular location. They are structurally complex and heterogeneous polyelectrolytes, (Schols and Voragen, 1994; Schols *et al.*, 1995; Ridley *et al.*, 2001) consisting of linear regions of (1 $\rightarrow$ 4)-α-D-galacturonosyl units and their methyl esters, interrupted in places by (1 $\rightarrow$ 2)-α-L-rhamnopyranosyl units. A fraction of these rhamnopyranosyl residues are branch points for neutral sugar side chains of (1 $\rightarrow$ 5)-α-L-arabinofuranosyl or (1 $\rightarrow$ 4)-β-D-galactopyranosyl residues. The rhamnose residues can be clustered together and the highly branched cluster, RG-1, may be isolated by treatment of cell walls with a polygalacturanase (Schols *et al.*, 1995). As well as having a heterogeneous structure, the distribution of pectic polysaccharides within the cell wall and middle lamella is not uniform (Willats *et al.*, 2001). For example it has been established that the pectic polysaccharides at the load-bearing corners of adjacent cells and in the middle lamella have a relatively low degree of esterification and are relatively depleted in rhamnogalacturonans RG-I and RG-II. Antibodies to RG-II localize it and its dimer in the primary cell wall next to the plasma membrane (Lord and Mollet, 2002).

From a physicochemical perspective, a major feature of the pectin matrix's structure of the plant cell wall and middle lamella is that it is formed from a polyelectrolyte. As a result of the range and strength of the charge interactions between polyelectrolytes and their counterions, the interactions often have a profound effect on the material properties of both synthetic polyelectrolyte and biopolyelectrolyte materials, including the extracellular matrix of connective tissues (Jin and Grodzinsky, 2001; Seog *et al.*, 2002; Dean *et al.*, 2003). For example, intermolecular repulsion forces between negatively charged glycosoaminoglycans are responsible for a large part of the resistance of cartilage to compression (Jin and Grodzinsky, 2001). More generally, the hydration of polyelectrolyte networks shows a strong dependence on charge density along the polymer chain (Rubinstein *et al.*, 1996; Manning and Ray, 1998). At low charge densities, the counterions required for electrical neutrality can generate large osmotic pressures which swell the network until balanced by the elastic restorative force provided by network crosslinking (Flory, 1953). At high charge densities, counterions can condense on the polymer backbone (Manning and Ray, 1998), reduce swelling and provoke a reduction in average dielectric constant which leads to further counterion condensation and eventually network collapse (Kramarenko *et al.*, 1997; Kokufuta *et al.*, 1998). The observed swelling behaviour of pectin networks is broadly comparable to that of polyelectrolyte networks (Tibbits *et al.*, 1998). The charged residues of pectins are often not uniformly distributed and as a consequence can contribute both to molecular association and also play a major role in determining network hydration. These roles are in opposition. Molecular association increases network stability and provides elastic resistance to network expansion, whereas hydration forces generated by ionized groups, and their counterions, drive network swelling. By controlling charge and its distribution along the pectin chain it is possible to produce a range of behaviour from highly crosslinked systems with minimal swelling, to lightly crosslinked systems with a high swelling potential. As pectin is a weak anionic polyelectrolyte, its charge is also sensitive to pH. The driving force for the polyelectrolyte swelling of the pectin network *in vivo* will depend on the magnitude of the charge repulsion between polymer chains; the charge density along the pectin backbone – whether it favours collapse or a Donnan-type swelling; and in the latter case the valency of the charge on counterion, with multivalent counterions producing a reduced osmotic effect. The forces opposing swelling *in vivo* include the osmotic stress provided by the cell contents and the crosslinking of the network.

The crosslinking of pectin networks *in vivo* may involve both non-covalent interactions and covalent crosslinks (Jarvis *et al.*, 2003). Of the non-covalent interactions, the association of blocks of charged galacturonic acid residues with inorganic cations, particularly Ca^{2+}, has been the most studied. The focus on the role of Ca^{2+} derives from studies on the calcium-mediated crosslinking of pectin gels. While a gel has a typical polymer concentration of a few percent there is a need to consider more fully the effect of changing pectin concentration to the > 30% w/w which is found in the intact cell wall (MacDougall *et al.*, 2001b). At these concentrations the inorganic cations K^{+} and Mg^{2+} can crosslink pectin networks (Zsivanovits *et al.*, 2004). Polyvalent organic cations, including fragments of the cell wall protein extensin (MacDougall *et al.*, 2001a), can also crosslink pectin networks. As well as functioning as crosslinkers, these polyvalent counterions also function to reduce Donnan-type swelling. The pectin environment *in vivo* is provided by the apoplastic sap which may contain a range of species which affect charge interactions involving pectin, including inorganic cations (K^{+}, Mg^{2+}, Ca^{2+}); organic cations (spermidine, putrescine); organic anions (citrate, malate); basic proteins (expansin, extensin). The pectin network can also be covalently crosslinked. Of the structurally characterised crosslinks, the borate diester crosslink of RG-II has a widespread distribution (Matsunaga *et al.*, 2004). Mutations which affect the glycosyl composition of RG-II lead to dwarfism in *Arabidopsis thaliana* (O'Neill *et al.*, 2001), a reduced tensile strength of the wall (Ryden *et al.*, 2003) and a reduced cell adhesion in callus cultures of *Nicotiana plumbaginifolia* (Iwai *et al.*, 2002). Boron deficiency results in cell wall swelling (Ishii *et al.*, 2001) and a large reduction in the amount of borate diester in the cell wall. The borate diester crosslink can therefore have an important impact on the mechanical properties and swelling behaviour of the cell wall. Its localisation suggests that it is important for the attachment of the cell wall to the plasma membrane and may help form a loose scaffold on which other pectic polysaccharides are assembled (Lord and Mollet, 2002). Another well-characterised crosslink, of a more limited distribution (Jarvis *et al.*, 2003) involves dimeric feruloyl esters between pectic polysaccharide side-chains (Levigne *et al.*, 2004).

It is therefore still an open question whether the properties of the cell wall are regulated through the control of cell wall synthesis and breakdown, or through the regulation of the environment of the cell wall.

In addition to the general relevance of the physical chemistry of polyelectrolyte networks to the behaviour of the polyelectrolyte network of the cell wall and the insights which are provided, there is growing research on the assembly of polyelectrolyte multilayers, or 'fuzzy nanoassemblies' from oppositely charged biopolymers (Decher, 1997). The multilayers may be formed on flat surfaces through alternate dipping of the surface in solutions of oppositely charged polymer. It is possible to form such structures with pectin. Multilayer assembly can be probed using a range of techniques including surface plasmon resonance (SPR) (Lahiri *et al.*, 1999); quartz crystal microbalance with dissipation monitoring (QCMD) (Picart *et al.*, 2002); fourier transform infrared spectroscopy with attenuated total reflection (FTIR-ATR); and atomic force microscopy (AFM). The various techniques give different insights into network assembly. SPR gives information on the masses deposited and the kinetics of binding; QCMD gives information on the hydrated mass which is deposited and the viscoelastic properties of the hydrated layer; FTIR-ATR (Xie and Granick, 2002) will give information on the mass and chemical characteristics of the layer including insight into how the local environment of the layer affects the ionization of charged groups; AFM can give information on the homogeneity of the deposited layer (Lavalle *et al.*, 2002). The combination of the various techniques is a potentially powerful way of examining the environmental responsiveness of pectin networks *in vitro*, and for exploring how the network responds to enzymic challenge.

Overall these biophysical studies on plant cell wall polymers demonstrate the tremendous importance of weak molecular forces in guiding the assembly of biological structures, and determining their functional role *in vivo.*

STARCH BIOSYNTHESIS

Amylose synthesis represents clearly the best known elongation of starch chains. A single enzyme, granule-bound starch synthase I (GBSSI), takes on this function and many mutants for this protein have been described in the literature. The absence of GBSSI suppresses amylose synthesis in the plant studied, without modifying the level of starch in it: mutants *waxy* of maize, *amf* of potato, *sta2* of *Chlamydomonas*, *lam* of pea, or mutants of type *waxy* of amaranthus, rice, barley, and wheat. The same behaviour has been observed in expressing an antisense RNA of GBSSII in potato.

The linkage of GBSSI with starch is a close interaction, non-covalent, which can be prevented by the presence of SDS and a strong reducing agent. Most of the biochemical characterizations of GBSSI result from *in vitro* incubation of native starch granules in the presence of ADP-glucose: the glucosyl residues are mainly incorporated in external chains of amylopectin, except when malto-oligosaccharides (MOS) are added in the incubation medium: then a massive synthesis of amylose takes place (Fig. 31.5). The role of MOS in the process of amylose synthesis remains unexplained. Do they serve as primers? Do they simply stimulate amylose synthesis at the cost of amylopectin synthesis? Are both mechanisms at work? This model, although elegant, comes up against an experimental fact: the starch granules are poorly porous, and oligosaccharides higher than maltohexase cannot penetrate them (Planchot *et al.*, 2000).

Amylopectin synthesis *in planta* is much more complex than amylose synthesis. It involves soluble starch-synthases (SS), branching enzymes (BE), debranching enzymes

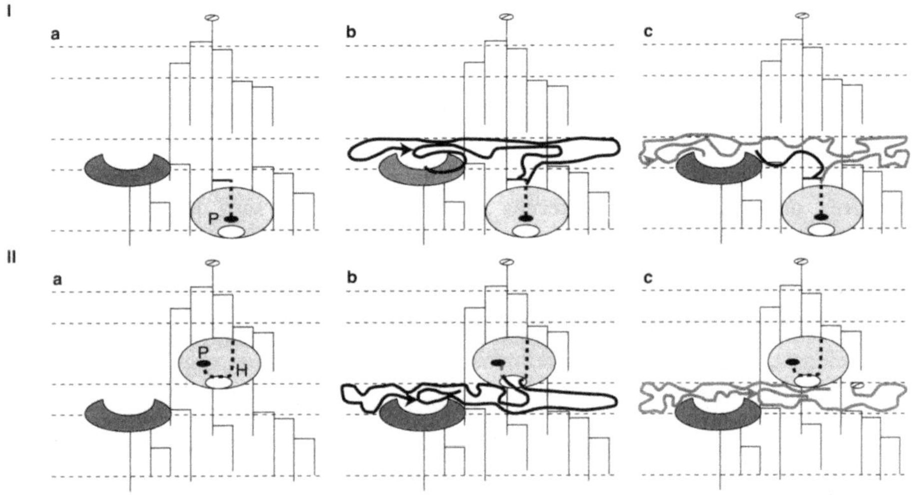

Fig. 32.5 Models of amylose biosynthesis by primers with amylopectin. The amylopectin matrix of origin is represented by fine vertical lines, whereas the external chains elongated by GBSSI are indicated by thick black lines. The newly synthesized amylose molecules correspond to thick gray lines. The white ovals indicate the reducing ends of amylopectin, the black ovals correspond to non-reducing ends. The gray ovals symbolize GBSSI and the crescent shapes represent a branching enzyme present inside the starch granule. The letter P indicates the position of the catalytic site of GBSSI and the letter H represents the position of a hypothetical site of hydrolysis. Parts I and II show two possible mechanisms of amylose synthesis. In part I, the release of the amylose molecule by cleavage is ensured by the branching enzyme or by another type of hydrolytic enzyme (not shown). Part II models the different steps of amylose synthesis as the GBSSI itself liberates the glucan newly synthesized by hydrolysis. In this last case, the presence of the branching enzyme in the starch granule explains the presence of branching points in the amylose.

(DBE) (Buleon *et al.*, 1998). The study of mutants, even if it proves very effective, sometimes comes up against limitations especially in the following cases: the first one is when the phenotypes become complex and the second one is when the functional redundancy linked to multiplication of the number of enzymatic isoforms able to catalyse the same reaction, but with slightly different specificities, no longer allows a particular function to be attributed to one isoform or another. At least five starch-synthases (elongation enzymes), three branching enzymes, four debranching enzymes, two α-1,4-glucanotransferases, two starch-phosphorylases, three α-amylases, or eight β-amylases are usually detected in plant genomes. The study of mutants allows an approach through the role of the enzyme in overall metabolism. However, the simple one gene-one protein relationship is unable to explain the complexity of the highly complicated biosynthesis pathways for a polysaccharide such as starch. In this context, mutant-derived starch cannot be used to obtain a direct relationship between a deficient enzymatic activity and a type of molecule synthesized or a type of structural organization. Therefore, other supplementary methodologies must be applied in order to be able to identify the biological role *in planta* of each of the proteins implicated in the biosynthesis process and to understand the physicochemical pathways leading to the granular and condensed form of the deposit that the starch granule represents.

Present knowledge is still based on a few mutants, allowing us to better understand the implication of one or another enzyme in processes of amylopectin synthesis (Fig. 31.6).

A first model was proposed by Ball *et al.* (1996) proposed an amylopectin synthesis model (Fig. 31.6a) where the implication of a debranching enzyme of the isoamylase type is absolutely required. The synthesis of an intermediate structure called "pre-amylopectin" would result from the joint action of the SS and the BE. The isoamylase being able to split α(1,6) supernumeraries, allowing so the implementation of the cluster's structure. When this activity is lacking, pre- amylopectin molecule would continue to be a substrate for SS and BE leading to production of phytoglycogène. Nevertheless, even if this model explains quite well the role that isoamylase plays in amylopectin synthesis, it does not explain how the other enzymes play and especially what their functions in the overall process are. Moreover, despite its shortcomings, this

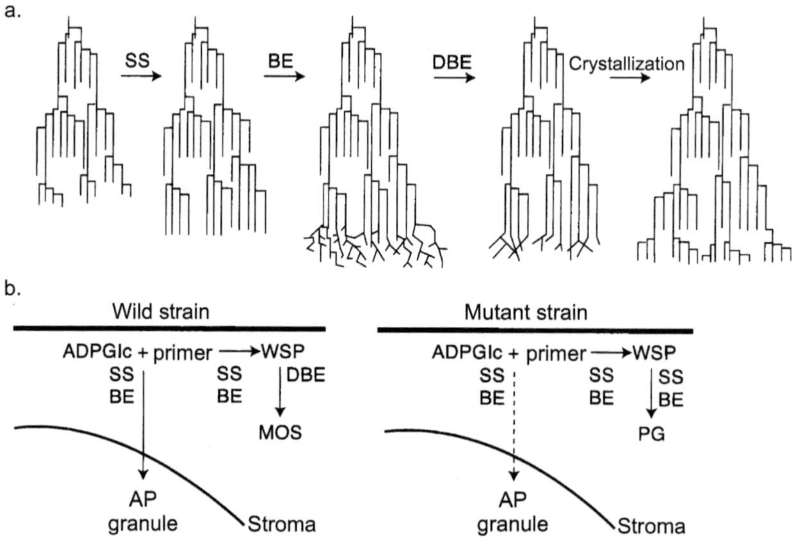

Fig. 31.6 The two models of amylopectin synthesis. a, model of glucane splicing (Ball *et al.*, 1996). The synthesis of the amylopectin cluster begins with elongation of external chains by the action of soluble starch-synthases (SS). When the chains become long enough, the branching enzymes (BE) introduce the α-1,6 linkages in the molecule. Finally, amylopectin undergoes a stage of splicing over the course of which the debranching enzymes of the isoamylase type (DBE) will suppress certain ramifications in α-1,6 incompatible with the crystallization of the molecule. b, model of elimination of water-soluble polysaccharides (WSP) (Zeeman *et al.*, 1998). The synthesis of amylopectin and that of WSP compete within the cell. In the wild cells, the WSP synthesized by the SS and the BE are specifically degraded by the DBE to limit the competition between the two reactions. In a mutant strain, lacking isoamylase activity, WSP are no longer degraded and can thus serve as primers for the synthesis of molecules of the phytoglycogen (PG) type, which occurs at the cost of amylopectin synthesis.

model has been confirmed by more recent works (Nielsen *et al.*, 2002, Wattebled *et al.*, 2005), contrary to the contradictory model proposed by Zeeman (Zeeman *et al.*, 1998, Fig. 31.6b). In Zeeman model, the originality lies in the specific fate of phytoglycogen, an unavoidable product of anabolism, which is completely degraded in normal genotypes.

The number of isoformes for each enzymatic activities involved in amylopectin synthesis makes particularly difficult the phenotype interpretation for obtained mutants. Even though study of the mutants remains a tool indispensable for understanding the synthesis of amylopectin (or of amylose). The studies of waxy mutants (without amylose) plants allowed to demonstrate that only the GBSSI is responsible for the synthesis of this fraction of starch. However no information about the mechanism of action of the GBSSI was obtained by the study of the mutants.

The analysis of pleitropic effects observed for some mutants led to the hypothesis of the coordination of the numerous activities of biosynthesis and degradation by the implication of these numerous proteins within multi-protein complexes (Colléoni *et al.*, 2003, Morell *et al.*, 2003). This hypothesis was confirmed with the analysis of heterologue expression of BE in yeast (Seo *et al.*, 2002). More recently Tetlow *et al.* (2004) indicated the important role that the phosphorylation of proteins could play on such complexes.

The influence of the genetic control is done by the modulation of the amylose content (GBSS), or the modulation of the fine structure of the amylopectin. Key parameters for a type kind of polymorphism are the clusters chains length (Hizukuri *et al.*, 2000) and cluster branching denisty (Bertoft *et al.*, 2000; Gérard *et al.*, 2000). The branch points density and the chains length cannot be resolved by studies in planta in which the polydispersity of molecular structures does not give enough experimental points to establish structure – function relationships.

The fine structure of cluster heads of amylopectin is the structural element very closely linked to the crystalline type (Gerard *et al.*, 2000). *In vitro* synthesis techniques (biomimetic systems) need to be developed in order to explain the determinism of the crystalline type of native starch.

Assays of *in vitro* biosynthesis systems from glycogen serving as acceptor of an elongation enzyme (amylosucrase of *Neisseria polysaccharea*) did not make it possible to obtain a collapse of gelled structures (Rolland-Sabate *et al.*, 2003), unlike the starch granule, which is a dense structure throughout its biosynthesis. It is a difficult task to break down the biosynthesis into elementary transition mechanisms.

PERSPECTIVES FOR MASTERY OF PLANT PRODUCTS

The aim of biological studies about organs of agronomic interest is to understand these process of organization in order to master the implementation of these complex assemblages for an adequate manipulation of the molecular recognition's phenomena.

Self-assembly behaviour brings us to reconsider the search for new properties when they are based on supramolecular organization. The idea is surely to combine molecules with specific properties, so that the structures are specific. From there, choosing molecules able to form systems corresponding to the desired properties is a logical step.

REFERENCES

Agashe, V.R., Hartl, F.U. 2000. Roles of molecular chaperones in cytoplasmic protein folding. *Cell Dev. Biol.*, 11, 15-25.

Azelos, M.A.V., Kolb, M. 1990. Crosslinked biopolymers: experimental evidence for scalar percolation theory. *Phys. Rev. Lett.*, 64, 1457-1460.

Ball, S., Guan, H.P., James, A., Myers, A., Keeling, P., Mouille, G., Buleon, A., Colonna, P., Preiss, J. 1996. From glycogen to amylopectin: a challenging model explaining the biogenesis of the plant starch granule. *Cell*, 86, 349-352.

Bernard-Lugo, I., Leopold, A.C. 1995. Seed stability during storage: raffinose content and seed glassy state. *Plant Physiol.*, 98, 1207-1210.

Buleon, A., Colonna, P., Planchot, V., Ball, S. 1998. Starch granules: structure and biosynthesis. *Int. J. Biol. Macromol.*, 23, 85-112.

Campbell, J.A., Davies, G.J., Bulone, V., Henrissat, B. 1997. A classification of nucleotide-diphospho-sugar glycosyltransferases based on amino acid sequence similarities. *Biochem. J.*, 316, 929-939.

Carpita, N.C. Gibeaut, D.M. 1993. Structural models of primary cell walls in flowering plants: consistency of molecular structure with the physical properties of the walls during growth. *Plant J.*, 3, 1-30.

Colléoni, C., Myers, A.M.., James, M.G. 2003 One- and two-dimensional Native PAGE activity of maize endosperm proteins reveal functional interactions between specific starch metabolizing enzymes. *J of Applied Glycoscience*, 50, 207-212.

Cosgrove, D.J. 2000. Loosening of plant cell walls by expansins. *Nature* 407, 321-326.

Cosgrove, D.J. 2001. Wall structure and wall loosening. A look backwards and forwards. *Plant Physiol.*, 125, 131-134.

Cosgrove, D.J., Li, L.C., Cho, H.T., Hoffmann-Benning, S., Moore, R.C., Blecker, D. 2002. The growing world of expansins. *Plant Cell Physiol.*, 43, 1436-1444.

Daoud, M., Williams, C.E. 1999. *Soft Matter Physics*. Springer Verlag, Berlin, 87.

Dean, D., Seog, J., Ortiz, C., Grodzinsky, A.J. 2003. Molecular-level theoretical model for electrostatic interactions within polyelectrolyte brushes: Applications to charged glycosaminoglycans. *Langmuir* 19, 5526-5539.

Decher, G. 1997. Fuzzy nanoassemblies: Toward layered polymeric multicomposites. *Science*, 277, 1232-1237.

Dubois, M., Deme, B., Gulik-Krzywicki, T., Dedieu, J.C., Vautrin, C., Desert, S., Perez, E., Zemb, T. 2001. Self-assembly of regular hollow icosahedra in salt-free cationic solutions. *Nature*, 411, 672-675.

Esau, K. 1977. *Anatomy of Seed Plants*. John Wiley and Sons Ltd., p. 550.

Feldman, D.E., Frydman, J. 2000. Protein folding *in vivo*: the importance of molecular chaperones. *Curr. Opin. Struct. Biol.*, 10, 26-33.

Flory, P.J. 1953. Principles of Polymer Chemistry. Cornell University Press.

Geli, M.L., Torrent, M., Ludevid, M.D. 1994. Two domains structural mediate two sequential events in gamma zein targeting: protein endoplasmic reticulum retention and protein body formation. *Plant Cell*, 6, 1191-1192.

Gerard, C., Planchot, V., Colonna, P., Bertoft, E. 2000. Relationships between branching density in amylopectin and crystalline structure of A- and B-type in maize mutant starches. *Carbohydrate Res*, 326, 130-144.

Henrissat, B. 2003. Carbohydrate acting enzymes in completely sequenced genomes: a step towards glycogenomics. 5th Carbohydrate Bioengineering Meeting, Groningen, April 6-9.

Henrissat, B., Bairoch, A. 1996. Updating the sequence based classification of glycosyl hydrolases. *Biochem. J.*, 316, 695-696.

Henry, F., Gaudillat, M., Costa, L.C., Lakkis, F. 2003. Free and/or bound water by dielectric measurements. *Food Chem.*, 82, 29-34.

Herman, E.M., Larkins, B.A. 1999. Protein storage bodies and vacuoles. *Plant Cell*, 11, 601-613.

Ishii T., Matsunaga T., Hayashi N., 2001. Formation of rhamnogalacturonan II-borate dimer in pectin determines cell wall thickness of pumpkin tissue. *Plant Physiol.*, 126, 1698-1705.

Iwai H., Masaoka N., Ishii T., Satoh S., 2002. A pectin glucuronyltransferase gene is essential for intercellular attachment in the plant meristem. *Proc. Nat. Acad. Sci.* USA, 99, 16319-16324.

Jarvis, M.C., Briggs, S.P.H., Knox, J.P. 2003. Intercellular adhesion and cell separation in plants. *Plant Cell Environ.*, 26, 977-989.

Jin, M.S., Grodzinsky, A.J. 2001. Effect of electrostatic interactions between glycosaminoglycans on the shear stiffness of cartilage: A molecular model and experiments. *Macromolecules*, 34, 8330-8339.

Kogan, M.J., Dalcol, I., Goristiza, P., Lopez-Iglesias, C., Pons, R., Pons, M., Sanz, F., Ludevid, D., Giralt, E. 2001. Self-assembly of the amphipathic helix (VHLPPP)8. A mechanism for zein protein body formation. *J. Mol. Biol.*, 312, 907-913.

Kogan, M.J., Dalcol, I., Goristiza, P., Lopez-Iglesias, C., Pons, R., Pons, M., Sanz, F., Giralt, E. 2002. Supramolecular properties of the proline-rich γ-zein N-terminal domain. *Biophys. J.*, 83, 1194-1204.

Kokufuta, E., Wang, B.L., Yoshida, R., Khokhlov A.R., Hirata, M. 1998. Volume phase transition of polyelectrolyte gels with different charge distributions. *Macromolecules*, 31, 6878-6884.

Kramarenko, E.Y., Khokhlov, A.R., Yoshikawa, K. 1997. Collapse of polyelectrolyte macromolecules revisited. *Macromolecules*, 30, 3383-3388.

Lahiri, J., Isaacs, L., Grzybowski, B., Carbeck, J.D., Whitesides, G.M. 1999. Biospecific binding of carbonic anhydrase to mixed SAMs presenting benzenesulfonamide ligands: A model system for studying lateral steric effects. *Langmuir*, 15, 7186-7198.

Lavalle, P., Gergely, C., Cuisinier, F.J.G., Decher, G., Schaaf, P., Voegel, J.C., Picart, C. 2002. Comparison of the structure of polyelectrolyte multilayer films exhibiting a linear and an exponential growth regime: An in situ atomic force microscopy study. *Macromolecules*, 35, 4458-4465.

Leckband, D., Israelachvili, J. 2001. Intermolecular forces in biology. *Q. Rev. Biophys.*, 34, 2, 105-267.

Leckband, D., Schmitt, F.J., Israelachvili, J.N., Knoll, W. 1994. Direct force measurements of specific and non specific protein interactions. *Biochemistry*, 33, 4611-4624.

Levigne, S.V., Ralet, M.C.J., Quemener, B.C., Pollet, B.N.L., Lapierre C., Thibault J-F.J., 2004. Isolation from sugar beet cell walls of Arabinan oligosaccharides esterified by two ferulic acid monomers. *Plant Physiol.*, 134, 1173-1180.

Lord, E.M., Mollet, J.C. 2002. Plant cell adhesion: A bioassay facilitates discovery of the first pectin biosynthetic gene. *Proc. Nat. Acad. Sci.* USA, 99, 15843-15845.

MacDougall, A.J., Brett, G.M., Morris, V.J., Rigby, N.M. Ridout, M.J., Ring, S.G. 2001a. The effect of peptide behaviour-pectin interactions on the gelation of a plant cell wall pectin. *Carbohydr. Res.*, 335, 115-126.

MacDougall, A.J., Rigby, N.M., Ryden, P., Tibbits, C.W., Ring, S.G. 2001b. Swelling behavior of the tomato cell wall network. *Biomacromolecules*, 2, 450-455.

Manning, G.S., Ray, J. 1998. Counterion condensation revisited. *J. Biomol. Structure Dynamics*, 16, 461-476.

Marchi-Artzner, V., Gulik-Krzywicki, T., Guedeau-Boudeville, M.A., Gosse, C., Sandersson, J.S., Dedieu, J.C., Lehn, J.M. 2001. Selective adhesion, lipid exchange and membrane fusion processes between vesicles of various sizes bearing complementary molecular recognition groups. *Chem. Phys. Chem.*, 2, 367-376.

Marzo, E.A. 1999. The ten classes of polymeric phase transitions: their use as models for self-assembly. *Prog. Polym. Sci.*, 24, 329-377.

Matsunaga, T., Ishii, T., Matsumoto, S., Higuchi, M., Darvill, A., Albersheim, P., O'Neill, M.A., 2004. Occurrence of the primary cell wall polysaccharide rhamnogalacturonan II in pteridophytes, lycophytes, and bryophytes. Implications for the evolution of vascular plants. *Plant Physiol.*, 134, 339-351.

Monod, J. 1970. *Le Hasard et la Necessite*. Seuil, Paris, 80, 118.

Morell, M.K., Kosar-Hashemi, B., Cmiel, M., Samuel, M.S., Chandler, P., Rahman, S., Buléon, A., Batey, I.L., Li, Z. 2003 Barley se6 mutants lack starch synthase Iia activity and contain a starch with novel properties. The Plant J. 34, 173-185.

Nielsen, H.T., Baunsgaard, L., Blennow, A. 2002. Intermediary glucan structures formed during starch granule biosynthesis are enriched in short side chains, a dynamic pulse labeling approach. *The J of Biological Chem.*, 277 (23), 20249-20255.

O'Neill, M.A., Eberhard, S., Albersheim, P., Darvill, A.G. 2001. Requirement of borate cross-linking of cell wall rhamnogalacturonan II for *Arabidopsis* growth. *Science*, 294, 846-849.

Picart, C., Mutterer, J., Richert, L., Luo, Y., Prestwich, G.D., Schaaf, P., Voegel, J.C., Lavalle, P. 2002. Molecular basis for the explanation of the exponential growth of polyelectrolyte multilayers. *Proc. Nat. Acad. Sci.* USA, 99, 12531-12535.

Planchot, V., Colonna, P., Roger, P. 2000. Suitability of starch granule porosity with regard to starch biosynthesis and susceptibility towards amylolysis. *Starch/Sarke*, 52, 333-339.

Ridley, B.L., O'Neill, M.A., Mohnen, D.A. 2001. Pectins: structure, biosynthesis, and oligogalacturonide- related signaling. *Phytochem.*, 57, 929-967.

Rolland-Sabate, A., Planchot, V., Veronese, G., Monsan, P., Colonna, P. 2004. Elongation and insolubilization of á-glucans by action of *Neisseria polysaccharea* amylosucrase. *J. Cereal Sci.* 40, 17-30.

Roos, Y. 1995. *Phase Transitions in Foods*. Academic Press Ltd., London.

Rubinstein, M., Colby, R.H., Dobrynin, A.V., Joanny, J-F. 1996. Elastic modulus and equilibrium swelling of polyelectrolyte gels. *Macromolecules*, 29:398-406.

Ryden, P., Sugimoto-Shirasu, K., Smith, A.C., Findlay, K., Reiter, W.D., McCann, M.C. 2003. Tensile properties of Arabidopsis cell walls depend on both a xyloglucan cross-linked microfibrillar network and rhamnogalacturonan II-borate complexes. *Plant Physiol.*, 132, 1033-1040.

Schols, H.A., Vierhuis, E., Bakx, E.J., Voragen, A.G.J. 1995. Different populations of pectic hairy regions occur in apple cell walls. *Carbohydr. Res.*, 275, 343-360.

Schols, H.A., Voragen, A.G.J. 1994. Occurrence of pectic hairy regions in various plant cell wall materials and their degradability by rhamnogalacturonase. *Carbohydr. Res.* 256: 83-95.

Schowalter, A.M. 1993. Structure and function of plant cell wall proteins. *Plant Cell*, 5, 9-23.

Seog, J., Dean, D., Plaas, A.H.K., Wong-Palms, S., Grodzinsky, A.J., Ortiz, C. 2002. Direct measurement of glycosaminoglycan intermolecular interactions via high-resolution force spectroscopy. *Macromolecules*, 35:5601-5615.

Shah, J.V., Flanagan, L.A., Janmey, P.A., Leterrier, J.F. 2000. Bidirectional translocation of neurofilaments along microtubules mediated in part by dynein/dynactin. *Mol. Biol. Cell*, 11, 3495-3508.

Staub, O., Rotin, D. 1996. WW domains. *Structure*, 4, 495-499.

Tetlow, I.J., Wait, R., Lu, Z., Akkasaeng, R., Bowsher, C.G., Esposito, S., Kosar-Hashemi, B., Morell, M.K., Emes, M.J. 2004. Protein phosphorylation in amyloplastes regulates starch branching enzyme activity and protein-protein interactions. *The Plant Cell,* 16, 694-708.

Tibbits, C.W., MacDougall, A.J., Ring, S.G. 1998. Calcium binding and swelling behaviour of a high methoxyl pectin gel. *Carbohydr. Res.,* 310, 101-107.

Walters, C. 1998. Understanding the mechanisms and kinetics of seed ageing. *Seed Sci. Res.,* 8, 223-244.

Wattebled, F., Dong, Y., Dumez, S., Delvallé, D., Planchot, V., Berbezy, P., Vyas, D., Colonna, P., Chatterjee, M., Ball, S., D'Hulst, C. 2005. Mutants of Arabidopsis lacking a chloroplastic isoamylase accumulate phytoglycogen and an abnormal form of amylopectin, *Plant Physiology,* 138, 184-195.

Willats, W.G.T., Orfila, C., Limberg, G., Buchholt, H.C., van Alebeek, G.J.W.M., Voragen, A.G.J., Marcus, S.E., Christensen, T.M.I.E., Mikkelsen, J.D., Murray, B.S., Knox, J.P. 2001. Modulation of the degree and pattern of methyl-esterification of pectic homogalacturonan in plant cell walls - Implications for pectin methyl esterase action, matrix properties, and cell adhesion. *J. Biol. Chem.* 276, 19404-19413.

Wolfram, S. 2002. *A New Kind of Science.* Wolfram Media Inc., Champaign, Illinois, 169-221.

Wolkers, W.F. 1998. The role of macromolecular stability in desiccation tolerance. Thesis, Wageningen Agricultural University, Wageningen.

Wu, X., Knudsen, B., Feller, S.M., Zheng, J., Sali, A., Cowburn, D. 1995. Structural basis for the specific interaction of lysine-containing proline-rich peptides with the N-terminal SH3 domain of c-Crk. *Structure,* 3, 215-216.

Xie, A.F., Granick, S. 2002. Local electrostatics within a polyelectrolyte multilayer with embedded weak polyelectrolyte. *Macromolecules,* 35, 1805-1813.

Zeeman, S.C., Umemoto, T., Lue, W.L., Au-Ying, P., Martin, C., Smith, A., Chen, J. 1998. A mutant of *Arabidopsis* lacking a chloroplastic isoamylase accumulates both starch and phytoglycogen. *Plant Cell*, 10, 1699-1711.

Zsivanovits, G., MacDougall, A.J., Smith, A.C., Ring, S.G. 2004. Material properties of concentrated pectin networks. *Carbohydr. Res.,* 339: 1317-1322.

32

CHAPTER

Metabolic Fingerprinting and Profiling by MID-FT-IR (Middle-Fourier Transformed-InfraRed)

G. Mouille and J.N. Johansen

The ability to obtain the unique biochemical identity of a given sample (the fingerprint) is of both scientific and commercial importance in the field of plant breeding, studies on the response of plants to environmental stresses, and for the characterization of genetic mutants. MID-FT-IR spectroscopy is a high-throughput approach that consists of a rapid classification of samples that gives a snapshot of the metabolic composition of (plant) tissues which have been subjected to any given experimental condition. Using this approach it is not initially necessary to identify all the metabolites in the tissue. MID-FT-IR spectroscopy is a method that measures the vibrations of the atomic molecular bonds within functional groups, and that generates a frequency spectrum that can be regarded as a biochemical or metabolic fingerprint. MID-FT-IR spectroscopy presents itself as an ideal candidate for high-throughput metabolic fingerprinting. In this chapter we will discuss the basic principles of the technique, and give examples where this method has been applied to metabolomics in plants.

PRINCIPLES OF MID-FT-IR

Molecular atomic bonds vibrate at specific frequencies depending on the elements (atoms), and the type of bonds. The energy of these molecular vibrations corresponds to the infrared (IR) region of the electromagnetic spectrum (wavelength from 400 cm^{-1} to 4000 cm^{-1}). Consequently, the wavelength of the light absorbed is characteristic of the chemical bond and nature of the bound atoms. For IR radiation to be absorbed by a molecule there must be an interaction of the radiation with an oscillating dipole (moment) associated with a vibrating bond. (Williams and Fleming, 1980).

In FT-IR spectroscopy, an interference wave that is produced in an interferometer (e.g. Michelson interferometer) interacts with the sample. The generated interference (pattern) is then converted into a spectrum through a Fourier transformation. The combination of this mathematical method with the use of an interferometer greatly decreases the time needed to acquire the data and increases the signal to noise ratio. In practice, the time needed for data acquisition is about one minute to a few minutes. Furthermore, the use of FT-IR leads to a high accuracy in frequency (cm^{-1}) due to the use of a laser to calibrate the wavelength.

Typically, the spectrum of an organic compound can be divided in two regions:

- The functional group region (4000 – ~1600 cm^{-1}) where the absorbance at a particular wavelength reveals the presence of the corresponding functional group (see http://www.muhlenberg.edu/depts/chemistry/irexer/IRDATATB.htm for characteristic peaks of common functional groups).
- The fingerprint region (~1600 – 600 cm^{-1}) is unique to a particular compound, but does not enable the identification of a particular chemical functional group.

The combined FT-IR spectra of the functional group and fingerprint regions can be considered as a unique "molecular fingerprint" of the sample, and it can be used for identification by comparison with a library of spectra of known components. However, it should be noted that the chemical environment (e.g. salts, pH, temperature) influences the frequency of the vibrations of the molecular bonds, and hence a change in the chemical environment can lead to a modification of the spectra obtained, generally through shifts of the wavelength of maximum of absorbances.

SAMPLE PREPARATION AND DATA ANALYSIS

There is no need for derivativization (needed for GC/MS analysis) or particular sample preparation prior to MID-FT-IR analysis. A wide range of samples, either in a solid, liquid or gas state can be analyzed using FT-IR spectroscopy and soluble as well as insoluble metabolites can be examined as the sample is usually dried before analysis. Samples for FT-IR can be prepared in several ways, and many accessories are now available for handling small amounts of material and for analyzing a large number of samples. This makes the technique particularly well suited for the purpose of metabolomics.

For liquid samples, the easiest way is to place one drop between two plates of barium fluoride (BaF_2 does not absorb the IR light). The drop will form a thin film between the plates.

Solid samples can either be analyzed using Attenuated Total Reflectance (ATR) accessories or the solid sample can be dissolved in a solvent, which is subsequently placed on a single BaF_2 plate. The solvent is then evaporated, leaving a thin film of the original material on the plate.

The sample can then be analyzed directly on this dried film.

Gas samples are analyzed through the use of specific cells or capillaries, but to our knowledge this kind of approach has not yet been used in metabolomics.

An interesting feature of MID-FT-IR spectroscopy is that it can be coupled to microscopy. It can then be used for organ or tissue specific metabolomics, without the need of any dissection steps. The size of the area/tissue to focus on should in this case be larger than 10 μm × 10 μm.

MID-FT-IR is highly sensitive, not only in terms of the amount of detectable material but also in the range of features that can be detected. In fact, differences in composition or structural differences between distinct samples can be detected. MID-FT-IR can then be considered as a quantitative and qualitative analytic method. This technique is qualitative because of the principle (see above) but also because of the need of only minimal preparation prior to analysis that preserves the native structure of the sample.

The spectrum of each sample generally contains a few hundred variables corresponding to absorbance values at different wave numbers, making visual analysis difficult. Furthermore, a MID-FT-IR spectrum cannot be considered as a "simple" multivariate set of variables (in contrast to a chromatogram, where each peak corresponds to one compound).

In plant biology the product to be analyzed often has a complex composition, resulting in a multifaceted spectrum, hence the analysis needs statistical tools that are known as chemometric methods. The most common chemometric methods are based on dimensionality reduction using principal component analysis (PCA) (Jollife, 1984), and discriminant analysis (Manly, 1994) often followed by cluster analysis. This statistical approach allows the classification of complex samples, however the original information consisting of the absorbance at different wave numbers, will be lost. Therefore, it is almost impossible to couple observed differences between analyzed samples to simple chemical modifications.

Furthermore, as discussed above, modification of the spectra could be due to structural change of the sample. These modifications often lead to a shift in wave numbers, which are very difficult to detect and interpret. Particular statistical tools need to be developed for this particular purpose.

METABOLIC FINGERPRINTING AND PROFILING BY MID-FT-IR IN PRACTICE

To date, few examples of the use of MID-FT-IR for metabolic fingerprinting or profiling in plants have been published. In the following section we will give examples of the use of MID-FT-IR to study the response of plants to environmental stresses, and to characterize and identify mutants.

MID-FT-IR was used in the study of the biochemical modification occurring in tomato fruits growing in saline conditions and to determine the metabolic discrimination between different tomato varieties (Johnson et al., 2003). The sample analyzed by the authors was the pulp of peripcarp and mesocarp of the tomato fruit. The FT-IR spectra obtained from these samples were considered as the metabolic fingerprint of the fruit. The data acquisition was followed by classical PCA and discriminant function analysis (DFA), and allowed the authors to distinguish the control from the salt-treated plants. Finally, the use of a genetic algorithm (GA) enabled the identification of the functional groups particularly affected during salt stress of the tomato plant, and also led to the identification of zones in the spectra important for discrimination of the two classes of fruit.

Johnson and colleagues (Johnson et al., 2004) used the same statistical approach to investigate legume silage fermentation through metabolic fingerprinting. In this particular case, the authors used high-throughput accessories. Samples were loaded onto a 400 well bespoke aluminium plate allowing kinetic investigation of the fermentation stages of different herbage types by distinct species of bacteria.

MID-FT-IR was also used to investigate plant-plant interference between *Arabidopsis thaliana* and *Brachypodium distachyion* (Gidman et al., 2003). MID-FT-IR spectra were acquired from ground whole leaf tissue. In this case, the authors detected changes in the metabolic profile of the leaves of plants that were grown alone or in competition. The authors suggested that MID-FT-IR spectroscopy is appropriate for rapid screening prior to targeted metabolite analyses. These studies show that MID-FT-IR is a very fast and sensitive tool to detect changes in global metabolite profiles.

A more specialized approach was developed for the analysis of the plant cell wall (Chen et al., 1998). This has been used for the characterization of cell wall mutants (Chen et al., 1998; Mouille et al., 2003). Here the plant samples were rinsed with water to remove the cytoplasm and then dried prior to the FT-IR analysis. Mouille et al. (2003) focused on the metabolic profile of the cell wall. Since the samples used in this study were small (150 µm width), the data was collected using a microscope coupled to the infrared spectrometer to be able to focus the IR beam on the selected tissue. Mouille and colleagues acquired spectra directly on squashed/rinsed four-day-old dark grown seedlings of *A. thaliana*. With this approach, they classified cell wall mutants and identified new mutants with features in common with previously identified cell wall mutants of *Arabidopsis thaliana* (Mouille et al., 2003; http://www-ijpb.versailles.inra.fr/en/bc/bc_cell_wall.htm).

Sibout and colleagues (Sibout et al., 2005) also took - advantage of a microscope coupled to MID-FT-IR to investigate tissue specific cell wall modification of *A. thaliana* stems (Figure 32.1A and 32.1B) (Sibout et al., 2005). The authors showed tissue specific modification in xylem and fibers of the secondary cell wall in *Atcad cd* (Cinnamyl Alcohol Dehydrogenase) double mutant (Figure 32.1C).

Another example of the use of microscopy coupled with MID-FT-IR is the study of pollen growth, where Wang et al (2005) acquired data directly on individually germinated pollen tubes of *Picea* and investigated the cell wall modification along the tube. This allowed analysis of cell wall modification during the growth of pollen tubes (Wang et al., 2005).

In conclusion, MID-FT-IR can be considered as a sensitive method for the qualitative and also quantitative detection of changes in highly complex biological samples. This technique, like NMR spectroscopy emerges as an interesting tool for metabolomics, not only because of its ability to detect and sometime identify compounds, but also for its ability to detect structural modification in the samples. These methods can then be considered as tools for what could be called "structuromics".

The main advantage of FT-IR coupled to microscopy technology, compared to other technical analysis methods used in metabolomics, such as NMR and GC/MS is that no dissection is needed prior to the analysis to achieve a similar spatial resolution. Thus the use of MID-FT-IR for metabolomics should be encouraged, as it is a very informative method owing to the large amount of information that IR spectra contain. The application of FT-IR should be done in parallel with the development of original statistical tools that are specifically designed for handling the complex data sets that IR spectra contain. In addition, the construction of accessories adapted for high-

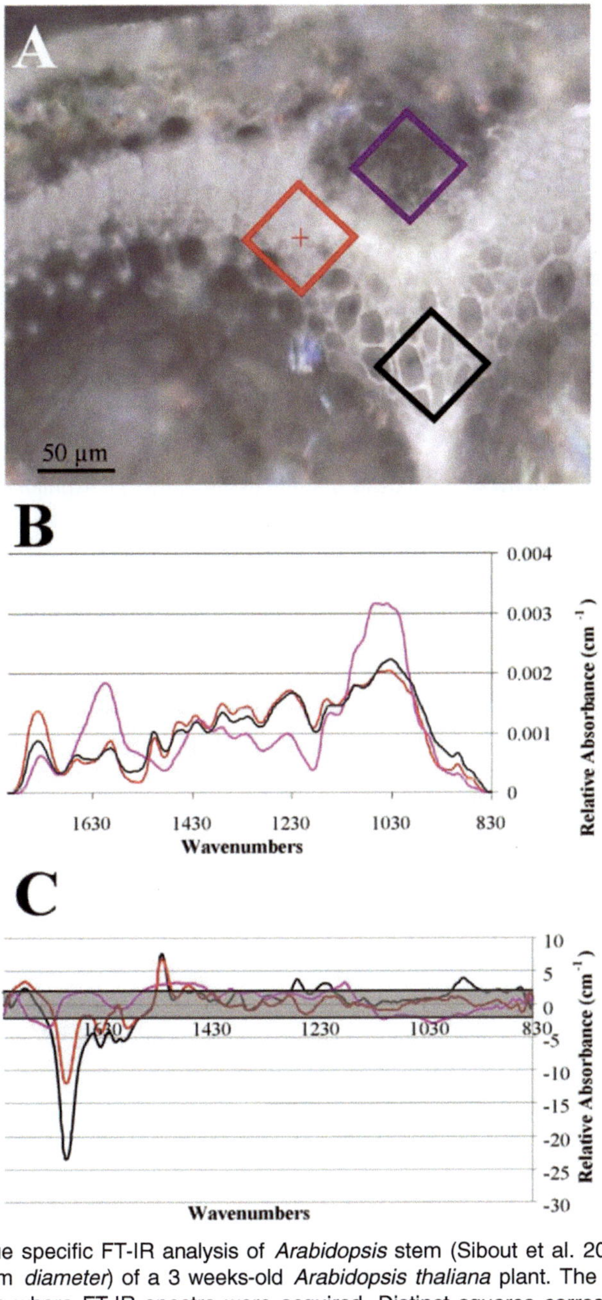

Fig. 32.1 Tissue specific FT-IR analysis of *Arabidopsis* stem (Sibout et al. 2005). (A) Section of a stem (600µm *diameter*) of a 3 weeks-old *Arabidopsis thaliana* plant. The colored squares define the zones where FT-IR spectra were acquired. Distinct squares correspond to specific tissues: phloem (purple), xylem (black), fiber (red). (B) Fourier transform mid-infrared (FT-IR) spectra of xylem area, fiber area and phloem in wild-type stems. (C) Identification of tissue specific cell wall alterations in the *Atcad cd* mutant. t value (Student's t-test) corresponding to the difference between *Atcad cd* mutants for the three distinct tissues investigated. Non-significant values ($p = 0.95$) are shaded.

throughput data acquisition, will make FT-IR a valuable and complementary tool for plant metabolic fingerprinting.

REFERENCES

Chen L., Carpita N.C, Reiter W-D, Wilson R.H, Jeffries C., McCann M.C. 1998. A rapid method to screen for cell-wall mutants using discriminant analysis Fourier transform infrared spectra. Plant Journal 16 (3) 385-392

Gidman E., Goodacre R., Emmett B., Smith A.R., Gwynn-Jones D. 2003. Investigating plant-plant interference by metabolic fingerprinting. Phytochemistry. 63(6):705-710.

Johnson, H.E., Broadhurst, D., Goodacre R., Smith A.R. (2003) Metabolic fingerprinting of salt-stressed tomatoes Phytochemistry 62 (6): 919-928

Johnson HE, Broadhurst D, Kell DB, Theodorou MK, Merry RJ, Griffith GW. 2004. High-throughput metabolic fingerprinting of legume silage fermentations via fourier transform infrared spectroscopy and chemometrics. Appl Environ Microbiol. 70(3): 1583-1592.

Jollife, I.T. 1986. Principal Component Analysis. Springer, New York, N.Y.

Manly, B.F.J. 1994. Multivariate Statistical Methods: A Primer., Chapman and Hall, London.

Mouille G, Robin S, Lecomte M, Pagant S, Hofte H. 2003. Classification and identification of Arabidopsis cell wall mutants using Fourier-Transform InfraRed (FT-IR) microspectroscopy. Plant J. 35(3):393-404.

Sibout R, Eudes A, Mouille G, Pollet B, Lapierre C, Jouanin L, Seguin A.2005. Cinnamyl alcohol dehydrogenase-C and -D are the primary genes involved in lignin biosynthesis in the floral stem of Arabidopsis. Plant Cell. 17(7):2059-2076.

Williams DH, Fleming I 1980. Infrared spectroscopy. *In* Spectroscopic Methods in Organic Chemistry, Ed. 3. McGraw Hill, New York, 35-73.

Wang Q, Kong L, Hao H, Wang X, Lin J, Samaj J, Baluska F. 2005. Effects of brefeldin A on pollen germination and tube growth. Antagonistic effects on endocytosis and secretion. Plant Physiol. Vol. 139(4):1692-1703.

Glossary

C. Meyer, J.F. Morot-Gaudry

Acrylamide gel: a polymer gel used for electrophoresis of DNA or protein, which can be used to measure their sizes (in base pairs for DNA, or in daltons for proteins).

Additive genetic effects: when the combined effects of alleles at different loci are equal to the sum of their individual effects.

AFLP™, Amplified Fragment Length Polymorphism: PCR amplification of genome DNA after digestion with two restriction enzymes and ligation of adaptors of approximately 20 nucleotide base pairs. PCR primers correspond to the adaptors plus two or three random bases at the 3' end. The profiles obtained are multibanded.

Agarose gel: a polysaccharide gel used for electrophoresis, to measure the size of nucleic acids (in bases or base pairs).

Algorithm: procedure of resolution of a problem, which can be translated into data-processing language in order to produce "automatically" a result. An algorithm is the logical sequence of steps by which a computational task can be performed. It is often essential to carry out a preliminary stage of algorithmic design, before any software development, in order to formally describe the sequences of programming prior to syntactic writing.

Allele: possible version of a locus. The locus can be a gene or a non-coding sequence of DNA. When in various individuals, different forms exist within a given locus, each one of these forms is an allele.

Alternative splicing: various ways of splicing out introns in eukaryotic pre-mRNAs, resulting in one gene producing several different mRNAs and protein products.

Amplification: an increase in the number of copies of a gene or a sequence of DNA. In the case of the PCR: *in vitro* synthesis of a large number of copies of a given DNA fragment.

Annotation: adding pertinent information such as description of the gene, amino acid sequence, or other commentary to the database entry of a raw sequence of DNA nucleotides.

Antisense: nucleic acid that has a sequence exactly opposite to a mRNA molecule made by plants. Antisense technology is used to express an antisense mRNA that decreases the expression of a given gene.

Backcross: a cross between a plant that is heterozygous for alleles obtained from two parental strains and a second plant from one of those parental strains; a backcross is also used to cross a mutant with its parents, or with an individual genetically identical to one of its parents.

BACs, Bacterial Artificial Chromosomes: are large segments of DNA, between 100,000 and 200,000 nucleotide bases long, cloned from another species into bacterial DNA. Once the foreign DNA has been cloned into the host bacteria, many copies of it can be made, and the BACs can then be stored in order in libraries. The BACs can be handled like plasmids of a very large size.

Bank of insertion mutants: collections of mutants in which one or several well identified elements (T-DNA, variable components, etc.) are inserted into genes.

BiFC, Bimolecular Complementation Fluorescence: microscopy technique which exploits the properties of the GFP.

Bioinformatics: the study of the application of computer and statistical techniques to the management of information. In genome projects, informatics includes the development of methods to search databases quickly, to analyze DNA sequence information, and to predict protein sequence and structure from DNA sequence data.

BLAST, Basic Local Alignment Search Tool: a computer program that identifies homologous (similar) genes in different organisms, such as human, nematode or plants. It is software which is used to compare sequences one by one and to determine their areas of homology.

Blotting: A technique for detecting one RNA within a mixture of RNAs (a Northern blot) or one type of DNA within a mixture of DNAs (a Southern blot). A blot can prove whether that one species of RNA or DNA is present, how much is there, and its approximate size. Blotting involves gel electrophoresis, transfer to a blotting membrane (typically nitrocellulose or activated nylon), and incubating with a radioactive probe. Exposing the membrane to X-ray film produces darkening at a location correlating with the position of the DNA or RNA of interest. See Northern, Southern and Western blotting.

BRET, Bioluminescence Resonance Energy Transfer: detection system based on proximity-dependent resonant energy transfer from a bioluminescent donor molecule to an acceptor. This technique was recently introduced in cell biology.

CALI, Chromophore Assisted Laser Inactivation: technique of microscopy which utilises an antibody to recognize a specific site and its derivation with a chromophore.

Candidate gene: a gene whose function suggests that it may be involved in the genetic variation observed for a particular trait.

CAPS, Cleaved Amplified Polymorphic Sequence (PCR/RFLP): digestion of the polymorphic products of amplification. A fragment of amplified genomic DNA (Sequence Tagged site, STS, or Sequence Characterised Amplified Region, SCAR) is digested by restriction enzymes in order to detect the polymorphic electrophoretic profiles of the restriction fragments.

cDNA, copy-DNA or complementary DNA: results from the synthesis of DNA by a reverse transcriptase enzyme using messenger-RNA (mRNA) as a template. Introns are not found in the cDNA.

cDNA library: collection of clones of cDNAs obtained after reverse transcription of a set of mRNAs from different organs, developmental stages, growth conditions, etc. .

Centromere: the constricted region near the centre of a chromosome. This is the region of the chromosome where the two sister chromatids are joined to one another. The majority of the DNA in the centromere contains highly repetitive sequences.

Chloroplast chromosome: circular DNA found in the photosynthesizing organelle (chloroplast) of plants, instead of the cell nucleus where most of the genetic material is located.

Chromatin: a complex of protein and DNA that makes up chromosomes.

Chromosomal deletion: the loss of part of the DNA of a chromosome.

Chromosome: the self-replicating genetic structure of cells containing DNA that bears in its nucleotide sequences, the linear array of genes. Plant genomes consist of a number of chromosomes, whose DNA is associated with different kinds of proteins.

Chromosome painting: attachment of certain fluorescent dyes to targeted parts of the chromosome. See FISH.

Chromosome walking: a technique for cloning sequences in the genome around a known piece of DNA (the starting probe). Typically used to "walk" from a starting point towards a specific nearby gene in order to clone that gene. Also used to obtain the full sequence of a gene, when only part was initially isolated.

Classical light microscopy: this involves passing light transmitted through or reflected from the subject through a series of lenses, to be detected directly by the eye, imaged on a photographic plate, or captured digitally.

Clone: an exact copy made of biological material such as a section of DNA (a gene or other region), a whole cell, or a complete organism.

Cloning Vector: a piece of DNA, such as a plasmid, into which a DNA segment can be inserted, transferred into an organism, and replicated or reproduced. A cloning vector is used to introduce foreign DNA into host cells, where the DNA can be reproduced in large quantities. Examples are plasmids, cosmids, and yeast artificial chromosomes; vectors are often recombinant molecules containing DNA sequences from several sources.

Codominance: situation in which two different alleles for a genetic trait are both expressed.

Colinearity: conservation of the order of genes along a chromosome.

Comparative genomics: the practice of comparing gene or protein sequences with each other, across the genomes of whole organisms, in the hope of elucidating functional and evolutionary significance.

Complementary sequence: nucleic acid base sequence that can form a double-stranded structure with another DNA fragment by following base-pairing rules (A pairs with T and C with G). The complementary sequence to 5′-GTAC-3′ for example, is 3′-CATG-5′.

Complex trait: trait that has a genetic component that does not follow strict Mendelian inheritance. May involve the interaction of two or more genes or with the environment.

Confocal microscopy: microscopy using an epifluorescence microscope to reduce the depth of field (providing a thinner focal slice) and to eliminate light from out-of-focus objects (reducing background). It is a valuable tool for obtaining high-resolution blur-free images of thick specimens at various depths.

Conserved sequence: a nucleotide base sequence in a DNA molecule (or an amino acid sequence in a protein) that has remained essentially unchanged throughout evolution.

Contig map: a map depicting the relative order of a linked library of overlapping clones representing a complete chromosomal segment.

COS, Conserved Ortholog Set: markers used in the approaches of comparative genomics (synteny).

Cosegregation: The tendency for closely linked genes and genetic markers to segregate (be inherited) together. Two genes (or a gene and a marker) cosegregate when they are not separable in the analysis of the descendants.

Co-suppression: phenomenon describing the reciprocal and co-ordinate inactivation of an endogenous gene and a homologous transgene. The co-suppression corresponds to – inactivation by TGS or PTGS–according to whether the transgene is homologous with the endogenous gene through the promoter or coding sequence.

Cot: term used for the kinetics of denaturation-renaturation of DNA. It is the product of the initial concentration in DNA (Co) expressed in mole of nucleotide/litre and the reaction time (t).

Crossing over: the breaking during meiosis of one maternal and one paternal chromosome, followed by the exchange of corresponding sections of DNA, and the rejoining of the chromosomes. This process can result in an exchange of alleles between chromosomes.

Cytogenetics: the study of the physical appearance of chromosomes.

Cytological map: a type of chromosome map whereby genes are located on the basis of cytological findings obtained with the aid of chromosome mutations.

Cytomic: cytological approach, which integrates — genomics and proteomics

Cytoplasmic trait: a genetic characteristic in which the corresponding genes are found outside the nucleus, e.g. in the chloroplasts or mitochondria.

DAF, DNA Fingerprinting Amplification: operates on the same principle as Random Amplified Polymorphic DNA,RAPD, but uses shorter starters (5 to 8 bases). The electrophoretic profiles of the amplified fragments of DNA are often complex.

Data warehouse: a collection of databases, data tables, and mechanisms to access the data on a single subject.

Deletion: a loss of part of the DNA from a chromosome, which can often lead to a disease or abnormality.

Deletion map: a description of a specific chromosome that uses defined mutations, specific deleted areas in the genome, as 'biochemical signposts,' or markers for specific areas.

Denaturation: In the case of DNA in particular, denaturation means the separation of the double strand into single strands.

DGGE, Denaturing Gradient Gel Electrophoresis: electrophoresis of DNA with a gradient of denaturation (urea, formamide). The DNA fragment, initially doublestranded, is partially denatured to form a Y shaped structure at a precise position along the gradient during electrophoresis. The electrophoretic mobility of the DNA is then modified. The sequence of the DNA determines the zone of the gradient where it will be denatured.

Differential display: PCR amplification of partial cDNAs by using an anchor 3′ primer and random primers. A given sample is compared to a control one and differential bands are identified utilising denaturing acrylamide gel electrophoresis.

Differential screening: technique used to identify cDNAs corresponding to mRNAs specific for a tissue, or a stage of development.

Directed mutagenesis: alteration of DNA sequence at a specific site and the reinsertion of the DNA into an organism to study any effects of the change.

DNA polymerase (DNA- or RNA-dependent): enzyme which lengthens the strands of nucleic acids by adding individual nucleotides through the formation of covalent chemical bonds. The order of the insertion of the nucleotides, is dictated by the order of the nucleotides in the complementary strand.

DNA Microarray: a dense arrangement of DNA fragments (PCR fragments or oligonucleotides) spotted or synthesized onto a solid support. See microarray.

Draft: uncorrected and/or partial genome sequence.

Ecotype: a locally adapted population of a widespread species. Such populations show minor changes of morphology and other characteristics, which are genetically regulated. They can still reproduce with other ecotypes of the same species.

Ectopic gene: a gene which is expressed in a tissue or organism, where it is not expressed normally.

Electrophoresis: a method of separating large molecules (such as DNA fragments or proteins) from a mixture of similar molecules. An electric current is passed through a medium containing the mixture, and each kind of molecule travels through the medium at a different rate, depending on its electrical charge and size. Agarose and acrylamide gels are the media commonly used for electrophoresis of nucleic acids and proteins.

EMS, Ethyl MethaneSulfonate: compound used to induce mutations (usually point mutations) in the genome (chemical mutagenesis).

Epigenetic modifications: modifications that can result either from a blocking of — transcription (transcriptional gene silencing, TGS), or of the specific degradation of mRNA (post-transcriptional gene silencing, PTGS).

Epigenetic phenomenon: heritable changes in gene expression that occur without a change in DNA sequence.

Epistasis: the masking of the phenotypic effect of alleles at one gene by alleles of another gene. A situation in which the phenotypic expression of a genotype at one locus, depends upon the genotype at another locus. A gene is said to be epistatic when its presence suppresses the effect of a gene at another locus. Epistatic genes are sometimes called inhibiting genes because of their effect on other genes, which are described as hypostatic.

ESI-MS/MS, Magnet-Spray Ionization MS/MS: technique of analysis of proteins by mass spectrometry.

EST, Expressed Sequence Tag: a short strand of DNA that is a part of a cDNA molecule and can act as the identifier of a gene. Used in locating and mapping genes.

Exon: section of a gene, which encodes part of a protein. Portion of a genomic DNA sequence, which is present in the final, mature mRNA after the introns have been removed. The term "exon" can also be used for the equivalent sections of the final mRNA.

Expression Profile: a snapshot of the genes expressed in specific tissues of an organism at a particular point in time. This snapshot tells an investigator where, when, and to what extent a particular gene is expressed in a cell, tissue, or organism and what physiological pathways may be active in the cell.

FISH Analyses, Fluorescent In-Situ Hybridization: fluorescent marked probes are hybridized directly with the chromosome in the cell nucleus, in order to determine specific gene locations.

Flanking region: a region of DNA, which is not transcribed into RNA but rather is adjacent to 5' end of the gene. The 5'-flanking region contains the promoter, and may also contain enhancers or other protein binding sites.

FLIP, Fluorescence Loss In Photobleaching: technique of microscopy based on the photo-instability of some fluorochromes.

Fluorochromes are essentially coloured dyes which accept light energy at a given wavelength and re-emit it at a higher wavelength. They can emit light in the visible wave length range when irradiated with light of another wave length.

Forward genetics: the traditional approach to genetics, which starts with a phenotype and then identifies the gene(s) or mutation that control or cause that trait. This is in contrast to reverse genetics, which starts from a gene/sequence in order to determine a phenotype.

FRAP, Fluorescence Recovery After Photobleaching: technique of microscopy whose principle rests on the phenomenon of photo-instability of some fluorochromes. It makes it possible to follow for example the diffusion of a protein into a subcellular compartment.

FRET, Fluorescence Resonance Energy Transfer: a technique of fluorescence microscopy, which makes it possible to visualize for example protein-protein interactions.

FST, Flanking Tag Sequence: DNA sequence adjacent to an insertion (T-DNA, transposon, etc.)

Full gene sequence: the complete order of nucleotides in a gene. This order determines which protein a gene will produce.

Functional genomics: an area of study aimed at determining the function of genes and the proteins they encode in determining traits, physiology or development of an organism. Generally the term is used for an experimental approach utilizing computational and high-throughput technologies at the level of whole genomes.

GenBank: a public database where DNA sequences are deposited and made freely available.

Gene: the fundamental physical and functional unit of heredity. A gene is an ordered sequence of nucleotides located in a particular position on a particular chromosome that encodes a specific functional product (i.e., a protein or RNA molecule). A gene is a transcription unit.

Gene expression: the process by which he coded information of a gene is converted into the structures present and operating in the cell. Expressed genes include those that are transcribed into mRNA and then translated into protein and those that are transcribed into RNA but not translated into protein (e.g., transfer and ribosomal RNAs).

Gene expression analysis: determination of the activity of the genes in a cell / tissue or an organism; currently mainly through determination of the quantity of mRNA of specific genes, also by quantification of proteins.

Gene mapping: determination of the relative positions of genes on a chromosome or plasmid and of the distance, in linkage units or physical units, between them.

Gene prediction: predictions of possible genes made by a computer program based on how well a stretch of DNA sequence matches known gene sequences.

Gene recombination: a process of chromosome exchange during cell division, resulting in new combinations of genes at specific locations on chromosomes.

Gene silencing: the process of inactivating a gene.

Gene transfer: incorporation of new DNA into the cell of an organism, usually by a vector such as a modified bacteria, *Agrobacterium tumefaciens* for example.

Gene transposition: transfer of a gene to a new position on the same or another chromosome or plasmid.

Genetic engineering: altering the genetic material of cells or organisms to enable them to make new substances or perform new functions.

Genetic map: a map of the relative positions of genetic loci on a chromosome, determined on the basis of how often the loci are inherited together.

Genetic marker: a gene or other identifiable portion of DNA, whose inheritance can be followed.

Genetic polymorphism: difference in DNA sequence among individuals, groups, or populations. See polymorphism.

Genome: the complete total amount of hereditary information of an organism (genes, genetic signal structures, as well as other DNA areas); hereditary material.

Genome sequence: sequence of nucleotides in the hereditary material of an organism. See full gene sequence.

Genomic bank or library: collection of DNA fragments representative of the totality of a genome, which is cloned in vectors (phage, YAC...) and propagated in host cells (bacteria, yeasts...).

Genomics: is the study of how genes and genetic information are organized within the genome, and how this organization determines their function.

Genotype: the combination of alleles located on homologous chromosomes that determines a specific characteristic or trait of the penotype.

Genotyping: determination of the genotype.

GFP, Green Fluorescent Protein: fluorescent protein, isolated from a jellyfish (*Aequorea victoria*). GFP is a fluorescent marker which have found numerous applications in plant biology for genomic and proteomic studies and in cell biology approaches upon living cells. GFP is used as tags for plant gene and protein location, or as indicator of distinct biological processes (protein diffusion, interactions, etc.).

GISH, Genomic *In Situ* Hybridization: *in situ* hybridization on a chromosome with a complex fluorescent probe (total or repeated fraction).

GUS: a reporter gene coding for the enzyme β-glucuronidase, whose activity can be monitored *in vivo* and *in vitro*, or visualized *in situ*.

Haploid: Having the same number of sets of chromosomes as a germ cell (pollen, egg) or half as many as a diploid somatic cell.

Heterozygote: an organism which is characterised by the presence of different alleles at one or more loci on homologous chromosomes.

High Throughput Biology: An experimental approach that generates massive amounts of raw data using highly automated technologies such as genome sequencing technology or microarray technology. The data is processed by a batch method using computational and other information management tools.

Homologous chromosome: chromosome containing the same linear gene sequences as another, each derived from one parent.

Homologous genes: genes derived from a common ancestor.

Homologous recombination: exchange of DNA fragments between paired chromosomes.

Homozygote: an organism that has two identical alleles of a gene.

HPLC, High Performance Liquid Chromatography: technique of separation and analysis of compounds by liquid chromatography on a column.

Hybridization: the process of joining two complementary strands of DNA or one each of DNA and RNA to form a double-stranded molecule; an important step in DNA microarray technology.

In situ **hybridization**: technique of microscopy, which allows the visualization of the tissue distribution of a RNA molecule by using labelled complementary probes.

Insert of DNA: fragment of DNA which is joined to a vector of DNA in order to be cloned.

Insertion mutagenesis: change in gene activity due to an insertion of DNA (T-DNA, transposons), that either changes the gene directly or disrupts control mechanisms.

Interactome: term of genomics used to define approaches, which seek to describe protein/protein interactions

Intellectual property rights: patents, copyrights, and trademarks.

Intron: a segment of a gene situated between exons, that is removed before translation of mRNA and does not function in coding for protein synthesis.

ISSR, Inter Simple Sequence Repeat: amplification of genomic DNA with primers that carry dinucleotide repeated patterns and some random nucleotides. The amplified fragments are thus located in the genome between two microsatellite loci. Other names: I-SSR PCR, Inter Simple Sequence Repeat PCR; IMA, Inter Microsatellite Amplification; IRA, Inter-Repeat Amplification; ISA, Inter-SSR Amplification. SSRs are among the most efficient markers for the detection of polymorphism.

Knockout: deactivation of a specific gene; used in laboratory organisms to study gene function.

Library: collection of a large number of different molecules (peptide library, protein library) or clones (gene library) or plant mutants (mutant library) for screening and/or selection.

Linkage disequilibrium: where alleles occur together more often than can be accounted for by chance. Indicates that the two alleles are physically close on the DNA strand or submitted to selection pressure.

Linkage map: a map of relative positions of genes on a chromosome. Genes inherited together are close to each other on the chromosome, and said to be linked.

Locus (plural loci): the position of a gene or other chromosome marker on a chromosome. The use of the term locus is sometimes restricted to mean expressed DNA regions.

Luciferase: an enzyme present in the cells of bioluminescent organisms that catalyzes the oxidation of luciferin and ATP to produce light.

MALDI-TOF/MS, Matrix Assisted Laser Desorption Ionisation-Time of Flight/Mass-Spectrometry; mass-spectrometric method for the determination of molecular mass (e.g. proteins) ; employed, among other things, for quality control in DNA synthesis.

MAR , Matrix Attachment Region: area of DNA binding to the chromosomal matrix.

MAS, Marker Assisted Selection: the use of DNA markers for the selection of desired individuals within a population, by linking the molecular marker to a specific trait. To be of maximal use, the markers should be closely linked to one or more target loci, which may often be quantitative trait loci.

Mass spectrometer: an instrument used to identify chemicals in a substance by their mass and charge.

Metabolic finger printing: an approach which aim to classify samples by global and fast analysis of a large number of metabolites.

Metabolite profiling: analysis of all metabolites of a metabolic pathway or a given chemical class of metabolites. Metabolite profiling aims at a quantitative assessment of a predefined number of target metabolites..

Microarray technology: a new approach to the study the expression of a large number of genes. This technology provides a quantitative assessment of how regulatory networks in a cell control extensive gene sets simultaneously. The method uses a robot to apply precisely very small (nano to picoliters) droplets containing functional DNA to glass slides. Researchers then attach fluorescent labels to DNA (derived from RNA) from the cell they are studying. The slides are put into a scanner that can measure the brightness of each fluorescent dot; the brightness reveals how much of a specific DNA fragment is present, an indicator of how active is the corresponding gene.

Microsatellite: a microsatellite is a simple sequence repeat (SSR). It might be a homopolymer ('...TTTTTTT...'), a dinucleotide repeat ('....CACACACACACACA.....'), trinucleotide repeat ('....AGTAGTAGTAGTAGT...') etc. Due to polymerase slip (polymerase chatter), during DNA replication there is a slight chance these repeat sequences may become altered; copies of the repeat unit can be created or removed. Consequently, the exact number of repeat units may differ between unrelated individuals. Considering all the known microsatellite markers, no two individuals are identical.

Mitochondrial DNA: the genetic material found in mitochondria, the organelles that generate energy for the cell. Not inherited in the same fashion as nuclear DNA.

Mutation: Change of the hereditary information resulting in a change in the sequence of the nucleotides in the DNA.

Northen blotting: hybridization of RNA molecules in a complex mixture separated by gel electrophoresis using a labelled specific probe.

Nuclear genome: the part of the genome of an organism that is found within the nucleus of the cell. It encodes most but not all of the structural and regulatory DNA sequences necessary for cellular function. See chloroplast and mitochondria.

Oligonucleotide: a molecule usually composed of 25 or fewer nucleotides; used as a DNA synthesis primer.

ORF, Open Reading Frame: the sequence of DNA or RNA located between the start-code sequence (initiation codon) and the stop-code sequence (termination codon).

Orthologous genes: are two genes coding for the same function, that are derived from the same ancestral gene. Two orthologous genes thus belong to two distinct species resulting from a common ancestor and these genes may indicate a position preserved in the genome of these two species.

Paralogous genes: are two homologous genes found at different chromosomal locations in the same organism, that have structural similarities indicating that they derived from a common ancestral gene.

Patent: in genetics, conferring the right or title to genes, gene variations, or identifiable portions of sequenced genetic material to an individual or organization.

PCR, Polymerase Chain Reaction: a process for amplifying DNA. A method for amplifying a DNA nucleotide sequence using a heat-stable polymerase and two oligonucleotide primers, one complementary to the (+) strand at one end of the sequence to be amplified and one complementary to the (-) strand at the other end. Because the newly synthesized DNA strands can subsequently serve as additional templates for the same primer sequences, successive rounds of primer annealing, strand elongation, and dissociation generate very rapidly a highly specific amplification of the desired sequence. PCR can be used to detect the existence of very small amounts the defined sequence in a DNA sample.

PCR-RFLP, PCR-Restriction Fragment Length Polymorphism: see CAPS.

Phenocopy: a trait, which although it appears to be identical to a genetic trait, is not caused by genetic inheritance but is in fact caused by an environmental change

Phenotype: the observable physical or biochemical characteristics of an organism, as determined by both genetic makeup and environmental influences.

Phenotyping: determination of the phenotype of an organism; determination of the observable physical or biochemical characteristics of an organism, as determined by both genetic makeup and environmental influences. A major challenge in phenotyping is balancing the need to gather sufficient information on each mutation against the need to screen large numbers of plants to obtain a broad coverage of the genome.

Physical map: a map of the physical locations of identifiable landmarks on DNA (e.g., restriction enzyme cutting sites, genes); distance is measured in nucleotide base pairs. The highest resolution map would be the complete nucleotide sequence of the chromosomes.

Plasmids: circular double-stranded DNA, which occur mainly in bacteria. Plasmids replicate independently of the chromosome and are used in genetic engineering for the cloning of foreign DNA sequences.

Pleiotropic gene: a gene which causes several different effects on the phenotype.

Polymorphism: genetic variations occurring in more than 1% of a population would be considered useful polymorphisms for genetic linkage analysis. See genetic polymorphism.

Population genetics: the study of variation in genes among a group of individuals.

Positional cloning: a technique used to identify genes, based on their location on a chromosome.

Post-transcriptional regulation: a process occurring after transcription, which affects the amount and type of protein synthesised from a gene. Includes mRNA processing, mRNA stability, translation efficiency and protein stability.

Post-translational regulation: a process, which alters the covalent structure of a protein after translation, e.g. phosphorylation, glycosylation and proteolytic cleavage.

PQL, Protein Quantity Locus: QTL of proteins.

Primary transcript: when a gene is transcribed in the nucleus, the initial product is the primary transcript, a RNA containing copies of all exons and introns. This primary transcript is then processed by the cell to remove the introns, and to polyadenylate the 3' end. The mature mRNA thus formed is then exported to the cytoplasm for translation.

Probe: single-stranded DNA or RNA molecules of specific nucleotide sequence, labelled either radioactively or immunologically, that are used to detect the complementary nucleotide sequence by hybridization.

Promoter: the part of a gene that contains the information to turn the gene on or off. The process of transcription (reading of DNA sequences to produce gene products) is initiated at the promoter.

Proteome: proteins expressed by a cell or organ at a particular time and under specific conditions.

Proteomics: analysis of the functions and interactions of proteins in organisms.

Pseudogene: a sequence of DNA similar to a gene but which is not transcribed; probably the remnant of a once-functional gene that accumulated mutations.

QTL, Quantitative Trait Locus: The location of a gene that affects a trait that is measured on a quantitative (linear) scale. These traits are typically affected by more than one gene, and also by the environment. Examples of quantitative traits are plant height and plant weight.

RAPD, Random Amplified Polymorphic DNA: PCR amplification of DNA fragments using a primer of 10 arbitrary bases.

Recombinant DNA molecules: an artificial combination of DNA molecules of different origin that are joined using recombinant DNA technologies.

Recombinant DNA technology: procedure used to join together sections of DNA in a cell-free system. Under appropriate conditions, a recombinant DNA molecule can enter a cell and replicate there, either autonomously or after it has become integrated into a chromosome.

Recombination: the process by which progeny derive a combination of genes different from that of either parent. In higher organisms, this can occur by crossing over. See Gene recombination.

Reporter gene: A gene whose phenotypic expression is easy to monitor; used to study promoter activity in different tissues or developmental stages. Recombinant DNA constructs are made in which the reporter gene is attached to a promoter region of particular interest and the construct introduced into a cell or organism. (e.g. GUS; luciferase, GFP,...)

Restriction enzyme: an endonuclease enzyme protein that recognizes specific, short nucleotide sequences and cuts DNA at those sites.

RFLP, Restriction fragment length polymorphism: variation between individuals in DNA fragment sizes generated by the digestion of DNA by specific restriction endonucleases. Polymorphic sequences that result in RFLPs are used as markers on both physical and genetic linkage maps. RFLPs are usually caused by mutations at a cutting site.

Reverse genetics: analysis that starts with a gene/DNA sequence and eventually leads to its biological function. Reverse genetic methods are more amenable to whole genome, high- throughput analysis than those used in forward genetics.

Reverse transcriptase: an enzyme used by retroviruses to form a complementary DNA sequence (cDNA), using RNA as a template.

RIL, Recombinant Inbred Line: an RIL is formed by crossing two inbred strains, followed by repeated selfing or sibling mating to create a new inbred line whose genome is a mosaic of the parental genomes.

RNAi, RNA Interference: this technique involves the transfer of double-stranded RNA into organisms, thereby specifically inactivating genes containing homologous sequences. RNAi has arisen from the observation that sense and antisense RNA are equally effective in suppressing specific gene expression. These results presented a paradox that was resolved by finding that small amounts of double-stranded (ds) RNA contaminating sense and antisense preparations, suppressed gene expression.

SAGE, Serial Analysis of Gene Expression: a method for the comprehensive analysis of gene expression patterns in a given tissue or treatment. It is based on the sequencing of short sections of 9-17 nucleotides ("tags") of cDNAs that have been ligated together into large concatemers. SAGE assumes that the abundance of a particular tag for a gene correlates with the abundance of that transcript in a sample.

Sequence analysis: analysis of the sequence of the elements of genes or proteins.

Sequencing: determination of the order of nucleotides (base sequences) in a DNA or RNA molecule, or the order of amino acids in a protein.

Shotgun method: a sequencing method that involves randomly sequencing cloned pieces of the genome, with no foreknowledge of where the piece originally came from. This can be contrasted with "directed" strategies, in which pieces of DNA from known chromosomal locations are sequenced. Because there are advantages to both strategies, researchers have used both random shotgun and directed strategies in combination to sequence genomes.

SNP: Single Nucleotide Polymorphism: nucleotide site in a genomic DNA sequence that varies from one individual to another. It is thought that the primary source of genetic difference between any two plants is due to the presence of single nucleotide polymorphisms in their DNA. Furthermore, these SNPs can be extremely useful in genetic mapping to follow the inheritance of specific sequences of DNA in a lineage.

SNP-typing is the process of determining the exact nucleotide at positions known to be polymorphic.

Southern blot: a technique for analyzing mixtures of DNA, whereby the presence and approximate size of one particular fragment of DNA can be ascertained by hybridization to a labelled specific probe. Named after its inventor, Dr E.M. Southern. See blotting.

SSR, Simple Sequence Repeat: See 'Microsatellite'.

Structural Genomics: studies of the structure and organization of the genome including DNA sequencing and physical and genetic mapping.

Synteny: the colinear relationships between the genomes of different organisms. Often chromosomal regions from related organisms, such as various grass species, contain corresponding genetic information and a similar gene order.

Tandem repeat sequences: multiple copies of the same nucleotide sequence on a chromosome; used as markers in physical mapping.

Targeted mutagenesis: deliberate change in the genetic structure directed at a specific site on the chromosome. Used in research to determine the targeted region's function.

T-DNA: The portion of the Ti (or Ri) plasmid of the bacterium, *Agrobacterium* that is transferred into the plant and integrated into the plant genome during infection by the bacterium. The wild-type T-DNA carries genes that cause the plant cells to grow as a tumor. T-DNA can be disarmed by removing these genes. The disarmed T-DNA can be used as a vector for genetic engineering in plants. .

Telomere: the end of a chromosome. This specialized structure is involved in the replication and stability of linear DNA molecules.

TILLING, Targeting Induced Local Lesions In Genomes: The end-product of the TILLING process is a plant (and its offspring) that has been identified with a change in a specific gene of interest. The plant line is then useful for determining the overall effect/role of that gene on the characteristics of the plant. TILLING is a high-throughput method for identifing specific gene knockouts in mutant populations and is useful as a tool of reverse genetics. TILLING utilizes PCR-based screening of plants generated through chemical mutagenesis (generally via EMS treatment), often resulting in the isolation of missense and nonsense mutant alleles of the targeted gene(s). TILLING permits the high-throughput identification of mutations in target genes without the production of genetically modified organisms and it can be an efficient way of identifying mutants in a specific gene that might not confer a strong phenotype by itself.

Tm: the melting point of a double-stranded nucleic acid. Technically, this is defined as the temperature at which 50% of the strands are in the double-stranded form and 50% are single-stranded, i.e. midway along the melting curve.

Transcription: the synthesis of a mRNA copy from a sequence of DNA (a gene) by a DNA polymerase; the first step in gene expression.

Transcription factor: protein assisting the transcription of a gene by a DNA polymerase.

Transcriptome: the full complement of activated genes, mRNAs, or transcripts in a tissue at a particular time, in response to a specific stimulus.

Transformation: change of the genetic status of a cell, e.g. by inserting–DNA.

Transgenic organism: an experimentally produced organism in which DNA has been artificially introduced.

Transposable element: mobile sequences of DNA present in the genome of the prokaryotes and the eukaryotes.

Two-dimensional polyacrylamide gel electrophoresis: a method of separating proteins as a function of their isoelectric point (pI) in the first dimension and as a function of their molecular mass in the second dimension. This technique, widely used in plant proteomics, combines isoelectric focusing (IEF) in the first dimension and SDS-PAGE (Sodium Dodecyl Sulfate-Poly Acrylamide Gel Electrophoresis) in the second dimension.

Two-hybrid system: a yeast-based method to test protein-protein interactions *in vivo* by reconstituting a functional GAL4 or LexA transcription factor.

VIGS, Virus Induced Gene Silencing: a method of transiently interrupting the function of a specific target gene through RNA interference. A plant virus is used as the vehicle for introducing the interfering RNA.

Western blot: a technique for analyzing proteins to show the presence, size and abundance of one particular type of protein. Similar to Southern or Northern blotting, except that a protein mixture is subject to electrophoresis on an acrylamide gel, and the "probe" is an antibody, which recognizes the protein of interest. See blotting.

YAC, Yeast Artificial Chromosome: a method used to clone and propagate large fragments of DNA in yeast

List of Contributors

Adam-Blondon Anne-Françoise, UMR INRA/CNRS/Université d'Evry Génomique Végétale, 2 rue Gaston Crémieux, BP5708, 91057 Evry cedex. **E-mail: adam@evry.inra.fr**

Al-Ghazi Yves, CSIRO Plant Industry, PO Box 1600, Canberra ACT 2601, Ausralia. E-mail: **ghaziy@ensam.inra.fr**

Alonso Ana Paula, Department of Plant Biology, Michigan State University, 166 Plant Biology Building, East Lansing, MI 48824. E-mail: **alonsoa@msu.edu**

Axelos Monique, INRA, Unité Physicochimie des macromolécules, BP 71 627, 44316 Nantes cédex 03. E-mail : **Monique.Axelos@nantes.inra.fr**

Bendahmane Abdelhafid, UMR Génomique végétale, INRA, 2 rue Gaston Crémieux, CP 5708, 91057 Évry. E-mail : **bendahm@evry.inra.fr**

Bergot Gaëlle, Institut de Biotechnologie des Plantes, CNRS-UMR 8618, Université de Paris XI, Bât. 630, 91405 Orsay cédex

Bernard Michel, UMR INRA / Univ. Clermont 2: Amélioration et santé des plantes INRA Site de Crouël 234 avenue du Brézet 63100 Clermont-Ferrand. E-mail : **Michel.Bernard@clermont.inra.fr**

Bolte Susanne, Plate-Forme d'Imagerie et de Biologie Cellulaire, IFR87 « La plante et son environnement », Institut des Sciences du Végétal, Avenue de la Terrasse, 91198 Gif-sur-Yvette cedex, France. E-mail : **susanne.bolte@isv.cnrs-gif.fr**

Bouchez David, INRA, Institut Jean-Pierre Bourgin, Génétique et amélioration des plantes, RD 10, 78026 Versailles cédex. E-mail: **bouchez@versailles.inra.fr**

Bouquet Alain, UMR INRA/ENSAM/CIRAD/IRD Biologie du Développement des Espèces Pérennes Cultivées, 2 Place Viala, 34060 Montpellier O1 Cedex. E-mail : **bouquet@ensam.inra.fr**

Boutet Stéphanie, INRA, Unité Nutrition azotée des plantes, RD 10, 78026 Versailles cédex. E-mail : **Stephanie.Boutet@versailles.inra.fr**

Boutté Yohann, Institut des sciences du végétal, Laboratoire de dynamique de la compartimentation cellulaire, bât. 23, avenue de la Terrasse, 91198 Gif-sur-Yvette cédex. E-mail : **bsj@isv.cnrs-gif.fr**

Bouzayen Mondher, UMR Génomique et biotechnologie des fruits,INRA/INPT/ ENSAT, BP 107 Auzeville, 31326 Castanet-Tolosan cédex. E-mail : **bouzayen@ensat.fr**

Brezellec Pierre, Laboratoire Génome et informatique, Tour Evry 2, 523 place des terrasses de l'agora, 91034 Évry cédex.

Briat Jean-François, UMR 5004, Biochimie et physiologie moléculaires des plantes, 2 place Viala, 34060 Montpellier cédex. E-mail: **briat@ensam.inra.fr**

Brown Spencer, Institut des sciences du végétal, Laboratoire de dynamique de la compartimentation cellulaire, bât. 23, avenue de la Terrasse, 91198 Gif-sur-Yvette cédex. E-mail: **spencer.brown@isv.cnrs-gif.fr**

Caboche Michel, UMR 1165, Génomique végétale, INRA, 2 rue Gaston Crémieux, CP 5708, 91057 Évry cédex. E-mail : **Michel.Caboche@versailles.inra.fr**

Carpentier Anne-Sophie, Laboratoire Génome et informatique, Tour Evry 2, 523 place des terrasses de l'agora, 91034 Évry cédex. E-mail : **carpentier@genopole.cnrs.fr**

Causse Mathilde, INRA, Génétique et amélioration des fruits et légumes, BP 94, 84143 Montfavet cédex. E-mail : **Mathilde.Causse@avignon.inra.fr**

Charcosset Alain, UMR INRA/INA-PG/Univ. Paris XI, Station de Génétique végétale, ferme du Moulon, 91190 Gif-sur-Yvette. E-mail: **charcosset@moulon.inra.fr**

Chiapello Hélène, INRA, Laboratoire Mathématique, informatique et génome, domaine de Vilvert, 78352 Jouy-en-Josas cédex.E-mail: **chiapell@diamant.jouy.inra.fr**

Choisne Nathalie, Génoscope, UMR 8030, 2 rue Gaston Crémieux, CP 5706, 91057 Évry cédex. E-mail : **choisne@genoscope.cns.fr**

Chupeau Yves, Institut Jean-Pierre Bourgin, INRA Centre de Versailles-Grignon, 78026 Versailles Cedex –France. E-mail: **chupeau@versailles.inra.fr**

Colonna Paul, INRA, Unité de recherche Polysaccharides, BP 71 627, 44316 Nantes cédex 03. E-mail: **colonna@nantes.inra.fr**

Daniel-Vedele Françoise, INRA, Unité de Nutrition azotée des plantes, RD 10, 78026 Versailles cédex. E-mail : **Francoise.Vedele@versailles.inra.fr**

Davey, Michael R. Plant Sciences Division, School of Biosciences, University of Nottingham, Sutton Bonington Campus, LE12 5RD, UK. E-mail: **mike.davey@nottingham.ac.uk**

Deborde Catherine, UMR Physiologie et Biotechnologie Végétales, INRA, Université Bordeaux 1, Université Victor Ségalen Bordeaux 2, BP 81, F-33883 Villenave d'Ornon, France. E-mail : **Catherine.Deborde@bordeaux.inra.fr**

Delseny Michel, UMR 5096, Laboratoire Génome et développement des plantes, université de Perpignan, 52 avenue de Villeneuve, 66860 Perpignan cédex. E-mail : **delseny@univ-perp.fr**

Demange Nadia, Génoscope, UMR 8030, 2 rue Gaston Crémieux, CP 5706, 91057 Évry cédex.

Diaz-Lazcoz Yolande, Laboratoire Génome et informatique, Tour Evry 2, 523 place des terrasses de l'agora, 91034 Évry cédex. E-mail : **diaz@genopole.cnrs.fr**

Dieuaide-Noubhani Martine, UMR 619 Physiologie et biotechnologie végétales, INRA, 33883 Villenave-d[1]Ornon cédex. E-mail : **dieuaide@bordeaux.inra.fr**

D'Hont Angélique, UMR 1096, PIA (Polymorphismes d'Intérêt Agronomique), CIRAD, TA40/03, Avenue Agropolis, 34398 Montpellier cédex 5, France. E-mail : **d'hont@cirad.fr**

Doumas Patrick, UMR 5004 Biochimie et physiologie moléculaires des plantes, INRA, 2 place Viala, 34060 Montpellier cédex. E-mail : **doumas@ensam.inra.fr**

Echeverria Manuel, UMR 5096 Laboratoire Génome et développement des plantes, Université de Perpignan, 52 avenue de Villeneuve, 66860 Perpignan cédex. E-mail : **Echeverria,manuel@univ-perp.fr**

Étienne Christelle, INRA, Station de Génétique et amélioration des fruits et légumes, BP 94, 84143 Montfavet cédex.

Falque Matthieu, INRA, Station de génétique végétale, ferme du Moulon, 91190 Gif-sur-Yvette. E-mail : **falque@moulon.inra.fr**

Gamas Pascal, Laboratoire des interactions plantes-microorganismes, CNRS-INRA, BP 52627, 31326 Castanet-Tolosan cédex. E-mail : **Pascal.Gamas@toulouse.inra.fr**

Gibrat Jean-François, INRA, Unité Mathématique, informatique et génome, domaine de Vilvert, 78352 Jouy-en-Josas cédex.E-mail: **gibrat@diamant.jouy.inra.fr**

Grandbastien Marie-Angèle, INRA, Laboratoire de biologie cellulaire, RD 10, 78026 Versailles cédex. E-mail : **Marie-Angele.Grandbastien@versailles.inra.fr**

Granier Fabienne, INRA, Unité de Génétique et amélioration des plantes, RD 10, 78026 Versailles cédex. E-mail : **granier@versailles.inra.fr**

Grivet Laurent, UMR 1096, PIA (Polymorphismes d'Intérêt Agronomique), CIRAD, TA40/03, Avenue Agropolis, 34398 Montpellier cedex 5, France.

Hénaut Alain, Laboratoire Génome et informatique, Tour Evry 2, 523 place des terrasses de l'agora, 91034 Évry cédex. E-mail : **henaut@genopole.cnrs.fr**

Hirel Bertrand, INRA, Unité de Nutrition azotée des plantes, RD 10, 78026 Versailles cédex. E-mail: **hirel@versailles.inra.fr**

Hoarau Jean Yves, UMR 1096, PIA (Polymorphismes d'Intérêt Agronomique), CIRAD, TA40/03, Avenue Agropolis, 34398 Montpellier cedex 5, France.

Joets Johann, UMR INRA / INA PG / Univ. Paris 11 / CNRS : Génétique Végétale INRA Ferme du Moulon 91190 Gif sur Yvette. E-mail : **joets@moulon.inra.fr**

Johansen Jorum N, Laboratoire de biologie cellulaire, RD 10, 78026 Versailles cédex.

Johnson Xenie, INRA, Institut JP Bourgin, Génétique et amélioration des plantes, RD 10, 78026 Versailles cédex. E-mail : **johnson @versailles.inra.fr**

Journet Étienne-Pascal, Laboratoire des interactions plantes-microorganismes, CNRS-INRA, BP 52627, 31326 Castanet-Tolosan cédex. E-mail : **journet@toulouse.inra.fr**

Joyard Jacques, UMR 5019, Laboratoire de Physiologie cellulaire végétale, CEA, avenue des Martyrs, 38054 Grenoble cédex 9.E-mail: **joyard@dsvsud.cea.fr**

Kluge Christoph, Institut des sciences du végétal, Laboratoire de dynamique de la compartimentation cellulaire, bât. 23, avenue de la Terrasse, 91198 Gif-sur-Yvette cédex. E-mail : **Kluge@isv.cnrs-gif.fr**

Krapp Annemarie, INRA, Unité Nutrition azotée des plantes, RD 10, 78026 Versailles cédex. E-mail: **krapp@versailles.inra.fr**

Kronenberger Jocelyne, INRA, Laboratoire de biologie cellulaire, RD 10, 78026 Versailles cédex. E-mail : **jocelyne.kronenberger@versailles.inra.fr**

Lasserre Éric, UMR 5096, Laboratoire Génome et développement des plantes, université de Perpignan, 52 avenue de Villeneuve, 66860 Perpignan cédex.

Lea Peter J., Department of Biological Sciences, Institute of Environmental and Natural Sciences, University of Lancaster, United Kingdom. E-mail: **p.lea@lancaster.ac.uk**

Le Gouis Jacques, INRA, Laboratoire de Génétique et amélioration des plantes, domaine de Brunehaut, 80200 Estrées-Mons. E-mail: **legouis@Mons.inra.fr**

Lefebvre Véronique, INRA, Unité de Génétique et amélioration des fruits et légumes, BP 94, 84143 Montfavet cédex. E-mail : **Veronique.Lefebvre@avignon.inra.fr**

Lelarge Caroline, Institut de Biotechnologie des Plantes, CNRS-UMR 8618, Université de Paris XI, Bât. 630, 91405 Orsay cédex. E-mail : **lelarge@ibp.u-psud.fr**

Loudet Olivier, INRA, Unité de Nutrition azotée des plantes, RD 10, 78026 Versailles cédex. E-mail: **loudet@versailles.inra.fr**

Louis Alexandra, Infobiogen, Tour Evry 2, 523 place des terrasses de L'agora, 91034 Évry cédex. E-mail : **louis @ infobiogen**

MacDougall Alistair J., BBSRC. Food Research Institute. Colney Lane. NR4 7UA. Norwich (UK). E-mail : **alistair.macdougall@bbsrc.ac.uk**

Manicacci Domenica, INRA, Station de génétique végétale, ferme du Moulon, 91190 Gif-sur-Yvette. E-mail : **manicacci@moulon.inra**

Marin Antoine, INRA, Unité Mathématique, informatique et génome, domaine de Vilvert, 78352 Jouy-en-Josas cédex. E-mail : **Antoine.Marin@jouy.inra.fr**

Martins Menck Carlos Frederico, Instituto de Ciências Biomédicas, Departamento de Microbiologia, Universidade de São Paulo. Av. Prof. Lineu Prestes No 1374 Cid. Universitária 05508-900, Sao Paulo, SP, Brasil. E-mail : **cfmmenck@usp.br**

Mazat Jean-Pierre, Inserm U688, Université Bordeaux 2, 146 rue Léo-Saignat, F 33076, Bordeaux-cedex France. **jpm@u-bordeaux2.fr.**

Menossi Marcelo, Centro de Biologia Molecular e Engenharia Genética and Departamento de Genética e Evolução, Instituto de Biologia, Universidade Estadual de Campinas. Caixa Postal 6010, CEP 13083-970, Campinas, SP, Brazil. E-mail: **menossi@unicamp.br**.

Meyer Christian, INRA, Unité Nutrition azotée des plantes, Institut Jean-Pierre Bourgin, RD 10, 78026 Versailles cédex. E-mail : **christian.meyer@versailles.inra.fr**

Mhiri Corinne, INRA, Laboratoire de biologie cellulaire, RD 10, 78026 Versailles cédex. E-mail : **Corinne.Mhiri@versailles.inra.fr**

Michelet Laure, Génoscope, UMR 8030, 2 rue Gaston Crémieux, CP 5706, 91057 Évry cédex.

Mohseni-Zadeh Sarah, Laboratoire Génome et informatique, Tour Évry 2, 523 place des terrasses de l'agora, 91034 Évry cédex.

Moing Annick, UMR Physiologie et Biotechnologie Végétales, INRA, Université Bordeaux 1, Université Victor Ségalen Bordeaux 2, BP 81, F-33883 Villenave d'Ornon, France. E-mail : **Annick.Moing@bordeaux.inra.fr**

Montalent Pierre, UMR INRA / INA PG / Univ. Paris 11 / CNRS : Génétique Végétale INRA Ferme du Moulon 91190 Gif sur Yvette. E-mail :**montalent@moulon.inra.fr**

Morot-Gaudry Jean-François, INRA, Unité Nutrition azotée des plantes, RD 10, 78026 Versailles cédex. E-mail : **morot@versailles.inra.fr**

Mouille Grégory, Laboratoire de biologie cellulaire, RD 10, 78026 Versailles cédex. E-mail : **Gregory.Mouille@versailles.inra.fr**

Murigneux Alain, Biogemma, Campus universitaire des Cézeaux, 24, avenue des Landais, 63170 Aubière. E-mail : **murigneux@biogemma.com**

Nicolas Pierre, INRA, Laboratoire Mathématique, informatique et génome, domaine de Vilvert, 78352 Jouy-en-Josas cédex. E-amil : **Pierre.Nicolas@jouy.inra.fr**

Noctor Graham, Institut de Biotechnologie des Plantes, CNRS-UMR 8618, Université de Paris XI, Bât. 630, 91405 Orsay Cedex. E-mail : **noctor@ibp.u-psud.fr**

Orjeda Gisela, Génoscope, 2 rue Gaston Crémieux, CP5708, 91057 Évry cédex.

Pelletier Éric, Génoscope, UMR 8030, 2 rue Gaston Crémieux, CP 5706, 91057 Évry cédex. E-mail : **gorjeda@evry.inra.fr**

Pasek Sophie, Laboratoire Génome et informatique, Tour Evry 2, 523 place des terrasses de l'agora, 91034 **pasek@genopole.cnrs.fr**

Pereira de Souza Anete, Centro de Biologia Molecular e Engenharia Genética and Departamento de Genética e Evolução, Instituto de Biologia, Universidade Estadual de Campinas. Caixa Postal 6010, CEP 13083-970, Campinas, SP, Brazil. E-mail: **ANETE@UNICAMP.BR**

Perez Pascual, Biogemma, Campus universitaire des Cézeaux, 24 avenue des

Landais, 63170 Aubière. E-mail : **pascual.perez@biogemma.com**

Planchot Véronique, INRA, Unité de recherche Polysaccharides, rue de la

Géraudière, BP 71 627, 44316 Nantes cédex 03. E-mail : **Veronique.Planchot @nantes.inra.fr**

Popineau Yves, INRA, Unité de recherche Protéines végétales, rue de la Géraudière, BP 71 627, 44316 Nantes cédex 03. E-mail : **Yves.Popineau@nantes.inra.fr**

Poujol-Talbot Christel, CNRS, Laboratoire de dynamique de la compartimentation cellulaire, avenue de la terrasse, 91198 Gif-sur-Yvette cédex.

Prioul Jean-Louis, Institut de Biotechnologie des Plantes, CNRS-UMR 8618, Université de Paris XI, Bât. 630, 91405 Orsay Cedex. E-mail : **Jean-Louis.Prioul@ibp.u-psud.fr**

Quétier Francis, Génoscope, UMR 8030, 2 rue Gaston Crémieux, CP 5706, 91057 Évry cédex. E-mail: **quetier@genoscope.cns.fr**

Raymond Philippe, UMR 619, Physiologie et biotechnologie végétales, INRA, 33883 Villenave d[1]Ornon cédex. E-mail : **raymond@bordeaux.inra.fr**

Ring Steve G., Food Research Institute. Colney Lane. NR4 7UA. Norwich (UK) steve.E-mail:**ring@bbsrc.ac.uk**

Risler Jean-Loup, Laboratoire Génome et informatique, Tour Evry 2, 523 place des terrasses de l'agora, 91034 Évry cédex. E-mail : **risler@genopole.cnrs.fr**

Riva Alessandra, Laboratoire Génome et informatique, Tour Evry 2, 523 place des terrasses de l'agora, 91034 Évry cédex. E-mail: **gucki@genopole.cnrs.fr**

Robin Stéphane, INA-PG-OMIP/ INRA-BIA, 16 rue Claude Bernard, 75231 Paris cédex 05. E-mail : **Stephane.Robin@inapg.inra.fr**

Roby Claude, UMR 5019 Physiologie cellulaire végétale, CEA, 17 rue des Martyrs 38054 Grenoble cédex. E-mail : **croby@cea.fr**

Rodolphe François, INRA, Laboratoire Mathématique, informatique et génome, domaine de Vilvert, 78352 Jouy-en-Josas cédex. E-mail : **Francois.Rodolphe@ jouy.inra.fr**

Rogowsky Peter, ENS, UMR 5667, 46 allée d[1]Italie, 69364 Lyon cédex 07. **peter.rogowsky@lyon.inra.fr**

Rolin Dominique, UMR Physiologie et Biotechnologie Végétales, INRA, Université Bordeaux 1, Université Victor Ségalen Bordeaux 2, BP 81, F-33883 Villenave d'Ornon, France. E-mail : **Dominique.Rolin@bordeaux.inra.fr**

Romieu Charles, UMR INRA/ENSAM/CIRAD/IRD Biologie du Développement des Espèces Pérennes Cultivées, 2 Place Viala, 34060 Montpellier Cedex 01. E-mail: **romieu@ensam.inra.fr**

Rossignol Michel, Unité de Recherche Protéomique INRA 2 place Viala 34060 Montpellier CEDEX 1 E-mail: **rossignol@ensam.inra.fr**

Rossini Pinto Luciana, Centro de Biologia Molecular e Engenharia Genética and Departamento de Genética e Evolução, Instituto de Biologia, Universidade Estadual de Campinas. Caixa Postal 6010, CEP 13083-970, Campinas, SP, Brazil

Rothan Christophe, UMR Physiologie et biotechnologie végétales, BP 81, 33883 Villenave-d[1]Ornon cédex. E-mail : **rothan@bordeaux.inra.fr**

Salanoubat Marcel, Génoscope, UMR 8030, 2 rue Gaston Crémieux, CP 5706, 91057 Évry cédex. E-mail : **salanou@genoscope.cns.fr**

Samain Sylvie, Génoscope, UMR 8030, 2 rue Gaston Crémieux, CP 5706, 91057 Évry cédex.

Santoni Sylvain, UMR Diversité et génomes des plantes cultivées, INRA,

Station d[1]amélioration des plantes, 2 place Viala, 34060 Montpellier cédex. E-mail: **santoni@ensam.inra.fr**

Satiat-Jeunemaitre Béatrice, CNRS, Laboratoire de dynamique de la compartimentation cellulaire, bât. 23, avenue de la terrasse, 91198 Gif-sur-Yvette cédex. E-mail : **beatrice.satiat-jeunemaitre@isv.cnrs-gif.fr**

Souza Glaucia, Instituto de Química - Departamento de Bioquímica, Universidade de São Paulo, Av. Prof. Lineu Prestes 748 - Bloco 12 T sala 1208 Cid. Universitária 05508-900, São Paulo, SP, Brazil.

Stevens Rebecca, INRA, Génétique et amélioration des fruits et légumes, BP 94, 84143 Montfavet cédex.

Tam Sheh May, Department of Plant Sciences, One Shields Avenue, University of California, Davis, 95616, USA. E-mail: **amshehmay@yahoo.com**

Torrésani Bruno, LATP, CMI, Université de Provence, 39 rue Joliot-Curie, 13453 Marseille Cedex 13. E-mail : **torresan@cmi.univ-mrs.fr**

Ulian Eugêno César, CTC - Centro de Tecnologia Canavieira. Bairro Santo Antonio S/N, Santo Antonio, Caixa-Postal: 162, 13400-970, Piracicaba, SP, Brazil.

Vedel Françoise, INRA, Unité Nutrition azotée des plantes, Institut Jean-Pierre bourgin, RD 10, 78026 Versailles Cedex. E-mail: **Francoise.Vedele@versailles.inra.fr**

Vincentz Michel, Pinto Luciana Rossini, Centro de Biologia Molecular e Engenharia Genética and Departamento de Genética e Evolução, Instituto de Biologia, Universidade Estadual de Campinas. Caixa Postal 6010, CEP 13083-970, Campinas, SP, Brazil. E-mail: **mgavince@unicamp.br**

Weissenbach Jean, Génoscope, UMR 8030, 2 rue Gaston Crémieux, CP 5706, 91057 Évry cédex. E-mail : **Jsbach@genoscope.cns.fr.**

Zivy Michel, UMR de Génétique végétale, INRA, ferme du Moulon, 91190 Gif-sur-Yvette. **E-mail: zivy@moulon.inra.fr**

Index

Q

R

Color Plate Section

Chapter 2

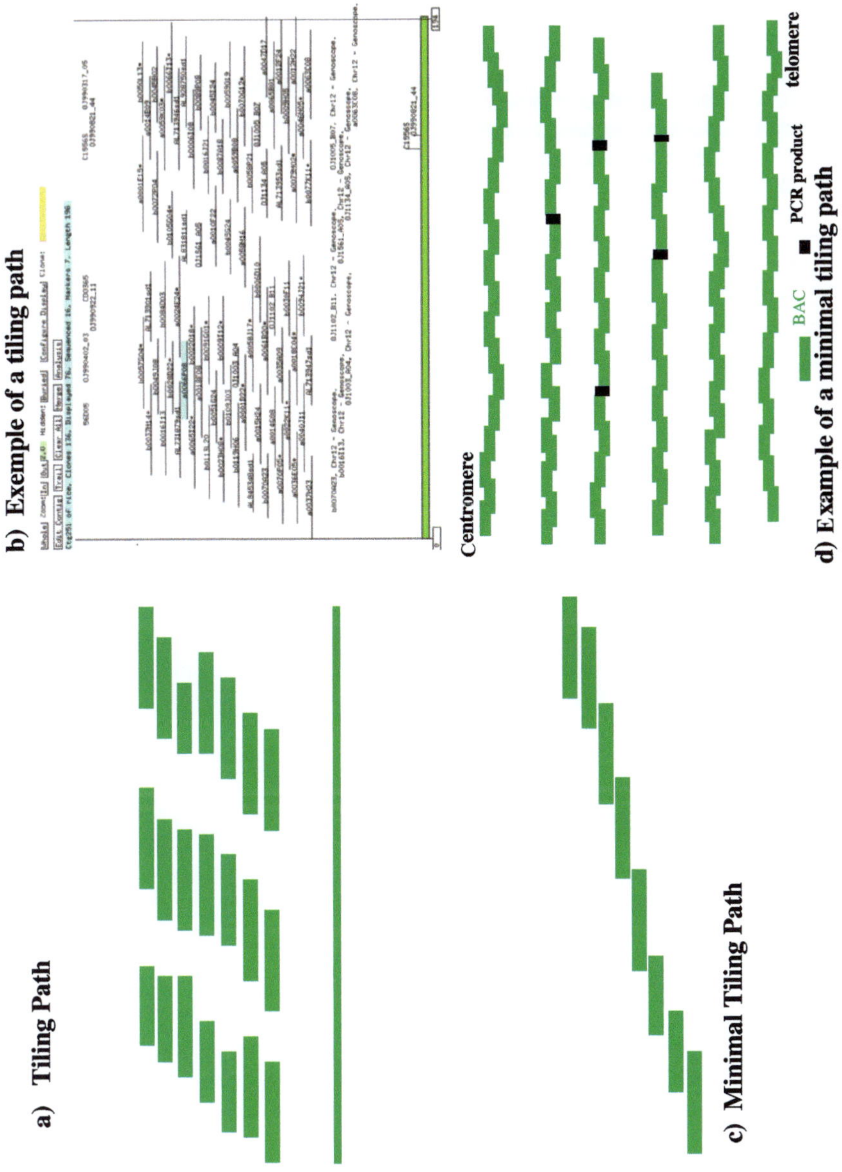

Fig. 2.3 Tiling path (TP) and minimum tiling path (MTP) : a) Basic scheme of the tiling path generated by the FPC software; b) screen display of a tiling path ; a part of the short arm of *Arabidopsis thaliana* chromosome 3; c) the minimal tiling path corresponding to the tiling path shown in a, reducing the number of BACs from 21 to 8 ; d) MTP covering completely the 10 Mb of the short arm of *A. thaliana* chromosome 3, with 102 BACs and 5 PCR products (for these later ones, the corresponding overlaping BACs were identified, but the sequencing of PCR products is cheaper and faster, while being sufficient to fill-in these short gaps).

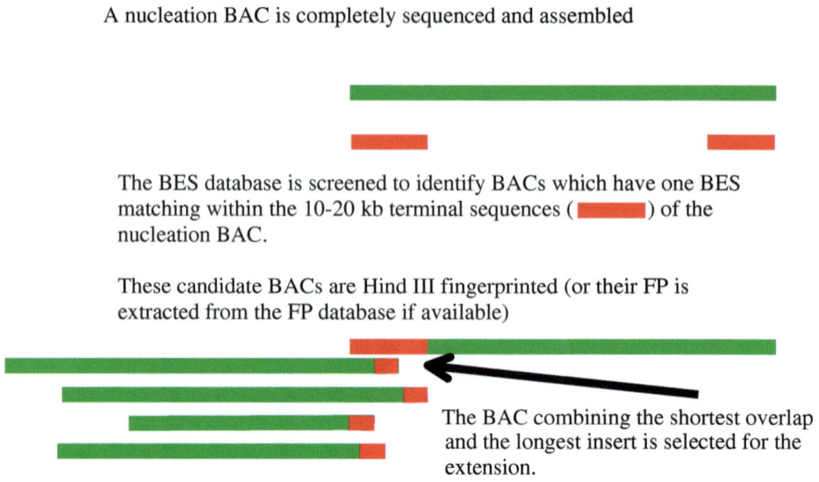

Fig. 2.4 The « sequence tag connectors » strategy (also called « sequence and go). The progression starts with "nucleation" BACs in a bi-directional process (for simplification, only the left extension is shown).

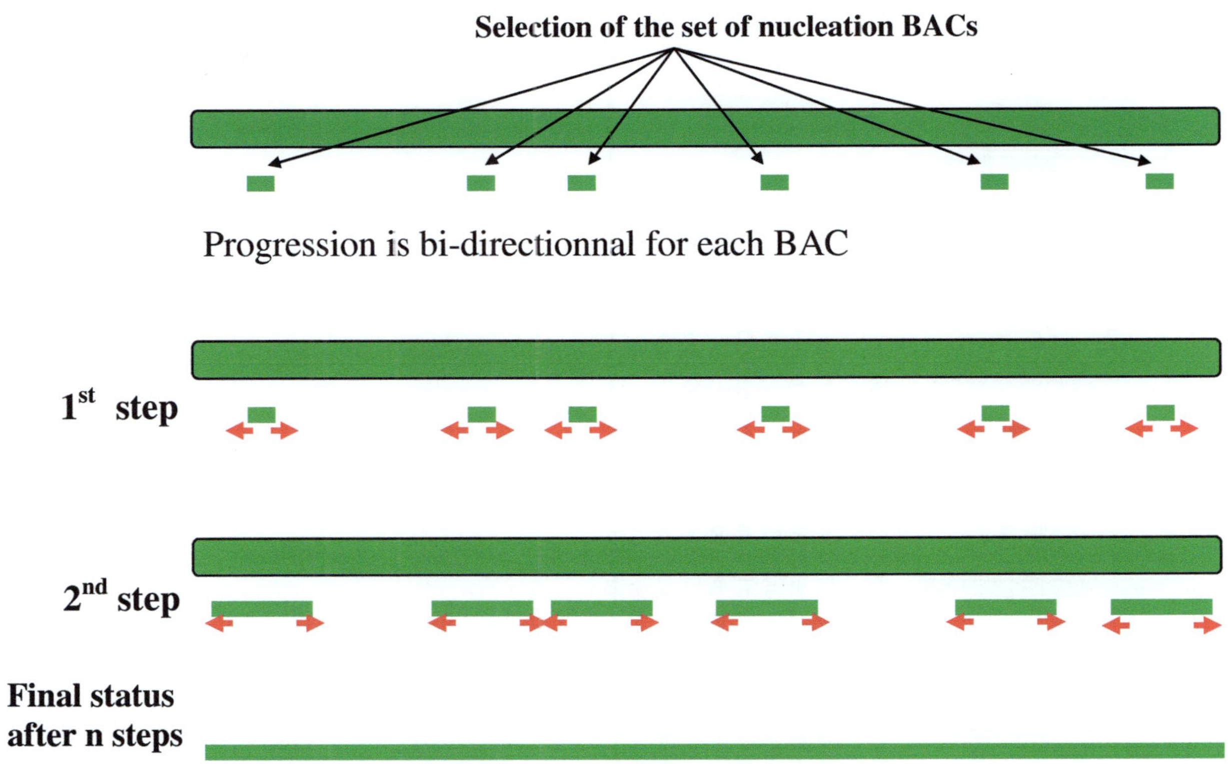

Fig. 2.5 STC progression to encompass whole chromosomes
The STC process is shown here as synchronous for the successive rounds of extension. In fact, the progression is independent and is linked to the rapidity at which each BAC is sequenced and its assembly is achieved.

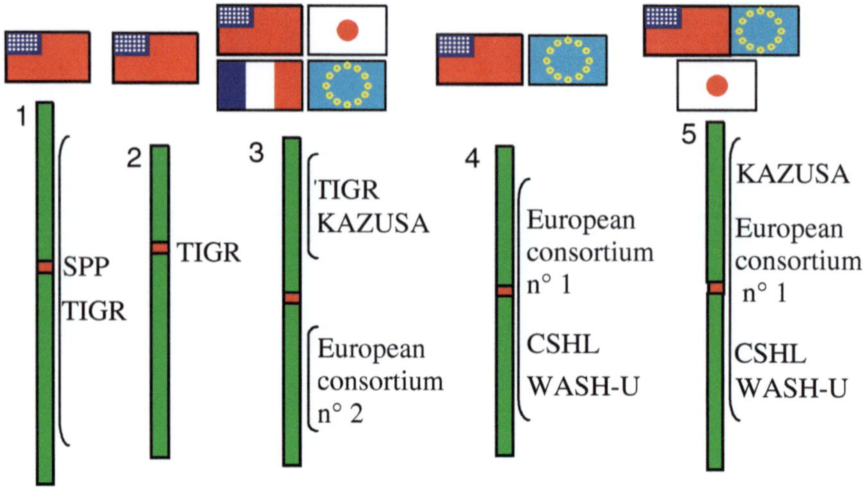

Fig. 2.6 Chromosome allocation for the *Arabidopsis thaliana* genome sequencing.
SPP Consortium
Stanford DNA Sequencing & Technology Center, Stanford University,
Arabidopsis thaliana Genome Center, University of Pennsylvannia
Plant Gene Expression Center, USDA
TIGR, The Institute for Genomic Research, Washington DC
KAZUSA DNA Research Institute, Japan
European Consortium n°1(coordinator : John Innes Center, Norwich)
European Consortium n°2 (coordinator : Genoscope, Evry)
Wash-U, Washington University, St Louis
CSHL, Cold Spring Harbor Laboratory, NY

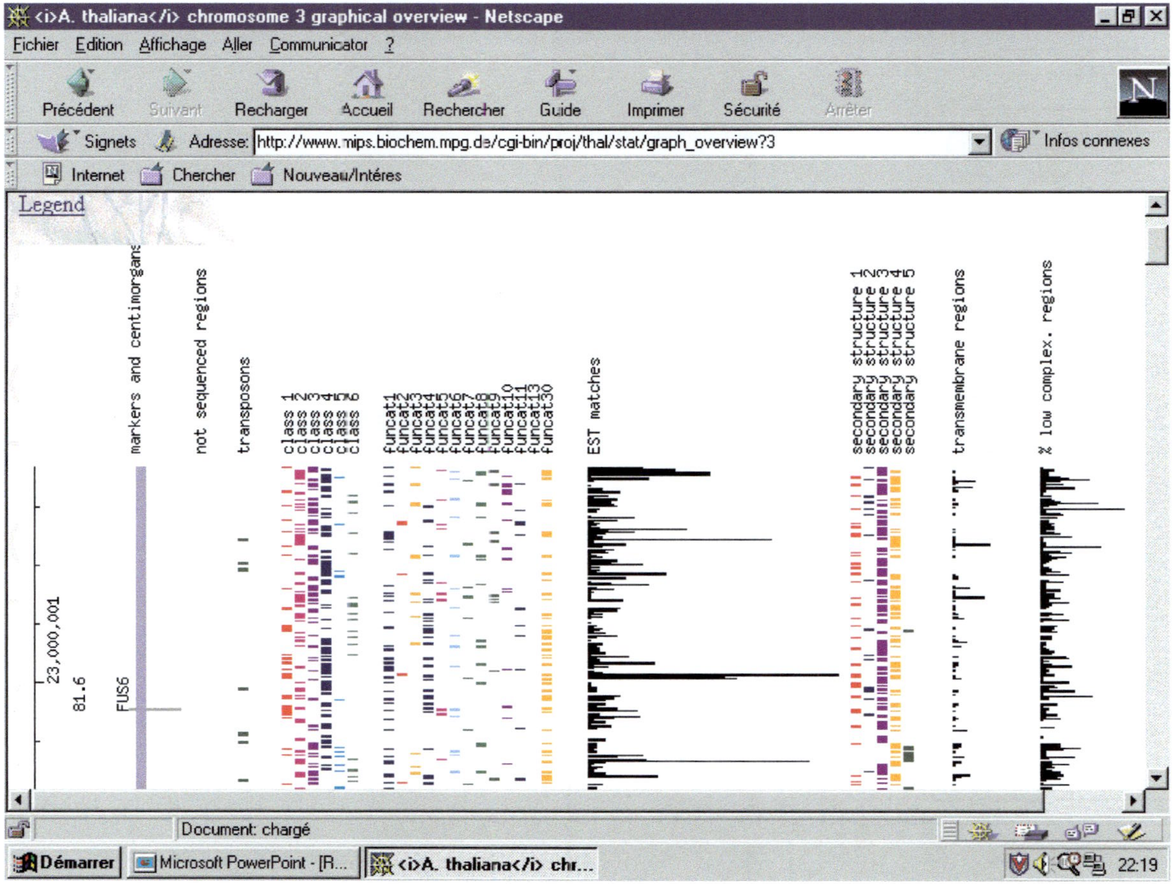

Fig. 2.7 An example in the *ab initio* annotation of *Arabidopsis thaliana* genome. Screen display (early 2000) of one end of *A. thaliana* chromosome 3 on the MIPS website (Münich, Germany).

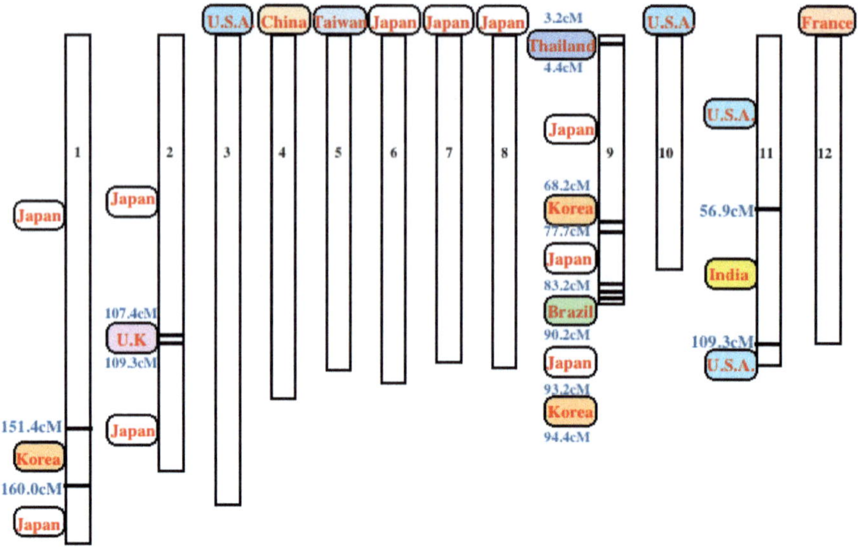

Fig. 2.8 Chromosome allocation for the IRGSP sequencing groups (November 2002)

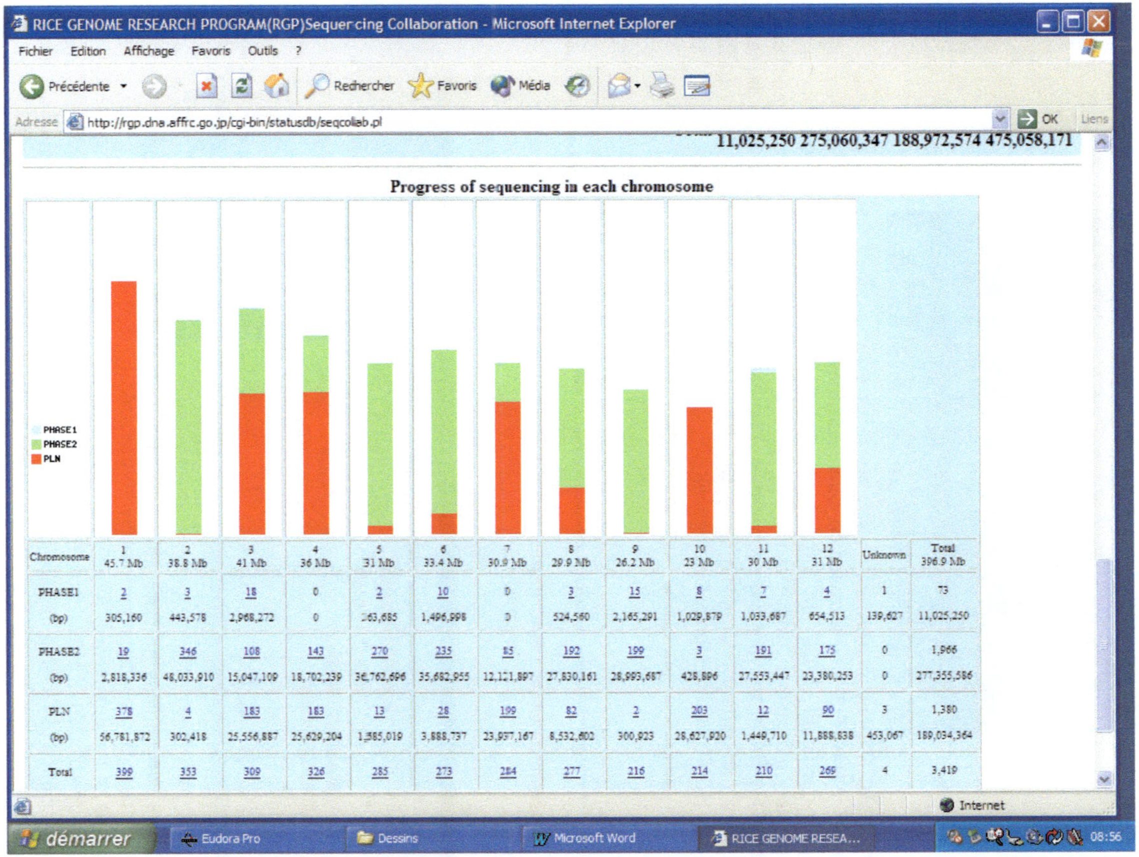

Fig. 2.9 Graphical view of the sequencing status for the Rice genome in March 2003
The sequencing status was regularly updated on the website http://rgp.dna.affrc.go.jp/cgi-bin/statusdb/irgsp-status.cgi. Phase 1 and phase 2 correspond to steps in partial sequencing, PLN is the GenBank section collecting finished sequences for plants, fungi and algae. The sequencing was completed in mid 2005.

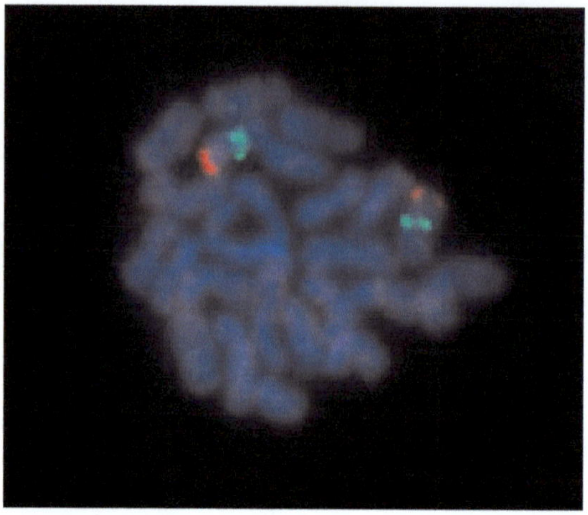

Fig. 2.10 BAC-FISH hybridizations
An example of BAC FISH hybridization on early pachytene rice chromosomes. Results confirm that the 2 candidate BACs are located on chromosome 12 and give several helpful complementary indications : which of the two arms, a rough estimate of the BAC positions, whether the two BACs are very close each other or far away, which one is the most proximal to the telomere (important information when two BACs are very close each other) (picture Angélique D'HONT, CIRAD, Montpellier, France).

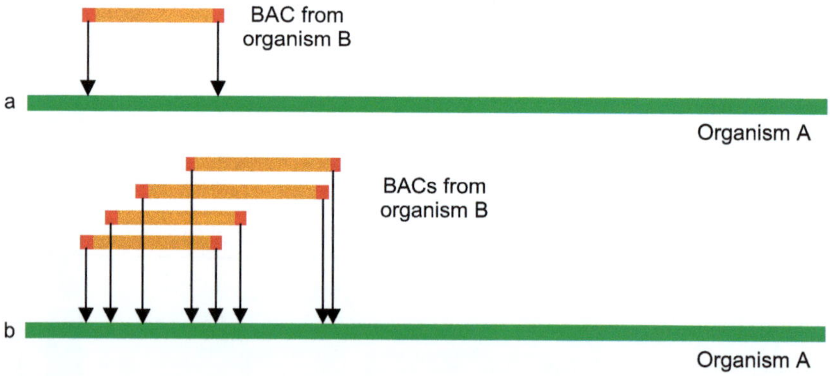

Fig. 2.11 A new strategy for the rapid construction of a physical map of a genome (when a complete sequence is available for a relative organism).
The genome of the organism A (green line) has been completely sequenced. a) the two end sequences (red blocks) of a BAC from the other organism B are positioned (BLAST alignment) on the A sequence ; b) even a medium coverage (6x) of a BAC library of the organism B allows the validation of the internal parts of this BAC from B (overlapping BACs of A).

Chapter 10

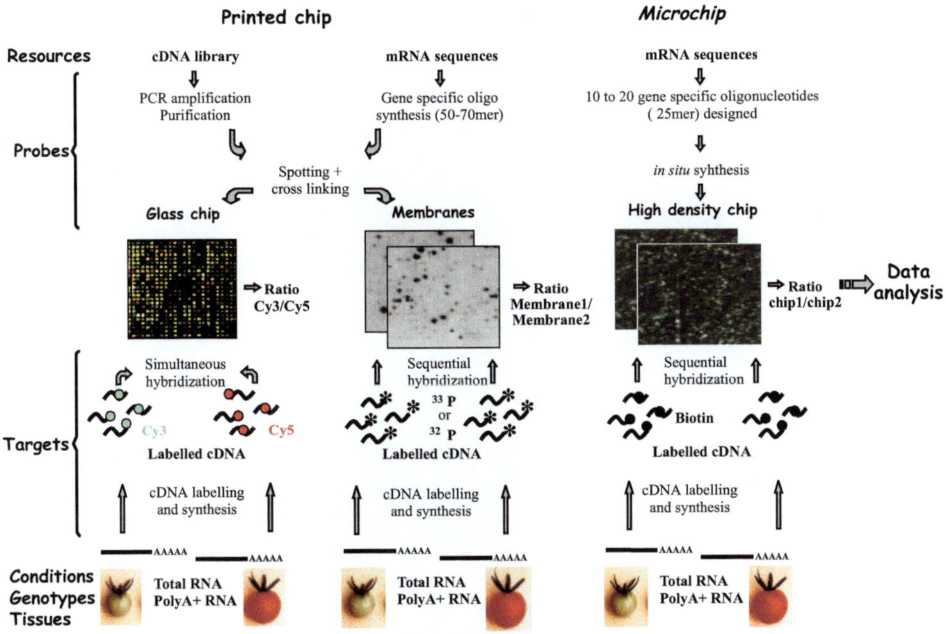

Fig. 10.1 (previous page): Scheme describing *in situ* (Affymetrix chips) and printed (micro- and macroarray) DNA array production and process.

Printed microarrays: (1) Probes are amplified from cDNA clones selected from libraries or from genomic DNA corresponding to ORFs (open reading frames) and then purified. Alternatively, 50 to 70 long oligonucleotides can be designed specifically to a given mRNA (cDNA clone or ORF) and synthesized. Probes can be then printed on the array on a slide or a nylon membrane by the use of a high density spotting robot. To prevent them from being stripped during hybridization, probes are light, heat or chemically cross linked to the support. During target processing (2), total RNA or PolyA+RNA is reverse transcribed (RT) and labeled either simultaneously or after transcription depending on the label used. Fluorochromes (Cy3 and Cy5) can be coupled to nucleotides for simultaneous labeling or bound after RT to the probe; radioactive compounds (33P or 32P) have to be incorporated during the RT only. The two fluorescent samples that are differently labeled are competitively hybridized to the slide. Each fluorescent label is excited by a specific energy laser and the fluorescence is recorded with a high resolution scanner. The relative signal intensity (ratio Cy3/Cy5) for both fluorochromes allows an estimate of the relative abundance of a given RNA species in the two samples. For radioactive labeling, targets are hybridized to different membranes or sequentially to the same membrane after stripping. The relative abundance of the transcript is estimated by comparison of the signals between the two membranes.

In situ synthesized probes (1) are 10 to 20 gene specific 25 base paired oligonucleotides synthesized directly on the chip by photolithographic methods. More than 250,000 elements can be arrayed on this kind of support (Affymetrix). Targets (2) are reverse transcribed mRNAs in which a T7 promoter is added. These cDNAs are transcribed into cRNA in the presence of biotinylated nucleotides and then fragmented before hybridization to the chips. Hybridized cRNAs are revealed upon interaction with a streptavidin/fluorophore complex. The relative abundance of the transcripts is estimated by comparison of the signals between the different hybridized chips.

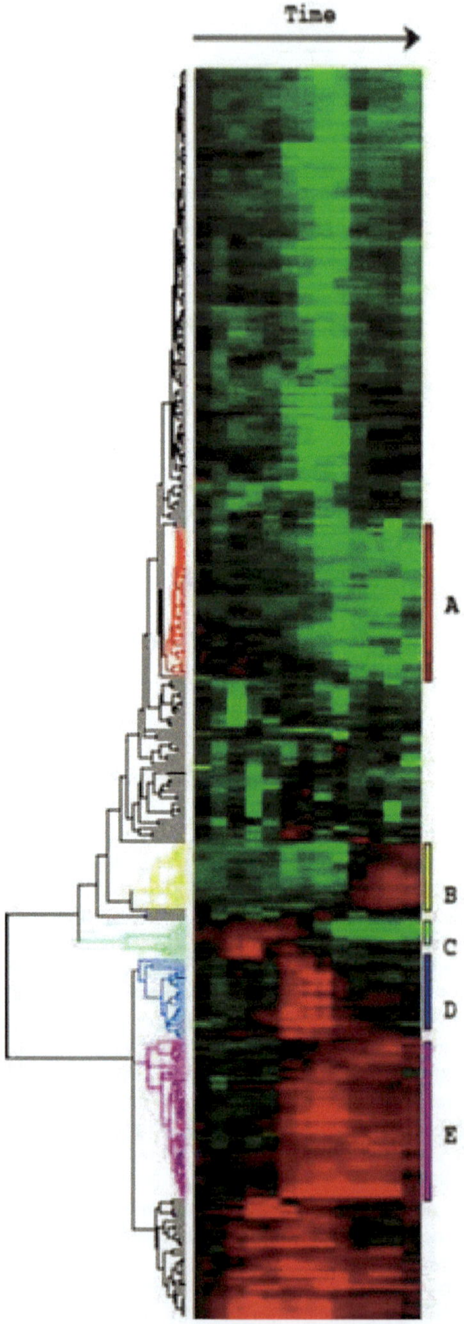

Fig. 10.3 Hierarchical clustering of time course expression profiles in yeast (Eisen *et al.*, 1998). Numeral values are coded in color (green = high expression level, red = low). Each row represents a gene and each column a condition (time). A functional interpretation of A-E groups is given in the original article.

Chapter 11

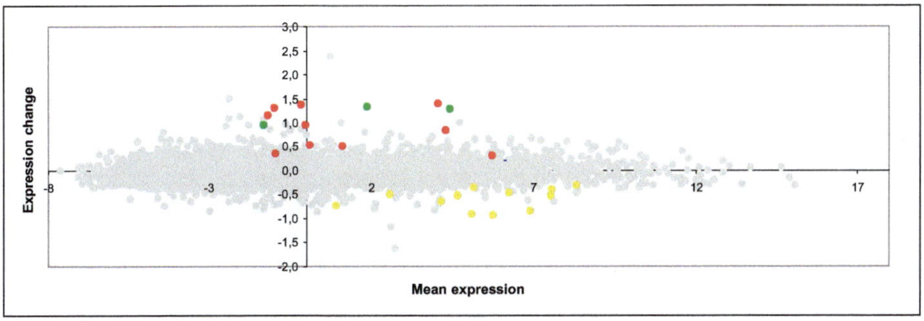

Fig. 11.4 The expression change in function of the factor sulphur as calculated by a spreadsheet (MS-Excel)

The figure shows the change of expression of the genes as a function of the sulphur source, against their mean expression. The potentially interesting genes are those away from the main body of the cloud. The highlighted genes are the ones which proved to be of particular interest for the problem investigated by Sekowska *et al.* (2001); the reader may refer to their work for a detailed discussion. Note that not all of these genes would have been detected using the spreadsheet.

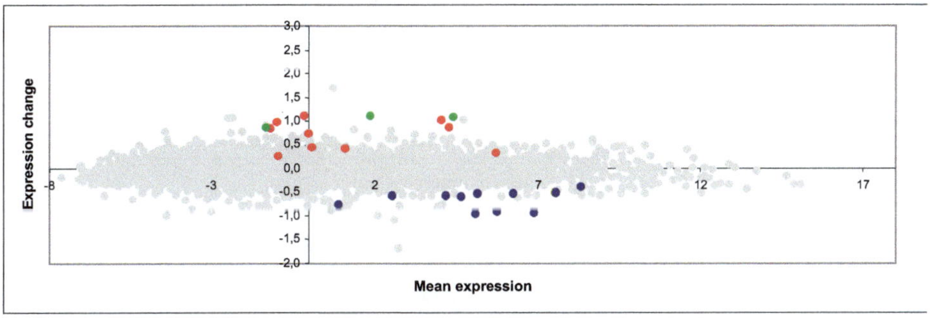

Fig. 11.5 The data cloud projected on the plane formed by axis 1 against axis 5 (PCA)

The figure shows the change of expression of the genes as a function of the sulphur source (axis 5), against their mean expression (axis 1), as calculated by PCA. The potentially interesting genes are those away from the main body of the cloud. The highlighted genes are the ones which proved to be of particular interest for the problem investigated by Sekowska *et al.* (2001); the reader may refer to their work for a detailed discussion. Note that not all of these genes would have been detected using PCA.

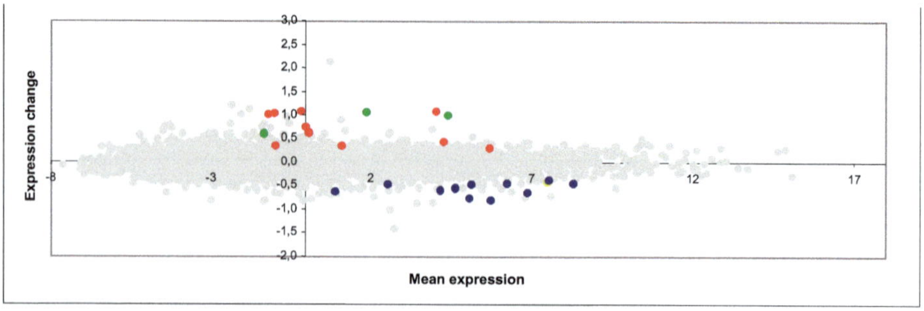

Fig. 11.6 The data cloud projected on the plane that separates well the sulphur source (ICA).

The figure shows the change of expression of the genes as a function of the sulphur source, as determined by ICA. The potentially interesting genes are those away from the main body of the cloud. The highlighted genes are the ones which proved to be of particular interest for the problem investigated by Sekowska *et al.* (2001); the reader may refer to their work for a detailed discussion. Note that not all of these genes would have been detected using ICA.

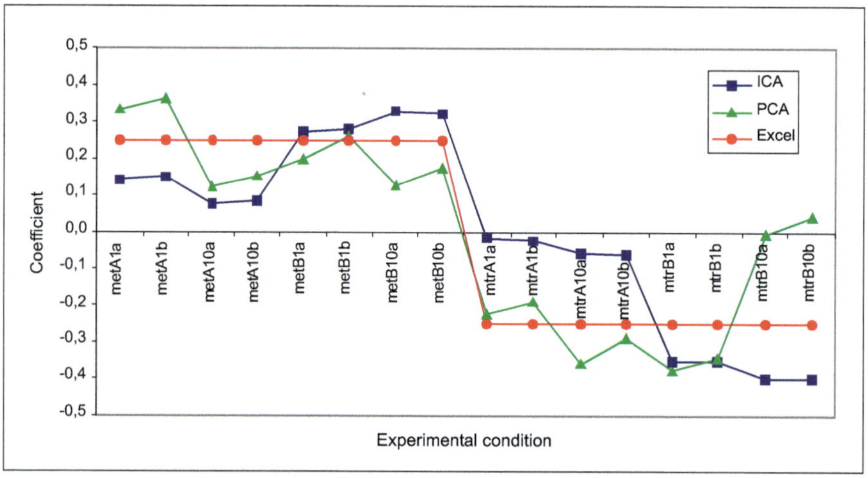

Fig. 11.7 The "weight" attributed to each experimental condition by a spreadsheet (MS-Excel), PCA and ICA.

The figure shows that a spreadsheet (MS-Excel) attributes the same "weight", or importance, to each experimental condition, whilst PCA and ICA do not.

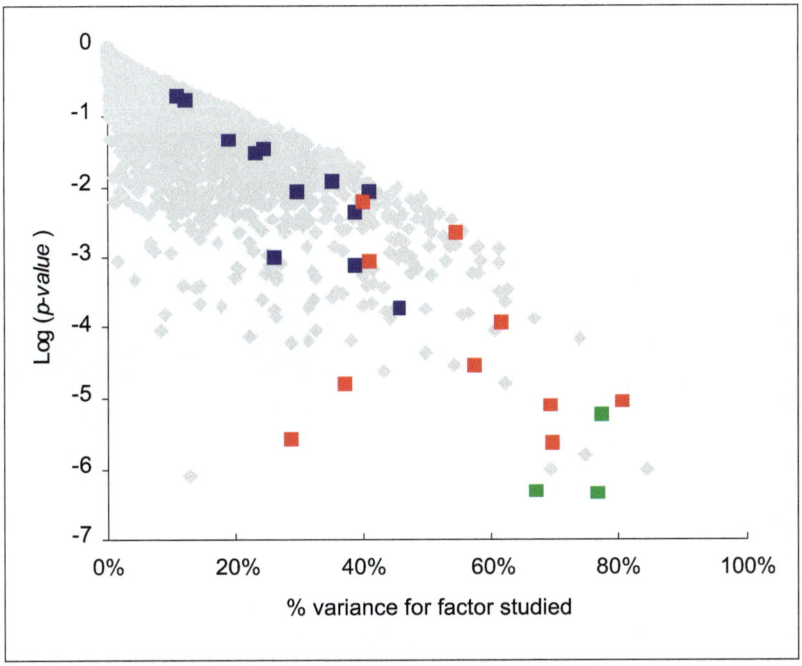

Fig. 11.8 The graphical representation of the results obtained with ANOVA for the factor sulphur.

The potentially interesting genes are those with a small *p-value* and a large variance, genes which are therefore in the bottom right part of the image. They are away from the main body of the cloud. The highlighted genes are the ones which proved to be of particular interest for the problem investigated by Sekowska *et al.* (2001); the reader may refer to their work for a detailed discussion. As can be seen, not all genes of interest would have been identified by the sole use of ANOVA.

Chapter 14

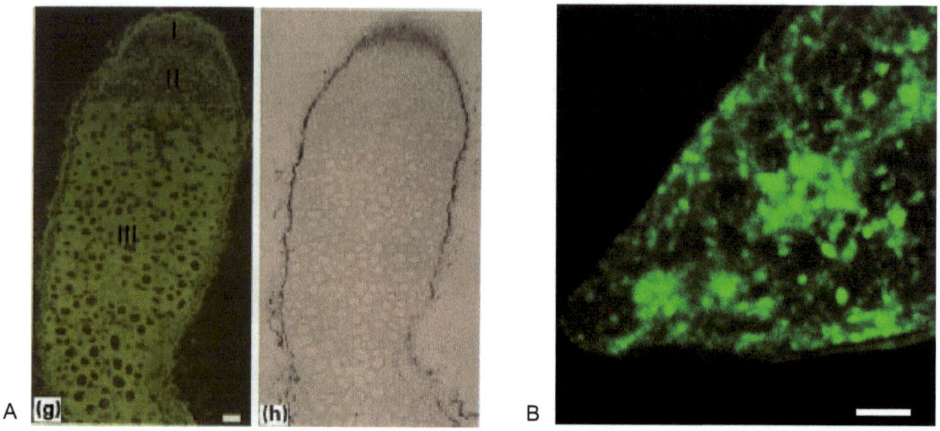

Fig. 14.1 RNA localization **A.** *In situ* hybridization with digoxigenin (DIG)-labelled probes. (g): Epifluorescence of a longitudinal section of a 20-days-old *Medicago sativa* symbiotic nodule, where characteristic zones can be seen: I: meristem; II invasion zone; III nitrogen-fixing zone enveloped by peripheral tissues. (h): Same section seen with bright field illumination indicating MsCA1 (coding for Carbonic Anhydrase) gene expression where a dark precipitate is detected with digoxigenin-labelled probes revealed by anti-DIG alkaline phosphatase antibody. For the colour reaction, nitro-blue tetrazolium salt (NTB) and 5-bromo-4-Chloro-3-indolyl-phosphate (BCTP) were used as substrate for the phosphatase (Coba de la Pena *et al.,* 1997). Bar 50 µm **B.** *In vivo* localization of enod40 RNA with GFP/MS2 bacteriophage in epidermal onion cells. Bar 20 µm (Campalans *et al.,* 2004)

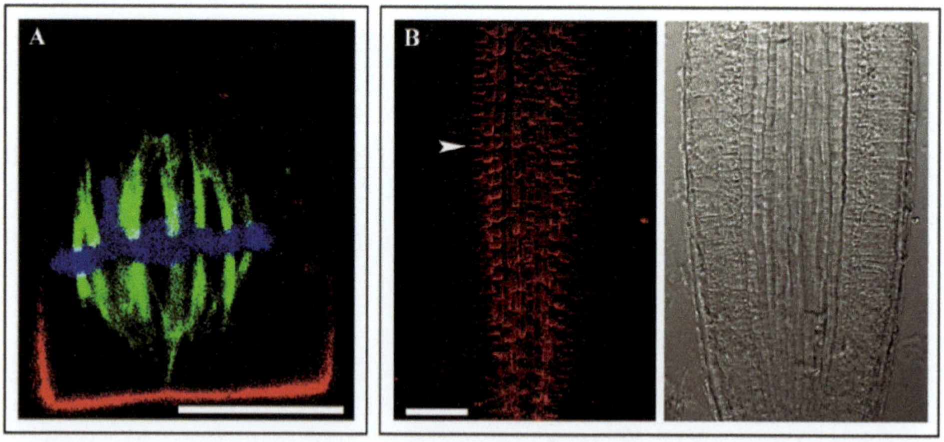

Fig. 14.2 Immunolabelling of plant cells: (A) Triple labelling in maize root cell. Immunostaining of PIN proteins i.e. plasma membrane associated proteins, facilitators of auxin transport (red channel reading Cy3 labelling), and microtubules (green channel reading the Fluorescein IsoThioCyanate (FITC) labelling). Hoechst staining of the nucleus (blue channel) (Bar: 8 µm). (B) Immunostaining of PIN proteins on thin sections of *Arabidopsis thaliana* root embedded in PEG. Left panel, PIN proteins label the basal membrane of cellular files (arrow). Right panel: differential interference contrast (Bar in B left: 40 µm). See also Boutté *et al.,* 2006.

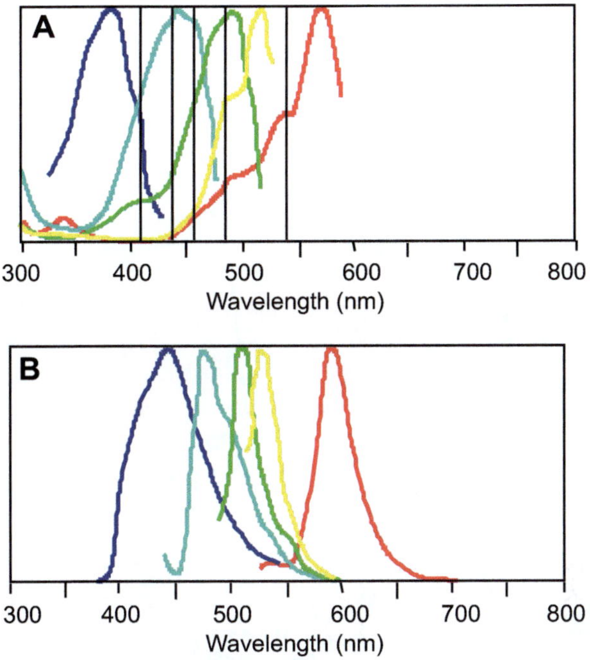

Fig. 14.5 Normalized excitation and emission spectra of EGFP and its spectral variants (modified from Clontech, Palo Alto) (From left to right: BFP, CFP, GFP, YFP, DsRed.) The positioning of laser lines often available on imaging devices is marked by vertical lines on the excitation panel (A). (B) Emission panel.

Chapter 20

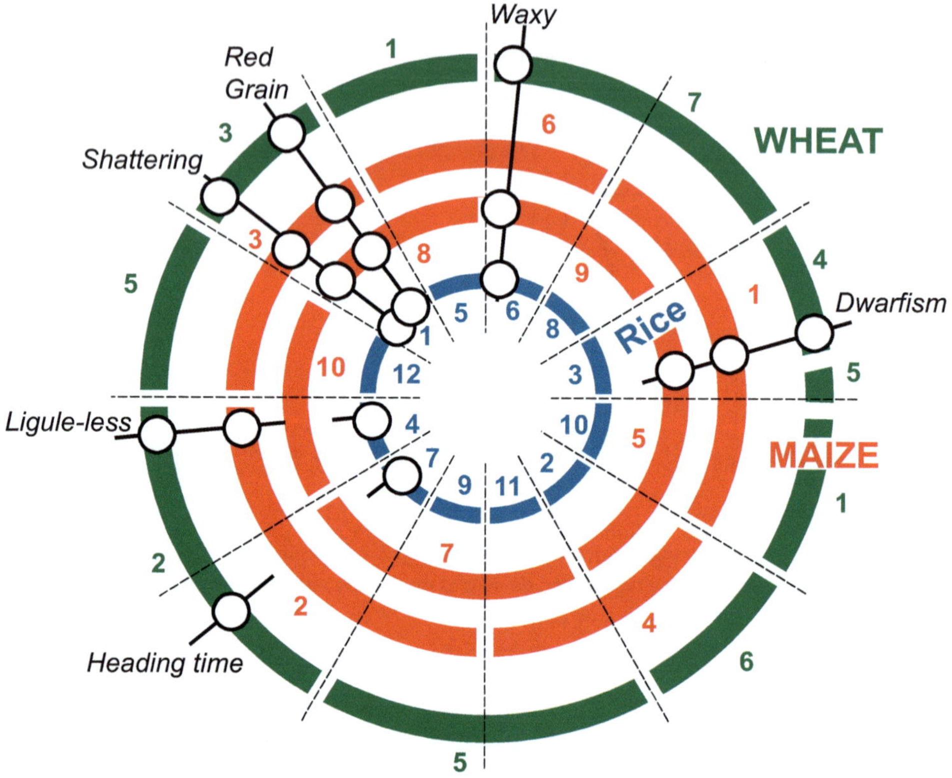

Fig. 20.2 Concentric circle model representing the rice, maize and wheat chromosomes. All maps are aligned relative to rice chromosomes. Numbers refer to chromosome number in each species nomenclature. Dashed radial lines represent the separation between rice chromosomes. A number of conserved traits have been positioned and are represented by open circles (O) along a radius (Devos and Gale, 1997).

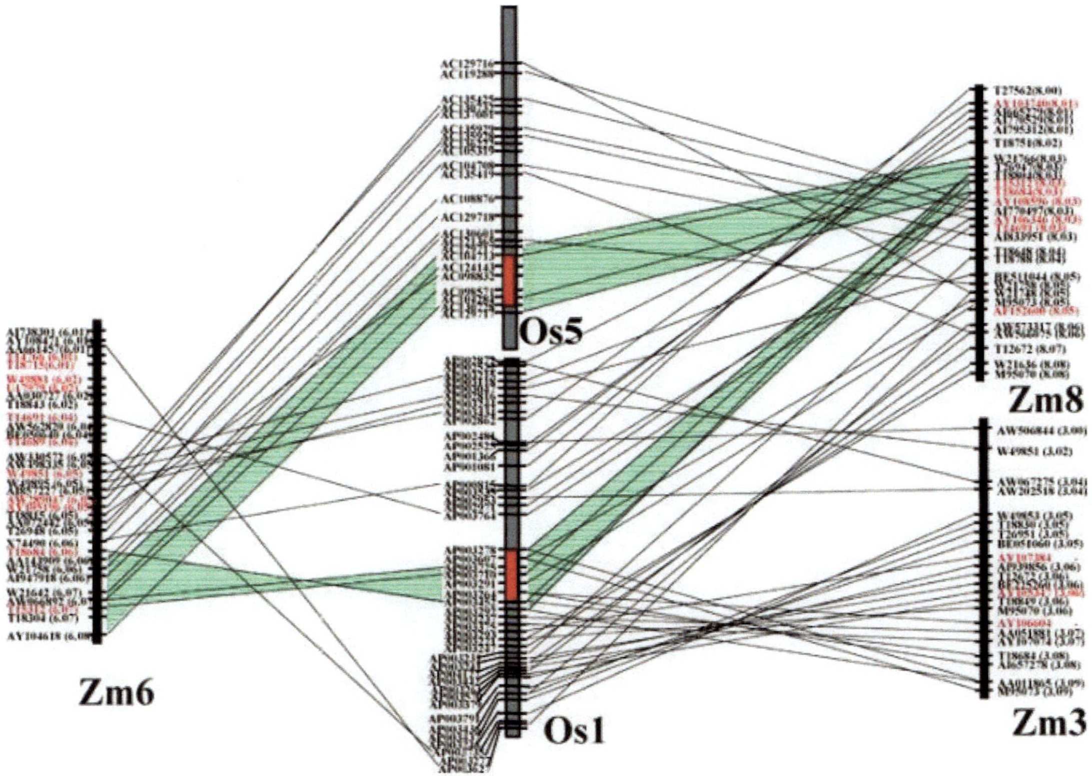

Fig. 20.3 Alignment between rice chromosomes Os 1 and Os 5 and maize chromosomes Zm 6, Zm 8 and Zm 3 (Salse *et al.*, 2004) Horizontal bars cn rice chromosomes indicate BAC clones and their accession numbers. Each bar, on maize chromosomes correspond to a mapped and sequenced marker with an homologue in rice. Markers in commcn between maize and rice are linked by straight lines. Shading illustrates collinear regions between rice and maize chromosomes and identifiy a region in rice Os 5 and Os 1 which is duplicated and represented in a different colour.

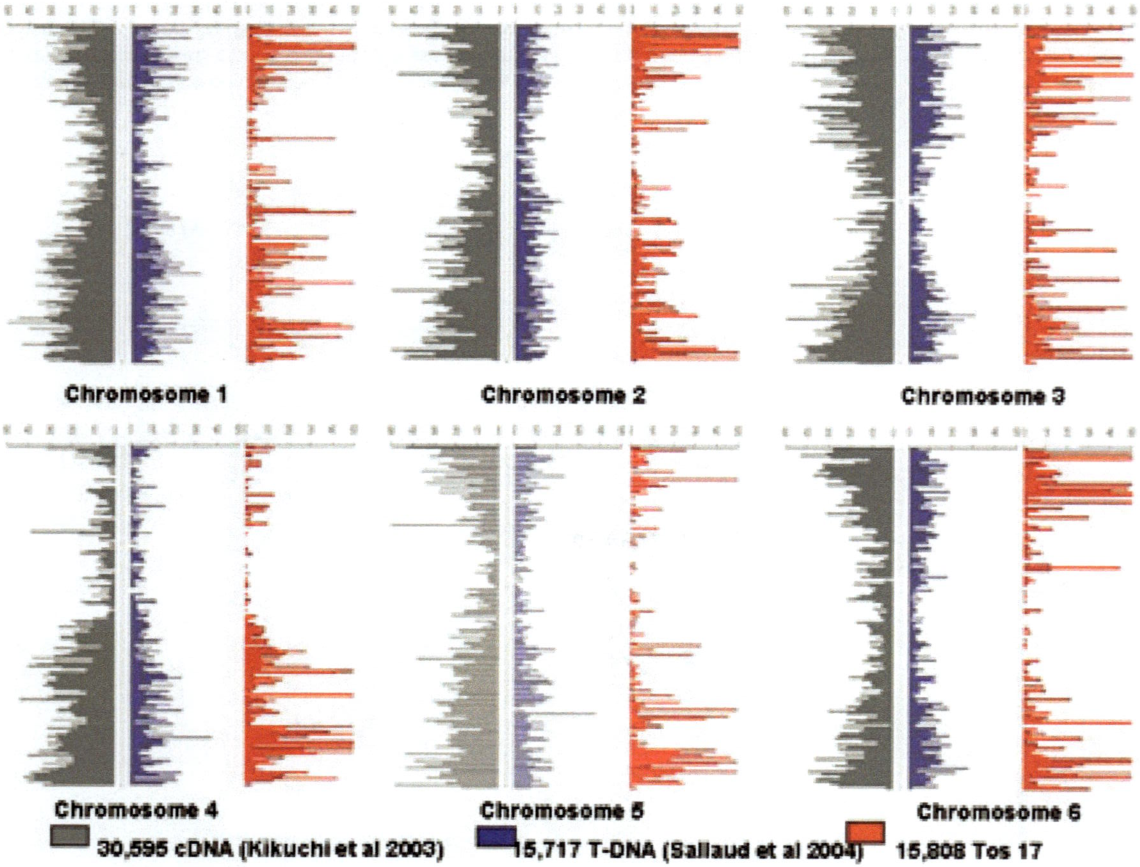

Fig. 20.5 Distribution of the full length cDNA, T-DNA insertions and *Tos 17* insertion along the rice chromosomes 1-6 (Sallaud *et al.*, 2004). Densities of T-DNA, and *Tos 17* FST and Full Length cDNA were plotted for each 200 kbp.

Chapter 21

Fig. 21.2 Aerial parts and endosymbiotic roots of *Medicago truncatula* (line Jemalong A17). a) Barrel-shaped, spiny pods; b) *Mt* seeds with truncated shape; c) *Mt* shoot bearing flowers / developing pods on each node; d) opening flower at the post-anthesis stage; e) 4 day-old seedling shoot bearing the first, monolobed leaf; f) young N-fixing, pink nodules on primary root 10 days post-inoculation with the microsymbiont *Sinorhizobium meliloti*; g) high-magnification view of nodule central zone showing infected cells packed with N-fixing bacteroids (photographed by Georges Truchet); h) root segment colonized by hyphae of the endosymbiotic mycorrhizal fungus *Glomus intraradices* (blue stain) that form arbuscules in numerous cells in the inner cortex (purple stain); i) magnified view of a fungal arbuscule (black stain) connected to the internal hyphae and located near xylem tissue; arbuscules are the major site of nutrient exchange between plant and fungal partners (h and i: courtesy of Mireille Chabaud).

Scale bars : a) 10 mm; b) 5 mm; c) 15 mm; d) 2 mm; e) 10 mm; f) 1 mm; g) 20 µm; h) 300 µm; i) 20 µm.

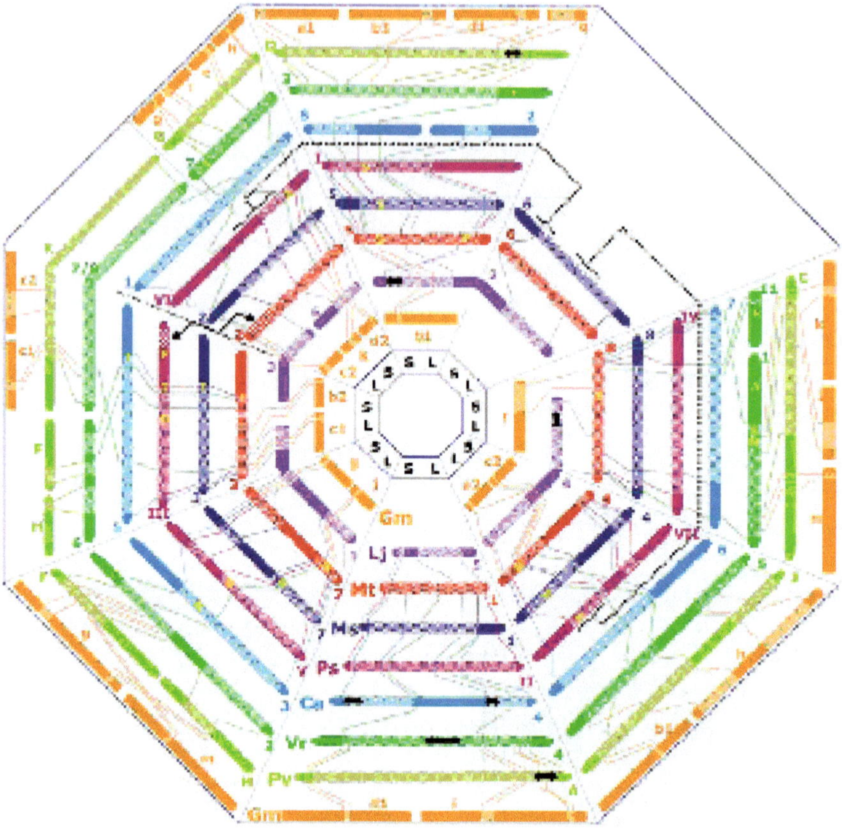

Figure 21.3 A simplified consensus map for eight legume species (reproduced, with permission, from Zhu *et al.,* 2005). Mt, *M. truncatula*; Ms, *Medicago sativa* (alfalfa); Lj, *L. japonicus*; Ps, *Pisum sativum* (pea); Ca, *Cicer arietinum* (chickpea); Vr, *Vigna radiata* (mungbean); Pv, *Phaseolus vulgaris* (common bean); Gm, *Glycine max* (soybean). S and L denote the short and long arms of each chromosome in *M. truncatula*. Syntenic blocks are drawn to scale based on genetic distance.

Chapter 24

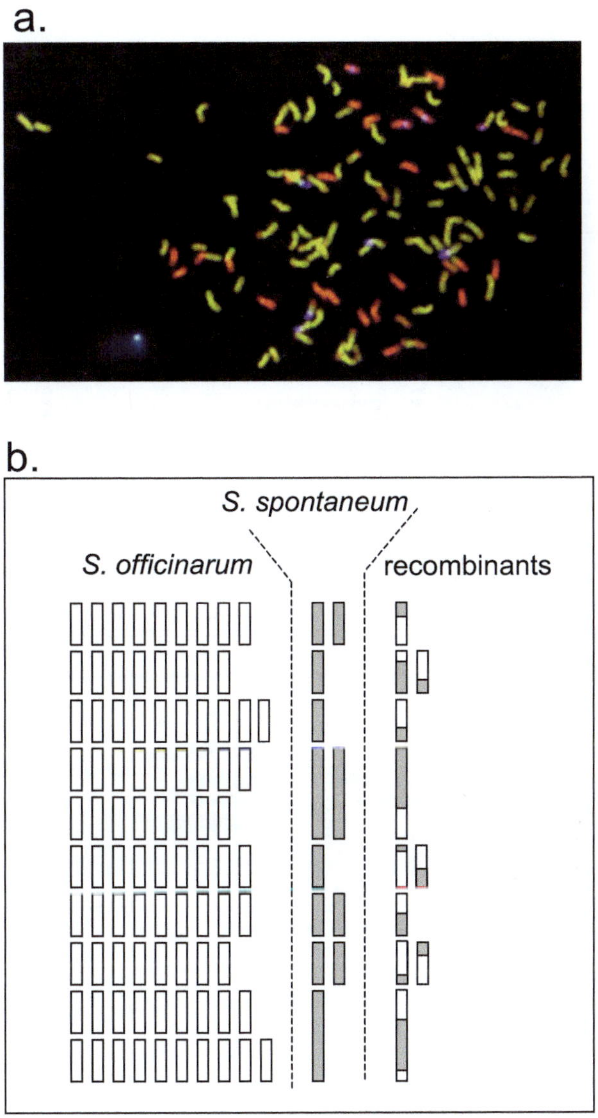

Fig. 24.1 **a.** Chromosome preparation of the sugarcane cultivar NCo 376 after GISH using *S. officinarum* genomic DNA (labeled with fluorescein isothiocyanate, green), *S. spontaneum* genomic DNA (labeled with Texas Red, red) and 18S-5.8S-25S rDNA (labeled with coumarin, blue). Around 70% of the chromosomes are inherited from *S. officinarum* (yellow), 20% from *S. spontaneum* (orange) and 10% are recombinant from both species. **b.** Schematic representation of the genome of current sugarcane cultivars as deduced from FISH and GISH experiments. Each bar represents a chromosome. White and grey bars correspond to *S. officinarum* and *S. spontaneum* chromosomes or chromosome segments, respectively. Chromosomes from the same row are homologous (or homeologous). The main characteristics of this genome are the high polyploidy, aneuploidy, the bi-specific origin of the chromosomes and the occurence of interspecific recombinant chromosomes. In addition, there are clear structural differences between the chromosomes of the two species.

Chapter 30

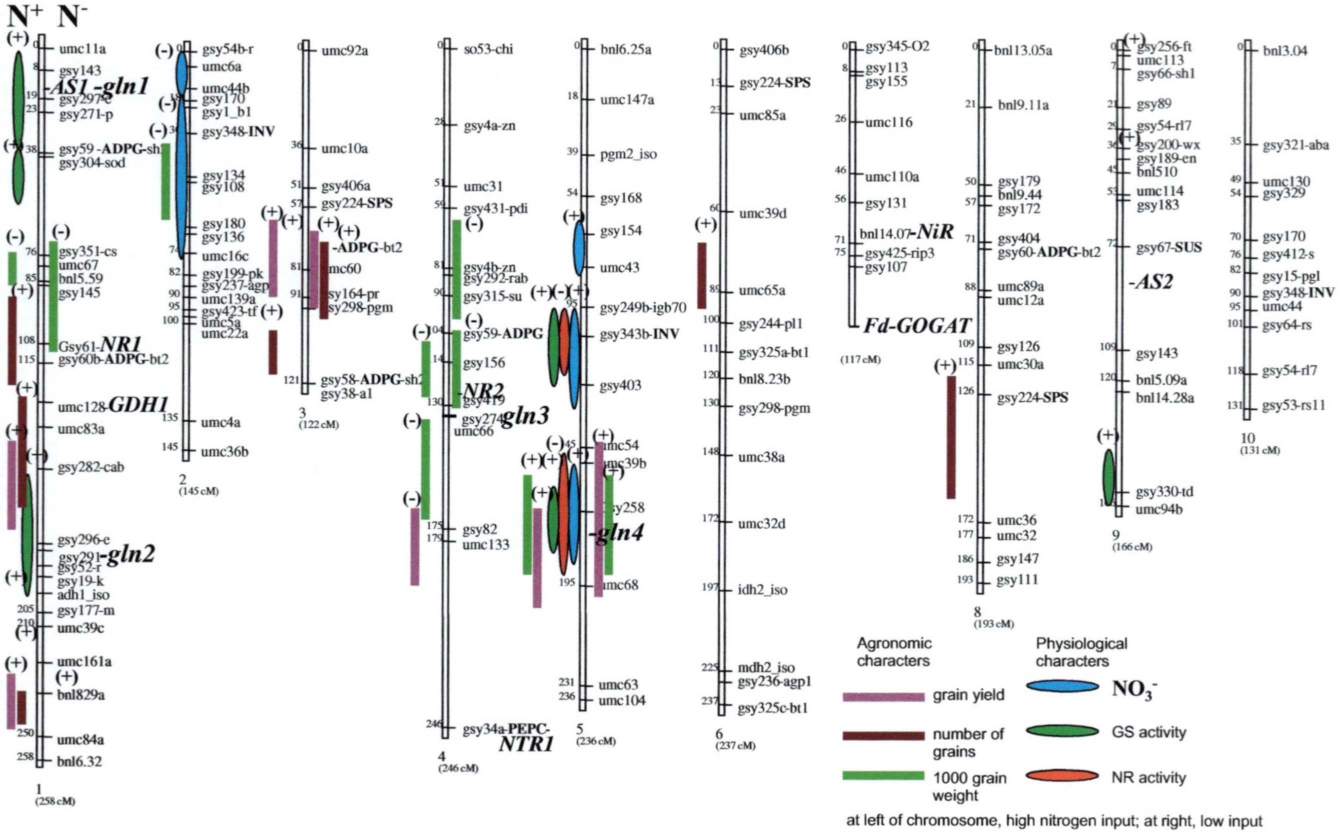

Fig. 30.1 Genetic map of maize (RFLP) and co-localization between the QTL for physiological characters and those linked to yield and to its components. Abbreviations of enzymes implicated in the nitrogen metabolism: ASN, asparagine synthetase; Fd-GOGAT, ferredoxin-dependent glutamate synthase; GS, glutamine synthetase; NR, nitrate reductase; NiR, nitrite reductase. Hirel *et al.*, 2001. © American Society of Plant Biologists.

Chapter 32

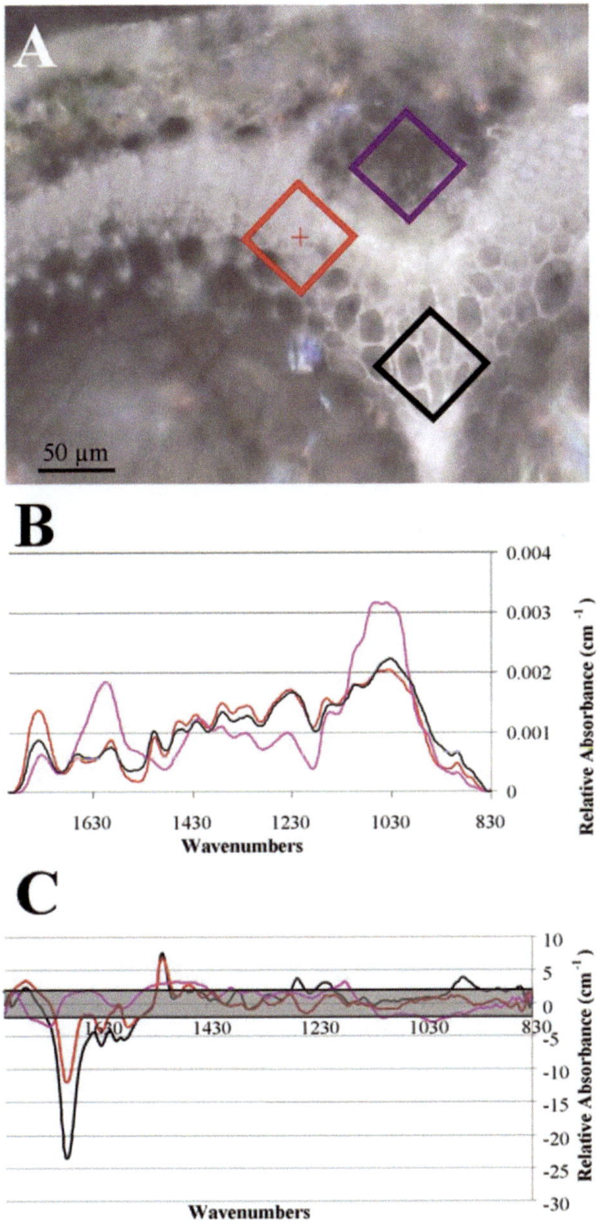

Fig. 32.1 Tissue specific FT-IR analysis of *Arabidopsis* stem (Sibout et al. 2005). (A) Section of a stem (600µm *diameter*) of a 3 weeks-old *Arabidopsis thaliana* plant. The colored squares define the zones where FT-IR spectra were acquired. Distinct squares correspond to specific tissues: phloem (purple), xylem (black), fiber (red). (B) Fourier transform mid-infrared (FT-IR) spectra of xylem area, fiber area and phloem in wild-type stems. (C) Identification of tissue specific cell wall alterations in the *Atcad cd* mutant. t value (Student's t-test) corresponding to the difference between *Atcad cd* mutants for the three distinct tissues investigated. Non-significant values (p = 0.95) are shaded.